HOSPITALITY AND TOURISM

PROCEEDINGS OF THE INTERNATIONAL HOSPITALITY & TOURISM POSTGRADUATE CONFERENCE 2013, SHAH ALAM, MALAYSIA, 2–3 SEPTEMBER 2013

Hospitality and Tourism

Synergizing creativity and innovation in research

Editors

Norzuwana Sumarjan, Mohd Salehuddin Mohd Zahari, Salleh Mohd Radzi, Zurinawati Mohi, Mohd Hafiz Mohd Hanafiah, Mohd Faeez Saiful Bakhtiar & Artinah Zainal

Faculty of Hotel & Tourism Management, Universiti Teknologi MARA, Malaysia

CRC Press
Taylor & Francis Group
Boca Raton London New York Leiden

CRC Press is an imprint of the
Taylor & Francis Group, an **informa** business
A BALKEMA BOOK

Organized by:

Faculty of Hotel & Tourism Management, Universiti Teknologi MARA, Malaysia

CRC Press/Balkema is an imprint of the Taylor & Francis Group, an informa business

Typeset by V Publishing Solutions Pvt Ltd., Chennai, India
Printed and bound in Great Britain by CPI Group (UK) Ltd, Croydon, CR0 4YY

Published by: CRC Press/Balkema
P.O. Box 11320, 2301 EH Leiden, The Netherlands
e-mail: Pub.NL@taylorandfrancis.com
www.crcpress.com – www.taylorandfrancis.com

ISBN: 978-1-138-00151-0 (Hbk)
ISBN: 978-1-315-79823-3 (eBook PDF)

Hospitality and Tourism – Sumarjan et al. (eds)
© 2014 Taylor & Francis Group, London, ISBN 978-1-138-00151-0

Table of contents

Hospitality and tourism marketing

Hospitality and Tourism – Sumarjan et al. (eds)
© 2014 Taylor & Francis Group, London, ISBN 978-1-138-00151-0

Preface

Dear Distinguished Delegates and Guests,

Welcome to the first *International Hospitality and Tourism Postgraduate Conference 2013*, organized by the Faculty of Hotel and Tourism Management, Universiti Teknologi MARA on September 2–3, 2013.

This conference mainly aims to provide an avenue for postgraduate students from all over the world to share, present and exchange their intellectual, creative, as well as innovative ideas or works revolving but not limited to the realm of hospitality and tourism. This conference also provide an opportunity for academia, hospitality practitioners and other relevant agencies or authorities to grasp the issues, the trend and practical ideas that can be useful to strengthen their current practices. The conference also serves as a platform for all the delegates and guests to develop network linkages for future collaboration.

A total of 142 papers (conceptual and completed) were received under a total of eight tracks covering the diverse segments of the hotel, tourism, gastronomy and foodservice industry. All submitted papers were then peer reviewed; the final number accepted is 116. The papers were selected based on the originality, significance and in align to the conference theme: "*Synergizing creativity and innovation in research.*"

We hope that this first conference can be a breakthrough effort to further inspire postgraduate students to produce more 'state of the art' and beneficial researches to enhance the overall hospitality and tourism education and best practices. We also hope that this proceeding can enthuse and benefit all delegates, readers and, especially future students who wish to embark into the challenging yet meaningful postgraduate kingdom.

We are very grateful and acknowledge the contribution of various parties who have been involved directly and indirectly—the organizing committees, students, reviewers, faculty members, speakers, sponsors, and the participants—for realizing the success of this conference.

Finally, we wish you success in your presentation, enjoy fruitful discussions and a pleasant stay in Shah Alam, Selangor, Malaysia.

Hospitality and Tourism – Sumarjan et al. (eds)
© 2014 Taylor & Francis Group, London, ISBN 978-1-138-00151-0

Acknowledgements

A. Parasuraman, *University of Miami, USA*
Ahmad Fitri Amir, *Universiti Teknologi MARA, Malaysia*
Artinah Zainal, *Universiti Teknologi MARA, Malaysia*
Azdel Abdul Aziz, *Universiti Teknologi MARA, Malaysia*
Azila Azmi, *Universiti Teknologi MARA, Malaysia*
Chemah Tamby Chik, *UniversitiTeknologi MARA, Malaysia*
Fadzilah Mohd Shariff, *UniversitiTeknologi MARA, Malaysia*
Faiz Izwan Anuar, *Universiti Teknologi MARA, Malaysia*
Flora Shumin Chang Abdullah, *Universiti Teknologi MARA, Malaysia*
Hamdin Mohd Salleh, *Universiti Teknologi MARA, Malaysia*
Hashim Fadzil Ariffin, *Universiti Teknologi MARA, Malaysia*
Hazmal Ismail, *Universiti Teknologi MARA, Malaysia*
Inoormaziah Azman, *Universiti Teknologi MARA, Malaysia*
Johanudin Lahap @ Wahab, *Universiti Teknologi MARA, Malaysia*
Kamarul Hasni Che Rose, *Universiti Teknologi MARA, Malaysia*
Khairun Najiah Ahmad, *Universiti Teknologi MARA, Malaysia*
Lim Lay Kian, *Universiti Teknologi MARA, Malaysia*
Mohd Faeez Saiful Bakhtiar, *Universiti Teknologi MARA, Malaysia*
Mohd Hafiz Mohd Hanafiah, *Universiti Teknologi MARA, Malaysia*
Mohd Raziff Jamaluddin, *Universiti Teknologi MARA, Malaysia*
Mohd Salehuddin Mohd Zahari, *Universiti Teknologi MARA, Malaysia*
Mohd Syaquif Yasin Kamaruddin, *Universiti Teknologi MARA, Malaysia*
Noor Azmi Ahmad, *Universiti Teknologi MARA, Malaysia*
Nor Azah Mustapha, *Universiti Teknologi MARA, Malaysia*
Norfezah Md Nor, *Universiti Teknologi MARA, Malaysia*
Norliza Aminudin, *Universiti Teknologi MARA, Malaysia*
Norzuwana Sumarjan, *Universiti Teknologi MARA, Malaysia*
Rafidah Aida Ramli, *Universiti Teknologi MARA, Malaysia*
Roslina Ahmad, *Universiti Teknologi MARA, Malaysia*
Rosmaliza Muhammad, *Universiti Teknologi MARA, Malaysia*
Salamiah A. Jamal, *Universiti Teknologi MARA, Malaysia*
Salleh Mohd Radzi, *Universiti Teknologi MARA, Malaysia*
Toh Poh See, *Universiti Teknologi MARA, Malaysia*
Zulhan Othman, *Universiti Teknologi MARA, Malaysia*
Zurinawati Mohi, *UniversitiTeknologi MARA, Malaysia*

Hospitality and Tourism – Sumarjan et al. (eds)
© 2014 Taylor & Francis Group, London, ISBN 978-1-138-00151-0

Organizing committee

PATRON

Tan Sri Dato' Prof. Ir. Dr. Sahol Hamid Abu Bakar
FASc, DPMS, SSAP, DJMK, DSPN, DJN, DSM, BCN

ADVISOR

Assoc. Professor Mohamad Abdullah Hemdi, PhD

CONFERENCE CHAIR

Assoc. Prof. Mohd Salehuddin Mohd Zahari, PhD

COMMITTEES

Assoc. Professor Salleh Mohd Radzi, PhD
Assoc. Professor Artinah Zainal, PhD
Zurinawati Mohi, PhD
Norzuwana Sumarjan, PhD, CHE
Nor Azah Mustapha
Mohd Hafiz Mohd Hanafiah
Mohd Raziff Jamaluddin
Azdel Abdul Aziz
Zulhan Othman
Hazmal Ismail
Inoormaziah Azman
Mohd Faeez Saiful Bakhtiar
Mohd Syaquif Yasin Kamaruddin
Mohd Noor Ismawi Ismail
Hamizad Abdul Hadi
Ahmad Fitri Amir
Mohd Zain Kutut
Siti Ruhanis Mohamad Nazer

GRADUATE ASSISTANTS

Amril Mu'idzuddin Amrahi
Shamsul Khairul Annuar Nordin
Nur Zalikha Othman
Mohd Daim Darson
Zulhilmi Suhaimi

Hospitality and tourism management

Hospitality and Tourism – Sumarjan et al. (eds)
© 2014 Taylor & Francis Group, London, ISBN 978-1-138-00151-0

Sustainable approach of fishing tourism in Kenyir Lake

M.A. Alias, N.A. Ahmad, M.A. Ahmad & F.H. Muhammad Safri
Universiti Teknologi MARA, Malaysia

ABSTRACT: Changing trends in travel pattern, recreational and adventure tourism which consist of various enjoyable activities are rapidly growing and becoming more popular. As such, Malaysia is promoting fishing activities as one of the divisions under recreational tourism due to the vast potential and selection of fishing locale which have diverse species of fish in accordance with the equatorial climate of the country. This study aimed to explore the sustainable approach of fishing tourism in Kenyir Lake. Qualitative approach with semi-structured interview with fishing/fisheries experts was undertaken to get a trustworthy data on the strategic sustainable approaches that need to be executed at the research area. Findings revealed a lot of strategies need to be implemented on managerial aspect of the lake, how to control the fish stocks and preserving natural resources.

Keywords: Fishing tourism, Kenyir Lake, sustainable approach

1 INTRODUCTION

Recreation associated with human life and it could be a restful day in a beautiful place with several activities such as bushwalking, orienteering, having a picnic, fishing, meditating or cycling (Mason, 2005). Gbadebo and Adedeji (2012) stated that recreation can be considered as a tourism when people are doing outdoor activities and spend on the tourism product such as accommodation, transportation and food (Koutoulas, 2004). This means recreation and tourism are related and there are many places offering this particular sector to cater the markets. In Malaysia, recreational tourism is becoming popular and these trends create a phenomenon where a lot of attraction places related to recreation are built (Nicholls, 2012).

Some places are catering niche market by offering specific interest such as fishing activities, mountain climbing, bird watching and hunting (Canadian Human Resource Council, 2012). Kenyir Lake, the largest man-made lake in South East Asia offers several attractions such as Kelah Sanctuary, National Park, river waterfalls, limestone caves, fishing, bird watching and jungle trekking (Kenyirlake.com, 2012; Virtualmalaysia.com, 2012). This place is the most popular fishing spot among anglers and fishing competitions are organised annually such as Boat On The Road (BOTR) which started since 2010 and Kenyir Cup 2012 International Fishing Competition (Pelancongkini. blogspot, 2012) (NST.com. 2012).

Kenyir Lake is well developed in terms of tourism and this place is one of the most visited places in Terengganu. Numbers of incoming visitors especially anglers are increasing and this will lead to degradation of this lake. There are some issues on degradation of places because of the unlimited people coming by did not practicing the "bag and limit size" or "catch and release" approach and it will lead to overfished at the area.This sport is still considered as a new attraction in Malaysia and there are still very few authorities or associations who are recording the growth of these sports, anglers' spending and fish stocks especially in this research area.

Therefore, in this study, the researcher wants to highlight the strategic sustainable approach that need to be implemented on this lake in order to preserve the natural resources as well as keep attracting people to come.

2 LITERATURE REVIEW

2.1 *Fishing tourism*

Since the creation of human in the world and the existence of settlements, food is one of the main needs for them to survive. During the ancient times, about 60,000 years ago, people started to hunt to get meat and fishing is one of the ways of it. They were mostly used nets by spinninggrasses, or spinning wool, or flax. It is more preferable for them to get as many fish as possible as quickly as they could (Carr, 2012).

Few centuries ago, due to the scarce of the fish stock, fishing has been divided into recreation and commercial (Alban & Boncoeur, 1999). Rashid (2011) supported that people began to make fishing as a hobby and more improvements in terms of fishing gear and their awareness on conserving the shortage of fish. The author also stated that in the current trends, fishing is considered as a sport to distinguish it from the commercial fishing. Furthermore, fishing is considered as a sport and it has been categorized as a recreational tourism (Marquez, 2010). According to Ditton et al. (2002) fishing can also be considered as a form of tourism when anglers cross state lines to go fishing.

Apparently, fishing activities was not only focusing on the rural area. Sort of payable, coast, deep sea, lake, mine, mountain river and common river are the targeted fishing spots by the anglers (Karim, 2010). Thus, it is an opportunity for the country to gain income from this activity. A study by Crossley et al. (2012) the commercial recreation and tourism field has a phenomenal economic, social, cultural, and personal impact which it contributed much to the country entirely.

2.2 *Sustainable approach*

Sustainable approach is broadly promoted as a comprehensive concept which targeted on social, economic and cultural (Reid and Schwab, 2006). According to Coccossis (1996), the sustainability concept can be assumed as it is a continuation of sustainable outcomes that come from the tourism contribution (Stabler, 1997). In response to the analytical impacts, strategic sustainable approach is crucial in order to conserve and sustain the attraction places (Moscardo, 2007). However Bressers (2004) mentioned there are three challenges to implement the new strategies: Some parties are unable to accept the policies as it differ with their societal and cultural norms, developing a new integration, and developing the power support to implement the new strategies.

Tourism sector which repeatedly associates with sustainable approach is a nature-based tourism (Luo & Deng, 2007). Developments done in nature-based tourism are closely related to natural environment (Maetzold, 2002). Turner et al. (1999) claimed that the impacts caused by the development in the natural environment compound are needed to be conserved and sustained. Fishing activities are normally deal with natural resources such as fishes, water, and vegetation (Navratil, Martinat, & Kallabova, 2009). Ditton, Holland, and Anderson (2002) stated that the increasing of popularity in the particular fishing area will leads to huge numbers of people attracted, while Turner et al. (1999) supported when demand on fishes are increase, it will affect the fish stocks and sustainable strategies need to done. The authors emphasized that habitat restoration and protective habitat management are reviewed as important to reduce the loss of habitat as well as maintaining healthy sustainable fisheries.

Interference of quality and quantity water faced by agriculture tourism has been sustained in order to increase agriculture productivity (Branca, Lipper, Neves, Lopa & Mwanyoka, 2011). Marine fisheries had implemented many sustainability approaches to preserve the fish stocks. There were artificial reefs (Pickering & Whitmarsh, 1996), and vegetation replanted such as sea grass and mangrove (Turner et al., 1999). All the efforts must being accompanied by the law enforcement to ensure all the strategies are succeeded (Malaysian Fisheries Society, 2013).

3 RESEARCH METHODOLOGY

This study began with intensive literature review to understand the attribute in sustainability of fishing. Feedbacks from tourism academician are sought to validate those attributes. In addition, interview with former fisheries officers was conducted to probe their insight about the current situation and the problem in Kenyir Lake. Semi-structured interview with the experts have been conducted within three months. The conversations have been recorded using a tape recorder. This study managed to capture 10 informants as the data had been saturated on the 10th.

4 FINDINGS

Tourism sector which repeatedly associates with sustainable approach is a nature-based tourism (Luo & Deng, 2007). Developments done in nature-based tourism closely related to natural environment (Maetzold, 2002). From the interview, informants gave their opinion based on their expertise on what should be done to conserve this lake. Informants' point of view is subsequently highlighted.

Informant 1 revealed some of the major approach that needs to be implemented in order to conserve and preserve the Kenyir Lake. First strategy is monitoring all development projects.

> *"Plan and formulate every development in Kenyir, and the location is the most importance. Need to have certain areas that are prohibited for any constructions".*

Second strategy is focusingon development program. They needto constantly do seminars and clinic to locals, boat and hotel operators with the aim to expose them about conservation, preservation and importance of cleanliness.

"Organize seminars to boat operators, hotel, resort and chalet staff. It is intended to reveals the importance of conserve and preserve the nature".

Third strategy is controlling and enforcing law to boat and hotel operators. All operating boats should pass the inspection and the boats must have all of the safety features such as life jacket and qualified boat interpreter.

"Before started their business, they have to pass several level of inspections such as size of boats, condition of boats, safety features and boat interpreter".

Fourth strategy is processing and filtering all the applications for programs or activities that wanted to be implemented in the Kenyir Lake area.

"Process and filter all application that related to activities that want to be done in Kenyir".

Next strategy focuses on sustaining the other attraction in Kenyir such as National Park, waterfalls, Kelah Sanctuary, caves, herbs farm and cage fish farming. Fishing activities in these compound are prohibited.

Informant 2 gave his opinion based on the marine strategic approaches to conserve the fish species. He suggested building the artificial reefs. It is in line with the literature by (Pickering and Whitmarsh, 1996) where artificial reefs are used as one of the sustainable approach. According to the informant, the reefs should be divided into two, which is for conservation and the other one is for recreation. Reefs for conservation are just for fish to breed and the location or the exact spot should not be exposed to public and should close for any fishing activities. Whereas the reefs for recreation is open to public, but for the fishing purposes only, netting is prohibited. He also mentioned that authorities should monitor and doing sport check to the locals and anglers.

"In marine area we had doing to build artificial reefs, I thing that this strategy also useful for the freshwater. But, emm.. The reefs must be divided. One is for conservation, the other one is for recreation. And.. emm.. Location of both reefs should be different. For the recreation reef, they can fish there but the other one is not. And one more thing..emm authority should regularly check the locals fishing gear. Some of them is too innovative and sometimes that gear will harmful the fish."

Informant 3, 5 and 6 emphasized on the monitoring system that should be implemented for anglers who are doing the fishing activity in this lake. Authorities who manage the lake should records all anglers/visitors information in order to observe and monitor their activity and their catches. It is also proposed by the informants to do "open-close season" for publics to fish at Sungai Terenggan and Sungai Pertang with terms and conditions. Other than that other informant stressed out that replant the trees and water treatment need to be done.

Informant 3

"To conserve the natural resources of the lake, they should implement the monitoring system whereby they should record the visitors' in and out. Any catches which is under protected species are not allowed to bring back. They should release it back, otherwise they will be fined".

Informant 6

"To sustain the natural resources, authority should do some program such as replant the trees and water treatment. They should have a department which assigned to check the water quality and cleaning up all the rubbish around the lake".

Informant 5

"Controlling system must be have so that the fish stock will maintain. Other than that, they should apply the restocking program. From what I know they do this program, but they released the barramundi (siakap). It is not wrong to release the "alien" species, but it should balance. They should review what this fish eat, and they should release their "food" also. If not, this barramundi will eat the local species because this fish also predator. And the other thing is, in Kenyir there have conservation area in emm...ha..in Sungai Pertang and Sungai Terenggan. To attract more people they can do the "open-close season" just like done by Tagal in Sabah whereby people can fish in the certain time. And just do the terms and condition such fees, catch and release concept and apply carrying capacity. Anglers willing to obey the requirements as long as they get their dream fish".

Informant 4 revealed that his organization has planning several strategies to boost the angling activity as one of the attraction in Malaysia. They are assuming that total numbers of anglers in Malaysia exceeds two million, but there are no official statistics on that. So he mentioned that their first strategy is to get the actual numbers of anglers in Malaysia. This is to calculate their total expenditure and when they get the data, this sports fishing can be categorized as an industry, just like European countries did.

"But from my side, we firstly want to get the actual statistics on total numbers of anglers in Malaysia. It is aimed err.. To make..emm.. Easier for us to record and calculate their expenditure. So that we can publish that our sports fishing industry is good. Just like European countries did".

Next strategy that needs to be implemented is licensing all the anglers in Malaysia. Through this approach, numbers of anglers can be calculated. Impact from this approach will be easy for other parties to record any data about fishing and also useful for any researcher.

"Next we are planning to license all the anglers. So if they want to travel or joining competition, they need to have license".

Informant 7 and 10 suggested that catch & release, bag & limit size and selected harvest practices must be enforced to sustain the fish stocks and for future generation. Besides that, informant added that "alien" species need to be taken out to save our local species from extinction.

Informant 7

"For me the most important thing is authority must enforce the practice of catch & release, bag & limit size and selected harvest. It is to ensure our future generation can feel our "golden treasure".

Informant 10

"Besides enforced those practices such as catch & release and so on. For me that's number two. Because fish are still many, just they don't want to eat our baits. The most important thing is they should bring out the "alien" species. Because of this fish, our local species become reduced".

Despite all this arguments and opinions, it is worth mentioning that developments in Kenyir Lake lead to several impacts. Therefore, strategies are needed in order to sustain the place and keep attracting people to come.

5 CONCLUSION

The findings of this study are anticipated to contribute towards a better understanding of the people towards the angling activities as a recreational tourism. The outcomes of this study are benefited to the government and tourism related bodies who are strive to develop the new tourism attractions in Malaysia by using the strategies from this study. Furthermore, from the findings, authorities, government bodies and any associations can utilize the strategies to be implemented to other potential places. Other than that, we are known Malaysia's fishing spots have a potential to be one of the important tourism division.

In conclusion, the above highlighted issues and suggestion could not be resolved without involvement and cooperation from all parties. Malaysian fishing sport has a big potential to be one of the famous and biggest contributor to the economy. Apart from that, in depth strategies are needed to ensure that our natural resources will not become extinct. All in all, from all of the efforts, it is hope that Malaysia will stated in a world travel guide as one of the famous fishing sport in the world.

REFERENCES

Alban, F. & Boncoeur, J. (1999). Commercial fishing, recreational fishing and tourism: Investigating the potential for developing a pluri-activity. Retrieved December 8, 2012, from http://www.eafe-fish.org

Branca, G., Lipper, L., Neves, B., Lopa, D., & Mwanyoka, I. (2011). Payments for watershed services supporting sustainable agricultural development in Tanzania.*The Journal of Environment & Development*, *20*(3), 278–302.

Bressers, H.T.A. (2004). *Governance for Sustainable Development: The Challenge of Adapting Form to Function.* Northampton, MA: Edward Elgar.

Canadian Human Resource Council (2012). Sectors in tourism. Retrieved November 8, 2012, from http://www.tc.gov.yk.ca/pdf/SectorsinTourism.pdf

Carr, D. (2012). The first civilizations and empires. Retrieved November 15, 2012, from http://myweb.unomaha.edu

Coccossis, H. (1996). *Tourism and Sustainability: Perspectives and Implications' in Priestley et al. (eds), Sustainable Tourism?* U.K: Cab International.

Crossley, J.C., Jamieson, L.M., & Brayley, R.E. (2012). Introduction to commercial recreation and tourism. Retrieved August 3, 2012, from http://www.sagamorepub.com/files/lookinside/218/pages-comrecebook.pdf

Ditton, R.B., Holland, S.M., & Anderson, D.K. (2002). Recreational fishing as tourism. *International Journal of Business and Management*, *5*(12), 41–50.

Gbadebo, A.M. & Adedeji, O.H. (2012). Tourism and recreation. Retrieved December 29, 2012, from http://www.unaab.edu.ng/attachments/464_Tourism_508_Notes.pdf

Karim, F. (2010, April). Jenis perairan/kawasan pancingan. *Pancing*, p. 4.

Kenyirlake.com (2012). Accommodation. Retrieved March 11, 2013, from http://www.kenyirlake.com/accommodation.htm

Koutoulas, D. (2004). Understanding the tourism product. Retrieved November 25, 2012, from http://www.academia.edu/1861406/Understanding_the_Tourism_Product

Luo, Y. & Deng, J. (2007). The new environmental paradigm and nature-based tourism motivation. *Journal of Travel Research*, 46, 392–402.

Maetzold, J.A. (2002). Nature-based tourism & agritourism trends: Unlimited opportunities. Retrieved December 15, 2012, from http://www.agmrc.org/media/cms/agritourism_E6794269B3FF6.pdf

Malaysian Fisheries Society (2013). Malaysian fisheries. Retrieved January 20, 2013, from http://mfs.org.my/

Marquez, J.V. (2010). Consumerism in developed countries causes destruction of wetlands in the tropics. Retrieved December 2, 2012, from http://cmsdata.iucn.org/downloads/wetland_destruction__art__para_revista__1.pdf

Mason, R. (2005). Recreation and tourism in the Australian alps. Retrieved November 28, 2012, from http://www.australianalps.environment.gov.au/learn/pubs/recreation.pdf

Moscardo, G. (2007). Sustainable tourism innovation: Challenging basic assumptions. *Tourism and Hospitality Research*, *8*(1), 4–13.

Navratil, J., Martinat, S., & Kallabova, E. (2009). Framework for utilizing angling as a tourism development tool in rural areas. Retrieved November 15, 2012, from http://www.agriculturejournals.cz/publicFiles/12079.pdf

Nicholls, S. (2012). Outdoor recreation and tourism. Retrieved May 10, 2013, from http://glisa.msu.edu/docs/NCA/MTIT_RecTourism.pdf

NST.com (2012, April 5). RM20,000 prize for giant fish in Lake Kenyir. *New Straits Times*. Retrieved March 11, 2013, from http://www.nst.com.my/nation/general/rm20-000-prize-for-giant-fish-in-lake-kenyir-1.71011

Pelancongkini.blogspot (2012). Kenyir cup 2012 international fishing competition. Retrieved March 11, 2013, from http://pelancongan-kini.blogspot.com/2012/10/kenyir-cup-2012-international-fishing.html

Pickering, H. & Whitmarsh, D. (1996). Artificial reefs and fisheries exploitation: A review of the 'attraction versus production' debate, the influence of design and its significance for policy. *Fisheries Research*, 31, 39–59.

Rashid, R. (2011, October). Family fun fishing. *Sirip*,. 28.

Reid, M. & Schwab, W. (2006). Barriers to sustainable development. *Journal of Asian and African Studies,* 41(5/6), 439–457.

Stabler, M.J. (1997). *Tourism and Sustainability: Principles to Practice*. Wallingford:CAB International.

Virtualmalaysia.com (2012). Kenyir lake. Retrieved March 11, 2013, from http://www.virtualmalaysia.com/destination/kenyir%20lake.html

Hospitality and Tourism – Sumarjan et al. (eds)
© 2014 Taylor & Francis Group, London, ISBN 978-1-138-00151-0

An overview of Yield Management practices among the front office employees in five-star hotels

D.A.A. Mohd Asri & N. Sumarjan
Universiti Teknologi MARA, Malaysia

ABSTRACT: Yield Management or YM is an integrated, continuous and systematic approach to maximize room revenue through the manipulation of room rates in response to forecast patterns of room demands. Numerous researches have been conducted about YM, however most of the researches involved the hotel revenue or yield managers. Limited research focuses on employee's perspective, specifically from the front office employees. The front office employees are responsible to sell the right product to the right customer, at the right time, for the right price via the right distribution channel with the ultimate goal to increase hotel revenues. It is important for them to know how YM function in order to maximize hotel revenue. Therefore, this study seeks to examine the YM practices among the five-star hotel front office employees in Malaysia. Additionally, the influence of the top management support on YM, and the effects on hotel performance will also be explored. This study is hoped to provide hoteliers with a better in-sights on how to improve their current YM practices to augment the hotel performance from the employees' perspectives. Finally, findings of this study will also add to the existing literature of YM that was dominantly targeted toward the management perspective.

Keywords: front office employees, revenue management, Yield Management

1 INTRODUCTION

The term Revenue Management (RM) and Yield Management (YM) was used interchangeably in the literature. Burgess & Bryant (2001) considered YM related with revenue derived from hotel rooms only, whereas RM may encompass all areas of hotel revenues. Hence, revenue management is seen as a development of YM and considered to be more strategic and broader. YM is "an integrated, continuous and systematic approach to maximizing room revenue through the manipulation of room rates in response to forecast patterns of room demands" (Jauncey, Mitchell & Slamet, 1995). Choi & Cho (2000) supported that YM applications were used mostly by hotel managers in order to maximize profits by offering seasonal demand for rooms.

However, hotel managers are not the only people who are responsible in YM. Without the existence of front office employees, reservationists and sales departments, the implementation can never be success (Jones & Hamilton, 1992). Thus, Donaghy, Beattie & McDowell (1997) corroborated that employees are required to have a thorough and precise understanding of YM if the hotels are to be successful.

2 LITERATURE REVIEW

2.1 *YM practices in the hotel industry*

The YM practices which are relevant to the front office employees consist of capacity management, overbooking controls, upgrading and up-selling.

2.1.1 *Capacity management*

Capacity management consists of selling a fixed perishable capacity within a given time horizon by controlling the availability of the products (Talluri & Van Ryzin, 2004). The two important aspects in managing capacity are average room rate and occupancy rate (Lockwood & Jones, 1990). Capacity management involves demand and supply forecasting (Gothesson & Riman, 2004; Vinod, 2004). Demand forecasting is an activity where an organization should have a detail data on past demand pattern of market segment and the methods to forecast current demand (Zeithaml, Bolton & Deighton, 2006).

2.1.2 *Overbooking control*

Overbooking control is one of the YM practices applicable to the front office employees. The overbooking process was conducted to reimburse cancellations and no-shows (Vinod, 2004). The no-show data is important in order to determine a sufficient overbooking policy (Donaghy et al., 1995). In line with that, oversales has resulted overbooking where it represent un-accommodated guest (walk out). The risks associated with walked guests are compensation, loss of goodwill and operational constraints.

2.1.3 *Upselling and upgrading*

Front office employees might have been empowered to up-sell or upgrade guest with various room types. Steinhardt & Gönsch (2012) stated upgrade is only fair if priority is given to customers who purchase higher quality products. In contrast, up-sell is a situation where customers are willing to pay for a higher price. Upgrading and up-selling can be beneficial if a hotel encounters a discrepancy between supply and demand.

2.2 *Top management support*

Top management support plays a crucial part in a hotel system. Tavitiyaman (2004) supported that employees developed positive attitudes, committed and will be satisfied to the organization when the top management are committed to their employees (Yuanlaie, 2011). There are several top management supports that are applicable to employees in making YM to be successful. Among those are training for employees, giving incentive and rewards, as well as developing yield culture (Gothesson & Riman, 2004).

2.2.1 *Training of employees*

The lack of employee training might be considered as a risk when using YM and therefore, training is an important aspect to make the system work (Zeithaml & Bitner, 2003). Thus, in ensuring a YM is successful, participation from the entire organization is required. Donaghy, et al. (1995) supported that extensive and specialized training will be required for the front line employees who deal directly with daily sales and inquiries.

2.2.2 *Incentives and reward schemes*

Babakus, Yavas, Karatepe & Avci (2003) argued that frontline employees perceived the effort of establishing and providing rewards for exceptional service performance as genuine commitments of the management. Berry, Parasuraman & Zeithaml (1991) also stated that appropriate reward policies is important to encourage frontline employees to deliver high-quality services. In the YM context, a productive incentive scheme incorporates incentive points that are directly related to the sales (Donaghy et al., 1995).

2.2.3 *Communication and interdepartmental co-operation*

Communication is an important aspect for the top level managers in order to disseminate the right instructions to the employees. With an effective communication, the risk of YM practices will be minimized. Some scholars have argued that communication may be the most critical determinant of damage control (Argenti, 2002; Reilly, 1993).

2.2.4 *Developing yield culture*

The term "YM culture" or "yield culture" is often used to encompass the importance of people in the YM systems. According to Jones & Hamilton (1992) and Brotherton & Mooney (1992), the first stage in implementing a YM culture is to ensure that it is fully understood by the organization. A yield culture should be practiced throughout the organization (Brotherton & Turner, 2001). Therefore, the entire organization, should have sufficient knowledge of the system and concept (Göthesson & Riman, 2004).

3 ISSUES

Majority of the previous studies had taken YM practices into account among top level managers or revenue managers. However, little is known about YM focusing on employee's perspectives. Therefore, this study proposes to identify the YM practices among the five-star hotel front office employees. It is important for these employees to have knowledge about YM as they are directly involve with the YM process. Emeksiz, Gursoy & Icoz (2006) stated that the front line employees had a limit authority in decision making related to YM and this could lead to delay in responding to guests.

Consequently, the employees must gain knowledge on YM and its related practices to avoid delays in decision making processes (Hansen & Eringa, 1998). Shoemaker (2003) stated if YM is practiced indiscriminately, it can have an adverse effect on customers' perceptions on loyalty and hotel performance. Thus, this study is proposed to examine the YM practices implemented by the front office employees and how it contributes toward hotel performance.

4 METHODOLOGY

This study will use a quantitative approach with semi-structured questionnaire as the means to obtain data. The sample will be the front office employees holding a non-managerial position working in Kuala Lumpur five-star hotels. All the

Kuala Lumpur hotels listed in the recent Malaysia Association of Hotels Directory will be invited to participate in the study. A drop-off method will be used. Data will be analyzed using the descriptive analysis and other relevant statistical tests.

5 CONCLUSION

Practically, this study will provide some insights to the hoteliers on how to better improve their YM practices and offer better solutions for the top management on how to succeed in implementing YM. Academically, this study will add to existing literature on current YM practices from the front office employee's perspectives.

REFERENCES

Argenti, P. (2002). Crisis communication: Lessons from 9/11. *Harvard Business Review, 80*(12), 103–109.

Babakus, E. Yavas, U. Karatepe, O.M. & Avci, T. (2003). The effect of management commitment to service quality on employees' affective and performance outcomes. *Journal of the Academy of Marketing science, 31*(3), 272–286.

Brotherton, B. & Mooney, S. (1992). Yield Management-progress and prospects. *International Journal Hospitality Management, 11*(1), 23–32.

Brotherton, B. & Turner, R. (2001), Introducing yield management systems in hotel: Getting the technical/human balance right. *Journal of service Research, 1*(2), 25–47.

Burgess, C. & Bryant, K. (2001). "Revenue Management—the Contribution of the Finance Function to Profitability". International *Journal of Contemporary Hospitality Management, 13*(3), 144–150.

Choi, T.Y. & Cho, V. (2000). Towards a knowledge discovery framework for yield management in the Hong Kong hotel industry. *International Journal of Hospitality Management, 19*(1), 17–31.

Donaghy, K. McMahon-Beattie, U. & McDowell, D. (1995). Yield Management: An overview. *International Journal of Hospitality Management, 14*(2), 139–50.

Donaghy, K. McMahon-Beattie, U. & McDowell, D. (1997). Implementing yield management: Lessons from hotel sector. *International Journal of Contemporary Hospitality Management, 9*(2), 50–54.

Emeksiz, M. Gursoy, D. & Icoz, O. (2006). A yield management model for five-star hotels: Computerized and non-computerized implementation. *International Journal of Hospitality Management, 25*, 536–551.

Göthesson, L. & Riman, S. (2004). Revenue management within Swedish Hotels. Tourism and Hospitality Management Master Thesis No 2004:61.

Hansen, C.N. & Eringa, K. (1998). Critical Success Factors in Yield Management: A development and analysis, progress in tourism and hospitality research, 4229–4244.

Jones, P. & Hamilton, D. (1992). Yield Management: Putting people in the big picture. *The Cornell HRA Quarterly, 33*(1), 88–95.

Jauncey, S. Mitchell, I. & Slamet P. (1995). The meaning and management of yield in hotels. *International Journal of Contemporary Hospitality Management,* 7(4), 23–26.

Parasuraman, A. Berry, L.L. & Zeithaml, V.A (1991). Perceived service quality as a customer-based performance measure. An empirical examination of organizational barriers using extended service quality models. *Human Resource Management, 30*(3), 335–364.

Reilly, A.H. (1993). Preparing for the worst: The process of effective crisis management. *Industrial & Environmental Crisis Quarterly, 7*(2), 115–143.

Shoemaker, S. (2003). Future of revenue management: The future of pricing in services. *Journal of Revenue and Pricing Management 2*(3), 271–279.

Steinhardt, C. & Gönsch, J. (2012). Integrated revenue management approaches for capacity control with planned upgrades. *European Journal of Operational Research, 223*(2), 380–391.

Talluri, K.T. & van Ryzin, G.J. (2004). The theory and practice of revenue management. Kluwer Academic Publishers.

Tavitiyaman, P. (2004). The effect of management commitment to service quality on employees' job satisfaction and prosocial service behaviors. (Order No 1424886, Oklahoma State University). *ProQuest Dissertations and Theses,* 87–87 p.

Vinod, B. (2004). Unlocking the value of revenue management in the hotel industry. *Journal of Revenue and Pricing Management*, *3*(3), 178.

Yuanlaie, S. (2011). The emotional exhaustion, motivation and job outcomes: A study of hotel frontline employees in phuket. (Order No 3498123, Oklahoma State University). *ProQuest Dissertations and Theses,* 110.

Zeithaml, V.A. & Bitner, M.J. (2003). Services Marketing. Integrating Customer Focus Across the Firm. New York: McGraw-Hill.

Zeithaml, V.A, Bolton, R.N. & Deighton, J. (2006). Forward-looking focus can firms have adaptive foresight? *Journal of Service Research, 9*(2), 168–183.

Hospitality and Tourism – Sumarjan et al. (eds)
© 2014 Taylor & Francis Group, London, ISBN 978-1-138-00151-0

Work motivation and job satisfaction of middle level management at five-star hotels in Kuala Lumpur

M.F. NurDalila, S.M. Radzi & Z. Othman
Faculty of Hotel and Tourism Management, Malaysia

ABSTRACT: This paper examines the relationship between work motivation (intrinsic, extrinsic, identified and introjected motivation) and job satisfaction of middle level management at five-star hotels in Kuala Lumpur. Self-administered questionnaires were used. Respondents are limited to the operational managers of five-star hotels in Kuala Lumpur. For data evaluation, SPSS program used, and regression analysis carried out to determine the relationship of the variables. The result revealed that there is a positive relationship between work motivation and job satisfaction. This study provides empirical findings on work motivation and job satisfaction from the participating managers. Intrinsic was the most agreeable motivation followed extrinsic motivation, introjected motivation, and identified motivation. Additionally, a moderate and positive correlations was found between work motivation and job satisfaction.

Keywords: extrinsic motivation, hotel, intrinsic motivation, work motivation, job satisfaction

1 INTRODUCTION

It is believed that the largest contributors for Malaysia are the hospitality and tourism industry. The rapid growth of the hotel industry has exaggerated the demand for employment of competent people. According to Travel and Tourism (2011), there will be a total of 1.6 billion jobs in the hotel industry during year 2020. Thus, there is a significant need in having a more qualified, motivated and skilled manpower to meet the challenges of this ever changing and demanding industry. This has become a challenge to the hotel industry due to the skilled and knowledgeable personnel shortage.

Malaysia is facing problems in attracting and retaining skilled and knowledgeable of high and low level employees due to low salary and rigid job traits. Hotel needs to encourage their employees especially among managers in terms of motivation to gain their confidence and loyalty. The hotel needs to ensure that managers are motivated and satisfied working in the organization as they handle top level management and low level management. It is proven that motivated and satisfied workforce can deliver powerfully and be more responsible to the bottom line.

There are four major factors of motivation that are intrinsic, extrinsic, introjected and identified. Numerous studies have been conducted about intrinsic and extrinsic types of motivation (Seiler, Lent, Pinkowska, & Pinazza, 2012), however limited study conducted on identified and introjected of extrinsic motivation (Moran, Diefendorff, Kim, & Liu, 2012). There are still limited studies on different types of motivational at work scale that lead to job satisfaction. Therefore, this study attempts to fill the gap by deeply investigating the vital role of work motivation and its factors that effects on job satisfaction.

2 LITERATURE REVIEW

2.1 *Work motivation*

Motivation can be understood as an individual's level of readiness to perform an action and has become an important factor for employees to remain in the organization. Work motivation can be divided into four dimensions that are intrinsic, extrinsic, identified and introjected motivation. According to Seiler et al. (2012), intrinsic motivation is related to internal or personal factor such as the opportunity to use one's ability, a sense of challenge and achievement, receiving appreciation, positive recognition and treated in a caring and considerate manner. Extrinsic motivation relates to tangible and external rewards such as salary, fringe benefits, security, promotion, contract of service, the work environment and conditions of work (Ryan & Deci, 2000). Gagne, Forest, Gilbert, Aube, Morin and Malorni (2010) asserted that identified motivation refers to someone that performs an activity to achieve a certain objectives and self-endorsement of goals. This also relates to personal values and identities. The last major dimension introjected motivation can be understood as the behavior involving self-worth and beliefs such as ego involvement (Koestner & Losier, 2002). These kinds of feelings can pressures people to behave in order to feel worthy in achieving success in the organization.

2.2 *Job satisfaction*

Job satisfaction refers to an employee's evaluation and the final feelings toward their jobs (Abdullah, Musa, Zahari, Rahman, & Khalid, 2011). Patah, Zain, Abdullah and Radzi (2009) stated that job satisfaction is the creation of job components that relate to the attitude of the employees in the organization such as good working environment, fairness of performance appraisal, enough authority and good relationship with the colleagues.

2.3 *Relationship between work motivation and job satisfaction*

Employee motivation and job satisfaction cannot be isolated, but they complement each other and respond to different organizational variables. Grenway (2007) informed that employees and managers work hard to receive the motivation and satisfaction to achieve success. Frye (2012) found that motivational factors may contribute to the job satisfaction. Job satisfaction depends on the employee behavior at the workplace which is determined by a set of intrinsic and extrinsic motivation. If employees satisfied with their jobs, they feel strongly motivated to improve their performance and make some positive changes.

3 RESEARCH METHODOLOGY

This study used the quantitative research design with a structured questionnaire as the research instrument. A descriptive research design and correlational approach were also used. Self-administered questionnaire was used. The Ministry of Tourism Malaysia (2013) listed twenty eight hotels categorized as five-star rated hotels in Kuala Lumpur. This study focused on the individual level of analysis involving the middle level management. To make sure the distribution of the questionnaires are equal between departments, at least two middle managers are required to answer the questionnaire, and this represents eight middle managers from operational department. The operational department involves front office, housekeeping, food production and food and beverage service departments.

A total of 224 questionnaires were distributed to the respondents via the Human Resource Department (HRD). Follow-up calls were done to make sure that the questionnaires received by the respondents. Within two weeks, the questionnaires collected from the HRD. A total of 158 questionnaires returned, and a total of 18 questionnaires found to be incomplete. Therefore, only 140 questionnaires coded for analysis. The questionnaires were pilot-tested to at least 30 managers and assistant managers in operational department at five-star hotels in Selangor. Data on the respondent demographic was also collected and included gender, age, qualification, designation, department and length of service. The questionnaire consists of three sections: Section A, B and C. Questions were prepared in English language as the sample of this study was among highly educated employees.

4 RESULTS AND DISCUSSIONS

Descriptive statistics such as mean, standard deviations, reliability coefficients and Pearson correlations computed. For the pilot study, the variables indicated good results above 0.70 of Cronbach's Alpha. The results for the pilot study considered acceptable and reliable. The results for the actual data represents high reliability standard above 0.80. Reported mean score shows that most of the respondents strongly agreed with the statements of intrinsic motivation ($M = 5.51$, $SD = 0.85$), extrinsic motivation ($M = 5.13$, $SD = 0.99$), introjected motivation ($M = 5.69$, $SD = 0.85$) and identified

Table 1. Measures of the study variables.

Study variables	No of items	Sources of scale	Type of scale
Work motivation – Intrinsic – Extrinsic – Identified – Introjected	30	Trembley et al. (2009); Gagne et al. (2010)	Seven-point Likert type scale from (1) "not at all" to (7) "exactly"
Job satisfaction	12	Ogaard (2006); Abdullah et al. (2011)	Five-point Likert type scale from (1) "strongly disagree" to (5) "strongly agree"

£ 129·00

motivation ($M = 5.52$, $SD = 0.96$). For the overall mean score of work motivation ($M = 5.46$, $SD = 0.84$) it can be seen that the values were quite high. The results indicate that the respondents acquired work motivation in the workplace. Reported mean score for job satisfaction depicts the highest mean score ($M = 4.34$, $SD = 0.57$) was on job responsibilities and the lowest mean score ($M = 3.61$, $SD = 1.10$) was on performance appraisal. Most of the mean score inclined to the scale of agree. It can be concluded that the respondents received satisfaction in the hotel. Table 2 exhibits the respondents' demographic profile that involved 140 respondents.

The Pearson correlations tested the intercorrelations of the study variables. Moderate and positive correlations were found between work motivation and job satisfaction ($r = 0.59^{**}$, $p < 0.01$). This study rejected the null hypotheses and accepted the alternate hypotheses.

Table 2. Frequencies of the respondents' demographic profile.

Characteristics	Frequency (N = 140)	Percentage (%)
Gender		
Male	76	54.3
Female	64	45.7
Age		
20–30 years	37	26.4
31–40 years	46	32.9
41–50 years	37	26.4
Above 50	20	14.3
Qualification		
SPM/STPM	13	9.3
Diploma	43	30.7
Degree	78	55.7
Master and above	6	4.3
Designation		
Manager	76	54.3
Assistant manager	64	45.7
Department		
Front office	56	40
Housekeeping	39	27.9
Food and beverage	45	32.1
Years in organization		
0–2 years	34	24.3
3–5 years	42	30
6–8 years	31	22.1
9 years and above	33	23.6
Years in position		
0–2 years	36	25.7
3–5 years	51	36.4
6–8 years	20	14.3
9 years and above	33	23.6

Previous study by Ayub and Rafif (2011) in a different setting reported that the correlation between work motivation and job satisfaction significantly related ($r = 0.56$, $p < 0.01$) on middle managers working in banks. This also supported by regression analysis on work motivation and job satisfaction where work motivation made a statistically significant contribution to the job satisfaction, and it explained 35% of the variation in the model.

5 STUDY IMPLICATIONS

This study provides information to hoteliers on how to recognize the critical factors that most of the managers wants and needs in the workplace. Manager plays an essential role in the hotel industry as they are going to handle the department and lower level employees. If the managers become demotivated, this will affect the hotel's performance. No doubt that work motivation is vital to everyone because it can create different positive effects such as loyalty, satisfaction, commitment and turnover towards organization. This study can obviously contribute in presenting a clear and precise understanding of work motivation and job satisfaction in the hotel industry. Not only that, the variable of work motivation can be used to define and to execute the training management practices in the hospitality industry. The human resource department will be able to identify highly motivated middle level managers that will be an asset to a particular hotel.

It is hoped that the findings of this study can be applied in the real situation and extending the existing body of literature in hospitality research. It will also form a basis to clarify the fact and findings in instilling more interest in future research. This research serves as resource materials for the management of hotels and provides benefit to others concerned with the managers' quality performance. In this study, insights into the possible predictors for job satisfaction are provided, which can serve as a tool for the human resource management in hospitality organizations. Besides that, this study contributes not only for five-star hotels in Kuala Lumpur but also for other star rating hotels because the results of this study will help the other hotels to understand the importance of these variables. It can be applied to the other employees because they faced the same environment and situation. The findings from this study should also provide an academic platform with valuable information and insights on the current situation of the industry, thus strengthening current literatures and findings. The students and academic profession may also be able to identify the appropriate ways of work motivation to adapt with the current trend.

6 LIMITATION

The first possible limitation related to the sample size, which only focuses on the operational departments. This involved a small number of respondents, and the demographic setting very limited. Besides that, quantitative approach provides limited information to this study. This study also limited to the four types of motivational factors.

7 SUGGESTIONS FOR FUTURE RESEARCH

Future research is suggested to include the employees from the back of the house. Other star ratings hotels involving three-star and four-star hotels could also be considered to get a larger population of managers. Besides that, this study could be extended to the other states in Malaysia involving managers so that later, the results could be generalized to the other country. Interview or qualitative approach could be used to gain opinion, information and better feedback from the respondents. It is recommended that the study explore other types of motivation such as external regulation and integrated regulation. Different types or dimensions of motivation provide different meaning that should be investigated to know more on employees' motivation. There will be more findings if the times provided are much longer. This study has proven that work motivation relates to job satisfaction and can affect turnover intention. It will be beneficial for future research to study the relationship of work motivation, job satisfaction and turnover intention.

8 CONCLUSION

As a conclusion, hotel industry need to improve and recognize potential employees to increase good performance that later will give a good image to the industry. The manager and assistant manager should be involved in decision making process to encourage them to use their own skills and abilities. By providing better working condition, flexible working hours and well paid, this will help them to have a sense of self pride, competence and self confidence that increases job satisfaction. This study has discussed on today's issues phenomenon and trends that the hotel should take a proactive action by providing necessary needs and wants of the employees. Manager have different needs and wants because they hold a big responsibility in the department. Hotels should find out the needs and wants of the managers even it is impossible to satisfy their needs and expectations. Discussions and events should be organized to share any problems, issues and opinions from the managers so that the hotel can improve their productivity and performance. Both employee and employer should try to generate better working conditions that will create a happy conducive and motivated atmosphere to reach the goals.

REFERENCES

Abdullah, R., Musa, M., Zahari, H., Rahman, R., & Khalid, K. (2011). The study of employee satisfaction and its effects towards loyalty in hotel industry in Klang Valley, Malaysia. *International Journal of Business and Social Science, 2*(3), 147–155.

Ayub, N. & Rafif, S. (2011). The relationship between work motivation and job satisfaction. *Pakistan Business Review*, 332–346.

Frye, W.D. (2012). An examination of job satisfaction of hotel front office managers according to extrinsic, intrinsic and general motivational factors. *International Journal of Business and Social Science, 3*(18).

Gagne, M., Forest, J., Gilbert, M., Aube, C., Morin, E.M., & Malorni, A. (2010). The motivation at work scale: Validation evidence in two languages. *Educational and Psychological Measurement, 70*, 628–646.

Grenway, B. (2007). The relationship between employee motivation and job satisfaction of African-American human service employees. Walden University.

Koestner, R. & Losier, G.F. (2002). Distinguishing three ways of being highly motivated: A closer look at introjection, identification, and intrinsic motivation. In E. Deci, & R. Ryan (Eds.). Handbook of self-determination research (pp. 101–121). Rochester, NY: University of Rochester Press.

Ministry of Tourism Malaysia (2013). Rated Tourist Accommodation Premises. Retrieved from http://www.motour.gov.my/en/check/registered-hotel.html.

Moran, C., Diefendorff, J.M., Kim, T.Y., & Liu, Z.Q. (2012). A profile approach to self-determination theory motivations at work. *Journal of Vocational Behavior, 81*, 354–363.

Ogaard, T. (2006). Do organizational practices matter for hotel industry employee's jobs? A study of organizational practice archetypical configurations and job outcomes. *International Journal of Hospitality Management, 25*, 647–661.

Patah, M.O.R.A., Zain, R.A., Abdullah, D., & Radzi, S. (2009). An empirical investigation into the influences of psychological empowerment and overall job satisfaction on employee loyalty: The case of malaysian front office receptionists. *Journal of Tourism, Hospitality and Culinary Arts, 3*(1), 43–62.

Ryan, R.M., & Deci, E.L. (2000). Intrinsic and Extrinsic Motivations: Classic Definitions and New Directions. *Contemporary Educational* Psychology, 25, 54–67.

Seiler, S., Lent, B., Pinkowska, M., & Pinazza, M. (2012). An integrated model of factors influencing project managers' motivation-Findings from a Swiss Survey. *International Journal of Project Management, 30*, 60–72.

Travel and Tourism Council (2011). The Authority on World Travel and Tourism. World Travel and Tourism Council, 35–40.

Trembley, M.A., Blanchard, C.M., Taylor, S., & Pelletier, L.G. (2009). Work extrinsic and intrinsic motivation scale: Its value for organizational psychology research. *Canadian Journal of Behavioral Science, 41*(4), 213–226.

Hospitality and Tourism – Sumarjan et al. (eds)
© 2014 Taylor & Francis Group, London, ISBN 978-1-138-00151-0

Corporate Social Responsibility (CSR) activities, brand image and hotel guest retention

N.Z. Othman & M.A. Hemdi
Universiti Teknologi MARA, Malaysia

ABSTRACT: Guest retention is a major consideration for hotel competitive advantage. Corporate Social Responsibility (CSR) activities in hotel industry have been found to have significant influence towards hotel profits, employee commitment and good organizational image. The purpose of this study is to investigate the relationship between CSR activities, hotel brand image and guest retention. Data for this study were gathered through self-administered questionnaires on hotel guests who experienced staying in Kuala Lumpur five-star rated hotels. Findings showed that CSR activities have significant positive influence toward hotel brand image and guest retention. Hotel brand image has also been found to be significant mediator between CSR activities and guest retention. Theoretical and practical implications of the findings were discussed. Limitations and future research approaches were forwarded.

Keywords: CSR activities, brand image, guest retention

1 INTRODUCTION

Tourism industry is the seventh largest contributor to the Malaysian economy with a Growth National Income (GNI) of RM37.4 billion in 2011 and is poised to enjoy consistent growth (Economy Transformation Programme Annual Report, 2011). Currently, Malaysia is the ninth most visited country in the world and received RM1 billion receipts per week from foreign visitors (Economy Transformation Programme Annual Report, 2011). By the year 2020, Malaysia is expected to receive 36 million tourist arrivals contributing to RM168 billion tourist receipts (Economic Transformation Programme Annual Report, 2011). The developments of tourism industry have increased hotel room supply. As more hotels are built, competitions for guests become stiffer. Hotels need to be innovative and proactive in attracting guest in order to compete effectively. Within this context, guest retention is a major consideration for competitive advantage. Thus, management of hotels needs to find ways and strategies to increase guest retention. Besides good services and excellent physical facilities, Corporate Social Responsibility (CSR) activities of the hotel towards environment and community are also important in influencing guest retention. Guest can then choose to evaluate a hotel based on the organization acts in a manner consistent with supporting the welfare of the community and society. Previous researchers agreed that guest expect organizations to match some of their social values as part of their contributions to the community (Marquis, Glynn, & Davis, 2007). Most studies on CSR were conducted in the Western countries. Limited studies have been conducted in Asian countries especially in Malaysia (Lines, 2004: Moon, 2005). Nevertheless, the focus of previous studies was not in hotel settings. Therefore, the problem in this study is to determine factors that influence guest retention in the hotel industry. Specifically, this study is to determine whether CSR activities have significant positive influenced on hotel brand image and guest retention.

2 LITERATURE REVIEW

2.1 *Guest retention*

Customer retention is defined as the process when customers continue to buy products and services within a determine time period (Narayandas, 1996; Prus & Brandt, 1995). Customer retention occurs when a customer is loyal to a company, a brand or a specific product or service, expressing long-term commitment and refusing to purchase from competitors. In the context of this study, guest retention is defined as intention to repurchase and intention to spread word of mouth about a particular hotel. This is in line with many previous studies that have used intention to repurchase

(Almohammad, 2010; Yen & Lu, 2008) and intention to spread word of mouth (Nadiri, Ekiz, & Erdagon, 2008) as elements of guest retention. Repurchase intention are usually obtained from surveys of current customers assessing their tendency to purchase the same brand, same product or service from the same company.

2.2 *CSR activities*

The CSR terms generally applies to company efforts that go beyond what may be required by regulators or environmental protection groups. According to Thomas and Nowak (2006), CSR involves the corporate sustainability, corporate citizenship, corporate social investment and corporate governance. CSR is one of the initiatives by the corporate to assess and take responsibility towards the environment and its impact on social welfare. Mohr, Webb and Harris (2001) defined CSR as company's commitment to minimizing or eliminating any harmful effects and maximizing its long-run beneficial impact on society. Whooley (2004) classified CSR into four categories namely environment, community, workplace and marketplace. For the purpose of this study, CSR dimensions pertaining to the environment and community were chosen since the subjects of the current study were the hotel guests.

2.3 *Hotel brand image*

Brand image is defined as perceptions about a brand as reflected by the brand associations held in consumer memory (Keller, 1993). Through this, we can see that a good brand images are instantly evoked, are positive and are almost always unique among competitive brands. In the context of hotel settings, Berry (1995) and Levitt (1983) argued that guests leave a hotel with only the memories of their experiences. Hence, the challenge for hotel operators is to ensure services provided to guests are able to create long lasting positive perceptions in guests' mind. One way to achieve that challenge is through creating a strong brand image which stands superior compared to the competitors in terms of tangible and intangible characteristics of the hotel products. The main idea behind CSR is that all activities undertaken by a company affect the way consumers perceive that company which formed brand image. Although all activities affect the reputation, the company must acknowledge that the activities deliberately undertaken to influence the brand must be accomplished in the name of the brand (Blumenthal & Bergstrom 2003).

Based from the above literature review, hypotheses of this study were offered.

H1: There is significant positive relationship between CSR activities (environment and community) and guest retention.

H2: There is significant positive relationship between CSR activities (environment and community) and hotel brand image.

H3: There is significant positive relationship between hotel brand image and guest retention.

H4: Hotel brand image mediates the relationship between CSR activities (environment and community) and guest retention.

3 METHODOLOGY

Data for this study were gathered through self-administered questionnaires. The survey questionnaire was personally distributed to the respondents which were the guest who had have experiences staying at 24 five-star rated hotels in Kuala Lumpur. From the 600 questionnaires distributed, a total of 300 questionnaires were returned, yielding a response rate of 50%. 17 questionnaires were found to be unusable because of incomplete information or the respondents were not the subject of focus in this study. Therefore, only 283 questionnaires (47.2%) were coded and analyzed. Descriptive statistics and regression analysis were used to analyze the data. Guidelines on mediation by Baron and Kenny (1986) were used to test mediation effect of brand image.

4 RESULTS

4.1 *Profile of respondents*

Out of the 283 respondents, 277 respondents (97.9%) were Malaysian. Majority of the respondents were female (54.1%) and the age ranges were between 21–30 years old (49.5%). On the frequency of staying at the hotel, most respondents (114 respondents or 40.3%) had stayed for the first time. Majority of the respondents (148 respondents) stayed at the hotel for 2 nights.

4.2 *Descriptive statistics*

From Table 1, mean for the study variables ranged from 3.53 to 3.99. Respondents in this study moderately perceived CSR activities practiced by the hotels. Similarly, hotel brand image was also moderately perceived by the respondents. Respondents in this study highly agreed that they would be retuning back to the hotel. In terms of inter correlations of the variables, Table 1 shows that all the study variables were significantly and positively correlated amongst each other. None of the observed correlation reached 0.90 which confirmed the non-existence of multi-collinerity. The Cronbach's Alpha for the study variables ranged from 0.61 to 0.92. Table 1 show the results of the mean and inter-correlation of the study variables.

4.3 *Hypothesis testing*

From Table 2, it was found that CSR activities were able to explain 38.0% ($R^2 = 0.38$, F-Change = 34.02, $p < 0.01$) of the observed variance in guest retention. Table 2 shows the results of the regression analysis of CSR activities on guest retention.

From Table 3, result it was found that all CSR activities except for energy supply all CSR activities except for energy supply was found significantly and positively related to hotel brand image. Table 3 shows the results of the regression analysis of CSR activities on hotel brand image.

Table 4 shows the results of the regression analysis of hotel brand image and guest retention. Result indicates that hotel brand image ($\beta = 0.54$, $p < 0.01$) able to explain the 29.0% ($R^2 = 0.29$, F-Change = 116.76, $p < 0.01$) of the observed variation on the guest retention.

Table 5 shows the summary of regression for the mediation effect of brand image. Results from Table 5 shows that, the effect of involvement in community project on guest retention became insignificant in the presence of hotel brand image, thus, full mediation occurred. Beta value and significant level of environmental practices and community involvement in quality of life decreasing in presence of hotel brand image. Therefore, this result implies partial mediation.

Table 1. Mean and correlation for study variables.

Variables	*M*	1	2	3	4	5	6	7
1. Sustainable development	3.71	–						
2. Environmental practices	3.62	0.62**	–					
3. Energy supply	3.42	0.42**	0.46**	–				
4. Involvement in community project	3.70	0.50**	0.46**	0.38**	–			
5. Community involvement in quality of life	3.53	0.21**	0.25**	0.40**	0.43**	–		
6. Hotel brand image	3.79	0.52**	0.54**	0.42**	0.66**	0.41**	–	
7. Guest retention	3.99	0.34**	0.42**	0.35**	0.44**	0.52**	0.54**	–

Note: $N = 283$, $^*p < 0.05$; $^{**}p < 0.01$, Five-point Likert Scale.

Table 2. Hierarchical regression of CSR activities and guest retention.

Predictors	Model 1/std. β
Step 1: CSR activities	
Environment	
• Sustainable development	0.03
• Environmental practices	0.24**
• Energy supply	0.02
Community	
• Involvement in community project	0.15*
• Community involvement in quality of life	0.39**
R^2	0.38
Adj. R^2	0.37
R^2 Change	0.38
F-Change	34.02**

Note: $N = 283$, $^*p < 0.05$; $^{**}p < 0.01$, Five-point Likert Scale.

Table 3. Hierarchical regression of CSR activities and hotel brand image.

Predictors	Model 1/std. β
Step 1: CSR activities	
Environment	
• Sustainable development	0.13*
• Environmental practices	0.20**
• Energy supply	0.60
Community	
• Involvement in community project	0.43**
• Community involvement in quality of life	0.12*
R^2	0.53
Adj. R^2	0.53
R^2 Change	0.53
F-Change	63.24**

Note: $N = 283$, $^*p < 0.05$; $^{**}p < 0.01$, Five-point Likert Scale.

Table 4. Hierarchical regression of hotel brand image and guest retention.

Predictors	Model 1/std. β
Step 1: Study variables	
Hotel brand image	0.54**
R^2	0.29
Adj. R^2	0.29
R^2 Change	0.29
F-Change	116.76**

Note: $N = 283$, $*p < 0.05$; $**p < 0.01$, Five-point Likert Scale.

Table 5. Summary of regression for the mediation effect of brand image on the relationship between CSR activities and guest retention.

Predictors	Model 1/ std. β	Model 2/ std. β
Step 1: CSR activities		
• Environmental practices	0.26**	0.17*
• Involvement in community project	0.16**	0.02
• Community involvement in quality of life	0.39**	0.35**
Step 2: Mediating variable		
Hotel brand image		0.30**
R^2	0.38	0.42
Adj. R^2	0.37	0.41
R^2 Change	0.38	0.04
F-Change	56.91**	20.55**

Note: $N = 283$, $*p < 0.05$; $**p < 0.01$, Five-point Likert Scale.

5 DISCUSSIONS

The objective of this study shows that CSR activities and hotel brand image have positive and significant influence on guest retention. Results of this study indicated that CSR activities pertaining to environmental practices, sustainable development, involvement in community projects and community involvement in quality of life have significant and positive influence on hotel brand image and guest retention. Hotel brand image was also found to be significant mediator on the relationship between CSR activities and guest retention.

Guest who has higher perception of the hotel involvement in improving local quality of life will express higher intentions to return. Specifically, if the hotel were perceived to have employed local workforce and bought products made by the local people, they would expressed higher intentions to comeback. These findings were in lined with Moir (2010) who found that hotel involvements in improving local quality of life have significant influenced on guest retention. Further, community projects such as sponsoring local events, donating for local programs and volunteering in local activities were significantly influenced guest intentions to return, in lined with Scofidio (2007) study. Respondents in this study also expressed positive brand image and higher intentions to comeback when they believed the hotel have implemented friendly environmental practices such as having water saving devices in bathroom, using recycle packaging for room amenities and using environmental friendly products. Similarly, if the guests perceived that the hotel shown concerns toward conserving the environment through saving the energy, using pleasant environmental signage in every room and using environmental friendly products, they will formed a high positive perception toward the hotel and consequently influenced them to return again. These findings were similar with previous researchers (Jones, Comfort, & Hillier, 2006; Lee & Park, 2009).

On theoretical implications this study study confirmed the linkage between CSR activities, hotel brand image and guest retention. This study also gave evidence to the importance of hotel brand image as the intervening variable from CSR activities predictor and guest retention. By demonstrating the existence of significant direct and indirect effects of CSR activities, hotel brand image and guest retention, this study provides clear evidence that CSR activities are important in fostering guest retention in hotel industry. Practically, the results from this study will help hotel managers and operators to plan and implement strategies pertaining to CSR activities, this study suggests that hotel operators should actively involve and organize community projects in order to improve the local community quality of life. Activities such as sponsoring public community facilities and events, hiring more local people to work, and buying products from local community would help to create a strong positive image about the hotel and subsequently would influenced guest to spread positive word of mouth about the hotel.

6 LIMITATIONS AND SUGGESTIONS

The CSR activities investigated in this study were limited in scope. Only CSR activities pertaining to environment and community were examined. Future researchers may need to widen the scope of investigations (for example, workplace and marketplace).

Secondly, this study was limited to guest who had experienced stayed at Five-star rated hotels. Future researchers may want to include guest from three- or four-star rated hotels from other locality.

REFERENCES

Almohammad, A. (2010). The role of brand equity in the effects of corporate social responsibility on consumer loyalty.

Baron, R.M., & Kenny, D.A. (1986). The moderator-mediator variable dinstiction in social psychological research: Conceptual, strategic amd statistical consideration. *Journal of Personality and Social Psychology*, *51*, 1173–1182.

Berry, L. (1995). Relationship marketing of services-growing interest, emerging perspective. *Journal of the Academy of Marketing Science*, *23*(4), 236–245.

Blumenthal, D., & Bergstorm, A. (2003). Brand councils that care: towards the convergence of branding and corporate social responsibility. *Journal of Brand Management*, *10,* 327–341.

Economy transformation programme annual report. (2011). Tourism Program. Retrived December 23, 2012, from http://www.epu.gov.my/tourism.

Keller, K. (2008). Strategic brand management: Building, measuring and managing brand equity (3rd Ed.): Pearson International Edition.

Levitt, T. (1983). The globalization of markets. *Harvard Business Review*, *61*, 92–102.

Lines, V.L. (2004). Corporate reputation in Asia: Looking beyond bottom-line performance. *Journal of Communication Management*, *8*(3), 233–245.

Marquis, C., Glynn, M.A., & Davis, G.F. (2007). Community Isomorphism and Corporate Social Action. *Academy of Management Review*, 1–48.

Mohr, L.A., Webb, D., & Harris, K. (2001). Do consumers expect companies to be socially responsible? The impact of corporate social responsibility on buying behaviour. *Journal of Consumer Affairs*, *35*(1), 45–71.

Moir, L. (2001). What do we mean by corporate social responsibility? *Corporate Governance*, *1*(2), 16–22.

Nadiri, H., Hussain, K., Ekiz, E.H., & Erdagon, S. (2008). An investigation on the factors influencing passengers' loyalty in the North Cyprus National Airline. *The TQM Journal*, *20*(3), 265–280.

Narayandas, N. (1996). The link between customer satisfaction and customer loyalty: An empirical investigation. *Harvard Business School*, 97–017.

Prus, A., & Brandt, D.R. (1995). Understanding your customers. *American Demographics*, 11.

Thomas, G. & Nowak, M. (2006). Corporate social responsibility: A definition. *Working Paper Series*, *62*, 1–20.

Whooley, N. (2004). Social Responsibility in Europe. Retrieved on August 22, 2012 from www.pwc.com/extweb/newcolth.nsf/0/503508DDA107A61885256F35005C1E35.

Yen, C.H., & Lu, H.P. (2008). Factors influencing online auction repurchase intention. *Internet Research*, *18*(1), 7–25.

Hospitality and Tourism – Sumarjan et al. (eds)
© 2014 Taylor & Francis Group, London, ISBN 978-1-138-00151-0

Strategies of travel agency in facing disintermediation

N.A.A. Nik Hashim, M.N.I. Ismail, Z. Awang & F.H. Muhammad Safri
Universiti Teknologi MARA, Malaysia

ABSTRACT: Tourism business has changed due to the usage of internet as it offers a new distribution channel whereby numerous people are doing their transaction through online and they have become passionate to this activity. New technologies have help the people to access rapid internet connection, make secure transactions, and provide easier and fast services which can be the causes of disintermediation in tourism industry. This is due to the fact that technology has allowed the people to communicate and contact directly with the suppliers. Seeing disintermediation as the main issue and problem here, this paper aim to identify strategies used by travel agencies in maintaining their business if they have been excluded by the tourism suppliers in Malaysia. A qualitative approach through face to face interview with managers of travel agent is considered to be the most suitable method for investigating this issue. The result revealed that, disintermediation has given positive and negative impact to travel agent. Travel agent need to take the threat of disintermediation as their opportunity to build up their business. The finding also shows that every travel agency has their own strategies in facing disintermediation. However, the implementation of website and e-commerce practices is the most useful strategy for travel agents to survive in future business.

Keywords: disintermediation, e-commerce, strategy, survive, technology, travel agent

1 INTRODUCTION

The internet is a technology that offers worldwide coverage and multimedia capability and becoming an important tool as means for promotion and distributing tourism services. The current statistics about the progress in e-tourism suggests that the tourism distribution chain is undergoing a technological revolution, which has brought a redefinition of the consumers, intermediaries and suppliers. (Abadzhiev, Euba, & Zeeshan, 2005). As a large number of the hospitality industry has use internet as their marketing tools and distribution channels, the relationship between the travel agencies and the hospitality suppliers has become complex. Since the tourism supplier has utilized the internet, the rivalry between the provider and also the travel agent has increased. The revolution of online booking websites like Expedia and Travelocity has changed the way on how consumers make a booking, and choose hotels and airlines (Carroll & Siguaw, 2003). As a result the business of travel agent has changed over the past ten years due to the drastic reduction in their income (Chung & Lam, 2009).

In the case of the relationship between hotels and travel agents, the hotels' initiatives to use new channel that allow booking at discounted rates would be perceived by its agent as threatening their survival in the business (Tse, 2003). In order to survive the business, most of the hotels cut the commissions that will be paid to the travel agents by building their own websites, and allow their guest to purchase room through online in a lower rate (Neuman, Hutzler, & Cohen, 2003). This is similar in the airlines industry where airliners paid travel agencies a commission on the issuance of a ticket. Nevertheless, nowadays most of the airlines have stopped paying commissions to travel agent in order to save costs. The same goes to other services such as buses, trains and restaurant. The importance of travel agents responsibility is reduced by the internet and ability of online booking which is called e-commerce.

Thus, this type of competition has directly led to a threatening situation known as disintermediation for the travel agent in tourism industry. Disintermediation has affected the revenue of travel agent since they are no longer being paid the commission from the suppliers such as airlines, hotel, cruise and ferry and bus operators (S. Joseph, personal communication, November 2, 2012). Airline companies encourage clients to call the company directly, cutting commissions and disintermediation of their travel agents. Hotels also develop their own websites to get their clients and other than that, they can now employ various price policies and loyalty programs that yield a

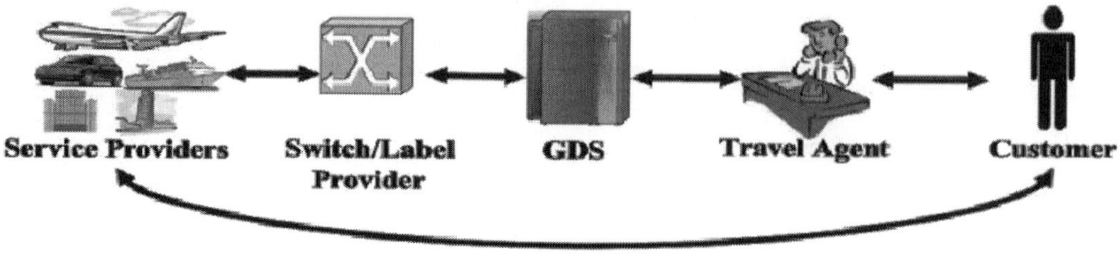

Figure 1. Traditional travel distribution (Amdekar and Padmanabhuni, 2006).

higher price transparency and more user-friendly interfaces (Patricia, n.d).

Since disintermediation is defined as the removal of supply chain intermediaries or third parties, expedited by the internet factors, (Kaewkitipong, 2011) therefore the travel agent cannot depend on the industries such as hotel, transportation and food and beverages anymore. They need to re-intermediate themselves by providing value-added services based on customer's needs and wants. Seeing disintermediation as the main issue and problem that has increased here, the researcher wants to identify the strategies used by travel agencies if they have been excluded by the tourism suppliers in Malaysia.

2 LITERATURE REVIEW

2.1 *Disintermediation*

Disintermediation was first well-known by this term and recognized as a problem in the second and third quarters of 1966. In the late 1990s when the introduction of internet-based shopping came, the term became a buzzword signifying the elimination of middlemen as a result of direct to consumere-commerce methods (King, 1999). In the travel and tourism industry, disintermediation has been known as the elimination of supply chain intermediaries or third parties, expedited by the internet factors (Kaewkitipong, 2011). Meanwhile, Bennett and Buhalis (2003) defined disintermediation as the removal of intermediaries (tour operator and travel agents) within the distribution channel, caused by electronic approach that allowed consumers to access and transact directly with suppliers (airlines, hotels, bus and cruise operators) and destinations.

2.2 *Business strategies*

Business strategy defined how particular business intends to succeed in its chosen market place against its competitors (Mathur & Kenyon, 1997). Porter (2001) defined strategy as the top management plan to achieve outcomes consistent with the organization mission and vision. A company's strategy consists of the competitive moves and business approaches that manager are employing to get better performance, expand their business and compete successfully (Thompson, Peteral, Gamble, & Strickland 2012). In the tourism industry, the travel agencies and tour operators need to develop their own strategies to stay alive in business. This is due to the increasing number of travel firms and threat of disintermediation and e-commerce (Abdul-Hamid, 2011). Based on Tourism Malaysia, in Jun 2012 there are about 3143 travel agents that have registered in Malaysia and this competitive environment is having a profound impact to the travel industry and need them to build up some strategies for their business.

2.3 *Travel agent strategies*

2.3.1 *Independent strategy*

Galbraith (1977) has mentioned that one of the strategies for travel agent to survive in business is to have an independent strategy. For example the travel agent may use their own creativity and resources by differentiating its services from the ones that their competitors offered. The traditional travel agent need to focus on personal service to ensure it can give value added services and professional advice to the customer (Bennett & Lai, 2005). They also need to take the threat of internet and other ICT as opportunity for them to build up their business.

2.3.2 *Technology investment*

Harris and Duckworth (2005) stated that, travel agent need to invest in new technology by hiring expert and creating competitive differentiation by focusing on certain market. The travel agent needs to adopt the ICT as their strategy in order to retain their position in the distribution chain (Garkavenko, Bremner, & Milne, n.d). Cheung and Lam (2009), suggested some of the approaches for traditional travel agents to get on internet. Firstly, an alliance with the technology experts could give an opportunity for the travel agent to learn all the electronic commerce application, especially for those travel agents who do not have any experience on that specific field. Second approach is through partnering. In this case, travel agent can have flexibility partnership with different technology provider to get all benefits of ICT because one technology provider may not suit with the needs and wants of travel agent.

2.3.3 *E-commerce*

The fast growing of the internet has increased the development of electronic commerce (Palmer & McCole, 1999). As mention by Turban et al. (2002), electronic commerce (e-Commerce) can be described as the secure trading of information, product and services via computer system and exchange of value on-line, as well as the support for any kind of business transactions over a digital infrastructure. The use of e-commerce as strategies in business had given a strong impact on e-commerce profitability and competitive advantage as well as development of company (Le, Thoung, & Anthony, 2002). Therefore because of that reason, travel agencies need to change their business quickly from traditional technique into e-commerce, in order to get new target market (Siebenaler & Groves, 2002). E-commerce is used to support and help the growth of the real and the virtual tourist market. It consists of business deals such as transactions, which are realized through telecommunication systems (Kazandzhieva, 2010). Thus, it will lead to long term achievement in term of business profitability and competitive advantage in the marketplace.

3 METHODOLOGY

3.1 *Population and sampling size*

The population of this study is the managers of travel agents in Selangor whereby it was conducted in qualitative approach. Fox, Hunn and Mathers (2009) suggested that in qualitative approach, the sample sizes used area normally very small amount will be enough until the researcher get the redundancies of information. As the information has reached the level of saturation, the researcher interviewed ten managers from different travel agencies in Selangor. Hair et al. (2007) stated the sample size using qualitative approach can be minimum as one of unit which Kumar (2005) also highlighted the amount of sample size is less important if the main focus of the study is to explore or describing situation, process, issues phenomenon in discovering new information.

3.2 *Data collection process*

The data was collected at travel agents that arelocated in Selangor area because Selangor has the second highest number of travel agents in Malaysia as stated by Tourism Malaysia. Therefore the chances of travel agents to compete with other are very high in order to survive in their business. Nevertheless, before conducting an interview with the travel agent manager, the potential travel agents has contacted to get permission and make an appointment with the travel agent manager for undertaking the interview session. Each of the interview session was expected to take about thirty minutes. Other than that, the interview session was conducted during weekdays but it depends on the availability of the potential travel agents' managers. Two months was required in collecting the data and gathers the information from all the respondents.

3.3 *Data analysis*

After the data been successfully collected from the respondents, the researcher analyzed and summarized the data by using Pen Portray Analysis. The data from the tape recording during the interview session was also being analyzed. Apart from that, the researcher had illustrated with diagrams on strategies that related with disintermediation based on information the researcher obtained from the interview session with travel agents manager.

4 RESULT AND FINDINGS

4.1 *Strategies in facing disintermediation*

There are a lot of strategies used by travel agencies to survive in future business. Some of the strategies are similar but some of it is not similar. As mentioned earlier in literature review, there are some of the negative impacts causes by disintermediation to travel agencies. However, through the finding it show that the travel agencies has re-intermediate their agencies by embracing those strategies used by the suppliers in order to survive in future business. In general, majority of travel agencies are focusing their strategies on marketing by implementing online booking or e-commerce services into their website to grab customer and survive in future business.

Table 1 show that 90 percent of informants are using e-commerce and website as their main strategies to survive in future business. Besides that, they also build a networking by collaborating with other travel agencies and company in line to sell their product. Meanwhile, the travel agencies also manipulate the price of their product by offering lower price than their competitors. Other than that,

Table 1. Percentage of strategies used by travel agencies.

No	Strategies	Informants Answer		Percentage (%)	
		Yes	No	Yes	No
1	Online Booking / E-Commerce	9	1	90%	10%
2	Website Implementation	9	1	90%	10%
3	Product Differentiation / Uniqueness	6	4	60%	40%
4	Travel Conferences	1	9	10%	90%
5	Joining MATTA Fair	4	6	40%	60%
6	Collaboration with Other Travel Agencies	8	2	80%	20%
7	Personal Selling	3	7	30%	70%
8	Public Relation With Clients	2	8	20%	80%
9	Best Service Quality	5	5	50%	50%
10	Collaboration with Corporate Companies	6	4	60%	40%
11	Service Fees Chargers	6	4	60%	40%
12	Price Manipulation	7	3	70%	30%
13	Handling Government Warrant	3	7	30%	70%
14	Direct Suppliers	4	6	40%	60%
15	Variety of Products	6	4	60%	40%
16	Social Media	7	3	70%	30%

as they are competing each other in business, most of the travel agencies had promote their product through social media such as Facebook, IP TV and developed a unique and different product or packages in attracting new customers purchase their product.

5 LIMITATION AND SUGGESTION FOR FUTURE RESEARCH

The first limitation is related to the demographic profiles. Due to time, logistics, and monetary constraints, this study was limited to only travel agencies in Selangor, Malaysia. As this study was looking on strategies they used in facing disintermediation, thus, generalization cannot be inferred to travel agencies in other locations in Malaysia, as the travel agencies might feature different impact and strategies to improve the future business. Hence, the future research could be undertaken in foreseeing the impact of disintermediation and strategies practices of all travel agencies in Malaysia.

Second limitation of this study is related to the research approach. Although this study reached-information on strategies by using qualitative approach but researcher do not know which impacts and strategies are the most influence. Therefore it would be suggested that future research can be undertaken in quantitative approach to see which impacts of disintermediation that bring effects to the travel agencies and which strategies are most useful for the travel agencies to survive in future business.

6 RECOMMENDATIONS

From the travel agencies perspective, as disintermediation given positive and negative impacts for them, they need to re-intermediate the agencies by offering some of unique and different product to their customers. In addition, they need to take the negative impacts of disintermediation as an opportunity to them in order to survive in business by creating and implementing strategies that used by the suppliers such airlines and hotels.

As for the suppliers side, seeing that disintermediation have given more opportunities and positive impacts to them, they need to cooperate with travel agencies in creating their product to look more attractive. Apart from that, airline such as Air Asia, travel agencies are given authority by them in setting their own service charge to the customer so that travel agencies can get some income. Therefore other suppliers like hotels and cruise operators should apply what have been practiced by airlines industry so that travel agencies can survive in business.

Last but not least, from the government side under Ministry of Tourism, the government should develop a national vision, a strategic plan and policy guidelines for travel agencies welfare and how to help the travel agencies in Malaysia to operate their business since they are being disintermediate by the industry. One of the methods is by setting and enforcing a ceiling price for certain products like what has been implemented before. This is to ensure that chances can be given to small travel agencies to cover their operation cost and give profits to survive in business. Other than that, the government can help travel agencies by helping them to install the e-commerce services so that they can sell the product in a lower price and customer can easily do the payment with them without them having to come to the office. The development of e-commerce strategies should involve all tourism stakeholders so that all of them can enjoy the facilities available.

7 CONCLUSION

In conclusion the above highlighted issues and suggestion could not be resolved without involvement and cooperation from all parties. In addition, clearly understanding on this issue of disintermediation, the travel agencies need to think some strategies on how to survive in a future business. This is due to that technology has conquered responsibilities of traditional travel agencies in the tourism industry.

REFERENCES

Abadzhiev, A., Euba, O. & Zeeshan, S. (2005). Internet and Intermediaries in the Tourism Distribution Channel-Study of Swedish, Bulgarian and Online Travel Agencies (Master's thesis, Goteborg University).

Abdul-Hamid, I. (2011). Travel Agency Strategies for Managing the Current Dynamic Environment. *SEGI Review*, 4(1), 71–79.

Amdekar, J. & Padmanabhuni, S. (2006). Future of Travel and Tourism Industry with the Adoption of Web Services in Electronic Distribution.

Bennett, M. & Buhalis, D. (2003). Tourism and Travel Distribution: The Travel Agent's Perspective. *Centre for e-tourism Research, School of Management, University of Surrey*.

Bennett, M.M. & Lai, C.W.K. (2005). The Impact of the Internet on Travel Agencies in Taiwan. *Tourism and Hospitality Research*, 6(1), 8–23.

Carroll, B. & Siguawa, J. (2003). The evolution of electronic distribution: Effect on Hotel and Intermediaries, 44(4), 38–50. Retrieved on November 7, 2012.

Cheung, R. & Lam, P. (2009). How travel agency survive in e-business world? *Communication of the IBIMA*, 10, 85–92.

Fox, N., Hunn, A. & Mathers, N. (2009). Sampling and sample size calculation. *National Institute for Health Research*, 2–41.

Galbraith, J.R. (1977). *Organization design, read and mass*: Addison Nesley.

Garkavenko, V., Bremmer, H. & Milne, S. (n.d). Travel Agents in the Information Age: New Zealand Experiences of Disintermediation. *New Tourism Research Institute, Auckland University of Technology.*

Hair, J.F., Money, A.H., Samouel, P. & Page, M. (2007). *Research Methods for Business.* England: John Wiley & Sons, Ltd, Publication.

Harris, L. & Duckworth, K. (2005). The future of the independent travel agent: The need for strategic choice. Strategic Change, *4*(1), 71–79.

Holma, A. (n.d). Triadic business relationships—the changing role of intermediaries in the business travel industry.

Kaewkitipong, L. (2011). Disintermediation in the tourismindustry: An investigation of Thaitourism SMEs. *International Journal Electronic Business*, 9, 516–517.

Kazandzhieva, V. (2010). The electronic commerce in thetravel agencies' activities in Bulgaria and terms of reference for its elaboration. *Journal of Economic*, 1(2), 67–76.

King, J. (1999). Disintermediation/Reintermediation. *Computerworld*, 33(50), 49–54.

Kumar, R. (2005). Research methodology: A step by step guide for beginners. Malaysia: SAGE Publication.

Le, T., Thoung, K. & Anthony, C. (2002). Amanagerial perspective on electronic commerce development in malaysia. Electronic commerce research.

Mathur, S.S. & Kenyon, A. (1997). *Creating value: Shaping tomorrow's business*. Oxford: Butterworth Heinemann.

Neuman, S., Hutzler, H., Cohen, M., (2003). Asia tourism hurt by war, deadly pneumonia outbreak. *Wall Street Journal*. Retrieved from http://online.wsj.com/article/SB104844775624006 300.html.

Palmer, A. & McCole, P. (1999). The virtual re-intermediation of travel services: A conceptual framework and empirical investigation. *Journal of Vacation Marketing*, 6 (1), 33–47.

Patricia, D.S. (n.d). The internet, threat or tools for travel agencies? 106–111.

Porter, M. (2001). Strategy and the internet. *Havard Business Review*, 63–78.

Siebenaler, T.C. & Groves, D.L. (2003). Travel agents and their survival. *Journal of Human Resources in Hospitality and Tourism*, 1(1), 1–17.

Thompson, A.A., Peteral, M.A., Gamble, J.E. & Strickland, A.J. (2012). *Crafting and executing strategy: Concept and cases*. New York: Mc Graw-Hill.

Tse A.,C. (2003). Disintermediation of travel agents in the hotel industry. *International Journal of Hospitality Management*, 453–460. Retrieved on October 6, 2012 from http://www.sciencedirect.com.

Turban, E., Lee, J., King, D., & Chung, H.M (2002). *Electronic commerce: A managerial perspective*. Upper Saddle River New Jersey: Prentice Hall.

Hospitality and Tourism – Sumarjan et al. (eds)
 ISBN 978-1-138-00151-0

Responsible tourism: Linking perceived benefits and practices in Pangkor Island

M.D. Darson, M.F. Wahab, E.R. Kassim & M.H. Hanafiah
Universiti Teknologi MARA, Malaysia

ABSTRACT: Island tourism sector regarded as a vital economic generator, creating business and job offers and improving income. However, this industry also affects the economic, socio-cultural and environment particularly on the local community quality of life. To prolong the development with constructive effects, many policy makers introduced the concept of responsible tourism. In comparison with development practices, responsible tourism is the most effective policy in maintaining the sustainability of the tourism development. This study focuses on the nexus between community's perceived benefits on tourism development and responsible tourism practice. Specific research objectives are to investigate: (1) the perceived benefits of tourism development; (2) the community's responsible tourism practice and; (3) the relationship between perceived benefits and responsible tourism practices. The Pangkor Island community was surveyed usinga quantitative approach. Data was analyzed using descriptive and inferential analysis. The study shows that most of the respondents view their responsible tourism practices from multiple perspectives and benefits in economic, environmental and social-cultural influenced their practice intention.

Keywords: responsible tourism, perceived benefits, island tourism, responsible practices

1 INTRODUCTION

Tourism becomes economically noteworthy as an income and employment source (Hamzah & Hampton, 2011). It affects the quality of life among the stakeholders particularly local resident where it should be put in significant concern as that the benefits may be communicated properly to the origin people who perceived direct development in a destination (Kim, 2002). It is due to the significant impact towards the local residents' life and eventually may influence the successfulness of development, planning, operation and sustainability of tourism (Jurowski, 1994).

However, tourism development resulted with two important events that may bear upon the island area. Besides delivering positive impacts such as enhancing local economies, being a source of new employment opportunities, additional tax receipts, foreign exchange benefits and revenue, tourism development has the potential towards negative outcomes (Ko and Steward, 2002). Therefore, it is necessary to develop the tourism industry with consensus among other stakeholder to minimize the negative impact. It is vital in ensuring the natural and local community resilience in terms of quality of life through life satisfaction.

Few have addressed the benefits of responsible tourism on enhancing the practice from the multiple stakeholders especially on tour operator and tourist rather than local residents. Most tourism studies stress more on tourists rather than permanent residents of the area in which tourism takes place and seemed to set priority to satisfying the needs of tourists rather than the population in the area. (Adanan et al., 2010).

Consequently, these issues also raise the fundamental question of what is the finest solution to minimize the impact of tourism. In other words, do perceived benefits of responsible tourism practice may influence the local residents' to practice? As the responsible tourism is a new robust paradigm for developing and managing tourism, further consideration should be done to investigate the practice (Settachai, 2009). Thus, this paper aims to look at the perspective on how perceived benefits affect the responsible tourism practice in Pangkor Island, Malaysia.

2 LITERATURE REVIEW

2.1 *Responsible tourism practices and perceived benefits*

It is widely believed that the tourism industry may impose different impacts where responsible tourism practices became the finest solution to minimize the impact and channel out the benefits towards the local resident. (Raviv, 2010; Settachai, 2009; Carasuk & Fisher, 2008; World Tourism Organization, 2009). The "green tourism" concept began in the early of 1980's and became the fundamental feature and also emerged the theory of sustainable tourism (Krippendorf, 1987).

However, green tourism has been argued by researchers as this concept is nothing but a "green wash" where many practitioners do not support the guidelines rather than follow the trend in the tourism (Stanford, 2008). Further consideration is needed to investigate the responsible tourism practices as it is still in the exploration stages. Therefore, the World Tourism Organization together with other organizations has agreed to align the entire tourism industry within responsible tourism practices (WTO, 2009; France, 1999).

The responsible tourism practice offers a comprehensive perspective on the practical implication of tourism that cut across many aspects of the community and involves multiple stakeholders (Settachai, 2009). Moreover, responsible tourism may be understood as any tourism activities that is being performed in a responsible manner (practices) by local residents, authority, private sector, tourist and other stakeholder that may increase (benefit) the value of economic, environmental and socio-cultural resilience.

Many agreed that this concept will maximize the perceived positive benefits than negative ones through social and ethic participation especially among local residents (Carasuk & Fisher, 2008; Stanfod, 2006; Goodwin and Spenceley, 2001). Recycling, ethics, water conservation, crime prevention, promoting local culture, air quality and pollution, community and social participation, energy management, health and awareness, waste and pollution and water conservation were suggested to become the responsible tourism practices that are implemented by the local residents (Stanfod, 2006; Federated Hospitality Association of South Africa, 2002; Spenceley et al., 2001).

However, perceived benefits of responsible tourism are the outcomes associated with integration that are valued by individuals where may be obtained through responsible tourism practices which include economic, environment and socio-cultural (Settachai, 2008). The practice may minimizes damage to the environment (flora, fauna, habitats, water, living marine resources, energy use, contamination, etc.) and ideally tries to improve the environment (Knot et al., 2004; Settachai, 2009). Economically, responsible tourism may benefit the host communities, the business sector and tourist themselves through economic resilience through increasing linkages, reducing leakages and ensuring that communities are involved in tourism development (Knot et al., 2004; International Ecotourism Society, 2004; Spenceley, 2001; Boothroyd & Davis 1993).

Further, responsible tourism practices are associated with the benefits of socio-cultural. Local cultures need to be presented in an authentic process by allowing host or local residents to control the conduct of their presentation. It therefore ensures that tour operator do not disrupt or lead to the displacement of local people, and ensure that the nature and scope of tourism are related to the local conditions and operate within the limits set by the appropriate infrastructure (Knot et al., 2004; Spenceley, 2001).

Due to these benefits of responsible tourism practices, it is believed to enhance the local residents' responsible behavior (Van de Merwe and Wocke, 2007; Knot et al., 2004; Spenceley, 2001). Therefore, in order to measure how perceived benefit affect the responsible tourism practice, it is important to identify what are the dimensions that may influence the responsible tourism practices.

3 METHODOLOGY

Pangkor Jetty, Pasir Bogak, Teluk Dalam and Teluk Nipah in Pangkor Island were selected as the scope of analysis. The areas are located within the tourism development areas and the respondents are directly involved in the tourism industry. Furthermore, the areas reported with numerous environmental issues in terms of coastal extension, improper planning and natural degradation.

Several dimensions were pre-identified to determine the influence factor of responsible tourism practices which are 1) economic benefits, 2) environmental benefits, 3) socio-cultural benefits and 4) responsible tourism practices. A self-administered questionnaire using 5-point Likert scale was used in assessing the perceive benefit of responsible tourism practices among local residents. A total 143 questionnaires were returned, usable and coded.

The survey and scale measurement was adapted from previous research (Spenceley et al., 2001). The items were analyzed using Exploratory Factor Analysis (EFA) analysis as to validate the newly developed items (Pallant, 2005).

4 FINDINGS

4.1 *Perception on economics benefit from responsible tourism development*

The most outstanding result in the sub-section analysis is related to the employment opportunity (m = 4.12) and standard of living (m = 4.07). This could be true as tourism businesses do hire local residents. As a result, the residents' standard of living increase and improved their perceived benefits on the development. In addition, the respondent also slightly agreed with the fact that recycling benefited them (m = 3.84) and environmental-sound technology improved their economic condition (m = 3.56). This could be supported as the responsible practices strive for cost saving through sustainable energy management and technology.

4.2 *Perception on environment benefit from responsible tourism development*

Most of the respondents agreed that responsible tourism practices controls the littering activities through recycling and reused concept (m = 4.06) and the concept has contributed to the preservation of the natural environment and protection of the wildlife in the community (m = 4.02). Besides that, the respondent believed that responsible tourism educates local residents to be accountable and alert towards the development changes and sustainability (m = 3.87) and this practice increase residents' knowledge towards development and sustainability (m = 3.95).

4.3 *Perception on socio-cultural benefit from responsible tourism development*

Most of the respondents viewed responsible tourism benefited them in terms of socio-cultural. They agreed that responsible tourism increased their pride (m = 4.02) and encourage them to involved in tourism entrepreneurial activities (m = 3.97).

Table 1. Perception on economics benefit from responsible tourism development (n = 143).

No.	Item	Mean	S.D
1.	Responsible tourism increase the standard of living	4.07	0.811
2.	Responsible tourism improves economic situation through the application of environmental sound technologies	3.56	0.964
3.	Responsible tourism creates employment opportunities to the local communities through their involvement and participation in the tourism industry	4.12	0.783
4.	Responsible tourism can save cost through environmental-saving activities such as recycling	3.84	0.885

Table 2. Perception on environment benefit from responsible tourism development (n = 143).

No.	Item	Mean	S.D
1.	Responsible tourism controls the littering activities by providing recycle and reused concept	4.06	0.882
2.	Responsible tourism has contributed to the preservation of the natural environment and protection of the wildlife in the community	4.02	0.773
3.	Responsible tourism educates local residents to be responsible and alert towards the development changes in Pangkor Island	3.87	0.846
4.	Responsible tourism practices increase local residents' knowledge towards tourism development and sustainability	3.95	0.816

Table 3. Perception on socio-cultural benefit from responsible tourism development (n = 143).

No.	Item	Mean	S.D
1.	Responsible tourism encourages a variety of cultural activities for local residents through residents participation	3.78	0.907
2.	Responsible tourism has encourages local residents to involve in entrepreneur activities	3.97	0.847
3.	Responsible tourism has increased residents' pride in the local culture towards residents through their participation in the decision making	4.02	0.773
4.	Responsible tourism helps to keep local residents alive and maintain cultural identity through promotion and resident participation	3.76	0.858

Table 4. Overall mean scores for perceived benefits (n = 143).

Minimum	Maximum	Mean	S.D
2.33	5.00	3.9195	0.47620

Table 5. Correlation between perceived benefits and responsible tourism practices.

Variables	Pearson correlation	Sig. (2-tailed)
Responsible tourism practices	1.000	0.000
Perceived benefits	0.632***	0.000

In addition, they also believed that responsible tourism practice helps maintaining their cultural identity (m = 3.76).

4.4 *Perceived benefits of responsible tourism practices*

Table 4 illustrates that based on the overall mean score. Most of the respondents were slightly agreed that responsible tourism practices bring benefits towards multiple dimensions. This is also interrelated with the intention to practice among the local residents.

4.5 *Perceived benefits towards responsible tourism practices*

The relationship between perceived benefits and responsible tourism practices was investigated using Pearson correlation coefficient. From Table 5, it shows that there was a moderate positive correlation between perceived benefits and responsible tourism practices. It shows that perceived benefits have ($r = 0.632$, $p < 0.000$). This point out that the notion of there is a relationship between perceived benefits and responsible tourism practices are accepted.

5 CONCLUSION

The result of the study revealed that perceived benefits have a positive impact towards responsible tourism practices from the local residents' perspective. This finding supports the argument from previous researchers who reported that perceived benefits may influence the responsible tourism practices towards local residents (Raviv, 2010; Settachai, 2008; Carasuk & Fisher, 2008; World Tourism Organization, 2009; Van de Merwe and Wocke, 2007; Knot et al., 2004; Spenceley, 2001). The finding of this research is also consistent with the previous study by Van de Merwe & Wocke, 2007; Knot et al., 2004; Spenceley, 2001 who contended that, benefit received by the residents through the practice will influence their intention to practice.

Moreover, the findings demonstrated that factors that influence responsible tourism practices are multi-dimensional and dynamic. As an addition, conclusive evidence has recognized the positive relationship between perceived benefits and responsible tourism practices. The discovery of the direct effect among benefits from responsible tourism practices has provided new information concerning the degree of strength which influences the residents' practices.

The result of this research is describing the opportunity and problems in many ways. Failure to take a cohesive approach to promote this practice towards the islanders in Malaysia will expose these islands into future degradation in terms of natural beauty, tourist advent and poor quality of life. Finally, this paper provides an explanation of perceived benefits towards responsible tourism practices that may be promoted in different area of study. Even this study has its own limitation; therefore, the future study is needed to contribute to better understanding of responsible tourism practices especially in Malaysia.

ACKNOWLEDGEMENTS

The work described in this study was funded by the Fundamental Research Grants Scheme, Ministry of Higher Education of Malaysia and Universiti Teknologi MARA Malaysia.

REFERENCES

Akmal, A., Radzi, S.M., Hanafiah, M.H. & Hamid, Z.A. (2010). Tourism development and its impact on residents' quality of life. Proceedings of 2010, 4th Tourism Outlook & 3rd ITSA Conference. Malaysia.

Boothroyd P. & Davis, H.C. (1993). Community economic development: Three.

Carasuk, R. & Fisher, D. (2008). Staycation: How global warming becomes a tourism constraint. Paper presented at the New Zealand Tourism and Hospitality Research Conference, Hanmer Springs, New Zealand.

Federated Hospitality Association of South Africa (2002). Statement of Intent to adhere to the Responsible Tourism Guidelines for the Hospitality Sector FEDHASA.

France, L. (1999). Sustainable tourism. In Pacione, M. (Ed.), Applied geography principles and practice: an introduction to useful research in physical, environmental and human geography, pp. 321.

Goodwin, H. & Spenceley, A. (2001/2002). National Responsible Tourism Guidelines For South Africa and Explanatory letter to accompany them. DEAT-November 2001. Finalized March, 2002 subsequently published as Guidelines for Responsible Tourism by DEAT, May 2002.

Hamzah, A. & Hampton, P.M. (2011). Tourism Development and Change in Small Island: Lesson from Perhentian Kecil, Malaysia. Centre for Innovative Planning and Development. Universiti Teknologi Malaysia. Monograph No. 11.

Jurowski, C. (1994). The interplay of elements affecting host community resident attitudes toward tourism: a path analytic approach. Unpublished doctoral dissertation. Virginia Polytechnic Institute and State University.

Kim, K. (2002). The effects of tourism impact upon quality of life of residents in the community. Unpublished doctoral dissertation, Virginia Polytechnic Institute and State University.

Knott, T., Burr, K. & Settachai, N. (2004). Responsible Tourism: An Introduction into Concepts for Achieving Responsible Tourism Development. Term Paper, Spring 2004. Department of Urban and Regional Planning, University of Hawaii at Manoa.

Ko, D.W. & Stewart, W.P. (2002). A structural equation model of residents' attitudes for tourism development. *Tourism Management, 23*(5), 521–530.

Krippendorf, J (1987). The Holiday Makers. Understanding the Impact of Leisure and Travel. Oxford: Butterworth Heinemann.

Raviv, C. (2010). Responsible tourism Qualmark accreditation: a comparative evaluation of tourism businesses and tourists' perceptions. Unpublished doctoral dissertation. Lincoln University.

Settachai, N. (2009). Global Tourism and Community Life, Saarbrucken: VDM Verlag.

Singletary, L.A. (2003). Empirical study of attributes and perceived benefits of application integration for enterprise systems. The Interdepartmental Program in Business Administration.

Spenceley, A. (2001). Responsible tourism guidelines for the South African tourism industry: Draft guidelines for discussion, Report to DfID/DEAT, October 2001 "Draft Guidelines Rept".

Stanfod, D. (2006). Responsible tourism, responsible tourists: what makes a responsible tourist in New Zealand? Unpublished doctoral dissertation, Victoria University of Wellington.

Stanford, D. (2008). Exceptional visitors: dimensions of tourist responsibility in the context of New Zealand. *Journal of Sustainable Tourism, 16*(3), 258–275.

Van der Merwe, M. & Wocke, A. (2007). An investigation into responsible tourism practices in the South African hotel industry. *South African Journal of Business Management, 38*(2), 1–15.

Webster, M. (1996). Webster's Encyclopedic Unabridged Dictionary of the English Language. New York: Random House.

World Tourism Organization. (2009). Sustainable Development of Tourism—Mission Statement. Retrieved 14 October 2010, from the World Wide Web: http://www.unwto.org/sdt/mission/en/mission.pp.

Hospitality and Tourism – Sumarjan et al. (eds)
© 2014 Taylor & Francis Group, London, ISBN 978-1-138-00151-0

Langkawi Island as a tourist friendly destination

A.Q. Abd Jalil, A.A. Azdel, F.I. Basri & M.I. Mustapha
Universiti Teknologi MARA, Malaysia

ABSTRACT: Tourism destination is seen as a combination of products and experience which influencing tourists to travel. The evolution of tourist tastes and demands as well as an increasing of travel potential, have called for a tourism destination to be more creative and innovative in offering niche tourism or more importantly, tourist friendly destination. Tourist friendly destination is a concept that provides tourists with satisfaction through connection and maximum use of space, activities, and products without any interference or problems. This study was done in Langkawi Island, Malaysia most popular island. This empirical study was done to identify whether Langkawi Island met with all the elements in the tourist friendly destination concept. Using a whole island area as the contextual setting, 302 domestic tourists were surveyed. Through a series of analyses, the majority of the respondents agreed that Langkawi Island can be considered as a tourist friendly destination. Although some improvements need to be done, this study showed that Langkawi Island had fulfill almost all the elements in the concept. The result also proved that satisfied respondents will lead them to revisit and recommend this island to others.

Keywords: Langkawi Island, revisit intention, recommend intention, satisfaction, space, tourist friendly destination

1 INTRODUCTION

The tourist friendly destination is a concept, which satisfies tourists through the utilization between the elements of activity, space and product without any interruption and difficulties at the destination (Anuar, Ahmad, Jusoh, & Hussain, 2012). It has been seen as a customer-oriented concept where tourists are regarded as the customers at the tourist friendly destination. The most concerning parts in tourist friendly destination is the concept will give a positive impact so that all experience and expectation are met and finally earned the right distribution to the target group of entrepreneur, such as the local community and the respective stakeholder. The importance of the tourist friendly destination perception can be seen through the study by HSBC Bank (2011) since the perception has been used by the relevant stakeholders to enhance the competition between each tourist destination. Besides, according to Swarbrooke (1995), tourists have a tendency to respond a tourist friendly destination wherein the facilities and services are easily accessible.

Langkawi Island has gone through many changes since it changed its status from being 'Mahsuri Legendary Island,' 'Langkawi 99 Magical Islands,' 'Langkawi Tax Free Island' (1987) and to 'Langkawi Tourism City' (2000). Its recognition as the 'Langkawi Geopark' (2007) has put the island on the international arena (Hashim and Aziz, 2010) and currently has been rebranding as the 'Naturally Langkawi' (2012). The popularity of Langkawi Island has spread all over this region after it has been chosen as the fourth best islands in Asia by a poll conducted by Conde Nast Traveller in 2012 (Bernama, 2012). The island has experienced tremendous development and many construction projects are carried out with the reason to accommodate tourism development. A big amount of investments by the government and private sector had drastically turned Langkawi into a popular destination and a shopping paradise for local and foreign tourists (Marzuki, 2011). For example, from 1988 to the end of 1992, about 106 tourism projects by private sectors valued at more than RM 1 billion had been permitted by the authorities. The influx of domestic and international arrivals in Langkawi also increased the number of hotel rooms from 859 in 1988 to 7072 in 2005 (Langkawi Development Authority, 2006).

However, a destination which is compelling or popular does not make it a tourist friendly destination and cannot guarantee that locals will be friendly to foreigners (World Economic Forum, 2010). Besides, Langkawi Island facesan increasing competition from other tourist destinations in South East Asia such as Bali, Phuket and Cebu. Langkawi falls behind other island destinations and international destinations such as Hawaii,

Mauritius and the Seychelles in terms of spending per day, length of stay and seasonality (The Langkawi Tourism Blueprint, 2011). Nevertheless, the Langkawi Lab Analysis (2011) reported there are still a lot of issues regarding the development of infrastructures, products and community involvement that can certainly affect the major plan of Langkawi development as a whole. Therefore, this study proposed to identify whether Langkawi Island nowadays fit the concept of the tourist friendly destination—a concept that provides satisfaction by fulfilling tourists' wants and needs through the maximization aspects of space, activity and products without interference or problems (Anuar et al., 2012).

2 LITERATURE REVIEW

Tourist friendly destinations are created from a mixture aspect of space, activities and products within the tourism system approach.

2.1 *The elements of space*

Space has been defined as a destination with complete facilities to accommodate the tourists' needs and sufficient space to fulfil primary motivations encourages tourists to spend time in the area and view the space as a holiday spot or tourist resort (Hitrec, 1995; Kaiser & Helber, 1978). Anuar et al. (2012) considered space element as a combination of infrastructure and superstructure, location, security, cost and value and finally accessibility. Yet, Vukonic (1997) added that that the more space used, it will increase tourists' satisfaction and attendance to the tourism destination. Hudson (1998) stated that a space is equipped with the required facilities and infrastructures needed by tourists in tourism destination, especially in tourist friendly destination. Therefore to ensure a tourist friendly destination concept fulfils the supply and demand of tourist, a space element with an infrastructure, facilities, accommodation, attraction and others is important (Anuar et al., 2012).

2.2 *The elements of activities*

According to Gunn (1994), activity is one of the thing that will be done when tourist travel to a specific attraction far from the normal life of work and home by satisfying their needs and wants. The role of activity element is one of the most important keys in the tourism industry. Without any activity or program, a particular tourism attraction or destination will become a doom as the environment becomes so dull without any activity to be organized. Thus, the importance of activity at such destinations must be taken care (Anuar et al., 2012). To guarantee that the activities fulfil tourist needs and wants, the element of activities should be measured and studied comprehensively. There are four activity elements that are very important in a particular tourism destination: culture and history, entertainment, special events and mix activities (Ritchie & Crouch, 2003). These elements are essential to design activities that are tourist friendly destination and can persuade tourists' motivation to decide their destination.

2.3 *The elements of products*

Kotler (1994) stated that product includes physical objects, services, places, organizations and so on. According to Swanson and Bridge (2006), product consists of attributes that can influence tourists decision making. Smith (1994) proposed a model of five element that represent a product, which includes physical characteristics, service, hospitality, freedom of choice and involvement in creating products at tourism destination. This model was later utilized by Xu (2010) to observe tourism products in other tourism sectors like theme parks, resorts, restaurants and natural environments.

2.4 *The relationship of tourist satisfaction and tourist behavioral intentions*

According to Zeithaml and Bitner (2003), customer satisfaction is a main contributor to guaranteeing a company's long-term profitability, customer retention and loyalty. Thus, an excellent delivery of customer satisfaction is the most vital determinant for an organization's survival and long-term profitability (Bolton & Drew, 1991). Meanwhile, behavioral intentions can be defined as a kind of tendency to behave such as recommend, purchase, travel, visit and return. Behavioral intention has been utilized as a variable to predict consumers' actual behaviors (Bigne, 2001; Castro, 2007). Behavior intention can be predicted by several antecedent such as past experiences, satisfaction and perceived value (Azjen & Fishbein, 1980).

According to Lin (2006), satisfaction level has a positive impact to behavioral intention. Meanwhile, Yeh (2000), Yoon and Usal (2005), Awadzi and Panda (2007) stated that there was a strong relationship between satisfaction level and willingness to revisit. When a tourist receives a good experience during their actual visit at a certain area, the likelyhood of revisit is high. Additionally, there is an agreement among scholars that satisfaction provides a ground for revisiting and positive word of mouth recommendations (Chi & Qu, 2008; Kozak, 2001; Yoon & Usal, 2005).

3 RESEARCH METHODOLOGY

This study aims to examine the relationship between the tourist friendly destination concept, tourist satisfaction and tourist behavioral intention. A descriptive study design using a quantitative approach was conducted. This research was conducted cross-sectionally and focused on the individual level of analysis (Hair, Money, Samouel, & Page, 2007). This study took place in Langkawi Island. Domestic tourists were the targeted respondents as they were the highest number of tourist received (2.85 million) in the year 2012 (Langkawi Development Authority, 2013).

4 FINDINGS

4.1 *Pearson correlation of Tourist Friendly Destination Concept (TFDC), Tourist Satisfaction (TS) and Tourist Behavioral Intention (TBI)*

The correlation between TFDC and TS showed the highest degree of relationship with a correlation (r = 0.789, p > 0.01). Meanwhile, the correlation between TS and TBI was also found significant (r = 0.614, p > 0.01). There was a moderate correlation (r = 0.598, p > 0.01) between TFDC and TBI. Overall, the results showed that relationships exist between all the dimensions in this study.

4.2 *The mediating effect of Tourist Satisfaction (TS) with Tourist Friendly Destination Concept (TFDC) and Tourist Behavioral Intention (TBI)*

Looking at Table 2, TFDC was able to explain 36 percent of the variation on the tourist behavioral intentions (R^2 = 0.357, F = 166.74, p < 0.001). The results demonstrated that TFDC has significant impact on tourist behavioural intentions (β = 0.598, p < 0.001).

Result of Step 2, the tourist satisfaction explain the additional 5.4% (R^2 change = 0.054) as a mediator to for tourist friendly destination concept to influence the tourist behavioral intentions (β = 0.598, p < 0.001). Therefore, it can be concluded that there is a partial mediation of tourist satisfaction towards tourist behavioral intentions to influence the revisit and recommend.

Table 1. The result of the Pearson correlation (N = 302).

Dimensions	Test significant of	TFDC	TS	TBI
TFDC	Pearson correlation	1		
	Sig. (2-tailed)			
TS	Pearson correlation	0.789**	1	
	Sig. (2-tailed)	0.000		
TBI	Pearson correlation	0.598**	0.614**	1
	Sig. (2-tailed)	0.000	0.000	

**Correlation is significant at the 0.01 level (2-tailed).

Table 2. The standard multiple regression.

Predictors	Model 1/std. β
Step 1: Model variables	
TFDC	0.598**
R^2	0.357
Adj. R^2	0.355
R^2 change	0.357
F-change	166.74

Note: *p < 0.05, **p < 0.01, ***p < 0.001.

Table 3. The standard multiple regression.

Predictors	Model 1/std. β
Step 2: Model variables	
TFDC	0.300
TS	0.377
R^2	0.411
Adj. R^2	0.407
R^2 change	0.054
F-change	27.28

Note: *p < 0.05, **p < 0.01, ***p < 0.001.

5 CONCLUSION

5.1 *The relationship between tourist friendly destination concept and tourist satisfaction*

The relationship among the variables were proven to exist as the respondents were satisfied with the tourist friendly destination concept in the Langkawi Island. This shows that Langkawi Island fulfills the tourist friendly destination concept elements. The respondents participated in this study were satisfied with the facilities, activities, programs and products provided in the island. The relationship was supported by the correlation analysis and the presented score are (r = 0.789, p > 0.01), the highest score among the others.

5.2 *The relationship between tourist satisfaction and tourist behavioral intentions*

This study showed that a relationship exist between these variables ($r = 0.614$, $p > 0.01$). Satisfied respondents are likely to revisit and recommend the destination to others. This is in line with most of the previous studies investigating the similar situation (Chi & Qu, 2008; Kozak, 2001; Lin, 2006; Yeh, 2000; Yoon & Usal, 2005).

5.3 *The relationship between tourist friendly destination concept and tourist behavioral intentions*

This study found a moderate relationship between these two variables ($r = 0.598$, $p > 0.01$). It is worth noted that the tourist friendly destination is an importance concept that should be implemented in order for the destination to be revisited and recommended to others. This is supported by Swarbrooke (1995) who stated tourists have a tendency to respond a tourist friendly destination with facilities and services that are easily reachable. Additionally, Vukonic (1997) stated that the space aspect can increase satisfaction and attendance of tourist to the tourism destination.

5.4 *The mediating affects of tourist satisfaction with tourist friendly destination concept and tourist behavioral intentions*

The value of R^2 increased about 5.4% after the mediating variable was introduced ($R^2 = 0.357$ to $R^2 = 0.411$). This showed that tourist satisfaction as the mediator played an important role for the respondents to revisit and recommend the destination that provide tourist friendly facilities to others. This is in support to Lin (2006) who found that satisfaction with tourist friendly destination concept has a positive impact to the behavioral intention. Thus, when a tourist receives a good experience in their actual visit at certain areas, the tendency of the tourist to revisit is high.

5.5 *Implications of the study*

This study shown that Langkawi Island can fulfill all the elements in the tourist friendly destination concept and provide all the needs and wants for the tourist. It also confirms that Langkawi Island is a tourist friendly destination as the island matched with the concept. All the three elements consisted in the concept are very important aspect to be provided by any destinations globally, especially Malaysia. Therefore, it is suggested that the government, private sectors and other related individual in the tourism industry to apply the concept in attracting the tourist and promoting their places so that they will achieve an increment in their output and profitability. Finally, the researchers hoped that findings of the study could contribute to the academicians and researchers theoretically.

REFERENCES

Anuar, A.N.A., Ahmad, H., Jusoh, H., & Hussain, M.Y. (2012). Understanding the factors influencing formation of tourist friendly destination concept. *Journal of Management and Sustainability*, *2*(1), 106–114.

Awadzi, W. & Panda, D. (2007). Relationship Marketing in the tourism Industry: Towards an integrated model for research. *Consortium Journal of Hospitality and Tourism*, *12*(1), 47–56.

Azjen, I. & Fishbein, M. (1980). Understanding attitudes and predicting social behavior. New Jersey: Prentice-Hall.

Bernama. (2012). PM Unveils 'Naturally Langkawi' Branding. Retrieved January 10th, from http://www.theborneopost.com/2012/11/01/pm-unveilsnaturally-langkawi-branding/.

Bigne, J.E. (2001). Tourism Image, evaluation variables and after purchase behavior: Inter-relationship. *Tourism Management*, *1*(22), 607–616.

Castro, C.B. (2007). The influence of market heterogeneity on the relationship between a destination's image and tourists' future behaviour. *Tourism Management*, *28*(1), 175–187.

Chi, C.G.Q. & Qu, H. (2008). Examining the structural relationships of destination image, tourist satisfaction and destination post-purchase behavior intention: An integrated approach. *Tourism Management*, *29*(4), 624–636.

Gunn, C.A. (1994). Tourism Planning (3rd Edition). London: Taylor and Francis.

Hair, J.F., Money, A.H., Samouel, P., & Page, M. (2007). Research Method for Business. West Sussex, United Kingdom: John Wiley @ Sons Ltd.

Hashim, H.S. & Aziz, R.A. (2010). Governance for Geopark: Langkawi Geopark as a showcase. Akademika 80 (Disember) 2010, 39–54.

Hitrec, T. (1995). Tourism Destination: Meaning, Development, Concept. In Wahab, S. & Pigram, J.J. (Eds.), Tourism, Development and growth: The challenge of sustainability (pp. 96). Great Britain: Routledge.

HSBC Bank. (2011). Malaysia antara 10 negara paling mesra. Kuala Lumpur: The New Straits Times Press (M'sia) Berhad.

Hudson, B.J. (1998). Waterfalls. Resources for tourism. *Annals of Tourism Research*, *25*(4), 958–973.

Kaiser, C., & Helber, L. (1978). Tourism: Planning and development. In Wahab, S. & Pigram, J.J. (Eds.), Tourism, development and growth: The challenge of sustainability (pp. 97). Great Britain: Routledge.

Kotler, P. (1994). Principles of marketing (6th ed.). Englewood Clifs: Prentice Hall.

Kozak, M. (2001). Repeaters' behavior at two distinict destinations. *Annals of Tourism Research*, *28*(3), 784–807.

Langkawi Development Authority, (2006). Tourism Statistic. Retrieved September 17th, 2006, from http://www.lada.gov.my/English/stat.htm.

Langkawi Development Authority, (2013). Visitors Arrival Statistics To Langkawi 2012. Retrieved June 4th, 2013, from http://www.lada.gov.my/v2/files/KemasukanPelancong2012.pdf.

Lin, M.M. (2006). A Study of the relationship among tourism image, tourist's satisfaction and behavior intention of Chi-Chi area. Nanhua University, Taiwan.

Marzuki, A. (2011). Resident attitudes towards impacts from tourism development in Langkawi Islands, Malaysia. *World Applied Sciences Journal*, 12, 25–34.

Ritchie, J.R.B., & Crouch, G.I. (2003). The competitive destination: A sustainable tourism perspective. UK: CABI International.

Smith, S.L.J. (1994). The tourism product. *Annals of Tourism Research*, *21*(3), 582–595.

Swanson, K., & Bridge. (2006). Travel motivations as souvenir purchase indicators. *Tourism Management*, 27, 671–683.

Swarbrooke, J. (1995). The development and management of visitor attractions. London: Butterworth Heinemann.

The Langkawi Tourism Blueprint 2011–2015. (2012). Retrieved November 10th, 2012, from http://www.epu.gov.my/html/themes/epu/images/common/pdf/blueprint/Blueprint-Cover.pdf.

Vukonic, B. (1997). Selective tourism growth: Targeted tourism destination. In Wahab, S., & Pigrim, J.J. (Eds.), Tourism, development and growth: The challenge of sustainability (pp. 95–108). London: Routledge.

World Economic Forum. (2010). The global competitiveness report 2010–2011. Retrieved May 5th, 2013, from http://www3.weforum.org/docs/WEF_GlobalCompetitivenessReport_2010-11.pdf.

Xu, J.B. (2010). Perception of tourism products. *Tourism Management*, 31, 607–610.

Yeh, M.S. (2000). A Study on tourist satisfaction of coastal recreational activities in the Jyibey Island of Penghu. Shin Hsin University, Taipei, Taiwan.

Yoon, Y. & Usal, M. (2005). An examination of the effects of motivation and satisfaction on destination loyalty: A structural model. *Tourism Management*, *26*(1), 45–56.

Zeithaml, V.A., & Bitner, M.J. (2003). Service marketing: Integrating customer focus across the firm (3rd ed.). Boston, MA: McGraw-Hill/Irwin.

Hospitality and Tourism – Sumarjan et al. (eds)
 ISBN 978-1-138-00151-0

Tourists' perceptions toward the authenticity or unique features of Melaka traditional Malay houses as cultural heritage tourism product

M.A. Ismail, F.I. Basri, N.A.A. Nik Hashim & A.A. Abd Majid
Universiti Teknologi MARA, Malaysia

ABSTRACT: Instead of become shelters for the people who live in, the Melaka traditional Malay houses reflect the owners' social structure, traditions, history and culture livelihood of Merlimau. The houses are recognized as an authentic and unique architectural legacy of the Malays in Peninsular Malaysia where most of them are still maintaining the authenticity based on the original structure during the colonial era of the Dutch and British. The houses are constructed based on history, socioeconomic and geography of the Merlimau area. This paper reports the potential of the houses as cultural heritage tourism products with regard to the authenticity or unique features as well as perceptions of the tourists. The result revealed that authenticity or the unique features of the houses as destination attributes have significant influence on the tourists' perceptions as the tourism product. The trends of building modern and concrete houses will affect the original structure of the traditional houses and it will result them to be demolished. Therefore, adaptation and transformation of the houses as tourism products is an effective strategy to improve the sustainability of cultural and historical heritage for future generation.

Keywords: cultural heritage tourism, Malay houses, Melaka traditional house, tourist perception

1 INTRODUCTION

Melaka Museum Corporation (2008) stated that one of the most unique and outstanding tourism products of cultural heritage tourism previously is the Melaka traditional Malay houses. The houses have unique and outstanding architectural legacy which may be preserved for the sake of tourism industry development in Merlimau. Merlimau is a small town, geographically located to the southeast of Melaka, about 23 km from the Melaka Historical City and 22 km from Muar by travelling through Jasin Town. This town blends the culture and heritage of Melaka and the southern state of Johor.

Merlimau area (4,674 hectare) is currently managed by the Jasin District. It consists of 30 traditional Malay villages where most of the villages have outstanding and unique traditional houses built as early as 18th and 19th century (Malaysia National Heritage Department [MNHD], 2011). The MHND stated that Melaka Traditional Malay Houses in Merlimau have their own unique features and Malay architectural authenticity of Melaka people and are highly potential to become a resource for cultural heritage tourism. The department also explained that the architectural styles of the houses inherited the Malays, Chinese, Dutch, British and others' characteristics of interior and exterior features (such as ventilation system, roof, porch, decorated brick stairways or staircases, doors and windows, fence, verandah) and many other elements rich with embroidered works and high quality wooden workmanship as well as decorations that are very unique from other traditional houses in Malaysia.

Unfortunately, modernization and change in lifestyles of the local communities in Merlimau causing the traditional houses to be transformed and demolished to modern houses with new structures. This situation is quite critical as the unique features might not be experienced by the younger generations.

1.1 *Authenticity or unique features*

World Heritage Conference (2008) noted that authenticity is the degree to which one is true to one's own personality, spirit, or character, despite the pressures of external and internal factors which also can be defined as a term of "originality, self". Authenticity is a central issue or key concept in the World Heritage Convention. The convention entails that "in order to be deemed of outstanding universal value, assets must meet the conditions of authenticity and must have a sufficient protection and management system to guarantee its safeguarding" (World Heritage Conference, 2008).

UNESCO (2005) and Stovel (2007) argued that subject on the type of cultural heritage and its cultural context, assets may be assumed to meet the conditions of authenticity if their cultural value as acknowledged in the nomination standards proposed are truthfully and credibly articulated through a variety of aesthetic attributes or unique features including structure, form and design, materials and substance, use and function, traditions, techniques and management systems, location and setting, language and other forms of intangible heritage, spirit and feeling and other internal and external factors.

The conception of authenticity is not easy to define. Specifically, authenticity is socially constructed, so it has different meanings in different cultural contexts as proposed by Labadi, (2007). Indeed, this carries certain complications associated with enforcing common standards. It also offers flexibility in taking the characteristics of each individual site into consideration when creating preservation decisions temporal change is not the only influence on authenticity. Diverse cultures interpret the concept in different ways (Labadi, 2007). The first and most important is the Nara Conference on Authenticity which was held in 1994. The conference gave rise to The Nara Document on Authenticity which summarizes how the term "authenticity" should be interpreted. Cultural heritage diversity is present in time and space, and demands respect for other cultures and all aspects of their belief systems. Authenticity of cultural heritage essential therefore is judged in its own cultural context (World Heritage Conference, 2008).

Several researchers agreed with the idea that authenticity is the key element of cultural heritage tourism. However, authenticity itself is a complex concept with countless definitions. In a broad sense, authenticity refers to an approach to communicate the real story of a place and consequently attach appeal to that place. Wang (1999) further classified authenticity in tourism into three categories: objective, constructive, and existential authenticity. Objective authenticity is concerned with tourists' perception of the toured objects determined by whether they respect the toured objects as authentic or not. Although the toured objects are supposed to present the real past, they do not necessarily have to be authentic as long as the tourists believe they have gained an authentic experience. The constructive authenticity sets importance in the construction of authenticity in terms of discourse as an alternative of the toured objects themselves, and the authenticity can be formed either by the tourists or tourism producers. Existential authenticity refers to a potential existential state stimulated by tourist activities. In facts, the tourists experience authenticity not because of the toured objects but because they are involved in some uncommon activities which signify a departure from their routines (Wang, 1999).

Based on the MNHD (2011) and PERZIM, the characteristics and unique features of Melaka Traditional Malay houses include "tunjuklangit, tanggabatu, alas tiang, kekisitingkap, tebarlayar, and ataptinggi," which some of them could not be found in other typical traditional Malay houses in other parts of Malaysia. The *Penghulu* (chieftain) Abdul Ghani bin Abdul Majid House in Merlimau, is among the oldest Melaka houses which still stands tall and is one of prettiest Traditional Malay houses in the country MHND (2011). It consists of authentic architecture which is very unique with a combination of Malay, Chinese and Dutch style. The scale of the house is much larger than the usual Melaka house because it was occupied by the head of the society holding the post of Demang or *Ketua Penghulu* (head chieftain). All the furniture in this house was made according to the European concepts where the table, cupboard, chairs and bed are slightly different from the traditional Malays who usually sat cross-legged on the floor (MNHD, 2011).

Based on aforementioned information, it is argued that authenticity is critical element of cultural heritage tourism product. This shows that the trend and tourist intention to visit or revisit the traditional houses as part of their travel itinerary are interrelated with the authenticity or aesthetic value of the houses and unique features which they might not find in any other traditional houses in the world. In order to assess the potential of Melaka traditional Malay houses to be developed as a cultural heritage tourism product, this study adopted the attributes of traditional houses as proposed by Gültekin (2011) in his study conducted in Akseki area, Turkey. According to his study, the authenticity or unique features of the traditional houses consist of three major elements which are described as follows:

1.1.1 *Great architectural*

The Melaka traditional Malay houses, also called long-roofed Melaka house can be found in all districts in Melaka, which were built with special architectural styles. The houses are popular with its brick staircases built with colorful tiles imported from China and India. The original houses are built from 12 to 16 main pillars in the main house or '*rumahibu*'. The house component consists of the main house, middle house and kitchen (MNHD, 2011). The division of the house consists of verandah or 'serambi', and porch or 'anjung' with staircase sometimes found in front of the house.

The middle space consists of a room between main house and kitchen with a space called 'selang'.

There is also a hidden room at the roof called 'peran' or attic for storing and sometimes used as a bedroom for daughters. Basically, the Melaka Traditional Malay houses were built from cengal, meranti and damarlaut wood. The magnificent and aesthetic value of the traditional houses in Merlimau was argued by the Melaka Museum Corporation (PERZIM) and other tourism agencies or authorities in Melaka. The characteristics or unique features of the traditional houses together with proper promotional approaches can attract tourists either domestic or international.

The houses to be studied and promoted for the survey were selected from the PERZIM conservation list and using certain criteria endorsed by the MNHD. Most of the selected houses have maintained their design originality (Melaka Museum Corporation, 2011). In addition, Poimiroo (1996) and Gültekin (2011) stated that the great and unique architectural styles of heritage buildings will phenomenally attract tourists and potentially become most visited attractions.

1.1.2 *Material*

Melaka traditional houses are famous for their uniqueness in architectural style or design and each of them has its own aesthetic value based on the status of the owners. The houses are mainly constructed with timber or wood materials. There is a pressing need to conserve and preserve this valuable architectural heritage, as these timber house structures and elements are exposed to decay and deterioration caused by environmental factors. Nor Haniza, Zuraini, Yacoband Helena, (2003) and Gültekin (2011) also mentioned that local materials from either natural or other resources such as wood, soil, stone, plaster and tiles of traditional houses with their own unique functions are becoming a tourism attraction. The MNHD (2011) described that the Melaka Malay Traditional Houses consists of two components which joined into one, very long-roofed and built with sixteen—pillars. Some of the house design was inspired by the owners built by the Chinese artisans during the 19th Century. As the result, it is not unusual that Chinese motifs adorned the walls, doors, windows and fence of the Malay traditional houses in Merlimau.

Most of the Melaka houses welcomed visitors with a very elegant brick stairway or staircase made from multi colored decorated tiles imported from China and India (Melaka Museum Corporation, 2008). The styles of the stairways and houses shape reflect the social status of the owners. The richer built bigger and prettier houses and the pillars decorated with a pineapple as a symbol of wealth on the left and right side of the stairs. The unique door or locally known as 'pintukoboi' situated between the verandah and the porch are embroidery decorated and carved. Basically, the traditional houses of rich Malay people in Melaka consist of three parts; the Rear House, Mother House, and Kitchen House (MNHD, 2011).

Some of the houses in Merlimau are different because the Rear House and the Kitchen House are separated from the Mother House by a courtyard. This was influenced by the Peranakan style houses with Chinese design and a tiled Dutch roof. Generally, there is a huge square water tank used to wash clothes in the courtyard. There was also a roofless bathroom that has a huge water tank. The roof of the Porch House, Mother House and the Rear House are made of zinc or Indian tiles, an exclusive material for a roof of that time. Some houses were painted with multi-colored motifs which influenced by the Chinese architecture. The traditional Malay houses usually maintained the color of wood (MNHD, 2011).

1.1.3 *Embroidery*

The houses were built with authentic Malay architectural, high and long-roofed, and brick staircases decorated with imported colorful tiles. The brick staircases with wooden embroidery works for 'kekisi' in verandah and porch are the unique features of the house that is hardly found in other traditional houses in Malaysia (MNHD, 2011). According to Ahmad, Jamaludin and Ani, (2010), the Melaka traditional Malay houses were built with full embroidery's work which are very attractive and have unique features such as the ventilation system called 'kekisi.' It was originally used to ventilate the house with as well as become one of the important decorations.

Associate Professor Dr. Anuar Talib, an architectural lecturer from Univesiti Teknologi MARA (UiTM) stated that traditional Malay house craftsman gets the inspiration of embroidery works and carving ornaments from the al-Quran and nature that blend with local cultural beliefs to become unique and artistic embroidery works (NurAzfahani, Nazhatulzalkis, & Ahmad, 2009). The woodworking is critically important previously where it can be widely seen on the exterior and interior of the traditional houses with different techniques used compared to the modern houses (Gültekin, 2011).

2 METHODOLOGY

A self-completed bilingual questionnaire (English and Bahasa Melayu) was designed to obtain the required information for the study from the domestic and international tourist. The data were collected in Demang Abdul Ghani Gallery.

The respondents were approached during their visit to the gallery; briefed on the purpose of the survey, and then approached by showing the pictures of other Melaka traditional Malay houses in Merlimau and the authenticity or unique features of the selected houses. A total of 384 completed questionnaires were collected.

3 RESULTS AND DISCUSSION

The coefficient alpha for overall dimensions of authenticity or unique features (great architectural, material and embroidery) was high (0.819). An alpha value above 0.60 considered acceptable and more than appropriate to proceed with further testing. The highest percentage of overall respondents were Malaysian with 88.3 percent (n = 339) compared to non-Malaysian with only 11.7 percent (n – 45).

3.1 *Analysis of authenticity or unique features*

The mean score was obtained from a descriptive statistic with regard to authenticity or unique features of the Melaka traditional Malay houses. The patterns of the mean score indicated that most of the respondents were having the same feeling toward all items in the authenticity or unique features dimensions. The mean score that were above 4.0 indicated the authenticity or unique features of the Melaka traditional Malay houses succeed to attract tourists' attention and attractive.

It is interesting to note that the highest mean score was reported regarding on the tourists' feelings whether they were impressed with the architectural styles and authentic design of the house (M = 4.55, A6), followed by the original structures of the house (M = 4.52, A1), the difficulty to find interesting and attractive traditional house like they have been visited in Merlimau (M = 4.52, A5) and the house was built with beautiful and interesting unique elements (M = 4.52, A8). Tourists agreed that the unique features of the house could not be found in other traditional houses (M = 4.51, A3) and the house has unique colorful decorated staircases (M = 4.51, A12).

Tourists agreed that the house was built by unique local and imported materials (M = 4.50, A7), the embroidery works and woodworking of the house are interesting, artistic and unique to be explored (M = 4.50, A10), the house has authentic architectural influence styles of the Malays, Chinese and Dutch (M = 4.46, A2), most of the decoration and embroidery works are influenced from various architectural styles and motifs (M = 4.39, A11), each of the elements of the house has their own special functions which are

Table 1. Overall mean score for each item under authenticity or unique features (N = 384).

Variables	Min	Max	Mean (M)	Std. deviation (SD)
Great architectural	2.33	5.00	4.4297	0.42918
Material	2.33	5.00	4.4618	0.50897
Embroidery	3.00	5.00	4.4688	0.41042
Authenticity or unique features	2.75	5.00	4.4475	0.39358

unique and interesting (M = 4.37, A9), this kind of Melaka traditional Malay house with special functions could only be found in Merlimau (M = 4.03, A4). The highest report on these items came from international tourists and tourists who were first time visited the house. The overall tourists' feelings on the house authenticity or unique features are summarized in Table 1. The result indicated the Melaka traditional Malay houses were authentic, interesting, impressing and attractive.

4 CONCLUSION

This research on Melaka traditional Malay houses has opened the doors to the awareness of the potentiality of the architectural legacy to be transformed, adapted and innovated into cultural heritage tourism product in Merlimau. The houses with special architectural styles and history need to be preserved for future generations as well as to be turned into a most beneficial asset to the tourism sector in Melaka. Each of the traditional houses has their own uniqueness, history, authenticity and interesting woodworking to be explored by tourists who loved the cultural heritage tourism activity or special interest tourism. The houses could potentially generate much more income for the state government as well as helping the local community to involve and generate income either directly or indirectly in the tourism sector by providing extra services such as foods or culinary services, spa or retreat center, transportation services, curators or on-site guides, accommodations, craft souvenirs with proper signboards and multilingual informational tools such as printed brochure or information boards. The authority should also consider providing the translation tools for international tourists.

The transformation of the traditional houses into cultural heritage tourism product should be a primary goal of the Melaka State government as well as focus by private sectors include travel agencies tourism authority that provide services to tourists who are interested to experience the visit to the 'living museum'. In order to achieve the goal, the traditional houses in Merlimau need to be critically inspected to assess the aesthetic and authentic unique features, evaluated, renovated or preserved. The owners must be afforded the opportunity to sustain the authenticity and unique features of the house structure, including monetary incentives, education and awareness to protect the house for tourism activity in the area.

Melaka tourism authority or Tourism Melaka should grab this opportunity to prosper the area with potential tourism products to be developed in order to increase the income from the tourism sector in the future as there is highly increasing demand for cultural heritage tourism.

The state government and private agencies need to make an extensive effort to promote the traditional houses to either international or domestic tourist aggressively. In addition, more interesting or interactive activities are needed to be planned and promoted including the cultural heritage trails to all the traditional houses and tourism products in Merlimau, lunch or dinner package, live cultural show, diorama and demonstration of the neighborhood's traditional activities, photography session with traditional Malay garments, traditional games, wedding ceremony and staying overnight in the house like chalets in order to make the sites and program lively as proposed by the National Trust for Historic Preservation. Therefore, it is possible to transform the traditional houses in selected villages in Merlimau into a cultural village consists of museum, art gallery, lodge and tourism research center for the area.

Printed and electronic brochures with effective map and direction of the trails need to be developed and published in the tourism authorities' websites in order to market the tourism product. The issues of demolishing the traditional houses into modern structured house could be reduced and solved if the control and transformation into tourism products by the authority are actively done. Instead of developing or building new tourism products, the authority could take into consideration to preserve and sustain the old and historical houses for future generation and for the sake of tourism development in Merlimau.

As a conclusion, if the local authorities and political leaders realize that the cultural heritage tourism is a potential economic development tool for the local communities in Merlimau, then it will be important to lay a solid framework and efforts need to be integrated from all relevant parties.

REFERENCES

Ahmad, N.A., Jamaludin, N., & Ani, A.I.C. (2010). Keberkesananpengudaraansemulajadibagi dalaman rumah:kajianimpakkeatasTelapakNaning, Melaka. Isu-isukontemporari Malaysia. *Dapatan inisiatif penyelidikan, 4*(93).

Gültekin, E.R. (2011). Traditional Akseki houses and cultural heritage tourism. *International Journal of Energy and Environment, 5*(5), 661–667.

Malaysia national heritage department (2011). The Penghulu Abdul Ghani house. Retrieved October 6, 2012 from https://www.heritage.gov.my/english/index.php?option=com_content&view=article&id=137&Itemid=217.

Melaka Museum Corporation (2008). rumahtradisional di Merlimau. Retrieved October 17, 2012 from http://www.perzim.gov.my/rumah_merlimau.htm.

Mohamad Bakri, D. (2008, March 11). Growth in Malacca tourism. The Star Online. Retrieved October 18, 2012from http://thestar.com.my/metro/story.asp?file=/2008/3/11/southneast/20579177&sec=southneast.

Nor Haniza, I., Zuraini, M.A., Yacob, O., & Helena, A.H. (2003). Case studies on timber defects of selected traditional houses in Malacca. *Journal of Design and the Built Environment, 3*(7), 81–90.

NurAzfahani, A., Nazhatulzalkis, J., & Ahmad, R. (2009). The effectiveness of cross ventilation for traditional Malay houses in Malaysia: Case study of 'Kutai' House in Perak. International Conference on Engineering, Environment, Economic, Safety and Health (CONVEEESH '09), Manado, Sulawesi, Indonesia, 26–27th October 2009 (ISBN 978-979-17509-7-4).

Poimiroo, J. (1996). The architectural attraction: How great architecture attracts tourism. Retrieved October 18, 2012 from http://hotelexecutive.com/business_review/1390/the-architectural-attraction-how-great-architecture-attracts-tourism.

Stovel, H. (2007). Effective use of authenticity and integrity as world heritage qualifying conditions. *City and Time, 2*(3), 3.

UNESCO, (2005). Operational guidelines for the implementation of the world heritage convention. UNESCO World Heritage Centre.

Wang, N. (1999). Rethinking Authenticity in Tourism Experience. *Annals of Tourism Research, 26*(2), 349–370.

World Heritage Convention. (2008). Operational guidelines for the implementation of the world heritage convention. Retrieved October 19, 2012 from http://whc.unesco.org/en/guidelines/.

World Tourism Organization. (2012). Tourism 2020 Vision. Retrieved October 17, 2012 from http://unwto.org/facts/eng/vision.htm.

Hospitality and Tourism – Sumarjan et al. (eds)
© 2014 Taylor & Francis Group, London, ISBN 978-1-138-00151-0

Service delay of cook to order food: How do Malaysia *Gerai* customers react?

N. Zainol, A.R. Ahmad Rozali, S.B. Usman & M.S. Mohd Zahari
Universiti Teknologi MARA, Malaysia

ABSTRACT: Every individual would agree that waiting for service of food in the restaurant increasingly is becoming common in peoples' daily life. Some may experience having to wait before service is rendered, some during its delivery and some after the core service has completed. The wait for service is denoted as the time from which a customer or consumers are ready to receive the service until the time the service commences or to the state of readiness felt by that customer during the wait. However, service delay beyond expected time affects customers dining mood, overall satisfaction and closely related to complain behavior and implicate their future re-visit intention. This study empirically investigates the attitudes and reaction of Malaysian customers toward service delay of cook to order food at one of the small scale types of restaurant popularly known by the Malaysian as *Gerai*.

Keywords: cook to order food, customer reaction, Malaysia *Gerai*, service delay

1 INTRODUCTION

Restaurant industry is one of the fastest growing industries and continuously booming along with other industries in Malaysia (Euromonitor, 2007). An array of restaurants is mushrooming in all subsectors of the foodservice industry. This sector is driven by the profit motive and the aspiration to provide customers with good quality of food and drinks at a reasonable price. Restaurant industry is directly contributed to local, national and regional economy and classified as the fourth largest consumer market with more than RM 37,386.2 billion in sales every year from more than 57,670 restaurants outlets nationwide include upscale, medium, ethnic and stalls (Euromonitor, 2010). Despite many types and unlimited choice of restaurants in this country, the present study only focused on one of the small type scale restaurant, which popularly known by the Malaysian as *Gerai*.

According to Winarno and Allain (2001) and Holt (2008), *Gerai* is a type of restaurant that sells "ready-to-eat food and beverages" or "cooked to—order food and beverages" either in the street or in the public eating places and normally operated by a family or individual. Subratty et al. (2004) classified *Gerai* as those restaurants that operate only at night and selling mostly cooked to order food and beverages. This type of restaurant is normally run without air-conditioning and menu prices are usually inexpensive compared to full service restaurants which attract the majority of the low to middle income customers. *Gerai* offers a range of local dishes, where the dishes are cooked quickly and can be taken away or consumed therein and the work stopped once all the products were sold (Holt, 2008).

There are number of reasons of looking at this type of restaurant. First, *Gerai* is found besides others to be one of the fastest growing types of restaurant in the country and remain popular places to eat for most people (Euromonitor, 2007). Second, it is well recognized and offer range of menu particularly on cook to order foods (ala carte) which is inexpensive compare to full service and medium restaurants and attract frequently visit by majority of low, middle to high income customers. Third, most of these restaurants operated during dinner time, customer therefore are patronizing for the purpose of dining.

With that hallmark, it is argued that service delay, complaint behaviour and the issue of repatronization are even worst at the *Gerai* compared to other types of restaurants. Based on the preliminary survey with 30 *Gerai* customers in Seberang Jaya, Pulau Pinang, majority of them admitted that they had experienced more service delay of cook to order food at the *Gerai* compared to other types of restaurants. These occurrences raises critical questions; what are the customers expected times for one portion of common cook to order food at *Gerai* to be delivered? What are the general attitudes and reaction of *Gerai* customers toward service delay beyond expected time?

Is there any difference on reaction toward service delay between age group? Thus, this paper is to answer such questions.

2 LITERATURE REVIEW

2.1 *Service delay and its perspective*

Service delays have received significant attention among the researchers from differing perspectives such as economics, marketing, management science and service industry including the restaurant industry. In the management science and marketing, service delays are a major source of consumer dissatisfaction (Hoffman & Kelley, 2000; Tom & Lucey, 1999; Tse, 2001). Tom and Lucey (1999), for instance looked on the waiting time or service delay and consumer satisfaction in supermarket. Time seems to be the factor most critical to consumers shopping experiences, not just in grocery stores but also in the retail outlet in general. According to Ittig (2002) some of retail and supermarket customers dislike waiting in line so much that they hire people to wait in lines for them. There was also evidence indicating service delay in the retail store that affect both overall customers satisfaction on the services rendered and future intention of repurchased or revisit.

Pruyn and Smidts (1998) argued negative experiences from service delay in many purchase situations and it seems to be a decisive element, not only with regard to the customer's level of satisfaction with service but their loyalty to the service company. In fact, there are relationships between perceived waiting time, the attributions of delays of service cause anger, satisfaction or dissatisfaction with service, consumer behavioral intention, propensity to complain and repurchase intentions (Cameron et al., 2003; Diaz & Ruiz, 2002) and customers perception of waiting time in the retail stores may differ from one individual to another (Cameron et al., 2003; Whiting & Donthu, 2009). Jones and Dent (1994) identified three types of customers. "Watchers" is a type who enjoy the bustle and do not mind waiting while "Neutrals" display neither enjoyment or frustration and "impatiens" who hate waiting and will always queue hop and complains.

Economists see service delay as form of price. Cameron et al. (2003) noted that customers incur high waiting costs during service delays and these could include financial costs, opportunity costs and social or emotional costs. They argued that the full cost of waiting activities equals the sum of market prices and the foregone value of the time used up. Zeithaml (2000) discussed the concept of perceived price and included time cost as part of perceived price. From a customer perspective, price is what is given up or sacrificed to obtain the products. Ittig (2002) attempted to incorporate the impact of waiting time on customer demand. He suggested that it may be appropriate for the researchers and service firms to devote more attention to the relationship between time and consumer behavior. He added that there are difficulties in measuring perceived price and in accounting for a price that may differ for each customer. However, service delay may costs a firm five to seven times more from new customer's compared to the existing customers (Chiang, 2007).

In the restaurant industry, delay for service typically represents the first direct interaction between customers and most delivery processes (Davis & Heineke, 1998). Many migrant based ethnic restaurants are criticized for providing inconsistent or poor service delivery to their customers compared to native ethnic restaurants (Josiam & Monteiro, 2004; Verbeke & Lopéz, 2005). Josiam and Monteiro (2004) reported that ethnic restaurants in the United States were having difficulty in attracting repeat customers due to the poor service delivery, service quality, poor service encounters, and poor employee interaction with customers. Pang (2002) stated that the positive acceptance of Chinese restaurants in Belgium, specifically in Antwerp has been tarnished by the poor service provided by the restaurant operators. The same scenario has occurred in Chinese restaurants in Germany where customers demand not just great food but also expect efficient service in delivering the food (Leung, 2002). Verbeke and Lopez (2005) argued that ethnic restaurants are continuing to develop rapidly but are still having problems in terms of service, particularly in waiting times for food to be delivered to the customers. In the United Kingdom, Jamal (1996) found that despite positive acceptance by British customers of ethnic restaurants, most customers commented that the service provided by the restaurant operators was not consistent and the quality of the service received was not up to their expectation.

3 METHODOLOGY

A questionnaire survey through a self reported experience was used to gain *Gerai* customers reaction of service delay of cook to order food and their re-patronization intention. It involved customers who had experienced two or more incidences of service delay at the *Gerai*. However, owing to large number of *Gerai* customers in the country only one contextual data collection setting which is Seberang Jaya, Pulau Pinang was chosen. To avoid disturbing customers mood of dining experiences, service delay was measured by selecting *Gerai* customers randomly at three popular supermarkets in Seberang Jaya city rather than at the *Gerai* itself. These shopping complexes and supermarkets were chosen because they are among the most popular in the city neighborhood.

The survey instrument used was developed and modified after review of literatures. The questionnaire contained five sections. Respondents were asked to report their views based on the five-point Likert scale ranging from 1, which represents "Strongly Disagree" to 5, "Strongly Agree" in all sections except the demographic profiles.

The surveys were conducted on four weekends at three selected shopping complexes, Carrefour, Pacific and Sunway, Seberang Jaya City, Penang. Respondents were randomly approached before entering the shopping malls with a screening question whether they had previously experienced of service delays at *Gerai*. Those met the criteria's were invited to take part in the study. The respondents were informed that all information provided by them was strictly confidential and that no individual would be identified. With positive response, 332 questionnaires were successfully distributed. The reliability test was than undertaken for Section B, C and D separately. The result showed that the instrument and items used was reliable with coefficient alpha value at 0.72 for Section B, 0.66 for Section C, 0.89 for Section D and 0.72 for Section E.

4 RESULT AND DISCUSSION

4.1 *Informant profiles*

Male exceeded the female with 65.4 percent ($n = 217$) against 34.6 percent ($n = 115$). The majority of respondents were between 18 to 35 years old made up 45.8 percent of the total sample ($n = 152$) followed by 36 to 45 years old which represented 30.1 percent ($n = 100$) and 46 years and above represented around 24.1 percent ($n = 80$). 40.7 percent ($n = 134$) of respondents were among the students compared to 22.3 percent ($n = 74$) of government servants, 21.0 percent ($n = 71$) were self employed and 16.0 percent ($n = 53$) the private sectors employees. On the frequency of visits, 37.4 percent ($n = 124$) of respondents dined between 2 and 3 times weekly, followed by 31.6 percent ($n = 105$) dined only once a week, 22.9 percent ($n = 76$) dined between 4 and 5 times, and 8.1 percent ($n = 27$) dined more than 7 times a weekly.

4.2 *Expected cook to order food to be delivered*

This section analysis looked on the times expected for one portion of cook to order food to be delivered. Majority of the respondents (79.5 percent, $n = 264$) expected that varieties of fried noodles and those similar to be delivered between 5 to 15 minutes. Similar time was expected for the fried rice (80.1 percent, $n = 266$), soups and those similar (75.6 percent, $n = 251$). With regard to roti canai and those similar, respondents expected between 5 to 10 minutes of these products to reach their dining table (79.5 percent, $n = 252$). Lastly, majority of the respondents (90.1 percent, $n = 299$) anticipated 5 to 10 minutes of freshly prepared drinks to be delivered. All these results not only in line with the result obtained from random interviewed with *Gerai* customers but strengthened the result revealed by Fraser et al. (2009).

4.3 *General attitudes about waiting and service delay*

The descriptive statistic looking at the mean score was used in examining the customer general attitudes and reaction about waiting and service delay beyond expected time in the *Gerai*. The respondents' scores are presented in Table 1.

Majority of the respondents agreed that waiting is a bad experience. This can clearly be seen from a greater level of agreement ($M = 3.57$, item 1) given to this item. They were more in agreement that good meal will spoil with too long waiting ($M = 3.38$, item 6) and therefore affects their mood

Table 1. Showing the mean score of the general attitudes about service delay ($n = 332$).

Items	*M*	S.D
Waiting too long is a bad experience	3.57	1.019
If I see a lot customers I would not order cook to order food	3.12	0.992
I do not mind waiting if I know the price is reasonable	3.53	0.849
Waiting too long affects my mood to dine	3.08	1.117
I do not mind waiting if I know for how long	3.14	1.053
A good meal spoil with too long waiting	3.38	0.933
I do not mind waiting if the staff give me an explanation	3.22	0.996
I do not feel like spending after I have been waiting	2.73	1.045
I would rather pay more for not to have long wait	2.71	0.993
I do not mind waiting in the weekends	3.11	0.879
I do not mind waiting if I know the quality of food suit my taste	3.52	0.891
I do mind waiting for good food despite I know service is inefficient	2.68	1.100

Scales: 1 = strongly disagree, 2 = disagree, 3 = slightly agree, 4 = agree, 5 = strongly agree.

to dine (M = 3.08, item 4). This notion is further supported when they do not feel like spending (M = 2.73, item 8) if they had to wait a long time, would not order or choose cook to order food (M = 3.12, item 2) if they see a lot of customers and would prefer to pay more for not to have long wait (M = 2.71, item 9).

Results also show that respondents were only willing to wait long with certain conditions. As such, they do not mind waiting if they know how long that they have to wait for the food to be delivered (M = 3.14, item 5) with an explanation from the staff (M = 3.22, item 7), good food is served (M = 2.68, item 12), quality and taste of the food suit their palate (M = 3.52, item 11), the price is reasonable t (M = 3.53, item 3) and waiting in the weekends (M = 3.11, item 10).

Together, these indicated that majority of the respondent were very concerned about waiting times of cook to order food and too long waiting do affects their mood to dine despite slightly tolerated on a few attributes. This result supported previous studies that revealed consumers dislike and tend to switch to other service provider if they have to wait longer than they expected (Dawes & Rowley, 1996; Jones & Dent, 1994; Tom & Lucey, 1995). In fact, Colgate and Norris (2001) contended that customers react in two ways when they encounter a service delay either remain with that service provider or leave and find a new provider.

4.4 *Comparison reaction of service delay based age group*

One-Way Analysis of Variance (ANOVA) test with the Scheffé *post hoc* was undertaken to see the differences between customer age groups reaction towards service delay. Ten out of seventeen items were found to have statistically significant differences between the age groups. By means of a Scheffé multiple comparison test, the differences found were mostly between the Young Adult (YA) and the Mature Adult (MA) groups compared to Middle Age Adults (MAA). Result showed that a long wait is a bad experience and young customers are found to be more aggressive than the mature adult. This is evident when Young Adults (YA) expressed that they will straight away complaint if they have to wait long than the Mature Adult (MA) and Middle Age Adults (MAA) (YA M = 3.83, p = 0.011 compared to M = 3.24, MAA M = 3.37) and will start grumbling if they have to wait long (M = 4.32, p = 0.002 compared to M = 4.04 and M = 4.05, respectively), especially when dine with their family (M = 3.95, p = 0.020, opposed to M = 3.51 and M = 3.70 respectively).

Additionally, the young adults also expressed that they will absolutely get angry if they have to wait longer for a simple food as opposed to mature and middle age adults (M = 4.32, p = 0.000, compared to M = 3.46 and M = 4.10, respectively). Further, they will definitely complaint to the service staff if they have to wait long (M = 3.99, p = 0.000 compared to M = 3.31 and M = 3.33) especially during slow period (M = 4.49, p = 0.000, opposed to M = 3.41 and M = 3.65) and receive bad quality of food (M = 4.51, p = 0.001, compared to M = 3.79 and M = 4.06).

It is interesting to note that this young adults admitted that they will definitely lose their mood to dine if they have to wait too long than the mature adults and middle age adults (M = 4.38, p = 0.000 compared to M = 4.23 and M = 4.07). Further, they would definitely force demand for compensation (M = 3.55, p = 0.000, compared to M = 2.88 and M = 3.21) and definitely walk out from the *Gerai* (M = 4.09, p = 0.038, compared to M = 3.54 and M = 3.76).

What can be said from sub section analysis results is that all age groups dislike the service delay. However, the young and middle aged groups are more aggressive and easily get angry if they had to wait long for service beyond expectation compared to a more tolerable of the older group. This result is in fact strengthening and supports Ngai et al. (2007) and Walker (2001) who noted that the tendencies of complaining are more obvious to younger people compare to the older people.

5 IMPLICATIONS AND CONCLUSION

The evidence from this study has reconfirmed previous investigations related to restaurant service that service delay has a negative impact on customer mood dining. Some of the views expressed are universal, while others are more dominant among particular segments of the customer population. Overall customers do not like waiting and will take various steps to moderate a potentially negative experience. Such steps include not ordering cook to order food when they see a lot of people waiting or even going to another restaurant. As reported in other studies; an explanation of any delay, the restaurant is obviously really busy, having staff apologize for delays, are things that reduce the customer's sense of dissatisfaction.

What is new in this study is perhaps more useful to the restaurateur. All age groups dislike the service delay and the young and middle aged groups are more aggressive and easily get angry if they had to wait long for service beyond expectation compared to the older group. These groups will react negatively by not revisiting or not allowing the delay to influence their dining experience. The older customers although might be less likely to react negatively on slow service but they are least likely to be tolerant of service delay. *Gerai* operators therefore should not overlook on this issue but rather improving the level of service and staff attitudes to an acceptable level. This can be done through improving their service by delivering more prompt and efficient service consistently to the customers. This could be achieved by properly training the staff and educate them to be more sensitive on this aspects. Even though, this kind of approach may add a cost in both resources and money but it would help the staff to better understands the importance service delivery in restaurant business operation.

It is hoped that the recommendation and information from this study could facilitate *Gerai* operators to better understand customer needs therefore can well positioning themselves and be more competitive in the fast growing trend of restaurant businesses in Malaysia. In fact, a better understanding of the *Gerai* customers experience in waiting cook to order food would also provide *Gerai* operators with valuable information to be more sensitive and effective toward operation management techniques and improving customer satisfaction and repeat patronization.

REFERENCES

Cameron, A.M., Baker, J., Peterson, M., & Braunsberger, K. (2003). The effects of music, wait-length evaluation, and mood on a low-cost wait experience. *Journal of Business Research,* 56, 421–430.

Chiang, C.C. (2007). *The effect of experiencing outcome versus process service failures on Taiwan and U.S restaurant customers' service ratings, a case of college students*. Ph.D. dissertation, Alliant International University, San Diego.

Colgate, M. & Norris, M. (2001). Developing a comprehensive picture of service failure. *International Journal of Service Industry Management,* 12(3), 215–233.

Davis, M.M. & Heineke, J. (1998). How disconfirmation, perception and actual waiting times impact customer satisfaction. *International Journal of Service Industry Management,* 9(1), 64–73.

Dawes, J. & Rowley, J. (1996). The waiting experience: Towards service quality in the leisure industry. *International Journal of Contemporary Hospitality Management*, 8(1), 16–21.

Diaz, A.B. & Ruiz, F.J. (2002). The consumer's reaction to delays in service. *International Journal of Service Industry Management*, 13(2), 118–140.

Euromonitor. (2007). Consumer foodservice in Malaysia. *Euromonitor, July.*

Euromonitor. (2010), Consumer foodservice in Malaysia, *Euromonitor, August.*

Fraser, R.A., Zahari, M.S., & Othman, Z. (2008). Customer reaction to service delays in Malaysian ethnic restaurants. *South Asian Journal of Tourism and Heritage,* 1(1), 35–45.

Goffin, K. (1999). Customer supports a cross-industry study of distribution channels and strategies. *International Journal of Physical Distribution and Logistics Management,* 29, 374–397.

Hoffman, D.K. & Kelley, S.W. (2000). Perceived justice needs a recovery evaluation: A contingency approach. *European Journal of Marketing*, 34, 418–429.

Holt, P. (2008). World: a waterside restaurant in Malaysia. Retrieved March 4, 2008 from nlpnow.com/2007/12/29/a-waterside-restaurant-in-malaysia.aspx.

Ittig, T.P. (2002). The real cost of making customers wait. *International Journal of Service Industry Management,* 13(1), 231–241.

Jamal, A. (1996). Acculturation: the symbolism of ethnic eating among contemporary British consumers. *British Food Journal,* 98(10), 14–28.

Jones, P. & Dent, M. (1994). *Improving service: managing response time in hotel and restaurant operation.* In Johnston, R. & Slack, N.D.C (Eds). Service Superiority, Operations Management Association, Warwick, 331–337.

Josiam, B.M. & Monteiro, P.A. (2004). Tandoori tastes: Perceptions of Indian restaurants in America. *International Journal of Contemporary Hospitality Management,* 16(1), 18–26.

Pang, C.L. (2002). Business opportunity or food pornography? Chinese restaurant ventures in Antwerp. *International Journal of Entrepreneurial Behaviours and Research,* 8, 148–161.

Pruyn, A. & Smidts, A. (1998). Effects of waiting on the satisfaction with the service: beyond objective time measures. *International Journal of Research in Marketing,* 15(4), 321–334.

Subratty, A.H., Beeharry, P., & Chan, S.M. (2004). A survey of hygiene practices among food vendors in rural areas in Mauritius. *Nutrition and Food Science*, 34(5), 203–205.

Tse, A.C. (2001). How much more are consumers willing to pay for a higher level of service? A preliminary survey. *Journal of Service Marketing,* 15(1), 11–17.

Verbeke, W. & Lopéz, G.P. (2005). Ethnic food attitudes and behavior among Belgians and Hispanics living in Belgium. *British Food Journal,* 107(11), 823–840.

Whiting, A. & Donthu, N. (2009). Closing the gap between perceived and actual waiting times in a call center: Results from a field study. *Journal of Services Marketing,* 23(5), 279–288.

Winnarno, F.G. & Allain, A. (2001). Street foods developing countries: Lessons from Asia. *Food, Nutrition and Agriculture*, 1, FAO.

Zeithaml, V.A. & Bitner, M. (2000). *Services marketing: integrating customer focus across the firm* (2nd Ed.). Boston: McGraw-Hill.

Hospitality and Tourism – Sumarjan et al. (eds)
© 2014 Taylor & Francis Group, London, ISBN 978-1-138-00151-0

The influence of career path elements and mentoring towards women employability in the Iran tourism industry

S. Simin, A. Zainal & N.H. Nazlan
Universiti Teknologi MARA, Malaysia

ABSTRACT: One of the important criteria in evaluating a country's development is the attention given towards women's welfare, development and also recognizing their active and creative powers. Iran's ten-year plan showed an indication of various amount of barriers for women's employability such as low level of literacy, low variety for job opportunities and high cost of labor particularly. The booming tourism industry has created vast economic opportunities, creating job prospects for women globally and it is also said to provide better opportunities for women's participation in the workforce, entrepreneurship and leadership as compared to other sectors of the economy. This study was conducted to examine the influence of career path elements and mentoring towards women employability in the Iran tourism industry. A total of 350 questionnaires were distributed to female workers in hotels, travel agencies and tour guide associations in the capital city of Iran with a response rate of 52 per cent. Results showed significant relationship between career path elements and mentoring towards women's employability.

Keywords: Iran tourism industry, career path elements, mentoring, women employability

1 INTRODUCTION

Nowadays women employment is considered to be one of the world's most important issues. As reported by the International Labor Organization (ILO), women accounted 50 percent of the population, 30 percent of the labor force, perform 60 percent of all working hours and receive 10 percent of the world's income (ILO, 2011). This indicates that women employment is currently not in a favorable position of the world's economy.

In recent years, tourism industry has been considered as job creator for women and presents both opportunities and challenges for gender equality and women empowerment (UNWTO, 2010). Numerous researchers of women employment studies in Iran agreed that the importance of women employability in Iran is contributed by factors such as religious, social, cultural, political limitations. The obstacles included gender inequality, lack of women's economic and social participation, unemployment, unequal job opportunities in Iran labor market, unfavorable interpretations of women occupational role and negative attitudes toward women activities (Poorghasem, 2002; Noroozi, 2003; Kristiansen, 2003; Hashemi, 2005; Rostami Povey, 2005; ICCIM, 2009; Rahmani, 2012).

The tourism industry could be a significant source of foreign exchange for Iran (Kaffashi, 2009). Tourism activities also could bring in lucrative source of income provided that it is appropriately managed. The Tourism sector is one of the few successful and expanding sectors in the world that is dominated by women (Obadić, 2009). Therefore, additional attention to female employment issues should be considered to improve the Iran's economy as a whole and also enhance the economic condition of the Iranian women. There has been some research on the general impact of tourism on female employment in Iran (Hashemi, 2005; Ziaee, 2007), but none of these studies provides a picture of the effect of career path elements and mentoring on female career in tourism industry, particularly in Iran. Due to limited information in this area, this current study will specifically discuss about the effects of education, mentoring and career move as elements for carrier path on women employability in the Iran tourism industry.

This study considers the following objectives:

I. To determine if career path elements influence women employability in Iran.
II. To evaluate if mentoring influence women employability in Iran.

2 LITERATURE REVIEW

With the advent of knowledge based economy, the economic structure of countries began to change very fast and employability began to be uncertain in the last decades of the 20th century (Tome, 2007). Employees have to stay marketable despite of the uncertain economic condition. Employability becomes very important nowadays as it facilitates them to move from one job to another, within and between organizations (Fugate, Kinicki & Ashforth, 2004). Employability has primarily benefited employees with high developed or high demanded skills (Clarke & Patrickson, 2008). Recent studies, distinguished four factors that influences employability: formal and informal career management practices, training, job experiences, education and tenure with current job (Juhdi, Pa'Wan, Othman, & Moksin, 2010). Ruddy (1995) identified career paths as one of the factors that influence employability. Ayres (2006) focused on items for career path emphasized on mobility (career moves), education and mentoring, and latest research are the three influences on tourism careers that provide current and future tourism employees with tools on which to build their careers.

Career path requires the employees to make a plan to obtain what is necessary for their advancement (Human Resource Management Glossary Index, 2011). Ruddy (1995) designed a "Career development model" of hotel managers and stated that career path elements are connected to age, pattern of mobility and education level.

Education Level as one of the elements for career path that has been considered as a type of learning undertaken by an individual (Ayres, 2006) and Baruch (2003) stated the positive role of education in career development is widely accepted. Education is also recognized as important factors in increasing women employment, particularly in Iran (Hashemi, 2005).

Career moves and *mobility* are used interchangeably throughout the results and discussion and refers to any movement between job, organization, industry or geographical location. Mobility refers to job moves, organizational moves or geographical moves are a key influence on employment and career.

Ayres (2006) stated that careers in the tourism industry are influenced by variety of mentoring relationships. Mentoring is most often defined as a professional relationship in which an experienced person (the mentor) assists another (the mentored—the one having a mentor) in developing specific skills and knowledge that will enhance the less-experienced person's professional and personal growth.

3 METHODOLOGY

3.1 *Participants and procedures*

The respondent of this study comprises of women workers in the hospitality and tourism industry (four- and five-stars hotels, travel agencies and tour guide association) in Tehran metropolitan area. This combination of different respondents was designed to obtain more meaningful information and represented the women workers in the Iran tourism. From the 350 questionnaires distributed, 186 questionnaires were returned (52% response rate) and analyzed.

3.2 *Measures*

The career path elements were measured by looking into educational level, pattern of mobility. 15 items was adopted from Ruddy (1995) measuring the overall career path. Educational level was measured by 7 items and pattern of mobility—8 items. Mentoring was measured using 4 items adopted from Ayres (2006).

This study opted for a 5 point Likert scale, ranging from 1—Strongly disagree to 5—Strongly agree. The questionnaire was made bilingual (English and Persian) to ensure clarity and avoid ambiguity among the respondents.

3.3 *Data analyses*

Descriptive statistic was used to provide an overview of respondents' profile. The regression analyses were conducted to assess the relationship between the independent variables of career path elements and mentoring toward the criterion variable of women employability.

4 RESULTS

4.1 *Respondents' profile*

A total of 186 usable questionnaires were analyzed and reported; 111 from travel agencies, 45 from hotel and 30 from tour guide association. The majority of respondents (69.4%) are between 21 and 29 years old. Four percent of the respondents ($n = 8$) have >10 years of experience, 43.5% ($n = 81$) have <3 years of experiences. The largest group 48.9% ($n = 91$) is those who work in organizations having between 5 to 20 employees. The income of 51.5% ($n = 96$) of the respondents is between 300 US$ to 500 US$ per month and only 4.8% ($n = 9$) earn more than 750 US$ per month. Majority of respondents ($n = 85$) have experienced a job turnover at least once while only 9 respondents (4.8%) have moved from one job to

Table 1. Result of regression analyses for testing for career path elements, mentoring on women employability.

Predictors	Model 1 Std. β
Career path elements	0.45***
Mentoring	0.19***
R2	0.33
Adj. R2	0.32
R2 Change	0.33
F Change	37.33

*$p < 0.005$; **$p < 0.001$; ***$p < 0.000$.

another job for more than 4 times in their career life.

4.2 *Hypotheses testing*

Table 1 provides the result of the regression analyses to test the influence of career path elements and mentoring on women employability. Table 1 shows the two variables (career path elements and mentoring) explains 32 percent (R2 = 0.32, F Change = 37.33, $p < 0.05$) of the changes in the variance of women employability. Career path elements ($\beta = 0.45$, $p < 0.05$) is a significant predictor of women employability, indicating that hypothesis 1 is supported. On the other hand, mentoring ($\beta = 0.19$, $p < 0.01$) also indicates a significant predictor of women employability, thus substantiating hypothesis 2.

5 DISCUSSION

Based on the finding, career path elements have an influence on women who are employed in the hospitality and tourism industry ($\beta = 0.45$, $p < 0.01$). This indicates that career path elements are indeed a significant predictor for female employees in Iran tourism industry. The finding for this research supports H1, and aligned with previous research. Ayres (2006) highlighted the role of formal and informal education, training pertaining and mobility to such careers. There is a moderate and positive relationship between mentoring and women employability. As reported in the findings, mentoring ($\beta = 0.19$, $p < 0.000$) significantly influence women employability thus supporting hypothesis 2. Although there is no widely accepted definition and agreement on the fundamental aspects, mentoring relationships have great ground for enhancement of an individual development in early and middle career stages (Kram, 1985; Ragins et al., 2000). Furthermore, Headlam-Wells (2004) suggested that mentorship enables women to access key resources and senior personnel, and helps increase their knowledge of organizational policies. Murphy and Ensher (2001) also highlighted the mentorship benefits in the organization.

6 IMPLICATIONS

The findings of this study also provide implications for the Ministry of Tourism of Iran. The conceptual link observed between elements such as career path elements and mentoring in organizational climate serves as determinants of women employability.

As such, a more in-depth concentration on such issue could potentially be a solution to improve Iran's economic position and simultaneously allowing women to also contribute to the economic progress.

This inferred the growing importance of working women to seek professional on skill and knowledge in enhancing their carrier path. Conducive working environment alone is not sufficient to entice women's career.

7 CONCLUSION

This study highlights some important women career issues in the Iran tourism industry. Tourism organizations need to acknowledge the issues of personal and professional development in order to maintain the flow of high employability for women into the industry. However, to make changes pertaining to female's career in tourism industry, the industry itself would need to encourage the development of a career path that is more diverse and dynamic. This will consequently serve as an avenue for future researchers to contribute a further theoretical development and formulation of new and unexpected research questions. Also it is essential for Iran tourism industry to pay much attention on establishing a clear set of career path elements (education and mobility) and mentoring that involve women employability.

REFERENCES

Ayres, H.J. (2006). "Influences On Career Development In Australian Tourism". PhD thesis. University of Canberra.

Baruch, Y. (2003). Transforming careers: from linear to multidirectional career paths: organizational and individual perspectives. *Career Development International*, 9(1), 58–73.

Clarke, M. & Patrickson, M. (2008). The new covenant of employability. *Employee Relations*, 30(2), 121–141.

Fugate, M., Kinicki, A.J. & Ashforth, B.E. (2004). Employability: A psycho-social Construct, its dimensions, and applications. *Journal of Vocational Behaviour,* 65(1), 14–38.

Hair J.F.J, Black W.C, Babin B.J, Anderson R.E, Tatham R.I (2006). Multivariate data analysis (6th ed.). New Jersey: Prentice Hall.

Hashemi, A. (2005). Female employment and the need. Abstract. From: www.farslabour.ir/cms2/Admin1256/img/1/women%5B1%5D.doc.

Headlam-Wells, J. (2004). E-mentoring for aspiring women managers. *Women in Management Review,* 19(4), 212–218.

ICCIM. (2009). Iran Chamber of commerce industries and mines report. http://www.iccim.ir/fa/index.php?option=com_content&view=article&catid=31&id=3362.

ILO Report. (2010). Women in labor markets: Measuring progress and identifying challenges, International Labour Office, Geneva.

Juhdi, N., Pa'wan, F., Othman, N.A. & Moksin, H. (2010). Internal and external employability of employees: a study among employees in Klang Valley, Malaysia.

Kaffashi, M. (2009). The Journal of Modern Thoughts in Education, *4*(4), 141–164.

Kram, K.E., & Isabella, L.A. (1985). Mentoring alternatives: The role of peer relationships in career development. *Academy of management Journal,* 110–132.

Kristiansen, N. (2003). Women's resources are wasted in Iran. published in online Kilden Magazine (information center for Gender Research in Norway).

Management Mentors (2011). http://www.managementmentors.com/resources/corporate-mentoring-programs-faqs/.

Moayedi. M. (1991). Statistical Glance at Women's Employment in Iran. Seaming monthly publication Kowsar Economic Organisation No. 39–40.

Murphy, S.E. & Ensher, E.A. (2001). The role of mentoring support and self-management strategies on reported career outcomes. *Journal of Career Development*, 27(4), 229–246.

Noroozi. L. (2013). Study for female labor market developments in Iran during (1997–2001) and its future prospects. http://www.lssi.ir/pages/pj/tarhaye%20tahghighati/66.htm.

Obadić, A. (2009). The significance of tourism as an employment generator of female labour force. Journal of Ekon. Misao Prakasa DBK. GOD XVIII. (1) 93–114. Retrieved May 2010.

Pallant, J. (2005). SPSS survivor manual. A Step-by-Step Guide to Data Analyses using SPSS version, 15(3).

Poorghasem. Sh. (2002). Research about Employment educated women. Ms.theses. Iran.

Ragins, B.R., Cotton, J.L., & Miller, J.S. (2000). Marginal mentoring: the effects of type of mentor, quality of relationship, and program design on work and career attitudes. *Academy of Management Journal,* 43(6), 1177–1194.

Rahmani. M. (2012). Iranian women committee. http://ir-women.com/spip.php?article10717.

Rostami, M. (2007). Research about women employability in handicraft. *Journal of Social Science*. Retrived on February 2009.

Ruddy, N.J. (1995). The career development of hotel managers: towards a model of individual career planning and management—PhD Thesis, University of Salford.

Tomé, E. (2007). Employability, skills and training in Portugal (1988–2000): Evidence from official data. *Journal of European Industrial Training*, 31(5), 336–357.

Ziaee, M. (2007). World Tourism Day slogan is based on make powerful women. Paper was presented at 2007 World Tourism Day celebration in Tabatabaee University, Iran. Retrieved in CHN. Available on:http://www.chn.ir/news/Print/?Section=1&id=24756.

Hospitality and Tourism – Sumarjan et al. (eds)

The Malay garden design concept opportunity to be developed as the tourism product

A.Z. Zakaria, I.H. Salleh & M.S. Abd Rashid
Universiti Teknologi MARA, Perak, Malaysia

ABSTRACT: The Malay garden design idea is a concept for the landscape design which deals with the characteristics of a private garden of the Malay community in Peninsular of Malaysia. The concept of cultural tourism shows its role by highlighting the activities of visitors and stock of products offered at a destination that classified as a cultural area. The purpose of this study is to explore the concept of Malay garden design in the eyes of tourists. The objectives of this research are; 1) to identify the relationship between cultural tourism and cultural landscape, 2) to convey the concept of the Malay garden as a new tourism product. This study used historical manuscripts on the Malay garden's existence and site visit to the old houses of the Malay community in peninsular Malaysia. This study expected to provide ideas on planning and development of new tourism products based on local culture especially the Malays.

Keywords: cultural tourism, Malay gardens, Malay landscape, national heritage, tourism products

1 INTRODUCTION

The Malays have many fascinating cultures and elements such as creativity and skills of weaving, carving, handicrafts and architecture which can be promoted to become a model of the concept of Malay gardens. However, the Malaysians have no personality to proud of landscape design by all, especially the Malays. The landscape architecture scenarios in Malaysia are identical with many modern designs. Currently, there are various design concepts in the field of landscape architecture. Among the well-known concept are the Japanese, Balinese and the English garden.

Religious influence distinguished Malay gardens against Japanese gardens and Balinese gardens, as 'Malay gardens' influenced by Islam, 'Japanese gardens' influenced by Chinese Buddhist and 'Balinese gardens' influenced by Hinduism. History has shown that, in the past, the Malays famously known as a civilized nation with prominent cultural value. Therefore, it is fascinating to discus on; Does civilized Malays, did not have their own garden design as the other races in the world? However, if we refer to the old manuscripts of the Malays, the Malays own gardens in their heyday, which does not prolong.

1.1 *The cultural landscape*

In the context of sociology, the term cultural landscape reflects the personality, culture, social and the local economics. The cultural landscape has a close association with human settlements in the area. Boehm (1996) argued that the cultural landscape closely related with the field of anthropology and archeology. Many researchers have concluded that the cultural landscape can be a basic overview of the social structure, economy and culture of a society who inhabit every corner of this earth.

1.2 *The cultural tourism*

The concept of cultural tourism is an extremely complex because it will take every result of artificial subjects. According to Smith (1979), cultural tourism refers to tourism activities that make the elements of culture as the main attraction. It includes the way of life, culture-based, food and drink, heritage, architecture, industrial products and handicrafts, way to entertain guests and all the elements that characterize the way of life in a place.

McIntosh and Goeldner (1986) argued cultural tourism encompasses all aspects of the adventure which also involves the traveler to learn about the history of a place and the foreign community heritage or way of life and their current thinking. In other words, cultural tourists will learn about the product and the culture of the places they have visited. According to Silberberg (1995), cultural tourism is a medium in an economic development which established in economic growth through the magnetism of tourists from the outside community or in part is tourists who fascinated to look at the historical, artistic, scientific or lifestyle heritage which is available in a community, region, group or an institution in the world.

2 PROMOTION STRATEGY

2.1 *Tourism promotion strategy*

Strategy in promoting a new tourism product should be strengthened with full involvement of various parties including governments, private agencies and the mass media. Promotion should not be done only in Malaysia but expanded globally. To encourage visits, it requires a variety of promotional methods, appropriate introduction to the concept of Malay gardens to the public and construction of public parks based on the concept of Malay garden design.

2.2 *A strong sense of place*

Referring to Harry (1985), an introduction to the unique character must continue to change and any changes supposed to design and cover as well as protect the uniqueness. The unique character or a strong sense of place often based on architectural styles particularly; climate quality and measure of light, rainfall, changes in temperature, uniqueness of the natural environment, memory and metaphor to show something happen that will give meaning to those who experienced it, use of local materials, craftsmanship, sensitivity in placing buildings and bridges are vital, the range of cultural and historical values, the values of the people, high quality public surroundings able to seen and accessed by all.

3 MALAY GARDEN CONCEPT

3.1 *The old Malay garden concept*

The aged Malay community has recorded a variety of gardens in their manuscripts. Among the well-known record are as follows: At the time of Majapahit Kingdom in the 14th century, there was a story about the city and the garden at the time as mentioned in the book of "Nagarakartagama". The interpretation; there is an extremely beautiful palace garden, red stonewalled, thick and high. In front of the west entrance, there is a square, and in the middle there was a circle. There is a Brahmana tree (*Ficus religiosa*) of row upon row, lined-row Bodhi tree (*Ficus rumphii*), Nagasari (*Mesuaferrea*) or Topen plant (*Mallotus floribunda*), Cempaka (*Micheliacampaka*) as well as flowering plants magnificently planted. Hikayat Hang Tuah also mentions about the garden; "... *bermula ada hampir kolam cenderamata itu sebuah balai gading bersendi-sendi dengan kayu arang tamar. Adapunbumi Taman ITUditambaknyadaripadatanahKoiDanditanamSerbiabagaijenisbunga-bungaan, air MawarMiraDan air MawarUnguDanBangui air Mawarputih, CempakaDanbungakenangaDan* ..." This shows us there are gardens with planned construction during those days.

In Sulalatus al-Salatin (*Sejarah Melayu*) written by Tun Seri Lanang, there was a legend about the Garden of Gunung Ledang. The garden has a collection of bamboo trees with a variety of fruit trees and flowers. There is also a fish pond in the garden. Garden of Gunung Ledang is a place for princesses and maidens revelers to enjoy the beautiful atmosphere. In Sulalatus al-Salatin (*Sejarah Melayu)* also mentions various plants which include *pohon ru, jambu laut, pinang gading, gandasuli, bakau, nyireh, tumu, api-api, kayu mahang, beringin, kuras, kempas, tui, ranggas, tulang daing, seri gading, ketirah, ibul* and *kulim*. The findings from the Hikayat Hang Tuah, written by Kassim Ahmad (1975) found that there were 52 different species of ornamental plants, 41 types of fruit trees and 7 types of plants can be classified as herbaceous plants/medicines.

Referring to the Bustan al-Salatin (*Taman Raja-raja*) written by Nuruddin al-Raniri, there are 48 different species of ornamental plants, 26 different species of fruit trees, 36 species of herbaceous plants and 4 different species of trees/logs. Based on the Hikayat Abdullah Bin Abdul Kadir by Amin Sweeney (2006), there are 3 types of ornamental plant species, 18 different species of fruit trees, 16 species of herbaceous plants and 4 different species of trees/logs. Furthermore, based on the Malay Medical Heritage refined by A. Samad Ahmad (1982), more than 200 species of plants can be used in traditional Malay medicine.

3.2 *Current Malay garden concept*

There were some differences in the concept of Malay community garden which include in terms of its geographical position aspects, cultural of societal aspects, and the terrain aspects. There are differences in terms of visible geographic distribution zones in peninsular Malaysia, the northern zone (Perlis, Kedah and Penang), middle zone (Perak, Selangor and Kuala Lumpur), the southern zone (Negeri Sembilan, Melaka and Johor) and the eastern zone (Kelantan, Terengganu and Pahang). It shows that culture has played an important role in the design and arrangement of a garden in the Malay community. Migration also influences the design and arrangement of the garden based on the origin civilization.

The southern states heavily influenced by the Javanese and Banjarese and the northern, and the eastern are prone to influence by the Siamese community. There are also similarities in the use of plant species in the Malay community. For example, the Malays also used the fragrant plants, salads, herbs and medicines as a major plant for their home surroundings.

The Malay people appreciate the beauty of house garden or the garden of the palace which planted with a variety of plants, particularly the fragrance plants. It is often depicted in old Malay manuscripts such as Hikayat Malim Deman, Hikayat Hang Tuah, The Malay Annals and others. Among the plants of choice of the Malay community are *Cocos nucifera* (Coconut), *Canangiumodoratum* (Kenanga), *Jasminum sambac* (Jasmine), *Vallarisglabra* (Kesidang), *Areca catechu* (Areca nut), *Cosmosviolaceace* (Ulam king), *Cymbopogon nardus* (Lemongrass), *Languas galanga* (Galangal), *Curcuma domestica* (Turmeric), *Lawsonia inermis* (Henna) and *Zingiber officinale* (Ginger).

4 THE MALAY GARDEN CONCEPT DESIGN MODEL

The Malay garden concept design model is a research initiative in creating a garden and landscape design concept exclusive to the Malays in Peninsular Malaysia. The data are the elements of the landscape and garden that will form the basis of The Malay garden design model (Table 1). From the old Malay manuscripts, types of plants divided into four main groups based on their functions and values; aesthetic aspect, herb and food plants, fruit plants and towering trees or hardwood trees.

According to historical studies, there were 98 species of plants for aesthetic purposes, 130 species of herb and food purposes, 68 species of fruit trees and 51 species of towering trees (hardwood trees) mentioned in old Malay manuscripts. Table 2 listed the seven most often cited plant species in the old Malay manuscripts. It is also an outcome of observation to the old Malay films specially directed by the late Tan Sri P. Ramlee (Tables 3 and 4).

Table 1. List of landscapefurniture observed by the author and recorded in the research papers done by researchers in the past about the Malay garden in the Malay Archipelago.

No.	Local name (as be called by local people)	English name
1.	Ampaian	Suspension
2.	Bangsal/Sulap	Barn
3.	Buaian	Tree swings
4.	Gerbang	Arch
5.	Guri	Guri (smaller than common traditional Malay water vessel)
6.	Jamban/tandas	Latrine
7.	Jamung andang/kandil	Torch
8.	Kepok padi/jelapang	Paddy store
9.	Kolah	Water tank
10.	Kolam	Pond
11.	Pagar	Fence
12.	Pangkin	Resting hut (usually without shelter)
13.	Pasu bunga	Flower pot
14.	Planter	Open timber platform (usually for washing clothes)
15.	Perigi	Well
16.	Perun	Dump site
17.	Reban	Hen coop
18.	Tempayan	Water vessel
19.	Titi	Log bridge (usually made from Areca nut trunk)
20.	Wakaf	Gazebo

(Source: Ahmad Zamil Zakaria, Ismail Hafiz Salleh Mohd Sabrizaa Abdul Rashid (2012), Landscape Furniture Present In The Ancient Malay Garden According To Old Manuscripts And Their Effects On The Formation Of Malay Garden Design Concept Model In Malaysia, PSU—USM International Conference on Art and Sciences).

Table 2. List of the most often cited plant species in the old Malay manuscripts.

No.	Local name (As it seems in the manuscript)	Botanical name	Description
1.	Bunga gandasuli/gandasuli	*Hedychium coronarium*	4 (BaS, TOHT, SaS, TMM)
2.	Kelapa/bunga kelapa/ nyiur/nyior	*Cocos nucifera*	5 (BaS, TOA, TOHT, TOI, TOMM)
3	Bunga delima/bunga delima wanta/delima	*Punica granatum*	6 (BaS, TOHT, TOI, TOMM, SaS, TMM)
4.	Pinang	*Areca catechu*	5 (BaS, TOA, TOHT, TOI, TOMM)
5.	Langsat	*Lansium domesticum*	4 (BaS, TOA, TOHT, TMM)
6.	Anggur	*Vitis spp.*	4 (BaS, TOA, TOHT, TOI)
7.	Gaharu	*Aquilaria malaccensis*	4 (BaS, TOA, TOI, TMM)

Legend; 1. Bustan al-Salatin—The Garden of Kings (BaS), 2. Tale of Abdullah (TOA), 3. Tale of Hang Tuah (TOHT), 4. Tale of Inderaputera (TOI), 5. Tale of Merong Mahawangsa (TOM), 6. Sulalatus al-Salatin—Malay Annal (SaS), 7. Traditional Malay Medicine (TMM).

Table 3. List of landscape furniture observed by the author from the Malay films.

Types of landscape furniture (as it appears in the film) / Malay film	Ampaian (suspension)	Bangsal/Sulap (barn)	Buaian (tree swing)	Gerbang (arch)	Guri	Jamban/tandas (latrine)	Jamung andang/ kandil	Kolam (pond)	Pagar (fence)	Pangkin	Pasu (flower pot)	Pelantar	Perigi (well)	Perun/perunan (dump site)	Reban (hen coop)	Tempayan (water vessel)	Titi	Wakaf (gazebo)
Hang Tuah (1956)				x			x	x	x		x							
Pendekar Bujang Lapok (1959)	x			x					x		x		x	x				
Batu Belah Batu Bertangkup (1959)					x	x	x	x	x		x				x	x		x
Nujum Pak Belalang (1959)	x				x	x	x	x	x	x	x		x		x			
Bawang Puteh Bawang Merah (1959)	x	x	x		x	x			x	x			x			x		
Seniman Bujang Lapok (1961)						x			x	x	x		x			x		
Ahmad Albab (1968)					x				x		x					x		
Laksamana Do Re Mi (1972)					x				x							x		
Puteri Gunung Ledang (2004)				x			x		x				x			x		x
Magika (2010)	x		x	x	x		x	x	x		x		x			x	x	
Hikayat Merong Mahawangsa (2011)				x	x		x		x	x		x				x	x	

(Source: Researcher observation, 2012).

Table 4. Among the plant species founded from the Malay films.

			Malay film										
No.	Local name	Botanical name	Hang Tuah (1956)	Pendekar Bujang Lapok (1959)	Batu Belah Batu Bertangkup (1959)	Nujum Pak Belalang (1959)	Bawang Puteh Bawang Merah (1959)	Seniman Bujang Lapok (1961)	Ahmad Albab (1968)	Laksamana Do Re Mi (1972)	Puteri Gunung Ledang (2004)	Magika (2010)	Hikayat Merong Mahawangsa (2011)
1.	Andung merah	*Cordyline fruticosa spp.*		x		x		x	x			x	
2.	Buluh/bunga buluh gading/rebung	*Bambusa vulgaris*		x	x		x		x	x	x	x	x
3.	Buluh perindu	*Bambusa magica*		x	x		x		x	x	x		
4.	Kelapa/bunga kelapa/nyiur/nyior	*Cocos nucifera*	x	x	x	x	x	x	x		x	x	x
5.	Pinang	*Areca catechu*	x	x	x	x			x		x		x
6.	Sirih	*Piper betle*	x	x	x		x				x		
7.	Pisang	*Musa spp.*	x	x	x	x	x	x	x	x	x	x	x
8.	Nipah	*Nypa fruticans*	x	x	x	x	x		x		x	x	x

(Source: Researcher observation, 2012).

5 CULTURAL TOURISM PRODUCT

There are several methods to promote the concept of the Malay garden design. There should be a complete guideline to landscape designers in terms of the principles and elements of design, philosophy and functions. This design model issued by the National Landscape Department (JLN) in collaboration with the various agencies involved with the heritage and culture of Malaysia. Next, the model of Malay garden concept design should be recognized by the local cultural experts and the presented concept is truly representative of the Malay community landscape. The traditional garden or the courtyard concept being built with garden landscape design Malay as an example and inspiration for landscape designers from inside and outside the country. Today, Taman Mini Malaysia in Melaka becomes as an information provider for Malay architecture, with showcasing models of several of the Malay. Furthermore, the Malay gardens must be promoted as the country's new tourism products. Ministry of Tourism and Culture (KPKM) and the Ministry of Communications and Multimedia (KMM) and the Ministry of Urban Wellbeing, Housing and Local Government (KPKT) need to work together in promoting landscape designers to use the concept of The Malay garden design in their landscape design. Lastly, National Landscape Department (JLN) and the Institute of Landscape Architects Malaysia (ILAM) is responsible to ensure the students majoring in landscape architecture at universities/colleges in Malaysia been introduced about the concept of The Malay design.

6 CONCLUSION

The concept of the Malay gardens is exclusive to the Malaysian Malay community. The concept will encourage the expansion of the landscape architecture field in Malaysia, particularly in areas of tourist attraction to showcase Malaysia's own uniqueness. Moreover, Malay gardens probably will become a national cultural heritage and widely used for public park construction in Malaysia.

REFERENCES

Abdul Rahman Al-Ahmadi (2003), Tamadun Rumpun Budaya Melayu, Kementerian Kebudayaan, Kesenian dan Pelancongan Malaysia.

Abdullah Sani Ahmad et al. (2006), Kajian Elemen Rekabentuk Landskap Melayu, Nilai Fungsi Komponen Dalam Landskap Melayu, Universiti Teknologi Malaysia.

Ahmad Zamil Zakaria, Ismail Hafiz Salleh Mohd Sabrizaa Abdul Rashid (2012), Landscape Furniture Present In The Ancient Malay Garden According To Old Manuscripts And Their Effects On The Formation Of Malay Garden Design Concept Model In Malaysia, PSU—USM International Conference on Art and Sciences.

Ali Ahmad (editor) (2000), Hikayat Inderaputera, Kuala Lumpur, Dewan Bahasa dan Pustaka.

Amin Sweeney (editor) (2006), Karya Lengkap Abdullah bin Abdul Kadir Munsyi, jilid 3: Hikayat Abdullah, Jakarta: Kepustakaan Populer Gramedia & École françaised Extreme-Orient.

Arkib Negara Malaysia (2008), Polemik Sejarah Malaysia—Jilid 1, Kementerian Penerangan Komunikasi dan Kebudayaan Malaysia, Malaysia.

Arkib Negara Malaysia (2009), Polemik Sejarah Malaysia—Jilid 2, Kementerian Penerangan Komunikasi dan Kebudayaan Malaysia, Malaysia.

Boehm, R.G (1996), Association of American Geographers: Cultural and Human Geography, (Online) retrieved on 15 March 2013 http://www.aag.org/Careers/Cultural_Human_Geography.html.

Bonnink, C. (1992). Cultural Tourism Development and Government Policy. MA Dissertation. Rijksuniversiteit Utrecht.

Hamzah Hamdani (2007), Hikayat Abdullah, PTS Publications, Malaysia.

Harry, L.G., (1985), Maintaining the spirit of place—A Process for the Preservation of Town Character, PDA Publishers Corporation, Mesa, Arizona.

Jelani Harun (editor) (2004), Bustan al-Salatin, Kuala Lumpur: Dewan Bahasa dan Pustaka.

Kasim Ahmad (1975), Hikayat Hang Tuah: Kota Melaka, Kuala Lumpur, Dewan Bahasa dan Pustaka.

McIntosh, R.W., Goeldner, C.R., & Ritchie, J.R.B. 1995. Tourism: Principles, Practices, Philosphies, 7th Edition, John Wiley & Sons, Inc., New York.

McKercher, B & Du Cros, H. (2002), The Partnership Between Tourism And Cultural Heritage Management, The Haworth Hospitality Press, London and Oxford, New York.

Perhimpunan Kebangsaan dan Serantau sempena 25 tahun Malaysia Merdeka dan Kongress Kebudayaan Melayu Pertama (1987), Pertemuan Dunia Melayu '82, Kuala Lumpur, Dewan Bahasa Dan Pustaka.

Samad Ahmad A. (editor) (1982), Warisan Perubatan Melayu, Kuala Lumpur, Dewan Bahasa dan Pustaka.

Samad Ahmad A. (editor) (1996), Sulalatus Salatin—Sejarah Melayu, Kuala Lumpur, Dewan Bahasa dan Pustaka.

Shellabear W.G. (1978), Sejarah Melayu, Kuala Lumpur, Penerbit Fajar Bakti Sdn.Bhd.

Silberberg, T. (1995). Cultural tourism and business opportunities for museums and heritage sites. Tourism Management, Elsevier.

Siti Hawa Salleh (1998), Hikayat Merong Mahawangsa, Kuala Lumpur, Yayasan Karyawan dan Penerbit Universiti Malaya.

Siti Hawa Salleh (editor) (1992), Bustan al-Salatin, Kuala Lumpur: Dewan Bahasa dan Pustaka.

Smith, Valene L. (1979), Hosts And Guests: The anthropology of tourism, University of Pennsylvania Press, Philadelphia.

Winstedt R.O (1969), A history of classical Malay literature, Oxford: Oxford University Press.

Zalina Abdul Aziz, Kamariah Jaafar (2012), Hikayat, From the Ancient Malay Kingdoms Retold by Ninotaziz, Utusan Publications & Distribution Sdn.Bhd.

Hospitality and Tourism – Sumarjan et al. (eds)
 ISBN 978-1-138-00151-0

Knowledge and business strategy model

S.K.A. Nordin, S.M. Radzi, H. Ismail & N. Ahmad
Universiti Teknologi MARA, Malaysia

ABSTRACT: Customers, competitors and technologies are the influences that make business firm change their strategic direction. Strategic management begins by answering how business firms could achieve competitive advantages in this kind of environment. One of the traditional solutions for business firm to adapt to this kind of environment is to execute business strategies in order to balance between the internal and external environmental factors. However, this does not ensure that the business firms will gain competitive advantages. Knowledge based view is believed to contribute to business firm in achieving competitive advantages and superior performance. Several researchers claimed that knowledge emerged as organization's crucial assets that play a vital role as competitive tool and direction for the company to survive and perform in turbulent environment of business. This encouraged business firms to acquire Knowledge Management (KM) and have the ability to create, disseminate, use and keep the knowledge asset from being imitated by the competitors. By just simply possess knowledge does not ensure firms to achieve competitive advantages and it must be aligned with the firms business strategy. Additionally, to attain superior performance and competitive advantages it is essential for business firm to fully utilize and integrate its knowledge resources in their business strategy. This paper discusses on the knowledge strategy and business strategy model that can be adapted in the Malaysia hotel business.

Keywords: Business strategy, competitive advantage, knowledge based view, knowledge strategy

1 INTRODUCTION

Most of the business firm having a hard time to remain competitive with current strategy execution due to the business environment factor such as new market attractiveness, emerging of multinational company and dynamic growth in global markets. The influence of customers, competitors and technologies are some reasons why business firm changed their strategic direction (Asoh, 2004). Therefore, strategic management team needs to find a new alternative in order to compete in the industry and to perform well. In the new era of knowledge-economy, knowledge based view is believed to contribute to business firm in achieving competitive advantages and superior performance. Knowledge economy is economy is driven by the knowledge and organization began to view knowledge as imperative strategic consideration like other assets such as land, equipment and capital.

2 RESEARCH ISSUES

In Malaysia hotel industry, the average occupancy rate can be considered as stable. In 2010, the average occupancy rate was at 60 percent (Tourism Malaysia, 2005, 2010). Kuala Lumpur and Selangor recorded the highest average daily room rate at RM260 in 2010 (Euro Monitor, 2012).

A review of the increment of hotel development in Malaysia showed that there is a high competition existed in the industry and this give implication to hotel performance. Hotel industry in Malaysia remained competitive with all players actively trying to capture the market share (Euro Monitor, 2012). Furthermore, customer recently has become sophisticated in finding the best deal for hotels, restaurants, travel agencies and tourist destination and the hospitality industry encounter a higher level of competition (Cheng, 2010). Additionally, the hotel chains in particular, also encounter the global pressure to improve their organization due to complexities such as promotional activities, pricing strategies, innovation activities, and control of capacity constraint, loans and fixed cost (Hallin, 2009).

Due to these issues, Knowledge Management (KM) in hotels is inevitable and the hotel management teams must consider knowledge as their asset to remain competitive in the turbulent environment. Rudež (2010) postulated that hotels are depending on knowledge development in their organization where KM is needed. Moreover, hotels need to acquire knowledge because knowledge can be accountable to tackle a specific problem in the industry, especially changes in tourist behavior (Sigala & Chalkiti, 2007). According to Zack (2002), KM is about managing and leveraging knowledge in the organization and therefore hotel that operate in an environmental uncertainty such as high competition needs to consider treating and managing knowledge as the strategic resource.

The hotel industry in Malaysia operated faces a high level of competition (Cheng, 2010) and global pressure (Hallin, 2009). Hence, the alignment of business strategy and knowledge strategy is expected to enhance business firm performance and attain competitive advantages (Al-Ammary, 2008; Asoh, 2004; Chen & Huang, 2012; Chew, 2008; Taxen, 2010). Therefore, this paper will discuss on how the alignment of knowledge strategy and business strategy could have an influence on Malaysia hotel industry performance.

2.1 *Knowledge strategy and business performance*

Knowledge strategy can be defined as "the overall approach an organization intends to take to align its knowledge resources and capabilities to the intellectual requirements of its strategy" (Zack, 1999a, p. 135). From this definition, it can be assumed that knowledge is an important asset in any organization. Bierly and Chakrabarti (1999) stated knowledge can contribute to firm performance if business strategy is formulated with difference knowledge development and deployment through knowledge strategy formulation. Knowledge strategy is a new concept in knowledge management field of research (Abdollahi et al., 2008). Due to this, many organizations faced difficulty in managing and fail to communicate its knowledge resource to its business strategy (Sadaghiyani & Tavallaee, 2011; Zack, 1999b). This has encouraged many business firm to recognized the need of knowledge strategy to manage its knowledge effectively and able to reallocate the resource in the organization.

Al-Ammary (2008), Asoh (2004), Chew (2008) and Zack (1999a) stated the conservatives knowledge strategy and aggressive knowledge strategy are well developed knowledge strategy in KM literature and influence business performance. Based on this argument, proposition one is developed:

P1: There is a positive relationship between knowledge strategy and business performance.

2.2 *Business strategy and business performance*

Mintzberg (1978) defined business strategy as a pattern in a stream of decisions. Business strategy concept can be viewed from two perspectives such as process and content (Ketchen, Thomas, & McDaniel, 1996). In process strategy it can be described as how the strategy is formed and while the content process focuses on competitive strategy of the business unit (Ketchen et al., 1996). However, Morgan and Strong (2003) mentioned process strategy reveals unclear relationship with business performance, while content strategy give firm a method to define its competitive advantages and influence business performance. They further indicated that strategy content can be viewed from three approaches: narrative approach, classificatory approach and comparative approach. Classificatory also known as typologies are the most notable in the management literature.

Many framework has been proposed regarding business strategy in organization. According to Asoh (2004, p. 33) though there are many framework for business strategy, "they can be classified as being generic." Among the generic business strategies included the one proposed by Miles and Snow (1978), and Porter (1980). Miles and Snow (1978) proposed defenders, prospectors, analyzers and reactors. Meanwhile, Porter (1980) proposed cost-leadership, differentiation and focus. These two generic business strategies are the most popular in the strategic management research.

Porter (1980) generic strategy seemed appropriate to be used in the hotel industry. According to Bordean, Borza, and Glaser-Segura (2011) hotel in the most part of the world compete with its competitors by pursuing low cost strategy, differentiation strategy and focus strategy because customer perceived price and values offer by the hotel. By applying this Porter's framework, hotel can attain favorable performance outcome. Analogous to this argument, Evans, Campbell, and Stonehouse (2003) reported Holiday Inn is one of the hotels that pursue cost-leadership while Mandarin-Oriental is pursuing the differentiation strategy. Meanwhile, Bordean et al. (2010) added Burj Al Arab hotel is exercising the focus strategy. This shows that Porter's (1980) generic strategy is widely used by most hotels worldwide. Hence, the following proposition is proposed:

P2: There is a positive relationship between business strategy and business performance.

2.3 *Knowledge strategy and business strategy alignment*

To sustain competitive advantages in turbulent environment for business firm, several researchers agreed to align knowledge strategy and business strategy in the business firm (Al-Ammary, 2008; Asoh, 2004; Shih & Chiang, 2005; Snyman & Kruger, 2004; Zack, 1999a).

Although many studies had proposed the alignment of Miles and Snow (1972) business strategy typology and knowledge strategy typology, there is lack of study that attempt to align Porter's business strategy typology with Zack's knowledge strategy typology. For example, in Al-Ammary (2008) study, he proposed alignment of prospector business strategy with aggressive knowledge strategy, analyzer business strategy with aggressive knowledge strategy, and defender business strategy with conservative knowledge strategy.

Porter's (1980) generic strategies resembled Miles and Snow (1978) generic strategies with defender and prospector are equivalent to cost-leadership and differentiation (Kim et al., 2004). Therefore, cost-leadership strategy can be aligned with conservative knowledge strategy while the differentiation strategy can be aligned with aggressive knowledge strategy.

Knowledge resources are derived from the logic of Resource Based View [RBV] (Nonaka, 2007). Knowledge resources in the organization will allow firm to compete on cost-leadership strategy or differentiation strategy (Ariyawardana & Bailey, 2002; Asoh, 2004; Snyman & Kruger, 2004). Additionally, Zack (1999a) postulated that in order to implement cost-leadership and differentiation successfully, it depends on the knowledge input that will deploy on the formulation of business strategy. From this statement the following proposition is developed:

P3: There is an alignment between knowledge strategy and business strategy.

2.4 *Alignment of knowledge strategy and business strategy and business performance*

Porter (1996) asserted that the essence of the strategy depend on the firm to create activities differently from its competitor by combining activities such as resource and core competencies and this is known as fit. He also argued that fit can create strong competitive advantages. Similarly, Miles et al. (1978) noted that organizational effectiveness depends on the ability to align internal and external domains.

In understanding the perspective of alignment, several researchers proposed different term such as "fit" (Porter, 1996), "strategic alignment" (Henderson & Venkatraman, 1993) and "bridge" (Ciborra, 1997). According to Knoll and Jarvenpaa (1994) "fit" is considered the same as alignment, linkage, top-down approach, congruence or match. Hence, whatever this word has been described, ultimately, organization try to match its internal and external resource (Asoh, 2004) and to integrate strategy with related contingency variables (Sun & Chen, 2008). The term "fit" is related with the firm as to match it with a set of component (Regev & Wegmann, 2004). This set of component consists of organization environment, strategy, structure, system, style, and culture (Van de Ven & Drazin, 1985).

From these perspectives of alignment, it can be assumed that in the turbulent environment of business, most of the business firm try to implement strategic alignment as to improve their performance (Asoh, 2004). There are many researches on alignment found in aligning information strategy and business strategy (Asoh, 2004; Zack, 1999a). Knowledge strategy and business strategy are also important element in alignment (Taxen, 2010) and act as basic foundation in proposing strategic alignment (Denford, 2009).

The alignment of knowledge strategy and business strategy gave a positive impact on performance (eg., Al-Ammary, 2008; Asoh, 2004; Chew, 2008; Denford, 2009; Massingham, 2004; Snyman & Kruger, 2004; Zack, 1999a). Snyman and Kruger (2004) stated business firm loss the opportunity to improve their performance if they fail to align knowledge strategy with business strategy. Al-Ammary (2008), Asoh (2004) and Shih and Chiang (2005) strongly argued firm that create alignment of knowledge strategy and business strategy lead to better performance than its rivals. Based on this argumentative, the following proposition is offered:

P4: Hotels that achieve the alignment between knowledge strategy and business strategy will perform better than those that are not.

3 CONCLUSION

One of the traditional solutions for business firm to adapt in volatile environment is that to execute business strategy in order to balance between internal and external environment factor (Green, Lisboa, & Yasin, 1993; Tse & Olsen, 1993). Researchers argued that if firms rely on traditional approach only to compete, chances to remain competitive are not guaranteed (Snyman & Kruger, 2004; Zack, 1999a, 2002). Similar to this statement, Massingham (2004) stressed most of the business firm having a hard time to remain competitive with current strategy execution because of business environment factors such as new market attractiveness, emerging of multinational company and dynamic growth in global markets.

Based on this argument, strategic management team needs to find alternatives in order to compete in the industry and to perform well. In the new era of knowledge-economy, knowledge based view is believed to contribute to business firm in achieving competitive advantages and superior performance. Alignment is one of the ways for business firm to figure how KM and business strategy can be incorporated in an organization. In a wider perspective, Porter (1996) indicated that alignment is part of the overall business strategy in firm organization. Meanwhile, in term of specific view of strategic alignment KM practices in the organization are seem appropriate to be linked with the business strategy (Al-Ammary, 2008; Asoh, 2004; Chew, 2008; Zack, 1999a, 2002) as they are the relevant elements for strategic alignment (Taxen, 2010).

REFERENCES

Abdollahi, A., Rezaeian, A., & Mohseni, M. (2008). Knowledge Strategy: Linking Knowledge Resources to Competitive Strategy. *World Applied Sciences Journal, 4*(2), 08–11.

Al-Ammary, J.H. (2008). *Knowledge Management Strategic Alignment in the Banking Sector at the Gulf Cooperation Council (GCC) Countries*. (Unpublished doctoral dissertation). Murdoch University Western Australia. Retrieved from http://researchrepository.murdoch.edu.au/433/2/02Whole.pdf.

Ariyawardana, A., & Bailey, W.C. (2002). The Relationship between Core Resources and Strategies of Firms: The Case of Sri Lankan Value-Added Tea Producers. *Sri Lankan Journal of Agricultural Economics, 4*(1), 01–18. doi: http://dx.doi.org/10.4038/sjae.v4i1.3481.

Asoh, D.A. (2004). *Business and Knowledge Strategies: Alignment and Performance Impact Analysis.* University at Albany State University of New York, ProQuest Dissertations and Theses.

Auzair, S.M. (2011). The Effect of Business Strategy and External Environment on Management Control System: A Study of Malaysian Hotels. *International Journal of Business and Social Science, 2*(13), 236–244.

Bierly, P., & Chakrabarti, A. (1999). Chapter 13—Generic Knowledge Strategies in the U.S. Pharmaceutical Industry. In H.Z. Michael (Ed.), *Knowledge and Strategy* (pp. 231–250). Boston: Butterworth-Heinemann.

Bordean, O.N., Borza, A.I., & Glaser-Segura, D. (2011). A Comparative Approach of the Generic Strategies within Hotel Industry: Romania vs. USA. *Management & Marketing Challenges for the Knowledge Society, 6*(4), 501–514. Retrieved from http://www.managementmarketing.ro/pdf/articole/239.pdf.

Bordean, O.N., Borza, A.I., Nistor, R.L., & Mitra, C.S. (2010). The Use of Michael Porter's Generic Strategies in the Romanian Hotel Industry. *International Journal of Trade, Economics and Finance, 1*(2), 173–178.

Chen, Y.-Y., & Huang, H.-L. (2012). Knowledge management fit and its implications for business performance: A profile deviation analysis. *Knowledge-Based Systems, 27*(0), 262–270. doi: 10.1016/j.knosys.2011.11.012.

Cheng, X. (2010). *A Systematic Review of Knowledge Management Research in the Hospitality and Tourism Industry.* University of Nevada, Las Vegas, ProQuest Dissertations and Theses. Retrieved from http://search.proquest.com.ezaccess.library.uitm.edu.my/docview/852514623?accountid=42518.

Chew, K.C. (2008). An Exploratory Case Study. Knowledge Management: Managing Organizational Knowledge Assets by Aligning Business Strategy, Knowledge Strategy, and Knowledge Management Strategy. Golden Gate University. Retrieved from http://search.proquest.com.ezaccess.library.uitm.edu.my/docview/304415288?accountid=42518. ProQuest Dissertations and Theses database.

Ciborra, C.U. (1997). De profundis? Deconstructing the concept of strategic alignment. *Scand. J. Inf. Syst., 9*(1), 67–82.

Denford, J.S. (2009). *The Alignment of Knowledge Strategies*. (Unpublished doctoral dissertation). Queen's University. Retrieved from http://qspace.library.queensu.ca/bitstream/1974/1738/1/Denford_James_Stephen_200901_PhD.pdf.

Euro Monitor. (2012). Travel Accommodation in Malaysia Retrieved 31 June, 2012, from http://www.portal.euromonitor.com.ezaccess.library.uitm.edu.my/Portal/Pages/Search/SearchResultsList.aspx.

Evans, N., Campbell, D., & Stonehouse, G. (2003). Chapter 10—Competitive strategy and strategic direction for travel and tourism organizations. *Strategic Management for Travel and Tourism* (pp. 209–240). Oxford: Butterworth-Heinemann.

Green, R.F., Lisboa, J., & Yasin, M.M. (1993). Porter's (1980) generic strategies in portugal. *European Business Review, 93*(2), 03–10. Retrieved from http://search.proquest.com.ezaccess.library.uitm.edu.my/docview/225428947?accountid=42518.

Hallin, C.A. (2009). Exploring the strategic impact of service employees' tacit knowledge: the development of an indicator for forecasting economic performance of hotel companies. University of Stavanger, University of Stavanger, Norway. Retrieved from http://idtjeneste.nb.no/URN:NBN:no-bibsys_brage_13374.

Hambrick, D.C., & Fredrickson, J.W. (2005). Are you sure you have strategy? *Academy of Management Executive, 19*(4), 51–62.

Henderson, J.C., & Venkatraman, N. (1993). Strategic Alignment: Leveraging Information Technology for Tranforming Organization. *IBM Systems Journal, 32*(1), 472–484. doi: 10.1147/SJ.1999.5387096.

Ketchen, D.J., Thomas, J.B., & McDaniel, R.R. (1996). Process, Content and Context: Synergistic Effects on Organizational Performance. *Journal of Management, 22*(2), 231–257. doi: 10.1177/014920639602200203.

Kim, E., Nam, D.-I., & Stimpert, J.L. (2004). The Applicability of Porter's Generic Strategies in the Digital Age: Assumptions, Conjectures, and Suggestions. *Journal of Management, 30*(5), 569–589. doi: 10.1016/j.jm.2003.12.001.

Knoll, K., & Jarvenpaa, S.L. (1994). *Information technology alignment or fit in highly turbulent environments: the concept of flexibility*. Paper presented at the Proceedings of the 1994 computer personnel research conference on Reinventing IS: managing information technology in changing organizations: managing information technology in changing organizations, Alexandria, Virginia, United States.

Massingham, P. (2004). Linking Business Level Strategy with Activities and Knowledge Resources. *Journal of Knowledge Management, 8*(6), 50–62.

Miles, R.E., & Snow, C.C. (1978). *Organizational Strategy, Structure and Process*. New York: McGraw-Hill.

Miles, R.E., Snow, C.C., Meyer, A.D., & Coleman, J.H.J. (1978). Organizational Strategy, Struture, and Process. *The Academy of Management Review, 3*(3), 546–562.

Mintzberg, H. (1978b). Patterns In Strategy Formation. [Article]. *Management Science, 24*(9), 934–948.

Morgan, R.E., & Strong, C.A. (2003). Business performance and dimensions of strategic orientation. *Journal of Business Research, 56*(3), 163–176. doi: 10.1016/s0148-2963(01)00218-1.

Nonaka, I. (2007). The Knowledge-Creating Company. *Harvard Business Review* Retrieved 5 May, 2012, from http://zonecours.hec.ca/documents/H2010-1-2312839.NONAKA-TheKnowledge-CreatingCompany.pdf.

Porter, M.E. (1980). Competitive Strategy Technique for Analyzing Industries and Competitors. New York: The Free Press.

Porter, M.E. (1996). What is Strategy. *Harvard Business Review* Retrieved 5 April, 2012, from http://www.ipocongress.ru/download/guide/article/what_is_strategy.pdf.

Regev, G., & Wegmann, A. (2004). Remaining Fit: On the Creation and Maintenance of Fit Retrieved 5 May, 2012, from http://www.ibissoft.se/events/bpmds04/regev.pdf.

Rudež, H.C. (2010). Knowledge Management in the Hotel Industry Before and After the Entry in the EU: The Case of Slovenia. *Organizacija, 43*(4), 197–202. doi: 10.2478/v10051-010-0019-9.

Sadaghiyani, J.S., & Tavallaee, R. (2011). *Review Different Knowledge Strategy Models and Designing a New Model.* Paper presented at the International Conference on Innovation, Management and Service, Singapore.

Shih, H.A., & Chiang, Y.H. (2005). Strategy alignment between HRM, KM, and corporate development. *International Journal of Manpower, 26*(6), 582–603. doi: http://dx.doi.org/10.1108/01437720510625476.

Sigala, M., & Chalkiti, K. (2007). Improving performance through tacit knowledge externalisation and utilisation: Preliminary findings from Greek hotels. *International Journal of Productivity and Performance Management, 56*(5), 456–483. doi: http://dx.doi.org/10.1108/17410400710757141.

Snyman, R., & Kruger, C.J. (2004). The Interdependency between Strategic Management and Strategic Knowledge Management. *Journal of Knowledge Management, 8*(1), 05–19. doi: http://dx.doi.org/10.1108/13673270410523871.

Sun, S.Y., & Chen, Y.Y. (2008). Consolidating the Strategic Alignment Model in Knowledge Management. *International Journal Innovation and Learning, 5*(1), 51–65. Retrieved from http://ejournal.narotama.ac.id/files/Consolidating%20the%20strategic%20alignment%20model%20in%20knowledge%20management.pdf.

Taxen, L. (2010). Aligning Business and Knowledge Strategies: A Practical Approach for Aligning Business and Knowledge Strategies. In M. Russ (Ed.), *Knowledge Management Strategies for Business Development* (pp. 277–308). Hershey PA: IGI Global.

Tourism Malaysia. (2005). Average Occupancy Rate Retrieved 5 May, 2012, from http://corporate.tourism.gov.my/images/research/pdf/2005/AOR/AOR_2004_2005.pdf.

Tourism Malaysia. (2010). Average Occupancy Rate Retrieved 5 May, 2012, from http://corporate.tourism.gov.my/images/research/pdf/2010/AOR/AOR_2009_2010.pdf.

Tsamenyi, M., Sahadev, S., & Qiao, Z.S. (2011). The relationship between business strategy, management control systems and performance: Evidence from China. *Advances in Accounting, 27*(1), 193–203. doi: 10.1016/j.adiac.2011.05.001.

Tse, E., & Olsen, M.D. (1993). Business Strategy and Organisational Structure: A Case of US Restaurant Firms. *International Journal of Contemporary Hospitality Management, 2*(3), 17–23. doi: http://dx.doi.org/10.1108/09596119010000595.

Van de Ven, A.H., & Drazin, R. (1985). The Concept of Fit in Contingency Theory. *Organizational Behavior* Retrieved 5 May, 2012, from http://www.freewebs.com/dangdinhtramen/302.pdf.

Zack, M.H. (1999a). Developing a Knowledge Strategy. *California Management Review, 41*(3), 125–145.

Zack, M.H. (1999b). Managing Codified Knowledge. *Sloan Management Review, 40*(4), 45–58.

Zack, M.H. (2002). A Strategic Pretext for Knowledge Management Retrieved 5 May, 2012, from http://citeseerx.ist.psu.edu/viewdoc/summary?doi=10.1.1.15.3266.

Hospitality and Tourism – Sumarjan et al. (eds)
© 2014 Taylor & Francis Group, London, ISBN 978-1-138-00151-0

Quality work life and occupational stress in hotel industry

M. Azinuddin & A. Zainal
Universiti Teknologi MARA, Malaysia

ABSTRACT: Competent and loyal employees are the strength of character for the organizational profits and apparently, no hotel can have outstanding operations without them. Although travel and tourism industry becomes increasingly central to the economies of Malaysia, it had been embittered by huge turnover rate that is deeply felt in the hospitality profession. Against this background, Quality Work Life (QWL) emerged as crucial for the service industry such as the hospitality organizations. Companies that offer better QWL and supportive work environment are likely to gain leverage in terms of hiring and retaining valuable personnel. Thus, there is a need to establish the extent to which the relationships of QWL and Occupational Stress are deem important and it will be a challenge to the hospitality organizations in Malaysia to be able to inaugurate a working environment that is able to retain employees.

Keywords: hotel industry, occupational stress, quality work life

1 INTRODUCTION

In this new era, people are the primary source for a company's competitive advantage and organizational prosperity and survival depends on how employees are treated (Kandasamy & Ancheri, 2009a). According to Tracey and Hinkin (2000), studies have confirmed that the competent and loyal employees are the backbone towards the organizational profits. This scenario is the same with the hotel industry. Apparently, no hotel can have outstanding operations without the excellent employees (Enz & Siguaw, 2000). Due to this, it is highly important to retain employees in the organization as they are considered as the greatest asset of an organization. Without employees, organizations will not be able to produce business results, achieve organization goals, or meet it's financial targets (Johari, Yahya, & Ahmad, 2012).

Looking into Malaysian hotel industry, it still shows sign of growth although it was heavily hit by the worldwide economic recession in 2009 (Hanan & Zainal, 2012a). Referring to Malaysia Tourism Satellite Account (2000–2010) by Malaysian Department of Statistics, employment for the tourism industry was recorded at 1.8 million persons in 2010, registering a marginal growth of 0.6 percent from 4.9 percent in 2009. In line with this notion, it is noted that the total number of hotels is increased to 3, 129 in 2010 from 2, 362 in 2009. This represents an increase of 32.5 percent (Economic Census 2011 Accommodation Services Report, Malaysia Department of Statistics). The report further stated that direct employment in the hotel industry shows an increment by 5.6% from 104, 645 in 2009 to 110, 535 in 2010. The profits from tourism industry increased 13.6% to RM100, 210.7 million from 2009 until 2010 (Malaysia Tourism Satellite Account, 2010).

In the midst of the current foregoing growth, there is hardly yet any research that elucidates the elements of QWL among personnel in the service sector (Kandasamy & Ancheri, 2009a) especially in Malaysian and Asian organizations (Daud, 2010; Bagtasos, 2011). Notably, studies have acknowledged the fact that improved Quality Work Life (QWL) is essential for organizations to attract and retain employees (Daud, 2010; May, Lau, & Johnson, 1999). Companies that are offering better QWL and supportive work environment are likely to gain leverage in terms of hiring and retaining valuable people (May et al., 1999) and enjoy exceptional growth and profitability (Lau & May, 1998). Moreover, in the last two decades, the issue of QWL has become critical owing to the increasing demands of today's business environment (Bagtasos, 2011). From this uprising concern, it gives an increased interest in QWL not only in business but also for many professions and fields (Akdere, 2006). Against this background, QWL has become an important indicator that the firms are able to offer suitable working environment, which eventually leads to high commitment from the employees (Muhamad-Noor & Abdullah, 2011).

2 DEVELOPMENT OF QWL

The development of QWL can be traced way back to the earlier management era (Muhamad-Noor & Abdullah, 2011) of where it was first introduced in 1972 during an International Labor Relations conference (Nayak & Joshi, n.d.). Ever since that, ranges of definitions and theoretical constructs have succeeded each other with the aim of mitigating the many problems facing the QWL concept (Martel and Dupuis, 2006). Owing to this, it has been observed that the method of defining QWL stands varied (Loscocco & Roschelle, 1991) and there is still yet to be acommonly accepted definition of the QWL concept (Daud, 2010). On the other hand, it has been well recognized that QWL stands as a multi-dimensional construct (Mortazavi, Yazdi, & Amini, 2012) and it can be strongly reiterated that different people have distinctive perspectives on defining QWL. Conceptually QWL has been defined as:

i. Confluence of two separate streams of thought. First, is to humanize the workplace and the second one is to improve the productivity (Walton, 1973).
ii. Promoting a work environment conducive to the satisfaction of employee's needs (Walker, 1992).
iii. Employee satisfaction with a variety of needs through resources, activities and outcomes stemming from the participation at the workplace (Sirgy, Efraty, Siegel, & Lee, 2001).
iv. A multi dimensional construct, made up of a number of interrelated factors that need careful consideration to conceptualize and measure (Rethinam & Ismail, 2008).

Basically, QWL has evolved over time and it is viewed as a wide-ranging concept that includes satisfaction towards work, participative management and improves work management (Nanjundeswaraswamy & Swamy, 2012). In a broad-spectrum, QWL is a comprehensive plan that enhances the employee's satisfaction by grooming their learning and in their environment and helps them in management and variation issues (Heidarie, Askary, Saedi, & Gorjian, 2012). Nowadays, the QWL concept has emerged as a main social subject in contemporary management on a global scale (Mirkamali & Thani, 2011).

3 PREVIOUS RESEARCHES IN QWL

Historically, most of the publications from the 1970s to 1980s highlight a number of attempts to empirically define the QWL concept (Martel & Dupuis, 2006). During this span of period, QWL had undergone through various evolution process that it grew to many distinctive theoretical concepts since it was first introduced. Consequently, there was a risk that these theoretical disparities would lead to the further ambiguity of the QWL concept as some studies defined the QWL concept based on the broad scope whereas some have defined it based on specific characteristics of work life. Stjernberg (1977) for instance, stressed that QWL concept should be linked with other life domains such as general life satisfaction, leisure and well being beyond the workplace. In other example, Davis (1983) defined QWL as the quality of relationship between employees and the total working environment with human dimensions added to the usual and technical economic consideration.

Interestingly, the lack of consensus among the scholars on the definition of QWL continued on to the 1990s as the concept is further extrapolated in becoming more dynamic. It can be seen that the focus of QWL is changing beyond the job satisfaction per se as more studies included factors such as job security, job involvement, reward system, training and career advancement opportunities, participation in decision-making and others (Danna and Griffin, 1999; Lau and May, 1998).

Entering into 2000s, the direction of studies in QWL is shifting once again as the new measurement of the concept were introduced. One significant example of this is a study conducted by Sirgy and et al. (2001) whereby they developed a new measurement of QWL based on need satisfaction and spillover theories. Martel and Dupuis (2006) also contributed to a new measurement that is based on suggested definition of Quality of Life.

Overall, the total body of studies has been substantial in QWL as it have been measured in a variety of settings using combinations of various questionnaires such as job satisfaction, organizational commitment, alienation, job stress, organizational identification, job involvement and finally work role ambiguity, conflict, and overload (Daud, 2010). On that note, most research and writings in QWL also has been linked with many personnel's job outcomes, such as productivity and increased retention (Koonmee & Virakul, 2007).

There have been also studies in the other industries such as banking, human resource, marketing managers, accountants and teachers whereby QWL have been examined as independent, intervening, moderating and dependent variables in various models offered to understand the employee behaviors better (Hong, Tan, & Bujang, 2010; Koonmee & Virakul, 2007; Sirgy et al., 2001; Talebi, Pokdel Bonab, Zemestani, & Aghdami, 2012).

Taking into accountall of the QWL studies so far, the approach and methodology mostly lying over general definitions of QWL in certain industries or organizations, with hotel industry without exception (Ma, Ma, Yu, & Hao, 2011; Manjunath & Kurian, 2011). For that reason, Kandasamy & Ancheri (2009b) designed and developed the attributes of QWL specifically for the hospitality employees as the service and tourism industry had been severely plagued by high turnover rate (Young & Corsun, 2010). In other words, the study is focusing on exploring hospitality employees' expected dimensions of QWL. The study yielded eight dimensions of QWL that consist of job characteristics, person-job fit, company image, human resource policies, physical working conditions, work-life balance, work group relationship and interaction with customers. Thus, these eight dimensions were intended to become as a foundation to a hotel organization in reducing the gap between employee expectations and the reality of the workplace attributes.

4 LITERATURE ON QWL AND OCCUPATIONAL STRESS

4.1 *Occupational stress*

Putting the perspectives onto the stress itself, it has been conceptually defined as existing in a person-environment relationship that can strain or threatens the personal resources, and as a mental state in response to pressure and daily hassles (Hong, Tan, & Bujang, 2010). It is further explained that lack of control over demands where the perceived demands in that particular person-environment relationship exceed the resources one can draw on to cope, distress will follow (Gibbons & Gibbons, 2007; Zohar, 1994). This is in line with the cognitive model of stress from Lazarus and Folkman (1984) and the most cited model in occupational stress (Gibbons & Gibbons, 2007), which is Karasek and Theorell's (1990) demand-control-support model. Moreover, there are increased evidences over the last 15–20 years that occupational stress is prevalent within the hospitality industry generally, compared to other occupational groups (Gibbons & Gibbons, 2007). This is partly down to the immediate impacts of unforeseen labor turnover that is clearly problematic in the hospitality industry that has also major recruitment difficulties (Abdullah et al., 2010; Rowley & Purcell, 2001). For example, Hanan and Zainal (2012b) carried out a study of occupational culture and stress of chefs on occupational commitment in five star hotels in Malaysia. The results suggested that there is a strong relationship as the occupational stress mediates the occupational culture and commitment. The other example is a survey on 40 chefs in Northern Ireland by Gibbons and Gibbons (2012). They identified other determinants of stress that includes poor communication in the workplace, threat of violence at the workplace, bullying at work, repetitive work, boring work, poor fixture and fittings, heavy workload and being short—staffed.

4.2 *Relationship between QWL and occupational stress*

It is already well documented that QWL and job satisfaction are two different concepts as job satisfaction is construed as one of many outcomes of QWL. Furthermore, it has been empirically proven that there is positive relationship between QWL and job satisfaction (Heidarie et al., 2012; Hong et al., 2010). As such, when the expectations of QWL are not fulfilled, it leads to the occupational stress (Daud, 2010; Denvir & McMahon, 1992; Gibbons, 2007; Kandasamy & Ancheri, 2009b; Rowley & Purcell, 2001; Young & Corsun, 2010) as it is likely that the job satisfaction is reduced or weakened.

In relating these two variables of QWL and occupational stress also, it is vital to note that studies have suggested that occupational stress and QWL are one of many important indicators that influenced the turnover intentions (Gibbons & Gibbons, 2007; Kandasamy & Ancheri, 2009a; Young & Corsun 2010). Thus, it is assumed that if an employee has a negative work experience, it is likely he or she would be stressed to the job, meaning to say the weaker the QWL, the stronger the occupational stress.

Conclusively, the effects of QWL towards the occupational stress are still further away from attention despite such importance of these variables as predictors of turnover intention in the hotel industry. As travel and tourism industry are becoming increasingly central to the economies of Malaysia, there is a real need to establish the extent to which these relationships of QWL and occupational stress is applicable to the country. This need is important to ensure further understanding of how Malaysian hospitality employees involve and participate in their work. Based on this premise, results from the study on these relationships were considered to be significant because they were intended to assist decision makers in identifying key workplace issues in order to develop strategies to address and improve the quality of working conditions in the Malaysian hospitality industry.

REFERENCES

Abdullah, R., Mohd-Alias, M.A., Zahari, H., Abdul-Karim, N., Abdullah, S.N., Salleh, H., & Musa, M.F. (2010). The study of factors contributing to chef turnover in hotels in Klang Valley, Malaysia. *Asian Social Science, 6*(1), 80–85.

Akdere, M. (2006). Improving QWL: Implications for human resources. *The Business Review, 6*(1), 173–177.
Bagtasos, M.R. (2011). Quality of work life: A review of literature. *De La Salle University Business and Economics Review, 20*(2), 1–8.
Danna, K. & Griffin, R.W. (1999). Health and well-being in the workplace: A review and synthesis of the literature. *Journal of Management, 25*(3): 357–384.
Daud, N. (2010). Investigating the relationship between quality of work life and organizational commitment amongst employees in Malaysian firms. *International Journal of Business and Management, 15*(10), 75–82.
Davis, L.E. (1983). Design of new organizations. In: Kolony, H., Beinum, H.V. (Eds.), *The quality of working life and the 1980s.* New York: Praeger Publisher.
Denvir, A. & McMahon, F. (1992). Labour turnover in london hotels and the cost effectiveness of preventative measures. *International Journal Hospitality Management, 11*(2), 143–154.
Economic Census (2011). Accommodation services (2012). Department of Statistics, Malaysia. Retrieved on 13 December 2012, from http://www.statistics.gov.my.
Enz, C.A. & Siguaw, J.A. (2000). Best practices in human resources. *Cornell Hotel and Restaurant Administration Quarterly, 41*(1), 48–61.
Hanan, F.A. & Zainal, A. (2012a). The relationship between occupational culture, organization tenure and occupational commitment of chefs in 4 and 5-star Hotels in Kuala Lumpur. In Zainal et al. (eds), *Current Issues in Hospitality and Tourism Research and Innovations.* Page Number 205–210. London: Taylor and Francis Group.
Hanan, F.A. & Zainal, A. (2012b). The relationship of occupational culture and stress of chefs on occupational commitment in Five star Hotels in Malaysia. Paper Presented at the Annual International Conference on Tourism and Hospitality Research (THoR 2012), Singapore.
Heidarie, A., Askary, P., Saedi, S., & Gorjian, B. (2012). Relationship between quality of work life, organizational health and commitment with job satisfaction. *Life Science Journal, 9*(3), 2300–2306.
Hong, K.S., Tan, K.W., & Bujang, S. (2010). Relationship between work life quality of teachers with work commitment, stress and satisfaction: A study in Kuching, Sarawak, Malaysia. *Jurnal Teknologi, 52*, 1–15.
Johari, J., Yahya, K.K., & Ahmad, M.N. (2012, March 12). Understanding the influence of human resource management practices on intention to stay: Evidence from Malaysia. Paper Presented at the 3rd International Conference on Economic and Business Research Proceeding, Bandung, Indonesia.
Kandasamy, I. & Ancheri, S. (2009a). WRKLFQUAL: A tool for measuring quality of work life. *Research and Practice in Human Resource Management, 17*(1), 59–70.
Kandasamy, I. & Ancheri, S. (2009b). Hotel Employees' Expectations of QWL: A qualitative study. *International Journal of Hospitality Management, 28*, 328–337.
Karasek, R. & Theorell, T. (1990). Healthy work: Stress, productivity and reconstruction of working life. New York: Basic Book.
Koonmee, K. & Virakul, B. (2007). Ethics, quality of work life, and employee job-related outcomes: A survey of HR and marketing managers in Thai businesses. *NIDA Development Journal, 47*(4): 67–97.
Lau, R.S.M. & May, B.E. (1998). A Win-win paradigm for quality of work life and business performance. *Human Resource Development Quarterly, 9*(3), 211–226.
Lazarus, R.S. & Folkman, S. (1984). *Stress, appraisal and coping.* New York: Springer Publications.
Lee, K., Carswell, J.J., & Allen, N. (2000). A meta-analytic review of occupational commitment: Relations with person—and work related variables. *Journal of Applied Psychology, 85*(5), 799–811.
Loscocco, K.A. & Roschelle, A.R. (1991). Influences on the quality of work and non-work life: Two decades in review. *Journal of Vocational Behavior, 39*(2), 182–225.
Ma, Y., Ma, Q., Yu, H., & Hao, J. (2011). Quality of work life and employee outcome: A Survey of employees in hotel business. *Advanced Materials Research,* 171–172, 433–436.
Malaysia tourism satellite account (2012). Department of Statistics, Malaysia. Retrieved on 13 December 2012, from http://www.statistics.gov.my.
Manjunath, S.J. & Kurian, S. (2011). Impact of quality work life of the hotel employees in customer satisfaction—A Study on star hotels in Bangalore. *South Asian Journal of Tourism and Heritage, 4*(2), 78–92.
Martel, J-P. & Dupuis, G. (2006). Quality of work life, theoretical and methodological problems, and presentation of a new model and measuring instrument. *Social Indicators Research, 77*, 333–368.
May, B.E., Lau, R.S.M., & Johnson, S.K. (1999). A longitudinal study of quality of quality work life and business performance. *South Dakota Business Review, 58*(2), 3–7.
Mirkamali, S.M. & Thani, F, N. (2011). A study on the quality of work life (QWL) Among faculty members of Tehran (UT) and Sharif University of Technology (SU). *Social and Behavioral Sciences, 29*, 179–187.
Mortazavi, S., Yazdi, S.V.S., & Amini, A. (2012). The role of the psychological capital on quality of work life and organization performance. *Interdisciplinary Journal of Contemporary Research in Business, 4*(2), 206–217.
Muhamad-Noor, S. & Abdullah, M.A. (2011, December 7). Quality work life among factory workers in Malaysia. Paper Presented at Asia Pacific International Conference on Environment-Behaviour Studies, Famagusta, North Cyprus.
Nayak, S. & Joshi, H.G. (n.d.). Quality of Work life among IT professionals in SME'S in select cities of India. *Journal of Law and Social Sciences. 1*(1).
Nanjundeswary, T.S. & Swamy, D.R. (2012). A Literature Review on Quality of Work Life and Leadership Styles. *International Journal of Engineering Research and Applications, 2*(3), 1053–1059.
Rowley, G. & Purcell, K. (2001). 'As cooks go, she went': Is labour churn inevitable? *International Journal of Hospitality Management, 20,* 163–185.
Rethinam, G.S. & Ismail, M. (2008). A Basic concept on QWL. *European Journal of Social Sciences, 7*(1), 60.

Stjenberg, T. (1977). Organizational Change and quality of life. Stockholm: EFI.

Tracey, J.B. & Hinkin, T.R. (2000). The cost of turnover: Putting a price on the learning curve. *Cornell Hotel and Restaurant Administration Quarterly, 49*(1), 12–27.

Tabassum, A. (2012). Interrelationships between quality of work life dimensions and faculty member job satisfaction in the private universities of Bangladesh. *European Journal of Business and Management, 4*(2), 78–90.

Talebi, B., Pakdel Bonab M., Zemestani, G., & Aghdami, N. (2012). Investigating the relationship between the employee's quality of work life (QWL) and Their Effectiveness in Banking. *European Journal of Experimental Biology, 2*(5), 1839–1842.

Walker, J.W. (1992). *Human resource strategy*. New York: McGraw Hill.

Walton, R.E. (1973). Quality work life: What is it? *11*, 11–21.

Young, C.A. & Corsun, D.L. (2010). Burned! The impact of work aspects, injury, and job satisfaction on unionized cooks' intentions to leave the cooking occupation. *Journal of Hospitality and Tourism Research, 34*(1), 78–102.

Zohar, D. (1994). Analysis of job stress profile in hotel industry. *International Journal of Hospitality Management, 13*(3), 219–231.

Hospitality and Tourism – Sumarjan et al. (eds)
© 2014 Taylor & Francis Group, London, ISBN 978-1-138-00151-0

Creating risk management awareness in Park Guiding—a case study of Bako National Park, Sarawak, Malaysia

N. Putit & M.K.Y. Chan
Universiti Teknologi MARA, Sarawak, Malaysia

H. Hanan
Kuala Lumpur Metropolitan University College (KLMUC), Malaysia

ABSTRACT: People travel to national parks to experience, anticipating a degree of adventure which ironically, the grandeur nature that attracts people to natural areas may also put them at risk. In Sarawak, there was no provision to evaluate the tour guides' awareness in risk management in the national parks and nature reserves before the implementation of licensing park guiding in 2007. This paper investigates the awareness of risk management among the candidates attending the training course in Bako National Park and their comprehension of their responsibilities during the practical examination conducted in relation to their experience in guiding. This study identifies strategic communication and demonstrative strategies employed, extends audience-centered communication to include both informal messages and demonstrative practices through observational data. The study revealed that before the Certificate in Park Guide training course was introduced in 2007, there were higher number of incidents that required emergency attention and accidents in Bako National Park. Since then, the number of emergency incidents and accidents had shown a decreasing trend.

Keywords: ecotourism, health and safety, licensed park guides, park guiding, risk management

1 INTRODUCTION

Over the last decade, tourism has grown tremendously and has become a popular global leisure activity. Traditional tourism is enjoyed by a large number of people in a short space of time to places of leisure interest. Recognized as Mass tourism, activities have contributed to undesired social and environment impact such as invasion of privacy and destruction of sensitive areas. This has then evolved to more responsible type of travel activities like ecotourism, green tourism, and sustainable tourism activities which meet the needs of the present without compromising the ability of the future generation.

Ecotourism activities require small scale tourists travelling to fragile and protected areas that should be of low impact, at the same time it helps educate tourists to be aware of conservation and protection of the area and benefit return to the local community. People travel to these areas to experience, anticipating a degree of adventure which ironically, the grandeur nature that attracts people to natural areas may also put them at risk. According to Page, Bently and Walker (2005), many of the risk factors are conceptualized as "human factors" and that that adventure operators have the greatest potential to control risk associated with latent errors. As such, risk management and safety and health precaution must be incorporated with the ecotourism activities as it is also considered as adventurous and dangerous activities that requires extra precaution. This is supported by Frangialli, (2003) as the well-being of tourists has become one of the primary concerns of the tourism industry.

1.1 *Incidences of risk mismanagement*

Extracts of several reports related to incidences of risk mismanagements are as listed below.

Glasgow Press and Journal, 1999:
In July 1999, 21 people were killed while canoeing in Switzerland when a flash flood swept down the valley. On recovering the bodies, rescue teams noted that many of the participants had lost their rubber boots and protective helmets.

The Christian Time Monitor Online, 2011:
Six people have died in water-related accidents around Yosemite. Two hikers drowned in the HetchHetchy Reservoir on June 29, and a hiker slipped and fell into the Merced River while on the

Mist Trail on May 13. Other tourist spots around the West have experienced accidents related to water and snow as well. In Oregon's Crater Lake National Park last week, a man jumped over a wall at the edge of the steep caldera, slipped on packed snow, and slid head-first 300 feet below the rim. It took rescuers five hours to bring him out. Four family members died in Wyoming earlier this week when a road collapsed and sent their car into a raging creek.

The Associated Press, 2012:
Recent incident in Halong Bay took the lives of 5 Taiwanese when their junk boat collided. The other 18 tourists were plunged into the water without life jackets fighting to survive and last year, 12 people from nine nations died when their boat sank in Halong Bay early in the morning as those on board slept. The captain of the boat was sentenced to 10 years in jail, the chief engineer to eight. An investigation into the accident found that crew had forgotten to close a valve that allowed the vessel to take on water after a pipe in the engine's cooling system broke. The accidents raised fresh questions over safety standards in one of Vietnam's most visited attractions.

1.2 *Licensing park guides in national parks and nature reserves in Sarawak*

In Sarawak, there was no provision to evaluate the tour guides' awareness in risk management in the national parks and nature reserves before the implementation of licensing park guiding in 2007. Regulations 8 of the National Parks and Nature Reserves Regulations, 1999 of Sarawak, Malaysia stipulates that a park guide must possess a license issued by the Controller of National Parks and Nature Reserves to provide guiding service in such areas. Thus a park guide has to be evaluated to obtain a certificate in Park Guiding, as a criterion to obtain the license. The Park Guiding Certificate training course provided by the Sarawak Forestry consists of a module in Risk Management, Health and Safety in Totally Protected Areas.

Therefore, this paper investigated the number of incidents requiring emergency attention and accidents. It also included the awareness of risk management of licensed park guides and their comprehension of their responsibilities in guided tours as perceived by their clients in Bako National Park.

2 METHODOLOGY

The data for this study were mainly obtained through a survey using administered questionnaires with 30 respondents on guided tours provided by licensed park guides in Bako National park. The survey instrument comprised of two parts. The first part obtained data from respondents regarding their demographics and their reasons for being in the Bako National Park. The second part of the questionnaires focused on the risk management awareness of the licensed park guide as perceived by the respondent. The respondent were to circle their response accordingly from "completely disagree to completely agree" based on the Likert Scale of: 1—I completely disagree; 2—disagree; 3—neutral; 4—agree 5—I completely agree and 6—I don't know. The completed questionnaires were collected in a period of two weekends during a low peak season in October 2012. After the data collection, the data from the questionnaires were reviewed, followed by coding procedures and tabulated by the SPSS/PC version 17.0.

3 RESULT AND DISCUSSION

The place of risk in society, drawing on work by Beck (1992), among others, is certainly important to an understanding of tourism practices. As Gidden (1991) has demonstrated, risk is an increasingly important factor in modern existence.

3.1 *Number of incidents requiring emergency attention and accidents in Bako National Park*

The number of incidents requiring emergency attention and accidents in Bako National Park incidences of accidents that occurred in Bako National Park over the period from January 2005 to September 2012 is shown in Figure 1.

Before the Certificate in Park Guide training course was introduced in 2007, there were higher number of incidents that required emergency attention and accidents in Bako National Park.

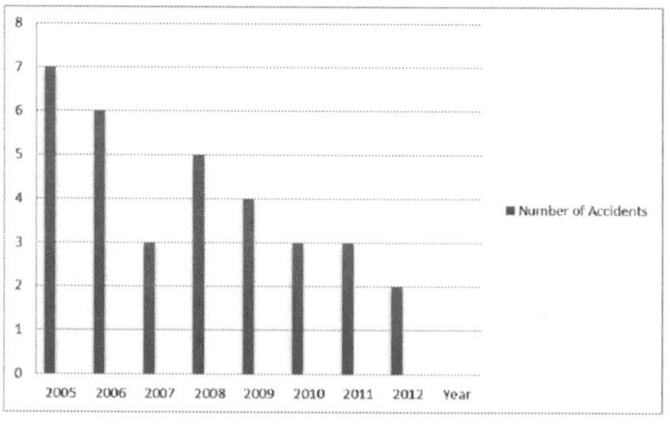

Figure 1. Statistic of number of incidents requiring emergency attention and accidents in Bako National Park from 2005–2012 (source: Bako National Park Record).

Since then, the number of emergency incidents and accidents had shown a decreasing trend. Although, there was no differentiation in incidence occurrence between guided and unguided tours and thus could be not be highlighted whether majority of the incidences were due to unguided tours. This outcome could still be deduced as attributed to the positive impact on the training module offered in the Certificate in Park Guiding Certificate training course creating awareness in risk management in the park guiding activities.

However, further detailed investigation should be conducted and if such cases were to have occurred, the management of the Park should consider imposing compulsory guided tour for all tourists. On a similar note, Queenstown in New Zealand, despite being marketed as the "adventure capital of the world" with active measure in place to ensure safety of the activities, accidents in the industry do occur. In the accident reports for all of fatalities of Queenstown, the overriding was clearly human error, with conclusions such a 'negligence', 'trip should not have taken place', 'pressure on guides due to late start', and 'failure to fully explain the trip' and 'what happens being when a problem arises' as being suggested by Page (1997). According to Cater (2006), adventure tourists cross boundaries in pursuit of thrills therefore; adventure guide must ensure that all procedures are followed to eliminate accidents or fatalities. He added participant in the adventure activities also must be aware of the risk and must apply the rules imposed by the adventure guides. The most successful ecotourism or adventure tourism operators are those that have reduced their actual risk levels whilst effectively commondifying the thrills within.

3.2 *Risk management awareness of licensed park guide as perceived by clients*

3.2.1 *Demographics of respondents*

There were about 60 visitors at the park at that period of time but only 30 respondents were willing to answer the questionnaires. From the 30 respondents, there was an almost equal representation of sexes: 50.0 percent male and 46.67 percent female with the majority accounting to 63 percent from age group of 30 year to 50 years of age. The percentage of the respondents for the age groups of above 50 years of age was about 73.33 percent. This indicated that the activities in relation to ecotourism are popular among those who are younger and considered physically fit.

From the data gathered, 60.0 percent of the respondents were single while 40 percent were married. Those respondents who were married either came together with their spouse or family while those who are single apparently came alone or with friends or associates.

International tourists from the western countries mainly from Germany, Netherland, Canada and United Kingdom made up of 83.3 percent of the respondents while 16.67 percent came from others states in Malaysia.

3.3 *Awareness of risk management of licensed park guides as perceived by their clients*

3.3.1 *Pre-tour risk management preparation by tourists*

The month of October was considered as a low peak season and therefore, not many tourists could be captured a within the span of two weeks. Based on information drawn from Questions 1 and 2 relating to the pre-tour risk management preparation by the tourists, 73.3 percent of the respondents were advised to take a travel insurance policy from their country of origin in case of an accident or emergency that might occurred while they travelled overseas either by their local agent or overseas agent. Most respondents agreed that they know about the place they are going and the activities they going to engage.

3.3.2 *Awareness of risk by tourists*

The awareness of risk by the tourists was reflected in Questions 3 to 5. 83.3 percent of the respondents agreed that risk management with 63.3 percent in the completely agreement was a very important issue indicating they would be very concern with activities which would jeopardise their safety. 76 percent of the respondent disagreed that they did not care about the risk they were facing while doing the ecotourism activities. There were 70 percent of the tourists who believed that they knew how to avoid risk.

3.3.3 *Awareness of risks by tourists in Bako National Park*

The awareness of risks that tourist would encountered in Bako National Park was drawn from Questions 6–7. It was shown that only 40 percent of the tourists thought they had prior knowledge of the level of risk they would be encountered in the park while 30 percent did not know about the level of risk and 20 percent responded with neutral answers. 53.3 percent of the tourist knew about the activities they would be engaging in the park. In the investigation of the Fox and Franz Josef glaciers, popular tourist attractions on the West Coast of New Zealand's South Island by Espiner (2001), it was discovered that many visitors especially international visitors have relatively poor awareness of natural hazards, and behave in ways which potentially compromise physical safety.

He argued that perceptions and behaviour are a consequence of diverse individual and situational factors including limited knowledge of the sites, beliefs about management, poor comprehension of hazard warning signs, and freedom from the normative constraints of everyday life.

3.3.4 *Awareness of risk management of licensed park guides and their practices as perceived by their clients*

The awareness of risk management of licensed park guides and their practices as perceived by the clients was drawn from Questions 8 to 17. More than 56.7 percent of the respondents agreed that the guide gave them a pre briefing while 33.3 percent was neutral. The later could be attributed to not understanding the park guides as 66.7 percent of the respondents agreed they were given sufficient information before going to the park. As 70 percent of the respondents agreed that they were asked to take a local insurance by their local tour operator, this would have led the park guide to request only 13 percent of the tourists to sign a waiver or contract. The 33.3 percent respondents did not know anything about it. However risk is argumentatively subjective as Cater (2006) suggested that while real risks are quantifiable and statistical in nature, perceived risks are more emotive and "profoundly qualitative". Individuals' perceptions of risk are influenced by a complex interaction of a number of factors that included an individual's past experiences, media presentations, vicarious experiences and a predisposition to anxiety (Davis-Berman et al., 2002).

33.0 percent of the respondents responded with "I don't know" upon being asked whether their guide informed them about risk management, while 30 percent said their guide did inform them about risk management. This could be attributed the meaning of risk management as understood by the tourists or the differences in the interpretation of risk management by the two parties i.e. the park guides and their clients. Thus, the comprehension of risk management by the park guide might be lacking.

The perception of only 46.0 percent of the respondents believing that the park guide knew about the risk they were facing in Bako National Park and 56.7 percent responded that they were reminded about being careful while trekking by their guide. Thus it indicated that either the park guide did not practice risk management or they were not able to comprehend the risk perceived by their clients as 60.0 percent of the respondents were sceptical about the park guide's knowledge of risk they were facing. This contrasted to the 66.6 percent of those guided tourists agreeing that their guide was very professional. In addition, 60.0 percent of the guided tourists would engage the same park guide again in their next trip to Bako National Park. However, this could be attributed to the personality of the park guide or their knowledge of the park products.

However, in any circumstances, according to Beeton (2001), the key factor is that no matter how potentially dangerous the crisis might have been, the operator must be in control and manage the situation without exposing guests to unnecessary danger. She further stressed that risk management and accident prevention were integral aspects of the preparatory phases of crisis management, in that they not only pre-empt so-called unpredictable circumstances, but also provide the operator with strategies to handle a crisis. The data collection process might have included respondents' emotional experience in their relationship established with their park guides with protectionism instinct. This could also be due to the need of projecting good image of themselves in front of researchers or they might feel obligated to answer the questionnaires even when they were in hurry.

4 CONCLUSION

Since the implementation of the Certificate in Park Guiding Certificate training course, there was a decreasing trend in incidents that required emergency attention and accidents in Bako National Park implicating the positive impact on the training module offered in creating awareness in risk management in the park guiding activities. The overriding factor of risk mismanagement was clearly human error and park guides must ensure that all procedures are followed to eliminate accidents or fatalities. Majority of the tourists accounting to 70.0 percent were aware of risk management and prepared for adventure activities however only 40.0 percent coming to Bako National Park thought they had prior knowledge of the level of risk they would be encountered in the park while 30.0 percent did not know about the level of risk and 20.0 percent responded with neutral answers. More than 50.0 percent of tourists' perceived that the licensed Park Guides' demonstrated risk management through both communication and practices in a professional manner. In contrast, more than 50.0 percent of the tourists perceived that the park guides did not know the knowledge of risk management. Risk is very subjective depending on individual perceptions of risk as influenced by a complex interaction of a number of factors that included an individual's past experiences, media presentations, vicarious experiences and a predisposition to anxiety. However, the park guides must be in control and manage the situation without exposing their clients to unnecessary danger. It is therefore, recommended the management of the Park should consider imposing compulsory guided tour for all tourists. Further research should be expanded to include non-guided tourists on their awareness of risk management. In addition, the differences in the perception of risk between the park guides and their clients should be investigated to enable designing training courses on risk management. Finally, the management should develop a tool to continual monitoring of the comprehension of risk management of the park guides and the practices.

REFERENCES

Bako National Park, (2005–2012). Annual Accident Report.

Beck, U. 1992. *Risky Society: towards a new modernity.* London. Sage.

Beeton, S, (2001). Horseback Tourism in Victoria, Australia: Cooperative, Proactive Crisis Management. *Current Issues in Tourism, 4*(5), 422–439.

Bently, T.A., Page, S.J & Laird, I.S. (2001). Accidents in the New Zealand adventure tourism Industry. *Safety Science, 38*, 31–38.

Berno, T., Moore, K., Simmons, D., & Hart, V. (1996). The nature of the adventure experience in *Queenstown. Australian Leisure, 8*(1), pp. 21–25.

Brown, I.D. (1989). Managing for Adventure Recreation. *Australian Parks and Recreation, 25*(4), 37–40.

Buckley, R.C. 2003. Adventure tourism and the clothing, fashion and entertainment industries. *Journal of Ecotourism, 2*, 126–134.

Cater, C.I. (2006). Playing with risk? Participant perceptions of risk and management implications in adventure tourism. *Tourism Management, 27*, 317–325.

Davis-Berman, J. & Berman, D. (2002). "Risk and Anxiety in Adventure Programming", the *Journal of Experiential Education, 25*, 2:305–310.

Espiner, S. (2001). The phenomenon of risk and its management in natural resource recreation and tourism settings: a case study of Fox and Franz Josef Glaciers, Westland National Park, New Zealand. Doctoral (PhD) Theses, Department of Social Science, Parks, Recreation, Tourism and Sport, Lincoln University, New Zealand.

Frangialli, F. (2003) Foreword. In Crisis Management in the Tourism Industry, D. Glaesser, ed., pp. xi–xii. Oxford: Butterworth Heinemann.

Glasgow Press and Journal. 1999. Swiss guides survivor face charges. (31st July), p. 17.

Gidden, A. 1991. Modernity and Self Identity. Cambridge. Polity.

Jansen-Verbeke, M. & Dietvorst, A. (1987). Leisure, Recreation, Tourism: A Geographic View on Integration. *Annals of Tourism Research, 14*, 361–376.

Mathieson, A. & Wall, G. (1982). Tourism: Economic, Physical and Social Impacts. London: Longman.

Page, S. (1997). Leisure travel: making it a growth market. again! New York: Wiley.

Page, S., Bentley, T., & Walker, L. (2005). Tourists Safety in New Zealand and Scotland. *Annals of Tourism Research, 32*(1), 150–166.

The Associated Press, 2012. "Tourist safety in Vietnam's famed Halong Bay questioned following latest deadly accident." http://www.canada.com/news.

The Christian Time Monitor Online, 2011. "Yosemite waterfall accident a cautionary tale for Yosemite visitors." http://www.csmonitor.com.

Hospitality and Tourism – Sumarjan et al. (eds)
© 2014 Taylor & Francis Group, London, ISBN 978-1-138-00151-0

The effects of POS, PSS, and OC on intention to leave in casual dining restaurants in Klang Valley area

N. Mohamad Nashuki, H. Ghazali & M. Othman
Universiti Putra, Malaysia

ABSTRACT: This study attempts to investigate the relationship between Perceived Organisational Support (POS), Perceived Supervisor Support (PSS), and Organisational Commitment (OC) with employees' intention to leave a job. There were a total of 717 questionnaires collected among casual dining restaurants employees in Klang Valley area. In order to determine the relationships, Pearson correlation and multiple linear regressions were run by using SPSS version 21. The results suggest that with higher level of POS, PSS, and OC would decrease the intention to leave a job as all of the predictors were found to be negatively correlated with the dependent variable. Furthermore, POS was also found to be the most influential factor in affecting employees' leaving intention between the antecedents. The finding is hoped to have important implications where the management can formulate strategies to reduce employees' turnover intention and ultimately retain them, as well as to monitor them better before they actually leave.

Keywords: intention to leave, KL restaurant industry, organizational commitment, perceived organizational support, perceived supervisor support

1 INTRODUCTION

Casual dining restaurant is a moderate-upscale dining and focuses on themes in menu, service, and décor (Rivera, DiPietro, Murphy & Muller, 2008). In Malaysia, casual dining restaurants are divided according to their price range. In outlets with higher price range, the guest cheque average is RM40–50 whereas outlet with lower price range received between RM10 to RM20 per person (Euromonitor International, 2012). It is a high energy and active workplace. The staffs are required to be knowledgeable on the menu items and products (Rivera et al., 2008).

As part of food service industry, casual dining restaurants are associated with having low pay, long hours, menial work, lack of benefits, hard work, physically exhausting, and under appreciated thus reflecting poor image to the public (Ingram & Jones, 1998; Wildes, 2005). Wildes (2005) then elaborated that psychologically, the job image that the restaurant industry suffers from is 'servitude' perception, which in restaurant industry it is pointed to the social stigma or stereotype attached to the job of serving others. These images contributed to the negative view of food service work. As a result, restaurants face difficulties in retaining their manpower and suffered from high rate of employee turnover.

1.1 *Employee turnover and turnover intention*

Turnover is defined as the movement of people in and out of employment within an organisation (Rahman et al., 2010). Employee turnover is a major cost for food service industry (Ingram & Jones, 1998; Kahumuza & Schlechter, 2008; Tuzun & Kalemci, 2012; Wildes, 2005). Malaysia experienced high rate of employee turnover both generally and in hospitality industry specifically. Based on a survey done by Malaysian Employers Federation (MEF) conducted on executive positions between June 2010 and July 2011, Hotel/Restaurant industry is placed third as the industry with the highest annual average turnover rate at 32.4 percent, just behind IT/Communication (75.72) and Associations/Societies (33). For example, Ryan, Ghazali, and Mohsin (2011) reported that turnover rate for non-managerial staff in fast food industry was 100 per cent. On the other hand, AON Hewitt, a human capital consulting and outsourcing firm, conducted a study shared that Malaysia is placed sixth in the Asia-Pacific in 2011 for staff turnover at the percentage of 15.9 (Goh, 2012).

Intention to leave is defined as the subjective estimation of an individual regarding the probability of leaving an organisation in the near future (Cho, Johanson, & Guchait, 2009). It is said to be a significant indicator of turnover action of an employee in the future (Tuzun & Kalemci, 2012). Tuzun and Kalemci (2012) added that since previous studies indicated that intention to leave is a better predictor to employees' actual turnover than other variables like overall job satisfaction and satisfaction with the work, therefore it became essential to understand the drive behind turnover intention and how to control them. Among the antecedents for intention to leave that has been studied extensively by scholars were Perceived Organisational Support (POS), Perceived Supervisor Support (PSS), and Organisational Commitment (OC). Therefore, this research will discussed how these predictors affect job leaving intention of employees of casual dining restaurants in Malaysia's key region, the Klang Valley area. The main objective of this study is to identify the relationship between POS, PSS, and OC on intention to leave a job.

2 LITERATURE REVIEW

2.1 *Perceived organisational support and intention to leave*

Employees do not know to what degree their employers reward their hard works. Therefore, they developed Perceived Organisational Support (POS), which holds that in order to meet socioemotional needs and to assess the benefits of increased work effort, employees form a general perception concerning the extent to which the organisation values their contributions and cares about their well-being (Robert Eisenberger, Huntington, Hutchison, & Sowa, 1986). Employees reciprocate the perceived support by increasing their performances, loyalties, and commitments (Kahumuza & Schlechter, 2008). Scholars have found negative relationships between POS and intention to leave (Allen, Shore, & Griffeth, 2003; Jawahar & Hemmasi, 2006; Kahumuza & Schlechter, 2008). In particular to hospitality industry, Cho et al. (2009) in their research done towards hotels employees in the USA found that employees with higher perception of POS have lower leaving intention. Blomme, Van Rheede, and Tromp (2010) reported that hospitality employees have greater propensity in leaving their organisation if they have lower POS, which showed similar pattern with employees outside the hospitality industry.

2.2 *Perceived supervisor support and intention to leave*

Perceived Supervisor Support (PSS) is the global perceptions that the employees have regarding the extent to which supervisor's value employees' contributions and care about their well-being (Eisenberger, Stringlhamber, Vandenberghe, Sucharski, & Rhoades, 2002). As employees work directly with supervisors, it is important for the latter to give fair treatments because they could affect the work performances of their subordinates. Moreover, supportive supervisor is effective in managing their subordinates' emotions, which would be essential in managing organisational commitment (Dawley, Andrews, & Bucklew, 2008).

Direct studies of PSS and intention to leave found negative relationships between the two variables (Basford, Offermann, & Wirtz, 2012; Dawley et al., 2008; Kahumuza & Schlechter, 2008). Report by Hay (2002) identified unsatisfactory with managers was the second highest contributor to the high number of turnover. In the survey, he found that 74 percent of employees who planned on staying at their organisation were satisfied with the management compared to 41 percent who planned to leave.

2.3 *Organisational commitment and intention to leave*

By definition, OC is described as the relative strength of an individual's identification with and involvement in a particular organisation (Mowday, Steers, & Porter, 1979). It is influenced by three dimensions—emotional attachment, perceived costs, and moral obligation. Meyer and Allen (1990) devised a model to describe the three dimensions as affective, continuance, and normative commitment. In their work, they have proposed that employees remain with the organisation for three different commitments. The commitments are between (1) affective commitment, where the employees themselves have the desire to remain in the organisation; (2) continuance commitment in which the employees recognition of the costs for leaving would be high; and (3) normative commitment where the employees have the feeling of obligation to stay and be loyal to the organisation. Regarding the relationship between OC and leaving intention, studies over the years have found that the two variables are negatively related (Kahumuza & Schlechter, 2008; Labatmediene, Endriulaitiene, & Gustainiene, 2007; Lee & Maurer, 1999; J.P. Meyer & Allen, 1991).

Based on the literature reviews, the following hypotheses were proposed:

- H_1: *POS will have a negative relationship with intention to leave a job.*
- H_2: *PSS will have a negative relationship with intention to leave a job.*
- H_3: *OC will have a negative relationship with intention to leave a job.*

3 METHODOLOGY

3.1 *Sample and data collection method*

The participants of this research comprised of the employees of casual dining restaurants in Klang Valley area. A self-administered questionnaire was distributed to 855 employees of casual dining restaurant outlets. The employees were advised that their participations are voluntary and their identities will be kept anonymous. Convenience sampling was used as sampling technique. Out of 855, a total of 717 responses were used for further analysis, which gives the response rate of 84%.

3.2 *Measures*

Responses to all the items for each variables were indicated on 7-point Likert scale (1 = strongly disagree to 7 = strongly agree).

3.2.1 *Perceived organisational support*

POS was measured using the 16-item scale developed by Eisenberger, Huntington, Hutchison, and Sowa (1986). The reliability of this construct measured by Cronbach's alpha was 0.88.

3.2.2 *Perceived supervisor support*

Assesment for PSS was done using 16-item scale developed by Kottke and Sharafinski (1988) ($\alpha = 0.89$).

3.2.3 *Organisational commitment*

This study used 18-item scale developed by Meyer, Allen, and Smith (1993) to measure OC ($\alpha = 0.77$).

3.2.4 *Intention to leave*

Intention to leave was assesed using 5-item scale developed by Wayne, Shore, and Liden (1997) ($\alpha = 0.63$).

3.2.5 *Demographic variables*

The respondents were also asked to specify their demographic profiles which were gender, age, marital status, educational level, working position, working hours per week, as well as their working experience in the current restaurant and in the restaurant industry.

3.3 *Data analysis*

Data analyses used for this study were based on the research objectives. Therefore, descriptive analysis, Pearson correlation, and multiple linear regression were used to test the hypotheses.

4 RESULTS

4.1 *Demographic profiles*

Male respondents were the dominant gender in this research, with 62.5 per cent. Respondents' average age was 24.08 years; 81.3 per cent were still single; more than half of them were either still studying or just finished their secondary school. Non-managerial employees made up 72.4 per cent of the respondent compared to managerial employees (23.4%). The respondents' average working hour per week was 54.3 hours, the average of 17.4 months of working experience in the restaurant they were working for, and the average of 35.3 months of restaurant industry's working experience.

4.2 *Interrelations between variables*

Pearson correlation was used to test hypotheses 1, 2, and 3 because it measures the association between variables. The results of correlations as well as the means and standard deviations are shown in Table 1. The correlation between intention to leave and POS showed negative relation at the level of 0.452. This association exhibits the highest magnitude compared to the other two relationships thus indicates that intention to leave-POS has the strongest connection. With the result showed negative association, it is therefore supported hypothesis 1 (H_1). In other words, if an employee had high level of POS, their tendency to leave the restaurant they currently employed is lessen and vice versa.

In regards to intention to leave-PSS relationship, it showed negative association at the level of 0.425. Therefore, hypothesis 2 (H_2) which proposed that intention to leave and PSS were negatively related is accepted. It means that employees who perceive to have greater support from their immediate supervisors were less likely to have leaving intention.

Table 1. Means, standard deviations, and correlations of variables.

	Mean	SD	1	2	3	4
1. Intention to leave	3.94	1.22	1.000			
2. POS	4.70	0.94	–0.452**	1.000		
3. PSS	4.82	0.97	–0.425**	0.679**	1.000	
4. OC	4.39	0.78	–0.181**	0.385**	0.382**	1.000

Table 2. Multiple linear regression.

Model	R	R^2		ANOVA F	Sig.
Regression	0.480[a]	0.230		71.080	0.000[a]
	Unstandardised coefficients				
Variable/predictor	B	Std. error	Standardised coefficient (beta)	t	Sig.
Coefficients[a]					
(Constant)	35.051	1.330		26.353	0.000
POS	–0.124	0.018	–0.309	–6.773	0.000
PSS	–0.088	0.018	–0.224	–4.919	0.000
OC	0.010	0.018	0.023	0.635	0.526

Lastly, OC also showed negative relation with intention to leave ($r = -0.181$). Employees with higher organisational commitment tend to not consider leaving the restaurants they were working for and vice versa. Since the result showed negative correlation between intention to leave and OC, hypothesis 3 (H_3) is therefore accepted.

4.3 *Multiple linear regression*

Multiple linear regression was also done as part of the analyses. POS, PSS, and OC were employed as joint predictors of intention to leave. From Table 2, it can be seen that about 23% of intention to leave can be explained by POS, PSS, and OC.

Standardised coefficient also suggested that POS is the variable that is most likely to predict intention to leave among employees of casual dining restaurants in Klang Valley area.

5 DISCUSSION

Overall, the results of this research contribute to the theoretical understanding of the role of POS, PSS, and OC had on casual dining restaurants employees' job leaving intention. In relation to POS, intention to leave a job by the employees showed significance and relate in negative manner. It means that employee with higher POS has lower intention to leave their job as compared to their colleagues with lower POS level. This finding is in line with previous researches that supported the negative relationship between POS and intention to leave. Employees with high POS have better motivation in helping the organisation to achieve their goals because they feel the need to reciprocate the helps from their management.

Negative relationship was also found between PSS and intention to leave, which was in line with previous studies. Employees who perceive greater support and care from their immediate supervisors are suggested to have lower leaving intentions. Some employees tend to be more committed to their immediate supervisors compared to the organisation because they could be more personal to the former. Therefore, the employees might have stayed with the restaurants because they feel the need to repay their supervisors kindness.

OC was also found to be negatively related. Employees that have better commitment were less likely to have leaving intentions. This is because employees with higher level of OC have better motivation to perform in the organisation's favour. OC is favoured in achieving long-term organisational goal due to the broader reaching implications it possessed (Perryer, Jordan, Firns, & Travaglione, 2010).

In summary, to reduce the intention to leave a job, which consider as a predictor to actual turnover, POS, PSS, and OC needed to be improved. Specifically, in the case of employees of casual dining restaurants in Klang Valley, the upper managements need to fix or improve the company's policies to work in their employees' favour because it was found that POS is the most influential factor in determining employees' intention to leave. Cho et al. (2009) suggested that in order to increase POS, the management has to include employees in decision making and made them feel that the organisation recognised their contributions.

6 CONCLUSION

This research provides the managers with information and suggestion on what determining the employee's turnover. Since POS, PSS, OC had significant contribution to the leaving intentions, it is suggested that the managers focused on these predictors to improve employee's commitment to them. Moreover, this research contributes to the general issue of turnover in Malaysian restaurant industry. With the continuous expansion of restaurant industry in Malaysia, it is important for the management to retain their workforce and attract more employees to work in their expanding outlets.

REFERENCES

Allen, D.G., Shore, L.M., & Griffeth, R.W. (2003). The role of perceived organizational support and supportive human resource practices in the turnover process. *Journal of Management*, 29(1), 99–118.

Basford, T.E., Offermann, L.R., & Wirtz, P.W. (2012). Considering the source: The impact of leadership level on follower motivation and intent to stay. *Journal of Leadership & Organizational Studies*, 19(2), 202–214.

Blomme, R.J., Van Rheede, A., & Tromp, D.M. (2010). Work-family conflict as a cause for turnover intentions in the hospitality industry. *Tourism and Hospitality Research*, 10(4), 269–285.

Cho, S., Johanson, M.M., & Guchait, P. (2009). Employees intent to leave: A comparison of determinants of intent to leave versus intent to stay. *International Journal of Hospitality Management*, 28, 374–381.

Dawley, D.D., Andrews, M.C., & Bucklew, N.S. (2008). Mentoring, supervisor support, and perceived organizational support: what matters most? *Leadership & Organization Development Journal*, 29(3), 235–247.

Eisenberger, R., Stringlhamber, F., Vandenberghe, C., Sucharski, I., & Rhoades, L. (2002). Perceived supervisor support: Contributions to perceived organizational support and employee retention. *Journal of Applied Psychology*, 87(3), 565–573.

Eisenberger, R., Huntington, R., Hutchison, S., & Sowa, D. (1986). Perceived organizational support. *Journal of Applied Psychology*, 71(3), 500–507.

Euromonitor International. (2012), October 31. *Full-service restaurants in Malaysia*. Retrieved from Euromonitor Passport GMID database.

Ghiselli, R.F., Lopa, J.M., & Bai, B. (2001). Job satisfaction, life satisfaction, and turnover intent among food-service managers. *The Cornell Hotel and Restaurant Administration Quaterly*, 42(2), 28–37.

Goh, L. 2012, February. Why Job-hoppers Hop, pp. 24–25.

Hay, M. (2002). Strategies for survival in the war of talent. *Career Development International*, 7(1), 52–55.

Ingram, H., & Jones, S. (1998). Teamwork and the management of food service operations. *Team Performance Management*, 4(2), 67–73.

Jawahar, I.M., & Hemmasi, P. (2006). Perceived organizational support for women's advancement and turnover intentions: The mediating role of job and employer satisfaction. *Women In Management Review*, 21(8), 643–661.

Kahumuza, J., & Schlechter, A.F. (2008). Examining the direct and some mediated relationships between perceived support and intention to quit. *Management Dynamics*, 17(3), 2–19.

Kottke, J.L., & Sharafinski, C.E. 9(1988). Measuring Perceived Supervisory and Organizational Support. *Educational and Psychological Measurement*, 48(4), 1075–1079.

Labatmediene, L., Endriulaitiene, A., & Gustainiene, L. (2007). Individual correlates of organizational commitment and intention to leave the organization. *Baltic Journal of Management*, 2(2), 196–212.

Lee, T.W., & Maurer, S.D. (1999). The effects of family structure on organizational commitment, intention to leave and voluntary turnover. *Journal of Managerial Issues*, 11(4), 493–513.

Meyer, J.P., & Allen, N.J. (1990). The measurement and antecedents of affective, continuance and normative commitment to the organization. *Journal of Occupational Psychology*, 63, 1–18.

Meyer, J.P., & Allen, N.J. (1991). A three-component conceptualization of organizational commitment. *Human Resource Management Review*, 1(1), 61–89.

Meyer, J.P., Allen, N.J., & Smith, C.A. (1993). Commitment to organizations and occupation extension and test of a three component conceptualization. *Journal of Applied Psychology*, 78(4), 538–551.

Mowday, R.T., Steers, R.M., & Porter, L.W. (1979). The measurement of organizational commitment. *Journal of Vocational Behavior*, 14, 224–247.

Perryer, C., Jordan, C., Firns, I., & Travaglione, A. (2010). Predicting turnover intentions: The interactive effects of organizational commitment and perceived organizational support. *Management Research Review*, 33(9), 911–923.

Rahman, B.A., Mohd Azuan, M.A., Zahari, H., Noraida, A.K., Siti, N.A., Salleh, H., & Mohd, F.M. (2010). The study of factors contributing to chef turnover in hotels in Klang Valley, Malaysia. *Asian Social Science*, 6(1), 80–85.

Rivera, M., DiPietro, R.B., Murphy, K.S., & Muller, C.C. (2008). Multi-unit managers: Training needs and competencies for casual dining restaurants. *International Journal of Contemporary Hospitality Management*, 20(6), 616–630.

Ryan, C., Ghazali, H., & Mohsin, A. (2011). Determinants of intention to leave a non-managerial job in the fast-food industry of West Malaysia. *International Journal of Contemporary Hospitality Management*, 23(3), 344–360.

Tuzun, I.K., & Kalemci, R.A. (2012). Organizational and supervisory support in relation to employee turnover intentions. *Journal of Managerial Psychology*, 27(5), 518–534.

Wayne, S.J., Shore, L.M., & Liden, R.C. (1997). Perceived organizational support and leader-member exchange: A social exchanger perspective. *Academy of Management Journal*, 40(1), 82–111.

Wildes, V.J. (2005). Stigma in food service work: How it affects restaurant servers' intention to stay in the business or recommend a job to another [Abstract]. *Tourism and Hospitality Research*, 5(3), 213–233.

Hospitality and Tourism – Sumarjan et al. (eds)
© 2014 Taylor & Francis Group, London, ISBN 978-1-138-00151-0

Customer perception towards service quality of fast food restaurants in Klang Valley

H. Rasidah & Z. Bohari
Universiti Teknologi MARA, Malaysia

ABSTRACT: Due to its convenience, the fast food industry in Klang Valley, Malaysia has becoming one of the most popular places for eating out. Therefore, quality service has becoming one of the most concerned issues. To examine expectations and perceptions of customers, evaluation method of service quality was conducted using SERVQUAL instrument. A survey of 404 restaurant's customers showing that the restaurant's customers are significantly dissatisfied with the seven dimensions of service quality, including tangibles, reliability, assurance, responsiveness, empathy, price and perceived value. According to the current service quality level of fast food restaurants, this study suggests that fast food restaurants in Klang Valley develop its service quality improvement strategy objectively and accurately.

Keywords: customer perception, fast food restaurants, Klang Valley, service quality

1 INTRODUCTION

Malaysian fast food industry have developed and expanded rapidly in recent years. Malaysian government has promoted and encouraged western fast food franchises due to proven business records (10th Malaysian Plan). Competition in Malaysian fast food industry has forced the industry to pay greater attention to service quality and customer satisfaction. Research has demonstrated that service quality has a positive impact on customer satisfaction (Andaleeb & Conway, 2006; H.S. Kim, Joung, Yuan, Wu & Chen, 2009). Starting from 1980, service quality studies started to attract the attention of scholars. In which, (Parasuraman, Zeithaml & Berry, 1985) defined service quality as the overall difference between a customer's expectations and perceptions of the service experience.

SERVQUAL scale, proposed by (Parasuraman et al., 1985) which includes five dimensions tangible, reliability, responsiveness, assurance and empathy and 22 paired items, especially for measuring service quality. These five gaps are the factors that the service industry cannot satisfy regarding customer demand and expectation. If the industry wants to satisfy customer demand, the difference of these five gaps must be reduced. In these five gaps, the first four gaps are the most major obstacles for the service industry to provide service quality. The fifth gaps is formed from the difference between customer perceived service quality and expected service quality and the functions of the previous four gaps. (Parasuraman et al., 1985) defined service quality as the overall difference between a customer's expectations and perceptions of the service experience. In other word, service quality can be measured by comparing the expectations of customers with their perceptions of the delivered service. A gap is created when the perceptions of the delivered service is not as per the expectations of the customer. This gap is addressed by identifying and implementing strategies that affect perceptions, or expectations, or both.

The SERVQUAL scale became the most frequently used scale to measure the service quality of the general service quality. Therefore, the purpose of this study is to apply the SERVQUAL instruments developed in order to obtain a better understanding the service quality expectations and perceptions of fast food restaurant's customers. Secondly this study is to identify the weakness within the dimensions of service quality based on customer's quality evaluations.

2 LITERATURE REVIEW

2.1 *SERVQUAL instrument*

Several model was exist to measure the quality of service perceived by the customers. One service quality measurement model that has been extensively applied is the SERVQUAL model developed by Parasuraman, Zeithaml, & Berry, (1988 & 1994); Parasuraman et al., (1985); Parasuraman,

Zeithaml & Berry, (1993); Zeithaml, Berry & Parasuraman, (1993). SERVQUAL as the most often used approach for measuring service quality has been to compare customers' expectations before a service encounter and their perceptions of the actual service delivered (C. Gronroos, 1984; C. Gronroos, 1982; Parasuraman et al., 1985). The SERVQUAL instrument has been the predominant method used to measure consumers' perceptions of service quality. It has five generic dimensions or factors and are stated as follows (Van Iwaarden, van der Wiele, Ball, & Millen, 2003):

i. *Tangibles*. Physical facilities, equipment and appearance of personnel.
ii. *Reliability*. Ability to perform the promised service dependably and accurately.
iii. *Responsiveness*. Willingness to help customers and provide prompt service.
iv. *Assurance* (including competence, courtesy, credibility and security). Knowledge and courtesy of employees and their ability to inspire trust and confidence.
v. *Empathy* (including access, communication, understanding the customer). Caring and individualized attention that the firm provides to its customers.

SERVQUAL instrument has emerged as the most popular standardized questionnaire to measure service quality. Although, since its creation, the scale has been criticized by many researchers in several disciplines (Babakus & Boller, 1992; Carman, 1990; Peter, Churchill & Brown, 1993; van Dyke, Prybutok, & Kappelman, 1999) the relevant literature reveals that the SERVQUAL is still regarded as a leading measure of service quality (Caruana, Ewing & Ramaseshan, 2000). Therefore, the SERVQUAL instrument was used as a tool of analysis in this study.

2.2 *The relationship between service quality, customer satisfaction and loyalty*

A number of researches have studied service quality and customer satisfaction in recent years (Choi, Lee, Kim, & Lee, 2005; Kang, James, & Alexandris, 2002; Kouthouris & Alexandris, 2005; Rajpoot, 2010). The outcomes from these studies have produced several contributions in relation to understanding the dimensional structure of service quality. Consumer perception of service quality is critical for the success of a firm as it is linked with customer satisfaction and loyalty. Indeed, several authors claimed that service quality is an important antecedent of customer satisfaction (Y. Ekinci, 2002; Faullant, Matzler, & Füller, 2008; Parasuraman et al., 1985; Yu & Shen, 2011). Some proved that satisfaction acts as a mediator between service quality and loyalty and has an overall stronger effect on loyalty than service quality. However, satisfaction alone may not be enough to actually achieve loyalty (Cronin & Taylor, 1992; Faullant et al., 2008; González, Comesana, & Brea, 2007; Heung, Wong, Qu & 2002; Kandampully & Suhartanto, 2000).

As far as restaurant experience is concerned, several authors argued that food, physical environment, and employee services are major components and have an impact on customer satisfaction (Namkung & Jang, 2007; Ryu & Jang, 2008). In the fast food industry, several authors studied the impact of perceived service quality on customer satisfaction and loyalty (Cheng, Chen, Hsu, & Hu, 2012; W. G. Kim, Ng, & Kim, 2009; Stevens, Knudsen, & Patton, 1995). The factors such as the taste of food, the outlet cleanliness, the fast service, and the friendliness of staff seem to play a major role. Moreover, two values that are gaining importance are price and nutritional values (Chang, Chen, & Hsu, 2010; Y. Ekinci, Riley, & Fife-Schaw, 1998; Oyewole, 1999; Stevens et al., 1995). Convenient operating hours are a further significant point for consumers (Oyewole, 1999).

3 METHODOLOGY

3.1 *Data collection and respondents*

The field survey was conducted onsite in 25 fast food restaurants in Klang Valley, Malaysia. The field survey method was chosen to gain information directly from respondents at the fast food restaurants setting. A total of 450 questionnaires were distributed, and 404 of them completed effectively, yielding 80.8% response rate. As shown in Table 1, respondents varied greatly in terms of age, occupation and number of meals taken in three months times.

3.2 *Measures*

A self-administered questionnaire, an adapted and modified version of SERVQUAL, was used in this study to analyze the service quality expectations and perceptions of fast food restaurants' customers.

The 29 service quality attributes, designed to measure the respondents' expectations and perceptions regarding quality of services offered by the fast food restaurants, were grouped into seven dimensions, including tangibles, reliability, responsiveness, assurance, empathy, price and perceived value. The structured questionnaire contained the SERVQUAL attributes,

Table 1. Selected characteristics of the respondents.

Characteristics	Frequency	Percentage
Gender		
Male	170	42.1
Female	234	57.9
Age		
20	82	20.3
21 < age < 31	108	26.7
31 < age < 40	180	44.6
41 < age < 50	30	7.4
>51	4	1
Occupation		
Public sector	34	8.4
Professionals	88	21.8
Housewives or retirees	100	24.8
Student	140	34.7
Others	42	1
Number of meals within three months		
Below 2	42	53
3–5	42	17.8
6–8	34	9.4
More than 9	68	19.8
Total	404	100

and several extra attributes were added to the original questionnaire, including some specific quality-related attributes to make them more applicable to the fast food restaurant. The items in the questionnaire were measured on a five-point scale, (1) indicating 'very low' and (5) 'very high'. The five-point scale was also different from SERVQUAL which had seven-point scales.

A pilot test was undertaken to assess the reliability of the attributes, and to ensure that the wordings of the questionnaire were clear. Twenty questionnaires were completed by the customers in accompaniment of researcher. Some problems were identified with the wordings and implications of some questions, so some minor revisions were made to avoid confusion. Reliability analysis was also applied to test the internal consistency of each of the expectation and perception attributes. The results showed that the Cronbach's α coefficient for all the expectations and perception attributes, ranging from 0.9150 to 0.9486, were quite high, indicating the measurement has internal consistent and reliable.

The respondents were asked to rate their expectations and perceptions of the various items for the fast food restaurants. Customers' perceptions of overall service quality was assessed in the second part of the questionnaires on the same five-point scale. While the third part of the questionnaires contained questions relating to socio demographic data about the respondents.

4 DATA ANALYSIS AND RESULTS

4.1 *t-test analysis*

SSPS 20.0 was used to analyze the data. Descriptive statistics analysis was used to measure guests' expectation and perception scores. Paired *t*-test was carried out to test the significant difference between the two means of expectations and perceptions regarding quality of services offered by the fast food restaurants. Table 2 shows the means, difference scores, *t*-values and significance obtained through the evaluation of each service dimensions. Descriptive statistical methods were used to describe tourists' expectations and perceptions. The means were computed by adding up the scores allocated by respondents for each dimensions and dividing the total value by the number of respondents. The gap scores (PM-EM) for each dimension was calculated by subtracting the expectation means from the perception means. Positive scores show better than expected service while negative scores show poor quality.

A zero score implies that quality is satisfactory. Paired *t*-test was carried out to test the significant difference between the means of expectations and perceptions. The *t*-test results (see Table 2) show negative gap scores between expectation and perception regarding the seven quality attributes of fast food restaurants. This means that the fast food restaurants offers, relatively poor service quality. Moreover, the significance of T value of each quality dimensions is all under 0.01, which means significant discrepancy exists between customers' expectation and perception. It indicates that the customers are significantly dissatisfied with fast food restaurant service regarding seven service dimensions studied.

The results also shows that the expectations of quality attributes of the fast food restaurant's customers are all above 4 which shows that the customers believe that all the seven attributes are important and highest expectation is price. This indicates that price is the most important attributes to the customers. Tangibility is the only one whose perception is above 4, which shows that tangibility is the best quality attributes that the fast food restaurants offer. The biggest gap exists between the expectation and perception is assurance, which show customers are most dissatisfied with the assurance given by the fast food restaurants. Further, the second dissatisfied attributes is empathy. This suggests that

Table 2. t-Test for customers' perception and expectation in the fast food restaurant.

Attribute	Expectation mean	Perception mean	Gap	T value	Sig.
Tangibility	4.2236	4.458	−0.2344	−8.042	0.00**
Reliability	4.3456	3.9241	−0.4215	−6.243	0.00**
Responsibility	4.2651	3.8432	−0.4219	−7.618	0.00**
Assurance	4.2362	3.0371	−1.1991	−5.653	0.00**
Empathy	4.2378	3.7096	−0.5282	−9.068	0.00**
Price	4.6104	4.5743	−0.0356	−8.564	0.00**
Perceived value	4.2871	3.8432	−0.4439	−7.618	0.00**

managers of the fast food restaurants have to re evaluate their strategies on giving more assurance and empathy to the customers since these two attributes important and can significantly affect customers' satisfaction.

Moreover, among five basic service quality attributes, assurance, empathy and responsibility are the three most dissatisfied service attributes. This indicates that the fast food restaurant managers need to improve on the performance and completing their promised service, an extra addition to the level and quality of service which they could offer the customers.

5 CONCLUSIONS

The study of service quality of fast food restaurants in Klang Valley, Malaysia using SERVQUAL instrument identified the weakness within the seven dimensions of service quality. The t-test analysis results suggest that the fast food restaurants customers are significantly dissatisfied with all seven dimensions of service quality including tangibles, reliability, assurance, responsiveness, empathy, price and perceived value. Moreover, among the seven dimensions of service quality, assurance and empathy revealed to be the most dissatisfied dimensions. Overall, based the results of t-test, it is suggested that fast food restaurants should make more efforts to improve its service quality along all seven critical service dimensions.

REFERENCES

Andaleeb, S.S., & Conway, C. (2006). Customer satisfaction in the restaurant industry: an examination of the transcation-specific model. *Journal of Services Marketing*, 20(1), 3–11.

Babakus, E., & Boller, G.W. (1992). An empirical assessment of the SERVQUAL scale. *Journal of Business Research*, 24(3), 253–268.

Carman, J.M. (1990). Consumer perceptions of service quality: an assessment of SERVQUAL dimensions. *Journal of Retailing*, 66(1), 33–55.

Caruana, A., Ewing, M.T., & Ramaseshan, B. (2000). Assessment of the three-column format SERVQUAL: an experimental approach *Journal of Business Research*, 49(1), 57–65.

Chang, K.C., Chen, M.C., & Hsu, C.L. (2010). Applying loss aversion to assess the effect of customers' asymmetric responses to service quality on post-dining behavioral intentions: an empirical survey in the restaurant sector. *International Journal of Hospitality Management*, 29(4), 620–631.

Cheng, Ching-Chan, Chen, Cheng-Ta, Hsu, Fu-Sung, & Hu, Hsiu-Yuan. (2012). Enhancing service quality improvement strategies of fine-dining restaurants: New insights from integrating a two-phase decision-making model of IPGA and DEMATEL analysis. *International Journal of Hospitality Management*, 31(4), 1155–1166. doi: http://dx.doi.org/10.1016/j.ijhm.2012.02.003.

Choi, K.S., Lee, H., Kim, C., & Lee, S. (2005). The service quality dimensions and patient satisfaction relationships in South Korea: Comparison across gender, age and types of service. *Journal of Service Marketing*, 19(3), 140–149.

Cronin, J.J., Jr., & Taylor, S.A. (1992). Measuring service quality: A re-examination and extension. *Journal of Marketing*, 56, 55–68.

Ekinci, Y. (2002). A Review of Theoretical Debates on the Measurement of Service Quality: Implications for Hospitality Research. *Journal of Hospitality & Tourism Research*, 26(3), 199–216.

Ekinci, Y., Riley, M., & Fife-Schaw, C. (1998). Which school of thought? The dimensions of resort hotel quality. *International Journal of Contemporary Hospitality Management*, 10(2/3), 63–67.

Faullant, R., Matzler, K., & Füller, J. (2008). The impact of satisfaction and image on loyalty: the case of Alpineski resorts. *Managing Service Quality*, 18(2), 163–178.

González, M.E.A., Comesana, L.R., & Brea, J.A.F. (2007). Assessing tourist behavioral intentions through perceived service quality and customer satisfaction. *Journal of Business Research*, 60, 153–160.

Gronroos, C. (1984). A service quality model and its marketing implications. *European Journal of Marketing*, 36–44.

Gronroos, C. (1982). Strategy management and marketing in the service sector: Boston.

Heung, V.C.S., Wong, M.Y., Qu, H., & (2002). Airport-restaurant service quality in Hong Kong: An application of SERVQUAL. *The Cornell Hotel and Restaurant Administration Quarterly*, 41(3), 86–96.

Kandampully, J., & Suhartanto, D. (2000). Customer loyalty in the hotel industry: the role of customer satisfaction and image. *International Journal of Contemporary Hospitality Management*, 12(6), 346–351.

Kang, G.D., James, J., & Alexandris, K. (2002). Measurement of internal service quality: Application of the SERVQUAL battery to internal service quality, *Managing Service Quality*, 12(5), 278–291.

Kim, H.S., Joung, H.W., Yuan, Y.H.E., Wu, C.K., & Chen, J.J. (2009). Examination of the reliability and validity of an instrument for measuring service quality of restaurants. *Journal of Foodservice*, 20(6), 280–286.

Kim, W.G., Ng, C.Y.N., & Y.S., Kim. (2009). Influence of institutional DINESERV on customer satisfaction, return intention and word of mouth. *International Journal of Hospitality Management*, 28(1), 10–17.

Kim, W.G., Ng, C.Y.N., & Kim, Y.S. (2009). Influence of institutional DINESERV on customer satisfaction, return intention and word-of-mouth. *International Journal of Hospitality Management,* 28(1), 10–17.

Kouthouris, C., & Alexandris, K. (2005). Can service quality predict customer satisfaction and behavioural intentions in the sport tourism industry? An application of the SERVQUAL model in an outdoors setting. *Journal of Sport Tourism*, 10(2), 101–111.

Namkung, Y., & Jang, S.C. (2007). Does food quality really matter in restaurants? Its impact on customer satisfaction and behavioral intentions. *Journal of Hospitality & Tourism Research* 31(3), 387–409.

Oyewole, P. (1999). Multi-attribute dimensions of service quality in the fast food restaurant industry. *Journal of Restaurant & Foodservice Marketing*, 3(3/4), 65–91.

Parasuraman, A., Zeithaml, V.A., & Berry, L.L. (1988). SERVQUAL: a multiple item scale for measuring consumer perceptions of service quality. *Journal of Retailing*, 64(1), 12–40.

Parasuraman, A., Zeithaml, V.A., & Berry, L.L. (1994). Alternatives scales for mesuring service quality: A comparative assessment based on psychometric and diagnostic criteria. *Journal of Retailing*, 70, 201–230.

Parasuraman, A., Zeithaml, V.A., & Berry, L.L. (1985). A conceptual model of service quality and its implications for future research. *Journal of Marketing*, 49(4), 41–50.

Parasuraman, A., Zeithaml, V.A., & Berry, L.L. (1993). Research note: more on improving service quality measurement. *Journal of Retailing*, 69(1), 140–147.

Peter, P., Churchill, G.A. Jr, & Brown, T.J. (1993). Caution on the use of difference scores in consumer research. *Journal of Consumer Research*, 19, 655–662.

Rajpoot, Nusser A. (2010). A Multiple Item Scale for Measuring Tangible Quality in Foodservice Industry. *Journal of Foodservice Business Research* 5(2), 109–127. doi: DOI: 10.1300/J369v05n02_08.

Ryu, K., & Jang, S. (2008). DINESCAPE: A scale for customers' perception of dining environments. *Journal of Foodservice Business Research*, 11(1), 2–22.

Stevens, P., Knudsen, B., & Patton, M. (1995). DINESERV: tool for measuring service quality in restaurants. *Hotel and Restaurant Administration Quality*, 4, 56–60.

van Dyke, T.P., Prybutok, V.R., & Kappelman, L.A. (1999). Cautions on the use of the SERVQUAL measures to assess the quality of information systems services. *Decision Sciences*, 30(3), 877–891.

Van Iwaarden, J., van der Wiele, T., Ball, L., & Millen, R. (2003). Applying SERVQUAL to websites: An exploratory study. *International Journal of Quality & Reliability Management*, 20(8), 919–935.

Yu, KT, & Shen, CY. 2011. Service quality management by expanded traditional importance-performance analysis. Journal of the Chinese Institute of Industrial Engineers, 28: 146–154. (2011). Service quality management by expanded traditional importance-performance analysis. *Journal of the Chinese Institute of Industrial Engineers*, 28, 146–154.

Zeithaml, V.A., Berry, L.L., & Parasuraman, A. (1993). The nature and determinants of customer expectations of service. *Journal of The Academy of Marketing Science*, 21(1).

Hospitality and Tourism – Sumarjan et al. (eds)
 ISBN 978-1-138-00151-0

Revealing the power of Appreciative Inquiry approach for rural tourism research

R. Che Aziz, M. Abdul, Y. Abd Aziz & A. Abdul Rahman
Universiti Putra, Malaysia

ABSTRACT: Major focus of Appreciative Inquiry (AI) approach is to learn about the strengths and successes of peoples and it is perceived to be a significant recent innovation in action-research. AI has been acknowledged as a vital tool to create and empower local people in order to take advantage of the opportunities provided by tourism development. AI approach does not primarily focus on what is going *wrong*, but rather on what is going *right*. In other words, this approach prompts discovery of the root causes of success rather than failure. AI was developed to gain and build strong, enduring relationships between communities and the tourism industry, based on the assumption that each organization or community has strengths which can become starting points for positive change. Past studies found that there is significant potential for the adoption of AI approach as it is rarely used in tourism field and it may offer researchers another worldwide and methodology for framing and conducting tourism research. This approach has been proposed as providing the opportunity to close the gap between theory and practice, which AI addresses an organizational issues in a significantly different way. With that respect, the purpose of this paper is to review recent study into the potential uses of Appreciative Inquiry as one of the promising tourism research tool in conducting tourism research. It does not mean that AI may replace all research approaches, nor it is a panacea that can overcome all the challenges of conducting research. However, AI can be considered as a new method, approach and/or strategy for research, which initiates positive changes in tourism communities.

Keywords: Appreciative Inquiry, community development, rural tourism, tourism development

1 INTRODUCTION

Rural tourism is a fast-growing sector in the global tourism market. Improving and protecting both natural and man-made attractions, economic growth, cultural development, and the establishment of infrastructure are some of the most significant issues facing this sector (Lo, Mohamad, Songan, & Yeo, 2012). There is potential for rural tourism to be a major force in rural economic development and contribute to poverty alleviation, particularly for rural communities. This sector also generates a large-scale employment opportunities and labour incentives in rural areas (Bramwell & Lane, 2012). Rural tourism may be key to transforming rural economies, but relevant policies must be established to ensure the success of these developments (World Tourism Organization, 2001). There has been little discussion thus far regarding tourism's contribution to rural communities. Two gaps have been further identified: making rural tourism a more viable industry in resource-dependent communities, and embedding the industry within a community that seeks alternatives from a "crisis context" (Koster & Lemelin, 2009). Thus, with this in mind, Appreciative Inquiry approach has been identified asan alternative tourism research tool for evaluating the impacts of rural tourism on community development.

Appreciative Inquiry (AI) is a growing and developing field offering community development strategies based on regional strengths (Cooperrider & Whitney, 2005). Although the AI approach originated from academia, there is a considerable body of literature covering its practical application and impact. AI is rooted in organizational development and strategies which assist organizations in transforming themselves. This approach has often been used for organization analysis, learning and development uniquely intended for discovering, understanding and spurring innovations and transformation in human/social system (The Mountain Institute, 2000). Constructivist, strength-based participatory action research methods which follow a grounded theory procedure are one of the important outcomes of AI (Cooperrider & Whitney, 2005). It serves as framework for personnel development, coaching, partnerships, alliance building, and large-scale community or organization (Finegold, Holland, & Lingham, 2002; Raymond & Hall, 2008a). The AI approach is grounded in part on the idea that changing any human group should involve encouraging groups to inquire about, learn from and build on methods which work when they are at their best (Laszlo & Cooperrider, 2010). Thus, this paper attempts to review recent study into the potential uses of Appreciative Inquiry as one of the promising tourism research tools in conducting tourism research. It does not mean that AI may replace all research approaches, nor it is a panacea that can overcome all the challenges of conducting research. However, AI can be considered as a new method, approach and/or strategy for research, which initiates positive changes in tourism communities.

2 APPRECIATIVE INQUIRY APPROACH

2.1 *Background of Appreciative Inquiry approach*

Appreciative Inquiry (AI) was first established in 1986 by David Cooperrider in his doctoral thesis on "*Appreciative Inquiry: Toward a methodology for understanding and enhancing organizational innovation*" (Cooperrider, Whitney, & Stavros, 2008; Cooperrider & Whitney, 2005). For more than a decade, non-profit organizations, businesses, families, health care organizations, school as well as government have applied AI principles. AI has been promoted since the late 1980s in a wide variety of organizations and for various purposes. Applications have included strategic planning, culture transformation, increase customer satisfaction, organization redesign, and leadership development. More recently AI has been applied to such situations as integrating organizations after a merger, building alliances and union-management partnerships, peace building, and implementing educational reform and economic development efforts (Whitney & Trosten-Bloom, 2010).

The concept of Appreciative Inquiry is defined in different ways by different scholars. The term is generally understood as an alternative approach, an innovative process, a theory, knowledge or a philosophy (Koster & Lemelin, 2009; Lewis, Passmore, & Cantore, 2008; Raymond & Hall, 2008a). For Cooperrider et al., (2008) and Cooperrider and Whitney (2005), The words "appreciative" and "inquiry" refer to:

> *Appreciate, v., 1. Valuing; the act of recognizing the best in people or the world around us; affirming past and present strengths, successes and potentials; to perceive those things that give life (healthy, vitality, excellent) to living systems. 2. To increase in value, e.g., the economy has appreciated in value.*
>
> *Inquire, v., 1. The act of exploration and discovery. 2. To ask questions; to be open to seeing new potentials and possibilities.*

The literature is full of contributions to the AI approach. AI has been perceived as a significant development for such diverse fields as action research (Boyd & Bright, 2007; Dick, 2006), organizational management (Rockey & Webb, 2005), community planning (Morsillo & Fisher, 2007; The Mountain Institute, 2000) and providing a new evaluation technique for field research (Lewis et al., 2008). Up to now, AI has been addressed as a group of process that inquires, identifies and further develops the best of "what is" in organizations in order to create a better future (Preskill & Catsambas, 2006). They claimed that AI is typically used in the organizational development field as an approach to a large-scale change; the method provides a blueprint for addressing issues, challenges, changes and concerns of organizations in ways that build on the successful, effective and energizing experiences of its members.

AI methodology does not primarily focus on what is going *wrong*, but rather on what is going *right*. Organization members first discover what is working well and are then asked to imagine what their organization might look like if every participant worked to that standard. Members can become agents of positive change by imagining *what can be,* basing their perceptions on *the best of what* is. (Cooperrider & Whitney, 2005; Michael, 2005; Preskill & Catsambas, 2006). This approach prompts discovery of the root causes of success rather than failure. AI encourages a conscious and deliberate choice to ask only positive questions regarding a community's needs for making a better life. Appreciative Inquiry has been embraced as a powerful organization development philosophy building on past success to propel positive change. AI is a highly participative, holistic approach to change that values the wisdom of members of the organization and amplifies positive forces (Bellinger & Elliott, 2011).

3 THE APPLICATION OF APPRECIATIVE INQUIRY APPROACH

AI was developed to gain and build strong, enduring relationships between communities and the tourism industry, based on the assumption that each organization or community has strengths which can become starting points for positive change (Raymond & Hall, 2008a, 2008b). AI has been applied in such diverse fields as conservation planning and tourism program development, providing examples of positive, future-oriented planning and management (Koster & Lemelin, 2009; The Mountain Institute, 2000). Inviting people to participate in dialogues and share stories about their past and present achievements, assets, unexplored potentials, innovations, strengths, opportunities, benchmarks, high-point moments, tradition, lived value, core and distinctive competencies, expressions of wisdom and possible futures is referred to as "positive core" (Cooperrider et al., 2008; Cooperrider & Whitney, 2005). AI directly harnesses the energy of this positive core to change agendas; the energy and excitement impels a desire to move forward to a shared goal.

A much more systematic study of AI seeks out the exceptional best of "what is" (Discovery) to help ignite the collective imagination of "what might be" (Dream). The purpose of this is the generation of a collectively desired future (Cooperrider et al., 2008). A vision which at first is only ephemeral becomes successfully translated from possibility and intention, and belief into concrete reality and practices. The AI process has been presented in various models, such as the original Cooperrider Model, the 4-D model and the Mohr/Jacobsgaard 4-I Model (Nyaupane & Poudel, 2012). These studies identify AI as a new creation or grand experiment which brings out the best of human organization, beginning with a clear purpose. AI has also been called a "narrative-based process of positive change." Cycles of activity start by engaging all organizational or community stakeholders in qualitative interviews to discuss strengths, resources and capabilities (Cooperrider & Whitney, 2005.p.g 15). Participants are then ushered through a series of future-focused activities which help them imagine the promise that their lives may hold in the future. The AI process has no rigid or fixed format; options range from break room discussions over coffee between members of a small team to company-wide initiatives involving every member from the boardroom down.

The AI process engages entire communities and their stakeholders to create a future that works for everyone and can be used to guide conservation, large group meetings or whole-system effort (Finegold et al., 2002; Shariff, Gramberg, & Foley, 2009). No matter what the purpose, the Appreciative Inquiry 4-D Cycle is the foundation for change. The essence of all these models is based on a set of principles which generally follow the framework of the 4-D Cycle, (see Fig. 1). Each AI process is home-grown and designated to meet the unique challenges of the organization and industry involved (Cooperrider & Whitney, 2005). The "four D's"of AI take many forms.

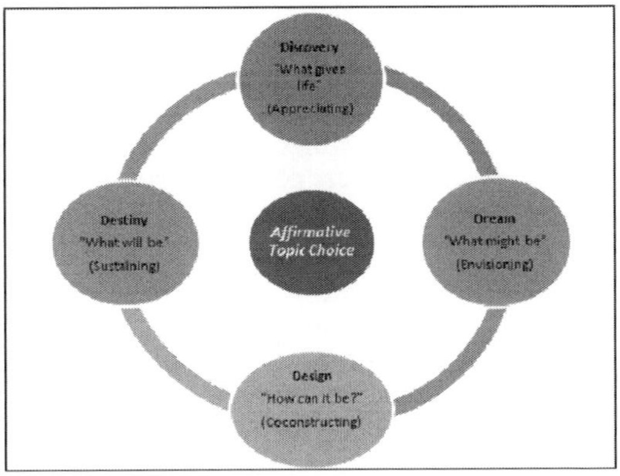

Figure 1. The Appreciative Inquiry 4-D cycle process. (Cooperrider et al., 2008).

DISCOVERY: Participants at this stage reflect and discuss the best of *what is* concerning the object of inquiry. During this stage, participants are interviewed about their own best experiences. This appears to be a key AI innovation. Personal and organizational highpoints are discovered, as well as what people value and how they hope and wish to enhance their organization's social, economic and environmentally vitality.

DREAM: At this stage, participants are asked to picture their group's ideal image and try to figure out the common aspirations of all stakeholders. They should then attempt to create a symbolic representation of that image. This stage also requires participants to carefully listen to stories about their group at its best and to share their hopes and dreams for their collective future (Cooperrider et al., 2008). Stories, analyses and maps of the "positive core" are essential resources for this process.

DESIGN: This stage will help enhance the organization by leveraging both its own past success and successes elsewhere with "strategic intent," signaling what the organization wants more of and recognizing that the future is built around what can be and what is (Laszlo & Cooperrider, 2010). These activities are conducted in large group forums or within small teams (Whitney & Trosten-Bloom, 2010). Because the dream of each individual, community or organization is different, this process provides a list of activities necessary to achieve these specific dreams. Stakeholders then select and prioritize activities and make plans according to their preferences (Judy & Hammond, 2006).

DESTINY: The destiny phase involves a series of actions inspired by the previous stages which support a continuous learning and innovation processes; this causes "what will be" to emerge

(Cooperrider et al., 2008). Building an "appreciative eye" into all of the organization's systems, procedures and ways of working is key to sustaining momentum (Cooperrider et al., 2008). The destiny phrase is ongoing and brings the organization back full-circle to the discovery phrase, in which continued application of the method may result in new topic choices, dialogues and continue learning.

The 4-D cycle acts as a basic guiding framework while conducting an AI process but it is not necessary that all Appreciative Inquiry proceed in the same way (Nyaupane & Poudel, 2012). The objective and the context most influence the form of engagement and the inquiry strategies. Cooperrider & Whitney (2005), demonstrated that AI may lead to the design of appreciative organizations. The transformation of the existing organizations into appreciative organizations and the creation of innovative organizations may follow a similar path through the 4-D Cycle, but may require different focus at each phase.

4 THE UNIQUES OF APPRECIATIVE INQUIRY APPROACH

What makes AI different from other approaches? This question can be discussed in various ways. Many other participatory change strategies exist and have been used in various settings with varying degrees of success. AI typically focuses more on local strengths and achievements, encourages local participation, and emphasizes local knowledge, rather than deficits and problems. Whereas other kind of approaches such as the top-down approach or participatory-led approach proceed by defining needs, problems, and obstacles in organizations or communities, these may lack appreciation of strengths and successes (Laszlo & Cooperrider, 2010). Additionally, these approaches may also fail to sustain participation and commitment of local people because they simply emphasize fixing problems in community (Judy & Hammond, 2006).

The uniqueness and power of AI interviews stem from a fundamentally affirmative focus involving appreciation for what gives life and energy to individuals, communities and organizations (Lewis et al., 2008). The initiators of AI also considered these "problem-solving" or "deficit-based" approaches in development and regard AI as an alternative. The problem-solving approaches as well as the deficit-based approaches may tend to leave people with an impression that their community is full of problems and needs which require the help of outside experts (Cooperrider & Whitney, 2005). The main theme of the deficit-based approach is to solve problems. Questions normally will begin with questions like "What is wrong?" and "What is the problem?"

Table 1. The power of AI approach.

Problem solving approach	Appreciative Inquiry approach
"Felt Need" Identification of Problem	Appreciating "Valuing the Best of What Is"
Analysis of Causes	Envisioning "What Might Be"
Analysis of Possible Solutions	Dialoguing "What Should Be"
Action Planning (Treatment)	Innovating "What will Be"
Organization is a problem to be solved	Organization is a mystery (infinite capacity) to be embraced

Adapted from Cooperrider & Whitney (2005. p. g. 13).

As a method of organizational analysis, AI also differs from conventional managerial problem solving. The basic underlying assumption of AI is that an organization is a "solution to be embraced" rather than a "problem to be solved" (Cooperrider et al., 2008). Table 1 highlighted the power of AI approach as compared to other traditional approach.

5 CONCLUSION

In the history of development research, Appreciative Inquiry (AI) has been thought of as a theoretical research perspective used in various field of studies as diverse as medicine, organizational development, human resources, engineering, and education. In several contexts it has been used to bring change into an organization or community, while facilitating a focus on organizational members appreciating and valuing one another (Raymond & Hall, 2008a). The approach is typically used to facilitate positive developments and may find use within the tourism industry. The AI 4-D Cycle serves as the foundation on which change is built, including Discover, Dream, Design and Destiny.

The unique core of AI is the involvement of all of community stakeholders throughout the inquiry process (Cooperrider & Whitney, 2005). AI allows the engagement of everyone who has interest or investment in the future, or information about and influence over an organization. Nyaupane & Poudel (2012) stated that AI acts as a research method alternative to conventional action research in the organizational development field and is applicable to tourism study. In addition, Raymond & Hall (2008) suggested that AI may be conducted regarding volunteer tourism processes in a similar way to which AI might be applied to commercial tourism. AI may find use as an alternative tourism research tool for community development in rural tourism. It is believed that there is significant potential for the adoption of the AI approach within tourism discipline (Nyaupane & Poudel, 2012; Raymond & Hall, 2008a). These researchers believe that understanding the rural populations' knowledge, needs and priority, without alienating them from research, is an important goal.

REFERENCES

Bellinger, A., & Elliott, T. (2011). What Are You Looking At? The Potential of Appreciative Inquiry as a Research Approach for Social Work. *British Journal of Social Work*, 41(4), 708–725. doi:10.1093/bjsw/bcr065.

Boyd, N.M., & Bright, D.S. (2007). Appreciative inquiry as a mode of action research for community psychology. *Journal of Community Psychology*, 35(8), 1019–1036. doi:10.1002/jcop.20208.

Bramwell, B., & Lane, B. (2012). Towards innovation in sustainable tourism research? *Journal of Sustainable Tourism*, 20(1), 1–7.

Cooperrider, D.L., & Whitney, D. (2005). *Appreciative inquiry: A Positive Revolution in Change*. San Francisco: Berrett-Koehler Publishers, Inc.

Cooperrider, D.L., Whitney, D., & Stavros, J.M. (2008). *Appreciative Inquiry Handbook-For Leaders of Change*. (R. Fry, Ed.) (2nd ed.). Brunswick, OH: Crown Custom Publishing, Inc & Berrett-Koehler Publishers, Inc.

Dick, B. (2006). Action research literature 2004–2006: Themes and trends. *Action Research*, 4(4), 439–458. doi:10.1177/1476750306070105.

Finegold, M.A., Holland, B.E.A.M.A.H., & Lingham, T. (2002). *Appreciative Inquiry and Public Dialogue: An Approach to Community Change. Public Organization Review* (Vol. 2, pp. 235–252). The Netherlands.

Judy, S., & Hammond, S. (2006). *An Introduction to Appreciative Inquiry* (pp. 1–12). Argenta.

Koster, R.L.P., & Lemelin, R.H. (2009). Appreciative Inquiry and Rural Tourism: A Case Study from Canada. *Tourism Geographies*, 11(2), 256–269. doi:10.1080/14616680902827209.

Laszlo, C., & Cooperrider, D.L. (2010). Creating Sustainable Value: A Strength-Based Whole System Approach. In T. Thatchenkery, D.L. Cooperrider, & M. Avital (Eds.), *Positive Design and Appreciative Construction: From Sustainbale Development to Sustainable Value* (pp. 17–33). United Kingdom: Emerald Group Publishing Limited.

Lewis, S., Passmore, J., & Cantore, S. (2008). *Appreciative Inquiry for Change Management—Using AI to facilitate Organizational Development*. London and Philadelphia: Kogan Page Limited.

Lo, M., Mohamad, A.A., Songan, P., & Yeo, A.W. (2012). Positioning Rural Tourism: Perspectives from the Local Communities. *International Journal of Trade, Economic and Finance*, 3(1).

Michael, S. (2005). The promise of Appreciative Inquiry as an interview tool for field research. *Development in Practice*, 15(2), 222–230.

Morsillo, J., & Fisher, A. (2007). Appreciative inquiry with youth to create meaningful community projects. *Journal of the Australian Psychological Society*, 19(1), 47–61.

Nyaupane, G.P., & Poudel, S. (2012). Application of appreciative inquiry in tourism research in rural communities. *Tourism Management*, 33(4), 978–987. doi:10.1016/j.tourman.2011.10.009.

Preskill, H., & Catsambas, T.T. (2006). *Reframing Evaluation Through Appreciative Inquiry*. United States of America: Sage Publications Inc.

Raymond, E.M., & Hall, C.M. (2008a). The Potential for Appreciative Inquiry in Tourism Research. *Current Issues in Tourism*, 11(3), 281–292. doi:10.2167/citMP022.0.

Raymond, E.M., & Hall, C.M. (2008b). The Development of Cross-Cultural (Mis) Understanding Through Volunteer. *Journal of Sustainable Tourism*, 16(5), 530–543. doi:10.2167/jost796.0.

Rockey, S., & Webb, L. (2005). Organizational Change Inside and Out: The Impact of an Appreciative Inquiry. *Journal for Nonprofit Management*, 9(1), 17–27. doi:10.1007/978-3-322-96514-1_1.

Shariff, Z., Gramberg, B. Van, & Foley, P. (2009). The Usefulness of Appreciative Inquiry as a Method to Identify Mass Sports Program Success. *Transylvanian Review of Administrative Sciences*, (30), 118–131.

The Mountain Institute. (2000). *Community-Based Tourism For Conservation and Development: A Resources Kit*. USA.

Whitney, D., & Trosten-Bloom, A. (2010). *The Power of Appreciative Inquiry—A Practical Guide to Positive Change*. (D. Cooperrider, Ed.) (2nd ed.). San Francisco: Berrett-Koehler Publisher, Inc.

World Tourism Organization. (2001). *Rural Tourism Master Plan for Malaysia* (pp. 1–180). Malaysia.

Hospitality and Tourism – Sumarjan et al. (eds)
© 2014 Taylor & Francis Group, London, ISBN 978-1-138-00151-0

Training transfer: The interaction of changed behavior and co-worker support on job performance

F. Mohd Shariff, A. Zainal & R. Hashim
Universiti Teknologi MARA, Malaysia

ABSTRACT: The present study examined the moderating effect of changed behavior on the relationship between co-workers support on frontline employees job performance. The authors hypothesize that co-workers support are more strongly related to performance among frontline employees reporting low than high level in changed behavior. Results of Structural Equation Modeling, gathered from 264 employees who held positions as food and beverage outlet servers in 4 and 5 star ratings hotels, support the hypothesis. The result indicates that when frontline employees' exhibit low changed behavior, the relationships between co-worker support and job performance are stronger. Implications for substantive research, strengths and limitations, and directions for future research are offered.

Keywords: changed behavior, employees, job performance training transfer

1 INTRODUCTION

Hospitality is one of the fastest growing business sectors, and in the ever-changing market conditions it is vitally important that it continues to build a skilled and competent workforce with a greater level of responsiveness. According to McKenzie (2009), skills training can never be cut as it is the lifeblood of the industry, and personal development for junior and senior staff alike keeps them motivated and challenged. The author added that, in a buoyant market, change would be seen as a driver for training needs and this environment is throwing up change at every turn.

Co-worker or peer support refers to co-workers assisting one another in their tasks when needed by sharing knowledge and expertise, as well as providing encouragement and support (Zhou & George, 2001). Peer support has been shown to increase the likelihood of learning transfer (Baldwin & Ford, 1998; Goldstein, 1980).

Facteau, Dobbins, Russell, Laddand Kudisch (1995) identified four major sources of social support: subordinate, peer, supervisor, and top management. Their study is designed to examine the independent and differential effects of subordinate support, peer support, supervisor support, and top management support on the training transfer of 967 managers and supervisors in departments within state government agencies. Their study concluded that, of four forms of social support, only subordinate and peer support are positively related to perceived transfer, supervisor support is negatively related to transfer, and top management support shows no significant relationship (Facteau et al., 1995). Similarly, study by Ford et al. (1992) concluded that those trainees who reported having colleagues and supervisors with positive attitudes towards training, reported that they had more opportunities to utilize new skills. Therefore, managers who believe that their subordinates and peers are supportive of their training efforts are more likely to perceive greater transfer of their training skills.

Clarke (2002) noted that the most consistent factor explaining the relationship between the work environment and transfer is the support trainees receive to use their new skills and knowledge in the workplace. Employees may also acquire task-relevant knowledge and expertise from supportive co-workers, which may make new ways of doing things possible (Woodman, Sawyer, & Griffin, 1993). In this context, working with helpful, supportive colleagues promotes an environment where new ideas can be discussed more openly and freely (Joiner, 2007).

1.1 *Behavior change*

Changes in job behavior or performance, involve studying the change in job behavior which takes place as a result of the training. This is to evaluate employees on whether the newly acquired skills, knowledge, or attitude are being used in the everyday environment of the learner. At this point, evaluation sees whether tasks are performed differently before and after the training. In line with Casper (2005), behavior change is a measure of the extent to which learners change their behavior and use the learned principles and techniques on the job.

Behavior change, or the actual utilization of what was learned in training on the job, was the theme of Kirkpatrick's step 3 (Kirkpatrick, 2005). Kirkpatrick (2005) listed five basic requirements that must exist if a person is going to change his/her job behavior. The basic requirements are: 1) he/she must want to improve; 2) he/she must recognize his own weaknesses; 3) he/she must work in a permissive climate; 4) he/she must have some help from someone who is interested and skilled; and 5) he/she must have an opportunity to try out the new ideas.

Green (2001) stated that the measurement of changes in job behavior as a result of participation in a training program is often referred to as the transfer of training or transfer of learning. Kirkpatrick (2005) concluded that a change in job behavior due to training is only possible if trainees acquire the new knowledge, skills and attitudes. In other words, the main antecedent of training transfer is the learning gained during the training. There is nothing more important than measuring the extent that on-the-job behavior has changed as a result of the training. If the trainees do not apply what they have learned, the program has been a failure even if learning has taken place. Therefore, measuring behavior change is necessary, not only to see if the behavior has changed, but also to determine the reasons why change has not occurred.

1.2 *Methodology*

In the present study, to identify whether or not there are any significant differences in the path coefficient, a two-group of path analysis, as suggested by Hair et al. (2006) was performed. The first step involved setting the parameter from co-workers support to employees' performance to be estimated freely. Secondly, the parameter was constrained to be equal across groups. A difference in chi squared ($\Delta\chi^2$) values between the two models could be used to identify whether changed behavior had a moderating effect on the relationship between co-workers support and employees performance (Hair et al., 2006). The procedure was applied to assess the moderating effect of changed behavior on the co-worker support and employees' job performance relationship.

1.3 *The moderating effect of changed behavior on co-workers support—performance relationship*

The path suggested that changed behavior moderates the relationship between co-worker support and job performance. The result reveals that the chi-squared differences were significant for the link between co-worker support and job performance (11.093, $p < 0.01$). The path coefficient of low CB was greater than that of high CB. Changed behavior significantly moderates the relationship between co-worker supports to employees' job performance.

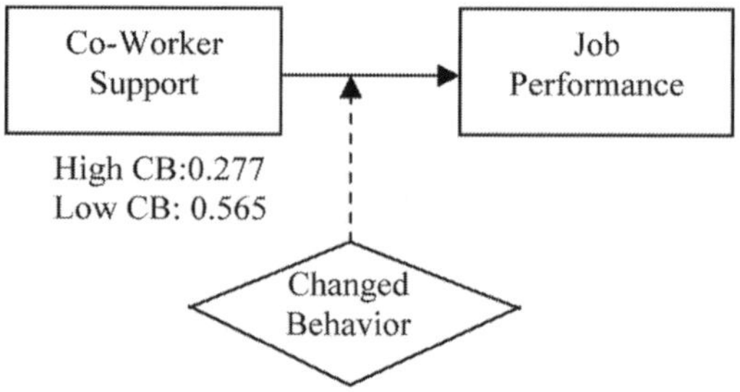

Figure 1. Changed behavior moderating effect between co-worker support and job performance.

This result indicates that when frontline employees' exhibit low changed behavior, the relationships between co-worker support and job performance are stronger. Thus, it seems that co-worker support is more effective for frontline employees with low changed behavior than those with high changed behavior characteristics. This result can be explained by referring to the characteristics of frontline employees with low changed behavior. There is a possibility that frontline employees with low changed behavior characteristics prefer to have less interaction or coordination with other work colleagues. Given that, it is possible that frontline employees with low changed behavior will prefer to seek advice or guidance from top management, in directing them to perform their job, instead of referring to their work colleagues. With sufficient guidance and direction from top management, the performance of frontline employees with low changed behavior would increase.

2 CONCLUSION

The conclusions drawn from the findings of the study are subject to several limitations that suggest a fruitful direction for future research. In evaluating the effectiveness of training, the most meaningful factor is the trainee's work performance and therefore, a better performance indicator may be in the knowledge and skills employees transfer from the training back to the work environment. Although employees may be able to apply the learned knowledge and skills during a training program, elements of the employees' characteristics and work environment may impact on their ability to transfer these learned knowledge and skills to the job. Individuals may be motivated to share knowledge and transfer these back to the job, but may be discouraged, inhibited, or prevented from doing so by circumstances in their work environment. This situation may occur due to the culture of Asian employees that present a particular group pattern of behavior and attitude which resulted from their cultural backgrounds, such as respect for authority, introvertness and self-restraint, other than experiencing communication barrier. In relation to this subject matter, hospitality business in Malaysia having diverse employees who are locals and foreign workers are experiencing the same situation. With the multi-nationality workforce, diversity will have a note-worthy impact on the hospitality industry.

The research was based on data collected from four and five stars rating hotel segments, primarily from frontline employees working in food and beverage departments in Klang Valley of Kuala Lumpur, Malaysia. Thus, the result may not be directly generalized to other segments of the hotel industry. Replication in other research settings is encouraged, as it will extend our knowledge across a broader range of the hospitality industry.

REFERENCES

Baldwin, T.T., & Ford, K.J. (1988). Transfer of training: A review and directions for future research. *Personnel Psychology,* 41(1), 63–105.

Casper, B. (2005). *The positive transfer of learned skills from training to changed behaviors at the job.* Unpublished Dissertation. Pepperdine University.

Clarke, N. (2002). Job/work environment factors influencing training effectiveness within a human service agency: Some indicative support for Baldwin and Fords' transfer climate construct. *International Journal of Training and Development,* 6(3), 146–162.

Facteau, J.D., Dobbins, G.H., Russell, J.E.A., Ladd, R.T., & Kudisch, J.D. (1995). The influence of general perceptions of the training environment on pre-training motivation and perceived training transfer. *Journal of Management* 21, 1–25.

Ford, J.K., Quinones, M.A., Sego, D.J., & Sorra, J.S. (1992). Factors affecting the opportunity to perform trained tasks on the job. *Personnel Psychology, 45,* 511–527.

Goldstein, I.L. (1980). Training in work organizations. *Annual Review of Psychology,* 31, 229–272.

Green, E.C. (2001). The influence of individual and work environment characteristics on trainee motivation and training effectiveness measures. UMI Microform 3036790.

Hair, J.F., Black, W.C., Babin, B.J., Anderson, R.E., & Tatham, R.L. (2006). *Multivariate Data Analysis* (6th ed.). New Jersey: Pearson Prentice Hall.

Joiner, T.A. (2007). Total quality management and performance: The role of organization support and co-worker support. *International Journal of Quality & Reliability Management,* 24(6), 617–627.

Kirkpatrick, D.L. (2005). Transferring learning to behavior: Using the four levels to improve performance. San Francisco, CA: Berrett-Koehler Publishers.

McKenzie, A. (2009). Can hospitality business afford to cut training spend? *Caterer & Hotelkeeper,* 199.

Woodman, R., Sawyer, J, & Griffin, R. (1993). Toward a theory of organizational creativity. *Academy of Management Review,* 18, 293–321.

Zhou, J., & George, J.M. (2001). When job dissatisfaction leads to creativity: encouraging the expression of voice. *Academy of Management Journal,* 44(4), 683–696.

Hospitality and Tourism – Sumarjan et al. (eds)
© 2014 Taylor & Francis Group, London, ISBN 978-1-138-00151-0

Quality of work life among hotel executives in Malaysia

I.M. Ghazali, S.M. Radzi, M.O.R.A. Patah & R.A. Aziz
Universiti Teknologi MARA, Malaysia

ABSTRACT: This paper fills the gap in understanding the Work Variable (WV), Non-Work Variable (NWV) towards Quality of Work Life (QWL) among hotel executives in Malaysia. The objectives of this study are to examine whether work variable has an influence on hotel executives' quality of work life and to examine whether non-work variable has an influence on hotel executives' quality of work life. This research utilized self-administered questionnaires and result revealed that the work variable had a positive influence on hotel executives' quality of work life while non-work variable shows insignificant relationship towards quality of work life.

Keywords: hotel industry, Non-Work Variable, Quality of Work Life, Work Variable

1 INTRODUCTION

The hotel industry needs to provide a good Quality of Work Life (QWL) in order to attract and retain capable employees. As explained by Sirgy et al., (2001), QWL does not only affect job satisfaction but also satisfaction in other life domains such as family life, leisure life, social life, financial life, and so on. Therefore, the focus of QWL is actually beyond the job satisfaction itself. It does involve the effect of the workplace on satisfaction with the job, satisfaction in non-work life domains, and satisfaction with overall life, personal happiness, and subjective well being. There is some evidence showing that a happy employee is a productive employee, dedicated and loyal employee (Greenhaus et al., 1987).

Studies also have affirmed the contribution of competent and loyal employees towards organizational profits (Hinkin and Tracey, 2000). In fact, no hotel can have excellent operations without excellent employees (Enz and Siguaw, 2000). These inferences are suggestive of the fact that hospitality employees are the most significant resource within the organization. Hotel managers are supposed to ensure that the attitudes and behaviors of employees are consistent with the expectations of their customers and the organization as a key to be successful in this global competitive environment.

Whitely and England (1977) rose in their study that work and family are the key domains of life to many people. The issue of work-family balance and conflict has continued to attract popular media and serious research attention worldwide. Kandasamy and Ancheri (2009) observed that despite acknowledging employees' start job with expectations that has been found to influence their QWL, a definition founded and formulated on the basis of employee expectation is conspicuously lacking, in the hospitality context.

On the other hand, this industry is finding it difficult to attract and retain their human resources. As explained by Yu (1999), in the light of the foregoing concerns in the hotel industry, it is a challenge for the management to motivate employees to stay on the job, by ensuring measures at enhancing the working condition to counter turnover. Therefore, it is essential and inevitable to induct a quality then into the work lives of hospitality employees. Research finding has acknowledged the impact of improved Quality of Work Life (QWL) on reduced turnover (Ference, 1982).

Based from the above mentioned statement, this study aims to investigate the executives' of five star hotel on whether work variable and non-work variable have an influence on quality of work life.

2 LITERATURE REVIEW

2.1 *Introduction*

An extensive review will focus on the variables under study mainly work variable, non-work variable and quality of work life to further define and support a theoretical framework.

2.2 *Work Variable (WV)*

Anderson and Ungemah (1999) mentioned that variable work hours programs are becoming more and more popular with companies both large and small, as employers realize they are good for employee morale and good for business. They explained that due to some reason such as the increase in two-career families, the rise in single-parent families and the huge number of working women including mothers of school-age children have forced companies to reconsider their work schedules. Employers have been pleasantly surprised to find the variable work hours have many benefits. Companies are realizing that everyone wins by providing the flexibility to accommodate family needs, leisure activities and other obligations.

According to Kim and Ling (2001), work variables are number of hours worked, work schedule inflexibility and work stressor. They explained that time pressure can be measured by the number of hours worked and schedule flexibility. They also discovered that past research studies have shown that long hours worked and work schedule inflexibility are related to high work-family conflict. On the other hand, job characteristics and work role pressure have a positive relationship with work-family conflict (Kim and Ling (2001). Work stressors can produce strain symptoms such as anxiety, frustration, tension and irritability (Greenhaus and Beutell, 1985).

Ballout (2008) noted that job involvement is one potential work-related variable that may cause work-family conflict. Moreover, individuals who are highly involved in their jobs or careers may devote more time and effort to the work role than to family role. According to Adams et al. (1996), job involvement has got to do with the degree of importance people assign to work involvement. Another work-related variable that relates to work-family conflict is job social support (Ballout, 2008). Adams et al. (1996) and Bernas and Major (2000) pointed out that social support from the non-work domain, such as the partners or family members is of great importance in reducing work-family conflict.

2.3 *Non-Work Variable (NWV)*

The non-work variables are gender, parental demands, and working spouse. Gender is perhaps the socio-demographic characteristic most often included in studies of work-family conflict and career success (Ballout, 2008). The evidence suggests that there are gender differences in work-family conflict due to social role differences between working men and working women. Traditional gender roles assumed that men are primarily responsible for family financial support by working, while women are primarily responsible for childcare and household duties (Loscocco, 2000).

In his research, Ballout (2008) indicated that parental demand can affect work-family conflict and subsequently success in careers. Due to the increased demands of spending a great amount of time at home taking care of their children, investing the required time and energy at work to support their career advancement and success, employed parents tend to experience more family interfere work conflict and work interfere family conflict than employed non-parents. On the other hand, non-parents often have more flexibility in managing their time and personal life, and fewer familial responsibilities, which less likely to experience family-to-work conflict.

2.4 *Quality of Work Life (QWL)*

Sirgy et al. (2001) mentioned that a new measure of QWL was developed based on need satisfaction and spillover theories. The measure was designed to meet the needs of an employee to capture the extent to which the work environment, job requirements, supervisory behavior, and ancillary programs in an organization. They further explained that QWL differs from job satisfaction whereby job satisfaction is construed as one of many outcomes of QWL.

Abo-Znadh and Carty (1999) noted that QWL efforts not only on how people can do work better, but also on how work may cause people to be better. Moreover, QWL in an organization also concerns about participation of workers regarding problem solving and decision making. Higher quality of work life would then correlate with lower work-to-family interference (Cheung and Tang, 2009, & Aziz et al. 2011).

2.5 *Relationship between work variable and quality of work life*

Time pressure which can be measured by the number of hours worked and schedule flexibility is one of the work characteristics that lead to work-family conflict. As time is a limited resource, the more time a women entrepreneur spends on her business, the less time she will have with her family (Kim and Ling, 2001). Jones and Fletcher (1996) confirmed in their research that work stressors affect family life in the form of unpleasant moods that have spilled over from work to family. Job involvement may lead to work interfering with family, which in turn leads to less time and energy devoted to family roles, thereby making it more difficult to comply with pressures associated with family roles (Ballout, 2008). Ballout (2008) also mentioned that supervisor and perceived organizational support could cause employees to report less spillover from work to family and to show greater job satisfaction and better performance. He further added that both work-specific variables and work-family conflict affect individuals' perceptions of career success.

2.6 *Relationship between non-work variable and quality of work life*

Ballout (2008) proposed that non-work variables to affect individuals' perceived career aspiration and success. In particular, he expected that such demographic variables do affect the way individuals make allocation investments to family responsibilities and career aspirations. According to Ashforth et al. (2000), social identity theory focuses on the choice made by individuals in role transitions involving home, work, and other places and postulates that roles can be arrayed on a continuum, ranging from high segmentation to high integration. While Ruderman et al. (2002) explored the benefits of multiple roles for managerial women and found that women managers who integrated careers with family and other non-work roles were successful and satisfied in engaging in both managerial and personal roles.

2.7 *Relationship between work variable, non-work variable and quality of work life*

As suggested by Ballout (2008), organizations that have supportive cultures can reduce work-family conflict of their employees by offering work-family support programs built around flexibility, teamwork, and cooperation. Further, when employees do experience greater conflict in the workplace that intrudes them into the home, their disposition and attitudes relating to career roles will be affected. Then employees may become preoccupied by managing the imbalance between work and home, if this is the case. Lack of balance will cause a detrimental effect on how committed employees feel toward their career competences.

3 RESEARCH METHODOLOGY

This research utilized self-administered questionnaires. The samples of the study were from 133 executives of five stars hotels in Kuala Lumpur. The data were later analyzed to test several hypotheses formulated for the study.

4 FINDING

4.1 *Response rate*

Out of 24 five star rated hotels, only 19 hotels have agreed to participate. A total of 51 questionnaires were returned out of 133 questionnaires and finally coded. This represents 38 percent response rate. According to Baldauf et al. (1999) and Tomaskovic-Devey et al. (1994), surveys of organizations typically receive substantially lower return rates than surveys of individuals, which is with 15% return rates sometimes reaching a level of acceptability for organizational surveys. Besides, Visser et al. (1996) mentioned that some studies with low response rates, even as low as 20%, are able to yield more accurate results than studies with response rates of 60% to 70%.

4.2 *Cronbach's alpha reliability analysis of study variables*

In order to ensure that the gathered data were reliable, the measurement of each variable and its sub scales were assessed for reliability. The Cronbach's alpha coefficient value for all variables in the study revealed a range of coefficient value from 0.63 to 0.66 accordingly. The dependent variable of quality of work life had a reliability coefficient of 0.65 constant with the number of independent variables. The independent variable of work variable showed that its dimensions of meaningfulness had a coefficient value of 0.66, and non-work variable had a reliability coefficient value of 0.63. The results also indicated that the reliabilities of the scales used in this study were commonly accepted and considered to be the criterion for demonstrating the internal consistency of a reliable scale (Hinton, Brownlow, McMurray and Cozens, 2004).

4.3 *Bootstrap/Indirect analysis*

The objectives of this study are to examine the relationships among work variable, non work variable, and quality of work life. The result from bootstrap analysis is used to examine all the relationships outlined in research objectives and to answer the rest of hypotheses in the study.

The Bootstrap/Indirect analysis was conducted to test the relationship between the independent and dependent variable. It was found that by simply looking to the bootstrap result on path (a) in Table 1, the result indicates that only work variable has positive and significant relationship with quality of work life ($B = 1.194$, $p < 0.01$). The analyses support the following hypothesis:

H1: Work variable significantly influence hotel executives' quality of work life.

Results gained in the study revealed that the work variable had a positive influence on hotel executives' quality of work life. It was earlier proposed that there was a positive influence of work variable on executives' quality of work life and based on the findings, it shows that there was indeed a

Table 1. Summary of bootstrap results (5000 bootstrap samples, N = 51).

Indepedent Variable (IV)	Depedent Variable (DV)	Direct effect (c')	Indirect effect (a*b)	95% CI for a*b	Total effect (c)
Work variable	Quality of work life	1.194*	0.0003	–0.1236, 0.0728	1.194*
Non work variable	Quality of work life	0.437	–0.0034	–0.1749, 0.0721	0.400

IV, independent variable; DV, dependent variable. In both analyses, the effect of the other independent variable were controlled for $^{**}p < 0.001$. $^{*}p < 0.05$.

significant positive and significant relationship with scores ($B = 1.194$, $p < 0.01$).

However, non-work variable shows insignificant relationship towards quality of work life ($B = 0.400$, $p\ 0 = 0.35$). The analyses were different from the following hypothesis:

H2: Non-work variable significantly influence hotel executives' quality of work life.

It was found that there was no statistically significant difference at the 0.05 level in non-work variable on executives' quality of work life. According to Table 1, the scores of non-work variable towards quality of work life were $B = 0.400$, $p\ 0 = 0.35$.

As expected by Weer (2006), the supportiveness of an organization's environment for non-work role participation did mitigate the negative impact of non-work emotional energy demands on work engagement. In addition, Thomas and Ganster (1995) found that employees who worked for organizations that were supportive of their involvement in family life experienced a greater sense of control over their work and family responsibilities, which in turn reduced their perceptions of conflict and strain and increased their job satisfaction.

5 CONCLUSION

The main objective of this study focused on the influence of Work Variable (WV) and Non-Work Variable (NWV) on hotel executives' Quality of Work Life (QWL).

Findings of this study suggest some important implications for hotels' operator in Malaysia particularly in Kuala Lumpur. From the study, the result indicates that only work variable has positive and significant relationship with quality of work life whereby non work variable shows insignificant relationship towards quality of work life. This indicates that executives did depend on their work life such as number of hours work, work schedule inflexibility, work stressor, job involvement and job social support to enhance their quality of work life. They show that they are passionately interested in dealing with things that are related to their jobs. On the other hand, non-work role (parental demands and working spouse) did not influence the executives' quality of work life as they can separate their personal life from work. For the industry, when the employees are aware of their responsibilities and could deliver good service, indirectly it will enhance the customers' satisfaction. Thus, this will not affect performance of executives towards the organization they served. Furthermore, the industry could retain their talented executives in the workforce longer.

The current study has some limitations that deserve to be mentioned. The study was conducted in Kuala Lumpur focusing on 5 star hotels only. Hence, the findings could not be generalized to the whole industry in Malaysia. The current study focused on work variable, non-work variable and its implication towards quality of work life. It is suggested that future studies to look into inter-role conflicts and how it influence QWL.

REFERENCES

Abo-Znadh, S., & Carty, R. (1999). An exploration of selected staff and job characteristics, and their relationship to quality of work life, among staff nurses in medical/surgical units in two tertiary care hospitals in Saudi Arabia. Doctoral Dissertation, George Mason University. Dissertation Abstract International, (UMI No: AAT 9908082).

Adams, G., King, L. & King, D.W. (1996). Relationships of job and family involvement, family social support, and work-family conflict with job and life satisfaction. *Journal of Applied Psychology*, 81 (4), 411–420.

Anderson, S. & Ungemah, D. (1999). Varible work hours: an implementation guide for employers. Oregon Department of Environmental Quality.

Aryee, S. (1992). Antecedents and outcomes of work-family conflict among married professional women: evidence from Singapore. *Human Relations,* 45 (2), 813–837.

Ashforth, B., Kreiner, G. & Fugate, M. (2000). All in a day's work: boundaries and micro role transitions. *Academy of Management Review,* 25 (3), 472–491.

Aziz, R.A., Nadzar, F.M., Hussaini, H., Maarof, A., Radzi, S.M. & Ismail, I. (2011). Quality of work life of librarians in government academic libraries in the Klang Valley, Malaysia. *The International Information & Library Review,* 43 (3), 149–158.

Baldauf, A., Reisinger, H., & Moncrief, W.C. (1999). Examining motivations to refuse in industrial mail surveys. *Journal of the Market Research Society,* 41, 345–353.

Ballout, H.I. (2008). Work-family conflict and career success: the effects of domain-specific determinants. *Journal of Management Development,* 27 (5), 437–466.

Barnett, R.C. & Baruch, G.K. (1987). Determinants of fathers' participations in family work. *Journal of Marriage and The Family,* 49 (1), 29–40.

Bernas, K.H. & Major, D.A. (2000). Contributors to stress resistance: testing a model of women's work-family conflict. *Psychology of Women Quarterly,* 24 (2), 170–178.

Cheung, F.Y., & Tang, C.S. (2009). Quality of work life as a mediator between emotional labor and work family interference. *Journal of Business Psychology, 24,* 245–255.

Enz, C.A., & Siguaw, J.A., (2000). Best practices in human resources. *Cornell Hotel and Restaurant Administration Quarterly,* 41 (1), 48–61.

Ference, E.A., (1982). Human resources development: toward a definition of training. *Cornell Hospitality Quarterly,* 23 (3), 25–31.

Greenhaus, J.H. & Beutell, N.J. (1985), Sources of conflict between work and family roles. *Academy of Management Review,* 10 (1), 76–88.

Greenhaus, J.H., A.G. Bedian & K.W. Mossholder: (1987). Work experiences, job performances, and feelings of personal and family well being. *Journal of Vocational Behavior,* 31, 200–215.

Hinkin, T.R., & Tracey, B.J., (2000). The cost of turnover: putting a price on the learning curve. *Cornell Hotel and Restaurant Administration Quarterly,* 41 (3), 14–21.

Hinton, P.R., Brownlow, C., McMurray, I., & Cozens, B. (2004). *SPSS Explained.* New York: Routledge Inc.

Jones, F. & Fletcher, B.C. (1996). Taking work home: a study of daily fluctuations in work stressors, effects on moods, and impact on marital partners. *Journal of Occupational and Organizational Psychology,* 69, 89–106.

Kandasamy. I, & Ancheri. S. (2009). Hotel employees' expectations of QWL: A qualitative study. *International Journal of Hospitality Management,* 28, 328–337.

Kim, J.L.S., & Ling, C.S. (2001). Work-family conflict of women entrepreneurs in Singapore. *Women in Management Review,* 16 (5), 204–221.

Loscocco, K. (2000). Age integration as a solution to work-family conflict. *The Gerontologist,* 40 (3), 292–300.

Martel, J.P., & Dupuis, G. (2006). Quality of work life: Theoretical and methodological problems, and presentation of a new model and measuring instrument. *Social Indicators Research,* 77, 333–368.

Martire, L.M., Stephens, M.P. & Townsend, A.L. (2000). Centrality of women's multiple roles: beneficial and detrimental consequences for psychological well-being. *Psychology and Aging,* 15 (1), 148–156.

Md-Sidin, S., Sambasivan, M. & Ismail, I. (2010). Relationship between work-family conflict and quality of life. *Journal of Managerial Psychology,25* (1), 58–81.

Morton, S.M.B., Bandara, D.K, Robinson, E.M., & Carr, P.E.A. (2003). In the 21st Century, what is an acceptable response rate? *Australiam and New Zealand Journal of Public Health,* 36 (2), 106–108.

Ruderman, M., Ohlott, P., Panter, K. & King, S. (2002). Benefits of multiple roles for managerial women. *Academy of Management Journal,* 45 (2), 369–386.

Sirgy, M.J., Efraty, D., Siegel, P., & Lee, D. (2001). A new measure of quality of work life (QWL) based on satisfaction and spillover theories. *Social Indicators Research,* 55, 241–302.

Thomas, L., & Ganster, D. (1995). Impact of family-supportive work variables on work-family conflict and strain: A control perspective. *Journal of Applied Psychology,* 80, 6–15.

Tomaskovic-Devey, D., Leiter, J., & Thompson, S. (1994). Organizational survey non response. *Administrative Science Quarterly,* 39, 439–457.

Weer, C.H. (2006). The Impact of Non-Work Role Commitment on Employees' Career Growth Prospects. Unpublished Doctor of Philosophy Dissertation. Drexel University.

Whitely, W. & England, G.W. (1977). Managerial values as a reflection of culture and the process of industrialization. *Academy of Management Journal,* 20, 439–453.

Yu, L., (1999). *The International Hospitality Business.* The Haworth Hospitality Press, New York.

Hospitality and Tourism – Sumarjan et al. (eds)

Restaurant patrons' satisfaction towards interaction quality of front-of-the-house employees

N.T. Jamaludin, S.N.A. Ayob, K.N. Sabri & Z. Mohi
Universiti Teknologi MARA, Malaysia

ABSTRACT: Service quality is an important element in determining restaurant patrons' satisfaction. The service quality field has raised major concern among marketing researchers. Many theory, measurements and models been deformed in recent years. Besides that, several service quality primary dimensions and sub dimensions developed when measuring the restaurant patrons' satisfaction. In addition, there are arisen issues on interaction quality among the front-of-the-house F&B employees in the restaurant industry. Many researchers and the practitioners claimed that although with a systematic human resource management, there is still a missing value in interaction quality of the front-of-the-house F&B employees' in the restaurant industry. Therefore, the researchers attempt to identify the sub dimensions for Interaction Quality of front-of-the-house F&B employees in Pendeta Coffee House's outlet in UiTM Shah Alam, Malaysia by adapting the multidimensional Model and Hierarchical Model. Pendeta Coffee House's outlets are one of UiTM business company which cater to the university's population by providing restaurant style services and generating income foodservice business.

Keywords: hierarchical model, interaction quality, attitude, professionalism skills, problem solving skills

1 INTRODUCTION

In any service organization as well as in the restaurant industry, achieving a high level of customer satisfaction and potential loyal patron has been a vital to improve competitiveness (Parasuraman, Zeithaml, & Berry, 1985, 1988). Most of the restaurant's patrons including in Malaysia have a claim for high standards and demands for excellent restaurant service (Wong & Lim, 2011). For example, the perfectly cooked food could be badly delivered to the patrons, irrespective types of restaurant. Regrettably, when the service suffered the whole dining experience will be just an average or sometimes turns into a horrible nightmare. Therefore, the restaurant operators have increased their competitive advantages, and now instead of having only a delicious and variety of foods to persuade patrons to dine in, restaurant operators offer first class employee as a great advantage (Liang & Zhang, 2012).

In the restaurant business, employees are ranging from management employees to Food and Beverage (F&B) employees. The F&B employee especially front-of-the-house employees considered a very important antecedent to boost up customer satisfaction later influencing the patrons' decision making whether to return, to recommend the restaurant, or in demonstrating loyalty (Douglas, 2003; Winsted, 2000).

As mentioned earlier, the service quality plays an important role in ensuring the restaurant patrons satisfactions. However, many front-of-the-house employees do not realize the importance of interaction quality towards the restaurant patrons' satisfaction. Based on discussions with the management team and the restaurant's patrons of Pendeta's Coffee House outlets, there are several issues pertaining to the interaction quality of the front-of-the-house F&B employees towards the restaurant patrons thus raised a concern. The interaction quality failure also occurred during researchers' personal experience when one of the F&B chewing bubble gum upon taking an order.

This study conducted in three Pendeta Coffee House's outlets in Universiti Teknologi MARA (UiTM) Shah Alam. Pendeta is a brand of Intekma Resort and Convention Centre (IRCC) and registered under UiTM Hospitality Management Services Sdn. Bhd. Previously known as ITM Resort and Convention Centre and started its operation since 28th September1997 until today.

2 RESEARCH FRAMEWORK AND HYPOTHESES

2.1 *Development of research framework*

The researchers adapted the dimensions of interaction quality from the combination of the multi-dimensional model and the hierarchical model as proposed by Dabholkar, Thorpe, and Rentz (1996) and Brady and Cronin (2001b) respectively.

2.2 *Development of hypotheses*

Service quality studied over last three decades and has been the most debated in service marketing areas including in the hospitality disciplines. In addition, many theories, measurement and models being developed (Brady & Cronin, 2001b). Similar to Brady and Cronin (2001b), several researchers (e.g., Ibrahim, Dana, Mahdi, Zin, Ramli, & Nor, 2013; Mohi, 2012; Mohi, Raja Abdullah, Benedict Jipiu, Abd. Majid, & Wu, 2012) claimed that the concepts and measurements of service quality, the dimensions of service quality, and the content of the dimensions.

Additionally, research regarding to the construct considered "vague" (Caruana, Ramaseshan, & Ewing, 2000). Based on the several views by Brady and Cronin (2001b); (Dabholkar et al. 1996; Mohi, 2012) through an extensive literature review, the potential sub-dimension of interaction quality for Pendeta Coffee House's outlets in UiTM Shah Alam, Malaysia discussed in the following subsection.

2.3 *Interaction quality*

From the restaurant patron's point of view, the most abrupt indication of service quality happens in the service encounter (Bitner, Booms, & Mohr, 1994; Malhotra, Mavondo, Mukherjee, & Hooley, 2012). According to (Ahmed & Al-Qudah, 2011; Surprenant & Solomon, 1987), interaction quality is a dyadic interaction between restaurant patron and the service staff. Surprenant and Solomon (1987) highlighted that the importance of service encounter beneath the concept of interaction quality is increasingly recognized in service industries. Therefore, this study proposed four sub-dimensions of interaction quality; attitude, behaviour, professionalism skills (in exchange of expertise proposed by Brady & Cronin, 2001b) and problem solving skills. Thus, the following proposed sub-dimensions of interaction quality identified:

i. Attitude (Brady & Cronin, 2001a; Ekinci, Dawes, & Massey, 2008; Mohi, Clemes, & Hu, 2011; Mohi et al. 2012; Peters & Mazdarani, 2008)
ii. Behavior (Kim & Moon, 2009; Mohi, 2012; Mohi et al. 2011; Mohi et al. 2012)
iii. Professionalism (Abu-ELSamen, Amjad, Akroush, Al Khawaldeh, & Al-Shbily, 2011; Mohi, 2012; Mohi et al. 2011; Mohi et al. 2012)
iv. Problem solving (Dabholkar et al. 1996; Huang, 2009; Lu, Zhang, & Wang, 2009; Wu & Hsu, 2012).

3 METHODOLOGY

3.1 *Research design*

Following Churchill (1979), the first step in questionnaire development by identifying the constructs as mentioned in Subsection 2.2, followed by adapting the items from the previous study relevant to this study. Once the items are finalized, prototype questionnaire will be developed. Next, after the questionnaires approved by the supervisor, pilot test will be conducted. Soon after, the relevancy of the items and questions in the questionnaire are then testified according to the results of the pilot test that shows the respondents' feedback and Cronbach alpha scores of $\alpha = 0.60$ (Churchill, 1979). Once the modification of the questionnaire completed, the final draft questionnaire will be distributed to the respondents.

3.2 *Sampling frame*

This study is a cross sectional study. The target respondents for this study will be UiTM's students and employees at during breakfast and lunch time. This study plan to implement an intercept procedure (mall or street intercept) since the researcher the researcher does not allowed to enter the restaurant (Andaleeb & Conway, 2012). Hair, Black, Babin, and Anderson (2010) recommended sample size in Exploratory Factor Analysis (EFA) should be 100 or larger, with the minimum amount is at least five times as variables to be analyzed.

3.3 *Development of questionnaires*

In designing the questionnaire, questions format, measurement and evaluation should take into consideration as it highly related to the issue of reliability and validity of the final set of questionnaire. This study will use closed-ended questionnaire for this study is descriptive in nature. The close-ended questionnaire chosen for tools used to collect the data because is easier for the data input and computer analysis for closed questions is relatively easy and less expensive Hair, Money, Samouel, and Page (2007).

3.4 *Plan for data collection*

The researchers plan to collect the data in three Pendeta Coffee House's outlets in (UiTM) Shah

Alam during the month of July to August at the outside of Pendeta Coffee House after the patrons had finished their meal.

The questionnaire plan for gathering the data based on the performance only. This approach will be employed considering the disadvantages of both the expectation (before) and perceptions (after) measurement approaches designed by Parasuraman et al. (1988). There are some criticism for expectation and perceptions approach; (1) psychometric problems have been identified with respect to the use of different scores, including issues of reliability, discriminate validity and variance restricted problems (Cronin & Taylor, 1994; Dabholkar et al. 1996; Zeithaml, Berry, & Parasuraman, 1996); (2) service quality stems from perceptions of performance compared to ideal standards (Teas, 1993). Based on the suggestion by Cronin and Taylor (1994), the measurement only look on the performance consistent with other studies (e.g., Andaleeb & Conway, 2012).

3.5 *Plan for data analysis*

Two types of data analysis procedures: preliminary data analysis (e.g., Z-score, skewness and kurtosis, and respondent sample profile) and data analysis EFA. In order to satisfy research objective 1, addressing a research question and testing Hypothesis 1 (H_{1a} to H_{1d}) will be answered through the EFA analysis. The SPSS 21.0 will be used to compute the statistical analyses of the quantitative data obtained from the research instrument such as perform paired t-tests at a significance level of $\alpha = 0.05$ to see if differences existed between mean satisfaction scores for various interaction quality attributes.

4 SIGNIFICANCE OF THE STUDY

Interaction quality is one of the important aspects in influencing patrons' satisfactions and positive feedbacks in the service industry. Undoubtedly, examining the interaction quality within the university coffee house setting, will contribute in both academic and practical aspects. Moreover, from the perspectives of practitioners, the scrutiny of the findings will provide the least and most important sub-dimensions of interaction quality as perceived by patrons in Pendeta Coffee House's outlets. Therefore, the findings of this study also can give a guideline for the management to have a clear view towards patrons' perceived satisfactions and their intention to return to the coffee house.

REFERENCES

Abu-ELSamen, Amjad, Akroush, M., Al Khawaldeh, F., & Al-Shbily, M. (2011). Towards An Integrated Model of Customer Service Skills and Customer Loyalty: The Mediating Role of Customer Satisfaction. *International Journal of Commerce Management, 21*(4), 349–380.

Ahmed, M., & Al-Qudah, H.S. (2011). The Relationship Between Employee Interaction And Service Quality In The Private Sector: A Jordanian Case Study Approach. *Asian Social Science, 7*(9), 43–51.

Andaleeb, S.S., & Conway, C. (2012). An Analysis of Restaurant Patrons' Experiences in Malaysia: A Comprehensive Hierarchical Modelling Approach. *Lincoln University*.

Bitner, M.J., Booms, B.H., & Mohr, L.A. (1994). Critical service encounters: The employee's viewpoint. *Journal of Marketing, 58*, 95–106.

Brady, M.K., & Cronin, J.J. (2001a). Customer Orientation: Effects on Customer Service Perceptions and Outcome Behaviors. *Journal of Service Research, 3*(3), 241–251. doi: 10.1177/109467050133005

Brady, M.K., & Cronin, J.J. Jr. (2001b). Some new thoughts on conceptualizing perceived service quality: a hierarchical approach. *Journal of Marketing, 65*, 34–49.

Caruana, A., Ramaseshan, B., & Ewing, M.T. (2000). The effect of anomie on academic dishonesty among university students. *International Journal of Educational Management, 14*(1), 23–30.

Churchill, G.A.J. (1979). A Paradigm for Developing Better Measures of Marketing Constructs. *Journal of Marketing Research, xvi*, 64–73.

Cronin, J.J., & Taylor, S.A. (1994). "SERVPERF versus SERVQUAL: reconciling performance based and perceptions-minus-expectations measurement of service quality". *Journal of Marketing, 58* (No. 1), 125–131.

Dabholkar, P.A., Thorpe, D.I., & Rentz, J.O. (1996). A measure of service quality for retail stores: scale development and validation. *Journal of the Academy of Marketing Science, 24*(1), 3–16.

Douglas, P.F. (2003). Restaurant Staff Service Tips. Canada.

Ekinci, Y., Dawes, P.L., & Massey, G.R. (2008). An extended model of antecedents and consequences of consumer satisfaction for hospitality services. *European Journal of Marketing, 42*(1/2), 35–68.

Hair, J.F., Money, A.H., Samouel, P., & Page, M. (2007). *Research Methods for Business*. UK: John Wiley & Son Ltd.

Hair, J.F. Jr., Black, W.C., Babin, B.C., & Anderson, R.E. (2010). Multivariate Data Analysis. *Seventh Edition*.

Huang, Q. (2009). The relationship between service quality and student satisfaction in higher education sector: A case study on the undergraduate sector of Xiamen University of China. *16–21*(30), 38–60.

Ibrahim, Zalina, Dana, Linda, Mahdi, Ahmad Faisal, Zin, Mohamad Zaid Mohd, Ramli, Mohd Anuar, & Nor, Mohd Roslan Mohd. (2013). Evaluation of Malaysian Retail Service Quality. *Asian Social Science, 9*(4), 14–26. doi: 10.5539/ass.v9n4p14

Kim, W.G., & Moon, Y.J. (2009). Customers' cognitive, emotional, and actionable response to the service scape: A test of the moderating effect of the restaurant type. *International Journal of Hospitality Management, 28*(1), 144–156.

Liang, R.D., & Zhang, J.S. (2012). The Effect Of Service Interaction Orientation On Customer Satisfaction And Behavioral Intention: The Moderating Effect Of Dining Frequency. *Asia Pacific Journal of Marketing and Logistics, 24*(21), 153–170.

Lu, Y., Zhang, L., & Wang, B. (2009). A multidimensional and hierarchical model of mobile service quality. *Electronic Commerce Research & Applications, 8*(8), 228–240.

Malhotra, N., Mavondo, F., Mukherjee, A., & Hooley, G. (2012). Service Quality Of Frontline Employees: A Profile Deviation Analysis. *Journal of Business Research*, 1–7.

Mohi, Z. (2012). An Analysis of Restaurant Patrons' Experiences in Malaysia: A Comprehensive Hierarchical Modelling Approach. *Lincoln University*.

Mohi, Z., Clemes, M.D., & Hu, B. (2011). The development of service quality dimensions for moderate upscale restaurant in Malaysia. *Paper presented at the 9th Asia-Pacific Council on Hotel, Restaurant, and Institutional Education (APacCHRIE) Conference, Hong Kong*.

Mohi, Z., Raja Abdullah, R.P.S., Benedict Jipiu, L., Abd. Majid, A.A., & Wu, H.C. (2012). Assessment of service quality dimensions in the fast food restaurants. *Paper presented at the International Hospitality and Tourism Conference, Kuala Lumpur, Malaysia*.

Parasuraman, A., Zeithaml, V.A., & Berry, L.L. (1985). A conceptual model of service quality and its implications for future research. *Journal of Marketing, 49*(4), 41–50.

Parasuraman, A., Zeithaml, V.A., & Berry, L.L. (1988). SERVQUAL: A multiple-item scale for measuring consumer perceptions of service quality. *Journal of Retailing, 64*(1), 12–40.

Peters, C., & Mazdarani, E. (2008). The impact of employee empowerment on service quality and customer satisfaction in service organizations: A case study of Länsförsäkringar Bank AB.

Surprenant, C.F., & Solomon, M.R. (1987). Predictability and personalization in the service encounter. *Journal of Marketing, 51*(2), 86–96.

Teas, R.K. (1993). Expectations, performance evaluation, and consumers' perceptions of quality. *Journal of Marketing, 57*, 18–34.

Winsted, K. (2000). Service behaviors that lead to satisfied customers. *European Journal of Marketing, 34*, 13.

Wong, K., & Lim, K.M.Y. (2011). The State of Customer Service in Malaysia.

Wu, H.C., & Hsu, F.S. (2012). A Multi-Dimensional and Hierarchical Model of Service Quality in the Gaming Industry. *International Journal of Tourism Sciences, Volume, 12*(3), 90–118.

Zeithaml, V.A., Berry, L.L., & Parasuraman, A. (1996). The behavioral consequences of service quality. *Journal of Marketing, 60*, 31–46.

Hospitality and Tourism – Sumarjan et al. (eds)
© 2014 Taylor & Francis Group, London, ISBN 978-1-138-00151-0

Customer acceptance and satisfaction towards foreign workers' service quality in casual dining restaurant

A.A. Sukur & A.L. Mohd Rasdi
Universiti Teknologi MARA, Malaysia

A.S. Sukur
Education Institution of Temenggong Ibrahim, Malaysia

ABSTRACT: Using the Service Quality (SERVQUAL) criteria, the purpose of this study is to investigate, to what extent the Malaysian customers willing to consume and satisfied with the service provided by foreign workers in a casual dining restaurant. A quantitative approach will be used to collect data. Using simple random sampling, structured questionnaires will be distributed to the customer who are dining in the casual restaurants and pay for services provided by foreign workers in Shah Alam. This study will benefit the government and local authority by analyzing back the needs of hiring foreign workers in Malaysia and finds the solutions to reduce or increase the foreign workers phenomena. Therefore, it only focuses at casual dining restaurant in Malaysia; future research utilizing the other types of restaurant and different country may be beneficial.

Keywords: casual dining restaurant, customer acceptance, customer satisfaction, foreign workers, SERVQUAL, service quality

1 INTRODUCTION

Malaysia's economy has grown in a few decades parallel due to developments in the country. Malaysia's economy has recorded an increased in Gross Domestic Product (GDP) through the year. According to the statistic provided by the Economic Planning Unit, Prime Minister's Department of Malaysia (2012), the changes in growth percentages of Malaysia's economic rate and GDP were based on economic sectors that available in Malaysia. There are five main sectors available in Malaysia which are agriculture, mining and quarrying, construction, manufacturing and services.

The growth of Malaysia's economic rate and GDP show that there were productivity and economy activities running in the country which led to the creation of employment opportunities. As a country that is rapidly developing and has development projects in almost all economic sectors, Malaysia has experienced the labor shortages, particularly in the agriculture, construction, factories and services sectors. Therefore, the demands for the foreign workers are expected to overcome the shortage of manpower in the various economic sectors of the country.

According to Economic Planning Unit, Prime Minister's Department of Malaysia (2012), the numbers of employment has increased from 11,959,000 in 2010 to 12,505,000 in 2012. This shows that there are increasing in the employment opportunities each year in Malaysia. Job availability creates opportunities to foreign workers. Low cost salary for foreign workers makes Malaysian employers willing to hire the foreign workers in reducing the cost of productions by minimizing the labor cost. This factor makes the numbers of foreign workers increased each year. The rose presence of foreign workers in Malaysia can be explained by excess demand for labor merge with rapid economic growth, as and a cheaper cost of foreign workers. In addition, industrial also bring foreign workers into Malaysia (Noor, Isa, Said, & Jalil, 2011). As the result now, Malaysia has faced the influx of foreign workers, as recorded by the Department of Immigration, Malaysia has roughly a total of 1.9 million foreign workers (2011) worked in economic sectors (Malaysian Labor Force, 2012). This implicated foreign workers from Indonesia (50.9 percent), Bangladesh (17.4 percent), Nepal (9.7 percent), Myanmar (7.8 percent), India (6.3 percent), and Vietnam (7.9 percent).

Therefore, the influxes of foreign workers entering the labor force bring lots of possible outcomes in the labor market and productivity issues in the visited country. Foreign worker's service quality is a part of the issues occur according to the situation. Numerous studies have been conducted to investigate the foreign worker phenomenon. But yet, there is only limited study that has been conducted in the hospitality industry.

2 LITERATURE REVIEW

2.1 *Casual dining restaurant*

There are various types of restaurant that available in Malaysia. In the industry, the classifications of restaurants are based on menu style, preparation methods, pricing and services provided. The varieties of restaurants crop up in Malaysia is to convenience the different categories of consumers (Jangga, Sahari, & Basir, 2012). The changes in Malaysian life style currently which like to dine outside make the numbers of restaurants lift up developed in Malaysia. Casual restaurant is a part of restaurant types available in Malaysia.

The growths of outlets in casual dining restaurant are due to increase in urbanization. This leads to changes in lifestyle amongst urbanities, with more people willing to dine out instead of having meals at home.

Jangga et al. (2012) have claimed the meaning of casual restaurant is a restaurant that serves moderately-priced food with a casual atmosphere and comprises a market segment between fast food establishments and fine dining restaurants. The casual atmosphere in casual dining makes Malaysian put intentions to visit the restaurant according to the price that offers by the service provider influence and attract Malaysian customer to come to the restaurant.

Not only the type of restaurant will influence the customer to come to the restaurant, but the service quality provide by the restaurant itself also important. Parasuraman et al. (1985) define service quality (SERVQUAL) as an overall evaluation that results from comparing a firm's performance with the customer's general expectations of how the industry should perform. If service expectations are not met, service failure will occur.

2.2 *Service quality*

The growth in service sectors has caught the pervious researchers' intentions to conduct the studies on service quality. For example, Parasuraman, Berry and Zeithaml (1988) have conducted a research to analyze the importance of service quality which can cause the business to success or failure. This is because in the service sector, the people in the industry need to interact and deal with the customers to sell their products and services. Parasuraman et al. (1985) has investigated the service quality (SERVQUAL) have five dimensions which were tangibility, reliability, responsiveness, assurance, and empathy of products and services.

According to Parasuraman et al. (1985), the quality of services is difficult to evaluate rather than products or goods that being purchased by the customers. Service quality perceptions result from a comparison of customer expectations with actual service performance (Parasuraman et al., 1985). This show that to measure the quality of services by a firm, customers will evaluate the service performance that delivered by the service provider either the services achieve customer's expectation or not. Service quality is important for a firm to make sure their customer will satisfied with the service provided which will lead to behavioral intention, where it looks as a benefit to the firm. SERVQUAL has been used widely in measuring quality of services provide including in casual dining where it has been used by Seyanont (2007) which analyze SERVQUAL in a casual dining restaurant in Phuket.

The influxes of foreign workers involved in Malaysia's service sectors especially in restaurants and accommodations, give an impact to service quality in Malaysian restaurants. Quality evaluations are not made only on the outcome of a service, but it also includes evaluation of the process of service delivery (Parasuraman et al., 1985). SERVQUAL dimension is an important tool for restaurant in understanding the customer's value and to analyze the organization's performance either they are meeting the needs and expectations of the customers or not. From the analysis, the organization can improve their performance in service quality to build customers' satisfactions towards their restaurants.

2.3 *Customer acceptance and satisfactions*

Behavioral intention towards service quality (SERVQUAL) by a foreign worker is used to measure the acceptance towards foreign workers by the customers in Malaysia. As defined by Fishbein and Ajzen (1975) behavior intent is "the strength of one's intention to perform a specified behavior". In this research, customer acceptance relates to respondents` receptiveness and intentions to engage in activities from the stretch of service provides such as being greeted, communications in the food's menu information and receiving product that's being ordered for the foreign workers. In addition, the study is to measure intentions such as readiness to receive service provided by foreign workers, willingness to use the service provided and willingness to loyal to the restaurant.

Therefore, customer acceptance towards service quality by foreign workers is important to drive the customer satisfaction which will lead to customer behavioral intentions. When the customer accepts the service is provided by foreign workers, the customer will analyze the SERVQUAL provided by the foreign workers and will decide either they were satisfied or not. Customer satisfactions are important for any firm. This is because, according to Seyanont (2007), the higher the levels of customer satisfaction, will lead to higher levels of repurchase intentions.

3 ISSUES

Hamid, Singh, Yusof and Abdullah (2011) claimed that local Malaysian workers did not interest in working as a laborer. Especially for younger generations who have higher levels of education, they prefer to work as professionals in the industry. These practices will lead to shortages in semi-skilled labor force in the industry. The continued growths in Malaysia's economy forced the country to seek for labor force to fulfill the shortages in maximizing the productivity activity.

Due to this factor, Malaysia has open up the country for foreign workers looking for job opportunities and fulfills the labor shortages in Malaysia's economy sectors. Therefore, the influxes of foreign workers entering the labor force bring lots of possible outcomes in the labor market and productivity issues in the visited country. Thus, in the service sector especially in the hospitality industry, service quality is the most important element to be cared. In addition, the level of customer acceptance towards service provided by foreign workers also important to be analyzed either the customer are willing to accept or they have to accept the service provide by foreign workers.

According to foreign workers' issues, numerous studies have been conducted to investigate the foreign worker phenomenon. But to be believed that, the originality of this study will contribute to a new body of knowledge in Malaysia's service sector as there is little research in the area of foreign workers in the hospitality industry where this study will be conducted to analyze on how the service quality applied by foreign workers in Malaysia.

4 CONCLUSION

The increasing in the total numbers of foreign workers is giving the positive and negative impact to Malaysia's economy. It is believed to be beneficial to invest extra attention on the foreign worker phenomenon. So, using the service quality (SERVQUAL) tools, this study aimed to examine the level of customer acceptance towards foreign worker's service quality (SERVQUAL), and the impacts to customer satisfaction in restaurants. By conducting this research, it will help in determining and examined the others impact of foreign workers in the other economy sector in Malaysia. Thus, it is believe that will help the government and local authority, especially the Ministry of Labor Force to analyze back the needs of hiring foreign workers in Malaysia and find the solutions to reduce or increase the foreign workers phenomena.

REFERENCES

Economic Planning Unit (2012). The Malaysian economic in figures 2012. Retrieved on 10th November 2012, from http://www.epu.gov.my/html/themes/epu/images/common/pdf/chart/The-Malaysian-Economic-In-Figures-2012.pdf.

Euromonitor (2012). Retrieved on 28th of December 2012, from http://www.portal.euromonitor.com.ezaccess.library.uitm.edu.my/Portal/Pages/Search/SearchResultsList.aspx.

Fishbein, M. & Ajzen, I. (1975), Belief, attitude, intention, and behavior: An introduction to theory and research, Addison-Wesley, Reading, MA. Retrieved on 28th December 2012, from http://people.umass.edu/aizen/f&a1975.html.

Hamid, A.R.A. (2011). The Employment Of Foreign Workers at Construction Sites. *2nd International Conference on Construction and Project Management* IPEDR, 15 (2011), 126–130.

Jangga R. et al. (2012). Factors Determining the Level of Satisfaction Experienced by Customers Who Visit Family Chain Restaurants. *3rd International Conference on Business and Economic Research,* 2762–2774.

Malaysian Labor Force (2012). Labor force survey report first half 2011. Department of Statistics, Malaysia, 2(10), 1–24. Retrieved on 10th of November 2012, from http://www.statistics.gov.my/portal/download_Labour/files/BPTMS/Laporan_PTB_H111_270911.pdf.

Malaysia Productivity Corporation (2012). Productivity report 2011/2012. Retrieved on 10th of November 2012, from http://bpap.mpc.gov.my/APR/APR2012.pdf.

Noor, Z.M. et al. (2011). The Impact of Foreign Workers on Labor Productivity in Malaysian Manufacturing sector. *Int Journal of Economic and Management*, 5(1), 169–178.

Parasuraman, A., Zeithaml, V.A., & Berry, L.L. (1985). A Conceptual Model of Service Quality and its Implication. *Journal of Marketing*, 41–50.

Parasuraman, A., Zeithaml, V.A., & Berry, L.L. (1988). The Service-Quality Puzzle. *Business Horizons*, 35–43.

Parasuraman, A., Zeithaml, V.A., & Berry, L.L. (1988). SERVQUAL: A Multi-Item Scale for Measuring Consumer Perceptions of The Service Quality. *Journal of Retailing,* 64(1), 12–40.

Seyanont, A. (2007). A comparative study of the service quality of casual dining restaurants in Phuket: Perspective of Thai and international customers. Oklahoma State University). ProQuest Dissertations and Theses, 225-n/a. Retrieved from http://search.proquest.com.ezaccess.library.uitm.edu.my/docview/304743942?accountid=42518. (304743942).

Hospitality and Tourism – Sumarjan et al. (eds)
 ISBN 978-1-138-00151-0

Local community outlook on tourism development in Tioman Island

M.H. Hanafiah, S.A. Abas, M.R. Jamaluddin & M.I. Zulkifly
Universiti Teknologi MARA, Malaysia

ABSTRACT: The tourism industry has been recognized as the symbol to help the local community especially in changing the economic and atmosphere due to the ability to generate income, employment as well improve the standard of living. The attitudes of the local community can directly affect the development of the tourism industry. Positive attitudes from the local residents help in boosting tourists' satisfaction levels and contribute to the word-of-mouth promotion among international tourists. In order to ensure the success of the tourism planning, the involvement and the involvement of the local communities in the area is pertinent. This study endeavors on local community attitudes' toward tourism development in Tioman Island. Local residents with personal benefit and positive perception of tourism development will support for future tourism development. However, the finding also revealed that the local community who perceived negative impact of tourism activities also supports the tourism development in the future due to the reliance on this sector as their main income. The role of perceived tourism impact as the moderator between the relationship between personal benefit from tourism and the future attitudes from tourism development become the contribution of this study.

Keywords: attitudes, future attitudes, perceived tourism impact, personal benefit

1 INTRODUCTION

The tourism industry has faced a rapid growth with the support from local communities who were involved in this industry. The growth of the tourism industry is crucial to the economic growth as well as the related industry such as transportation, leisure services and hospitality. Generally, tourism is acknowledged as one of the major attributes for cultural and economic nowadays and it gives the opportunity to the local communities as well. On the other hand, tourism also becomes the symbol to help the communities especially in changing the economics and atmosphere. This is because the ability of the tourism industry to generate income, currency exchanges and provides the employment opportunity (Hanafiah, Harun & Jamaludin, 2010).

However, Kayat (2002) pointed out that some of the developing countries utilized this tourism industry for economic stimulation purposes by gaining the foreign investment and capital. The local community is the main stakeholders of tourism developments (Harrill, R., 2004). Currently, the number of studies on local community attitudes towards future tourism development is increasing due to the importance of tourism appropriateness. However, limited research was embarked on the roles of personal benefit from tourism development will influence the future support towards future tourism development. As stated by Jafari (2001), it is suggested that research on tourism development should consider the roles of perceived of personal benefit from tourism development towards their support for future development in tourism destination.

This paper look into the personal benefit that local community gain from tourism development and how their response from the development of tourism as well as their support for future development. This study assessed local community attitudes towards tourism development in Tioman Island and particularly underscores the personal benefit from tourism development, perceived tourism impact as well as the attitudes for future tourism development. Furthermore, this research looks into the local community participation and non participation in the tourism development. By understanding local community attitudes towards tourism development, the planning for future tourism development will get support from the local community who are one of the stakeholders as well as involve in the decision making process.

1.1 *Tourism development in Tioman Island*

Tioman Island is the largest island off the South-East Coast of Peninsular Malaysia and principal target for promotion as a tourist destination. This island situated at 40 kilometers from the coast, its modest area of 138 KM is nevertheless bigger than such well-known island. Its location had imbued the island with some strategic importance in the maritime trade between China and Southeast Asia from about the beginning of the second millennium. Tioman probably served as an important landmark and watering place for the ancient mariners. The island is rugged; its mountainous backbone constitutes a formidable barrier to overland movements and restricts the area of habitable space. Climatically, the northeast monsoon from November to march, often accompanied by gale force winds, produces a marked seasonality in the tourism industry (Tioman Development Authority, 2011). In 2011, the number of tourist arrivals reached to 126,660 tourists domestically as well as internationally. This is the increasing of 8.8% in 2010 which recorded the number of tourists is 213,158 people (Tioman Development Authority, 2011).

The traditional economy of the island is one of farming, fishing and village cultivation associated with spontaneous settlement since the 1920. Until the early 1980s the principle means of livelihood were coconut cultivation and fishing involving 70% to 80% of the population. Radical changes in land use began with the Pahang State Government approval to build a beach resort in 1965 of seafront land originally occupied by a Malay Kampung. Ownership was later transferred to a major Malaysian business corporation with diversified interest in the hotel, travel and tours, development and management of golf courses and operation of resort in Malaysian, Fiji and the Mauritius (Azaruddin, 1994). Under the new management, the original 70 room resort was extended to 373 rooms and international championship golf course was established. The package tours provide a captive market in this resort which an effect operates a typical gateway destination where tourist spending hardly spills over to nearby villages (Tioman Development Authority, 2011).

2 LITERATURE REVIEW

Local communities played a vital role in tourism development as they are important to provide a good situation especially for tourists in order to boost up the number of tourist's arrival as well as improve their quality of life. They are sometimes known as 'service suppliers' as well as a focal point to cater the accommodation, information, transformation, facilities and services for the sake of tourism development in their area (Andereck & Nyaupane, 2011). Previous researchers mentioned about the term of community in different tourism setting and reflect to what their common interest.

Local community influenced the tourism development activities through organizing themselves at all levels to act more effectively in development and good in sense of contact with government. According to Latkova and Vogt (2012), local communities must have sense to help their community in order to ensure they can gain the positive benefit from tourism development. As suggested from previous research, the communities have to work closely to the non government organizations as the effort to educate people about tourism development.

The importance of gaining local community support for the tourism development has been discussed by tourism scholars as vital components in achieving successful tourism industry. Support for future tourism development is a key factor in developing and implementing successful initiatives. Some of local community perceptions of tourism may positively whereas others are negative (Srikaya, Teye, & Sonmez, 2002). Previously, research conducted by Wang and Pfister (2008) on the role personal benefit that local community obtained from tourism development discovered that, benefit and cost are assumed to be more carefully evaluated and benefits exceed the cost, the residents' be more positive on tourism but then as converse from this, when cost exceed the benefit, residents' tent to be more negative toward tourism development.

Furthermore, the significant role played by the personal benefit from tourism development particularly the indirect social values in the process of exchange for residents' favorable attitudes towards tourism development. In addition, Nunkoo and Ramkissoon (2012) found out that even though residents' did not enjoy a direct economic benefit from tourism, the majority of them will be neutral in their attitudes towards tourism. But then, the findings revealed that most of residents' believed that there will be an economic benefit from tourism development.

Residents' perception of social and cultural impacts of tourism development has been studied extensively. However, the findings of those studies have produced contradictory results. While, several studies reported that residents tend to perceive the social and cultural impact of tourism development negatively (Dyer et al., 2007; Johnson et al., 1994; Jurowski et al., 1997; Tosun, 2002). More than that, some of researchers argue that host residents view tourism as providing various social and cultural benefits to the community. However, other effects may be less welcomed, such as changes in social and family structure. As suggested by Hovinen (2002), tourism provides new opportunities and instigates social change. But then, because of the flexible working patterns and new opportunities tourism erodes gender segregation (Cooper, 1995).

Previous research examined the relationship between residents' attitudes and level of tourism development. Allen et al. (1988) found that residents' perception of tourism impact become less positive as level of tourism in community is increased. Perdue et al. (1990) also summarized that residents' initial attitudes towards tourism were enthusiastic but as a cost outweighed benefit of tourism development, attitudes achieve a threshold after which residents support for tourism declined. Meanwhile, Allen et al. (1993) discovered that, communities with low tourism and high total economic activity, look tourism as development favorable than communities with low tourism and high economic activity and communities with high tourism development and low economic activity.

3 METHOD

The population of this research is the local community on Tioman Island. There are five villages on this island which are Kampung Salang, Kampung Tekek, Kampung Air Batang, Kampung Paya Genting, Kampung Mukat and Kampung Juara. In 2011, the total number of local communities in Tioman Island is 3314 residents. The list of the population out of the five villages was obtained from the Tioman Development Authority.

The sample included the local community from five different villages on Tioman Island and was selected by using the convenience sampling. The sample size of 248 respondents is based on Krescjie and Morgan (1970). Variables are adopted from the Latkova and Vogt (2012) and Lankford and Howard (1994). The questionnaires developed from a review of existing literature dealing with local community attitudes toward tourism development. The measurement of the attitudes used in this study applied the Tourism Impact Attitude Scales (TIAS) developed by Lankford and Howard (1994) as the standardized of measurement of resident attitudes toward tourism developed. This measurement has been tested in various settings of tourism area and destination (Latkova & Vogt, 2012). The number of questionnaires distributed is 250 however, only 129 responses received.

Table 1. Population of local community in Tioman Island.

Name of village	Total house	Total residents
Kampung Salang	36	279
Kampung Air Batang	459	2,092
Kampung Paya, Genting	86	435
Kampung Mukut	41	225
Kampung Juara	62	283
Total	687	3,314

4 RESULTS

4.1 *Respondents' Personal benefit from tourism development*

Majority of respondents' perceives themselves benefitted from tourism activity in Tioman Island. This particularly in safety and security (M = 4.50), feel safe in their current job (M = 4.34) and level of social life (M = 4.31) were identified as three major perspectives of personal benefit that respondents' obtained from tourism development. This result is consistent with the establishment of police booth at every village in Tioman Island in order to control and coordinate the safety and security purposes especially for both local communities and tourists as well. Moreover, most of respondents' feel that they are safe in their current job which is in line with the 'Human Capital Program' that organized by the Tioman Development Authority.

However, the respondents' perceived that the cost of the basic necessity in Tioman Island is expensive (M = 1.20). This might due to the high transportation cost and retail operators tend to increase the price of the product in order to cover the cost of transportation. The majority of respondents in Tioman Island perceived lots of benefits from more tourism development (M = 4.32) as well as current tourism development (M = 3.9070). Overall, respondents' were happy lives in Tioman Island.

Table 2. Respondents' Personal benefit from tourism development.

Items	N	Mean	Std. deviation
Generate extra income	129	3.67	1.35
Secure in my current job	129	4.34	0.834
Extra public facilities	129	3.46	1.17
Condition of the environment (air, water & land)	129	4.15	0.761
Social life	129	4.31	0.716
Safety and security	129	4.50	0.575
Cost of the basic necessity product	129	1.20	0.564
Benefited from tourism development	129	3.90	1.07
Benefited from more tourism development	129	4.32	0.901
I am happy to live here	129	4.74	0.534

Table 3. Respondents' by perceived positive tourism impact.

Items	N	Mean	Std. deviation
Improves the local economy	129	4.61	0.489
Shopping, restaurants and entertainment	129	3.78	1.134
Public development (road, public facilities)	129	3.82	1.162
Generate income and improvement in standard of living	129	4.42	0.681
Jobs for local homeowners	129	4.49	0.574
Recreational opportunities for local homeowners	129	4.10	0.900
Incentives for the protection and conservation of natural resources	129	4.07	1.017
Increases the quality of life	129	4.39	0.938

4.2 *Respondents' perceived positive impact*

The increasing number of tourists in Tioman Island had improved the local community economy (M = 4.61), contributes to the income and standard of living (M = 4.42) and at the same time, provides job opportunity (M = 4.49). It is in line with the declaration of Tioman Island as a duty free zone in 2002 which boosted up the number of tourist arrivals. Due the influx of tourist, local communities gain extra income, especially those who are operating the chalets, resorts or rooms, sea taxi providers, shops as well as the handicraft shops. Based on Gursory and Rutherford (2004), residents tend to view tourism as a tool to generate income as well as create job opportunities and as the number of tourists increased, it will also improve their level of economy.

Furthermore, the respondents agreed tourism development increases recreational facilities in Tioman Island (M = 4.10). Recently, the Tioman Development Authority (TDA) installed the 'pontoon' at Kampung Juara, Kampung Genting and Kampung Paya as the effort to make the water recreation activities more systematic and secure to all communities and tourists. More than that, the facilities for scuba also introduced especially for children. The authority had engaged with Scuba School International to carry out the program namely as Kids Scuba Camp at Kampung Salang. This shows that the authority really concerns in terms of providing the recreational activities to the local communities.

4.3 *Respondents' perceived negative impacts*

Most of respondents perceived that tourism resulted in increasing the cost of living (M = 3.37)

Table 4. Respondents' by perceived negative tourism impact.

Items	N	Mean	Std. deviation
Traffic problem	129	2.97	1.526
Litter problem	129	3.22	1.358
Cost of living	129	3.37	1.649
Low salary	129	3.18	1.406
Overcrowding problem	129	3.06	1.345
Unfair property taxes	129	2.74	1.253
Friction between community and tourist	129	2.22	1.134
Number of crimes	129	1.81	1.144

and previous studies on tourism impacts found that, cost of living increase because of the tourism activity (Conlin and Baum, 1996). Furthermore, respondents agreed that tourism sector pays the lowest salary to their employee (M = 3.18). In addition, respondents perceived that overcrowding problems occurred because of the tourism development in Tioman Island. The declaration of the duty free zone boosted up the arrival of tourists and at the same time some of local communities feel inconvenient of the overcrowding of tourists. The arrival of more tourists has always been linked with many social problems such as frictions between tourist and local community as well as the crime resulted from the tourism activities. However, in this case, the respondents do not agree that tourism brings the friction and crime problem to their community.

5 CONCLUSION

The signs of personal benefit from tourism development show that most of respondents obtained benefit from the tourism activity in Tioman Island. In other word, tourism brings improvement of their quality of life. The more benefit respondents' perceived gaining from tourism personally, the more likely they are to increase the support for tourism development. Personal benefit and costs should be evaluated carefully and when benefits exceed the cost the perception tends to be positive attitudes towards tourism development. Personal benefits from tourism development had been identified as a significant aspect associated with the attitudes towards tourism development. Respondents' perceived safe and security as well as feel secure in their current job are two personal benefit items that they're obtained from the tourism development in Tioman Island.

The major concern for residents in Tioman Island was the potential of tourism to increase the cost of living. Most of previous study on tourism impact discovered that the cost of living is increasing because of tourism activity. More than that, litter and overcrowding problems occurred because of tourism development on Tioman Island. Conversely, tourism activity in Tioman Island brings opportunity especially for the local community in terms of providing the employment in tourism related business as well as improving their quality of life. Local community perceived positive impact of tourism with the number of facilities is increasing for the community purposes.

Even though the tourism activity is extensively promoted, the authority also considered about the incentives for the protection and conservation of the Tioman Marine Park. Local community in Tioman Island believes that, the tourism sector will continue to play a major role in the economy of the community. In line with that, the Tioman Development Authority should manage the growth of tourism activities wisely with limitation on the spending leakages. In addition, the role of the government and authority is important to enhance the promotion in order to boost up the tourist arrival to Tioman Island.

ACKNOWLEDGEMENTS

The work described in this study was funded by the Fundamental Research Grants Scheme, Ministry of Higher Education of Malaysia and Universiti Teknologi MARA Malaysia.

REFERENCES

Allen, L.R., H.R. Hafer, P.T. Long, & R.R. Perdue. (1993). "Rural Residents' Attitudes toward Recreation and Tourism Development." *Journal of Travel Research, 31* (4): 27–33.

Allen, L.R., P.T. Long, R.R. Perdue, & S. Kieselbach. (1988). "The Impact of Tourism Development on Residents' Perceptions of Community Life." *Journal of Travel Research, 27* (1): 16–21.

Andereck, K.L., & Nyaupane, G.P. (2011). Exploring the nature of tourism and quality of life perceptions among residents. *Journal of Travel Research, 50*, 248–260.

Azaruddin, O. (1994). Tourist industry in Tioman Island. Malaysia. *Journal of Geography, 25*(2): 59–67.

Conlin, M.V., & T. Baum. (1996). Island Tourism: Management Principles and Practice. New York: John Wiley.

Cooper, C. (1995). Strategic planning for sustainable tourism: The case of the offshore islands of the UK. *Journal of Sustainable Tourism, 3*(4): 191–209.

Dyer, P., Aberdeen, L., & Schuler, S. (2007). Tourism impacts on an Australian indigenous community. *Tourism Management, 24*, 83–95.

Hanafiah, M.H., M.F. Harun & Jamaluddin M.R., (2010). Bilateral Trade and Tourism Demand. World. *Applied Sciences Journal 10* (Special Issue of Tourism & Hospitality), 110–114.

Harrill, R. (2004). Residents' attitudes toward tourism development: A literature review with implications for planning. *Journal of Planning Literature, 18*(3): 251–266.

Jafari, J. (2001). The socio-economic costs of tourism to developing countries. *Annals of Tourism Research, 1*(2): 227–234.

Johnson, K.M., & C.L. Beale. (2002). Nonmetro Recreation Counties: Their Identification and Rapid Growth. *Rural America, 17*(4): 12–19.

Jurowski, C., Uysal, M., & Williams, D.R. (1997). A theoretical analysis of host community resident reactions to tourism. *Journal of Travel Research, 34*(2): 3–11.

Kayat, K. (2002). Power, social exchanges and tourism in Langkawi: Rethinking residents' perceptions. *International Journal of Tourism Research, 4*, 171–191.

Kepe, T. (2004). Poverty alleviation and biodiversity conservation: a South African Perspective. Oryx: *The International Journal of Conservation, 38* (2): 143–145.

Krejcie, R.V., & Morgan, D.W. (1970). Determining sample size for research activities. *Educational and Psychological Measurement, 30*, 607–610.

Lankford, S.V., & D.R. Howard. (1994). Developing a Tourism Impact Attitude Scale. *Annals of Tourism Research, 21*(1): 121–139.

Latkova, P., & Vogt, C.A. (2012). Residents' attitudes toward existing and future tourism development in rural communities. *Journal of Travel Research, 51*(1): 50–67.

Nunkoo, R., & Ramkissoon, H. (2012). Power, trust, social exchange and community support. *Annals of Tourism Research, 39*(3): 997–1023.

Perdue, R.R., P.T. Long, & L.R. Allen. (1990). Resident Support for Tourism Development. *Annals of Tourism Research, 17* (4): 586–599.

Sirakaya, E., Teye V., & Sonmez, S. (2002). "Understanding residents" support for tourism development in the Central Region of Ghana. *Journal of Travel Research, 41*, 57–67.

Tosun, C. (2002). Host perceptions of impacts. *Annals of Tourism Research, 29*, 231–253.

Tioman Development Authority (2011). Annual Report 2011. Retrieved June 22, 2012, from http://www.tioman.gov.my/Laporan%20Tahunan%202011.pdf.

Wang, Y., & R.E. Pfister. (2008). "Residents' Attitudes toward Tourism and Perceived Personal Benefits in a Rural Community." *Journal of Travel Research, 47*(1): 84–93.

Hospitality and Tourism – Sumarjan et al. (eds)

The perception of minimum wage towards employee job satisfaction

N.A. Talib, N.A. Ahmad, A.A. Azdel & M.S.Y. Kamaruddin
Universiti Teknologi MARA, Malaysia

ABSTRACT: Investigation between job satisfaction and work motivation in the hotel industry is not a new scenario. Numerous researches have been conducted focusing in the area, however only few studies focusing on minimum wage as part of motivation factor. Based on the government's new guideline on the minimum wage for all sectors in 2012, this study is proposed to examine the effect of minimum wage towards employees' job satisfaction. The hotel employee's perception about this policy will also be explored. This case study will be conducted using a quantitative approach with the employees as the future respondents. Through this study, employees can express their opinion with regards to the minimum wage policy thus can help the hotel to better develop their employees' salary scheme. Additionally, the finding is also hoped to serve as a platform to regenerate more ideas regarding minimum wage particularly from Malaysian context.

Keywords: extrinsic motivation, job satisfaction, minimum wage

1 INTRODUCTION

Asian country starts to increase the minimum wage to tackle the wealth gaps in the region. Bank Negara, (2012) reported that Malaysia joined more than 150 other countries in the world in standardizing the minimum wage policy. As Malaysian companies began to implement the new minimum wage in January 2013, the hotel industry is trying its best to establish ways to roll out new salaries to its employees (Bank Negara, 2012). Hidayu (2012) stated that when the minimum wage to all sectors of employees was enforced, hotel employees new total salaries including additional income gained from the service charge could be more than a hotelier could handle.

Other than that, operation costs in the hotel business will increase and for this reason hoteliers are reluctant to comply with the minimum wage policy. It is undoubtedly that many firms will have a profound effect on the costs when the policies of labor in the state change (Heriot, Lang, & Thomson, 2010). For this study, a wage (salary) system is the focal point that researcher tries to stress that as the policies of labor that currently has some changes in Malaysia.

Heriot, Lang and Thomson (2010) explained that this industry is particularly relevant because it contains a large number of businesses and these companies have many employees earning minimum wage with other benefits (points, meal allowance, etc.).

Furthermore, the effort and contribution of employees are the key competitive advantage and is crucial to the success of the firm (Bent & Freathy, 1997). Therefore, this study is proposed to examine the effect of minimum wage towards employees' job satisfaction. The hotel employee's perception about this policy will also be explored. Moreover, it can highlight and further improve our understanding of this important issue.

2 LITERATURE REVIEW

2.1 *Wage policy in Malaysia 2013*

Hidayu (2012) highlighted that the minimum wage policy is a mandatory in Malaysia and considering some potential of long term benefit for the organization, most of the hotel organizations in Malaysia are currently looking into the best ways to implement the mandatory order to reach a compromise between the employees, hotel owners and management. In addition, she reported that Human Resources practitioners' share and exchange ideas on the type of approach every hotel could take in compliance with the order and to avoid any consequential negative effects.

Bank Negara (2012) reported that Malaysia joined more than 150 other countries in the world in standardizing the minimum wage policy. Hence, the minimum wage set at RM900 per month (RM4.33 per hour) for Peninsular Malaysia and RM800 per month (RM3.85 per hour) for Sabah,

Sarawak and the Federal Territory of Labuan. This minimum wage applies to both local and foreign workers, except domestic workers such as domestic helpers and gardeners (Bank Negara, 2012).

Meer and West (2012) argued that the question of how a minimum wage affects employees remain one of the most widely studied and most controversial topics in labor economics. Thus, they suggest that a more conclusive answer is needed to the question of the effects of the minimum wage if higher minimum wages are to be evaluated alongside alternative policy instruments as a method of increasing the standard of living for low-income households.

2.2 *Motivation*

According to Ayub and Rafif (2011) in their research they have mentioned that most of the motivation literature suggests that employee motivation can come from other tangible factors such as money or/and intangible factors such as sense of achievement. To gain more understanding on the motivation aspect, it is advisable for researchers to refer a variety of motivation theories.

Motivators involve factors built into the job itself, such as achievement, recognition, responsibility and advancement. Hygiene factors are extrinsic to the job, such as interpersonal relationships, salary, supervision and company policy (Herzberg, 1966). Bogardus (2007) suggested that, when hygiene factors are maintained, dissatisfaction can be avoided. When opposite, dissatisfaction is most probable to occur and motivation can't take place. These factors relate to the content of the work and if they are in proper form, it tends to eliminate job dissatisfaction (Ellis, 2005).

Table 1 below is presented in this study to make readers more clear what exactly Hezberg try to explain in his Two-Factor Theory.

Due to the framework of study where salary: minimum wage is the main independent variable that researcher looking for, researcher are more concern to look into the Hygiene's (extrinsic motivation) were in this group Hezberg has come out that salary is one of the factors of employee motivation. He also has mentioned that salary is one of the factors that can lead to dissatisfies of the employee if organization or management are not given them an appropriate amount.

Maslow (1943) said that people work to survive and live through financial compensation, to make new friends, to have job security, for a sense of achievement and to feel important in the society, to have a sense of identity, and most especially to have job satisfaction. All employees that have job satisfaction are high performers in their respective workplaces. Morgan (1997) suggested to the industry on how an organization can satisfy needs of an employee's base on the different levels of Maslow Hierarchy. Maslow proposed that there are five different levels of needs people have to seek for satisfaction of their basic needs (Mullins, 2007).

Table 1. Hezberg Two-Factor Theory.

Motivators (intrinsic motivation)—leading to satisfaction	Hygiene's (extrinsic moti = vation)—leading to dissatisfaction
i. Achievement	i. Company policy
ii. Recognition	ii. Supervision
iii. Work itself	iii. Relationship with boss
iv. Responsibility	iv. Work conditions
v. Advancement	v. Salary
vi. Growth	vi. Relationship with peers
	vii. Security

2.3 *Job satisfaction*

Most of the literature has mentioned that job satisfaction is another name for employee satisfaction. It is often being used in a process to understand the psychology of the industrial-organizational (Dupnock, 2010). No matter how previous researches define job satisfaction, at some points they are agreed that job satisfaction bring the definition of the attitudes and feelings of employee have about their work (Funmilola, Sola, &

Table 2. Example on how an organization can satisfy needs of an employee.

Self-actualizing
- Encouragement of complete employee commitment
- Job a major expressive dimension of employee's life

Ego
- Creation of job with scope for achievement, autonomy, responsibility, and personal control
- Work enhancing personal identity
- Feedback and recognition for good performance

Social
- Work organization that permits interaction with colleagues
- Social and sport facilities
- Office parties and outings

Security
- Pension and health care plans
- Job tenure
- Emphasis on career paths within the organization

Physiological
- Salaries and wages
- Safe and pleasant working conditions

Source: Investigation and analysis of motivation factors at work by Maren Bassy, 2002.

Olusola, 2013). In this study, the researcher uses job satisfaction as the dependent variable in a framework in measuring the reflection of applying the new minimum wage policy in the organization.

According to Funmilola et al. (2013), job satisfaction or dissatisfaction is indicated based on the positive and negative of employee attitudes. They also mentioned that job satisfaction attributed not only depend on one but many factors. Many researchers have agreed that the attributes of job satisfaction consist of career path, pay package, work environment and any other factors that can relate to the human psychology. Researcher agreed with the statement wherein creating a job satisfying, there a many attributes has to rely on, yet minimum wage is only the focus for this moment which it is a new policy that been implemented in Malaysia.

Rehman, Saif, Khan, & Nawaz (2013) explained that in any organizations with regardless of the type and at any region, job satisfaction is one of the crucial problems that appear. Wherefore, degree of employee satisfaction will bring over their commitment and this is an indication for organizational output and effectual operations (Rehamn et al., 2013).

Referring the studied by Gebremichel and Rao (2013), they has explained the importance of job satisfaction to an organization's success. This matter widely studied which the findings have come out with job satisfaction influence employee behavior and subsequently influence others organizational variables such as productivity, turnover and absenteeism. Therefore, it was suggested that most of the organization has to take an initiative and effort in creating job satisfaction to gain the positive feedback for the organization as what's been discussed in most of studied.

In line with the above statement, the assessment of job satisfaction is a common activity in many organizations where management feels that employee well-being is important (Spector 1997). Based on the review of literature, a hypothesis been formulated:

H1: Extrinsic motivation (minimum wage) influence employee job satisfaction in Grand Continental Hotel.

3 METHODOLOGY

This study will be conducted through self—administered questionnaire and 105 questionnaires will be distributed to all non—management levels the employee of Grand Continental Hotel. Reliability and validity of the research instruments will be tested through Cronbach Alpha coefficient test, expert review and pilot testing to ensure instruments measures the right variables. In measuring the effect of minimum wage implementation towards job satisfaction regression analysis will be used.

4 EXPECTED FINDING

Commerce International Merchant Bankers Berhad (CIMB) 2012 reported that Malaysia's minimum wage is set in Malaysia will benefit some 3.2 million private sector workers and arguably, boost productivity by increasing worker satisfactions.

For a simple understanding, researchers can say that CIMB (2012), has clearly mentioned that by implementing minimum wage employee will be more satisfied and is likely to vary depending on time, place and the type of job. According to Juanis (2013), wage or salary in a hotel industry is a sensitive and crucial issue where the management or employers have to give serious attention and due consideration because hotel industry is a service oriented industry. Therefore, hopefully through regression analysis it can show to what extent minimum wage salary is set by the government and implemented by the hotel influence employee job satisfaction.

Meer and West (2012) suggested in their studies which an effect of the minimum wage should be more transparent in employee perspectives. However, despite extensive empirical research, debates on the labor market especially employees' satisfaction that impact of minimum wage increases is still unsettled, and the results conflict (Cui, 2012). Therefore, the finding is also hoped to serve as a platform to regenerate more ideas regarding minimum wage particularly from Malaysian context.

5 CONCLUSION

In the preferential culture of hotel industries, it is not easy to implement the minimum wage policy as set up by the government. With this study, researcher hopes that it can benefit the Grand Continental hotel itself to understand their employee perception about implantation effect of minimum wage policy. In addition to that, this paper contributes to understand how Malaysia minimum wage is adjusting the dynamic of wage setting behavior under a binding wage. In line with the current issues this study hopefully clarifies and provides an answer pertaining to the minimum wage policy and employee motivation effect. It is believed that this study will benefit to both academia and industry, while at the same time show the significance of this issue in Malaysia.

REFERENCES

Allison, S.J. (2012). The eroding Hukou system and the New Minimum Wage in China: The impacts of economic inequality, labor shortages and social unrest. Senior theses, Trinity College, Hartford, CT 2012. Trinity College Digital Repository, http://digitalrepository.trincoll.edu/theses/176.

Ayu, N. & Rafif, S. (2011). The relationship between work motivation and job satisfaction. Pakistan Business review July 2011. Retrieved May 05, 2013 from http://www.iobm.edu.pk/pbr/pbr_1107/110708_workmotivation.pdf.

Azeem, S.M. (2010). Job satisfaction and organizational commitment among employees in the Sultanate of Oman. Psychology, 2010, 1, 295–299. doi:10.4236/psych.2010/14038 published Online October 2010.

Bank Negara (2012). Outlook and policy in 2013. Retrieved April 30, 2013 from http://www.bnm.gov.my.

Bent, R., & Freathy, P. (1997). Motivating the employee in the independent retail sector. *Journal of Retailing and Consumer Services*, 4(3), 201–208.

Bogardus, A.M. (2007). PHR/SPHR: Professional in Human Resources certification study guide. Indianapolis. Wiley Publishing, Inc.

CIMB (2012). Economic update: Minimum wage policy—Curse or cure? Retrieved April 23, 2013 from, http://www.etp.pemandu.gov.my.

Dupnock, L.M. (2010). Independent restaurant employee job satisfaction in college towns. Retrieved April 25, 2013 from http://www.dspace.iup.edu/bitstream/handle/2069/313/Lisa%20DupnockCorrected.pdf?sequence=1.

Ellis, C.W. (2005). Management skills for new managers. USA. AMACOM.

Funmilola, O.F., Sola, K.T., & Olusola, A.G. (2013). Impact on job satisfaction dimensions on job performance in a small and medium enterprise in Ibadan, Soouth Western, Nigeria. Institute of Interdisciplinary Business Research, March 2013, Vol 4, No 11.

Gebremichel, H. & Rao, B.V.P. (2013). Job satisfaction and organizational commitment between academic staff and supporting staff (Wolaita Sodo University—Ethiopia as a case). Far East Journal of Psychology and Business, Vol. 11 No1 April 2013. 11–32.

Heriot, K., Lang, T., & Thomson, N. (2010). Evaluating the impact of Federal Legislation on Small Business: An Exploratory Study of the New Minimum Wage Rate and the Health Insurance Tax. *Small Business Institute Journal*, 6, 63–90.

Herzberg, F. (1966). Work and the nature of man. Cleveland, OH: World Publishing Company.

Keijzers, B. (2010). Employee motivational related to employee performance in the organization. Retrieved on 12 May 2013. Retrieved May 13, 2013 from http://www.arno.uvt.nl/show.cgi?fid=121057.

Kingir, S. & Mesci, M. (2010). Factors that affect hotel employees motivation the case of Bodrum. Serbian Journal of Management 5 (1), 59–76.

Lawler, E.E. (1973). Employee reactions to job characteristics. *Journal of Applied Psychology*, 55, 259–286.

Maslow, A.H. (1943). A theory of human motivation. *Psychological Review*, 50, 370–396.

Meer, J. & West, J. (2012, August 9). Effects of the minimum wage on employment dynamics. Texas. Retrieved April 23, 2013, from http://www.econweb.com.tamu.edu/jmeer/Meer_West_Minimum_Wage.pdf.

Michel, D.D. (2012). The mediating role of job satisfaction in the relationship between reward satisfaction, turnover intention and work effort across occupational status. Retrieved May 13, 2013 from http://www.s-d.be/assets/Talent-Award-2013/Theses-2013/VUB.pdf.

Morgan, G. (1997). Images in organizations (2nd Ed.). Thousand Oaks, London: Sage Publications.

Mullins, L.J. (2007). Management and Organisational behaviour. Eighth edition. Prentice Hall.

Panwar, S. & Gupta, S. (2012). Money: Does it really affect hotel employee's performance as a motivational factor!!!. VSRD International *Journal of Business & Management Research*, 2 (2), 38–46.

Pavelka, T. (2012). The Minimum Wage in the Czech Republic—the instrument for motivation to work? The 6th International Days of Statistics and Economins, Prague, September 13–15, 2012. Retrieved April 29, 2013 from http://www.msed.vse.cz/files/2012/Pavelka_2012.pdf.

Re'em, Y. (2010). Motivating public sector employees: an application-oriented analysis of possibilities and practical tools. Herti School of Governance—Working Papers, No 60, July 2011. Retrieved April 25, 2013 from http://www.hertie-school.org/fleadmin/images/Downloads/working_papers/60.pdf.

Salanova, A. & Kirmanen, S. (2010). Employee Satisfaction and work motivation. Retrieved May 13, 2013 from http://www.publications.theseus.fi/bitstream/handle/10024/10979/Salanova_Anna_Kirmanen_Sanni.pdf?sequense=1.

Siira, H.J. (2012). Employee motivation and its effects on job performance. Turku University of Applied Sciences. Retrieved May 15, 2013 from http://publications.theseus.fi/bitstream/handle/10024/46767/Siira_Heidi.pdf?sequence=1.

Spector, P.E. (1997). Job satisfaction: application, assessment, cause, and consequences. USA. SAGE Publications, Inc.

Stringer, C., Didham, J., & Theivanathampillai, P. (2011). Motivation, pay satisfaction, and job satisfaction of front-line employees. *Qualitative Research in Accounting & Management*, 8 (2), 161–179.

Swaffield, J.K. (2012). Minimum wage hikes and the wage growth of low-wage workers. Discussion papers in economics. No 12/10. Retrieved April 23, 2013 from http://www.york.ac.uk/media/economics/documents/discussionpapers/2012/1210.pdf.

Tang, T.L.P., Luna-Arocas, R., Sutarso, T., & Tang, D.S.H. (2004). Does the love of money moderate and mediate the income-pay satisfaction relationship? *Journal of Managerial Psychology*, 2, 111–135.

Ukandu, N.E. & Ukpere, W.I. (2011). Strategies to improve the level of employee motivation in the fast food outlets in Cape Town, South Africa.

Hospitality and Tourism – Sumarjan et al. (eds)

Tourist information center service quality: A case of Malaysia Tourism Centre (MaTiC)

M.I. Izham, N. Baba & M.H. Hanafiah
Universiti Teknologi MARA, Malaysia

ABSTRACT: This study aims to assess the visitor's perception on the tourism information centre quality standard in Malaysia. It is a challenge to MaTiC in developing a high standard service quality and thus sustaining the level of consistency. The study's outcomes obtained through surveys on visitor's satisfaction. Service quality considered as a philosophy for any tourism information centre and tourism related services. As the Malaysian tourism appears to have suffered mainly due to inadequate information and promoting activities, it is necessary to determine the strategic customer service approaches of tourism information centre. This study provides useful information for tourism service quality improvement. MaTiC should consider other possible way to use in delivering the information and develop a methodical customer service coordination.

Keywords: service quality, tourist information centre, satisfaction

1 INTRODUCTION

Travel abroad has become a new trend among tourist mostly from the Middle East, Europe and South East Asia. The Malaysia tourism report showed that the numbers of tourist arrival in 2011 are 24,714,324 which are 0.6% growth than tourist arrival in 2010 which are 24 million (Tourism Malaysia, 2011). According to Hanafiah & Harun (2010), many countries strive to develop the tourism sector and increase the number of incoming visitors because of several reasons: international tourists bring foreign currency to the host country; tourism sector is much more merciful toward the environment than many industries producing goods, and growth in the tourism industry accelerates other related industries (retail, entertainment and transportation).

Psarrors (2009) classify a Tourist Information Center (TIC) as a physical location that provides tourist information to the visitors who tour the place or area locally. The objectives of tourist information center are to provide quality information to visitors and identify additional ways to distribute information in order to reach wider audiences. In Malaysia, one of the responsible agencies to disseminate the tourism is Malaysia Tourism Centre (MaTiC). MaTiC aims at promoting tourism-related activities organized by the Ministry as well as other state-run activities through its information counter, tourism websites, printed materials and the electronic board. Furthermore, MaTiC also manages events such as cultural and tourism-related events through shows, exhibitions, briefings and demonstrations to attract local and foreign tourist.

Globally, hospitality organizations attempt to assess their visitor satisfaction and explore practical tools for developing and maintaining a high standard of service quality and guest satisfaction. Thus, it creates a challenge to MaTiC in developing a high standard service quality and thus sustaining the level of consistency. Consistency is one of the important matters in delivery service and can directly affect the level of perceived service and thus customer satisfaction (Bashirian & Majdpour, 2011).

The tourism industry in Malaysia has succeeded to undertake reasonable and effective promotional activities due to positive policy guidelines, resource availability, and exclusive marketing orientation. Therefore, to maintain consistency, the front staff is the first point of contact that is supposed to set a positive tone on the perception of the entire TIC operation. Thus, developing an adequate image of TIC through a standard service level is one of the main challenges of TIC operation.

With the widespread stiff competition, intangible nature of TIC services and demand of prepurchase information for prospective tourists, promotional measures in the tourism industry play a key role in the performance and progress of this industry. As Malaysia tourism appears to have suffered mainly due to inadequate information and promoting activities, it is necessary to identify the strategic customer service approaches of TIC and to find out the remedies to overcome the problems or limitations identified.

2 LITERATURE REVIEW

2.1 *Service quality in delivering information*

The subject of service quality is immensely rich in the context of definitions, models and measurement issue (Seth & Deshmukh, 2004). Hence, service quality enable management to identify quality problems and thus help in planning for the launch of a quality improvement program thereby improving the efficiency, profitability and overall performance. Therefore, the tourist information centre should give the best services when dealing with tourist. Zeithaml & Bitner (2003) defined perceived service quality as a global judgment or attitude relating to the superiority of a service. It is widely accepted today that service quality is a multidimensional concept. However, Kothari & Sharma (2009) argued that measuring service quality is extremely challenging task for any service industry. Hersh (2010) said that the service quality dimension is a useful management tool since it aims to identify the gaps between customer's expectations and customers' perceptions of the services. The measurement of visitor's perceptions and expectations has been a disputable issue in the literature.

Moreover, Hersh (2010) mentions that owing to the customer-oriented service endeavors, tourism enterprises, either private or public, need to improve service offerings by determining the needs of their target groups. Exploring the current ratings of customer expectations and customer perceptions on service attributes provides a tool for managers in order to improve the service quality of the firm. Since the development of SERVQUAL by Parasuraman, Zeithaml, & Berry (1985), service quality has been widely researched and applied in different types of industries. Moreover, service quality is the consumers' consideration to the perceived quality, instead of their satisfaction toward an incident or conduct. In SERVQUAL multiple-item scale, five dimensions are identified to measure service quality, including tangibles, reliability, responsiveness, assurance, and empathy.

2.2 *Visitor satisfaction*

Johnston (1995) revealed that service quality satisfaction determinants are being attentiveness, responsiveness, care and affection. Meanwhile, the main sources of dissatisfaction are integrity, reliability, responsiveness, availability and functionality. Johnston (1997) illustrates that certain actions, such as increasing the speed of processing information and customers, are likely to have a significant effect in terms of pleasing customers; however other activities, such as improving the reliability of equipment, will lessen dissatisfaction rather than delight customers and suggests that it is more valuable to ensure that the dissatisfaction are dealt with. It appears to be a consensus emerging that satisfaction refers to the outcome of personalized service transactions and the overall service encounter, whereas service quality is the customer's overall impression of the relative inferiority/superiority of the organization and its services (Johnston, 1995). In fact, some empirical studies show that the two constructs are conceptually distinct (Boulding, Kalra, Staelin, & Zeithaml, 1993). The idea of linking service quality with customer satisfaction has existed for a long time.

Comparing on the two related constructs, Parasuraman, Zeithaml, & Berry (1988) conceptualized perceived service quality as a long-run overall evaluation about a service, whereas satisfaction is a transaction-specific evaluation. Based on these conceptualizations, they posited that incident of satisfaction over time result in perceptions of service quality. Other researchers supported the argument that customer satisfaction leads to service quality. For example, Bitner (1990) developed a model of service encounter analysis and empirically supported the effect of satisfaction on service quality. Bolton & Drew (1991) also proposed that satisfaction leads to service quality.

In addition, Lee et al. (2000) discuss on the contrast to this perspective, some other researchers argued and empirically supported that perceived service quality is a predecessor of customer satisfaction. Cronin & Taylor (1992) reported that overall service quality, and purchase intention, the coefficients of path for service quality satisfaction purchase intention appeared to be all significant while the coefficients of path to satisfaction—service quality—purchase intention were insignificant. Spreng & MacKoy (1996) also studied the relationship between service quality and satisfaction based on their modified Oliver's (1993) satisfaction/service quality design.

Broderick & Vachirapornpuk (2002) found that the extent and type of customer participation had the greatest impact on the quality of the service experience and issues such as customers' zone of tolerance, the degree of understanding and emotional response potentially determined, expected and perceived service quality. Similarly,

Lassar, Manolis, & Winsor (2000), demonstrated that a functional-quality based model did a better job of predicting customer satisfaction than a SERVQUAL instrument for those customers actively involved or particularly involved in service delivery. Thus, Churchill & Suprenant (1982), Cronin and Taylor (1992) & Kozak (2001) indicate that service quality performance could be an appropriate procedure to evaluate customer satisfaction because the customer is likely to be satisfied when a product or service performance is at a desired level.

3 METHOD

A descriptive research method used for this study. The sampling frame for this paper is convenience. Marshall (1996) describes the convenience sampling as the least rigorous approach, involving the selection of the most accessible subjects. Research Division of Tourism Malaysia (2012) reported that the total MaTiC visitors in 2011 is 178,552 visitors. According to Krejcie, & Morgan, (1970) & Sekaran (2003), it is suggested that the sample size on 15,000 populations is 250 units. This study adapted existing scale for measuring the variables such as information on counter service quality and satisfaction on information counter service quality. The data were analyzed by Statistical Package for the Social Sciences (SPSS).

4 RESULTS

218 respondents are able to participate in the study with 132 of them were foreigners. The respondents who participated in the survey come from various countries such as Indonesia, Finland, UK, Belgium, USA, Japan, Spain, Pakistan and Malaysia itself. Descriptive statistic used to assess the effectiveness of information counter service quality in delivering the information. Respondents were asked to provide ratings based on Five-point Likert Scale, in which one (1) indicated "Strongly Disagree" and seven (5) indicated "Strongly Agree".

Table 1. Variables measurement.

Variables	Measurement	Sources
Information counter service quality	Likert scale (1–5) From 1 = "strongly disagree" to 5 = "strongly agree"	Kothari and Sharma (2009)
Satisfaction on information counter service quality	Likert scale (1–5) From 1 "strongly dissatisfied" to 5 = "strongly satisfied"	Dmitrović et al. (2009)

4.1 *Information counter service quality*

The Table 2 shows that tangibility dimension positively perceived by the visitors (m = 4.18) followed by the assurance dimension (m = 4.15), reliability dimension (m = 4.13), responsiveness (m = 4.12) and empathy dimension (m = 4.08). This finding supported by Hersh (2010) where he argued that the service quality aspect is a valuable management tool since it aims to identify the gaps between customer's expectations and customers' perceptions of the services. The range of perceptions and expectations has been a contentious issue in the literature. Identifying the gaps is the best way to determine quality, identify potential problems, and predict loyalty.

4.2 *Satisfaction on information counter service quality level*

The level of satisfaction on information counter offered at the Malaysia Tourism Centre (MaTiC) is pleasing (m = 4.09). The highest satisfaction is on the service offered by this organization and the information given by the staff (m = 4.11). Further, they were satisfied with the information given by the counter staff helped in planning their trip (m = 4.10). This is in line with Dmitrović T., Cvelbar L. K., Kolar T., Brenčič M.M, Ograjenšek I. & Žabkar V. (2009) who insist on the necessity of recognizing and understanding the cognitive and behavioral consequences of satisfaction has serious implications for destination management. Moreover, visitor satisfaction has a vital role in planning marketable tourism products and services for destinations and its evaluation must be a critical framework used to evaluate the performance of destination products and services (Yoon and Uysal, 2005).

Table 2. Mean scores for information counter service quality.

No.	Dimensions	Means	S.D
1.	Tangibles	4.18	0.69
2.	Reliability	4.13	0.729
3.	Responsiveness	4.12	0.708
4.	Assurance	4.15	0.688
5.	Empathy	4.08	0.715
	Overall mean score	4.13	0.706

Table 3. Mean scores for the level of satisfaction on tourist information counter.

No.	Tourist satisfaction	Mean	S.D
1.	The information given by the staff helps me to make my travel choice.	4.11	0.714
2.	I am satisfied with the service offered by the information counter staff.	4.11	0.690
3.	The information counter staff helps me in planning my trip.	4.10	0.707
4.	The information given meets my requirement.	4.03	0.727
5.	Overall, I am satisfied with the service offered by this organization.	4.11	0.716
	Overall mean score	4.09	0.710

Table 4. Regression analysis on the relationship between service quality and tourist satisfaction.

Predictors	Model 1
Dependent variable: Satisfaction	
Service quality	0.925*
R^2	0.683
Adj. R^2	0.681
F	465.195*

Note: *$p < 0.05$.

4.3 *Analysis of the relationship between Malaysia Tourism Centre (MaTiC) service quality and tourist satisfaction*

Regression analysis was conducted to verify the relationship between service quality and visitor satisfaction. The Table 4 summarizes the outcomes of the analysis.

The above result interpretation can be explained ($R^2 = 0.683$, Adj. $R^2 = 0.681$, $F = 465.195$, $p < 0.05$). The results showed that there is a strong significance relationship between MaTiC service quality and visitor satisfaction. MaTiC service quality ($\beta = 0.925$, $p < 0.05$) shows a significant positive impact on visitor satisfaction. The hypothesis is strongly supported. Hence, the researchers reject the null hypothesis and accept the alternate hypothesis.

Visitor satisfaction refers to the outcome of personalized service transactions and the overall service encounter, whereas service quality is the customer's overall impression of the relative inferiority/superiority of the organization and its services (Johnston, 1995). In fact, some empirical studies show that the two constructs are conceptually distinct (Boulding et al., 1993). The idea of linking service quality with customer satisfaction has existed for a long time.

5 CONCLUSION

In order to maintain their level of satisfaction, the service quality should be improved from time to time. Moreover, the study helps to identify all the lacking factors that can develop towards disapproval of tourist. By analyzing and understanding the findings, Malaysia Tourism Centre (MaTiC) may improve their performance and maintain the quality of the services offered. This is advisable to be practiced to ensure the visitor assess the right information from valid resources and may motivate them to travel which trigger national economy in tourism industries. MaTiC should consider other possible way to use in delivering the information. Now days, most of the tourists prefer to browse the information through the internet and online software. Therefore, MaTiC should develop a customer service system that the visitor can use to communicate virtually. The examples of virtual communication are such Yahoo Messenger, Skype and Facebook. By developing its own virtual communication, MaTiC can directly promote Malaysia in the eyes of the world.

ACKNOWLEDGEMENTS

The work described in this study was funded by the Research Intensive Funds (RIF), Universiti Teknologi MARA Malaysia.

REFERENCES

Bashirian N. & Majdpour T. (2011). "The role of standard service quality in front office of five star hotels in Tehran in order to satisfy customers". Luleå University of Technology Department of Business Administration, Technology and Social Sciences.

Bitner, M.J. (1990). "Evaluating Service Encounters: The effect of physical surroundings and employees responses". Journal of Marketing. Vol. 54 (April 1990). Pp 69–82.

Boulding, W., Kalra, A., Staelin, R. & Zeithaml, V.A. (1993). "A dynamic process model of service quality: from expectations to behavioral intentions", Journal of Marketing Research, Vol. 30, February, pp. 7–27.

Broderick, A.J. & Vachirapornpuk, S. (2002). "Service quality in internet banking: the importance of customer role", Marketing Intelligence & Planning, Vol. 20 No. 6, pp. 327–335.

Churchill, G.A. & Suprenant, C. (1992). An Investigation into the Determinant of Customer Satisfaction. Journal of Marketing Research, Vol. 19, pp. 491–504.

Dmitrović T., Cvelbar L.K., Kolar T., Brenčič M.M, Ograjenšek I. & Žabkar V. (2009). "Conceptualizing tourist satisfaction at the destination level". International Journal Of Culture, Tourism And Hospitality Research. Vol. 3 No. 2, 2009, pp. 116–126.

Hanafiah, M.H. & Harun. M.F (2010). "Tourism Demand in Malaysia: A cross-sectional pool time-series analysis". International Journal of Trade, Economics and Finance, Vol. 1, No. 1, June, pp. 80–83.

Hersh (2010). "Evaluate the impact of tourism service quality on customer satisfaction". Institute of Interdisciplinary Business Research. October 2010. Vol. 2, No. 6.

Johnston, R. (1997). "Identifying the critical determinants of service quality in retail banking: importance and effect", The International Journal of Bank Marketing, Vol. 15 No. 4, pp. 111–116.

Johnston, R. (1995). "The determinants of service quality: satisfiers and dissatisfiers" International Journal of Service Industry Management, Vol. 6 No. 5, pp. 53–71.

Kothari & Sharma (2009). Defining and Measuring the Effect of Service Quality in Selection of a Mutual Fund in Indian Context. Journal of Services Research, Vol. 9, No. 2 (October 2009–March 2010).

Kozak, M. (2001). "Comparative assessment of tourist satisfaction with destinations across two nationalities", Tourism Management, Vol. 22 No. 4, pp. 391–401.

Krejcie, R., & Morgan, D. (1970). Determining sample size for research activities. Educational and Psychological Measurement, 30, 607–610.

Lassar, W.M., Manolis, C. & Winsor, R.D. (2000). "Service quality perspectives and satisfaction in private banking", International Journal of Bank Marketing, Vol. 18 No. 4, pp. 244–271.

Lee H., Lee Y., & Yoo D., (2000). "The determinants of perceived service quality and its relationship with satisfaction", Journal of Services Marketing, Vol. 14 Iss. 3, pp. 217–231.

Malaysia Tourism Centre (MaTiC) (2011). "Malaysia Tourism Centre objective and function". Retrieved on Mei, 2011 from world wide web http://www.matic.gov.my/en/about-matic/corporate-information/objective-function.html.

Parasuraman, A., Zeithaml, V.A. & Berry, L.L. (1985). "A Conceptual Model of Service Quality and its Implications for Future Research", Journal of Marketing, Vol. 49, Fall, pp. 41–50.

Parasuraman, A., Zeithaml, V.A., & Berry, L.L. (1988). SERVQUAL: A multiple-item scale for measuring consumer perceptions of service quality. Journal of Retailing. 64 (1): 12–40.

Psarrors. (2009). "Tourist Information Centers as a vital component for providing quality visitors services. The case of & Manchester, UK, Cape Town, SA & Athens, GR", Destination Development & Marketing Consultant.

Research Division of Tourism Malaysia (2012). "Malaysia Tourist Arrival Data by 2010–2011". Retrieved on February, 2012 from world wide web http://corporate.tourism.gov.my/images/research/pdf/2011/TouristArrivals_JanDec_2011.pdf.

Sekaran (2003). "Research methods for business. A skill building approach". John Wiley & Sons.

Seth & Deshmukh (2004). Service quality models: a review, International Journal of Quality & Reliability Management Vol. 22 No. 9, 2005 pp. 913–949.

Spreng, R.A. & MacKoy, R.D. (1996). "An empirical examination of a model of perceived service quality and satisfaction", Journal of Retailing, Vol. 72 No. 2, pp. 201–214.

Yoon, Y., & Uysal, M. (2005). An examination of the effects of motivation and satisfaction on destination loyalty: a structural model. *Tourism Management*, 26 (1), 45–56.

Zeithaml, V.A., & Bitner, M.J. (2003). Services Marketing: Integrating Customer Focus Across the Firm. New York: McGraw-Hill.

Hospitality and tourism marketing

Hospitality and Tourism – Sumarjan et al. (eds)
© 2014 Taylor & Francis Group, London, ISBN 978-1-138-00151-0

Uptown night market and tourists' satisfaction

N. Kamaruddin & M.F.S. Bakhtiar
Universiti Teknologi MARA, Malaysia

ABSTRACT: Night markets are popular destination among local communities and domestic tourists typical to Malaysian way of life. The vibrant surrounding of night market together with its variety and affordable products available is making it a must-visit destination by locals. Through understanding of these unique night market's attributes, tourist's satisfaction can be assessed to determine the viability of transforming night markets into a popular destination not just for local as well as international tourists. Through quantitative research approach, this paper examined tourist's satisfaction of Uptown night markets. In accessing the outcome, self reported questionnaire was applied and 384 respondents were gathered using convenience sampling technique. The findings of this study are hopeful to contribute an in-depth understanding about tourist's satisfaction toward Uptown Night Market attributes and determining the opportunities and constrains in exposing Uptown night markets as one of the tourism destinations in Malaysia.

Keywords: behavioral, tourist's satisfaction, Uptown night market

1 INTRODUCTION

Uptown night markets in Malaysia are popular nighttime destination among locals offering the excitement of 'strolling, shopping and eating' all at the same place (Rohaizah and Roni, 2011). This concept of nighttime destination has existed long enough and continues to be perceived by locals offering variety of products, convenient to access and affordable (Khursiah and Yeng, 2011). Night markets were remarked as a way of integrating various ethnic groups in the country and aimed towards becoming a tourists' attraction destination since it provides a glimpse into the unique Malaysian culture and way of life (Kuala Lumpur Structure Plan 2020, 2000).

In the viewpoint to transform night market as a tourist attraction, Taiwan has been successful in implementing the idea (Nor, Khusiah and Rohaizah, 2011) as it is regarded to be one of the most popular tourism attractions in Taiwan (Taiwan Bureau, 2006). Nonetheless, in Malaysia Uptown night markets are only popular among the locals and domestic tourist; it is uncommonto international tourists (Khursiah and Yeng, 2011). Hsieh and Chang (2006) pointed out that by understanding the attributes of night markets, tourist experience and satisfaction about night markets can be analyzed hence the result may be used as a benchmark to indicate the viability of transforming this spot as a popular local and international tourist destination. In addition, it is also beneficial to study the aspect influencing the strengths and weaknesses of the destination (Ibrahim and Leng, 2003). Nor et al. (2011) mentioned that the success criteria of a destination would also depend on its ability to satisfy its customers, tourists as well as the local community. Therefore, this study pursue to the above importance by empirically valuing tourist's satisfaction of Uptown night markets.

2 LITERATURE REVIEW

2.1 *Night market*

Night markets are generally known as Pasar Malam in Malaysia and Indonesia. "Pasar malam" is a Malay and Indonesian word that literally means night market, "pasar" being related to "bazaar" in Persian (Nor, Khusiah and Rohaizah, 2011). Meanwhile, Lee, Chang, Hou and Lin, (2008) defined night market as street markets operating during night time mainly in urban and suburban areas that generally tend to have more leisure, shopping and eating areas. Night markets in Malaysia are a common destination among locals and it can be found in almost all small towns and cities in every state (Khursiah and Yeng, 2011). Additionally, (Huang, Liau and Tzeng, 2009), define night market as a trading place comprising of small businesses gathered together during evening and offers a variety of products and food to visitors.

Abu (2004) claimed Uptown night markets to be a modernization of the night market activities which has now become a symbol of the lively urban lifestyle in Malaysia. Abu (2004) noted that prices of certain products may be relatively higher as compared to traditional night markets, still Uptown night markets were preferred by visitors due to its overall attributes including the variety of products and food offered, physical settings and facilities available. Apparently, Nor et al. (2011) stated that the night markets environment and physical settings could also influence visitor's preference towards specific night market.

2.2 *Tourist satisfaction*

Satisfaction in regards to this subject, describes the level of enjoyment or disappointment, originating from expectations of the product (Kotler, 1994). More precisely satisfaction defines the level of a person's feeling of pleasure resulting from comparing a product or a service outcome in relation to his or her expectation (Kotler, 1994). Meanwhile, Ai Cham (2011) mentioned that satisfaction is generally seen as post-purchase construct that is related to how much a consumer likes or dislikes a service or product after purchasing and experiencing it. From a tourism perspective, satisfaction is primarily referred to as a function of pre-travel expectations and post travel experiences (Chi and Qu, 2008).

When experiences are compatible to expectations it results in the feeling of gratification, the tourist is then satisfied and leaves the destination with good memories. However, when a tourist feels displeasure, it means that the tourist is dissatisfied (Reisinger and Turner, 2003). As claimed by Baker and Crompton (2000), the higher a tourist's satisfaction increases, the more it affects their behavioral intention.

3 METHODOLOGY

In accessing tourists' satisfaction of Uptown night market, a quantitative research approach was employed using self administered questionnaire. Domestic tourists who visited two Uptown night markets in Klang Valley area (Shah Alam and Kota Damansara) were chosen for data collection using convenience sampling technique. A total of 384 valid respondents were gathered and analyzed using descriptive analysis approach.

4 RESULT

4.1 *Uptown night market attributes*

Before commencing into statistical analysis, all data were checked for irregularities, missing item, or unrealistic responses. In response to the objective of the study, mean score of four Uptown night markets attributes relating to products and food, night market atmosphere, night market physical settings and public facilities were undertaken and shown in multiple tables.

A review of Table 1 showed that majority of respondents were pleased with the choices of products and food offered at Uptown night markets ($M = 4.22$). Meanwhile, respondents were also satisfied with the price of products and food offered ($M = 4.04$). Moving to quality of products and food, both were perceived moderately, ($M = 3.80$) and ($M = 3.67$) respectively with minimal inclination towards satisfaction. Likewise, most of respondents rated the condition of products and food displays, labeling, appearances as well as packaging moderately with mean score between ($M = 3.57$) to ($M = 3.53$).

Table 2 depicted the descriptive summary of eight items of night market environment. Findings indicated that most respondents perceive the scenery of Uptown night markets moderately with the

Table 1. Attribute of Uptown night market (product and food).

Product and food		
Item	(M)	SD
Choices of products and food	4.22	0.604
Price of product and food	4.04	0.637
Quality of products	3.80	0.635
Taste of food	3.67	0.625
Product and food display	3.57	0.631
Product and food labeling	3.57	0.626
Product and food appearance	3.54	0.638
Product packaging	3.53	0.604

Scale: 1 = Strongly dissatisfied, 2 = Dissatisfied, 3 = Moderate, 4 = Satisfied, 5 = Totally satisfied.

Table 2. Attribute of Uptown night market (environment).

Night market environment		
Item	(M)	SD
Scenery of the area	3.76	0.596
Friendliness of traders	3.54	0.810
Noise	3.16	0.519
Safety	3.04	0.657
Relaxing	3.00	0.576
Crowd	2.73	0.599
Cleanliness	2.72	0.584
Traffic flow	2.68	0.583

Scale: 1 = Strongly dissatisfied, 2 = Dissatisfied, 3 = Moderate, 4 = Satisfied, 5 = Totally satisfied.

mean score of 3.76. Same result is achieved for friendliness of traders with ($M = 3.54$). Contrarily, they were minimally dissatisfied with the noise of surrounding, safety elements and relaxing facilities at Uptown night markets. It is indicated through lower mean score for the items between ($M = 3.16$) to ($M = 3.00$). Next, they were clearly dissatisfied with the crowd level, cleanliness as well as the traffic flow at Uptown night market. Their dissatisfaction was reflected through even lower mean score achieved for those items.

Table 3 consists of three items to measure night market physical settings. Most respondents rated a moderate score the entrance and exit point with mean score of ($M = 3.83$), Moving to the size of shop area and its arrangement, both were also perceived moderately. Despite the average moderate score, an in-depth review shows that respondents were still inclined toward satisfaction with the overall physical settings of Uptown night market.

Table 4 provides a descriptive summary of six items measuring night market's public facility. None of the facilities offered reach a total satisfactory score. Only the parking space was rated moderately with slight inclination towards satisfaction. The rest was perceived unsatisfactory. As such, rest area, rest room, trash cans and public

Table 3. Attributes of Uptown night market (physical setting).

Night market physical setting		
Item	(M)	SD
Overall night market layout (entrance and exit point)	3.83	0.665
Size of shop area	3.56	0.675
Shop arrangement	3.53	0.604

Scale: 1 = Strongly dissatisfied, 2 = Dissatisfied, 3 = Moderate, 4 = Satisfied, 5 = Totally satisfied.

Table 4. Attributes of Uptown night market (public facilities).

Public facilities		
Item	(M)	SD
Parking space	3.62	0.727
Sign age	3.09	0.547
Rest area	2.95	0.573
Rest room	2.89	0.539
Trash cans	2.84	0.547
Public transportation	2.78	0.571

Scale: 1 = Strongly dissatisfied, 2 = Dissatisfied, 3 = Moderate, 4 = Satisfied, 5 = Totally satisfied.

Table 5. Overall mean score.

Variable	(M)	SD
Product and food	3.85	0.430
Night market environment	3.08	0.362
Physical setting	3.64	0.526
Public facilities	2.61	0.332
Overall attributes	3.38	0.316

Scale: 1 = Strongly dissatisfied, 2 = Dissatisfied, 3 = Moderate, 4 = Satisfied, 5 = Totally satisfied.

transportation at night market were rated dissatisfactory between ($M = 2.78$) and ($M = 2.95$).

Table 5 details the total mean score of all four Uptown night markets attributes. The mean score for products and food offered in Uptown night markets was ($M = 3.85$). Hence it is indicated that this attribute met their minimal expectation. Another attribute with slightly lower score but at minimal satisfaction score was the physical setting with ($M = 3.64$). Night market environment with ($M = 3.08$) revealed that majority of respondents felt that this attribute did not meet their satisfaction. The lowest score was achieved for public facilities provided with ($M = 2.61$), it shows that the condition of public facilities provided in Uptown night markets would need the most attention for improvement. Since the overall mean score of all the attributes of Uptown night markets was ($M = 3.38$), it can be concluded that all attributes of Uptown night market would require improvement to better satisfy visitors or tourists with public facilities needing the most attention. On the contrary, products and food offered are important elements that may encourage patronage to Uptown night market.

4.2 *Tourists' satisfaction*

Table 6 depicted the mean score of six items measuring tourists' satisfaction. The findings revealed that most respondents were satisfied with the overall Uptown night market attributes which has offered them a good overall shopping experience with mean score ($M = 4.05$). They also enjoyed their visit and inclined to havea positive feeling towards Uptown night market ($M = 3.97$). In the meantime, most respondents were slightly satisfied with their decision to visit and shopping at Uptown night market ($M = 3.90$), they thought that it was a pleasant visit ($M = 3.85$) while exceeding their minimal expectation ($M = 3.80$).

The table above clearly revealed that most of the respondents were satisfied with their visit to Uptown night markets and agreed that Uptown night market has given them a good shopping and

Table 6. Tourists' satisfaction.

Item	(M)	SD
Overall Uptown night market has given a good shopping experience (quality of product and food, the environment, physical setting and public facilities)	4.05	0.611
I really enjoyed the visit at Uptown night market	3.97	0.635
I have a positive feeling regarding Uptown night market	3.97	0.535
I am satisfied with my decision to visit and shopping at Uptown night market	3.90	0.579
This was a pleasant visit to Uptown night market	3.85	0.500
The visit to Uptown night market exceeded my expectations	3.80	0.561

Scale: 1 = Strongly dissatisfied, 2 = Dissatisfied, 3 = Moderate, 4 = Satisfied, 5 = Totally satisfied.

Table 7. Overall mean score of Tourists' satisfaction.

Variable	(M)	SD
Tourists' satisfaction	4.0321	0.38760

Scale: 1 = Strongly dissatisfied, 2 = Dissatisfied, 3 = Moderate, 4 = Satisfied, 5 = Totally satisfied.

visit experience. This is due to the variety choices of products and food offered in Uptown night markets at a reasonable price with a comfortable environment and calculated physical settings at the Uptown night markets, which led to the satisfaction of respondents.

5 DISCUSSION

From the analysis on the attributes of Uptown night markets, it is interesting to note that most of the respondents revealed that the products and food as well as the price as one of the important attributes that encouraged them to visit Uptown night markets. This empirical finding was supported by Rohaizah and Roni (2011) who stated that the variety of products and food offered in night markets at a reasonable price influenced the visitors to patronize and shop at night markets. Meanwhile, an analysis on Uptown night markets environmental attributes revealed that most of the respondents had a moderate overall satisfaction towards the scenery, arrangement of the shops, landscape and physical atmosphere at Uptown night markets.

Besides, most of the respondents claimed that they were slightly satisfied with the friendliness of the traders, noise level, safety and rest area at Uptown night markets. Unfortunately, most of them expressed that they were not pleased with the cleanliness, crowd level as well as the traffic flow. This finding was also supported by Rohaizah and Roni (2011) who mentioned that the traffic flow and crowded situation in night markets are critical during peak hours, so it is recommended that additional parking area or at least official traffic controller may be available to counter the situation.

Next, the analysis of Uptown night market physical settings shows that most of respondents were satisfied and pleased, especially with the entrance and exit points as well as the size of shop areas and arrangements. The final attribute analyzed was the public facilities; it shows that most respondents were slightly satisfied with the parking space and sign age provided at the Uptown night market area. Contrarily, most respondents expressed their dissatisfaction with rest rooms condition, trash cans and public transportation provided. Therefore, it is important to provide sufficient and functional public facilities including rest room, seating area, and covered rooftop surroundings. For conclusion, products and food were considered the most important attributes encouraging visit to Uptown night market, followed with the physical settings, night market atmosphere and finally public facilities available.

Moving into analyses on tourists' satisfaction, it was discovered that most tourists were generally satisfied with their visit even though several Uptown night market attributes were perceived poorly; this was discussed in previous section. They also agreed to having good shopping and visit experience; hence they perceived this destination positively. For them, visiting Uptown night market was a good decision; it was a pleasant visit and overall exceeding their minimal expectation. From the above facts, there is a potential to transform Uptown night market into a popular destination for both local and international tourists once the main attributes are addressed significantly.

REFERENCES

Abu, N.K. (2004). Service quality dimension: a study on various sizes of grocery retailers a conceptual paper. *Transportration Research Record, 2*, 12–20.

Ai Cam, T.T. (2011). Explaining Tourists Satisfaction and Intention to Revisit Nha Trang, Vietnam. *Journal of Travel Research, 85*, 15–45.

Anderson, W. & Sullivan, M. (1998). The antecedents and consequences of customer satisfaction for firms. *Journal of Marketing Science, 12*, 125–143.

Baker, D. & Crompton, J. (2000). Quality, Satisfaction and Behavioural Intention. *Annal of Tourism Research, 27*, 785–804.

Chi, C. & Qu, H. (2008). Examining the structural relation ship of destination image, tourist satisfaction and destination loyalty: an integrated approach. *Journal of Tourism Management, 29*, 624–63.

Hsieh, A-T. & Chang, J. (2006). Shopping and tourist night market in Taiwan. *Tourism Management, 27*, 138–145.

Huang, S.O., Liou, Y.H., & Tzeng, G.H. (2009). Development strategies for improving the service of tourist night markets through Hybird MCDM technique. *International Journal of Culture, Tourism and Hospitality Research*.

Ibrahim, M.F., & Leng, S.K. (2003). Shoppers' perceptions of retail developments: Suburban shopping centres and night markets in Singapore. *Journal of Retail & Leisure Property, 3*, 176–189.

Khursiah, A.A. & Avini, S. (2011). Customer buying behaviour at night market. *Journal of Contemporary Research in Business, 3*, 377–394.

Khursiah, A.A., & Yeng, L.W. (2011). Exploring the potential of a night market as a tourist attraction. *International Journal of Business and Social Science, 2*, 146–157.

Kuala Lumpur Structure Plan 2020 (2000). Percetakan Nasional Malaysia Berhad. Kuala Lumpur.

Kotler, P. (1994). Marketing Management: Analysis, Planning, Implementation and Control, 8th Edition. Englewood Cliffs NJ: Prentice-Hall International.

Lee, S-H., Chang, S-C., Hou, J-S., & Lin, C-H. (2008). Night market experience and image of temporary residents and foreign visitors. *International Journal of Culture, Tourism and Hospitality Research, 2*, 217–233.

Nor, K.I., Khusiah, A.A., & Rohaizah, A.A (2011). Typology of night markets in Malaysia. *Journal of Case Research in Business and Economics*.

Rohaizah, A.L., & Roni, H. (2011). Understanding the night market's environment in Selangor, Malaysia. *Interdisciplinary Journal of Contemporary Research in Business, 3*, 668–682.

Reisinger, Y. & Turner, L.W. (2003). Cross-cultural behavior in tourism: Concepts and analysis. Oxford: Butterworth-Heinemann.

Taiwan Tourism Bureau, (2006). 2005 Tourism Survey on Tourist Spending Trends and Satisfaction Levels. Retrieved from www.taiwan.net.tw/lan/index/index.asp.

Hospitality and Tourism – Sumarjan et al. (eds)
© 2014 Taylor & Francis Group, London, ISBN 978-1-138-00151-0

Craft souvenirs: Perceived authenticity and tourist purchase behaviour

F.I. Basri & M.F.S. Bakhtiar
Universiti Teknologi MARA, Malaysia

ABSTRACT: Malaysia craft industry is experiencing incredible growth over the years. Unfortunately, the increasing demand for craft souvenirs has resulted in imitation of craft products and even some imported from other countries. Due to this, tourists have misled the authentic craft products as they may think that it is locally made. Moreover, these products also have been sold together with the local products in the market. Through quantitative research approach, this paper empirically examined tourist's perceived authenticity of craft souvenirs and influenced their purchasing behaviour. Through self-reported questionnaire, 384 respondents were approached using the convenience sampling method. The result indicates that most of the respondents agreed on the uniqueness and originality as well as cultural and historic integrity were the most important attributes that portray the craft's authenticity. Overall, understanding the authenticity attributes of Malaysia's craft souvenirs is crucial as it influences tourists purchasing behaviour.

Keywords: authenticity, craft souvenirs, originality, purchase behaviour

1 INTRODUCTION

The development of Malaysia's craft industrial sector helps strengthen our economic growth and it is due to the rapid expansion of the Small and Medium scale Enterprise (SME) which includes in the craft industry (Awang Teh, 1998). In order to promote this particular local product, various craft centres including Craft Complex have been developed by local state governments for tourists that visit Malaysia. According to Bernama (2012), Malaysia's government has a strong vision and objectives in order to lead the development as well as the promotion of the competitive Malaysia crafts industry.

According to Kraftangan Malaysia (2012), the government vision for the upcoming year is to increase sales of local craft products in both domestic and international levels as well as to be the dynamic force in producing a competitive craft product in the market. Hence, a stable market demand and supply is one of the important factors that can lead to their vision being achieved. Based from the scenario, it is very important to know how the tourist's perceived authenticity of crafts souvenirs because it will influence their purchase behaviour. Anderson and Littrell (1995) explained in their studies that tourists not only investing time by shopping during their travel, but also spend approximately one-third of their total expenditure on retail purchase on crafts.

Beliefs about the craft product's features and its authenticity had an effect on the attitude towards their shopping experiences, which in return it will affect the purchase intention of the tourists (Yu and Littrell, 2003). Hence, it is important to examine how tourists perceive the authenticity of craft souvenirs found in the market. Moreover, it is worth determining whether tourists are still buying the product even if they know that it is not authentic and locally made. Therefore, this study particularly examined tourists perceived the authenticity of Malaysia's craft souvenirs attributes towards their purchasing behaviour.

2 LITERATURE REVIEW

2.1 *Uniqueness and originality*

A craft can be defined as an item that has been produced by artisans or craft maker, either completely by hand or with the help of tools and machines, as long as the direct manual contribution of the artisan remains the mainly significant component of the finished product (Chudasri, Walker and Evans, 2012). Craft souvenirs can be produced handmade or through the use of the machine as long as the manual contribution by the crafts person remains as the main components of the final product to (Swanson, 2004).

According to Etsa and Adi (2011), craft souvenirs are usually unique and different which it will represent the destination to where it belongs and also brings the artistic value and importance towards the travelling period. The uniqueness and originality are related to colours or design, availability of resources, unique and different from available product and whether it is copying, duplicating or mass production (Littrell, Anderson and Brown, 1993). For example, the uniqueness and originality can be seen from its one of a kind from the wood item and not look alike with other wooden crafts product.

Adamson (2010) also mentioned that a definition of craft originality is that a craft is almost always a subject of triangulation between craft sperson, the tool being used and material to make a craft souvenirs, moreover the hand that is being used might be considered as a tool, and there is no particular reason why any specific type of tool should be considered prohibited for this relation between the handmade crafts as long as it is not by a mass production and used a machine technology.

2.2 *Cultural and historic integrity*

Tourists are concerned about the cultural and historical value in a craft product. They are keen to know where the product come from are and what are the story behind the specific product. According to Reisenger and Steiner (2006), the crafts person's knowledge on historical value behind each product is undeniable and really valuable because it draws on an inferred and appreciation towards the craft work. The cultural and historical value is important in order to perceive authenticity and the pureness of a craft product.

Similarly, Etsa and Adi (2011) agreed that cultural and historical value is one of the important attributes on perceiving authenticity of the crafts. Based on their study, the result indicates that the tourists strongly agreed that the authenticities of craft products are closely related with the traditional culture that is related to the visited place and the story behind the product. Moreover, cultural and historical value can satisfy the tourists when buying the crafts product because they feel that each of the products have their own value that cannot be duplicated and buy from other places. The originality of the product belongs to the specific area or visited places and it is their place identity. Meanwhile, there are also other attributes that are affecting the tourist's understanding of craft's authenticity and their purchase behaviour.

2.3 *Material*

The material that was being selected and used in a craft making is very important in order to determine the quality of the products. According to Yu and Littrell (2003), most of the tourists believed that the authenticity of crafts product was based on the materials being used. The specialty and uniqueness of the crafts can be seen by the material chosen, the process in producing the craft and also the value of authenticity (Zulaikha and Brereton, 2011).

Craftsmen's usually used the best materials and exhibited fine workmanship and knowledge gained across a lifetime of training. In contrast, some tourists defined the authentic as from the used of indigenous materials, such as local woods, clays or grasses, however, the craft's form and content may be very contemporary (Littrell, Anderson and Brown, 1993). In short, only those crafts that convey indigenous material in form and content are eligible for consideration as authentic crafts.

2.4 *Price*

The price of the crafts product is closely related to the quality and competitiveness of the crafts seller and producers. Some of the price-sensitive tourists might be reluctant to buy the desire product due to the higher price in the market. Most of the price displayed is quite competitive and sometimes it does not value for money. As the craft products are ranging from a variety of materials and quality, the price also will be different according to the craft making process. The more difficult the crafts to be produced, the higher the price it would be.

According to Etsa and Adi (2011), some of the tourists that were price sensitive may be likely needed to sacrifice their intention to buy authentic craft souvenirs, as they preferred cheaper price and there is a possibility for tourists who love the crafts products will consider on buying even though it is expensive. As long as it is value for money and portray its own authenticity attributes.

2.5 *Perceived authenticity*

The authenticity of craft products can be perceived through its unique quality, originality and specialty of craft product. Craft souvenirs are related to tourist activities which part of their shopping experience, such as seeing a craft being made or demonstrated in a shop and they looked for unique and original products, colours, and designs (Littrell, Anderson and Brown, 1993). The qualities of craft products in some places where significance to tourists perceived authenticity based on the type of product. Meanwhile, due to the imitations products that can be purchased openly in the market, tourists are encouraged to aware of perceiving the authenticity of the craft products.

There are basically several motivators where tourists often extend their searches for authenticity in their purchase of craft souvenirs. However, although tourist demand for authentic craft souvenirs has often been mentioned in numerous researches, the meaning of craft souvenir authenticity has not been satisfactorily defined, and this has given the opportunity of the term to be used in different ways to imply different meanings.

2.6 *Tourist's purchase behaviour*

Tourist purchase behaviour can be defined as the actions a person act and takes towards buying and using the products and their decision making process that precedes and determine those actions (Choibamroong, 2006). Several studies have estimated that tourist spending on souvenir purchases can be counted for approximately one third of their total expenditure (Yu and Littrell, 2003).

According to Peck and Childers (2006), products with a specific nature and characteristics such as cultural elements and authentic value will tend to induce instantaneous on buying the product. Therefore, the tourists purchase behaviour is a key foundation of all marketing or promoting activity which is carried out to develop, promote, as well as to sell the tourism products.

3 METHODOLOGY

In accessing respondents' perceive about authenticity of Malaysia's craft souvenirs, a quantitative research approach was employed using self-administered questionnaire. International tourists at Craft Complex, Jalan Conlay, Kuala Lumpur were chosen for data collection using the convenience sampling technique. A total of 384 valid respondents was gathered and analysed using descriptive analysis approach.

4 FINDINGS AND DISCUSSION

4.1 *Demographic profile of the respondents*

Before commencing into statistical analysis, all data were checked for irregularities, missing item, or unrealistic responses. Frequency test was used against all demographic variables (gender, age, marital status, monthly income and country of origin). Out of the 384 respondents, 43.2 percent were male as opposed to 56.8 percent female. About half of the respondents which 41.1 percent were between 41 to 50 years old, followed by 33.1 percent between 41 to 50 years old and another 25.8 percent from 20 to 30 years old. Moving to their marital status, there are about 47.7 percent of the respondents was married and another 52.3 percent was single. Meanwhile, a review of respondents' income revealed that the majority (53.1%) respondent's monthly income were between USD 3,001 and USD 5,000. Followed by 37.2 percent of the respondent's income were between USD 1001 and USD 3000. The remaining percentage is 9.1 percent which their income below USD 1,000 and only 0.5 percent of respondent's income were USD 5,001 and above. Based on the result, a majority of the tourists earns above USD 3,000. The last demographic component analyses was respondents' country of origin. Based from the analysis, 80.2 percent respondents were from the Asia region (China, Singapore, Brunei, Vietnam, South Korea, India, Thailand and U.A.E) and another 19.8 percent from European region (United Kingdom). Hence, it can be generated that the findings of this study depict the Asian perspective of the subject matter.

4.2 *Overall analysis of craft attributes*

The analysis conducted clearly shows the mean value of the overall attributes on the craft souvenirs, uniqueness and originality, material, cultural and historic integrity as well as price (refer Table 1). The results have inclined towards agreement of all statements from the four attributes. However, the result indicates that the cultural and historic integrity has given the highest mean as the majority of the respondents agreed on the important of this attribute as compared to uniqueness and originality, material and price. Therefore, it can be concluded that this attribute were seen as contributing attributes to the authenticity of craft souvenirs.

The result was supported by Yu and Littrell (2003), whereby the researchers result findings were shown that the cultural and historic integrity has most influenced tourists to perceive the authenticity value of the craft products. Most of the tourists were concerned about the cultural and historic value particularly about the histories behind the craft product. Moreover, it was also supported by Etsa and Adi (2011) whereby as stated in their findings, most of the tourists agreed that the reasons they were attracted to the craft souvenir product was because of the unique stories and cultural portrays from each of the products.

Interestingly, the majority of tourists agreed that local's craft souvenir products are unique and

Table 1. Overall mean score of craft attributes.

Items	n	(M)	(SD)
Uniqueness and originality	384	4.0573	0.20671
Material	384	3.7847	0.61549
Cultural and historic integrity	384	4.0781	0.34526
Price	384	3.9512	0.60356

Scale: 1 = Strongly Disagree, 2 = Disagree, 3 = Neutral, 4 = Agree, 5 = Strongly Agree.

original as they were inclined towards agreement of this statement. Hence it can be summarized that Malaysia's craft products are generally unique and portray its authenticity attributes.

The result also indicated a moderate mean score of the price attributes which majority of tourists has agreed that the price factor also contributes to the craft souvenir authenticity. In other word, the reasonable price with a clear price tag and the provision of bargaining opportunities will gain tourists' attention and interests on purchasing the crafts product in the market.

Moreover, the mean score of material is slightly lower than other dimensions with minimal inclination towards agreement hence it can be concluded that the majority of tourists slightly agreed that the materials being used generally portray the authenticity of crafts product.

4.3 *Implications*

This study revealed that the cultural and historic integrity as well as uniqueness and originality are the most important elements on portrays the authenticity of crafts souvenirs. With such indicator, craftsmen or Small Medium Enterprise (SME) should produce better products that according to the tourists' needs. For the practical perspectives, this study provides the significant insight into the craft industries as well as Small Medium Enterprise (SME). The craft industry has given a huge contribution towards the country economies because the local craft products have gained interest from the international tourists. Therefore the individual or organization that involved in the craft industries should take more consideration in producing an authentic craft product up to the tourists' expectation. Therefore, any craftsman should follow tourist's expectation on local craft products especially become more sensitive with the prices offered and select the most quality and unique materials to maintain the authenticity of local craft products.

5 CONCLUSION

5.1 *Summary of the findings*

Result revealed that female outweighs male respondents because this is probably due to the fact that female tourists are generally more attracted to shopping for souvenir product as compared to the male tourists. Moreover, the middle aged group (31–50 years old) are more interested in buying craft souvenirs as compared to younger generation. Moving to their marital status, the percentage differences between single and married respondents are very small, it can be concluded that tourist's interest of buying craft souvenirs is not influenced by their marital status. The results also revealed that the majority of the respondents earns above USD 3,000, hence they can be categorized as tourists with high purchasing power.

Moreover, the result of the analyses revealed that the majority of the respondents agreed with the overall cultural and historic integrity as well as the uniqueness and originality attributes of Malaysia's craft souvenirs. As opposed to cultural and historic integrity as well as uniqueness and originality, respondents were only slightly agreed with the overall material used and price of Malaysia's craft souvenirs.

Based from the findings, it was clear that respondents were inclined to the agreement of the authenticity of Malaysia's craft souvenirs has attracted them to purchase the craft souvenirs. This study has achieved the objective specifically in determining tourists perception of the authenticity of craft attributes. In short, the study findings hoped to offer some opportunities to the craft industries in order to produce authentic craft products that meet tourists' expectations in the future, as well as in terms of promoting the local craft products towards internationally.

Thus, understanding the tourist's need is vital for the craft industries as to ensure tourist's interest in buying the craft products with regard to its authenticity attributes. Hence, this will not only improve tourists purchase decision but encourage repeat purchase behaviour and creating awareness towards Malaysia crafts products.

REFERENCES

Adamson, G. (2010). The craft reader. (Rev. ed.). O ford, U.K: Berg.

Anderson, L.F., & Littrell, M.A. (1995). Souvenir-purchase behavior of women tourists. *Annals of Tourism Research*, 22(2), 328–48.

Awang Teh, M. (1998). Usahawan kraftangan oleh bekas-bekas pelatih perbadanan kemajuan kraftangan malaysia. *Journal of Social Science,* 11–18.

Bernama (2012, April 3). Local arts and crafts reflect Malaysia's rich cultural legacy. Retrieved October 20, 2012, from http://www.theborneopost.comb/2012/04/03/local-arts-and-crafts-reflect-malaysias-rich-cultural-legacy.

Choibamroong, T. (2006). Knowledge of tourists' behavior: A key success factor for managers in tourism business. *International Journal of Tourism Research*, 1–8.

Chudasri, D., Walker, S., & Evans, M. (2012). An overview of the issues facing the craft industry and the potential for design, with a case study in Upper Northern Thailand. Paper presented at the DRS 2012 Bangkok Conference. Chulalongkorn University, Bangkok.

Etsa, A.S., & Adi, I. (2011). Outsourced souvenirs: An investigation towards authenticity anxiety and tourists purchase behavior. Proceeding of the International Conference on Social Science, Economics and Art 2011, 196–201.

Kraftangan Malaysia (2012). Malaysian Handicraft Development Corporation (MHDC). Retrieved October 2, 2012 from http://www.malaysiacraft.com.my/index.php.file.

Littrell, M.A., Anderson, L.F., & Brown, P.J. (1993). What makes a craft souvenir authentic. *Annals of Tourism Research*, 20(1), 197–215.

Swanson, K. (2004). Tourists' and retailers' perceptions of souvenirs. *Journal of Vacation Marketing*, 10(4), 363–377.

Yu, H., & Littrell, M.A. (2003). Product and process orientations to tourism shopping. *Journal of Travel Research,* 42, 140–150.

Zulaikha, E., & Brereton, M. (2011). Innovation strategies for developing the traditional souvenir craft industry. The First International Postgraduate Conference on Engineering, Designing and Developing the Built Environment for Sustainable Wellbeing, 53–58.

Hospitality and Tourism – Sumarjan et al. (eds)
© 2014 Taylor & Francis Group, London, ISBN 978-1-138-00151-0

Shopping mall attractiveness: Factors affecting consumer decision towards visitation

Z. Awang, N. Aminudin, N.A.A. Nik Hashim & N.A. Mustapha
Universiti Teknologi MARA, Malaysia

ABSTRACT: Retail industry has been recognized as one of the industries that contributed to the rapid expansion of economic in Malaysia. As shopping malls is one of the sectors of the retail industry, it has played a prominent role in making Kuala Lumpur as the second shopper paradise in Asia Pacific. The number of visitors has grown each year and in order to cater a large number of visitors, the development of the shopping mall in Klang Valley area has rapidly increased and it has lead to the oversupply situation. Besides, the management and the developers are not aware of the attributes and the attractiveness factors of the shopping mall that have attracted consumers to visit the shopping mall. Thus, this study aimed at identifying the attributes of the shopping mall that gives most effect to the consumer decision making in choosing shopping mall. The objective focused on three attributes which are accessibility and convenient, entertainment and environment. Self administered questionnaires were distributed to the consumers of top five regional shopping malls in Klang Valley. A total of two hundred and thirty questionnaires were obtained and analyzed using SPSS version 20. A descriptive and inferential statistic was used to obtain the results. The finding shows that those three attractiveness factors have affected consumer decision making but as to compare between those three attributes, the convenience attributes have given the most influence and effect towards consumers' decision. Further implication and recommendations are also been given.

Keywords: attributes, consumer, decision making, mall attractiveness, shopping mall

1 INTRODUCTION

One of the preferred tourism activities as stated by Paige and Littrell (2003) is shopping and has been known as the primary resources in generating tourism income. Turner and Reisinger (2001) said that shopping is one of the most crucial aspects of tourism which several researchers have concluded that no visit is complete without spending time for shopping. Timothy (2005) agreed that shopping has been seen beyond the conventional practices. That is why shopping mall have played crucial roles in contributing towards the economic growth as the shopping experience will enhance the consumer to visits the place again. Nowadays, people are likely to seek for comfort, easiness and pleasantness in dealing with shopping experiences. There had been rising demands for shopping malls which can offer consumers preferences. Shopping mall even can be a lovely place for tourist and traveller to come and visit.

All shopping malls have their own attractiveness. Attractiveness can be the quality that specific shopping malls have which can attract consumers to visit the mall. Wong, Yu and Yuan (2001) stated that, the growing number of malls has lead shoppers to select their destination. They are likely to patronize malls that are more attractive and exhibit a large variety of stores and merchandise that suit with their preferences. These days, people with more money and time are willing to travel further to shop and by the increasing number of retail areas, consumers have a growing selection of shopping destination. Shopping malls have lots of aspects that make it attractive to consumers which may both encourage and discourage shoppers to shop and visits certain shopping malls. Every consumer know themselves what they need and want when it comes in selecting the malls that will be their favourite.

The launch of 20 malls in the Klang Valley with a total net floor area of 4.4 million square feet this year makes the retail property market is likely to face an oversupply situation with pressure on rental rates. They have been facing problems in securing sufficient tenants and shoppers. (Mahalingam, 2010). The president of Malaysian shopping mall association stated that there is a signs that some of the malls are failing and running on deficit. This is not good for developers, owners, retailers and consumers. Most of the problems are due to market saturation and not the financial crisis.

Even though the numbers of shopping malls are increasing, and most of every each shopping mall has almost the same attributes, but it is not easy to pull the shoppers or consumers to actually get into the shopping malls. Certain design qualities of malls can encourage or discourage the shoppers and non-shoppers, such as lighting, ceiling height, colour scheme, existence of public transportation, parking facilities and many others.

These are the matter that authority needs to be alert at because consumers nowadays are not seeking for new and big places to visit at, but there are attributes and circumstances that make them shop, stay and hang out at the shopping malls. This study was conducted to identify the attractiveness factors or factors enhancing consumer visits to the shopping mall in Klang Valley, so it can be adapted into the other shopping malls in Malaysia especially at other cities in Malaysia so, they can be excel as a shopping destination like Kuala Lumpur.

2 LITERATURE REVIEW

2.1 *Environment*

According to Zhang (2011), the mall environment is considered as an influence factor of the consumer behavior when their perception of a shopping mall environment becomes more favorable. Consumer will see products quality to be higher which can influence consumer to exhibit greater approach behavior in the mall environment and influence them to shop in the particular mall. The environment may influence the consumer behaviour because it may change and affect the customer's feeling. This is supported by Moye (2000) who stated that the major asset of stores retailers is their ability to plead the consumer using physical environment. There are several environmental dimensions which are related with the mall environment such as quality, lighting, layout, carpeting and aisle width and placement are mall attribute used to project image and influence mall choice. The researcher also believed that dimension of the environment can be categorized as ambient condition for example temperature, airquality, noise, music and scent. Other than that is the space or function which include the layout, equipment and furnishing of the mall. The ambient in the mall may affect the perception and human response towards the environment. Consumers view the environment and develop perception of the sore particularly the mall as effective and a good place to shop.

2.2 *Entertainment*

Clow and Baack (2009) stated that managers of shopping mall employ various strategies and tactics to make the shopping experience as enjoyable as possible and the best example is through entertainment. Entertainment is a big factor for shopping mall because it causes consumer's shopping experience to be exciting or delight which thus can build loyalty among customer on that shopping mall. In addition, shopping mall that offers all in one roof will minimize time spent on travelling for the shoppers due to shopping mall management that have combined all stores with entertainment for shopping, socializing, walking, eating and leisure. It shows that they are able to plan their tenant mix and ensure that the styles of stores are balanced and complete to meet their shoppers' demand and expectations.

According to Kim, Christiansen, Choi and Feinberg (2005), malls are now built with large entertainment centers. As the competition between malls increases, enhancing the entertainment value for the consumer has becomes an important way of differentiating the mall product and give a positive relationship on excitements for them to stay and loyal to the shopping mall. They have also mentioned that entertainment is "something that amuses, pleases or diverts especially a performance or a show". There is an undeniable overlap between the two concepts. However, they do not appear to be interchangeable. An entertaining mall may make shopping more pleasurable or a welcome diversion from daily life without reaching what may be termed a high level. It is possible that exciting malls may be malls with intensive levels of entertainment.

2.3 *Convenience*

Pinnington (2004) said that consumers nowadays want a product in a convenient place and location due to the busy lifestyle and defined convenience as a need that is generated by peoples. Convenient factors include parking spaces, accessibility through public transportation and also operating hours. According to Hanzaae, Kheiry and Abghari (2011) parking seems to be a huge factor because nowadays, cars appear as a vital mode of transport for people nowadays especially shoppers. Therefore, the availability of parking spaces and the parking facilities that are offered by the shopping mall management can be assumed as a major factor that can enhanced the shopping mall attractiveness. Ibrahim and McGoldrick (2003) have stated that the availability of public transportation near to the shopping mall are able to enhance consumer to choose the mall due to the factor that they do not have to face the traffic congestion problem at the same time it can save money for fuel consumption and also time. Tiwari and Abraham (2010) believed that operating hours and time consumption to reach the shopping centre are one of the main criteria which the consumers seek for while deciding the best shopping centre to visits. The fact that convenience has often been ignored on the other side of the value equation suggests its potential has gone un-recognized. Developed and empirically tested by Clulow and Reimers (2009), the 13 attributes of retail centre convenience are parking availability, parking proximity, pedestrian areas, proximity to home or work, one-stop shopping, store compatibility, trading hours, enclosure, access lanes, retail concentration, centre size, traffic controls and shopper services. Machleit et al. (2003) said that shopping hassles such as delays in parking, lack of staff assistance, congested sidewalks, poor signage and poor access can all act as barriers to hedonic value.

2.4 *Consumer decision making*

Sternberg and Grigorenko (1997) and Stone (1954) agreed that the literature of consumer decision-making have a fairly long but varied history in the context of theory and research in retailing. Sproles and Kendall (1986) defined a consumer decision making as a mental orientation characterizing a consumer's approach in making choices. All consumers' decisions generally fall along a range of three categories which are the routine decision-making, limited decision-making and extensive decision-making (Boyd & Walker 1990, Howard 1994, Lamb et al 2004, Lancaster & Jobber 1994, Levy & Weitz 1992, Schiffman & Kanuk 2000, Sheth & Mittal 2004, Solomon 2002). Schiffman et al. (2008) believe that in consumer decision making literature there are four types of decision making views which are economic view, passive view, emotional view and cognitive view.

Past researchers have tried to explore for critical attributes that consumers recognize generally when observing stimuli, such as price, advertising, design or odour. Wu and Delong (2006) stressed that consumer perception is important and serves as a background for the understanding of consumers and a necessary approach to locate consumers in different cultures. Consumers tend to learn by the flow of information through the five senses—sight, hearing, smell, touch and taste—although, each consumer receives, organizes and interprets the sensory information in an individual way (Strydom, Jooste & Cant, 2000). Gagliano and Hathcote (1994) found that a store's appearance, employees, atmosphere and equipment have more impact on a customer because it is their first impression of the store.

3 METHODOLOGY

This study aimed to discuss on the attractiveness attributes among the top regional shopping malls in Klang valley area that may influence consumer decision in visiting the shopping malls. The shopping malls that have been selected were the famous shopping mall among tourist and also among the top ten shopping malls in Malaysia. Researcher used quantitative method in collecting the data. A convenience sampling techniques was applied whereby the selection of the respondents was based on the visitors' arrivals as well as the availability of the respondents which researchers can reach them during the research being conducted. A structured and self administered questionnaire was applied, and it was adapted and adopted from several researchers. As this research focused on the consumers at selected shopping malls and all shopping malls are private properties, instead of just go and distribute the questionnaires at the shopping mall which it can become more complicated to get the data, the questionnaires were placed at the selected F&B outlet at the shopping mall and also been distributed at the public transportation stations.

4 RESULT AND FINDINGS

The relationship between the shopping malls attributes and customers' decision in visiting the shopping mall is depicted in Table 1. It clearly shows that there is a positive relationship between all the attributes and the consumer's decision. As been listed from the table, these three (3) attributes were able to explain 72 percent ($R2 = 0.72$, $F = 599.358$, $p < 0.001$) of the variance towards consumers decision making. The result shows that environment, entertainment and convenience attributes in shopping mall have significantly affected consumers' decision. Pallant (2005) stated that as to compare between all the attributes of independent variables, it is essential to look on the beta value. The outcomes shows that the convenience attributes ($\beta = 0.79$, $p < 0.001$) of the shopping mall have

Table 1. Results of multiple regressions of the shopping mall attributes on the consumers decision making.

Predictors	Model 1/std. β
Step 1: Model variables	
Shopping mall attributes	
Environment	0.684
Entertainment	0.779
Convenience	0.796
R^2	0.724
Adj. R^2	0.723
R^2 change	0.724
F-change	599.358

influence most towards the consumers' decision. It was closely followed by environment attribute ($\beta = 0.68$, $p < 0.001$) and entertainment attributes ($\beta = 0.77$, $p < 0.001$). As all three (3) attributes of the shopping mall was found to significantly and positively affecting consumers in making decision, it shows that the assumption on the decision made by the consumer base on the environment, entertainment and convenience factors are strongly supported.

Evidently it shows that between all this three (3) attributes, convenient had the most significant impact towards the consumers' decision making, followed by entertainment and environment. This result proves the convenient factors are the shopping mall attributes that consumer highly considered in deciding and selecting the shopping mall that their will visits.

5 LIMITATIONS AND SUGGESTION FOR FUTURE RESEARCH

As other research study accomplishments, this study also has its fair share of limitations that deserve further justifications. The first limitation is allied with the population or sample of the study. This study was focusing on the consumers of shopping mall in only top five selected shopping malls in Klang Valley area which the results only been made at the study contextual setting. As to make the research represent an ambitious and broad description of shopping mall consumers, the study can be carried out at other state's shopping mall which will also strengthening the result of the study. Further study might be conducted in other states in Malaysia to test whether the setting and attributes that have been used in this study can be successfully adapted at other shopping malls in Malaysia.

The second limitation is the research design itself. Most study conducted on the shopping mall is only focusing on the shopping mall consumers. Researchers did not address the difficulties that are faced by the management, and also the developers. There was no study that considered the developer and also the management point of view. A qualitative study might be conducted in reviewing and evaluating the developer and shopping mall management perspectives and their opinion.

6 RECOMMENDATIONS

Shopping mall attributes played a prominent role in affecting the consumers' decision making and it cannot be denied that the success of the shopping mall will not be achieved without the consumers. From the consumer point of view, it is necessary for the shopping mall to have all that will provide them with a delightful experience. So it is the developer and management of shopping mall responsibility to make their shopping mall stand out and different form the other. With this indication, the management might come with a new arrangement of strategy in how to attract the consumer in choosing the shopping mall. Every shopping mall have their own uniqueness, it depends on the role of the marketing team of the shopping mall management on how to make their shopping mall stand out like the other successful shopping mall and to be a brand new successful shopping mall.

Shopping malls in Malaysia specifically in Klang Valley area are facing oversupply situation and having a problem in securing sufficient consumer. This might be due to the fact that there are too many shopping malls built at the same area and near to each other. The authority should look into this matter where by maybe the authority can limit the construction of retail property and come out with a policy that the shopping malls can only be build up at a certain area that have attained a strict requirement that have been set up by the authority. The Malaysian shopping mall association should also play their roles in dodging this problem before it became more severe.

7 CONCLUSION

To conclude, the management of the shopping mall and also the authority should work together in helping the shopping mall that has been facing the insufficient consumer problem. Yes, we have enough property, but we did not have enough tenants and consumer. What can be done is the promotion towards shopping malls must not only focused to specific area and the successful shopping malls but also to the newly developed and the present shopping mall.

REFERENCES

Boyd, H.W., Jr. & Walker, O.C., Jr. (1990). Marketing management: A strategic Approach. Boston: R.R. Donnelley.

Clow, K. & Baack, D. (2009). Marketing management: A Customer-oriented approach. California: Sage Publication Inc.

Gagliano, K.B. & Hathcote, J. (1994). Customer expectations and perceptions of service quality in retail apparel specialty stores. *Journal of Service Marketing*, 8(1), 60–69.

Hanzaae, Kheiry & Abghari (2011), Investigation of the effects of stores' tenant mix and internal and external environment condition on customer satisfaction from shopping centre in Iran. *Asian Journal of Business Management Studies*, 2(2), 94–100.

Ibrahim, M. & P. McGoldrick (2003), Shopping choices with public transport options: An agenda for the 21st century. Ashgate, Hampshire UK.

Kim, I., Christiansen, T., Feinberg, R., & Choi, H. (2005). Mall entertainment and shopping behaviors: A graphical modeling approach. *Advances in Consumer Research*, 32(1), 487–492.

Lamb, C.W., Jr., Hair, J.F., Jr. & Mcdaniel, C. 2004. Marketing 7th ed. Oxford: Oxford University Press.

Lancaster, G. & Jobber, D. 1994. Selling and sales management 3rd ed. London: Clays.

Levy, M. & Weitz, B.A. 1992. Retailing management. Boston: Von Hoffmann Press.

Machleit, K., Meyer, T., & Eroglu, S. 2005, 'Evaluating the nature of hassles and uplifts in the Retail shopping context', *Journal of Business Research*, 58(5), 655–663.

Mahalingam,, E. (2010). Malls, more malls everywhere. The Star Online. Available at http://biz.thestar.com.my/news/story.asp?file=/2010/8/7/business/6796926&sec=business.

Moye, L.N. (2000). Influence of Shopping Orientations, Slected Environmental Dimensions with Apparel Shopping Scenarios, and Attitude on Store Patronage for Female Consumers.

Paige, R.C. and Littrell, M.A. (2003). Tourism activities and shopping preferences. Journal of Shopping Center Research, 10(2), 7–25.

Pallant, J. (2005). SPSS survival manual: A step by step guide to data analysis using SPSS. Sydney: Sabon Publishers.

Pinnington, D. (2004), Defining Convenience: A Shopper centric Report. Available at http://www.shopper centric.com/uploads/report/Short%20Report01%20%20Defining%20Convenience.pdf-6617.pdf.

Reimers, V., and Clulow, V. (2009B), Retail centres: It's time to make them convenient, *International Journal of Retail & Distribution Management*, 37(7), 541–562.

Schiffman, L.G. & Kannuk, L.L. (2008). Consumer Behaviour. 10th Ed. New Jersey: Pearson/Prentice Hall.

Sheth, J.N. & Mittal, B. (2004). Customer behaviour: A managerial perspective 2nd Ed. Ohio: South-Western.

Solomon, M.R. (2002). Consumer behaviour: Buying, having and being 5th ed. New Jersey: Prentice-Hall.

Sprolles, G.B. & Kendall, E.L. (1986). A Methodology For Profiling Consumer'S Decision-Making Styles. *The Journal Of Consumer Affairs*, 20(2), 267–279.

Sternberg R.J., Grigorenko E. Are cognitive styles still in style? *Am Psychol* 1997;52(7):700–712.

Stone GP. City shoppers and urban identification: observations on the social Psychology of city life. *Am J Sociol* 1954;60:36–45.

Strydom, J.W., Jooste, J. & Cant, M.C. (2000). Marketing Management. 4th Ed. Cape Town: Oxford University Press.

Timothy, D.J. (2005). Shopping Tourism, retailing, and leisure, Clevedon: Channel View Publications.

Tiwari, R.K. & Abraham, A. (2010), Understanding the Consumer Behaviour towards Shopping Malls in Raipur City. International Journal of Management & Strategy, 1(1), 114.

Turner, L.W. & Reisinger, Y. (2001). Shopping satisfaction for domestic tourists. *Journal of Retailing and Consumer Services,* 8(1), 15–27.

Wonderful Malaysia (2012). Top 10 Shoping Malls in Kuala Lumpur. Retrieved on December 12, 2012 from http://www.wonderfulmalaysia.com.

Wong, G., L. Yu & Yuan, L. (2001). SCATTR: An Instrument for Measuring Shopping Center Attractiveness. *International Journal Retail and Distribution Management*, 29 (2),76–86.

Wu, J. & Delong, M. (2006). Chinese Perceptions Of Western-Branded Denim Jeans: A Shanghai Case Study. *Journal Of Fashion Marketing And Management*, 10(2), 238–250.

Zhang, Y. (2011). The Influence of The Mall Environment on Shoppers Values and Consumer Behavior in China, 18(1):214–224.

Hospitality and Tourism – Sumarjan et al. (eds)
© 2014 Taylor & Francis Group, London, ISBN 978-1-138-00151-0

Destination attractiveness of a theme park: A case study in Malaysia

M.A. Ahmad, M.R. Jamaluddin, M.A. Alias & A.Q. Abd Jalil
Universiti Teknologi MARA, Malaysia

ABSTRACT: Tourism in Malaysia has evolved significantly, from providing nature and heritage tourism-related activities, to the creation of fun and leisure related experience to cater the needs of fun and energetic tourism activities. In line with the previous notion, this study aims to examine the effect of theme park destination attractiveness towards visitors' satisfaction. Quantitative approach was chosen hence 384 respondents were interviewed to examine the stipulated objective. Findings show that facilities and entertainment are the main determinants in predicting the variance in satisfaction and behavioural intention. Surprisingly, price is not a key determinant to promote behavioural intention. By looking at the facilities provided, the price is perceived as value for money. It is concluded that destination attractiveness are important to determine the satisfaction and behavioural intention. Therefore, theme park providers must pay attention to the attributes listed to ensure positive word-of-mouth.

Keywords: behavioral intention, destination attractiveness, satisfaction, theme park, tourism product

1 INTRODUCTION

In the past two decades, tourism in Malaysia emerged from a small industry to one of the biggest industry in Malaysia. Currently, Malaysia is ranked 9th in the most visited country in the world list by World Tourism Organization (Bernama, 2012). Malaysia is blessed with a variety of tourism products that has a potential to attract different segment of tourist to visit the country. Presently, the popular tourism products in Malaysia are cultural tourism, nature tourism, medical tourism, beach tourism, agriculture tourism, island tourism and the latest one, theme park tourism. Among the local destination, Penang and Melaka are the famous historical and cultural sites in Malaysia that offer festival, heritage buildings, variety of foods and retail shopping (Tourism Malaysia, 2012). The fact that both states are listed as United Nations Educational, Scientific and Cultural Organization (UNESCO) as heritage city, had enhanced the image of these destinations. Another Malaysia's signature tourism product is Tioman Island. It is listed in the top ten most popular tourist attractions offered to the tourist. Nonetheless, with the improvement of infrastructure and mode of transportation in Malaysia, theme park has gained attention by Malaysian as a holiday destination especially during public and school holiday. The trend of visiting theme park can be seen several years ago with the opening of Universal Studio in Singapore, where it has attracted Malaysian to visit Singapore.

The word "theme park" is interpreted differently by people. From personal interview, such dominant responds received are "amusement park", "park with special theme", "indoor and outdoor attraction", and "large park with games and fun". Raluca & Gina (2012) stated theme park as a subset of visitor attraction that plays an important role in generating demand and become the main motivators for tourism trips, and also one of the core elements of the tourism product to be offered to the tourist. Domenech (2011) mentioned over the last fifty years, theme parks have become a top tourism destination for many tourists. Malaysia is no exception with the rise of several theme parks namely Genting Highland, Sunway Lagoon, Bukit Merah Laketown Resort, Bukit Gambang Resort City and the latest one, Legoland. This is consistent with the trend in Asia that sees the development of many theme parks especially in the emerging markets like Thailand, Indonesia, Singapore, and Malaysia. For example, Thailand offers Dream World theme park in Bangkok, Indonesia offers Trans Studio in Bandung and Singapore offers Universal Studio in Sentosa.

In tourism study, numerous researches focused in other types of attractions such as nature, heritage, and city, but less study on theme park were undertaken (Domenech, 2011). Therefore, a research study to assess theme park destination attractiveness is timely to measure the rating towards the theme park attractiveness in Malaysia. This study aims to determine the most important attributes that affect the satisfaction level of visitor when visiting theme park. The relationship between satisfaction and behavioural intentions towards the theme park were also highlighted in this study. The underlying principle for these objectives is to provide an opportunity for this theme park to recognize visitor's perception towards their products.

2 LITERATURE REVIEW

2.1 *Theme park as tourist attraction*

Domenech (2011) defined theme park as a park that includes a bunch of theme attractions including beautiful architecture, landscape, rides, food, services, employees and retail shops. Milman (2010) mentioned the original concept of theme parks was implemented by Walt Disney's first theme park namely Disneyland. This theme park opened in 1955 in Anaheim, California. Since then theme park has become an influential tourism attractions to the most countries in the world because of the chain effect that it provides to other hospitality service provider and feeder.

2.2 *Destination attractiveness*

Attractiveness is the common word to describe popularity in the tourism industry. From the literature review, majority of the scholars use attractiveness to express the positive attributes that comprise of several elements like price, entertainment, accessibility, services, facilities, safety, location, and others. Destination attractiveness is the crème of the crop in the tourism study because of the important findings that it provides. As stated by Vengesayi (2003), the more a destination is able to meet the needs of tourist, the more it is perceived to be attractive and the more the destination is likely to be chosen.

2.3 *Price*

Price has been proven to be an important measurement for tangible goods but contradict can be said to the intangible goods. Domenech (2011) supported this notion by saying that price is not an important element when selecting a destination for holiday purposes. In searching for experience, price is secondary element that promotes satisfaction and behavioural intention. Nevertheless Zhou (2005) noted that price is one of the key attribute concerns by the tourist when choosing holiday destination.

2.3.1 *Entertainment*

Theme park provides a range of indoor and outdoor entertainment. Going to theme park means having fun. Theme park offers variety of entertainment but the essential one is ride. Domenech (2011) stated that thrill ride is one of the critical elements under the theme park attributes, with roller coaster be the top of the list. Zhou (2005) observed that tourist looks for outdoor activities, gambling, nightlife, museum and cultural sites for getting an entertainment.

2.4 *Accessibility*

Duration to-and-fro theme park is among the factor that worth research for. Pleasant journey by any mean of transportation can complement the overall satisfaction. Accessibility may influence other factors included in the destination attributes. This is in line with the findings from Mohamad, Abdullah and Mokhalis (2012) that concluded journey is the main priority in planning a holiday destination.

2.4.1 *Services*

Service is another secondary attributes that complement the core products of a theme park. If technology in the theme park provides the fun experience, service offered by the operator provides the pleasant feeling. Emphasize in service will create memory because the close interaction that it provides to the visitor. Poor service will reflect the images of the destinations as well as the intention for the next visit as proved by Haah and Yavas (2004).

2.5 *Facilities*

Facilities and service work in the same manner to complement the core products. Providing facilities to human could be a challenging task because of the difference in demand and behavior of a human. However standard facilities like accommodation, food services, and transportation are essential for a theme park. Research by Sangchoul et al, (2012) confirmed that facilities are crucial in enhancing experience. This is further supported by Yahya (2000) that said tourism destination could not stand alone without the presence of facilities because it is the heart of the theme park.

2.6 *Safety*

The issue of safety is always being a priority for a group visitors especially family and group tour. With the explosion of information communication technology, the report of any safety incident can be browsed easily. Domenech (2011) proved that safety is the main concern while spending time with a family in the theme park. Safety does not only enhance the theme park image but also the overall country's image. Pikkemaat & Schuckert (2012) proved that safety and security is significant in the future patronize intention as well as important attributes that the tourist concern when making a decision on selecting the destination.

2.7 *Satisfaction*

Satisfaction often derives after the overall experience. However in the social science study, the presence of satisfaction is influential in mediating the relationship with outcome variables like trust and behavioural intention. According to Cam (2012), perceptions of service quality and value, will affect satisfaction, loyalty and post-purchase behaviors of the visitor. He adds that visitors who experience low service quality have the possibility to not recommend and revisit the destinations in the future. Apart from that, Visitor Satisfaction Survey (2012) classifies satisfaction as occurrence of visitors' pre-visit expectations and their actual.

2.8 *Behavioural intentions*

In tourism study, the role of behavioural intention is the most powerful force for the sustainability of the products. The development of sustainability tourism has prompted researchers to include behavioural intention such as positive word-of-mouth and revisit intention in the outcome determinant. As founded by George and George (2004), behavioural intention shows how frequent visitors decided to revisit the destination in the near future. Positive feedback towards measurements of destination attractiveness will definitely enhance possibility of future patronization. Careful consideration must be emphasized to this issue as stated by Chen and Tsai (2006) that posited tourist decides to revisit based on the rating of pre, during and post visit tour experience. This is confirmed by Mohamad et al. (2012) that concluded positive tourism destination image will reflect the tourist satisfaction as well as the intention to return in the future, and willingness to speak positive, and recommend the destination to others. Cam (2011) added satisfied customers are the one who act as word-of-mouth advertising agent which will influence and recommend to family, friends or potential customer about the products.

3 METHODOLOGY

This study began with extensive literature review to understand the attributes that matter to theme park. Feedbacks from tourism academician are sought to validate those attributes. In addition, interviews with three sales managers of theme park were conducted to probe their insight about the measurements listed in the conceptual framework. One theme park was chosen as the setting for this study. This theme park is one of the prominent theme parks in Malaysia that offers modern technology on rides and entertainment. The permission from sales manager was granted in order to conduct the survey in the theme park. A self-administered questionnaire survey is used to tap the respondent's feedback towards destination attributes. The interview was conducted over two weekends. As suggested by the sales manager the field survey was conducted over the over weekend because these days are the peak time for visitors. This study managed to capture 384 visitors by using convenience sampling method.

4 FINDING AND ANALYSIS

In order to measure the strength of the relationship between destinations attributes, satisfaction and behavioural intention, Pearson correlation analysis was performed. Table 1 summarizes the result of the correlation analysis. Among the six attributes, entertainment ($r = 455$, $p < 0.01$) is found to be highest one rated by the respondents. It is anticipated because visitors look for the products available upon entering the theme park. Furthermore, the first thing that is available at the entrance is the brochure about entertainment information that definitely influences the score. As stated earlier, people go to theme park to feel the utmost entertainment and to spend time with family. From observation, the theme park offers variety of products that cater to the different needs of kid, adult and senior citizen. Respondents rated highly for the items listed for entertainment namely "good place for holiday", "variety of games and activities", "enjoyable atmosphere" and "place for fun entertainment" and "entertaining activities for kids and adult". This is in line with the findings from Domenech (2011) that proved types of entertainment offered such as roller coaster and games become the important reason visitors choose theme park for holiday destination. Next, facilities are rated the second highest relationship that drive satisfaction ($r = 0.349$, $p < 0.01$). Facilities scales that comprise of "Comfortable shelter from sunshine and rain", "convenient facilities" and "sufficient facilities" are significant towards determining behavioral intention. Accessibilities ($r = 0.219$, $p < 0.01$) make up the top three lists of the attributes that have strong explanatory power towards the behavioral intention. Accessibility attributes of "strategically located", "visibility of the signage", "easy to reach" and "ample parking space" must be treated significantly by theme park service providers. Although all attributes were rated significantly, price score the lowest. Price to visit

Table 1. Correlations analysis.

Items	1	2	3	4	5	6	7	8
1. Behavioural intention	–	0.282**	0.455**	0.057**	0.219**	0.349**	0.132**	0.700**
2. Accessibility		–	0.324**	0.200**	0.200**	0.112**	0.238**	0.357**
3. Entertainment			–	0.049**	0.255**	0.388**	0.267**	0.476**
4. Price				–	0.060**	0.240**	0.066*	0.195**
5. Safety					–	0.093**	0.241**	0.315**
6. Facilities						–	0.16**	0.449**
7. Service							–	0.284**
8. Satisfaction								–

Note: **Correlation is significant at the 0.01 level (2-tailed). *Correlation is significant at the 0.05 level (2-tailed).

theme park is very weak to explain the variance in behavioral intention. For satisfaction, as commonly studied, it is a strong determinant for future patronization ($r = 0.700$, $p < 0.01$). The strong relationship of satisfaction and behavioral intention is justified as people must get the feel of good factor at the end of the visit before reacting to the behavioral intention scale that consist of future revisit intention and positive word-of-mouth.

5 IMPLICATION

This study provides several important implications towards the theoretical and practical based on the results obtained. From this study, the overall satisfaction level of a theme park was discovered. Other than that, the important attributes which contribute the most towards the satisfaction level of a theme park has been identified. Therefore, the outcomes from this study provide a several implications which are essential for the future research particularly on the theme park industry. As for the theoretical implication, since less study regarding the theme park destination attributes were undertaken, the outcome from this study will provide new dimensions towards the theme park destination attractiveness attributes. In addition, numerous studies only focuses on theme park development, only few research was undertaken on which factors attract more visitors to visit theme park (Domenech, 2011). Since this theme park has a huge potential in attracting people to visit Malaysia, thus from the practical implication, the findings from this research will provide appropriate responses which are believed to be important for the related authorities especially to Tourism Malaysia, policy maker and other related authorities to continuously promote theme park as a tourism product.

Additionally, the outcome from this study is not only important to this theme park profit but also significant for enhancing the other tourism sites including the lodging industry, food and beverages industry, public transport including the taxi drivers, shuttle bus as well as the benefit to the local community. Importantly, if this theme park was highlighted as a successful tourism product and the visitors' behavioural intentions in visiting theme park were acknowledged, it will help the management of the theme park to provide a better improvement and it will attract more tourists and lead to better revenue, more employment, and business opportunities which will improve the quality of life. With that, this is one of the Tourism National Key Economic Area (NKEA) that will be achieved through this study.

6 CONCLUSION AND SUGGESTIONS FOR FUTURE STUDY

The limitation that needs to be highlighted in this study is related with the application of convenience sampling which is not represented by all visitors of theme park. Therefore, further research needs to be undertaken in order to examine a wider sample of visitors of theme park. As a conclusion, the outcome from this study will provide a guideline especially for theme park future planning and development. Findings from this study revealed that majority of the respondents are satisfied with the entertainment offered at the theme park and it influenced the most towards visitor satisfaction level as well as behavioural intentions. Nevertheless, the theme park management also needs to focus in implementing several improvements in terms of the price and facilities and also take proactive actions by implementing necessary strategies to overcome this issue.

REFERENCES

Bernama. (2012), January 30. Malaysia ranked among top 10 nations for tourists. *New Straits Time*. Retrieved November 8, 2012, from http://www.nst.com.my/opinion/columnist/malaysia-ranked-among-top-10-nations-for-tourists-1.39046.

Cam, T.T.A. (2011). Explaining tourists' satisfaction and intention to revisit NhaTrang, Viet Nam. Retrieved Mac 28, 2013 from http://munin.uit.no/bitstream/handle/10037/3782/thesis.pdf?sequence=1.

Chen, C.F., & Tsai, D.C. (2006). How destination image and evaluative factors affect behavioural intentions? Retrieved May 28, 2013 available at http://cht.tcm.ncku.edu.tw/.

Domenech, J.M. (2011). Determining factors of theme park attendance. Retrieved April 21, 2013 from http://dacc.coalliance.org/fedora/repository/.

George, P.B. & George, P.B. (2004). Past visits and the intention to revisit a destination. *The Journal of Tourism Studies*, 15(2).

Haahti, A. & Yavas, U. (2004). A multi-attribute approach to understanding image of a theme park: the case of Santa Park in Lapland. *European Business Review*. 16(4). 390–397.

Milman, A. (2010). The global theme park industry. *Worldwide Hospitality and Tourism Themes,* 2(3).

Mohamad, M. Abdullah, A.R., & Mokhlis, S. (2012). Tourists' evaluations of destination image and future behavioural intention: the case of Malaysia. *Journal of Management and Sustainability,* 2(1).

Pikkemaat, B. & Schuckert. (2012). Successful factors on theme park: an exploratory study. Retrieved December 20, 2012 from http://www.academia.edu/329316/Success_Factors_of_Theme_Parks_An_Exploratory_Study.

Raluca, D.C. & Gina, S. (2012). Theme park: the main concept of tourism industry development. Retrieved December 20, 2012 from http://steconomice.uoradea.ro/anale/volume/2008/v2-economy-and-business-administration/113.pdf.

Sangchoul, Y. Jonathan, D. & Liping A. Cai. (2012). Exploring Asian cruise traveler's travel experience and perception, 3. Retrieve December 9, 2012, available at http://scholarworks.umass.edu/gradconf_hospitality/2011/Presentation/34/.

Tourism Malaysia, (2012). Media Release: Tourist arrival to Malaysia continue to increase. Retrieved November 7, 2012 available at http://corporate.tourism.gov.my/mediacentre.asp.

Vengesayi, S. (2003). A conceptual model of tourism destination competitiveness and attractiveness. ANZMAC 2003 Conference Proceedings Adelaide 1–3 December 2003.

Visitor Survey Satisfaction. (2012). Impact Assessment. Retrieved October 6, 2012, from http://www.tba.co.nz/kete/PDF_files/ITP407_visitor_satisfaction_survey.pdf.

Yahya, S. (2000). Foreign consumer perception of tourism services, 8(1), 7–17. Retrieved on November 30, 2012 available at http://psasir.upm.edu.my/3286/1/Foreign_Consumer_Perception_of_Tourism_Services.pdf.

Zhou, L. (2005). Destination attributes that attract international tourist to Cape Town. Retrieved November 30, 2012 from http://etd.uwc.ac.za/usrfiles/modules/etd/docs/etd_init_9726_1173950974.pdf.

Hospitality and Tourism – Sumarjan et al. (eds)
© 2014 Taylor & Francis Group, London, ISBN 978-1-138-00151-0

Tourist perception on homestay programme: A case of Kuala Medang homestay

M.F. Ahmad, F.S. Chang & E.R. Kassim
Universiti Teknologi MARA, Malaysia

A. Husin
MAHSA University, Malaysia

ABSTRACT: Homestay programme in Malaysia is comprised of various ecotourism packages, offering agro-tourism, nature-tourism and/or cultural-tourism activities. However, homestay hosts need to improve their activities in order to attract more tourists in the future. There has been a stable decrease in the number of homestay tourist arrivals in Kuala Medang, which is chosen as the setting of this study. A questionnaire was developed to examine tourists' perception, preference and satisfaction level toward the homestay activities and how it affected post-behavioural intentions. A convenient sample of 346 local and international tourists were surveyed and the results showed majority of them were satisfied with the homestay activities which led to positive post-behavioural intentions. Activities which tourists most enjoyed were visit to the *Orang Asli* village, top-spin *"gasing"* game and demonstration, fishing, culture performance as well as bamboo water rafting.

Keywords: homestay activities, tourist satisfaction, post-behavioural intentions

1 INTRODUCTION

One of the most successful programmes that have been identified by the Ministry is the homestay programme which surpassed its target of 23% of occupancy rate and achieved a rate of 32% (The Star Online, 2012). Hosts of the homestay programme are able to earn a considerable income for the local people as the programme showcases Malaysia's rich and unique cultural heritage lifestyle (The Star Online, 2012).

The homestay program offers a great implication to the improvement and stabilization of the social economic conditions. However, problems arise as those who are engaged in homestay programmes lack marketing knowledge which is essential for successful business. Moreover, insufficient attention is given to meeting the requirements of the rural tourism consumers, the diversity of activities provided for them, their satisfaction and behavioural intentions (Morajab, 2010). The activities organized by the homestay hosts must really satisfy the tourists who participated in the programme (Thaothampitak and Weerakit, 2008). This is to ensure that these tourists will feel the utmost experience of traditional Malay lifestyle as to satisfy them in every way so that they will revisit the place and also recommend it to other people. According to Thaochalee, Laoakkha and Panthachai (2011), activities in homestay villages are very important to satisfy tourists' needs who take part in the homestay programme. Such activities include traditional offerings, planting of cotton, natural attractions (mountain, waterfall), handicrafts making, cooking of local dishes, and tribal culture houses.

The number of tourist arrivals from the year 2002 (397) until 2009 (4,918) has been increasing but it drops slightly in year 2010 (2,537). However, since 2010 until September 2012, the number of tourist arrivals in each year has never exceeded the number of tourist arrivals in 2009. The activities provided by the homestay hosts could be uninteresting thus need to be improved in order for them to achieve their tourist arrival records once more. This is in line with Thongma, Leelapattana and Hung (2011) who stated that the activities provided by the homestay hosts for tourists must be improved in order to attract tourists to come to their homestay villages.

Previous studies have explored and looked into adult learning in the homestay programme (Cyprian, 2012), the factors influencing communities' life of homestay hosts (Yusof, Ibrahim, Muda and Amin, 2012), discovering ecotourism products in potential homestay villages (Saaidin, 2012), homestay management and sustainable community development (Silparcha and Hannam, 2009), homestay programme and rural community development (Ibrahim and Razak, 2010), as well as tourist motivation to use homestays and satisfaction based on cultural and heritage-based attributes (Seubsamarn, 2009). While Ryan (1995) urged that the attitude, expectations, and perception of holiday maker are significant variables in setting goals, influencing behaviour and determining final satisfaction. Therefore this study will fill in the gap by looking at the perception of tourist towards the homestay activities provided in the Kuala Medang Homestay.

2 LITERATURE REVIEW

2.1 *Kuala Medang homestay*

Kampung Kuala Medang is a traditional Malay village situated at bank of Jelai River, 50 kilometres shy of Kuala Lipis. This village still retains its traditional lifestyle and characteristics. Rich with 'flora and fauna' this is the best place for people who seek for nature and adventure getaway. Kuala Medang was also nominated as the 'Best Homestay Program in Malaysia (Tourism Pahang, 2013). It can be reached by major road links either from Kuala Lipis or Raub. Initially, Kuala Medang is a traditional village located in remote areas, far from the city centre where the journey will go through rivers, rapids and lush mangroves. There are a few neighbouring attractions such as Changkah Pasir Rapid, Tema Waterfall Rapid, and Aborigines (Orang Asli) settlement in Pos Pantos. Among the activities offered are as such performances and cultural activities of the indigenous and Malay community, traditional games, traditional cuisines and a feast. Apart from that, rafting and kayaking in Changkah Pasir Rapid and Tema Waterfall Rapid, make crafts demonstrations, live experience as rubber tappers, oil palm harvesters and paddy relay, visit traditional villages, cave exploring, learning traditional games such as *gasing* (top-spin) and so on (Pahang Trekker, 2012).

Kampung Kuala Medang is known for its own products such as *dodol, kuih bahulu, serbuk kunyit* (turmeric powder) and noodles, made by residents as part of the village industry. Other attractions in the village include *jus salak* (juice extracted from salak fruits), *halwa, emping, bubur gula kabung* and mat-making using *mengkuang* (pandanus) leaves.

2.2 *Culture and heritage activities*

According to Rosenfeld (2008), a cultural or heritage attraction is defined as "a tool of economic development that achieves economic growth through attracting visitors from outside a host community, who are motivated wholly or in part by interest in the historical, artistic, scientific or lifestyle or heritage offerings of a community, region, group or institution". Silberberg (1995) argued that cultural and heritage tourism is a tool of economic development that achieves economic growth through attracting visitors from outside a host community, who are motivated wholly or in part by interest in the historical, artistic, scientific or lifestyle or heritage offerings of a community, region, group or institution which means such travel is focused upon experiencing cultural environments, including landscapes, the visual and performing arts and special lifestyles, traditions, values, and events. Cultural/heritage destination attributes have been studied by many other researchers. For instance, Sofield and Li (1998) studied the cultural tourism of China by selecting history, culture, traditional festivals, historical events, historical sites, beautiful scenic heritage, architecture, folk arts (music, dancing, craftwork) and folk culture villages as the attributes of significance. Janiskee (1996) emphasized the importance of events through several features such as festivals, traditional ceremonies, historic houses, music, dancing, craftwork, food, and the direct experience of traditional life.

2.3 *Aborigines' activities*

Jamak, Ali, Salleh and Abdullah (2010) defined *Orang Asli* as "original people" or "first people". They further added that the *Orang Asli* comprises of three main tribal groups, namely Negritos, Proto-Malays and Senoi, with all these three of which can be further separated into 18 dissimilar cultural-linguistic groups. In a study on "Assessing Entrepreneurship and Micro Business Enterprise of Orang Asli Aborigines in Pahang" by Jamak et al. (2010) found that these aborigines have strong desires to improve themselves, and they are willing to participate or apply for any government projects as long as it benefits their livelihood and also standard of living. This is supported by Jamak, Eusof and Sivapalan (2008) in their research on "Issues on Small Business Enterprise of Orang Asli Aborigines in Pahang" in which they found that about 70% of all of their respondents (aborigines) prefer doing business and regard being self-employed businesses as a preferred way of life as compared to being farmers, hunters, fishermen or forest harvesters which have been their usual way of life. Hu, Wang and Wang (2012) argued that places possess abundant natural resources, such as hot springs, and giant trees in Dabajian Mountain are having great potentials in developing tourism with their leisure farming and cultural tourism. The hosts of Kuala Medang Homestay provide activities such as visiting the

Orang Asli settlement in Pos Pantos, Kuala Lipis, Pahang, as well as watching and taking part in the traditional *Sewang* dance together with the aborigines (go2homestay.com, 2012).

2.4 *Water-based activities*

Basically, water based activities are on-water and water adjacent activities such as rafting, tubing, kayaking, swimming, wading, and fishing. The National Association of State Park Directors (2004) defined water based activities are as such: swimming, boating, and fishing. Exercising in water has become increasingly popular, and it has been reported that water exercise for older individuals is therapeutically beneficial (Robert, Jones, and Bobo, 1996; Takeshima, Nakata, Kobayashi, Tanaka, and Pollock, 1997). Arai and Pedlar (1997) argued that recreational activities have been found to help build social and familial infrastructures. Previous research on marine and coastal water based activities has focused on the value of free water based activities, including beach activities, angling, boating and swimming (Curtis, 2003). Through these activities, one can combat social exclusion (Curtis, 2003). Whereas in a study conducted by Wilks and Atherton (1994) on health and safety in Australian Marine Tourism, they suggested that the importance of gaining specific information about travellers' health and safety in marine settings is vital whereby the information can then be used by agents and operators to ensure their services are provided with the highest duty of care to the customers. In the homestay programme, there are a few water-based activities organized such as fishing, river cruising as well as bamboo water rafting.

3 METHODOLOGY

As this study sought to examine tourist perception towards the homestay activities provided in Kuala Medang Homestay, a descriptive research design using a quantitative approach with a self-structured questionnaire was used. The sample size was 346 in which used to represent the estimated population number of about 3,700. Convenience sampling was used where the selection of respondent was based on the tourists who could provide the information required. The survey questionnaire was divided into six main sections. The first section (Section A) was about the respondent's demographic profiles, with nominal scale items, such as age, gender, marital status, employment status, and monthly income were asked. In the second section (Section B), homestay tourist profiles' questions were developed which looked at the information about the respondent's trip, for examples domestic or international tourist, travelling companion, the number of time the respondent has taken part in any homestay programme, length of stay, the amount spent for each person during the homestay programme, as well as the source of information about the homestay programme. The third section (Section C) highlighted the homestay activities that the respondent participated as well as which of the activities that they liked and enjoyed the most. The fourth section (Section D) included the questions pertaining tourist perception towards the homestay activities. This section was divided into three categories which were cultural and heritage activities, aborigines' activities, and water-based activities. A five-point Likert scale was used in the survey which varies from 1 (strongly disagree) to 5 (strongly agree).

4 FINDINGS AND ANALYSES

4.1 *Respondents' profiles*

A total number of 295 questionnaires were usable and analyzed. The results showed that 51.8% of the respondents were male and the age groups with highest respondents were between the age of 25–34 followed by 20–24. 60% of them were single as the rest were married, and travelled together in organized groups. Nearly 80% of the respondents were domestic tourists and majority of them took the 3-days-and-2-nights homestay programme package. The respondents chose "visit to the *Orang Asli* settlement" as the most preferred activity. This was then followed by the "top-spin (*gasing*) game and demonstration", "culture performance", "fishing" as well as "bamboo water rafting" activity.

4.2 *Analysis of tourist perception on homestay activities*

Table 1 reported the overall tourists' perception on the culture and heritage activities provided in Kuala Medang Homestay. It is clear that the mean score of the overall tourist perception on the culture and heritage activities in Kuala Medang Homestay was at (M = 4.4793) which indicated that the respondents agreed with the culture and heritage activities provided in the homestay programme.

Table 2 described the overall tourists' perception on the aborigines' activities provided in Kuala Medang Homestay. It is clear that the mean score of the overall tourist perception on the aborigines' activities in Kuala Medang Homestay was at (M = 4.5915) which indicated that the respondents agreed with the aborigines' activities provided in the homestay programme.

Table 3 exhibited the overall tourists' perception on the water-based activities provided in Kuala

Table 1. Overall mean score of culture and heritage activities.

Variable	N	Minimum	Maximum	Mean	Std. deviation
Culture and heritage activities	295	1.60	5.00	4.4793	0.55262

Table 2. Overall mean score of aborigines' activities.

Variable	N	Minimum	Maximum	Mean	Std. deviation
Aborigines' activities	295	1.67	5.00	4.5915	0.60975

Table 3. Overall mean score of water-based activities.

Variable	N	Minimum	Maximum	Mean	Std. deviation
Water-based activities	295	1.40	5.00	4.3485	0.61527

Table 4. Overall mean score for homestay activities.

Variable	N	Minimum	Maximum	Mean	Std. deviation
Homestay activities	295	1.63	5.00	4.4805	0.52912

Medang Homestay. It is clear that the mean score of the overall tourist perception on the water-based activities in Kuala Medang Homestay was at (M = 4.3485) which indicated that the respondents agreed with the water-based activities provided in the homestay programme.

The mean score for the overall perceived homestay activities from Table 4 is (M = 4.4805), by which it can be seen that majority of the respondents agreed that all the three homestay activities, which are culture and heritage activities, aborigines' activities, and water-based activities, were provided in the homestay programme.

5 CONCLUSION

No doubt that homestay programme in Malaysia is a tourism product that has a big potential because tourists all over the world come and visit Malaysia to experience the real Malaysian lifestyle. The Malay culture and customs represent the colours of Malaysia which therefore attract tourists hence it leads to several aspects in the tourism industry. Revenue is generated in terms of foreign currency by international tourists and as for the community and the whole country, domestic consumption has stimulated the economic flows. The villagers gained supplementary income from the tourism benefits. There has been an increase of local employment in tourism and homestay activities, and the development of the village infrastructure. Furthermore, greater and better preservation of natural resources and attractions as well as local culture and traditions have been taken into account in looking to the cultural and environmental aspect, as pointed out by Silparcha and Hannam (2009).

Moreover, homestay operators can identify the latest trends that tourists seek for in homestay programme. This benefits both parties such as local communities and homestay operators through the studies of tourist satisfaction on homestay activities and their behavioural intentions. It is seen that tourists prefer homestay activities which are really connected with culture and heritage, aborigines which are one of the oldest of people of Malaysia, as well as water-based activities. Therefore, all homestay hosts throughout Malaysia should focus more on these activities because tourists are looking for authenticity and originality of Malaysia.

Subsequently, the Ministry of Tourism and private sectors should be concerned about the natural resources as well as the culture and customs of the Malays, by preserving and sustaining it at all costs. This could be one of the most important factors in attracting more tourists to visit Kuala Medang Homestay. In addition, the Ministry of Tourism should create appropriate tourism policy such as carrying capacity in order to preserve the natural environment. Thaothampitak and Weerakit (2008) also voiced out that natural environment must be sustained and preserved at all time in order for more tourists to visit again in the future.

Finally, among the three categorized homestay activities, tourists longed for aborigines' activities with the highest mean score followed by culture and heritage activities which is then followed by water-based activities. This indicates that tourists, be it local or international, found that aborigines' activities are the most satisfying and enjoyed activities, by which also supported by the earlier findings on the most preferred activity by tourists. Therefore, the homestay hosts should make sure that the natural environment and settlement of the aborigines are well preserved as well as making sure that they maintain their traditional lifestyle.

As a conclusion, homestay hosts must provide various kinds of activities which tourists demand in order to make them experience the best out of the activities thus leads to positive post-behavioural intentions. The hosts should also try to reduce activities which are common and can be found anywhere to make sure that the natural environment is being less used.

REFERENCES

Arai, S.M., & Pedlar, A.M. (1997). Building Communities Through Leisure: Citizen Participation in a Healthy Community Initiative, Journal of Leisure Research, 29(1), 167–182.

Curtis, J.A. (2003). Demand for Water-based Leisure Activity. Journal of Environmental Planning and Management, 46(1), 65–77.

Cyprian, O.E. (2012). Adult Learning Development From The Prism Of Homestay. Higher Education Studies. 2(3). 77–87. doi: 10.5539/hes.v2n3p77.

Homestays in Pahang (2012). Retrieved on 1 October 2012 from www.go2homestay.com/pahang.

Hu, Y.C., Wang, J.H. & Wang, R.Y. (2012). Evaluating the Performance of Taiwan Homestay Using Analytic Network Process. Mathematical Problems in Engineering, 2012, 1–24.

Ibrahim, Y. & Razzaq, A.A.R. (2010). Homestay Program and Rural Community Development in Malaysia.

Jamak, A., A.B.S., Eusof, Md., A.J., & Sivapalan, S. (2008). Issues on Small Business Enterprise of Orang Asli Aborigines in Pahang, Malaysia. 5th International Conference on Small and Medium Sized Enterprises: Management—Marketing—Economics Aspect, Athens, Greece, 10–14 August 2008.

Jamak, A., A.B.S., Ali, M, R.M., Salleh, R. & Abdullah, A. (2010). Assessing Entrepreneurship And Micro Business Enterprise Of Orang Asli Aborigines In Pahang, Malaysia. Oxford Business & Economics Conference Program 2010.

Janiskee, R.L. (1996). Historic Houses and Special Events. Annals of Tourism Research. 23(2), 398–414.

Morajab, L. (2010). International Tourist Satisfaction and Dissemination Information towards Kampung Banghuris Homestay Attributes. Page 5–6.

National Association of State Park Directors (2004). State Park Facts. Retrieved 12 November, 2012, from http://www.naspd.org/.

Pahangtrekker.malaysiatrekker,com/kuala_lipis_home stay.htm.

Robert, J.J., Jones, L. & Bobo, M. (1996). The Physiologic Response Of Exercising In The Water And On Land With And Without the X1000 Walk'N Tone Exercise Belt. Res Q Exercise Sport, 67(3), 310–315.

Ryan, C. (1995). Researching Tourist Satisfaction. West Street, New York: Routledge.

Rosenfeld, R.A. (2008). Cultural and Heritage Tourism. Municipal Economic Development Tool Kit. Page 1–2.

Saaidin, H.M. (2012). Discovering Ecotourism Products in the Community of Kampung Kempadang and Kampung Deraka. Interdisciplinary Journal of Contemporary Research in Business. 3(12), 83–91.

Seubsamarn, K. (2009). Tourist Motivation To Use Homestays in Thailand and Their Satisfaction Based on the Destination's Cultural and Heritage-Based Attribute.

Silberberg, T. (1995). Cultural Tourism and Business Opportunity for Museums and Heritage Sites. Tourism Management. 16(5), 361–365.

Silparcha, W., & Hannam, K. (2009). Homestay and Sustainable Community Development.

Sofield, T.H.B., & Li, F.M.S. (1998). Tourism Development and Cultural Policies in China. Annals of Tourism Research, 25(2), 362–392.

Takeshima, N., Nakata, M., Kobayashi, F., Tanaka, K. & Pollock, M.L. (1997). Oxygen Uptake And Heart Rate Differences Between Walking On Land And In Water In The Elderly. JAPA, 5(2), 126–134.

Thaochalee, P., Laoakkha, S., and Panthachai, T. (2011). Home-Stay Tourist Villages: The Development of Cultural Tourism Management among Phuthai Tribal Groups in Isan. European Journal of Social Sciences—Volume 22, Number 4, Page 567–578.

Thaothampitak, W. & Weerakit, N. (2008). Tourist Motivation and Satisfaction: The Case Study of Trang Province, Thailand.

The Star Online. Strict Rules For Homestays To Ensure True Malaysian Experience. Retrieved 13 September, 2012, from http://thestar.com.my/news/story.asp?sec=nation&file=/2009/6/9/nation/4079048.

The Star Online. Tourism To Give Malaysia A Boost. Retrieved 13 September, 2012, from http://thestar.com.my/news/story.asp?file=/2012/5/2/nation/11192349&sec=nation.

Thongma, W., Leelapattana, W., & Hung, J.-T. (2011). Tourists' Satisfaction Towards Tourism Activities Management of Maesa Community, Pongyang Sub-District, Maerim District, Chiang Mai Province, Thailand.

Tourism Pahang. "10 Tourism Icons". Retrieved 22 May, 2013, from http://www.pahangtourism.org.my/index.php/explore/pahang-tourism-icons.

Wilks, J., & Atherton, T., (1994). Health and Safety in Australian Marine Tourism: A Social, Medical and Legal Appraisal. The Journal of Tourism Studies, 5(2), 1–16.

Yusof, Y., Ibrahim, Y., Muda, M.S., Amin, W.M., W.A.A. (2012). Community Based Tourism and Quality of Life. Review of Integrative Business & Economics Research, 1(1).

Hospitality and Tourism – Sumarjan et al. (eds)
© 2014 Taylor & Francis Group, London, ISBN 978-1-138-00151-0

The relationship between experiential marketing and tourists' satisfaction on Kelantan food

H.N. Syahida & A. Zainal
Universiti Teknologi MARA, Malaysia

ABSTRACT: Known as a Cradle of Malay Culture, Kelantan is famous for its distinctive cultural heritage, natural environments, and unparalleled hospitality of the locals including their unique foods which have influenced the development of the tourism industry. In relation to that, Kelantan Tourism Malaysia together with the help of corporate organizations aggressively promotes Kelantan food products locally and internationally. However, it raises critical questions of how tourists perceived Kelantan food during their visits. Thus, this study was to offer an approach of understanding the relationship between experiential marketing and tourists' satisfaction on Kelantan food. A total of two hundred and eight (208) international tourists around Kota Bharu, Pengkalan Chepa and Kubang Kerian area had participated in this study. Data were gathered through self-administered survey questionnaires and the results of the five dimensions of experiential marketing namely sense, feel, think, act and relate recorded a significant relationship with the tourists' satisfaction ($R^2 = 0.761$, $\beta = 0.872$) at significant level $p < 0.001$. Implication and recommendations for future research are also provided.

Keywords: experiential marketing, Kelantan food, tourist satisfaction

1 INTRODUCTION

As revealed by UNWTO (2012), 89 countries were showing the escalation in the number of tourists and Malaysia is not an exception. Previously, there was a progressive growth in the number of tourist arrivals from 10.2 million in 2000 to 24.7 million in 2011 (Tourism Malaysia, 2012). In this respect, Malaysia has recognized the tourism industry as a major contributor to the National economic revenues compared to other sectors (Tenth Malaysia Plan 2011–2015).

Recent research showed that food has been measured as an important element of satisfactory among tourists during holiday as they spend roughly 40 percent of their resources for local food consumption (Karim, 2010). Malaysia is simply a gastronomic paradise in terms of the exceptionality of food, beverages and food cultures derived from multi races. Due to this, all states in Malaysia strived to promote their own distinctive and unique identities by promoting their local cuisines (Zainal, Zali & Kassim, 2010).

The increment of tourists arrival in Kelantan from 941 319 in 2011 to 1211 754 in 2012 (Tourism Malaysia Kelantan, 2013), undoubtedly has great importance for the development of Kelantan tourism in the future.

As food becomes a major part of satisfactory in tourism nowadays, it is important to know how Kelantan food is evaluated and appreciated especially by the foreign visitors. In view of the above, William (2006) stated that experiential marketing is a current trend used to satisfy the tourist because it takes a postmodern and views them as emotional beings and concerned with achieving pleasurable experiences since the orientation of the tourism and hospitality environment are increasingly dynamic, unpredictable, unstable and competitive. Hence, in the absence of comprehensive discussion on experiential marketing in predicting tourists' satisfaction on ethnic food food, this current study will investigate into this area as a focal point.

2 LITERATURE REVIEW

2.1 *Food in tourism*

Food is important in the context of tourism. Food provides energy and essential nutrients needed for body functions, and thus, eating is at the same time regarded as compulsory in tourist activity (Richards, 2002). As food is important in travelling, tourists are willing to spend their money for food. Because of this, food and beverage expenditures amount to one-third of overall tourist expenditures in the global tourism turnover (Meler & Cerović, 2003).

In line with the above, the significant implications of tourist food consumption on destinations have received a growing research interest recently. Food has become an integral part of tourist experience and important element in tourism marketing. This is due to the fact that a food product is a source of pleasure and an unavoidable element in creating authentic and unique traveling experience. Owing to this, food is declared as an important element in determining the level of tourist satisfaction with integral tourist product (Grbac & Milohanovic, 2008), hence it contributes to the local economic impact.

Having said above, and with the evolution of food consumption during travelling, domestic and international visitors are becoming more adventurous and receptive to new experiences. The knowledge of culinary through printed and electronic media seems to be growing and tourists are not neophobia to genuine and authentic food which can be found in local eating-places. In view of this, there is a need to explore factors that contribute to enhance the satisfactory on food consumption. Therefore, experiential marketing is highlighted in this study and explained further below.

2.2 *Experiential marketing*

Experiential marketing is vital for business, especially in the hospitality and tourism industry (Erin & Kenny, 2008) since the products of this industry are always experiential (William, 2006) consumers increasingly expected engaging experiences and not just a process to purchase goods and services (Pine & Gilmore, 1999). Thus, experiential marketing which consisted of five basic dimensions which are sense (sensory), feel (inner feeling and emotion), think (intellect), act (bodily affect), and relate (relating with self, other people and culture (Schmitt (1999)) are closely related to tourists. Through conceptualization of these dimensions, experiential marketing did offer customer on experiences base as a focal point. This involves a creation of inspiring thoughts about comfort and pleasure, as well as inspiring a sense of practicality.

Notably, tourists are continuously seeking for fantasies, feelings, and fun (Hirschman & Holbrook, 1982) and they appreciate the benefits of happy adventures and the joyful processes of consumption rather than functions and benefits. These include the capture of sensory pleasure, variety, and cognitive stimulation (Keller, 1993). Hence, it is stressed that an experience orientation had become a mainstream nowadays.

In line with this, the application of experiential marketing have been discovered in many contexts and industry such as retailing (Verhoef, Parasuraman, Roggeven, Tsiros & Schlesinger, 2009), arts (Joy & Sherry, 2003), and hospitality and tourism (Pine & Gilmore, 2002; Leighton, 2007). These researchers agreed that the used of experiential marketing was able to create a positive emotion and contribute to the high satisfaction. Furthermore, experiential satisfaction in turn relates positively to loyalty intentions.

2.3 *Satisfaction*

Customer satisfaction has occupied a central position in the service and marketing literature (Trianasari, Sparks & Butcher, 2012). Erin and Kenny (2008) defined customer satisfaction as the evaluation and emotional recognition of the completely purchasing process after purchase and customer satisfaction plays an important role in achieving customer loyalty and profitability (Barsky & Nash, 2003).

Due to that, many researchers and academia have highlighted the importance of customers' satisfaction in today's market. Due to the fact that tourism consumers are more demanding, there is a need to generate more qualitative tourist products (Bruno & Ana, 2008). Their evaluation is influenced by number of factors such as the consumer's emotional state, psychographics and social environment that influence the judgment. They tend to compare their expectations with current experience rather than measuring the absolute service performance (Scott, Laws & Bosberger, 2009). Therefore, customers become the most valuable resource in measuring standards, products performance and monitoring continuing changes in market expectations. This shows the power has now shifted to the customer.

Owing to that, customers' satisfaction contributes towards profound effects on destination. Satisfaction is vital for profitability and future behavioral of customers. Satisfaction may not only indicate loyalty in terms of actual repatronage behavior (Straube, 1997), yet it become a predictor of a customer's willingness to recommend to others (Oh, 2000). In contrast, dissatisfied tourists may not return to the same destination and may not recommend it to other tourists (Chen & Chen 2010). Even worse, dissatisfied tourists may express complaining behavior on a destination and damage its market reputation (Reisinger & Turner, 2003).

Based on the literature, the following hypothesis was formulated:

H1: Experiential marketing (sense, feel, think, act and relate) influence tourists' satisfaction.

3 METHODOLOGY

This study was conducted through self-administered questionnaire and 400 questionnaires were distributed to the tourists in Kelantan in the area of Kota Bharu, Pengkalan Chepa and Kubang Kerian. Yet, due to the restricted resources such as time, money and workforce, only 208 (52.26%) of respondents were able to be gathered. Data were analyzed using descriptive statistics and linear regression. The questionnaire comprised of three sections namely experiential marketing (independent variable), tourists' satisfaction (dependent variable) and demographic profile. The instruments for experiential marketing were adapted and modified accordingly to the local context. Section A was from Lin, Chang, Lin, Tseng and Lan (2009); Maghnati, Kwek and Nasermoadeli (2012) while tourists' satisfaction (Section B) from Lin et al., (2009) and Wang, Lo, Chi and Yang (2004). All items utilized a 5-point Likert scale from 1-strongly disagree to 5-strongly agree.

4 RESULTS

4.1 *Profile of sample*

50.5% (105 respondents) of the respondents were male while the remaining 49.5% (103 respondents) were female with mean age in the range of 29–39 years. With regards to the marital status of the respondents, majority of them are single which contributed to 57.2% (119 respondents) and holds at least a bachelor's degree. Additionally, 33.2% (69 respondents) are working in privates sectors and majority of the household incomes were between RM6000 to RM8999 (approximately 27.9% or 58 respondents). Almost 59.6% (124 respondents) visited Kelantan for the purpose of holiday and 57.7% (120 respondents) were first-time tourists in Kelantan.

Table 1. The overall mean score on each variable and dimension

Part	Dimension	Mean (M)	Std. deviation (DV)
I	Experiential marketing		
	• Sense	3.80	0.60
	• Feel	3.89	0.64
	• Think	3.95	0.65
	• Act	4.03	0.65
	• Relate	3.94	0.64
II	Tourists' satisfaction	4.06	0.69

N=208.

4.2 *Hypothesis testing*

Table 1 depicts the descriptive analysis of experiential marketing and tourists' satisfaction.

Table 2 provides the results of the linear regression analysis to test the relationship between experiential marketing on tourists' satisfaction. The test determines whether experiential marketing have significant influence on tourists' satisfaction.

At the significant level 0.001, experiential marketing overall have significant positive influence towards tourist's satisfaction where $\beta = 0.872$, $R^2 = 0.761$.

5 DISCUSSION

Based from the findings presented, five dimensions of experiential marketing has been identified as high influence on tourists' satisfaction with a significantly high beta value $\beta = 0.872$ ($p < 0.001$) and correlation $R^2 = 0.761$, indicating that experiential marketing is a significant predictor for tourists' satisfaction. Thus, accepting the research hypothesis.

The findings for this present study was similar with previous research, where experiential marketing are said contribute to the customers' satisfaction (Erin & Kenny, 2008). In addition, the result of this study is also consistent with other researchers (Hosamy & Witham, 2010; Oh, Fiore & Jeoung, 2007) that five realms of experiential marketing leads to a new frontier of knowledge to measure tourists' satisfaction. Notably, these researchers also agree that the well-staged of experiential marketing will create a positive memories which subsequently influences tourists' evaluations including overall satisfaction.

Table 2. Results of linear regression analysis for experiential marketing on tourists' satisfaction.

Hypothesis	Regression R^2	Regression correlation β	Regression statistical sig.
H1: Experiential marketing influence tourists' satisfaction	0.761	0.872	0.000**

CI = confidence interval, $^{**}p < 0.001$.

Hence, based on previous literatures and the findings of this study, it is safe to conclude that experiential marketing has taken place and properly delivered by the state of Kelantan as a travel and food heaven. Tourists found that the food in Kelantan is appealing and the uniqueness of the food gives a feeling of creation when consuming Kelantan food. The mildness (less spicy) and sweetness in taste of most food has led to the travelers' acceptance of the local food and thus, may influence them to spread good words-of-mouth to friends and relatives. Importantly, this dissemination of good words is able to bring up Kelantan's name as a food heaven.

6 IMPLICATION

The hypothesis result evidently shows that the experiential marketing influenced tourists' satisfaction and majority of them are able to accept the taste of Kelantan food when reasonably priced. Therefore, food can be one of the tourism products for Kelantan besides the fascinating destination. Hence, Kelantan will be able to promote food tourism product to the tourists and will contribute further to the local economy.

In view with the above suggestion, this study will give clarification on factors that influence tourists' satisfaction on Kelantan food. In attracting more tourists to experience Kelantan food, sense pleasurable play an important role. This includes aroma, taste and appeal. Furthermore, the findings presented can inform marketer by suggesting means for enhancing emotion driven marketing. It can be done by developing products and promotions that motivates tourists to imagine and create subjective meaning in the mind of tourists that can be constantly reinforced. It is important since the goal is to create a compilation of memory about Kelantan food which can be shared with family and friends once they return. Thus, tourism bodies and food operators can take a proactive action by focusing on the post-purchase stage. For example, they can send postcards and online newsletter updates to spur their memory and positive word of mouth besides creating a web platform to encourage tourists to share their experience regarding their visits to Kelantan especially on food with other viewers.

7 LIMITATIONS AND RECOMMENDATIONS

Several limitations existed in this current research. The empirical analyses are conducted based on data collected at a specific time due to constrain in resources such as time and money. In conjunction with that, the researcher had co-operated with tour guide and hired the questionnaire distributors in Kelantan for data collection. Hence, short briefings were conducted by the researcher in order to clearly explain the objective of this study. Another limitation that the present study faced was the lack of participation and cooperation from the respondents (international tourists). Some of the tourists rushed to check-in for departure (airport) and limited time to shop (shopping mall) thus, not giving full cooperation. They felt interrupted and implicitly or explicitly declined to participate.

Besides, convenience sampling method used in the current study which has no statistical methods for measuring the sampling error thus unable to generalize the findings to the target population with any measured degree of confidence (Hair, Money, Page, & Samouel, 2007). Therefore, probability sampling methods are suggested for future research to ensure that the sample is representative. Lastly, although intentionally this study would like to look into tourists satisfaction, other perspectives such as future behavioral intention could be done. Hence, the need to look at future behavioral intention is recommended for future research.

8 CONCLUSION

Looking at the high level of satisfaction highlighted by the respondents, it shows the current trend of tourism desired by tourist nowadays are seeking the experiential opportunity to enhance their experience with interesting activities. It can be concluded that tourism bodies, government and related sectors has to cooperate and support each other in revealing the significant influence of experiential marketing in predicting tourists' satisfaction on Kelantan food. Besides that, aggressive and continual quality improvement and promotional efforts are needed in order to ensure the potential of Kelantan food in attracting international tourist to visit Kelantan will be fully utilized. Thus this will embark and promote Kelantan as food destination to tourists.

REFERENCES

Barsky, J., & Nash, L. (2003). Customer satisfaction: Applying concepts to industry-wide measures. *Cornell Hotel and Restaurant Administration Quarterly, 44* (5/6), 173–183. doi: 10.1177/001088040304400523.

Bruno, G., & Ana, M. (2008). *Contribution of Food Products in Creating Cultural Identity of Tourist Destination.* WSEAS International Conference on Cultural Heritage and Tourism (CUHT'08), Heraklion, Crete Island, Greece. Retrieved from http://www.wseas.us/e-library/conferences/2008/crete/cuht/cuht14.pdf.

Chen, C.F. & Chen, F.S. (2010). Experience quality, perceived value, satisfaction and behavioral intentions for heritage tourists. *Tourism Management*, *31*(4), 29–35.

Erin, Y.Y., & Kenny W.C. (2008). Relationships among experiential marketing, experiential value and customer satisfaction. *Journal of Hospitality & Tourism Research*, *32*(3), 387–410. doi: 10.1177/1096348008317392.

Gilmore, J.H. & Pine, B.J. II (2002). Differentiating hospitality operations via experiences. *Cornell Hotel & Restaurant Administration Quarterly*, *43*(3), 87–96. doi: 10.1016/S0010-8804(02)80022-2.

Grbac, B. & Milohanovic, A. (2008). *Contribution of Food Products in Creating Cultural Identity of Tourist Destination.* Paper Presented at the WSEAS International Conference on Cultural Heritage and Tourism, Crete Island, Greece. Retrieved from http://www.wseas.us/e-library/conferences/2008/crete/cuht/cuht14.pdf.

Hair, J., Money, A., Page, M. & Samouel, P. (2007). *Research Methods for Business*. England: John Wiley & Sons Ltd.

Hosamy, S. & Witham, M. (2010). Dimension of cruisers' experiences, satisfaction and intention to recommend. *Journal of Travel Research*, *49*(3), 351–364. doi: 10.1177/0047287509346859.

Joy, A. & Sherry, J.F. Jr (2003). Speaking of arts as embodied imagination: A multisensory approach to understanding aesthetic experience. *Journal of Consumer Research*, *30*(2), 259–282.

Karim, A.S. & Chi, G.Q.C. (2010). Culinary tourism as a destination attraction: An empirical examination of the destination's food image. *Journal of Hospitality Marketing & Management, 19*(6), 531–555.doi: 10.1080/19368623. 2010.493064.

Leighton, D. (2007). Step back in time and live the legend: Experiential marketing and the heritage sector. *International Journal of Nonprofit and Voluntary Sector Marketing, 12*(2), 117–125.

Lin, K.M., Chang, C.M., Lin, Z.P., Tseng, M.L., & Lan. L.W. (2009). Application of experiential marketing strategy to identify factors affecting guests' leisure behaviour in Taiwan hot-spring hotel. *WSEAS Transaction on Business and Economics*, *6*(5), 229–240. Retrieved from http://www.wseas.us/e-library/transactions/economics/2009/31-814.pdf.

Maghnati, F., Ling, K.C., & Nasermoadeli, A. (2012). Exploring the Relationship between Experiential Marketing and Experiential Value in Smartphone Industry. *International Business Research*, *5*(11), 500–520. doi: 10.5539/ibr.v5n11p169.

Ministry of Tourism Malaysia (2007). Fabulous Food 1Malaysia. Retrieved on October 2012 from http://www.fabfood1malaysia.com/.

Oh, H. (2000). Diners' perceptions of quality, value and satisfaction. *Cornell Hotel and Restaurant Administration Quarterly*, *41*(3), 58–66.

Oh, H., Fiore, A.M., & Jeoung, M. (2007). Measuring experience economy concepts: Tourism applications. *Journal of Travel Research*, *46*(2), 119–132. doi: 10.1177/001088040004100317.

Reisinger, Y. & Turner, L.W. (2003). *Cross-cultural Behaviour in Tourism: Concepts and Analysis*. Butterworth-Heinemann: Oxford.

Richards, G. (2002). Gastronomy: An essential ingredient in tourism production and consumption?. In A.M. Hjalager & G. Richards (Ed.), *Tourism andgastronomy* (pp. 3–20). London: Routledge.

Scott, N., Laws, E., & Bocksberger, P. (2009). The marketing of hospitality and leisure experiences. *Journal of Hospitality Marketing and Management, 18*(2–3), 99–110. doi: 10.1080/19368620802590126.

Straube, P. (1997). Why customer satisfaction may not mean loyalty. *Nation's Restaurant News*, *31*(30), 32–33.

Tenth Malaysia Plan recognized tourism industry. (2010). Tourism industry. Retrieved on November 2012, from http://www.pmo.gov.my/dokumenattached/RMK/RMK10_Mds.pdf.

Tourism Malaysia, (2012). *Malaysia Economy Updates.* Retrieved on October 2012 from http://www.tourismmalaysia.gov.my.

Tourism Malaysia Kelantan (2013). International tourist arrival in Kelantan. Retrieved on April 2013 from http://kelantan.gov.my/tic/.

Trianasari, B. Sparks, B., & Butcher, K. (2012). The role of cultural familiarity on consumer satisfaction in the context of hotel service failure. *Current Issues in Hospitality and Tourism Research and Innovations*.

UNWTO (2012). International tourism receipts and expenditure. Retrieved on November 2012 from www.unwto.org/facts.

Verhoef, P.C., Lemon, K.N., Parasuraman, A., Roggeveen, A., Tsiros, M. & Schlesinger, L.A. (2009). Customer experience creation: Determinants, dynamics and management strategies. *Journal of Retailing*, *85*(1), 31–41. doi: 10.1016/j.jretai.2008.11.001.

Wang, Y., Lo, H.P., Chi, R. & Yang, Y. (2004). An integrated framework for customer value and customer-relationship-management performance: A customer-based perspective from China. Managing Service Quality, *2*(3), 169–182. doi: 10.1108/09604520410528590.

Williams, A. (2006). Tourism and hospitality marketing: fantasy, feeling and fun. *International Journal of Contemporary Hospitality Management*, *18*, 482–495. doi: 10.1108/09596110610681520Q.

Zainal, A., Zali, A.N., & Kassim, M.N. (2010). Malaysian gastronomy routes as a tourist destination. *Journal of Hospitality, Tourism and Culinary Arts*, *2*(1), 15–24.

Hospitality and Tourism – Sumarjan et al. (eds)
© 2014 Taylor & Francis Group, London, ISBN 978-1-138-00151-0

Tourist perception on island tourism: A case of Tunku Abdul Rahman Marine Parks

E.R. Kassim, S.A. Jamal, M.F. Ahmad & M.D. Darson
Universiti Teknologi MARA, Malaysia

ABSTARCT: This study is to explore in-depth understanding of the tourist perceptions towards the activities and services provided and to investigate the overall perception of tourist towards the activities and services provided by tour operator in Tunku Abdul Rahman Marine Park. However, if a customer's expectation is not met by the perceived performance, dissatisfaction experience will occur. Therefore, satisfaction, being positive or negative, can be determined by performance and regarded as a vital basis of competitive issues. The result from this study attempt to gain overall perception of tourist towards the activities, price, satisfaction and quality of services provided by tour operator in Tunku Abdul Rahman Marine Park and from the result researcher could determine the factor of behavioural intention toward the Marine Park, or island. The significance of this paper will soon help to enhance tour operator's services and give a wonderful experience to the tourist and meet the tourist needs and wants in future. Findings clearly revealed that tourist perception on island activities, service quality and price value are positively or negatively influenced tourist satisfaction in island tourism and their intention to revisit.

Keywords: island activities, mediating, revisit intention

1 INTRODUCTION

Currently in Malaysia there are six Marine Parks with a total of 41 islands which are the Tunku Abdul Rahman Marine Parks (5 islands), Payar Island Marine Park (4 islands), Redang Island Marine Park (9 islands), Tioman Island Marine Park (7 islands), Mersing Marine Park (13 islands) and Labuan Marine Park (3 islands) (Abdul Khalil Abdul Karim, 2003). All these Marine Parks are developed as tourist destinations. This is because tourism is not regarded as destructive towards the marine park's marine ecology and leisure activities if they are sustainably managed. In general, the level of tourism sector's development in the marine parks is very good due to the fact that almost all marine parks are fully equipped with various facilities for the tourists such as resorts, chalets, restaurants, water sport activities, souvenir shops, jetties and others. Thus, the influx of tourists from both local and abroad can be observed yearly. However lately, the robust development of the tourism industry and the high number of tourist arrivals in year 2006–2009 were 989,197 tourists visited the marine parks and seen as invitation to destroy the marine park's ecosystem. Thus, this research concentrates on tourist perception, island activities, and quality of service in Tunku Abdul Rahman Marine Park. Comprises of five island Tunku Abdul Rahman Marine Park is gazetted in 1974, as one of the unique features with high esthetical values as a natural heritage displays untouched quality of nature making it ideal for those seeking a more tranquil and deserted atmosphere. Previous studies have explored and looked into island development focusing the impacts on marine (William & Thomas, 1991), the illegal immigrants in Sabah (Hassan, Omar & Dollah, 2010), economic impacts of tourism within the island (Whiteoak, 2009), as well as sustainable tourism development in the Islands (Hamzah & Hampton, 2011). While Ryan (1995) urged that the attitude, expectations, and perception of holiday maker are significant variables in setting goals, influencing behavior and determining final satisfaction. Therefore this study will fill in the gap by looking at the perceptions of tourist towards the activities and services provided on the island. The purpose of the present research is to investigate the Tourist discernment and reputation towards the activities and services provided by tour operator in Tunku Abdul Rahman Marine Park which is one of the conserve marine parks in Sabah. The main objective isto investigate the overall tourist perception towards the activities and services provided by tour operator in Tunku Abdul Rahman Marine Parks.

2 LITERATURE REVIEW

2.1 *Tunku Abdul Rahman Marine Parks*

Tunku Abdul Rahman Marine Parks are circle around all that is to discover under water, Diving and snorkelling are the most popular activities besides sun bathing along the many public beaches that can be found on all of the 5 islands namely Manukan, Mamutik, Sapi, Gaya, and Sulug island that make up the marine park. Other activities include kayaking, sailing, picnicking, barbeque lunch buffet and beach water sports. Tourist can stay at the biggest island which is Pulau Gaya as they offer world class island resorts and managed by international well known company for example Gaya Island Resort, Bunga Raya Island Resort and Gayana Eco Resort, for tourist who are adventurous they can do some light jungle trekking, as there are over 20 km of trails throughout the island and tourist can experience the amazing flora and fauna along the trails. During these treks you might encounter bearded pigs, proboscis monkeys, long-tailed macaques, monitor lizards, hornbills and even a few snake species.

2.2 *Island activities*

Su, Jiang, & Tu (2004) urged that marine leisure activities require physical and active participation by the person involved. Island water sport activities focus on the basic and advanced level of water related activities which take place in the ocean with specific emphasis on the importance of adhering to safety regulation and proper usage of equipment, the development of each participant's ability, and provide a comprehensive understanding and appreciation of the environmental issues. Furthermore marine leisure implies a more passive form of recreational activity, such as taking advantage of the seaside resort, cruise ships, marine parks and tourist submarines (Yu, 2009). The types of island activities are banana boating, parasailling, fly fish, scuba diving, snorkelling, sea walking, sport fishing and others, While on the other hand, many researchers believed that this categorization of the island recreation market will help to facilitate the relative activities in each of these subsections (Bammel & Burrus-Bammel, 2009).

2.3 *Service quality on island activities*

Rahim and Nasir (2011) supported that Service quality refers to the highest possible delivery of service standard based on the elements of tangibility, reliability, responsiveness, assurance and empathy of service provided to the customer. In recent years, numerous studies have focused on customers' satisfaction and service quality in the hospitality industry. Several contributions have been made in relation to various mechanisms for improving customer satisfaction (Fitsmmons, 2006). Nevertheless service quality is a focused evaluation that reflects the customer's perception of specific dimensions of service: reliability, responsiveness, assurance, empathy, tangibles (Zeithaml, 2000). In line with the assertion Lori et al. (2003) stated that, the objective of the instrument is to provide a tool for assessing customer perception of service quality in service and retailing organization. Researcher added that the perceived quality is the degree and direction of discrepancy between customer's perceptions and expectations. Customer perceptions are subjective assessments of actual service experiences. The instrument was based on the principle that service quality is different between the customers' expectations and the service they received.

2.4 *Price value*

Delivering high quality service within the hospitality industry positively influences a customer's perception of value. The practice of the purchase decision originates from the trade-off between benefits and sacrifices (Al-Sabbahy, Ekinci, & Riley, 2004). Customers judging the value of the services to is worth the cost paid at the end of the sporting activities and will tend to be more satisfied with the service when they actually receive what they have expected. In other words perceived value pricing is an important price setting practice. Perceived value pricing indicates the importance of providing remuneration and functionalities to the consumer and the simultaneous need to price it effectively so that the firm can take appropriate value (Kanagal, 2010). Price related factors such as perceived value or money well spent may not be able to be evaluated until such time that the service has been performed (Kennett, Sneath, & Henson, 2001). Value may be evaluated based on factors such as cost of tickets, cost of activities, and cost of equipment rental, concessions and merchandise. Customers are not able to judge service quality before the service is performed.

3 METHODOLOGY

The population were the tourists' of Tunku Abdul Rahman Marine Parks consist of international and local visitors. The number of sample was 315 respondents based on the total visitors to the marine parks. Convenience sampling approach was used as the tourist perception data is to be collected from respondents in five islands in the marine parks. The element of convenience sampling was used where a selection of respondents will be based on the participants who are willing to participate in the survey. The survey questionnaire was divided into six main sections, The first section (Part A) was the respondents demographic profiles, (Part B) on perceived island activities, (Part C) on service quality, (Part D) price value, and lastly (Part E) on overall satisfaction. The type of instrumentation used in this research was a seven point summated rating scale, each scale has a minimum of "1" and a maximum of "7" which means a very strong attitude in favour of the statement while a score closed to "1" means a very strong attitude against the statement.

As the study was to access the perceptions of tourist towards the activities and services provided in the island, the questionnaires were distributed in the five islands in Tunku Abdul Rahman Marine Parks. Nevertheless before carrying out the survey, permission was requested from Sabah Park Department to enter the parks legally, and they had issued an annual permit for conducting research in the marine parks. Questionnaires were distributed to visitors who have experienced services and those who participated in at least one of the island activities.

4 FINDINGS & DISCUSSION

4.1 *Demographic*

Number figures consecutively in the order in which the total number of respondents who participated in this research was 315 respondents. The participants consisted of 43.2% male and 56.8% female. Most of the respondents were in the age range of 20 to 24 years old and only 17 respondents were 55 to 64 years old. Majority of the respondents were local (80.3%) and (19.7%) were international tourist. More than half respondents (60.3%) were single and (39.7%) were married. Most of the respondents (35.2%) were travelling with their friends and partner, (18.4%) with family, (17.5%) with spouse and group tour. While (83.5%) of respondent were travelling day trip to the island and only (6%) staying for one night or more. Majority of the respondents' choose snorkelling activities as their most

Table 1. Overall mean scores for perceived island activities variable.

Variable	N	Minimum	Maximum	Mean	Std. deviation
Activities	315	3.29	6.86	5.3605	0.66844
Valid N (listwise)	315				

Table 2. Overall mean scores for perceived service quality variable.

	N	Minimum	Maximum	Mean	Std. deviation
Service quality	315	3.89	6.44	5.3855	0.53671
Valid N (listwise)	315				

Table 3. Overall mean scores for perceived service quality variable.

Variable	N	Minimum	Maximum	Mean	Std. deviation
Price	315	3.00	6.14	5.1383	0.68870
Valid N (listwise)	315				

Table 4. Overall mean scores for tourist perception.

	N	Minimum	Maximum	Mean	Std. deviation
Descriptive statistics					
Perception	315	3.74	6.22	5.3141	0.51333
Valid N (listwise)	315				

preferred activity (70.8%) and the least preferred activity was glass bottom boat (0.6%).

4.2 *Result of overall mean score*

The overall visitors' perception on island activities provided is summarized in Table 1, from the table it shows that the overall perception on perceived island activities in Tunku Abdul Rahman Marine Parks were at (M = 5.360) which indicate that the respondents slightly agreed with the activities offered by tour operators. Therefore, this analysis answered objective one of the study.

The overall visitors' perception on service quality provided is summarized in Table 2. From the table it shows that the overall perception on perceived service quality in Tunku Abdul Rahman Marine Parks were at (M = 5.385) which indicate that the respondents slightly agreed with the services offered by tour operators, and just met their expectations, however tourist perception did not exceeded the mean 6.00 above to describe better perception on service quality.

This section reports the variable of perception respondents on price value in the island obtain during their trip. The descriptive statistics below show the average mean of (M = 5.138) which indicate that respondents slightly agreed with the price offered and the price of activities worth for its services perceived by tourists.

This section reveled the overall tourist perception on island activities. Based on pattern of the mean score in Table 4, most of the respondents were more inclined toward 'slightly agree' feelings with activities, services and price of activities offered in the marine parks.

5 CONCLUSION

From this study, tourists as main target market portrayed their attitudes toward perception on the island and their overall satisfaction in Tunku Abdul Rahman Marine Parks through various dimensions. Study outcomes clearly revealed that large fraction of tourists slightly agreed for most of the dimensions being asked. Most respondents expressed a moderate expectation as they have some clear or insufficient information about the marine parks. Besides, it is proven in this study that expectation having a significant effect to the tourists' satisfaction which consequently affects their loyalty to revisit and recommend to others. State government and responsible authorities should emphasize more on the policy planning in order to develop a good and clear image of the island tourism destination Sabah since Sabah is well known as an Eco-Tourism destination. A comprehensive approach through collaboration with various sector especially tourism service providers are needed to establish a recognizable image that portray the specialty of Sabah islands attraction and differentiate from other states in Malaysia. These findings have given some implications not only for the state government, but also the tourism service providers as a whole.

REFERENCES

Al-Shabbahy, H.Z., Ekinci, Y.. & Riley, M. (2004). An investigating of perceived value dimensions: Implications for hospitality research. *Journal Of Travel Research*, 42(3), 226.

Bammel, G., & Burrus-Bammel, L. (1996). *Leisure And Human Behavior* (3rd ed.). Dubuque, IA, USA: Brown & Benchmark.

Fitszmmon, J.A., & Fitszmmon, M.J. (2006). *Service Management: Operations, Strategy, Information Technology*. United States of America: McGraw-Hill Higher Education.

Hassan, W.S., Omar, M.A., & Dollah, R. (2010). The illegal immigrants in sabah: why do they come back? *Borneo Research Journal,* 4, 115–128.

Kanagal, N.B. (2010). Conceptualization of perceived value pricing in strategic marketing. *Journal of Management and Marketing Research,* 12, 1–17.

Karim, A.K (2003). Malaysian legislation on the management of marine protected areas and marine parks. *Report of the Regional Symposium on Marine Protected Areas and Their Management, Chennai.*

Kennett, P., Sneath, J., & Henson, S. (2001). Fan satisfaction and segmentation: A case study of minor league hockey spectators. *Journal of Targeting Measurement and Analysis for Marketing,* 2, 132–142.

Lori S.C. (2003). Exploring Linkages between Quality System, Service Quality, and Performance Excellences: Service Providers' Perspectives. *Journal of Quality Management,* 9(2), 44–56.

Rahim, R.C., & Nasir, N. (2011). The Measurement of Service Quality using SERVQUAL: The Case Study of Peladang Setiu Agro Resort, Terengganu, Malaysia. *Innovation, Development Sustainability, and Economic Growth*, 1112–1127.

Ryan, C. (1995). *Researching Tourist Satisfaction*. West Street, New York: Routledge.

Su, D.J., Jiang, M.H., & Tu, J. (2004). The study of using fish habors to develop marine recreation sport using the example of Taipei country. *A Report Was Entrusted And Studied By National Council On Physical Fitness And Sports.*

Whiteoak, K. (2009). The potential impacts of climate change on the phillip island little penguin colony-regional economic impacts. *Westernport Greenhouse Alliance*, 1–31.

Yu, J.H. (2009). Perception of tourist attraction and satisfaction levels toward marine leisure sports in Kenting Taiwan. Proquest Dissertation and Theses, 1–125.

Zeithaml, V.A. (1988): Consumer Perceptions of Price, Quality and Value: A Means-end Model and Synthesis of Evidence. *Journal of Marketing,* 52(1), 2–22.

Hospitality and Tourism – Sumarjan et al. (eds)
© 2014 Taylor & Francis Group, London, ISBN 978-1-138-00151-0

The impact of food quality and its attributes on customer satisfaction at Malay restaurants

M.A.A. Bashir, H.F. Ariffin, N. Baba & S. Mantihal
Universiti Teknologi MARA, Malaysia

ABSTRACT: The foodservice business today is a dynamic industry. This can be seen from the assortment of foodservice establishments escalating in all sub-sectors of the industry, including restaurants. Food quality appears to be acknowledged as elemental factor to satisfy restaurant customers; yet, most restaurant quality–related studies focused on atmospherics and service delivery, often neglecting the significance of food itself as the foundation of restaurant. The objectives of this study were to assess the effects of food quality on satisfaction and to further examine the contribution of individual food attributes on satisfaction. With regression analyses, the results demonstrated overall food quality significantly affects customer satisfaction; on top, food taste was the greatest contributor to it. Thus, restaurant operators should pay attention to the key food quality attributes that elicit customer satisfaction that eventually led to long term business success in terms of customer loyalty, organizational profits and return patronage.

Keywords: customer satisfaction, food quality, food quality attributes, Malay restaurants

1 INTRODUCTION

Looking at the phenomenon of increasing numbers of Malaysians dining out, entrepreneurs have taken this golden opportunity to operate businesses in foodservice especially restaurant industry. Consequently, with increasing competition between restaurants, attracting new customers can no longer guarantee profits and success, but retaining existing customers is of more importance. In fact, a competitive environment provides customers with more alternatives to choose from (Haghighi, Dorosti, Rahnama, & Hoseinpour, 2012). As a result, the key to sustainable competitive advantage lies in delivering high-quality service that will in turn lead to satisfied customers. Relating to restaurant industry, total foodservice encompasses both tangible and intangible components. With regard on tangible components, this is the time when food quality has taken its important role in sustaining restaurant business. Food quality has been considered as a fundamental element of customers' dining experiences that affects customer satisfaction and determines future behaviors (Ha & Jang, 2010). Namkung and Jang (2007) as well affirmed that quality of food is one of the best means to maximize success in the restaurant business. Despite the importance of food quality in restaurant businesses, according to Namkung and Jang (2007), a critical challenge facing restaurant industry these days is to offer quality food that is not only compelling for the customers but also can be better to business competitors. They further argued that although food quality has been viewed as important, most restaurant quality–related studies have focused on atmospherics and service delivery, often neglecting the importance of food itself as the core competency of a restaurant. Therefore, the objectives of this study were to assess the effect of food quality on satisfaction and to further examine the contribution of individual food attributes on satisfaction.

1.1 *Food quality and food quality attributes*

Among the vital elements in consumer food perceptions and food choice decisions is the quality (Grunert, 2005). Generally, consumers favor products of high quality including the choice of food that they are consuming. Hence, it is necessary to know consumers' own perceptions of quality as consumers typically will be making purchasing decisions on these beliefs (Rijswijk & Frewer, 2008). Food quality has been commonly accepted as a fundamental element of the overall restaurant experience (Ryu, Lee, & Kim, 2012; Ha & Jang, 2010; Namkung & Jang, 2007). According to Peri (2006), food quality is an essential condition to satisfy the needs and expectations of customers.

Despite the importance of food quality in restaurant business, there is no agreement on the individual attributes that represent food quality. Sulek and Hensley (2004) included all food attributes into only one variable, food quality, while Kivela, Inbakaran, and Reece (1999), who designed a model of dining satisfaction and return patronage, saw that food quality had numerous attributes: presentation, tastiness, menu item variety, and temperature. Yet, little has been done with the critical attributes of food quality in relation to customer satisfaction. A comprehensive review of the literature reveals that the general description of food quality among researchers focuses on (a) presentation, (b) variety, (c) healthy options, (d) taste, (e) freshness, and (f) temperature.

Presentation refers to how attractively food is presented and decorated as a tangible cue for customer perception of quality. Kivela et al. (1999) pointed out that the presentation of food is a key food attribute in modeling dining satisfaction and return patronage. In addition, Raajpoot (2002) described food presentation as one of the product and/or service factors in TANGSERV.

Variety entails the number or assortment of different menu items. Restaurants continuously develop new menus to attract diners, and several proactive restaurateurs have created a variety of food and beverage offerings. In previous studies, menu item variety was a crucial attribute of food quality in creating dining satisfaction (Raajpoot, 2002; Kivela et al., 1999).

Healthy options involve offering nutritious and healthy food. Johns and Tyas (1996) stated that healthy food could have a significant effect on customer perceived evaluation of the restaurant experience. Kivela et al. (1999) noted the importance of healthy foods in restaurants and suggested nutritious food as one of the core properties in dining satisfaction and return patronage. The more notable thing is that restaurant customers are increasingly interested in healthy menu items (Sulek & Hensley, 2004).

Taste is viewed as a key attribute in food in the dining experience (Kivela et al., 1999). Many customers have become knowledgeable about food; as a result the taste of food in restaurants has become ever more important (Cortese, 2003). Thus, it is not surprising that shabby restaurants with gourmet cooking are packed with customers. Thus, taste is usually believed to influence restaurant customer satisfaction and future behavior intentions (Kivela et al., 1999).

Freshness generally refers to the fresh state of food and appears to be correlated to crispness, juiciness, and aroma (Péneau, Hoehn, Roth, Escher, & Nuessli, 2006). In previous studies, freshness of food has been cited as a key fundamental quality indication of food (Kivela et al., 1999; Johns & Tyas, 1996).

Temperature is also a sensory aspect of food quality (Kivela et al., 1999; Johns & Tyas, 1996). According to Delwiche (2004), temperature influenced how the perceived flavor of food was evaluated, work together with other sensory properties such as taste, smell, and sight. Thus, temperature could be considered as one determinant enhancing enjoyment in the food experience.

1.2 *Customer satisfaction*

Customer satisfaction is an essential signal of a company's past, current, and future performance and, as a result, has long been a critical focus among marketing practitioners and scholars (Oliver, 1999). Many researches have been made on the topic of consumer satisfaction. Customer satisfaction is principally a customer's evaluation of a good or service in terms of whether it has met their needs and expectations or not (Lamb, Hair, & McDaniel, 2011).

In modeling satisfaction, two general conceptualizations of satisfaction exist in the literature: transaction-specific satisfaction and cumulative satisfaction (Anderson & Fornell, 1993). In the transaction-specific perspective, satisfaction is momentary: How happy is a customer with a product or service at a certain point in time (Cronin & Taylor, 1992). Instead, cumulative satisfaction is a customer's evaluation of the total consumption experience with a product or service to date, which directly affects post purchase phenomena such as attitude change, repeat purchase, and brand loyalty (Johnson & Fornell, 1991).

The main concept of customer satisfaction is based on Lewin's (1938) expectancy–disconfirmation theory, which speculates that satisfaction judgment is based on comparisons between expectations held a priori and the perceived performance post hoc by a customer with an evaluative norm. The expectancy–disconfirmation model emphasize that customer satisfaction is a function of subjective disconfirmation under various conditions (Churchill & Surprenant, 1982). More specifically, the consumer is satisfied if perceived performance exceeds consumer expectations (a positive disconfirmation). On the other hand, the consumer is dissatisfied if perceived performance falls short of his or her expectations (a negative disconfirmation). This subjective evaluation toward confirmation or disconfirmation causes satisfaction-related emotions.

Customer satisfaction covers a lot of elements whether it is satisfaction over the performance of a product, consumption of a product, a service rendered or an idea given. It is also a well established fact that satisfied customers led to long term business success in terms of customer loyalty, organizational profits and return patronage (Ryu & Han, 2010; Namkung & Jang, 2007).

1.3 *Relationship between quality and satisfaction*

In recent years, academic and managerial concern in the conceptualization and measurement of service quality and consumer satisfaction has amplified in the services marketing literature. Although these two constructs appear to be similar, researchers have questioned whether service quality and customer satisfaction are the same (Iacobucci, Grayson, & Ostrom, 1994; Dabholkar, 1993).

On the other hand, Dabholkar (1995) claimed that the two constructs are different for recent customers, although they overlap for long-term customers because customer satisfaction evaluations develop more and more cognitive over time. Assuming that the two constructs are distinct, theoretical and empirical arguments for the order of occurrence between quality and satisfaction have continued (Taylor & Baker, 1994). Conventionally, researchers have suggested that incidents of satisfaction over time generate perceptions of quality (Bitner, 1990; Parasuraman, Zeithaml, & Berry, 1988). However, more recently, the contrary view appears to be strongly favored. Oliver (1993) first suggested that service quality was the antecedent of customer satisfaction no matter whether these constructs were measured for a specific time or over time. Several researchers have provided empirical support for Oliver's argument that customer satisfaction is a consequence of service quality (Anderson & Sullivan, 1993). Recognizing that satisfaction is super-ordinate to quality, Taylor and Baker (1994) also observed quality as one of the key service dimensions factored into consumer satisfaction constructs. Generally, researchers in marketing agree that satisfaction and perceived quality are greatly interrelated (Bitner & Hubbert, 1994; Churchill & Surprenant, 1982) and have maintained that perceived quality is one of the foundation determinants of overall satisfaction (Oliver, 1997; Fornell, Johnson, Anderson, Cha, & Bryant, 1996).

2 METHOD

2.1 *Instrument development*

A questionnaire was developed based on customer perceptions of food quality attributes and customer satisfaction in relation to the restaurant experience. The six selected food quality attributes were presentation, menu item variety, healthy options, taste, freshness, and temperature. The perceived quality of food attributes and satisfaction was measured on a 7-point scale ranging from 1 (extremely disagree) to 7 (extremely agree).

Table 1. Reliability coefficient for each section of the questionnaire.

Sections	No. of items	Cronbach's alpha
Section A: Food quality	18	0.96
Section B: Satisfaction	6	0.96

*Note: No. of respondents = 30.

Before the questionnaire was finalized, two restaurant managers in full-service restaurants and two academic professionals in the hospitality industry reviewed the questionnaire to ensure content validity. Subsequently, a pilot study was conducted to ensure the reliability of each construct.

The reliability of measurements during pilot study was well above the recommended cutoff of 0.70, indicating internal consistency (Nunnally, 1978).

Care was taken to rephrase terms used in the questions to suit common usage in Malaysia. Feedbacks from respondents were considered and some changes were made before arriving at the final version of the instrument.

2.2 *Data gathering procedure*

The type of sampling was non-probability sampling. Data were collected based on convenience sampling since the respondents selected were mainly from Malay restaurants' customers in Shah Alam, Selangor. The selected Malay restaurants for the survey were chosen based on several characteristics, which are (1) offers Malay cuisines on the menu, (2) medium price range of RM10–RM20 per person, and (3) offer full-service to the customers. The researcher conducted the survey at five selected mid-scale Malay restaurants in Shah Alam area in the month of April 2013 during lunch and dinner time. Restaurant patrons were approached and requested to complete the survey and were informed that their individual responses were anonymous and confidential.

A total of 400 questionnaires were distributed. However, after performing data screening and cleaning procedures, only 330 usable questionnaires were obtained for further data analyses.

3 RESULTS

3.1 *Sample profile*

Among the respondents, 77.3% (n = 255) of them were in the range of 20–29 years old, 21.2%

($n = 70$) were 30–39 years old, while the remaining 1.5% ($n = 5$) was 40–49 years old. In regards to gender, there were a slightly high proportion of male customers with 56.4% (n = 186) as compared to female customers with 43.6% (n = 144). Based on respondents' ethnicity, a large proportion of 89.4% ($n = 295$) of the total respondents were Malay, followed by 9.1% (n = 30) was Chinese, and 1.5% ($n = 5$) was Indian. In terms of respondents' education level, a large proportion with 88.8% ($n = 293$) undergoes college or university education. This is followed by 7.9% (n = 26) from secondary school education, and the remaining 3.3% ($n = 11$) did not undergo any formal education. For respondents' reported monthly income, approximately 80.3% (n = 265) of respondents were having monthly income of RM3,000 and below, and the remaining respondents (19.7%, n = 65) were having monthly income of RM3,000 and above.

3.2 *The effect of food quality on satisfaction*

The relationship between food quality and satisfaction founded to be significant with standardized coefficient of 0.79 ($p < 0.001$).

In addition, 62.4% of the variation for satisfaction was explained by food quality, indicating the substantial effects of food quality on satisfaction.

3.3 *The contribution of individual food attributes on satisfaction*

Food quality attributes were able to explain 71% of the variation in satisfaction. Among all of the attributes, four (food presentation, food taste, food freshness, and food temperature) made statistically significant contributions and positive relationships with satisfaction.

In term of importance, food taste (β = 0.47, $p < 0.001$) made the largest unique contribution to satisfaction. Alternatively, food variety and healthy options were discovered to be not significant predictors to satisfaction.

Table 2. The effect of food quality on satisfaction.

Sections	B	SE B	β
Constant	0.869	0.176	
Food quality	0.834	0.036	0.790***

*Note: $R^2 = 0.624$, ***$p < 0.001$.

Table 3. The effects of food quality attributes on satisfaction.

Sections	B	SE B	β
Constant	0.663	0.183	
Food presentation	0.069	0.028	0.097*
Food variety	–0.020	0.039	–0.021
Healthy options	–0.024	0.032	–0.032
Food taste	0.458	0.042	0.469***
Food freshness	0.245	0.040	0.334***
Food temperature	0.119	0.033	0.138***

*Note: $R^2 = 0.709$, *$p < 0.05$, ***$p < 0.001$.

4 DISCUSSION AND CONCLUSIONS

Analyses for the relative importance of food attributes revealed that food presentation, food taste, food freshness, and food temperature were the most significant contributors to customer. It is interesting to note that temperature was one determinant of satisfaction. To make customers satisfied, it is important to serve food within the appropriate temperature range and to maintain it at the proper temperature during consumption. Thus, restaurant operators have to pay attention to food temperature at the time of service to customers. In addition, they need to use food plates with appropriate temperatures to maintain food quality. In addition, variety of food was not a significant predictor of satisfaction. The findings indicated that managers should pay more attention to popular menu offerings and allocate their limited resources to improve the quality of those items rather than developing various menu items.

When considering the predictive power of food quality as a fundamental element of the restaurant experience, restaurateurs should not underestimate quality food a restaurant has to offer. Overall, the findings suggested that restaurateurs can improve customer satisfaction by offering tasty and visually appealing foods at an appropriate temperature. Also, restaurateurs should recognize that customer place a high value on fresh state of food and this appears to be related to crispness, juiciness, and aroma of the food. In conclusion, from a managerial perspective, it might be useful to prioritize resources by focusing on the most important food quality attributes along with key attributes of service and atmospherics, and market the restaurant accordingly in order to sustain in the restaurant business.

5 LIMITATIONS AND FUTURE RESEARCH

This study does have some limitations. Initially, study findings may not be generalized. Data were collected from customers at five mid-scale Malay restaurants in Shah Alam. If the survey were expanded to include more states and countries, the magnitude and direction of the relationships among constructs may be different. Therefore, a more comprehensive sample, considering geographic dispersion, would ensure external validity. Similarly, the results may not directly apply to other segments of restaurant industry. Thus, future studies should target other segments of the restaurant industry.

Food quality was treated as a main independent variable in this study. However, many other factors may influence customer's restaurant experience, so it may be possible to provide deeper insight into the factors those restaurant owners and managers need to stress in their total offering. By considering these aspects, addition of more independent variables such as service and atmospherics may be desirable for future research to assess the relative influence of food quality compared to other factors for customer satisfaction. Moreover, future research could include a set of other variables that are related to one type of restaurant that would not be relevant for other types of restaurants. For instance, if the study were replicated for fast-food restaurants, location, accessibility to the restaurant, and convenience may also be critical in driving customer satisfaction.

REFERENCES

Anderson, E.W., & Fornell, C. (1993). A customer satisfaction research prospectus. In R.L. Oliver & T.R. Rust (Eds.), *Service quality: New directions in theory and practice*, 239–266. Newbury Park, CA: Sage.

Anderson, E.W., & Sullivan, M.W. (1993). The antecedents and consequences of customer satisfaction for firms. *Marketing Science,* 12(2), 125–143.

Becker, T. (2000). Consumer perception of fresh meat quality: a framework for analysis. *British Food Journal,* 102(3), 158–176.

Bitner, M.J. (1990). Evaluating service encounters: The effects of physical surroundings and employee responses. *Journal of Marketing,* 54(2), 69–82.

Bitner, M.J., & Hubbert, A.R. (1994). Encounter satisfaction versus overall satisfaction versus quality. In R.T. Rust & R.L. Oliver (Eds.), *Service quality: New directions intheory and practice*, 72–84. Thousand Oaks, CA: Sage.

Churchill, G.A., Jr., & Surprenant, C. (1982). An investigation into the determinants of customer satisfaction. *Journal of Marketing Research,* 19(4), 491–504.

Cortese, A. (2003, October 27). Fine dining? Just across the lobby: Some of the best new restaurants are popping up in hotels. *Business Week,* 130–131.

Cronin, J.J., Jr., & Taylor, S.A. (1992). Measuring service quality: A reexamination and extension. *Journal of Marketing,* 56(3), 55–68.

Dabholkar, P.A. (1993). Customer satisfaction and service quality: Two constructs or one? In D.W. Cravens & P.R. Dickson (Eds.), *Enhancing knowledge development inmarketing*, 10–18. Chicago: American Marketing Association.

Dabholkar, P.A. (1995). The convergence of customer satisfaction and service quality evaluations with increasing customer patronage. *Journal of Consumer Satisfaction, Dissatisfaction, and Complaining Behavior,* 8, 32–43.

Delwiche, J. (2004). The impact of perceptual interactions on perceived flavor. *Food Quality and Preference,* 15(2), 137–146.

Fornell, C., Johnson, M.D., Anderson, E.W., Cha, J., & Bryant, B.E. (1996). The American customer satisfaction index: Nature, purpose, and findings. *Journal of Marketing,* 60(4), 7–18.

Grunert, K.G. (2005). Food quality and safety: consumer perception and demand. *European Review of Agricultural Economics,* 32(3), 369–391.

Ha, J., & Jang, S.S. (2010). Effects of service quality and food quality: The moderating role of atmospherics in an ethnic restaurant segment. *International Journal of Hospitality Management,* 29(3), 520–529.

Haghighi, M., Dorosti, A., Rahnama, A., & Hoseinpour, A. (2012). Evaluation of factors affecting customer loyalty in the restaurant industry. African Journal of Business Management, 6(14), 5039–5046.

Iacobucci, D., Grayson, K.A., & Ostrom. A.L. (1994). The calculus of service quality and customer satisfaction: Theoretical and empirical differentiation and integration. In T.A. Swartz, D.E. Bowen, & S.W. Brown (Eds.), *Advances in services marketing and management,* 3, 1–67. Greenwich, CT: JAI.

Johns, N., & Tyas, P. (1996). Investigating the perceived components of the meal experience, using perceptual gap methodology. *Progress in Tourism and Hospitality Research*, 2(1), 15–26.

Johnson, M.D., & Fornell, C. (1991). A framework for comparing customer satisfaction across individuals and product categories. *Journal of Economic Psychology,* 12(2), 267–286.

Kivela, J., Inbakaran, R., & Reece, J. (1999). Consumer research in the restaurant environment, Part 1: a conceptual model of dining satisfaction and return patronage. *International Journal of Contemporary-Hospitality Management*, 11(5), 205–222.

Lamb, C.W., Hair, J.J.F., & McDaniel, C. (2011). *Marketing 11th edition*, South-Western Cengage Learning, Mason.

Lewin, K. (1938). *The conceptual representation and measurement of psychological forces*. Durham, NC: Duke University Press.

Namkung, Y., & Jang, S. (2007). Does Food Quality Really Matter in Restaurants? Its Impact On Customer Satisfaction and Behavioral Intentions. *Journal of Hospitality & Tourism Research,* 31(3), 387–409. doi: 10.1177/1096348007299924.

Nunnally, J.C. (1978). *Psychometric theory.* New York: McGraw-Hill.

Oliver, R.L. (1993). Cognitive, affective and attribute bases of the satisfaction response. *Journal of Consumer Research,* 20(3), 418–430.

Oliver, R.L. (1997). *Satisfaction: A behavioral perspective on the consumer*. New York: McGraw-Hill.

Oliver, R.L. (1999). Whence consumer loyalty? *Journal of Marketing,* 63(4), 33–44.

Parasuraman, A., Zeithaml, V.A., & Berry, L.L. (1988). SERVQUAL: A multiple-item scale for measuring consumer perceptions of service quality. *Journal of Retailing,* 64(1), 12–37.

Péneau, S., Hoehn, E., Roth, H.-R., Escher, F., & Nuessli, J. (2006). Importance and consumer perception of freshness of apples. *Food Quality and Preference,* 17(1/2), 9–19.

Peri, C. (2006). The universe of food quality. *Food Quality and Preference,* 17(1), 3–8.

Pfaff, M. (1977). The index of consumer satisfaction: Measurement problems and opportunities. In H.K. Hunt (Ed.), *Conceptualization and measurement of consumersatisfaction and dissatisfaction,* 36–71. Cambridge, MA: Marketing Science Institute.

Raajpoot, N.A. (2002). TANGSERV: A multiple item scale for measuring tangible quality in foodservice industry. *Journal of Foodservice Business Research,* 5(2), 109–127.

Rijswijk, W.V., & Frewer, L.J. (2008). Consumer perceptions of food quality and safety and their relation to traceability. *British Food Journal,* 110(10), 1034–1046.

Ryu, K., & Han, H. (2010). Influence of the Quality of Food, Service, and Physical Environment on Customer Satisfaction and Behavioral Intention in Quick-Casual Restaurants: Moderating Role of Perceived Price. *Journal of Hospitality & Tourism Research,* 34(3), 310–329.

Ryu, K., Lee, H.R., & Kim, W.G. (2012). The influence of the quality of the physical environment, food, and service on restaurant image, customer perceived value, customer satisfaction, and behavioral intentions. International *Journal of Contemporary Hospitality Management,* 24(2), 200–223.

Sulek, J.M., & Hensley, R.L. (2004). The Relative Importance of Food, Atmosphere, and Fairness of Wait The Case of a Full-service Restaurant. *Cornell Hotel and Restaurant Administration Quarterly,* 45(3), 235–247.

Taylor, S.A., & Baker, T.L. (1994). An assessment of the relationship between service quality and customer satisfaction in the formation of consumers' purchase intentions. *Journal of Retailing,* 70(2), 163–178.

Hospitality and Tourism – Sumarjan et al. (eds)
© 2014 Taylor & Francis Group, London, ISBN 978-1-138-00151-0

The relationship between perceived service performance and customer satisfaction in Malay upscale restaurants in Kuala Lumpur, Malaysia

A.R.M. Nazri
Universiti Pendidikan Sultan Idris, Malaysia

A. Zainal
Universiti Teknologi MARA, Malaysia

ABSTRACT: To date, popularity of Malay upscale restaurant is still outshined by other international full service restaurants that come from different parts of the world such as America, Mexico, Thailand, Spain, France, Korea, Japan and others. The highly demand towards international full service restaurant has made local restaurants particularly Malay full service restaurants trying to be a player and taking a share of this market. This study focuses on the relationship between perceived service performance and customer satisfaction in Malay upscale restaurants in Kuala Lumpur. It looks at the dimensions of physical environment quality, interactional quality and outcome quality. A total of 5 Malay upscale restaurants were chosen and 170 customers participated in this survey. Data were collected through self-administered questionnaires. Descriptive statistics, reliability analysis, and line arregression analysis were used in the data analysis. Outcome quality emerged to have the most impact on customer satisfaction ($\beta = 0.785$, $p < 0.01$), followed by interactional quality ($\beta = 0.660$, $p < 0.01$) and physical environment quality ($\beta = 0.579$, $p < 0.01$).

Keywords: customer satisfaction, Malay food, perceived service performance, upscale restaurant

1 INTRODUCTION

In the year of 2011, Euro monitor International (2012) reported, the overall consumer full service restaurants in Malaysia saw improvement due to recovery from the economic recession and as an outcome it can be seen a 5% increase in current value growth in 2011, reaching RM 10.2 billion. Owing to that, people are more willing to spend on eating out be it at full-service restaurants, fast food restaurants, fine dining or casual dining eateries depending on their budget and disposal income (Euro monitor International, 2011).

The growing trend of foreign food consumption among Malaysian consumers is one of the reasons that boosted the value growth of full service restaurant especially in Kuala Lumpur. Basically the highly demand towards international full service restaurant has made local restaurants particularly Malay full service restaurants trying to be a player and taking a share of this market. In order to win the competition in today's market, the operators will have to take efforts to offer the best value of their food and provide customer with a favorable ambience (Soriano, 2002). Therefore it is noteworthy to state customer satisfaction has become one of the most critical marketing priorities because it generally assumed to be significant determinant of repeat sales, positive word of mouth, and customer loyalty (Ryu & Han, 2008).

Hwang and Ok (2012) stated that customer's first experience the physical environment thus physical environment quality gradually affects interactional and outcome qualities. Hence, with limited literature discussion on perceived service performance that lead to customer satisfaction in experience through dining at upscale ethnic restaurants, and with the suggested views above this current study embarked through investigating the complex relationships in order to measure the level of customer satisfaction.

2 LITERATURE REVIEW

2.1 *Perceived service performance*

In an increasingly competitive environment, restaurants must be customer oriented using marketing concept that identifies customer needs thus leading to their satisfaction and increased retention (Baber, Goodman & Goh, 2011). The complexity of hospitality industry is one reason why restaurateurs should focus not only food quality but also other elements such as physical environment and employee service in order to increased customer satisfaction (Kivela, 1997). Marketing academics have generally agreed that service quality is a hierarchical and multidimensional construct (Brady & Cronin, 2001; Grönroos, 1984; Rust & Oliver, 1994). The hierarchical and multidimensional framework is believed to offer an improved and more systematic approach that explain the complexity of human reactions to a service experience leading some researchers to propose that perceptions of service quality are not only multidimensional but also occur at various levels (Brady & Cronin, 2001). The hierarchical model of service quality suggests that the restaurant patrons' form perceptions of each of the three primary dimensions: interaction quality, physical environment quality and outcome quality, in order to form an overall service quality perception (Brady & Cronin, 2001).

2.2 *Physical environment quality*

Physical environment quality can be defined as physical surroundings, not the natural or social environment (Bitner, 1992). The physical environment quality has a broader definition than the SERVQUAL tangibles dimension which refers to the physical aspects of the service. In a traditional service setting, the service environment relates to the physical ambience of the service encounter (Rust & Oliver, 1994). According to Ryu and Han (2011), eating away from home for a majority of restaurant patrons is more important than dining. Such patrons may seek a memorable dining experience away from home so that the atmosphere of the restaurant can play acritical role in creating a memorable experience (Ryu & Han, 2011). To measure customer's perception about the physical environment, DINESCAPE scale proposed by Ryu and Jang (2007) which is also widely used are ambient conditions (intangible aspect affecting individual responses to the environment), facility aesthetic (architectural design, décor and interior design, which customer view and evaluate), spatial layout (the ways in which machinery, equipment and furnishing are arranged) and seating comforts (the level of physical comfort derived from the seating quality).

2.3 *Interactional quality*

Interaction quality represents the interplay between the restaurant patrons and the service personnel (Surprenant & Solomon, 1987). In the restaurant industry, the front-of-the-house employees, for example waiter or waitress represent large numbers of employees who have direct contact with restaurant patrons (Yoo, Shin & Yang, 2006). The restaurant industry involves a high degree of interaction between the front-of the-house employees and restaurant patrons; therefore there are numerous opportunities for service failures to occur (Yoo et al., 2006). For example, front-of-the-house employees who are able to provide a prompt and courteous service are likely to enhance customer satisfaction. Normally, the more effort an employee demonstrates, the greater the level of service quality and the greater the level of customer satisfaction. The employees' actual attitude and behavior and service expertise may change a restaurant patron's assessment of the service (Brady & Cronin, 2001). Oh (2000) stated that delivering superior interactional quality enhance success in the restaurant business.

2.4 *Outcome quality*

Gronroos (1990) defined outcome quality as results of service transaction and the benefit the customer receives at the final stage of service transaction. It is evaluated after going through the process of quality (i.e physical environment and interactional qualities) and occurs after service delivery. In addition, Rust and Oliver (1994) regarded outcome quality as the outcome of the service act representing what the consumer gained from the service and whether the outcome fulfilled the customer's needs. The outcome quality focuses on the outcome of the service act and indicates the perceptions of quality (Fassnacht & Koese, 2006). Food is the main and the final stage of the overall dining experience and can be measured through using many attributes.

In the restaurant setting, food quality not only provides tangible evidence of outcome quality but is also an important component of outcome quality in the hospitality industry (Andersson & Mossberg, 2004; Johns & Tyas, 1996; Kim, Lee & Yoo, 2006). Even though many scholar used variety indicator to measure food quality, there are six widely accepted as outcome quality in the restaurant: food freshness, portion size, menu variety, food presentation, food taste and food temperature (Andaleeb & Conway, 2006; Raajpoot, 2002; Namkung & Jang, 2007).

2.5 *Customer satisfaction*

According to Oliver (1999), "satisfaction is the consumer's fulfillment response. It is a judgment that a product or service feature, or the product or service itself, provided (or is providing) a pleasurable level of consumption-related fulfillment, including levels of under- or over-fulfillment". Yang and Peterson (2004) added, loyal customers are more likely than nonloyal customers to engage in positive word of mouth behaviors and spend extra money in specific service operation. Patrons visit upscale restaurants not only to satisfy hunger, but to enjoy a high-quality dining experience and prestigious service in a luxurious setting (Walker & Lundberg, 2005). Indeed, total foodservice in the restaurant industry encompasses both tangible (food and physical facilities) and intangible (employee-customer interaction) components. A proper combination of the tangible and intangible aspects should result in a customer's perception of high restaurant service performance, which in turn should lead to attaining customer satisfaction in the restaurant industry.

Based on the literature, the following hypotheses were formulated:

H1: Perceived service performance significantly influence customer satisfaction.

H1a: Physical environment quality significantly influence customer satisfaction.

H1b: Interactional quality significantly influence customer satisfaction.

H1c: Outcome quality significantly influence customer satisfaction.

3 METHODOLOGY

A total of 5 Malay upscale restaurants (taken Malaysia Online Directory, 701panduan.com, 2013), in Kuala Lumpur were chosen and 170 customers participated in this survey. Data were gathered through self-administered survey questionnaire that were analyzed using descriptive statistics, reliability test and linear regression. The reliability of the scales used was also analyzed using Cronbach's Alpha coefficient score, and all scale registered moderately to fairly high reliability scores.

The questionnaire comprised of three sections namely perceived service performance (dependent variable, customer satisfaction (dependent variable) and demographic profile. The instrument for perceived service performance (Section A) was adapted from Brady and Cronin (2001). Section B measures the customer satisfaction dimensions that were adapted from Cronin, Brady and Hult (2000).

4 RESULTS

4.1 *Profile of sample*

61.8% (N = 105) of the respondents were female while the remaining 38.2% (N = 65) were male with mean age of 34.5 years. 34.1% of respondent were Malay and 32.4% of respondent were Chinese and majority of respondents were reported married, consisting of 54.1% (N = 92). 60% (N = 102) of the respondents have bachelor degree and 17.1% (N = 29) of respondents hold a master degree. Almost 73.5% (N = 125) of respondents has an income range between RM 30,000 and above. 5.3% (N = 9) has an annual income between RM 25,000 to RM 30,000 and majority of respondents works in private sector, consisting of 51.2% (N = 87). 72.4% (N = 123) of respondents indicates that the purpose of dining was for leisure.

Table 1. Results of regression analysis for perceived service performance on customer satisfaction.

Hypothesis	Regression, R^2	Regression correlation, β	Regression statistical sig.
H1	0.586	0.765	*0.000
H1a	0.335	0.579	*0.000
H1b	0.436	0.660	*0.000
H1c	0.616	0.785	*0.000

*p < 0.001.

4.2 *Hypotheses testing*

Table 1 provides the results of the linear regression analyses to test the relationship between perceived service performances on customer satisfaction. The test determines whether perceived service performance have significant relationship in developing customer satisfaction. Three (3) sub-hypothesis were also been tested to find out which dimension in perceived service performance give more influence to customer satisfaction.

Table 1 illustrates the result of the regression analysis and correlation on the relationship between perceived service performance and its dimension towards customer satisfaction. At the significant level 0.001, perceived service performance overall has positive significant influence to customer satisfaction with ($\beta = 0.586$, $R^2 = 0.765$). The three (3) sub-variables namely physical environment quality, interactional quality and outcome quality has significant unique contribution to customer satisfaction with value of ($\beta = 0.335$, $R^2 = 0.579$), ($\beta = 0.436$, $R^2 = 0.660$) and ($\beta = 0.616$ $R^2 = 0.785$). This indicated that, hypothesis 1 and the sub-hypotheses are supported.

5 DISCUSSION

Based from the findings presented, it was observed that the sub-hypothesis namely outcome quality (H1c) reported as most correlated with $R^2 = 0.616$ and have a high beta value, $\beta = 0.785$

($p < 0.001$). Thus, the finding therefore support the sub-hypothesis (H1c) and this is similar with previous study that revealed t food quality is the core of the overall restaurant dining experience (Raajpoot, 2002). Indeed, in the restaurant business, outcome (or food) quality delights customers, providing them with enjoyment (Hwang and Ok, 2012) and Namkung and Jang (2007) also suggested that food quality is important in pleasing customer. Outcome quality is the result of service transaction, the benefit the customer receives at the final stage of service transaction after going through physical environment and interactional qualities. Moreover, as claimed by the owner of the Malay upscale restaurants that they only used quality ingredients and pay more attention on the dishes in order to fulfill customer expectation.

Second sub-hypothesis for perceived service performance namely interactional quality (H1b), also reported significant influence on customer satisfaction with $R^2 = 0.436$ and beta value $\beta = 0.660$ ($p < 0.001$). This finding supported sub-hypothesis (H1b). From the previous study, delivering superior interactional quality enhance success in the restaurant business (Oh, 2000). Hence, service staff spends the most time with customer and with positive interaction between service staff and customer demonstrated to satisfaction and behavioral interaction.

Lastly, the sub-hypothesis for perceived service performance which is physical environment quality (H1a), reported significant influence on customer satisfaction with $R^2 = 0.335$ and beta value $\beta = 0.579$ ($p < 0.001$) and therefore this finding supported sub-hypothesis (H1a). This is in line with various studies that stated physical environment is critical to a patron's satisfaction and behavior in a restaurants (Bitner, 1990; Kim and Moon, 2009; Kivela 1997 and Robson, 1999). Hence, by portraying Malay Archipelago elements in the restaurants that can induce customers' emotion including pleasure (good or happy) and arousal (excited or active), this can lead customer satisfaction.

6 IMPLICATION

Perceived service performance effects customer satisfaction. The results have implication for determining how management should allocate their budget. Among the perceived service performance constructs in Malay upscale restaurants, the most significant factor was outcome quality. A crucial challenge facing restaurant industries today is to provide quality of food that is not only compelling for the customers by also can be superior to business competitors (Namkung and Jang, 2007). According to Peri (2006), food quality is a necessary condition to satisfy the needs and expectations of customers. Thus, it is important to recognize that the outcome quality elements (e.g., taste, variety of menu, freshness, healthy, portion size, presentation and temperature) that can be managed to a large extent by the restaurant owners. Restaurants must take full advantages of providing quality food by emphasizing critical food properties an elicit a positive customer dining experience.

Besides that, the results also indicated interactional quality effect customer satisfaction. This means that delivering comfortable, timely and dependent service might enhance customer satisfaction. Thus, training staffs may be expensive; restaurateur should consider systematic training to help increase interaction quality. In addition, customer of upscale restaurants often spends 2 hours or more, sensing physical surroundings consciously and unconsciously before, during and after their meal. Thus, physical environment also contribute to customer satisfaction. Although foods and services should be of acceptable quality, pleasing servicescape (e.g. lighting, décor and layout) may influence the degree of positive emotion and subsequent behavior.

7 LIMITATION AND RECOMMENDATION

The researcher acknowledges several limitations in this current research. First, the data for this study were collected for Malay upscale restaurants only, in which the result might not be generalized to other types of restaurants. Perhaps this study can be extended to all types of restaurant within the same setting. A non-probability sampling approach was used in the current study, which is considered to be simplest form of non-probability sampling (Cooper & Shinder, 2003). This also negatively influences the generalisability of the findings. Third limitation appeared was related to the location (Kuala Lumpur) of the data collection. Again, the barrier comes in term of personal budget constraints. It would be much more significant if the data collection were obtained in other available Malay upscale restaurants such as in Pulau Pinang, Johor, Perak and Melaka other than concentrating in Kuala Lumpur only. In line with the limitations, it is essential to expand this study by looking at broader scope and aspects especially to others traditional Malay food belonging to each state in Malaysia for an in-depth understanding about each traditional food. Lastly, the constrain found in this study was in the direction of the findings in the questionnaire due to the fact that this study is only looking at customer satisfaction levels without looking at behavioural intention.

Hence, the need to look at customer behavioural intention for this type of restaurants would be recommended for future research.

8 CONCLUSION

This study demonstrates that customer satisfaction influenced by perceived service performance, where outcome quality was seen to be the best predictor of customer satisfaction. This research will consequently serve as an avenue for future researchers to contribute in further theoretical development and formulation of new and unexplored research question, especially pertaining on customer satisfaction in restaurants business industry. This is because customer satisfaction will lead to determinant of repeat sales, positive word of mouth, and customer loyalty. By understanding customer behavior in experience dining at the upscale restaurants, so it can help the restaurant operator to achieve their marketing objectives at the same time accomplish organizational goals.

REFERENCES

Andaleeb, S.S., & Conway, C. (2006). Customer satisfaction in the restaurant industry: an examination of the transaction-specific model. *Journal of Service Marketing*, 5(7), 454–459.

Andersson, T.D., & Mossberg, L. (2004). *The dining experience: Do restaurants satisfy customer needs?* Food Service Technology.

Barber N., Goodman R.J., & Goh B.K. (2011). Restaurant consumer repeat patronage. *International Journal of Hospitality*, 30(2), 329–336.

Bitner, M.J. (1992). Evaluating service encounters: the effects of physical surroundings and employee responses. *Journal of Marketing*, 54 (2), 69–82.

Brady, M.K., & Cronin, J.J.J. (2001). Some New Thoughts on Conceptualizing Perceived service quality: A hierarchical Approach. *Journal of Marketing.*

Cooper, D.R., & Schindler, P.S. (2003). *Business research methods (8th ed.)*. New York, NY: McGraw-Hill.

Cronin, J.J.J., Brady, M.K., & Hult, G.T.M. (2000). Assessing the effects of quality, value and customer satisfaction on consumer behavioural intentions in service environments. *Journal of Retailing.*

Euromonitor International (2011). Full-Service Restaurant in Malaysia. Retrieved October, 2012 from http://www.euromonitor.com.

Euromonitor International (2012). Full-Service Restaurant in Malaysia. Retrieved October, 2012 from http://www.euromonitor.com.

Fassnacht, M., & Koese, I. (2006). Quality of electronic services: Conceptualizing and testing a hierarchical model. *Journal of Service Research.*

Grönroos, C. (1984). A service quality model and its marketing implications. *European Journal of Marketing.*

Gronroos, C. (1990). *Service Management and Marketing: Managing the Moment of Truth in Service Competition.* United States: Lexington Book.

Hwang, J., & Ok, C. (2012). The Antecedents and Consequence of Consumer Attitudes toward Restaurant Brand: A Comparative Study between Casual Dining and Fine Dining Restaurants. *International Journal of Hospitality Management.*

Johns, N., & Tyas, P. (1996). Use of service quality gap theory to differentiate between foodservice outlets. *Service Industries Journal.*

Kim, W.G., Lee, Y.K., & Yoo, Y.J. (2006). Predictors of relationship quality and relationship outcomes in luxury restaurants. *Journal of Hospitality & Tourism Research.*

Kim, W.G., & Moon, Y.J. (2009). Customers' cognitive, emotional, and actionable response to the servicescape: a test of the moderating effect of the restaurant type. *International. Journal of Contemporary Hospitality Management.*

Kivela J.J. (1997). Restaurant Marketing: Selection and Segmentation in Hong Kong. *International of Contemporary Hospitality Management.*

Malaysia Online Directory (2012*)*. Restaurants within Kuala Lumpur [Online] Retrieved August 28, 2012 from *http://www.701Panduan.com.*

Namkung, Y., & Jang, S.C.S. (2007*)*. Does Food Quality Really Matter in Restaurants? Its Impact on Customer Satisfaction and Behavioural Intentions. *Journal of Hospitality & Tourism Research.*

Oliver, R.L. (1999). When Consumer Loyalty. *The Journal of Marketing.*

Oh, H. (2000). Diners' Perceptions of Quality, Value, and Satisfaction: A practical viewpoint. *Cornell Hotel & Restaurant Administration Quarterly.*

Peri, C. (2006). The universe of food quality. *Food Quality and Preference.*

Raajpoot, N.A. (2002). TANGSERV: A multiple item scale for measuring tangible quality in foodservice industry. *Journal of Foodservice Business Research.*

Robson, S.K.A. (1999). Turning the tables: The psychology of design for high-volume restaurants. *Cornell Hotel & Restaurant Administration Quarterly.*

Rust, R.T., & Oliver, R.l. (1994). *Service Quality: New Directions in Theory and Practice*. Sage, London.

Ryu, K., Jang, S.S. (2007). The Effect of Environmental Perceptions on Behavioral Intentions Through Emotions: The Case of Upscale Restaurants. *Journal of Hospitality and Tourism Research.*

Ryu, K., Han, H., & Kim, T.H. (2008). The Relationships among Overall Quick-casual Restaurant Image, Perceived Value, Customer Satisfaction, and Behavioral Intentions. International. *Journal of Hospitality Management.*

Ryu, K.,& Han, H. (2011). New or repeat customers: How does physical environment influence their restaurant experience?. *International Journal of Hospitality Management.*

Soriano, D.R. (2002). Customers' Expectations Factors in Restaurants: The Situation in Spain. *International Journal of Quality & Reliability Management.*

Surprenant, C.F., & Solomon, M.R. (1987). Predictability and personalization in the service encounter. *Journal of Marketing.*

Walker, J.R., & Lundberg, D.E. (2005). *The restaurant: from concept to operation.* John Wiley & Sons, Hoboken, NJ (4th edition).

Yang, Z., & Peterson, R.T. (2004). *Customer perceived value, satisfaction, and loyalty: The role of switching costs. Wiley Interscience* (http://www.interscience.wiley.com).

Yoo, J.J.E., Shin, S.Y., & Yang, I.S. (2006). Key attributes of internal service recovery strategies as perceived by frontline food service employees. *International Journal of Hospitality Management.*

Hospitality and Tourism – Sumarjan et al. (eds)
© 2014 Taylor & Francis Group, London, ISBN 978-1-138-00151-0

Destination attractiveness of outlet mall in Malaysia

N.S. Shamsudin, M.R. Jamaluddin, N.A. Ahmad & M.D. Darson
Universiti Teknologi MARA, Malaysia

ABSTRACT: The retail industry in Malaysia is growing by leaps and bounds and has contributed significantly towards the economic development. The contribution of shopping towards overall holiday satisfaction is essential that required concerted effort by the Malaysian authorities to ensure Malaysia being among the top of the list for tourist shopping destination. Previously, research looked at the importance of price and location of the shopping mall but less attention were paid to the assortment, atmosphere and services especially in Malaysia. This study aims at examining the outlet mall attributes towards visitor's satisfaction and therefore correlational study was conducted to satisfy this objective. This study used convenience sampling method and managed to interview 250 respondents. Result shows that atmosphere, assortment and accessibility are the top three attributes that has strong relationship with visitor's satisfaction. Outlet mall management must properly manage these attributes and highlights them in the future marketing activities in order to improve satisfaction, and in turn increase profitability.

Keywords: destination attractiveness, outlet mall attributes, satisfaction

1 INTRODUCTION

Tourism contribution towards Malaysia's economic development is undeniably influential especially to the creation of jobs and strong currency exchange. Among the popular tourism products in Malaysia are cultural tourism, island tourism and, heritage tourism. Shopping tourism has started to gain popularity with the mushrooming development of shopping mall throughout Malaysia. The role of shopping in the holiday itinerary is important especially to the souvenir and textile items. Oh, Cheng, Lehto & O'Leary (2004) confirmed this and further noted that tour package also lists shopping in the tour itinerary. For Malaysia, the portion of shopping in the tourism receipts is significant. Tourism Malaysia (2012) states that, in 2010 tourism receipts reached RM 56.5 million and significantly contributed by accommodation, transportation, food, and shopping.

The rapid expansion of economic development over the years combines with external economic and social influences have led to a retail boom in Malaysia. In Malaysia, the phenomenon of outlet retail industry is still relatively new than in the US, which has the largest outlet mall in the world. In ASEAN, the rise of outlet mall can be seen in Singapore, Bandung (Indonesia) and Bangkok (Thailand). These shopping spots cater specifically for budget travellers who are concern and committed with lower price and generally terms as price sensitive buyer. To name a few, popular shopping outlets in ASEAN are Mustapha Centre in Singapore, Rumah Mode in Bandung, and Levi Factory Outlet in Bangkok.

In Malaysia, factory outlet mall industry is still new as only few outlet stores and malls in Malaysia. However the concept of outlet store is offered by Quicksilver, Esprit, Mango, and Armani Exchange that offer off-season merchandises at a lower price. These stores are located either at a shopping mall or at an outlet mall. Outlet malls in Malaysia are located at Iskandar, Johor and Equine Park, Kuala Lumpur. Johor Premium Outlet was the first to be established in South East Asia under the Simon Property Group. Another outlet mall is planned to be established at KLIA from Japanese developer. This outlet namely Mitsui Outlet Park KLIA will cater branded merchandises and will start the operation in 2014. Mitsui Outlet Park KLIA will be developed with food & beverages services and entertainment facilities with several themes like "Knowledge and Attractions", "Prime Time Complex" and "World Food Expo" (The Malaysian Insider, 2012).

Malaysia has the potential to be the next favourite outlet mall destination in ASEAN with the development of Iskandar Economic Zone in Johor that is able to attract foreign direct investment to develop more outlet malls. In line with this notion, Malaysia has planned to generate more income by tapping international tourist as their target market.

The demand for outlet centres is increasing in Central, Eastern and South European countries. However, within the Asia and Middle East regions, the outlet centre sector is only well established in China, Japan, Turkey and Korea, and as for other countries, they are considered underdeveloped (Colonial First State Global Asset Management, 2011).

In relating to the opportunities that the outlet mall may provides, lack of studies were done to diagnose the impact towards satisfaction in Malaysia. The overall dimension for outlet mall competitiveness will provide the opportunities to serve the visitors better and in turn, enhancing Malaysia's image as the the best place for shopping destination.

2 LITERATURE REVIEW

2.1 *Shopping*

The term "shopping center" has been evolving since the early 1950s (The International Council of Shopping Centre, 1999). Given the development of the industry, there are various types of centers currently existing that go beyond the standard definitions. Industry classification originally provided four basic terms which are neighborhood, community, regional, and super-regional centers. However, as the industry has developed and changed more types of centers have evolved and these four classifications are no longer adequate as The International Council of Shopping Centers (1999) has also defined eight principal shopping center types namely neighborhood center, community center, regional center, superregional center, fashion/specialty center, power center, theme/festival center, and outlet mall.

Previous literatures have recognized a variety of shopping attributes important to consumers. Retail and marketing literature suggests that stores must identify those attributes of importance to consumers and provide products and services demanded by their target markets. Consumers differ with respect to their preference for retail attributes and store patronage. A feature or attribute of importance to one consumer might be less influential for another consumer in determining store choice. Having said that, store attributes are influential in determining consumer's decision to patronage the shopping mall and drive consumer's future patronage intention.

Sit, Merrilees & Birch (2003) postulated that, there are a lot of past studies and its attributes on retail image of individual stores like supermarket, but very few have touched on the shopping mall image. Their study review the past researches and highlight the retail store concept as,

> *"the way in which the retail store is interpreted in the shoppers' mind partly on its functional qualities and partly by an aura of psychological attributes."*

2.2 *Satisfaction*

There are many interpretations of satisfaction but the interesting one defined by Brady & Robertson, (2001) stated that customer satisfaction as an individual's feeling of pleasure or disappointment resulting from comparing a product's perceived performance in relation to their expectations. By relating to this, from shopping point of view, satisfaction can be defined as individual's overall experience that meet the "big four" shopping centre attributes.

Past studies relate the shopping as one of the attributes in the destination competitiveness and prove that there is a positive relationship between excitement, desire to spend time in the mall, and involvement with re-patronage intention (Kinley, Forney, & Kim, 2012). Moreover, Huang and Hsu (2009) added that satisfaction in shopping is a major force to determine the overall travel satisfaction.

3 METHODOLOGY

This study employed a quantitative approach and cross-sectional study to describe the characteristics of the variables of interest at an outlet mall. This is in line with the objective of this study as to understand on the shopping attributes that influence the visitors' satisfaction of an outlet mall. Information was obtained through self-administered questionnaire from visitors who had experience visiting outlet center in Malaysia. Therefore purposive sampling was used for this study as the determined population was unknown. 250 questionnaires were distributed to visitors of outlet mall through email, online survey and by pesonal questionnaire distribution. The questionnaire was adapted from a review of existing literature dealing with shopping attributes and satisfaction. A seven point Likert scale was used ranging from "1 = strongly disagree" to "7 = strongly agree" that probably be the best rating for the sample population selected in the study. A pilot test targeting 30 people from diverse background was undertaken to ensure suitability of the question's wording and clarity. Each participant was asked to answer the survey and to interpret the questions to make sure that they understand and able to respond without confusion. They were also encouraged to give comments on the questionnaire and highlight points such formatting, numbering, spelling, grammar and so forth. Based from their comments, some questions have been rephrased and amended for a better comprehension in the real survey. As a result, 32 items were retained in the shopping attributes variable, and five items in the satisfaction. The 2-month field survey was conducted and generated 223 valid responsed or 89 per cent response rate.

4 RESULTS

By looking at Table 1, it shows that respondents agree with the items listed in the shopping mall attributes. The highest rating is given to "Assortment/Selection" (M = 5.80, S.D 0.734). It is indeed important for every shopping mall to offer wide selection of goods and services to maximize experience in the shopping mall. The next highest rating is "Atmospherics" (M = 5.74, S.D 0.681). Attention to the detail particularly to the layout, decoration and walking walkways are influential for shopping mall image. From these two ratings, it shows that the attributes during shopping is rated significantly by the respondents. However, the score for "Accessibility" (M = 5.58, S.D 0.605) and "Services" (M = 5.50, S.D 0.599) are rated slightly lower but are still important to maximize the overall shopping experience in the shopping mall.

Table 1. Overall mean score for each shopping mall attributres.

No.	Item	Mean	S.D
1.	Assortment/selection	5.80	0.734
2.	Accessibility/convenience	5.58	0.605
3.	Atmospherics	5.74	0.681
4.	Services	5.50	0.599

The next part of the analysis, Pearson correlation was conducted to measure the strength of relationship between the shopping mall attributes and satisfaction. Table 2 summarizes the result and it shows that "Atmospherics" is the strong predictor among four shopping attributes to explain the satisfaction during shopping ($r = 0.509$, $p < 0.01$). In this study, the high-end shopping mall experience were examined and therefore the result justify that it is significant to offer the great surrounding for better satisfaction. Apart from that, "Assortment/Selection" has moderate positive relationship ($r = 0.437$, $p < 0.01$) with satisfaction. Assortment or selection attributes are influential to the customer that demand variety of international brands especially for the price insensitive buyers. However, "Accessibility/Convenience"| ($r = 0.378$, $p < 0.01$) and "Services" ($r = 0.319$, $p < 0.01$) has a weak relationship with significant. All in all, the items listed in the shopping attributes are significant in determining the satisfaction towards shopping experience.

5 DISCUSSION

Based on the result, it is noted that shopping attributes are clearly seen to influence the visitors' satisfaction. Respondents agree that assortment leads to their satisfaction. Among the measurement that listed under this dimension are stores establishment, variety of stores, quality of merchandise, products selection and special promotion offered by the outlet.

Similar respond received to the accessibility that comprised of access road conditions to the outlet mall, accessibility to get around within the outlet,

Table 2. Correlation between shopping mall attributes and satisfaction.

Dimensions	Satisfaction	Assortment/selection	Accessibility/convenience	Atmospherics	Services
Satisfaction	1				
	223				
Assortment/ selection	0.437**	1			
	0.000				
	223	223			
Accessibility/ convenience	0.378**	0.413**	1		
	0.000	0.000			
	223	223	223		
Atmospherics	0.509**	0.618**	0.573	1	
	0.000	0.000	0.000		
	223	223	223	223	
Services	0.319**	0.367**	0.414**	0.456**	1
	0.000	0.000	0.000	0.000	
	223	223	223	223	223

convenience to find entrance and exit of the outlet, good public transportation, parking facilities and spaces as well as ease of navigation to reach the outlet. In addition, for the atmospheric measurement, items are shopping experience at the outlet center, spacious store environment, cleanliness, interior decoration, temperature, ambience and the colour of the building. All items listed recorded more than five Mean score that can be categorized as somewhat agree to most of the measurements.

Finally, the lowest score among the dimensions listed is "Services" that comprised of courtesy and friendliness of the outlet's sales assistants and the possession of product knowledge, sufficiency of the signboard, adequacy of restrooms, convenience of the washroom location, cleanliness, the adequacy of escalators, variety of food at the food court, and accessibility of Automated Teller Machines.

6 CONCLUSION

Based on the results recapped on the preceding section, interesting and significant findings were highlighted. However, attention is given to the limitation and claim for further discussion. The first possible limitation is related to the sample size. As the sample size for the study was only focusing on the past customer of an outlet mall or stores, the result may not be generalized at other segment of the outlet centre in Malaysia especially other countries as the respondents' experiences may differ. Therefore, in producing detail information pertaining to other future outlet centres in Malaysia, future research could be undertaken in foreseeing the shopping attributes and customer satisfaction.

As a result, this will provide more awareness and reflection on different point of view and feedback from the respondents. In getting clear result for this method, comparative study may be done through collaboration of larger sampling size. Furthermore, observation, interview and analysis of both oral and written feedback with different perspectives such as outlet developer, management, vendor, supplier and authorities could also be done. Qualitative approach can also be implemented to suit with the objectives of the future study.

The second limitation dealt with the external validity. Since the respondents were Malaysian, Singaporean, Indonesian and Brunei, this study does not classify the result based on the country of origin and most of the respondents were Malaysian. It creates the pattern of having opinions more on the Malaysian perspectives rather than multiple consensuses. These situations basically created multiple experiences for customers who have visited the outlet center. For future research, it is suggested that comparison between customers with different country of origins should be done especially from the Singaporean perspectives.

There should be future researches to be explored from this study. The research should focus on the satisfaction from diverse perspectives. Comparison between different stakeholder such as authority, supplier and vendor might also benefit the researcher.

In addition to that, future exploration related to facilities, promotion and marketing attributes that may influence the revisit intention could be done in order to get clearer picture on shopping tourism in Malaysia.

Shopping mall attributes undeniably may influence the customer satisfaction through assortment, accessibility, atmospherics and services. It is concluded that assortment and atmospheric are the most significant attributes that demanded by the visitors to maximize satisfaction.

As for the management, a standard procedure in filtering the quality of products to be sold in the outlet should be set to the vendor. It is due to the main definition of the outlet stores which sell off season or out-dated merchandises to other countries. Therefore, reasonable price should be determined that is equivalent to the quality of the product or merchandise. If this issue does not handled properly, it will definitely violate customers' right as the price of the products are being sold on the same price even though it is off-season item.

The abovementioned issue is related to the government objectives of outlet centre development. As for the authority aims to attract more international tourists to visit and spend at the outlet quality products and price should be monitored to create value for money in return to the visitors. To create customer satisfaction, the product and services must exceed or meet their expectation. Therefore, authority needs to ensure that the services and products offered are sold accordingly to the policy that they have set. The intervention of Ministry of Trade in this context is highly recommended to control the product quality.

REFERENCES

Brady, M.K., & Robertson, C.J. (2001). Searching for a consensus on the antecedent role of service quality and satisfaction: an exploratory cross-national study. Journal of Business Research, 51 (1), 53–60.

Colonial First State Global Asset Management (2011). Outlet Shopping Centres Are Making Their Mark Across The Globe. Retrieved January 26, 2013, from http://www.cfsgam.com.au/uploadedFiles/Microsites/Property/Research/Outlet_Centres_15%20Dec%202011.

Huang, J., & Hsu C.H. (2010). The impact of customer-to-customer interaction on cruise experience and vacation satisfaction. Journal of Travel Research, 49 (1), 79–92.

Kinley, T.R., Forney, J.A., & Kim, Y.K. (2012). Travel motivation as a determinant of shopping venue. International Journal of Culture, Tourism and Hospitality Research, 6 (3), 266–278.
Oh, J.Y., Cheng, C., Lehto, X.Y., & O'Leary, J.T. (2004). Predictors of tourists' shopping behaviour: examination of socio-demographic characteristics and trip typologies. Journal of Vacation Marketing, 10 (4), 308–319.
Sit, J., Merrilees, B., & Birch, D. (2003). Entertainment-seeking shopping centre patrons: the missing segment. International Journal of Retail and Distribution Management, 31 (2), 80–94.
The International Council of Shopping Centre (1999). ICSC Shopping Center Definitions Basic Configurations and Types. Retrieved May 25, 2013, from http://www.icsc.org/ICSC-shopping-center-definitions-basic-configurations-and-types.
The Malaysian Insider (2012). Malaysia Airports, Mitsui to Develop Upscale Factory Outlet Park at KLIA. Retrieved December 25, 2012, from http://www.themalaysianinsider.com/business/article/malaysia-airports-mitsui-to develop-upscale-factory-outlet-park-at-klia.
Tourism Malaysia (2012). Malaysia Tourists Profile 2010 by Selected Markets. Retrieved December 1, 2012 from http://corporate.tourism.gov.my.

Hospitality and Tourism – Sumarjan et al. (eds)
© 2014 Taylor & Francis Group, London, ISBN 978-1-138-00151-0

Service quality dimension of low cost airline in Malaysia

M.I. Mustapha, A.Q. Abd Jalil, F.I. Basri & Z. Mohi
Universiti Teknologi MARA, Malaysia

ABSTRACT: The purpose of this paper is to identify service quality dimension from passengers perceptive of the low-cost carriers for Malaysian Market. A sample size of 272 passengers who had travelled on low cost carrier in Malaysia and the data was analyzed using exploratory factor analysis. The study found the essential dimensions of service quality such as personnel, reservations and airline tangibles contribute to passenger satisfaction level of the low cost airline. Practically, this research benefits the airline's operators in understanding the passengers perceived importance of the services on low cost airline in Malaysia.

Keywords: low cost airline, Malaysia, service quality dimensions

1 INTRODUCTION

Low Cost Carriers (LCCs) have radically altered the structure and business model of the traditional airline industry. New approaches to operations, infrastructure usage, distribution and passenger service have been employed and costs are lowered across all categories. The growth of LCCs not only significantly increased competition, but provide useful examples for improving the efficiency of their own cost base (Smith & Pierce, 2006).

Whether for business purposes, official duties or for holidays, passengers travelling on airlines expect certain levels of service quality, and this applies to Malaysian air travelers as well. Consumers' overall impressions of service quality are linked to how efficient an organization renders its services, and this impression determines customers' satisfactions and future behavioral intentions to continuously using the airline (Abdullah, Abd. Manaf, & Mohd Noor, 2007).

1.1 *Research issues*

Service quality has been given considerable attention in the marketing literatures by both practitioners and academic researchers in the recent years (Caruana, 2002). The reason for the overwhelming interest in service quality is that both practitioners and academic researchers believe service quality is crucial to the success of any business organization because of it largely impacts on customer satisfaction, repeat purchase behavior, and ultimately, an organization's profitability (Zeithaml & Bitner, 2003).

Despite SERVQUAL's popularity, many researchers (Brady & Cronin 2001; Babakus & Mongold, Cronin & Taylor, 1994). Van Dyke, Kappelman, & Prybutok, (1997) argue that the SERVQUAL instrument has a serious problem that limits the instrument's usefulness. The difficulties with the SERVQUAL instrument identified in the literature can be grouped into two main categories: conceptual and empirical. Conceptual problems relate to the ambiguity of the expectations construct and using a single instrument across several industries (Brady & Cronin, 2001; Van Dyke et al., 1997). Several empirical studies show concern about the five SERVQUAL dimensions and still lack of consistency in different service environments (Babakus & Mongold, 1992; Buttle, 1996; Finn & Lamb, 1991). The empirical problems associated with SERVQUAL include reduced reliability, poor convergent validity, poor predictive validity, and unstable dimensionality (Van Dyke et al., 1997; Spreng, MacKenzie, & Olshavsky, 1996).

Parasuraman, Berry and Zeithmal (1988) strongly claim that service quality is a multi-dimensional concept consisting of five dimensions, and claim that the five dimensions of SERVQUAL are generic and universally applicable. Beside Parasuraman et al. (1988), several researchers (e.g., Brady & Cronin, 2001; Dabholkar, Thorpe, & Rentz 1996; Lehtinen & Lehtinen, 1991), especially in tourism discipline noted service quality is multi-dimensional in nature (Martinez & Martinez, 2008). In addition, researchers such as Alexandris, Dimitriadis, & Markata (2002) maintain that the dimensions are more likely industry-based and culturally-based. Several service quality multi-dimensional models have been developed for different service industries, such as hotels (Ingram & Daskalakis, 1999;

Oh, 1999), travel agencies (Martinez & Martinez, 2008), sports recreation (Ko & Pastore, 2005), and tertiary education (Clemes, Ozanne, & Laurensen, 2007).

In with the above notion, Malaysia a home grown of low lost airlines namely Air Asia, commence their operation at the new millennium and adopting no frills service in concept. The emergence undeniably converted the previously non customer group into a potential customer and expanded the usage rate of leisure air travelers thus increasing the size and growth rate of Malaysia's air travel market (Yeoh & Kim, 2011). Due to the emerging of Malaysians market, they have attracted many other airline companies to the advantage of increasing market demand in Malaysia such, Tiger Airways, Jet Star Airways, Fire Fly and recently Malindo Air. This study seeks to better understanding the depth of service quality of the airline industry in Malaysia focusing on low cost carrier based on passengers' perceptions.

2 LITERATURE REVIEW

The general importance of quality to the airline industry is recognized in the annual Airline Quality report. Bowen and Headley (2013) reported domestic airlines incorporate an internally-focused model that based on financial data, fares, passenger load factors and service related issues. As service quality strategies fundamentally focus externally on the customer and satisfaction (a market orientation) rather than internally on company profits and costs (a production orientation), a distinctive and clear management approach as well as organization commitment are required (Cronin & Taylor, 1994).

Zagat, an aviation research firm rates airline service on the basis of five criteria: (1) overall performance; (2) comfort; (3) service; (4) food; and (5) web site (Rhoades, 2006). The US Department of Transportation (DOT) also utilizes a multidimensional conception of service quality in its regular reports on airline service quality, which provide data on the following operational measures: (1) flight delays; (2) mishandled baggage; (3) oversold flights; and (4) consumer complaints (Saha & Theingi, 2009).

Rhoades and Waguespack (2005) compared the service quality of US airlines before and after the terrorist attacks of 11 September 2001 and found that service quality was at its optimum in 2002 as decreased utilization of seats facilitated on-time performance, reduced overbooking, and fewer customer complaints. Tiernan, Rhoades, and Waguespak (2008) offered a wider perspective on service-quality measures on the basis of data from the Association of European Airlines (AEA) when they found the important factors in determining positive customer perceptions of the industry included on-time flight arrivals and an absence of cancellations and baggage problems.

Ekiz, Hussain, and Bavik (2006) developed an AIRQUAL scale to overcome the psychometrical application problems of the existing quality scale to measure the quality perceptions of airline customers in North Cyprus. The results identified five dimensions: airline tangibles, terminal tangibles, personnel, empathy, and image. Nevertheless, according to Ekiz et al. (2006), the AIRQUAL scale was design to measure service quality for Cyprus National Airline only, and due to the generalizability of findings, there is a need to modify the scale in order to apply to others airline.

De Jager, Van Dyl and Toriola (2012) developed service quality dimensions for Airline in Italy and South Africa. They identified several service quality dimensions for airline passenger by taking account the industry players as well as airline passenger's perspective. The result from factor loading for airline in Italy identified the scale had five dimensions (cabin servicescape (food and films), convenience of booking, cabin service scape (cabin and cabin crew), timeliness of flight, country origin of Airline). For airline service quality dimensions in South Africa, De Jager et al. (2012) factored six dimensions: cabin service scape (food and films), convenience of booking, cabin service scape (inflight entertainment), timeliness of flight, country origin of airline, and special offers and freebies.

Jiang (2013) conducted a study on service quality of low-cost long-haul airlines, the case of Jetstar Airways and AirAsia X. The outcome from Exploratory Factor Analysis (EFA) for both airlines, Jiang (2013) factored five dimensions of service quality, which are ground service, flight experience, service reliability, airfare & schedule, and consistency.

After reviewing the past studies, the results supported claims by several researchers that the service quality dimensions need to be re-defined in each setting, from the customers' point of view and also from the culture-specific perspective (e.g., Cronin & Taylor, 1994; Martínez & Martínez, 2007; Ueltschy & Krampf, 2001).

3 METHODOLOGY

3.1 *Instruments*

This study sought to identify the LCC airline's dimensions for service quality perceived by passenger in Malaysia setting following the suggestion by several researchers (e.g. Cronin & Taylor, 1994; De Jager et al., 2012; Eqiz et al., 2006; Jiang, 2013; Martínez & Martínez, 2007; Ueltschy & Krampf, 2001). Based on review of literature seven dimensions was identified that tailored to the context of LCC services in Malaysia: (a) Assurance; (b) Airline Tangibles; (c) Personnel (d) Terminal Tangibles; (e) Ground Services; (f) Special offers and Freebies; (g) Reliability; and (f) Ground service.

Once the dimensions were identified, the prototype questionnaire was developed. Face and content validity of the questionnaire was undertaken to ensure the validity of the items used in the study. The Cronbach alpha for pilot test was between 0.6 and 0.8 indicating internal consistency (Hair et al., 2010). The data collection was undertaken in April 2013 after following recommendation from several academicians in hospitality field.

3.2 *Data collection*

A survey was conducted from low cost airline passengers in Malaysia. A convenience sampling method was used to select the potential respondents. Out of the 300 questionnaires distributed, 282 were able to retain within two weeks response period.

The data was analyzed using SPSS 21. Preliminary data analysis was performed before proceeding to Exploratory Factor Analysis (EFA) to ensure the "cleanliness" of data (Pallant, 2007). The tests conducted including the skewness, kurtosis, and z-scores. After the preliminary data process, 10 questionnaires were found not suitable for further analysis, resulted in a total of 272 (90%) usable responses rates. The usable responses rate was above the minimum sample size of 150 as suggested by Hair et al. (2010); hence was considered to be acceptable for the purpose of this study. Demographic sample profile is presented in Table 1.

Table 1. Demographic sample profile.

	Frequency	Percentage
Gender		
Male	155	57.0
Female	117	43.0
Age		
18–27 years old	92	33.8
28–37 years old	76	27.9
38–47 years old	61	22.4
48–57 years old	31	11.4
58 years old and above	12	4.4
Occupation		
Government	82	30.1
Private	85	31.3
Business	17	6.3
Student	76	27.9
Housewife	8	2.9
Retired	2	0.7
Unemployed	2	0.7
Purpose		
Holiday/pleasure	148	54.4
Business/meeting	29	10.7
Shopping	24	8.8
Visiting friends/relative	40	14.7
Food/culture	11	4.0
Entertainment/event	20	7.4

4 FINDING

In order to satisfy the research objective, this study employed EFA using VARIMAX rotation. 30 items were proposed to measure the seven dimensions. Overall test conducted for EFA revealed that the correlation matrix test value was large (5079.027). The correlation matrix revealed that most correlations were above 0.30 that the data shared common factors appropriate for factor analysis indicating the data was suitable for EFA (Hair et al., 2010). The Bartlett's Test of Sphericity result was statistically significant at the 0.001% level. The scree plot in Figure 1 shows that the extraction of seven factors was appropriate for this analysis.

As reported in Table 2, the service quality items loaded on seven separate factors. All items except the comfort of waiting hall" are highly loaded on a single factor, indicating an adequate unidimensionality among the items based on guideline by Hair et al., (2010). All of the factor loadings for the items retained are above 0.50. Factor loading values ranged from 0.518 to 0.852. Consequently, inline of achieving the research objective, seven factors were subsequently named in accordance with the construct that they represented (Hair et al., 2010; Pallant, 2007).

The remaining items were subjected to a reliability test. Reliability was measured with

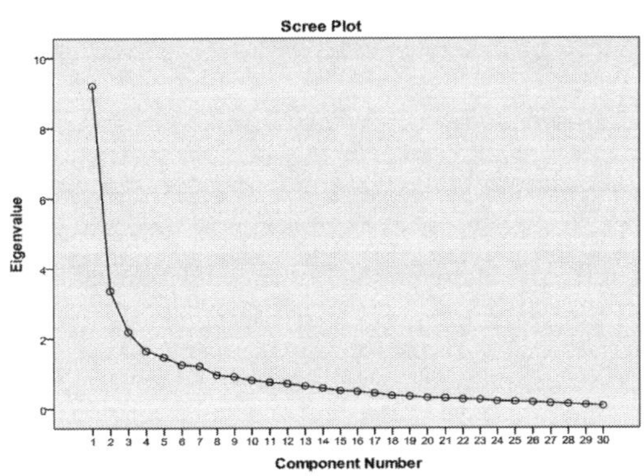

Figure 1. Scree plot.

Table 2. Factor loading.

Item	Factor loading	Alpha	% of total variance
Assurance			
Airline provides tasty foods and drinks	0.841	0.844	30.68
Airline are dependable in fulfilling my needs	0.821		
Airline performs service right at the first time	0.78		
Shows sincerity in solving problem	0.658		
Provides services as promised	0.562		
Personnel			
Personnel are dependable	0.76	0.786	11.18
Personnel has a good attitude towards the passenger	0.696		
Personnel gives exact answer	0.617		
Personnel gives consistent service	0.615		
Personnel shows personal care	0.518		
Reliability			
I didn't face any problem during check in	0.751	0.779	7.27
Most of my flights arrived on time	0.707		
I don't have any problem when dealing with this airline	0.633		
Most of my flights departed on time	0.61		
Services provided by airlines are consistent	0.573		
Terminal tangibles			
Trolleys are easy to find	0.818	0.826	5.47
Facilities provided by the airline are up-to-date	0.772		
Check-in information is helpful	0.652		
Comfortable waiting area hall	0.574		
Special offers and freebies			
Provide accumulative points for frequent flier	0.852	0.858	4.89
Provide discounts for frequent fliers	0.848		
Gives special offers for frequent fliers	0.673		
Ground service			
Booking the tickets online with ease	0.712	0.805	4.164
Check-in kiosk is easy to use	0.654		
Convenient web and mobile check-in	0.637		
Less problem when dealing the tickets online	0.552		
Website that is easily accessible	0.527		
Airline tangibles			
Aircraft looking modern and new	0.683	0.702	4.04
Air conditioning is working perfectly	0.664		
Cabin area clean and neat	0.65		
Overall scale		0.906	67.72

Cronbach alpha. Cronbach alpha scores were between 0.702 and 0.858, higher than the general criterion of 0.60 recommended by Churchill (1979), so indicating the internal consistency of the variables in the exploratory study.

5 CONCLUSION

This study has highlighted the proposed service quality dimensions for low cost Airline for Malaysia market. Having a clear insight on the customer's preferences on the service quality therefore the airline operator should take proactive actions in delivering a good service consequently generate more income and build good customer relationship. Resourcing the customer-important dimensions will improve airline's competitive positioning in the marketplace. The results of this study also provide an opportunity for airline management especially low cost to use the information to increase favorable behavioral intentions, which will in turn, help to increase the airline profit.

The ultimate success of an airline company often depends on its ability to integrate and coordinate the various customer touch points of the service value chain, and delivering it consistently (De Jager et al., 2012). An internal marketing strategy must also be developed, as each employee contributes towards creating value-in-use for the client.

REFERENCES

Abdullah, K., Abd. Manaf, N.H., & Mohd Noor, K. (2007). Measuring the service quality of airline services in Malaysia. *IIUM Journal of Economics and Management,* 15(1), 1–29.

Alexandris, K., Dimitriadis, N., & Markata, D. (2002). Can perceptions of service quality predict behavioral intentions? An exploratory study in the hotel sector in Greece. *Managing Service Quality,* 12(4), 224–232.

Babakus, E.& Mongold, W.G. (1992). Adapting the SERVQUAL scale hospital services: An empirical investigation. *Health Services Research,* 26(2), 767–786.

Bowen, D.B. & Headley, D.E. (2013). *Airline Quality Rating Report.* Advance Aviation analytics Institute of Research.

Brady, M.K. & Cronin, J.J.J. (2001). Some New Thoughts on Conceptualizing Perceived Service Quality: A hierarchical approach. *Journal of Marketing,* 65(3), 34–49.

Buttle, F. (1996). SERVQUAL: Review, critique, research agenda. *European Journal of Marketing*, 30 (1), 8–35.

Caruana, A. (2002). Service loyalty: The Effects of Service Quality and the Mediating Role of Customer Satisfaction. *European Journal of Marketing,* 36(7/8), 811–830.

Churchill, G.A. (1979). A paradigm for developing better measures of marketing constructs. *Journal of Marketing Research,* 16(1), 64–73.

Clemes, M.D., Ozanne, L.K., & Laurensen, W.L. (2001). Patients' perceptions of service quality dimensions: An empirical examination of health care in New Zealand. *Health Marketing Quarterly,* 19(1), 3–22.

Cronin, J.J. & Taylor, S.A. (1994). SERVPERF versus SERVQUAL: Reconciling Performance-Based and Perceptions-Minus-Expectations Measurement of Service Quality. *Journal of Marketing,* 58(1), 125–131.

Dabholkar, P.A., Thorpe, D.I., & Rentz, J.O. (1996). A measure of service quality for retail stores: Scale development and validation. *Academy of Marketing Science Journal,* 24(1), 3–16.

De Jager, W.J., Van Zyl, D., & Toriola, L.A. (2012). Airline service quality in South Africa and Italy. *Journal of Air Transport Management,* 25, 19–21.

Ekiz, H.E., Hussain, K., & Bavik, A. (2006). Perceptions of service quality in North Cyprus national airline, *Tourism and Hospitality Industry 2006—New Trends in Tourism and Hospitality Management, Proceedings of 18th Biennial International Conference, Croatia: Faculty of Tourism and Hospitality Management, Opatija,* May 3–5, (3–5), 778–790.

Finn, D.W., & Lamb, C.W. (1991). An evaluation of the SERVQUAL scales in a retailing setting. *Advances in Consumer Research*, 18(4), 483–490.

Hair, J.F., Black, W.C., Babin, B.J., & Anderson, R.E. (2010). *Multivariate data analysis* (7th ed.). Englewood Cliffs: PrenticeHall.

Ingram, H., & Daskalakis, G. (1999). Measuring quality gaps in hotels: The case of Crete. *International Journal of Contemporary Hospitality Management,* 11(1), 24–30.

Jiang, H. (2013). Service quality of low-cost long-haul airlines: The case of Jetstar Airways and AirAsia X. *Journal of Air Transport Management,* 26, 20–24.

Ko, Y.J., & Pastore, D.I. (2005). A hierarchical model of service quality for the recreational sport industry. *Sport Marketing Quarterly,* 14(2), 84–97.

Lehtinen, U., & Lehtinen, J. (1991). Two approaches to service quality dimensions. *Service Industries Journal,* 11(3), 278–303.

Martinez, C.L. & Martinez, G.J. (2008). Developing a multi-dimensional and hierarchical service quality model for the travel agency industry. *Tourism Management,* 29(4), 706–720.

Oh, H. (1999). Service quality, customer satisfaction and customer value: A holistic perspective. *Hospitality Management*, 18(1), 67–82.

Pallant, J. (2007). *SPSS Survival Manual.* Australia: Allen & Unwin.

Parasuraman, A., Zeithaml, V.A., & Berry, L.L. (1988). SERVQUAL: A multiple-item scale for measuring consumer perceptions of service quality. *Journal of Retailing,* 64(1), 12–40.

Rhoades, D. (2006). Growth, customer service and profitability Southwest style. *Managing Service Quality,* 16(5), 538–47.

Rhoades, D.L. & Waguespack, B. Jr. (2005). Strategic imperatives and the pursuit of quality in the US airline industry, *Managing Service Quality,* 15(4), 344–356.

Saha, G.C. & Theingi. (2009). Service Quality, Satisfaction and Behavioural Intentions: A Study on Low-Cost Airline Carriers in Thailand. *Managing Service Quality,* 12(3), 350–372.

Smith, M.& Pierce, S. (2013). *Airline Cost Performance.* IATA Economics Briefing, 5.

Spreng, R.A., MacKenzie, S.B., & Olshavsky, R.W. (1996). A re-examination of the determinants of consumer satisfaction. *Journal of Marketing,* 60(3), 15–32.

Tiernan, S., Rhoades, D.L. & Waguespack, B. Jr. (2008), Aireline service quality—Exploratory analysis of consumer perceptions and operational performance in the USA and EU, *Managing Service Quality,* 18(3), 212–224.

Ueltschy, L.C., & Krampf, R F. (2001). Cultural sensitivity to satisfaction and service quality measures. *Journal of Marketing Theory and Practice*, 14–31.

Van Dyke, T.P., Kappelman, L.A., & Prybutok, V.R. (1997). Measuring information systems service quality: Concerns on the use of the SERVQUAL questionnaire. *MIS Quarterly,* 21(2), 195–208.

Yeoh, E., & Kim, J.L.C. (2011). Malaysian low cost airlines: key influencing factors on customers' repeat purchase intention. *World Applied Sciences Journal (Special Issue of Tourism & Hospitality,* (12), 35–43.

Zeithaml, V.A. & Bitner, M.J. (2003). *Service marketing: Integrating customer focus across the firm* (3rd Ed.). Boston, MA: McGraw-Hill/Irwin.

Hospitality and Tourism – Sumarjan et al. (eds)
 ISBN 978-1-138-00151-0

Destination attributes, tourist satisfaction and behavioral intentions toward Penang

F.H. Muhammad Safri, H. Ismail, M.A. Alias & N.A.A. Nik Hashim
Universiti Teknologi MARA, Malaysia

ABSTRACT: Penang is not a newcomer to the international tourism market; in fact it is one of the most visited states in Malaysia. Its abundance in heritage, historical sites, multi-culture and diversity of people and other attributes make Penang very attractive to international and domestic tourists. However, when considering the competitive market, no destination can succeed without putting in some effort. Therefore, the identification of Penang's important attributes and a comprehensive evaluation of these attributes in terms of their effectiveness in impacting both domestic and international tourists on their decision when choosing their prospective destination would be of great value to Penang's tourism planners. This paper report the data from the perception of tourist regarding their knowledge in destination attributes of Penang and how the destination is likely to be chosen.

Keywords: behavioral intentions, destination attributes, Penang, tourist satisfactions

1 INTRODUCTION

Cities and countries increasingly are turning to tourism as an important element in generating their economic. Some destinations appear to be more successful and more attractive than others in offering tourism activities and drawing tourists (Formica, 2000). The attractions are the primary elements of destination appeals. They are the key motivators for visitation to a destination. This is generally what attract tourists from one destination to another.

Fishbein, and Ajzen (1975) mentioned that the choice of goods and services are the result of the comparison of its perceived attributes in the person's set of preferences. Additionally, by looking at some benefits that the consumer are looking for, only then they will evaluate attributes of the destination that have potential to satisfy them (Mayo and Jarvis, 1981).

Penang is not a newcomer to the international tourism market; in fact it is one of the most visited destinations in Malaysia. Its abundance heritage, historical sites, multi-culture and diversity of people and other attributes make Penang very attractive to international tourists. When considering the competitive market, no destination can succeed without putting in some effort.

However, according to Penang Institute (2013) a policy advisory body to the state government of Penang, even though the conferment of the Unesco World Heritage status to George Town in 2008 gave the industry a major boost, they still (1) lack of information on tourist profiles which inhibits targeted promotional activities; inadequate promotion in traditional tourist markets such as Europe, (2) lack of focus in tourism efforts; insufficient expertise in marketing and branding of existing tourist attractions and (3) lack of international recognition as a tourist destination.

Therefore, the identification of Penang's important attributes and a comprehensive evaluation of these attributes in terms of their effectiveness in impacting both domestic and international tourists on their decision when choosing their prospective destination would be of great value to Penang's tourism planners. It will help them to formulate appropriate marketing strategies and it is essential for Penang to continue being successful in the more competitive international tourism market.

2 LITERATURE REVIEW

Destination attractiveness is a frequent tourism research issue. Destination attractiveness can be defined as the sum of the perceived capability of a destination to deliver satisfaction and benefits to tourists (Hu and Ritchie, 1993). Victor (1989) also emphasizes that "attractiveness" is the initial motivation for tourists to choose a destination, according to their interests and preferences (Benckendorff & Pearce, 2003). Attractiveness is seen as an important perception of tourists when they participate in a tourism activity (Funk, Ridinger, & Moorman, 2004) and is also defined as the main generator in the construction of a tourism destination. Some studies have been carried out to identify the attractiveness of a tourist destination on the basis of destination attribute analysis (Gearing, Swart & Var 1974; Ritchie & Zins, 1978; Tang & Rochananond, 1990).

There are many characteristics that can be used to assess the attractiveness of tourist destinations. These criteria may differ from or similar to other locations. Cam (2011) defined the destination image as perceived quality evaluation of a combination of five different attributes/facets such as environment, infrastructure and accessibility, culture and social, tourist leisure and entertainment and local food. A research done by Zhou (2005) found that the most commonly used destination attributes to measure the attractiveness are price, culture and history, nightlife and entertainment, shopping, landscape, weather, accessibility, safety, local people's attitude toward tourists, and service. The statement is supported by several authors that used the same destination attributes such as Javalghi, Thomas and Rao (1992); Calantone, Benedetto, Hakem and Bojanic (1989); Goodrich (1977); Goodrich (1978); Haahti and Yavas (1983); Haahti (1986); Driscoll, Lawson and Niven (1994).

3 METHODOLOGY

This study involved respondents that have experience visiting the tourist attractions around Penang. A screening question was used to potential respondents to ensure that they are eligible to participate in the study.

According to the website of Tourism Penang, the number of tourists that visited Penang in year 2011 is 5.96 million. And based on the figures, the researcher had taken a total of 384 respondents in order to get a better result by the end of this study (Krejcie & Morgan's 1970).

The convenience sampling design was used in the study. The respondents are selected just because they are willing to participate in the study and the researcher did not have to consider selecting subjects that are representative of the entire population. This is the reason why most researchers rely on sampling techniques like convenience sampling, the most common of all sampling techniques. Many researchers prefer this sampling technique because it is fast, inexpensive, easy and the respondents are readily available.

With regard to research instruments used in this study, five sections were created: Section A looking at the demographic profile of the respondents, Section B is dealing with evaluation of destination image attributes, Section C is focusing on measuring the satisfaction level, Section D investigating on the behavioral intention of the tourist, and lastly Section E assessed the importance level of destination attributes.

Several top destinations in Penang were chosen to distribute the questionnaire such as Padang Kota Lama, Georgetown, Quensbay Mall. These areas were selected because it is known for the high number of tourists lingering around the area plus international tourist can be easily seen and approached.

4 RESULTS AND FINDINGS

This section discusses the results of the research gained from the tourists perspectives in Penang. The total of 420 surveys was distributed to the tourists around the Penang top tourist destination.

4.1 *Destination attributes*

Looking at Table 1, the majority of the respondents rated entertainment as the most satisfied in Penang. It is then followed by local foods, culture and social, environment and, lastly infrastructure and accessibility.

4.2 *Satisfaction*

As for tourist satisfaction level, Table 2 shows that most of the tourists are quite satisfied with their visit in Penang. This indicates tourism in Penang is on the right track.

4.3 *Behavioral intentions*

In Table 3, most of the respondents are willing to recommend Penang to others. They also plan to

Table 1. Destination attributes.

Dimensions	N	Mean score
Entertainment	384	4.20
Local foods	384	4.08
Culture and social	384	4.01
Environment	384	3.99
Infrastructure and accessibility	384	3.79

Table 2. Satisfaction level.

Item	N	Mean score
Satisfaction	384	4.17

Table 3. Intention to recommend.

Item	N	Mean score
Recommend	384	4.20
Revisit	384	4.25

Table 4. Importance level.

No	Attributes	Mean score
1	Services (shopping, accommodation, food, and transportation)	4.58
2	Accessibility (information available)	4.52
3	Safety (personal safety in trip)	4.47
4	Landscape (beautiful scenery and natural attractions)	4.35
5	Local people' attitude toward visitors	4.33
6	Relaxation (e.g., sun, beaches, fishing)	4.31
7	Special events and activities (e.g., Chinese New Year, Thaipusam, music festival, water festival and etc)	4.30
8	Culture & history (history, monument, heritage, arts, handcraft and ways of life of local people	4.29
9	Nightlife and entertainment	3.96
10	Climate (e.g., pleasant weather)	3.74
11	Price (e.g., cost, good value for money)	3.70

revisit Penang again in the future hence showing why Penang deserves to be in the top 3 most visited destination in Malaysia.

4.4 *Important attributes*

Based on Table 4, Services, Accessibility and Safety considered as the most important factor when visiting a tourist destination. If Penang desire to successfully attracting the domestic and international tourist, the three important attributes mentioned earlier must be improved significantly.

5 CONCLUSION

The results gained from this study can be used by authorities or government bodies to improve any drawback toward the responds from tourist. For example, tourist perceives services, accessibility and security to be the most important attributes, thus the authority should focus on these attributes in order to increase the satisfaction level among the tourists. In addition to that, by concentrating on the development of innovative marketing and promote tourism, Penang may sustain their title as among the top destination to visit and could surpasses Pahang and Kuala Lumpur, which currently on top.

Other than that, marketer can use appropriate media to portray the uniqueness of Penang. Based on the result, people satisfied the most with the entertainment provided in Penang followed by local foods as well as cultural and social. Thus the marketer can really use the information to emphasize on the attractions mentioned earlier in order to attract more tourists which eventually will increase their overall profit.

In general, this study provided support for a strong relationship between destination image and satisfaction, which in turn would lead to behavioral intention. More importantly, this research sends out the message that program, activities and promotional campaigns, should focus and appeal towards the tourists. Perhaps, loyalty programs could be created to produce repeat businesses. By concentrating on fulfilling the needs of the tourist as well as educating the local citizen to appreciate tourist, then only satisfaction as well as behavioral intention would significantly increase to the desired state.

REFERENCES

Benckendorff, P., & Pearce, P.L. (2003). Australian Tourist Attractions: The Links Between Organizational Characteristics And Planning. *Journal of Travel Research*, 42, 24–35.

Calantone, R.J., Benedetto, C.A., Halam, A., & Bojanic, D.C. (1989) Multiple Multinational Tourism Positioning Using Correspondence Analysis. *Journal of Travel Research*, 28, 25–32.

Cam, T.T.A. (2011). Explaining Tourists Satisfaction And Intention To Revisit NhaTrang, Viet Nam. Retrieved from http://munin.uit.no/bitstream/handle/10037/3782/thesis.pdf?sequence=1.

Driscoll, A., Lawson, R., & Niven, B. (1994). Measuring Tourists' Destination Perceptions. *Annals of Tourism Research,* 21(3), 499–51.

Fishbein, M., & Ajzen, I. (1975). *Understanding attitudes and predicting social behaviour*. New Jersey: Prentice Hall.

Formica, S. (2000). Destination Attractiveness As A Function Of Supply And Demand Interaction. Retrieved from http://www.scholar.lib.vt.edu/theses/available/etd-11142000-15560052.

Funk, D.C., Ridinger, L.L., & Moorman, A.M. (2004). Exploring Origins of Involvement: Understanding the Relationship between Consumer Motives and Involvement with Professional Sport Teams. *Leisure Sciences*, 26(1), 35–61.

Gearing, C.E., Swart, W.W., & Var, T. (1974). Establishing a measure of touristic attractiveness. *Journal of Travel Research*, 12(4), 1–8.

Goodrich J.N. (1978). The Relationship between Preferences for and Perceptions of Vacation Destinations: Application of a Choice Model. *Journal of Travel Research*. 8–13.

Goodrich, J.N. (1977). Differences in perceived similarity of tourism regions: A spatial analysis. *Journal of Travel Research*, 16, 10–13.

Haahti, A. & U. Yavas (1983) Tourists' perceptions of Finland and selected European countries as travel destinations. *European Journal of Marketing,* 17(2), 34–42.

Haahti, A.J. (1986) Finland's competitive position as a destination. *Annals of Tourism Research*, 13, 11–35.

Hu, Y., & Ritchie, B.J.R. (1993). Measuring Destination Attractiveness: A Contextual Approach. *Journal of Travel Research,* 32(2), 25–34.

Javalgi, R.G., Thomas, E.G., & Rao, S.R. (1992). US Pleasure Travellers' Perceptions of Selected European Destinations. *European Journal of Marketing*, 26(7), 45–64.

Mayo, E, J., & Jarvis, L.P. (1981). The psychology of leisure travel: Effective marketing and selling of travel services. Boston: CBI publishing company.

Penang Institute (2013). Penang Economic Outlook 2013: SWOT—An Analysis of Penang's Economy. Retrieved from http://penangmonthly.com/penang-economic-outlook-2013-swot-an-analysis-of-penangs-economy/.

Ritchie, J.R.B. & Zins, M. (1978). Culture As Determinant Of The Attractiveness Of A Tourism Region. *Annals of Tourism Research,* 5(2), 252–267.

Tang, John C.S. & Rochananond, N. (1990). Attractiveness As A Tourist Destination: A Comparative Study Of Thailand And Selected Countries. *Socio-Economic Planning Sciences,* 24(3), 229–236.

Victor, T.C. (1989). Marketing Implications For Attractions. *Tourism Management,* 10(3), 229–232.

Zhou, L. (2005). Destination Attributes That Attract International Tourists To Cape Town. Unpublished manuscript, University of the Western Cape, Cape Town, South Africa.

Hospitality and Tourism – Sumarjan et al. (eds)

The effects of place, performance and personality (3P's) Pull Factor of screen tourism and viewers' travel intention

N. Abd Hadi Khan, N. Sumarjan, M.F. Hassan & M.A. Ahmad
Universiti Teknologi MARA, Malaysia

ABSTRACT: The powerful impact of airing popular media products such as films or television shows on destination choice had increased tourist visitation numbers to film destination. Grounded with the 3P's (Place, Performance and Personality) Pull Factor theory, this study examines how the effects of "Place", "Performance" and "Personality" Pull Factor lead to trigger the viewers' travel intention. Data was collected from Malaysian viewers' using self-administered questionnaire. Findings from the Malaysian perspectives revealed that the 3P's Pull Factor theory is recognized as the contributor towards viewers' travel intention to film destinations. The significant effects of the 3P's Pull Factor theory towards viewers' travel intention have given some implication not only to the academic perspectives, but also practical perspectives, namely film makers, destination developers and promoter as well as relevant tourism authorities.

Keywords: 3P's Pull Factor, screen tourism, viewers travel intention

1 INTRODUCTION

Area of research relating to cinematic tourism is relatively small but Kraaijenzank (2009) acknowledged that film tourism is an interesting subject to be studied. The use of movies as product and destination promotion began since late nineteenth century (Horrigan, 2009). Rewtrakunphaiboon (2009) further acknowledged that the number of tourists visiting destinations featured in films and television series are exponentially increasing. During the early studies of screen tourism phenomenon, Riley, Baker and Van Doren (1998) stated that a larger body of data was missing in the discussion to support the idea that films could attract tourists to visit the actual film locations. Beeton (2005) emphasized that many questions need to be addressed in screen tourism research including determination of psychological drivers in a film that encourage tourist visitation. Particularly notable is the dearth in a cross-disciplinary working between tourism and media studies in Malaysia context. It is considered as a losing point for Malaysia, because the country could share its cultural characteristics through film and television, as these mediums could provide information and excite the audiences with our local language, history, literature, society, landscapes and personalities (Borneo Post, 2012). Therefore more researches are required in order to study these gaps and other factors relating to film induced tourism. Although numerous studies have been conducted investigating the screen tourism phenomenon in several countries (Connell, 2012), either on certain television shows (Kim and Wang, 2012; Wearing et al., 2011; Connell and Meyer, 2009; Kraaijenzank, 2009) or with diverse subject matters and themes (Croy, 2011; Macionis and O'Connor, 2011; Butler, 2011; Kim, Robinson and Long, 2006).

With this idea, an empirical investigation is needed, particularly to examine the effects of 3P's (Place, Performance and Personality) Pull Factor (Gjorgievski & Trpkova, 2012) from the perspectives of Malaysian viewers towards their travel intention after media consumption. The referred Pull Factor theory actually involves three main attributes due to which a film-induced tourist decides on a certain destination. Through this theory and understanding the fundamentals of screen tourism, it could have an impact on future postmodern tourist trends (Gjorgievski and Trpkova, 2012). This is considered as an opportunity for the Malaysia film industry and Malaysian destination developers to understand the drivers of screen tourism and investigate the effects.

2 LITERATURE REVIEW

2.1 *Screen tourism*

Connell (2012) and Lin (2012) identified the variation of terms used to define the concept of film tourism. It has been referred in the literatures with the terms "Film Induced Tourism", "Media

Induced Tourism", "Movie Induced Tourism", "Cinematographic Tourism", "Screen Tourism", "Media Pilgrim" or seen as a media pilgrimage (Connell, 2012; Lin, 2012; Horrigan, 2009). Beeton (2005) argued that by using these terms, it can be expanded to include television, video and Digital Video Disc (DVD). The study of screen tourism is complex. It is because of the inter-related aspects of disciplines such as sociology and psychology, as well as industry-based sectors such as film making industry, destination marketing, community development and strategic planning of a destination (Beeton, 2005).

2.2 *3P's Pull Factor theory*

The "Push and Pull" model is one popular typology for understanding travel motivation (Crompton, 1979). Most Push Factors logically originated from intangible or intrinsic desires of human beings, while pull factors are attracting people to a specific destination once the decision to travel has been made (Lam and Hsu, 2006). Some characteristics such as the perfect pictures of the film which portray wonderful locations (Place), the engagement of audience to personalize the story (Performance) and their adoration towards celebrities who took part in the acting (Personality) highlights the pre-visit role audiences to the film sites (Croy, 2012).

Gjorgievski and Trpkova (2012) agreed that a person desires to visit a particular place which is attractive because it was seen on a movie. However, the film-induced tourism does not merely occur at the on-locations, it also involves off-locations shooting (Beeton, 2005). It is also interesting to note on how films or TV show could help in promoting business places featured in it; that hotels, guesthouses and dining places featured in the film could be promoted and later become income generator and cause tourism to happen (Berić, Kovačević, Simat and Božić, 2013).

Tooke and Baker (1996) suggested that tourists may want to experience at least part of what was portrayed in the film as not to merely gazing at the film locations. This happens because the plot and storyline in many films induced the audiences to sense that they are involved in a real story as it is happening near them (Kim, Robinson & Long, 2006).

Additionally, fascination with the famous and infamous is not new in tourism industry where in the late 20th and early 21st century; contact with famous person makes an individual feel more worthy (Beeton, 2005). Visit to film destinations also allows the tourists to get in touch with places closely associated with admired persons, and perhaps the opportunity to touch the artifacts and memorabilia (Busby & Klug, 2001). Therefore Kim, Robinson and Long (2006) declared that it is not unreasonable to believe that celebrities do have the pulling power of attracting audiences.

Lam and Hsu (2006) asserted that it is beneficial to understand the reason why people travel and identifying the factors which influence their behavioral intention of choosing a travel destination especially to the assistance of tourism planning and marketing. As familiarization of the destination was developed among the tourists, they seem to have precise awareness about holiday opportunities (Kim & Richardson, 2003). Croy (2011) pointed out that film-induced tourism plays the major role as an attractor, motivator, and as a demanded experience for the tourist to experience.

3 RESEARCH METHODOLOGY

3.1 *Sampling and instrumentation*

The researcher applied convenience sampling method, and the samples chosen were Malaysian viewers who had watched any television programs or movies within six months period. To ensure that the samples' watching experience is still fresh in mind, a screening question ("Have you watch any television programs or movies within six months period?") was asked before the researcher solicits potential respondent permission to participate in the study. A structured questionnaire written in dual language (English and Malay language) was used as a tool to collect responses from the samples involving Malaysian audiences.

With regards to the sample size, the researcher referred to Salkind (2003), where the rule of thumb for an unknown population or sample size should be more than 30 but less than 500. Additionally, the researcher also considered the common rule for multiple regression analysis as emphasized by Hair, Anderson, Tatham, and Black (1998), that it is to have the sample size at least five (5) times more than the items to be analyzed. In this study, there are three (3) independent variables with 20 items to be analyzed; hence a minimum number of 100 completed questionnaires (20 items × 5) are needed to test the regression coefficient.

4 RESULTS AND DISCUSSION

There are 44 males (25 percent) and 132 females (75 percent) in the sample, giving a total of 176 respondents. The Standard Multiple Regressions analysis was conducted in order to answer research questions as follows:

RQ1: How the effects of "Place" Pull Factor of screen tourism could induce travel intention of viewers?

Table 1. Results of multiple regressions of 3P's Pull Factor on viewers' travel intention.

	R	R^2	Adj. R^2	Sig.
RQ1	0.611	0.374	0.370	0.000
RQ2	0.590	0.348	0.344	0.000
RQ3	0.653	0.426	0.423	0.000

RQ2: How the effects of "Performance" Pull Factor of screen tourism could induce travel intention of viewers?

RQ3: How the effects of "Personality" Pull Factor of screen tourism could induce travel intention of viewers?

Looking at Table 1, the three (3) attributes of Pull Factor theory namely Place, Performance and Personality were able to explain 37 percent (adj. $R^2 = 0.370$, sig. = 0.000), 34.4 percent (adj. $R^2 = 0.344$, sig. = 0.000), and 42.3 percent (adj. $R^2 = 0.423$, sig. = 0.000), respectively of the variance in viewers' travel intention. As each of the attributes are significantly and positively influence the viewers' travel intention ($p < 0.01$), it can be said that the assumption on the effects of 3P's Pull Factor of screen tourism in inducing viewers' travel intention are supported.

The "Place" Pull Factor is contributing towards viewers' travel intention perhaps because they could directly witness the physical environment of the filming locations from their television screens. Thus, it can be said that different people would have their own point of view in defining the "attractiveness" of the locations portrayed in films or TV shows which could induce their travel intention. In relation to the effects of "Performance" Pull Factor, The researcher predicted that emotions of viewers were developed by the "Performance" Pull Factor. Therefore it would increase the sentiment of being at the film location and indirectly induce their travel intention. Lastly, "Personality" Pull Factor was found as the most significant contributor towards viewers' travel intention, compared to other Pull Factors stipulated in the study, perhaps the celebrities and famous person who involved in films or TV shows act as spokesperson which might induce the viewers travel intention to film locations.

5 FUTURE RESEARCH AND CONCLUSION

Replication of this study, conducted based on local films or TV shows is recommended to further expand the local screen tourism knowledge. The future studies should study the effects of 3P's Pull Factor on local films or TV shows. Besides, it is recommended to use different sampling technique, approach the respondents at specific research setting, improvised the survey instrument or using mix method approach in developing the Push Factor-Travel intention model for future research, to see whether the outcome is similar or different.

As a conclusion, The findings of the study have indicated that the study framework developed based on the 3P's Pull Factor theory by Gjorgievski and Trpkova (2012) was successfully validated. It was found that the effects of 3P's Pull Factor on viewers' travel intention were somewhat supported based on statistical analysis conducted. Last but not least, all Pull Factors involved (Place, Performance and Personality) were found to be significant predictors of viewers' travel intention.

REFERENCES

Beeton, S. (2005). *Film-induced Tourism*. Great Britain: Cromwell Press.

Berić, D., Kovačević, M., Simat, K. & Božić, S. (2013). Film Tourism: A Contemporary Resource for Promoting Serbia. *Turizam,* 17(1), 18–28.

Borneo Post.(2012). *State to be Promoted as Filming Location*. Retrieved on 28th October 2012, from http://www.theborneopost.com/2012/09/17/state-to-be-promoted-as-filming-location/.

Busby, G. & Klug, J. (2001). Movie-induced tourism: The challenge of measurement and other issues. *Journal of Vacation Marketing*, 7(4), 316–332.

Butler, R. (2011). It's Only Make Believe: The Implications of Fictional and Authentic Locations in Films. *Worldwide Hospitality and Tourism Themes,* 3(2), 91–101.

Connell, J. & Meyer, D. (2009). 'Balamory Revisited': An Evaluation of the Screen Tourism Destination-Tourist Nexus. *Tourism Management,* 30(2), 194–207.

Connell, J. (2012). Progress in Tourism Management: Film Tourism—Evolution, Progress and Prospects. *Tourism Management,* 33, 1007–1029.

Crompton, J. (1979). Motivation for Pleasure Vacation. *Annals of Tourism Research,* 6(4), 408–424.

Croy, W.J. (2011). Film Tourism: Sustained Economic Contributions to Destinations. *Worldwide Hospitality and Tourism Theme,* 3(2), 159–164.

Gjorgievski, M. & Trpkova, S.M. (2012). Movie Induced Tourism: A New Tourism Phenomenon. *UTMS Journal of Economics,* 3(1), 97–104.

Hair, J.F., Anderson, R.E., Tatham, R.L. & Black, W.C. (1998). *Multivariate Data Analysis* (5th ed.). Upper Saddle River, N.J: Prentice-Hall.

Horrigan, D. (2009). Branded Content: A New Model For Driving Tourism Via Film and Branding Strategies. *Tourismos: An International Multidisciplinary Journal of Tourism,* 4(3), 51–65.

Kim, H. & Richardson, S.L. (2003). Motion Picture Impacts on Destination Images. *Annals of Tourism Research,* 30(1), 216–237.

Kim, S. & Wang, H. (2012). From Television to the Film Set: Korean Drama Daejanggeum Drives Chinese, Taiwanese, Japanese and Thai Audiences to Screen-Tourism. *The International Communication Gazette,* 74(5).

Kim, S., Robinson, M., & Long, P. (2006). *Understanding popular media production and potential tourist consumption: A methodological agenda.* Paper presented at the International Tourism and Media Conference. Retrieved from http://dspace.flinders.edu.au/xmlui/bitstream/handle/2328/26062/Kim%20Understanding.pdf?sequence=1.

Kraaijenzank, M. (2009). *Movie-Induced Tourism.* Unpublished Master's Thesis, Aalborg University.

Lam, T. & Hsu, C.H.C. (2006). Predicting Behavioral Intention of Choosing a Travel Destination. *Tourism Management,* 27, 589–599.

Lin, X. (2012). *On Development Way of Movie-Induced Tourism Products that Embedded Ethnic Feature: The Analysis Based on Industrial Convergence Development Characteristic.* Paper presented at the International Conference on Future Information Technology and Management Science & Engineering Lecture Notes in Information Technology. Retrieved from http://www.ier-institute.org/2070-1918/lnit14/v14/471.pdf.

Macionis, N. & O'Connor, N. (2011). How Can the Film-Induced Tourism Phenomenon Be Sustainably Managed?. *Worldwide Hospitality and Tourism Themes,* 3(2), 173–178.

Rewtrakunphaiboon, W. (2009). *Film-induced Tourism: Inventing a Vacation to a Location.* Retrieved on 28th October 2012, from http://www.bu.ac.th/knowledgecenter/epaper/jan_june2009/pdf/Walaiporn.pdf.

Riley, R.W., Baker, D. & Van Doren, C.S. (1998). Movie Induced Tourism. *Annals of Tourism Research,* 25(4), 919–935.

Salkind, N.J. (2003). *Exploring Research.* New Jersey: Pearson Education.

Tooke, N. & Baker, M. (1996). Seeing is Believing: The Effect of Film on Visitor Numbers to Screened Locations. *Tourism Management,* 17(2), 87–94.

Wearing, S., Buchmann, A. & Jobberns, C. (2011). Free Willy: The Whale-Watching Legacy. *Worldwide Hospitality and Tourism Themes,* 3(2), 127–140.

Hospitality and Tourism – Sumarjan et al. (eds)
© 2014 Taylor & Francis Group, London, ISBN 978-1-138-00151-0

Assessing Customer-Based Brand Equity theory of repurchase intention in the family restaurant

M.A. Abdul Majid & C. Tamby Chik
Universiti Teknologi MARA, Malaysia

ABSTRACT: Brand equity has become a very important factor that influences consumer's perceptions of a brand, regardless of industry. This study applied four of five Aaker's components of brand equity i.e. brand loyalty, perceived quality, brand awareness and brand associations that defined as Customer-Based Brand Equity (CBBE). This study investigates the relationships between the four dimensions of CBBE and repurchases intention towards a family restaurant in Klang Valley, Malaysia and to determine the key determinant of CBBE on repurchase intention. A total of 450 questionnaires were distributed to customers at the participating restaurants and 349 usable questionnaires were analysed. The results showed that only brand loyalty and brand association had significant relationships ($p < 0.001$) towards repurchase intention and brand association was found to be the strongest determinant ($\beta = 0.55$) on repurchase intention. Furthermore, the findings established that perceived quality was the most important component in how customers perceived CBBE in the family restaurant.

Keywords: Customer-Based Brand Equity, repurchase intention, family restaurant

1 INTRODUCTION

Due to customer's changing demand and increased competition among restaurants, brand has become vital for the business to succeed. One of the strategies is to focus on improving their brand to add value to the restaurants. It is crucial to the success of restaurants because strong brands offer the key elements of differentiation among competitors (Kim & Kim, 2004). The brand value perceived by consumers and the company owning the brand has been conceptualized with the term "brand equity" (Randall, 2000), and many companies regard this as their most valuable asset (Ambler, 2003).

Due to the fact that the restaurant industry is both tangible and intangible in nature, the industry becomes more and more competitive in order to gain sustainable competitive advantage of the brand. Furthermore, the key asset for many restaurant businesses is the brand name and what that brand signifies (Kim & Kim, 2004). Williams (2002) stressed that customers would buy brands they trust and familiar in order to reduce the perceived risk. Moreover, people choose brands, not just because of their function, but consumers feel that the particular brands are reflecting their personalities (Mariotti, 2002). Besides that, brand equity and discounts are the potential ways in which customers can differentiate the brands. Certainly, promotion e.g. coupon and package meal has commonly been used as the marketing strategy by most restaurants. This leads to price war and could reduce profit and weaken customers' loyalty.

It is contended that customers choose one restaurant because of price instead of brand. This in turn has affected the marketing strategy and efforts of restaurants to promote their brand to customers. Hence, there is a need for researchers to conclude and resolve this argument in the Malaysian context, particularly in the family restaurant. Therefore, this study aims to investigate the relationships between the four dimensions of CBBE and repurchase intention towards family restaurants in Klang Valley. In addition, the purposes were also to determine the key determinant of CBBE on repurchase intention and to examine the overall perceptions of customers towards CBBE in the family restaurant.

1.1 *Research hypotheses*

The CBBE dimensions namely brand loyalty, perceived quality, brand awareness and brand association are treated as dependent variables while repurchase intention is treated as the dependent variable. The hypotheses are set forth.

H1: There is a relationship between CBBE dimensions and repurchase intention.

H1a: There is a relationship between brand loyalty and repurchase intention.

H1b: There is a relationship between perceived quality and repurchase intention.
H1c: There is a relationship between brand awareness and repurchase intention.
H1d: There is a relationship between brand association and repurchase intention.

2 LITERATURE REVIEW

2.1 *Brand equity*

Brand equity has been the interest among researchers in marketing management since the 1990's (Keller, 1993; Cobb-Walgren, Ruble, & Donthu, 1995; Lassar, Mittal & Sharma, 1995; Aaker, 1996). Previous researchers have different concepts in brand equity and found that it affects many perspectives such as value added to the products (Aaker, 1996; Keller, 1993); value of the customer (Aaker, 1991; Martensen & Grønholdt, 2003); value of the firm (Aaker, 1991; Kim & Kim, 2005); brand preference and purchase intention (Cobb-Walgren et al., 1995); response of consumer to the marketing of brand (Keller, 1993); incremental utility (Simon & Sullivan, 1993) and consequence of marketing efforts (Villarejo-Ramos & Sánchez-Franco, 2005).

Several researchers have highlighted the significance of building brand equity as it brings the benefit to the firm of more consumer brand preference and purchase intentions (Cobb-Walgren et al., 1995; Myers, 2003). According to Aaker (1991), brand equity stipulates value to customers by enhancing their interpretation and processing of information, confidence in the purchase decision, and satisfaction. Brand equity also provides value to the firm by enhancing efficiency and effectiveness of marketing programs, prices and profits, brand extensions, trade leverage, and competitive advantage (Yoo & Donthu, 2001).

2.2 *Conceptualization of CBBE*

Generally, the operationalization of Customer-Based Brand Equity (CBBE) can be categorized into two types, which are consumer perception and consumer behaviour. Cobb-Walgren et al. (1995) and Yoo and Donthu (2001) claimed that consumer perception consists of perceived quality, brand awareness and brand associations while consumer behaviour consists of brand loyalty and willingness to pay a premium price. Lassar et al. (1995) defined the operationalization of brand equity as "the enhancement in the perceived utility and desirability a brand name confers on a product."

Aaker (1991, 1996) proposed that CBBE consists of four dimensions, which are brand loyalty, perceived quality, brand awareness and brand association. His theory has been accepted and adopted by many scholars (Keller, 1993; Low & Lamb, 2000; Prasad & Dev, 2000; Yoo & Donthu, 2001). Moreover, Yoo, Donthu and Lee (2000) have tested a model by using three dimensions of Aaker's (1991) components, which are brand loyalty, perceived quality and brand association. They found that these dimensions contribute to brand equity. Moreover, in a study by Kim and Kim (2004), they confirmed that brand equity should consist of the four components.

2.3 *Relationship between brand equity and repurchase intention*

Based on Aaker's (1991) theory, there were several researchers have conducted the study of CBBE with repurchase intention. Washburn and Plank (2002) tested the relationship between the dimensions of brand equity including brand loyalty, perceived quality, brand awareness and brand association and repurchase intention and found that there was a significant correlation between the measured dimensions. Similarly, in a study by Zhou and Jiang (2011), they discovered that brand loyalty and perceived quality had a positive effect on the hotel revisit intentions. However, it was inconsistent with the findings of Kim, Jin-Sun and Kim (2008). They found that, except brand loyalty, the three components of brand equity did not influence revisit intention. Revisit intention and repurchase intention are two similar terms, except for in hotel study, repurchase intention is changed to revisit intention. Indeed, Butcher (2005) stated that the brand equity influenced the customers' repurchase intention as one service outcome.

The literature review has shown that there is a positive relationship between perceived quality and intentions after the purchase (Castro, Armario & Ruiz, 2007). Besides that, good location selection ensures that the customer will use the same brand at future purchases due to convenience thus leveraging brand equity (Yoo et al., 2000; Bucklin & Silva-Risso, 2008). Good location is part of brand association. For that reason, it implied that brand association is associated with repurchase intention. This also validated by the findings of Zhou and Jiang (2011). They combined brand awareness and brand association as one dimension as suggested by Yoo and Donthu (2001), and found that the dimension had positive effects on revisit intentions.

3 METHODOLOGY

3.1 *Research design*

This study employed the use of a quantitative research design with a structured questionnaire as the research instrument. A descriptive research design and correlational approach have also been used. A cross sectional study was adopted where data has been gathered just once at the same point of time for each participated restaurant namely Rasamas, The Chicken Rice Shop and Pizza Hut.

3.2 *Pilot study*

Before the researcher proceeded with the main survey, a pilot study was conducted between March 19 and March 21, 2013 among 30 Master's degree students of Universiti Teknologi MARA, Shah Alam who were familiar with the restaurant brands. A preliminary list of 33-items scale was constructed based on previous studies in CBBE and repurchase intention to indicate the importance of brand equity from the consumer's perspective on restaurant brands. From the pilot study, a reliability analysis was conducted to measure the internal consistency of all study variables.

3.3 *Sample size and sampling technique*

The sample size of this study was 300. However, according to Hair, Black, Babin and Anderson (2010), 100 percent of usable questionnaires is unlikely to achieve. Therefore, 450 questionnaires were distributed in order to guarantee that at least 300 usable questionnaires to be collected. The family restaurants were selected within Petaling, Gombak and Klang districts of Klang Valley. Permissions were requested from each participating restaurant and "drop-off/pick up" method was used. Questionnaires were distributed by the managers of the restaurants using a convenience sampling method to the respondents who were readily available at the restaurant. According to Sekaran and Bougie (2010), convenience sampling is quick, convenient and less expensive.

3.4 *Response rate*

Between April 14 and May 1, 2013, all questionnaires were collected from 9 participating restaurants. From the 450 questionnaires that were handed to the restaurant's managers, 407 questionnaires were collected for a response rate of 90.4%. However, only 349 questionnaires (77.6%) were analyzed because 58 questionnaires were unusable due to incomplete and invalid responses.

4 RESULTS AND ANALYSIS

4.1 *Profile of sample*

Among the 349 respondents, more than half (54.2%) of them are male while 45.8% were female. The majority (45.3%) of the respondents indicated that they were in their twenties, with the second largest age group (22.9%) being in their thirties. The third largest age group (13.8%) indicated they were under age 20. Lastly, more than 10% (10.3%) of the sample were in their forties while less than 10% (7.7%) were age 50 and above. The samples included with more than 70% (71.9%) were Malay while more than 10% (13.5%) were Chinese, followed by Indian (9.2%) and others (5.4%). More than half (51.9%) of the respondents were married and 42.1% were single while less than 10% (6.0%) were divorced.

Concerning the education level, more than half of the respondents were either having a Bachelor's Degree or Diploma (36.4% and 32.7% respectively). About 15% (14.0%) of the respondents have SPM while 13.7% have a Master's Degree and only 3.2% have a Doctorate Degree. The respondents were also asked about their occupation. About 40% (38.7%) of them work in the private sector and 31.5% work in the public sector. 15.2% of them were students while others were either self-employed (7.7%) or unemployed (6.9%). Besides that, respondents were asked about their monthly income. The monthly income that most frequently (32.1%) reported by respondents was below RM2000 and 21.8% were between RM3000 to RM3499 while 18.1% of the respondents earned RM3500 and above. More than 15% (16.3%) of the sample earned monthly income between RM2000 to RM2499 and more than 10% (11.7%) earned between RM2500 to RM2999.

4.2 *Reliability coefficient*

Internal reliability of the 33-items scale of 30 respondents from the pilot study was examined by using Cronbach's alpha. The result of alpha was between 0.794 and 0.931. Among the independent variables, perceived quality with 0.931 is the highest value. Generalizing from the results, the alpha values for the internal consistency of all study variables were considered as acceptable. Table 1 shows the reliability values for the measured variables of this study.

Table 1. Reliability statistics.

Variables	Alpha value	Items
Brand loyalty	0.921	6
Perceived quality	0.931	11
Brand awareness	0.794	5
Brand association	0.892	8
Repurchase intention	0.876	3

4.3 *Relationships between CBBE and repurchase intention*

The first objective was assessed through hypothesis testing and multiple linear regression was employed to test the assumption of this study. Table 2 shows the model summary of multiple regression analysis.

The model as a whole was significant $[F(4,344) = 102.1, p < 0.001]$. This model made a statistically significant contribution towards dependent variable and it explained 54% of the variation in repurchase intention ratings. The remaining 46% is explained by other predictor variables.

Based on Table 2, the first and fourth hypotheses demonstrated that the brand loyalty and brand association were significantly related to repurchase intention at the significant level of 0.05. It can be concluded that H1a and H1d hypotheses are supported. However, the second and third hypotheses established that the perceived quality and brand awareness were not significantly related to repurchase intention. Therefore, it can be concluded that H1b and H1c are not supported. Table 3 shows the summary of hypotheses testing results.

4.4 *Key determinant of CBBE*

Among the four predictors, in terms of importance, *brand association* ($\beta = 0.55$) made the largest unique contribution as shown in Table 2. It followed by *brand loyalty*, *brand awareness* and *perceived quality* ($\beta = 0.29$, $\beta = -0.07$ and $\beta = 0.04$ respectively). This means that the brand association was the key determinant in predicting the customers' repurchase intention towards family restaurants in Klang Valley.

Table 2. Summary of multiple regression analysis.

	B	SE B	β
Constant	0.22	0.25	
Brand loyalty	0.31	0.05	0.29*
Perceived quality	0.06	0.08	0.04
Brand awareness	–0.08	0.06	–0.07
Brand association	0.65	0.07	0.55*

Note: $R^2 = 0.54$, $*p < 0.001$.

Table 3. Summary of hypotheses testing.

Indicator	Remarks
H1a	Supported
H1b	Not supported
H1c	Not supported
H1d	Supported

Table 4. Overall mean scores of CBBE dimensions.

Construct	Minimum	Maximum	*M*	*SD*
Brand loyalty	1.00	7.00	4.36	1.363
Perceived quality	1.64	7.00	4.82	1.055
Brand awareness	1.00	7.00	4.76	1.378
Brand association	1.00	7.00	4.63	1.225

4.5 *Level of customer's perception towards CBBE*

Table 4 reports the mean scores of CBBE dimensions that were obtained from customers of family restaurants in Klang Valley by using descriptive statistic. The result of mean scores and standard deviation from the descriptive statistic has provided the evidence of degree of importance of customer's overall perceptions towards CBBE in family restaurant.

Based on four dimensions of CBBE, perceived quality ($M = 4.82$, $SD = 1.055$) was the most important component in how customers perceived the CBBE of family restaurants in relative to other dimensions. It followed by brand awareness ($M = 4.76$, $SD = 1.378$), brand association ($M = 4.63$, $SD = 1.225$) and brand loyalty ($M = 4.63$, $SD = 1.363$). In addition, the dimension had standard deviations of more than 1.000, which means the customers' perception of brand association was widely scattered.

5 IMPLICATIONS, RECOMMENDATIONS AND CONCLUSION

It is worth noting that although perceived quality was not significantly related to repurchase intention, it was found to be the most important element affecting customer's perception on CBBE in family restaurants. From that finding, researchers suggest that family restaurant should provide the best quality of products and services. It should consistently deliver products and services such as confidence and knowledgeable staff regarding the menu, neat and clean staff appearance, gives customers individual attention, handles customer's complaints effectively, serving food at the promised time and convenient operating hours. All the criteria mentioned above are also considered as the factors to enhance the image of the restaurant. However, the finding that perceived quality did not significantly related to repurchase intention was an unexpected result because customers who experience the quality of products and services at the restaurants are more likely to show their potential intent to repurchase at the same restaurant.

Besides that, the result showed that brand loyalty obtained the lowest score among the four dimensions. In general, this indicated that family restaurants in Klang Valley did not gain loyal customers. One of the ways to attract customers is through loyalty program by using promotion especially pricing discounts. Although it could reduce profit and customers might choose the restaurant based on price, Zhou and Jiang (2011) suggested that it is an effective way to attract customers to experience the brand. Therefore, it is imperative to form customers' brand loyalty in order to improve family restaurant repurchase intentions because as expected, brand loyalty and repurchase intention were significantly related. It indicates that brand loyalty is an important dimension of brand equity but it was difficult to manage. It is vital in order to ensure the customers to consider the restaurant as the first choice and recommend the restaurant to others.

In contrast, brand awareness and brand association are measured by recognition and recall of a restaurant's logo or physical appearance. It could occur in people's mind without any experience with specific brands (Kim et al., 2008). In addition, Zhou and Jiang stated that building customers' brand awareness/brand association encourages revisit intentions. In this study, it was found that relationship between brand association and repurchase intention was significant and brand association was the key determinant of CBBE on repurchases intention. However, there was no significant relationship between brand awareness and repurchase intention.

In conclusion, the results of this study showed that CBBE leads to repurchase intention and lack of brand equity can affect customers' intention to purchase again at the restaurants. It is certain that if the restaurant operators do not make an effort to improve and develop CBBE, they should expect decreasing in numbers of customers. Therefore, restaurants are required to work continuously to build and develop their restaurant brand and earn advantages over other competing restaurants.

REFERENCES

Aaker, D.A. (1996). *Building Strong Brands*. New York: The Free Press.

Aaker, D.A. (1991). *Managing Brand Equity: Capitalizing on the Value of the Brand Name*. New York: The Free Press.

Ambler, T. (2003). *Marketing and the Bottom Line*. Harlow: Pearson Education.

Butcher, K. (2005). Differential Impact of Social Influence in the Hospitality Encounter. *International Journal of Contemporary Hospitality Management, 17*(2), 125–135.

Castro, C.B., Armario, E.M., & Ruiz, D.M. (2007). The influence of market heterogeneity on the relationship between a destination's image and tourists' future behavior. *Tourism Management, 28,* 175–187.

Cobb-Walgren, C.J., Ruble, C.A., & Donthu, N. (1995). Brand equity, brand preference, and purchase intent, *Journal of Advertising, 24*(3), 25–40.

Hair, J.F.J., Black, W.C., Babin, B.J., & Anderson, R.E. (2010). *Multivariate data analysis: A global perspective (7th ed.)*. New Jersey: Pearson.

Keller, K.L. (1993). Conceptualizing, Measuring, and Managing Consumer-Based Brand Equity. *Journal of Marketing, 57*, 1–22.

Kim, W.G., Jin-Sun, B., & Kim, H.J. (2008). Multidimensional Customer-Based Brand Equity and Its Consequences in Mid priced Hotels. *Journal of Hospitality & Tourism Research, 32*(2), 235–254.

Kim, H.B., & Kim, W.G. (2005). The relationship between brand equity and firms' performance in luxury hotels and chain restaurants. *Tourism Management, 26,* 549–560.

Kim, W.G., & Kim, H.B. (2004). Measuring customer-based restaurant brand equity: Investigating the relationship between brand equity and firms' performance. *Cornell H.R.A. Quarterly, 45*(2), 115–131.

Lassar, W., Mittal, B., & Sharma, A. (1995). Measuring Customer-Based Brand Equity. *Journal of Consumer Marketing, 12*(4). 11–19.

Low, G.S., & Lamb, C.W.J. (2000). The measurement and dimensionality of brand associations. *Journal of Product and Brand Management, 9*(6), 350–368.

Mariotti, J. (2002). *Smart Branding.* NY: New York Press.

Martensen, A., & Grønholdt, L. (2003). Improving Library Users' Perceived Quality, Satisfaction and Loyalty: An Integrated Measurement and Management System. *The Journal of Academic Librarianship, 29*(3), 140–147.

Myers, C.A. (2003). Managing brand equity: A look at the impact of attributes. *Journal of Product and Brand Management, 12*(1), 39–51.

Prasad, K., & Dev, C.S. (2000). Managing hotel brand equity: A customer-centric framework for assessing performance. *Cornell Hotel and Restaurant Administration Quarterly, 41*(3), 22–31.

Randall, G. (2000). *Branding*. London: Kogan Page.

Sekaran, U., & Bougie, R. (2010). *Research Methods for Business: A Skill Building Approach (5th ed.)*. West Sussex: John Wiley & Sons Ltd.

Siddarth, S., Bucklin, R., & Silva-Risso, J. (2008). Distribution Intensity and New Car Choice. *Journal of Marketing Research, 45*(4), 473–486.

Simon, C.J., & Sullivan, M.W. (1993). *The measurement and determinants of brand equity: A financial approach. Marketing Science, 12*(1), 28–52.

Villarejo-Ramos, A.F., & Sánchez-Franco, M.J. (2005). The impact of marketing communication and price promotion on brand equity. *Journal of Brand Management, 12*, 431–444.

Washburn, J.H., & Plank, R.E. (2002). Measuring brand equity: An evaluation of a consumer-based brand equity scale. *Journal of Marketing Theory and Practice, 10*(1), 46–61.
Williams, A. (2002). *Understanding the Hospitality Consumer.* Oxford: Butterworth-Heinemann.
Yoo, B., & Donthu, N. (2001). Developing and validating multidimensional consumer-based brand equity scale. *Journal of Business Research, 52*(1), 1–14.
Yoo, B., Donthu, N., & Lee, S. (2000). An examination of selected marketing mix elements and brand equity. *Journal of the Academy of Marketing Science, 28*(2), 195–211.
Zhou, Y., & Jiang, J. (2011). The Impacts of Customer-Based Brand Equity on Revisit Intentions: An Empirical Study on Business and Leisure Traveler at Five Shanghai Budget Hotels. *Research Journal of International Studies, 22,* 110–119.

Hospitality and Tourism – Sumarjan et al. (eds)
© 2014 Taylor & Francis Group, London, ISBN 978-1-138-00151-0

The effects of brand personality dimensions on customer satisfaction in Kuala Lumpur hotels

N.A. Zaireen, M. Sulaiman & Z. Mohi
Universiti Teknologi MARA, Malaysia

ABSTRACT: Sincerity, excitement, competence, sophistication and ruggedness are the dimensions of brand personality. It is the aim of this study to apply the same dimensions on hotel brands as to test whether the dimensions would reflect the hotel's personality and eventually affect customers' satisfaction. Using a quantitative approach, data gathered through surveys on the guests of five-star hotels in Kuala Lumpur. Descriptive statistics, correlation analysis and regression analysis were used in this study. From the five dimensions, sincerity had higher scores, and it is believed that brand personality is adaptable to hotel brands. From this study, the hospitality industry will be benefited with the essential understanding of the brand personality impact on customers' satisfaction.

Keywords: brand personality, customer satisfaction, five-star hotels

1 INTRODUCTION

The tourism industry in Malaysia, especially the lodging industry, is expanding rapidly because of the increasing number of tourist arrivals in Malaysia. As the numbers of hotels in Kuala Lumpur are increasing, this encourages the competition among existing hotels and basically the competition is in term of market share amongst the hotels. Matovic (2002) supported that the competition among the hotels will bring benefit to the hotels such as improving the hotels' market share and the hotels' brands.

Kotler & Armtrong (2001) defined brand as "a seller's promise to deliver a specific set of features, benefits and services consistent to the buyers". Hotel brand will make it easier for customers to choose their preferred hotel and due to branding; hotel guests can easily differentiate one hotel to another. Thus, branding is crucial to the hotel industry. Biel (1993) mentioned "a unique brand personality in consumer minds can lead to an emotional bond to the brand". Therefore, a hotel with a unique brand personality will gain positive perception from their guests and this will lead to customer satisfaction and create loyalty towards the hotel itself.

However, Ambroise, Ferrandi, Jolivot, Merunka, Valette-Florence and ZineDanguir (2003) argued that brand personality dimensions' structural and validity is questioned, especially in the contexts of different culture. Many studies on brand personality dimensions conducted in Western countries and limited studies were conducted in Asian countries. In addition, Aaker (1997) mentioned that even though ruggedness revealed as one of the human personalities, it cannot be applied to Brand Personality because the result may vary due to the different locations and cultures.

2 LITERATURE REVIEW

2.1 *Brand personality*

The definition of brand personality by Aaker (1997) is "the set of human characteristics associated with the brand." Whereby, Keller (1993) cited that the brand personality will serve a "symbolic or self-expressive function." From the perspective of customers, when they feel that the brand of a hotel is relevant to them, and they decide to stay loyal is because they feel that it is a right personality for them (Donahay & Rosenberger, 2007). The dimensions of brand personality, namely sincerity, excitement, competence and sophistication are not only reflects the human personality, but the characteristics are also applicable to the hotel industry.

Hypothesis 1: There is a positive relationship between the hotel's brand personality and customer satisfaction.

2.2 *Brand personality dimensions*

The first dimension of brand personality is sincerity which according to Aaker (1997) are domestic, honest, genuine and cheerful. However, Lee and Back (2010) mentioned that the items of sincerity are wholesome, down-to-earth, honest and cheerful. These components resemble the attitude of the employees. As an example, McDonalds was rated high on sincerity based on the employees' attitude towards the customers.

Hypothesis 1a: Sincerity has a significant effect towards the customer's satisfaction.

The second dimension of brand personality is excitement and Aaker (1997) mentioned that this component consists of characteristics such as daring, spirited, imaginative, and up-to-date. This will be most suitable to the hotel's environment. Aaker, Bennet-Martinez and Garolera (2001) argued that excitement is the dimension that creates the hotel's theme.

Hypothesis 1b: Excitement has a significant effect towards the customer's satisfaction.

The third dimension is competence. Aaker (1997) mentioned that this dimension has characteristics that include reliable, responsible, dependable and efficient. This shall apply to the systems and technology being used in the hotel. Lee, Back and Kim (2009) justified that in competence dimension, there are items such as corporate, successful, technical, hard working and also confident. There are all applicable to the hotels' employees including the superior employees.

Hypothesis 1c: Competence has a significant effect towards the customer's satisfaction.

Next dimension would be sophistication, and the elements is glamorous, pretentious, charming and romantic, (Aaker, 1997). These characteristics associated with the customers that would like to have a prestigious moment when staying at the hotel.

Hypothesis 1d: Sophistication has a significant effect towards the customer's satisfaction.

Ruggedness is the last dimension in brand personality. This dimension resembles strong, masculine and western and reflected by the restaurant, cafe or even the facilities of the hotel itself (Aaker, 1997). The facilities that resemble ruggedness would give satisfaction to the guests that seek for something that is different from what they have or even something similar to their hometown as was mentioned by Norman (1963).

Hypothesis 1e: Ruggedness has a significant effect towards the customer's satisfaction.

Customer satisfaction was used to measure the post-purchase behavior of the customers. Jones, Mothersbaugh and Beatty (2000) defined satisfaction is defined as "an overall evaluation of performance based on all prior experiences with a firm." In addition, customer satisfaction can also be defined as a post-purchase evaluation and judgment (Homburt & Giering, 2001). In this study, customers' satisfaction is based on their experience of staying in the hotels in Kuala Lumpur. Their level of satisfaction rated not only based on the hotels' services but also on the product attributes.

3 METHODOLOGY

3.1 *Research instrument*

The scale used to describe the hotels' brand personality was a five-point scale ranging from 1 (not at all descriptive) to 5 (extremely descriptive) and for customer satisfaction, it was described by using a five-point Likert scale ranging from one to five, denoting very dissatisfied to very satisfied. Similar with most rating scales, the Likert-type scale often measures attitude and opinion on a five or seven categories (Preston & Colman, 2000), thus indicating that a five-point scale is widely used.

3.2 *Data collection*

Convenience sampling been used by selecting the people who experienced staying in one of the hotels of upscale hotels in Kuala Lumpur. The questionnaires that been distributed were 566 whereby the respondents that have answered the questionnaires altogether were 283 respondents that generate 50% of response rate. Table 2 shows the highest number of respondents of each variable in the demographic profile section.

4 RESULTS AND DISCUSSION

Assessments of internal reliability using the Cronbach's Alpha method were undertaken for brand personality dimensions' and customer satisfaction's instrument. According to Pallant (2005), the reliability test of Cronbach Alpha should be above 0.70. Based on the result of the reliability test, for brand personality variable, The highest Cronbach Alpha value is for sincerity at 0.874 for six items and whereby, ruggedness was at 0.750 for five items.

Based on the Pearson Correlation analysis, the result shows that sincerity with the score of (r = 0.54**; $p < 0.001$) has a strong correlation towards customer satisfaction. The lowest score and weak correlation are between ruggedness

Table 1. Development of questionnaire.

Study variables	Sources of scale	Number of items	Type of scale
Brand personality	Asperin (2007); Lee & Back (2010)	4 to 5 items for each dimension	Five-point scale ranging from 5 denoting (extremely descriptive) to 1 denoting (not at all descriptive)
Customer satisfaction	Skogland & Siguaw (2004); Choi, OK, & Hyun (2010) Asperin (2007)	14 items 5 items 4 items	Five-point scale ranging from 5 denoting (very satisfied) to 1 denoting (very dissatisfied)

Table 2. Respondents' demographic profile.

Variables	Indicator	Frequency	Percentage (%)
Gender	Male	106	37.5
	Female	177	62.5
Residency	Malaysia	268	95.0
	International	15	5.0
Age	26–35	111	39.2
Education level	College or University	244	86.2
Employment status	Professional	139	49.1
Income level	RM3,003–RM9,000	139	49.1
Hotel	Equatorial hotel	70	12.0

Table 3. Results of reliability test (N = 283).

Items	Cronbach's alpha	No. of items
Brand personality		
Sincerity	0.874	6
Excitement	0.858	4
Competence	0.742	5
Sophistication	0.770	5
Ruggedness	0.750	5
Customer satisfaction		
Satisfaction and personal opinion	0.884	14

dimension and customer satisfaction which is (r = 0.08; p = 0.19). Even though there was an insignificant correlation between ruggedness dimension and customer satisfaction, ruggedness did have correlation towards customer satisfaction. Therefore, it can be concluded that there are positive relationship between the hotel's brand personality and customer satisfaction. In order to answer the hypothesis 1a to 1e and also to satisfy research objective 2, a Regression Analysis was undertaken.

Table 4. Results of regression of brand personality on customer satisfaction.

Predictors	Model 1/Std. β
Step 1: Model variables	
Brand personality attributes	
Sincerity	0.54***
Excitement	0.47***
Competence	0.44***
Sophistication	0.45***
Ruggedness	0.08
R^2	0.32
Adj. R^2	0.32
R^2 Change	0.32
F-Change	133.950***

From Table 4, the result shows that out of the five dimensions, sincerity impacted the most to customer's satisfaction with the score of ($\beta = 0.54$; $p < 0.001$). Then, followed by excitement ($\beta = 0.47$; $p < 0.001$), sophistication with the score of ($\beta = 0.45$; p < 0.001) and also competence ($\beta = 0.44$; p < 0.001). As for ruggedness, it does not significantly contribute to the satisfaction of the hotel guests and the score is ($\beta = 0.08$; $p > 0.05$). Therefore, the analyses supported hypothesis 1 (1a to 1d). However, hypothesis 1e is not supported due to the insignificant result of ruggedness dimension to customer satisfaction.

5 THE IMPLICATION AND CONCLUSION

As was supported by the research results, it is strongly agreed that brand personality is one of the marketing approaches to enhance the satisfaction of the hotel guests. Brand personality and its dimensions do contribute to customer satisfaction. The study findings clearly showed that sincerity contributed the most to customer satisfaction. It is crucial for the hotels train their employees to practice the best sincere attitude in front of their guests. A brand personality resembles to what the hotel truly is and the hotel will serve their guests according to what the hotel's personality is. Based on the hotel's personality, the hotel guests may easily describe the products and services of the hotels and even its employees as the brand personality of the hotels. As for ruggedness, the result showed that it is not significantly effect on customer satisfaction. Thus, this dimension did not have any positive or negative influence on the feeling of satisfaction of the hotel guests. However, the ruggedness is more suitable to reflect the theme hotels for example, Hard Rock Hotel.

For future research, it is recommended an investigation between the brand personality and the post-purchase behavior such as repeat purchase intention, loyalty or the intention to recommend that will provide a wide array of knowledge on the brand personality for the lodging industry. As a conclusion, based on the discussion above, it is strongly supported that brand personality is not only crucial to the clothing or the automotive industry, it is also crucial to the hospitality industry namely the lodging industry. The hotels acknowledge the importance of guests experience and brand personality dimensions, which will create a great influence in the development of favorable attitude and behavioral brand-related outcomes.

REFERENCES

Aaker, J. (1997). Dimensions of brand personality. *Journal of Marketing Research,* 34(3), 342–352.

Aaker, J., Bennet-Martines, V. & Garolera, J. (2001). Consumption symbols as carriers of culture: A study of Japanese and Spanish. *Advances in Consumer Personality Research, 22*(1), 391–395.

Ambroise, L., Ferrandi, J.-M., Jolivot, A.-G., Merunka, D., Valette-Florence, P. & ZineDanguir, S. (2003). Modelling and measuring Brand Personality: A cross-cultural application, Proceedings of the Ninth Cross-Cultural Research Conference, Rose Hall.

Biel, A.L. (1993). Converting image into equity in brand equity and advertising: Advertising's role in building strong brands, 67–82.

Donahay, B. & Rosenberger, P.J. III. (2007). Using Brand Personality to measure the effectiveness of image transfer in Formula One racing. Marketing Bulletin, 18, Article 1.

Homburg, C. & Giering, A. (2001). Personal characteristics as moderators of the relationship between customer satisfaction and loyalty: An empirical analysis. *Psychology and Marketing Journal, 18*(1), 43–66.

Jones, M.A., Mothersbaugh, D.L., & Beatty, S.E. (2000). Switching barriers and repurchase intentions in services. *Journal of Retailing, 76*(2), 259–274.

Kotler, P. & Armstrong, G. (2001). *Principles of marketing* (9th ed.). New Jersey: Prentice Hall.

Keller, K.D. (1993). Conceptualizing, measuring and managing customer-based brand equity. *Journal of Marketing, 57*(1), 1–22.

Lee, Y.K., Back, K.J., & Kim, J.Y. (2009). Family restaurant brand personality and its impact on customer's emotion, satisfaction and brand loyalty. *Journal of Hospitality & Tourism Research, 33*(3), 305–328.

Lee, J.S. & Back, K.J. (2010). Examining antecedents and consequences of brand personality in the upper-upscale business hotel industry. *Journal of Travel and Tourism* Marketing, 27(2), 132–145.

Matovic, D. (2002). The competitive market structure of the U.S. lodging industry and its impact on the financial performances of hotel brands. Faculty of the Virginia Polytechnic Institute and State University, Blacksburg, Virginia.

Norman, W.T. (1963). Toward an adequate taxonomy of personality attribute: Replicated factor structure in peer nomination personality ratings. *Journal of Abnormal and Social Psychology,* 66, 83–574.

Hospitality and Tourism – Sumarjan et al. (eds)
© 2014 Taylor & Francis Group, London, ISBN 978-1-138-00151-0

Destination image of Perak: Perspective and loyalty among Singaporean tourists

A.M.F. Wahab, M.D. Darson, M.H. Zamri & L.K. Lim
Universiti Teknologi MARA, Malaysia

ABSTRACT: This study investigates the destination image of Perak from Singaporean tourists' perspective. It is also looking their loyalty toward Perak as a tourist destination. Using Perak as contextual study setting, 150 Singaporean tourists were successfully surveyed. Findings revealed that Singaporean tourists perceived a negative image on destination in Perak. Shopping facilities has recorded the lowest rating among all images listed while safety condition image made up the highest score. Apart from that, image was found significantly affecting the loyalty toward destination. Finally, the implications and recommendations for state government and the local tourism service providers were also emphasized.

Keywords: destination image, loyalty, Perak

1 INTRODUCTION

Destinations primarily compete based on their perceived images relative to other competitors in the marketplace (Baloglu & Mangaloglu, 2001). This situation significantly gearing the effort to develop a positive image of tourist destination and to reach its competitive advantage. Images are significantly affecting tourists' behavior, starting from their mental constructions about destination attributes to the travel decision making process (Bigné, Sanchez & Sanchez, 2001; Chen & Tsai, 2007). With that, it is argued that investigating the role of destination images is critical as it help to promote tourist destinations as holiday resources (Nicoletta & Servidio, 2012).

Viewing from tourist arrivals, Singaporean remain the leading group of international tourist to Malaysia when the number of tourist arrivals from this country in 2011 are recorded at 13.37 million (Tourism Malaysia, 2012). In the year 2012, tourists from this country made up 52 percent from 25.03 million total international tourist arrivals in Malaysia. With this positive indication, many states in Malaysia including Perak are competing each other in attract the neighboring country tourists by strengthening their destination images.

The intention of State Government of Perak to grab the Singaporean tourists market were clear stipulated when Datuk Hamidah Othman as Perak State Executive Councilor for Tourism said that, "*We want to explore many ways that we can all work together to increase tourist arrivals from Singapore to Perak*" (The Borneo Post, 2012).

With a mission to promote tourism and foreign investment especially from Singapore, State Government of Perak has signed a five-year strategic collaboration with Firefly, and one of the focuses is to enable international air connection between Sultan Azlan Shah Airport in Ipoh, Perak and Changi International Airport in Singapore (Firefly, 2010). This action were driven by the fact that Singapore is the biggest tourism spender from South East Asia region, sitting in the 11th place in the world ranking of international tourism spenders in the year 2012, jumped two places from 2011 (United Nation World Tourism Organization, 2013). According to the same source, an astonishing RM 69.4 billion total expenditure worldwide has been recorded by tourists from this island country in the year 2012. It indicated that this targeted group of tourists has strong capability in spending their money during travelling and some consideration is continuously undertaking to tap this valuable market.

In line with this notion, although Perak state is having lots of invaluable attractions and facilities, questionsarise whether it has developed a strong image among the Singaporean and what is Singaporean tourists' loyalty toward Perak as their destination. To date, there is still lack of studies per se looking at this issue. Understanding the perspective of Singaporean tourists towards Perak as their destination is therefore consider important to strategize the marketing approach and maximizing the use of available resources among this tiny island tourists. With that, one hypothesis is aims to be tested.

1. *There is a significant relationship between destination image and tourist loyalty*

2 LITERATURE REVIEW

2.1 *Destination image*

Early 1970s is the era when destination image concept was identifiable in the tourism studies (Stepchenkova, 2005). Hunt (1971, 1975), Mayo (1973), and Gunn (1972) have started the studies related to destination image and many definitions of the destination image have been discussed. For some, destination image is simply described as overall impression. Hunt (1971) stated that state tourism image is the impression that an individual has about a state in which they do not stay.

Crompton (1979) defined destination image as the summation of beliefs and ideas as well as impressions that a person has about the destination. While Fridgen (1987) denote destination image as mental representation of an object or place which is not physically appeared in front of the observer. Echtner & Ritchie (1991) noted destination image as the individual perceptions toward destination attributes and the holistic impression made by the destination. Destination image also defined as the tourists' total impression about a destination (Kotler, Haider & Rein, 1993).

2.2 *Loyalty*

Loyalty can be defined and measured in various and multiples aspects (Jacoby & Chestnut, 1978). Loyal customers tend to spend more with the service provider in the short term (O'Brien & Jones, 1995) while, in the long term, the loyal customers will attract new customers by word of mouth (Reichheld & Teal, 1996).

Oppermann (2000) contended that the degree of tourists' loyalty to a destination can be portrayed in their intentions to revisit the destination and in their willingness to propose it to others. By having positive experiences towards the service, products, and other resources provided by tourism destinations, repeat visits among tourists as well as good word of mouth to friends and/or relatives could be assured (Chi & Qu, 2008). It is no doubt that loyalty has significant present and future value for company profits and survival (Valazques, Saura, & Molina, 2011).

2.3 *Relationship between destination image and loyalty*

From previous studies, two significant roles of the destination image towards behaviors have been identified. First, destination image influencing the selection of destination in the decision-making process and second, its influences in forming after-decision-making behaviors include participation (on-site experience), evaluation (satisfaction) and future behavioral intentions such as intention to revisit and willingness to recommend (Bigne et al., 2001; Lee, Lee, & Lee, 2005; Mansfeld, 1992).

Chen & Tsai (2007) investigate the destination image and factors that affect behavioral intentions. They proposed a integrated tourist behavior model by including destination image and perceived value into the "quality—satisfaction—behavioral intentions" paradigm. The structural relationship analysis revealed that destination image appeared to have a significant relationship with behavioral intentions such as intention to revisit and willingness to recommend. Destination image influence not only the decision-making process but also condition-safter-decision-making behaviors of tourists.

3 METHODOLOGY

The sample populations for this study were among Singaporean tourists who have visited Perak and stay at least one night in the state. The convenience sampling method was used. The data was collected at four hotels in Ipoh who were frequent stayed by majority of Singaporean when they visited Perak capital city. As questionnaire surveyed is undertaken during low period or not during schools holiday only 150 questionnaires were successfully distributed. The data set were coded and keyed for analysis using a Statistical Package of Social Science (SPSS), Version 20.0.

4 RESULT AND ANALYSES

4.1 *Demographic profiles*

The majority of the respondents were male with 64.0 percent compared to female (36 percent). Most of the respondents' age is between 25–44 years (52.7 percent as opposed to 32.0 percent which over 44 years old. 78.0 percent graduated from university with 28 percent obtained the high school certificate.

4.2 *Destination image*

The magnitude of mean score between 2.5 to below 3.5 indicate that majority of Singaporean tourists felt slightly disagree toward most of the items related to Perak destination image. Slightly disagree was expressed to the beautifulness of the scenery (M = 3.21) and the beauty of parks image i (M = 3.30). Slightly disagree were also expressed with regard to the variety of cultural attraction (M = 3.26), interesting cultural activities (M = 3.37) and local custom image (M = 3.38).

Similar feeling goes to facilities for shopping (M = 3.01), accommodations' quality (M = 3.45) and value destination for money (M = 3.33), relaxing destination (M = 3.49), having a pleasant environment (M = 3.43) and possess exciting places (M = 3.29).

Besides having slightly disagree feeling to most of the items, tourists also portrayed the inclination toward neither disagree nor agree to safety condition image (M = 4.33) in infrastructure, diversity of flora and fauna (M = 3.97), food, culture and cuisine (M = 3.87) and quality food (M = 3.56).

4.3 *Loyalty*

This descriptive analysis looking at Singaporean tourists' loyalty toward Perak as a tourism destination is presented in Table 1. Magnitude between 2.5 until to below 3.5 in most of the items signify the tendency toward slightly disagree while only one item recorded the value in the range between 1.5 until below 2.5 which represents the inclination towards disagree.

For revisit intention, tourists expressed slightly disagreed feeling through their willingness to return to Perak (M = 3.04) and readiness to plan the next trip to Perak in the future (M = 2.95). Similar predisposition was recorded in willingness to recommend Perak as a good destination (M = 2.94).

Tourists portrayed the disagreed feeling in encouraging their friends and relatives to visit Perak (M = 2.33). It is proven when the mean score recorded is the lowest among all items in the loyalty variable.

Table 1. Reported mean scores for destination image.

Variables	Mean (M)	Std. Deviation (SD)
Destination image		
1. Great variety of flora and fauna	3.95	1.965
2. Beautiful scenery	3.21	1.615
3. Beautiful parks	3.30	1.678
4. Variety of cultural attraction	3.26	1.681
5. Interesting cultural activities	3.37	1.628
6. Appealing local custom	3.38	1.771
7. Shopping facilities	3.01	1.614
8. Quality accommodation	3.45	1.709
9. Good value for money	3.33	1.645
10. Safety	4.33	1.665
11. A relaxing destination	3.49	1.666
12. Pleasant environment	3.43	1.607
13. Exciting places	3.29	1.688
14. Different cuisine	3.87	1.792
15. Quality food	3.56	1.603

Note: i. N = 150; ii. 1 = Strongly Disagree, 2 = Disagree, 3 = Slightly Disagree, 4 = Neither Disagree nor Agree, 5 = Slightly Agree, 6 = Agree, 7 = Strongly Agree.

Table 2. Reported mean scores for loyalty.

Variables	Mean (M)	Std. Deviation (SD)
Loyalty		
1. I will try to return to Perak after this visit	3.04	1.399
2. I will plan next trip to Perak in the future	2.95	1.289
3. I will encourages relatives and friends to visit Perak	2.33	1.422
4. I will recommend Perak for those looking for a good destination	2.94	1.439

Note: i. N = 150; ii. 1 = Strongly Disagree, 2 = Disagree, 3 = Slightly Disagree, 4 = Neither Disagree nor Agree, 5 = Slightly Agree, 6 = Agree, 7 = Strongly Agree.

4.4 *Destination image of versus loyalty*

A simple linear regression analysis was used to approximate the coefficient of the linear equation between destination image and loyalty. Result denotes that destination image were able to explain 50.2 percent (R^2 = 0.50.2, F-Change = 148.983, $p < 0.01$**) of the variance in the loyalty dimensions. The outcomes showed that the image of destination significantly contributes to the tourist loyalty to revisit and recommend the destination to others. It shows that the destination image ($\beta = 0.708$, $p < 0.01$**) was found positively and significantly affecting the loyalty which can be said that the hypothesis is supported.

The result of the hypothesis is consistent with Chen & Tsai (2007) and Rodri'guez del Bosque & San Martin's (2008) which discovered significant role of the destination image in influencing the behavioral intentions to revisit and willingness to recommend. Faullant, Matzler & Füller (2008) also supported the findings by noted that once tourists having an optimistic image about the destination, they will not take serious consideration on the performance evaluation of the visited destination as they have the inclination to give more room of improvement if failures happen.

5 DISCUSSION AND CONCLUSION

Referring to the perspective of Singaporean tourists toward destination image of Perak, the analysis revealed that tourists were slightly disagreed toward most images listed with shopping facilities recorded the lowest score compare to other images. Viewing the relationship between destination image and loyalty, image appears to have significant effect on loyalty through intention to revisit and willingness to recommend the destination to others. Therefore, efforts should be directed to further strengthen the destination image of Perak in destination marketing strategies in order to facilitate the state effort to grab more Singaporean tourists market.

Theoretically, the findings have contributed to the literature and provided valuable information regarding the current situation of the tourism industry in a specific region from destination image aspect. On the other hand, the findings will become a base reference to further clarify more deeply regarding which aspects of destination segment that should be investigate more in the future research. Besides, it could assist the state government of Perak and other related authorities to reshaping their strategy in a promotional campaign to tap the Singaporean tourists market. Apart from that, the information from the research will potentially identify the strengths and weaknesses of destinations in Perak and all the input will help those authorities to plan in advance the suitable approach so that the potentially negative image in tourists' mind about the destination before can be reduced. This action crucially needs to be done as images either positive or negative can be shared easily by groups of people. Apart from that, by having a fresh approach and strategy toward establishing excellent destination image, it undoubtedly will become a pull factor for a destination as many scholars have found a relationship that link the destination image and visitation intent in their studies.

In conclusion, the above highlighted issues need a comprehensive action and collective involvement from all related parties. State government through responsible authorities should lead the action to rejuvenate and refurbish the image of destination in Perak and local tourism service providers should support the effort effectively. Therefore, the state governments' mission to grab the Singaporean tourist market would be able to achieve, and it could be translated through the rising number of their arrival and repeat visit to the state.

REFERENCES

Baloglu, S. (1999). A path analytic model of visitation intention involving information sources, socio-psychological motivations, and destination image. *Journal of Travel and Tourism Marketing*, 8(3), 81–91.

Baloglu, S., & Mangaloglu, M. (2001). Tourism destinations images of Turkey, Egypt, Greece, and Italy as perceived by US-based tour operators and travel agents. *Tourism Management,* 22(1), 1–9.

Bigné, J.E., Sánchez, M.I., & Sánchez, J. (2001). Tourism image, evaluation variables and after purchase behavior: Inter-relationship. *Tourism Management*, 22(6), 607–616.

Chen, C.F., & Tsai, D. (2007). How destination image and evaluative factors affect behavioral intentions?. *Tourism Management*, 28, 1115–1122.

Chi, C.G-Q., & Qu, H. (2008). Examining the structural relationships of destination image, tourist satisfaction and destination loyalty: An integrated approach. *Tourism Management*, 29, 624–636.

Chon, K.S. (1990). The Role of Destination Image in Tourism: A Review and Discussion. *Revue du Tourisme*, 2, 2–9.

Crompton, J.L. (1979). An assessment of the image of Mexico as a vacation destination and the influence geographical location upon that image. *Journal of Travel Research,* 17(Spring), 18–23.

Echtner, C.M, & Ritchie, J.R.B. (1991) The meaning and measurement of destination image. *The Journal of Tourism Studies*, 2(2), 2–12.

Faullant, R., Matzler, K., & Füller, J. (2008). The impact of satisfaction and image on loyalty: The case of Alpine ski resorts. *Managing Service Quality,* 18(2), 163–178.

Firefly (2010). Strategic Collaboration between Firefly and Perak. Retrieved on December 10, 2012 from http://firefly.com.my/news-releases/strategic-collaboration-between-firefly-and-perak.

Fridgen, J.D. (1987). Use of cognitive maps to determine perceived tourism region. *Leisure Sciences*, 9, 101–117.

Gallarza, M.G., Saura, I.G., & Garcia, H.C. (2002). Destination image: Towards a conceptual framework. *Annals of Tourism Research*, 29(1), 56–78.

Goeldner, C.R. & Ritchies, B.J. (2006). *Tourism: Principles, practices and philosophies (10th Ed.)*. New Jersey: John Wiley and Sons, Inc.

Gunn, C.A. (1972). *Vacationscape: Designing tourist regions*. Austin: Bureau of Business Research, University of Texas.

Hunt, J. (1975). Image as a factor in tourism development. *Journal of Travel Research,* 13(3), 1–17.

Jacoby, J. & Chestnut, R.W. (1978). *Brand Loyalty: Measurement and Management*. New York: Wiley.

Ko, D-W. & Park, S-H. (2000). Five aspects of tourism image: A review. *International Journal of Tourism Sciences*, 1(1), 79–92.

Kotler, P., Haider, D.H., & Rein, I. (1993). *Marketing Places: Attracting Investment, Industry and Tourism to Cities, States and Nations*. The Free Press: New York.

Krejcie, R., & Morgan, D. (1970). Determining sample size for research activities. *Educational and Psychological Measurement*, 30, 607–610.

Lee, C., Lee, Y., & Lee, B. (2005). Korea's destination image formed by the 2002 world cup. *Annals of Tourism Research*, 32(4), 839–858.

Mansfeld, Y. (1992). From motivation to actual travel. *Annals of Tourism Research*, 19, 399–419.

Mayo, E.J. (1973). Regional images and regional travel behavior. Research for Changing Travel Patterns: Interpretation and Utilization. In Proceedings of the Travel Research Association, Fourth Annual Conference, 211–218.

Ministry of Tourism Malaysia (2012). Rated Tourists Accommodation Premises. Retrieved on December 12, 2012 from http://www.motour.gov.my.

Nicoletta, R. & Servidio, R. (2012). Tourists' opinions and their selection of tourism destination images: An affective and motivational evaluation. *Tourism Management Perspectives*, 4, 19–27.

O'Brien, L. & Jones, C. (1995). Do rewards really create loyalty?. *Harvard Business Review*, May/June, 75–82.

Oppermann, M. (2000). Tourism destination loyalty. *Journal of Travel Research*, 39(1), 78–84.

Pike, S. (2002). Destination image analysis—a review of 142 papers from 1973 to 2000. *Tourism Management*, 23, 541–549.

Reichheld, F. & Teal, T. (1996). *The loyalty effect: The hidden force behind growth, profits and lasting value*. Boston, MA: Harvard Business School Press.

Rodri'guez del Bosque, I. & San Martin (2008). Tourist Satisfaction: A Cognitive-Affective Model. *Annals of Tourism Research*, 35(2), 551–573.

Sgtravellers.com (2010). Perak seeks bigger share of Singapore tourist dollar. Retrieved on December 11, 2012 from http://sgtravellers.com/.

Singapore Budget (2013). Budget Highlights. Retrieved on May 12, 2013 from http://www.singaporebudget.gov.sg.

Stepchenkova, S. (2005). *Russia's destination image among American pleasure travelers*. Master dissertation, Purdue University.

The Borneo Post (2012). Singaporean tourist arrivals to Perak soar 700 percent between Jan to Aug, says State Exco. Retrieved on December 7, 2012 from http://www.theborneopost.com.

Tourism Malaysia (2012). Tourist arrivals to Malaysia for Jan–Dec 2012. Retrieved on December 10, 2012 from http://corporate.tourism.gov.my.

Valazques, Saura & Molina, (2011). Conceptualizing and measuring loyalty: Towards a conceptual model of tourist loyalty antecedents. *Journal of Vacation Marketing*, 17(1), 65–81.

Hospitality and Tourism – Sumarjan et al. (eds)
© 2014 Taylor & Francis Group, London, ISBN 978-1-138-00151-0

The influence of food neophilia and neophobia towards tourists consumption of Malay food in Malaysia

M.N. Mohd Hafiz, A. Zainal, Z. Ahmad Nizan & I. Shahariah
Universiti Teknologi MARA, Malaysia

ABSTRACT: It is known that physiological factors are the focal reasons why majority of the people nowadays are into travelling. Practically, they are pushed and pulled by internal or external attributes before deciding where to visit and for what reasons to visit certain places of interest. In studying human attitudes there are many unexpected behaviours and different reactions involved and food is found to be one of the reasons why some people travel. This present study examines the physiological factors of tourists towards food while in Malaysia. In analyzing tourist's consumption of Malay food, Food Neophobia Scale (FNS) and Variety-seeking tendency scale (VERSEEK Scale) were adapted with a minor modification made before analyzing and examining the food neophilia and food neophobia characteristics. 200 questionnaires were analysed using a series of statistical analyses. The results indicated that food was one of the contributors towards attracting tourists to visit Malaysia. It was found that although the Malay food is unfamiliar to tourists but they tend to have the tendency to try the variety of Malay food.

Keywords: food consumption, food neophilia, food neophobia, Malay food, physiological factor

1 INTRODUCTION

From the global perspective, the tourism industry has become an essential component in the economic development agenda and it has continuously presented a positive impact to the worldwide economy scenario (Harrill, 2004; Goeldner & Ritchie, 2006). These movements have lent a hand to the local economic expansion such as local expertise extended, job opportunities and others sectors related (UNWTO, 2012). This is in line with Hjalager and Richard, 2002; Quan and Wang, 2004 and Hjalager, 2004 which stated that, to be presented as a destination preferred myriad, ongoing investment and planning need to be taken care of in order to sustain within this industry.

Datuk Seri Utama Mohd Najib Tun Abdul Razak, Malaysia's Prime Minister announced that the tourism industry has become the second largest industry contributor to the National wealth (Malaysia Planning Economic Unit, 2011). It has been proven in 2011 when Malaysia successfully attracts 24.7 million International tourists and earned tourism receipts RM 58.3 billion which is an increment 0.6 per cent comparing 2010 tourism receipts. (Tourism Malaysia, 2012).

Consistent to that Malaysia as a multiracial and multicultural country with many variations of authentic food, each has many local creations in terms of culinary delight and become a valuable resource to attract tourists to engage with the local culture (Zainal, Zali & Kasim, 2010). Okumus, Okumus and McKercher (2007) acknowledged the trends nowadays where people begin travelling for the reasons of gastronomy. Thus, this is the opportunity for Malaysia to promote their local food creation especially the authentic Malay food which is unique in the usage of ingredients

The research objectives are to analyse the diversity of tourists' behaviour who would like to try or curious to consume different food and those who are reluctant to taste unfamiliar food during tour holiday; an important topic to be discovered through analysing the characteristic of physiological factors.

i. *To examine if food neophilia significantly influence tourists consumption of Malay food.*
ii. *To examine if food neophobia significantly influence tourists consumption of Malay food.*

2 LITERATURE REVIEW

2.1 *Physiological factors*

Physiological factors are state that people can be influence on how they react at any specific moment (Mishra & Mishra, 2010). According to Sajna (2005), food acts as multiple roles in motivating human eating behaviour. In developed countries such as the United States of America, people are not just eating to fulfil their physiological needs but eating to represent who they are (Pliner & Hobden, 1992). Hall and Mitchell (2000) stated that food is not only for eating purposes but it is connected with the personal identity, production, tradition, sustainability and consumption. Besides that, the food is adopted as an important tool in the tourism industry apart from the destination attraction.

2.2 *Food neophilia*

Kim, Eves and Scarles (2009) defined food neophilia as a tendency to taste something new. Veeck (2004) viewed food neophilia is an adventure behavior and personality that is eager to experience new food. Pliner and Hobden (1992) stated that, food neophilia is associated with acquiring a different taste physiology, which allows individuals to experience food with more pleasure. Therefore, food neophilia was identified as proficiency to discriminate food items in their taste evaluation. (Pliner and Hobden, 1992; Ritchey, Frank, Hursti & Touorila, 2003).

2.3 *Food neophobia*

Food neophobia can be defined as a person who is reluctant to try novel food such as food products, dishes and cuisines (Kim et al., 2009). Veeck (2004) viewed food neophobia as unfamiliar food that people purposely avoid and might be expected not to taste unusual food or a person plainly who avoids unknown food. It can be said that, food neophobia is one of the individual trait and have been adopted as a barrier to experiencing the local food (Cohen & Avieli, 2004). They added as consequences, local food is getting lack of interest from foreign people thus, it gives a vast effect to the national food tourism industry (Mitchell & Hall, 2003).

2.4 *Food consumption*

Food consumption is related with the understanding of various behaviors before deciding to consume food such as preference, liking, intake and choice (Mak, Lumbers, Eves & Chang, 2012). According to Koster, (2009) and Sobal, Bisogni, Devine and Jastran (2006), food consumption is a complex behaviour (cultural, social, psychological and sensory acceptance factor) and play a role in the decision making process. Fields (2002) suggested that the motivator for consumption of food and beverages in a tourist destination can be categorized into physical, cultural, interpersonal and status and prestige motivators. In line with that, food consumption and tourists are considered as closely related since during travel, tourists need food to eat when they are away from home in order to fulfil their hunger (Jang and Wu, 2006).

Ritchey et al. (2003) and Cohen and Avieli, (2004) recommended that, in order to investigate an unfamiliar food choice and food consumption among tourists, it is necessary to engage food neophilia and food neophobia elements in order to understand the tourists' behaviour.

2.5 *Malay food*

Local food can be defined as food and beverages that is produced or grown in the local area or local specialty food that has a local identity (Nummedal & Hall, 2006). Local food does not rely on the local grown but also includes raw material from outside of the area that contribute to the local or region's identity.

Malay food has been said to vary from state to state, though they have similar names. Historically the traditional Malaysian culinary has been greatly influenced by traders from the neighbouring countries such as Indonesia, India, the Middle-East, China and others. Thus, in offering visitors many different experiences, indeed the local cuisines should become apparent, obvious and visible as a destination's intangible heritage; as experiencing the delectable taste allow tourists to gain an authentic cultural experience (Zainal et al, 2010).

Rozin (2006) posited that there are three components of Malay food that differentiates it from other ethnics like Chinese and Indian (a) the staple ingredients, (b) the flavour principles (lemon grass, ginger, chillies, dried herbs and spices) and (c) cooking methods.

2.6 *The relationship between physiological factors toward tourists consumption of food*

Ryu and Jang (2006) examined the tourist's experience of local food. They found that opinion and suggestion from other people do not significantly influence on taste the local food. But the most significant is the personal attitude influencing the consumption of the local food. It was proven by Cohen and Avieli, (2004) and Chang, Kivela and Mak., (2010) that people who are food neophobia can significantly affect the food consumption during holiday.

According to Pelchat and Pliner, 1995; McFarlane and Pliner, 1997; Tourila, Lahteenmaki, Pohjalainen and Lotti, 2001, the more they (people) are exposed to different culture the more they contribute to food neophilia and slowly reducing the percentage of as a food neophobia. They added, exposure, nutritional information, familiarity and positive experience factors would increase food neophilia and decreasing food neophobia.

3 METHODOLOGY

This study employed a correlational approach by using self-administered questionnaire to collect research data. The samples are international tourists that spend their vacation trip in Malaysia with an age range from 20 years old and above. The survey was carried out at the KLIA, LCCT, KLIA Transit, KLIA express and LCCT shuttle bus transportation services. 450 sets of questionnaires have been distributed and a total of 200 sets of questionnaires yielding approximately 44.5% was returned.

The instruments were adapted from Pliner and Hobden, 1992; Van Triojjp and Steenkamp, 1992; Sajna, 2005 and Kim et al., 2009 with minor modifications made to six (6) food neophobia items that are related to willingness or reluctant to try unfamiliar food. On the other hand, seven (7) items of food neophilia having similarity with tendency to consume new food are adapted with minor modifications made to match this study. This was based on the VERSEEK scale (variety-seeking tendency with respect to food) which consisted of eight (8) items created by Van Trijp and Steenkamp (1992).

Tourist's consumption of the Malay food was adapted from Sajna, 2005 and Kim et al., (2009) studies. Five (5) items were registered in order to comprehend the overall conditions focusing on consumption of Malay food during a holiday trip in Malaysia.

4 RESULTS

4.1 *Respondents profile*

The majority of the respondents is from the Asian continent (45%) followed by Europe (22%), Middle East (12.5%) and the rest of Africa, America and others. A total of 104 respondents were males (52%) while another 96 were females (48%). The age range is between 20 years old until above 60 years old. Almost 39% of the respondents visited Malaysia for holidaying with the average length of stay in Malaysia was 4–6 nights. 42.5% (85 respondents) can be grouped into food neophilia category recorded having a Malay food more than three times while in Malaysia.

4.2 *Hypothesis testing*

In this study Cronbach's alpha is used in measuring the reliability of the data. Hence, food neophilia, $\alpha = 0.83$, food neophobia $\alpha = 0.77$ and tourists consumption of Malay food $\alpha = 0.89$. It is in line with Chandon, Pierre-Yves and Philippe (1996) and Tabachnick and Fidell (2001) indicated that, value 0.50 is generally accepted when measuring attitudes of people.

Table 1. Regression analysis for physiological factors on tourists consumption of Malay food.

Independent variable (control variables)	Std. β
Physiological factors	
Food neophilia	0.670***
Food neophobia	0.08
R^2	0.42
Adj. R^2	0.41
R^2 Change	0.42
F Change	71.05***
Sig. F Change	0.000

Scale: $p < 0.05$*, $p < 0.01$**, $p < 0.001$***.

Meanwhile, Table 1, shows that the two control variables (food neophilia and food neophobia) explains 42% ($R^2 = 0.42$, F Change = 71.05, $p < 0.001$) of the variance in tourist consumption of the Malay food in Malaysia. Food neophilia ($\beta = 0.67$, $p < 0.001$) is a statistically significant predictor of tourists' consumption while food neophobia recorded insignificant predictor of tourists' consumption ($\beta = 0.08$, $p = 0.15$).

5 IMPLICATION

The physiological factors toward tourists consumption of Malay food among the respondents is moderate and majority of them are able to accept the taste of the Malay food and Malay traditional culture (eating with fingers). As such by accommodating to the palate of visitors and providing more varieties of traditional Malay food which are good in quality, hygienic and well presented will definitely be an advantage to the tourism industry.

At the same times, for agencies and non-governmental organizations, they can increase their profits by enhancing their current tour packages with the implementation of food based activities and food trails available in the country. This measure fitted with Kivela and Crotts's research (2006) where nowadays many international tour agencies and tour organizers around the world have included the gastronomic products as part of their tour itineraries.

In line with the above suggestions, the government authorities such as Ministry of Tourism can also gain benefits from this study whereby they can implement many more attractive gastronomy tourism activities that seriously involve authentic Malay food and Malay culture so as to gain competitive advantages in promoting Malaysia as a tourist destination.

6 LIMITATIONS AND RECOMMENDATION FOR FUTURE RESEARCH

There are a few limitations that appeared while the research was undertaken. Firstly, although the present study's findings gathered with high response rate, it could be noted that, the ratio of the tourists was unbalanced, representing 90 respondents (Asia), 44 respondents (Europe), 10 respondents (America), Africa (11 respondents), Middle East (25 respondents) and 8 respondents from Australia. With such proportion, the result has indeed showed a positive significant toward Malay food since majority of the respondents are from Asia. Secondly, tourists are reluctant to be engaged with strangers when they are in the airport.

Third limitation was the mixed up of menus in the tourists' minds. Although several examples of Malay food were described in the questionnaire but tourists have problems of recalling the exact dish. As such researcher had to provide some well-known Malay food pictures and personal explanation regarding the food before tourists could answer the questions.

In expressing their thoughts on food, this study allows the respondents to use personal descriptors about the Malay food such 'delicious', 'creamy' 'oily', 'sweetness', 'delectable', 'hot and spicy' and 'variety of choices'. This has assisted the researcher in understanding more about tourists views of Malay food in particular and will help other researchers to embark on future studies. Thus, it is recommended that it is essential that future studies to expand this study by looking at broader scope and aspects especially to other traditional food belonging to each state in Malaysia for an in-depth understanding about each traditional food.

7 CONCLUSIONS

The result of the findings indicates that the majority of the respondents accepted and admitted experiencing the Malay food while in Malaysia. Although Malay food used a variety of aromatic ingredients in the preparation of the cuisine, which may be unpleasant at times but tourists do not hesitate to taste and enjoy the authentic taste of Malay food such as rendang, satay and peanut sauce and others.

Hence, tourism bodies and related government sectors should dynamically co-operate and support each others in attracting international tourists to explore further the authentic Malay cuisines that could truly reflect Malaysia to the world just like many European countries such as France, Italy and Spain have done to their food tourism.

REFERENCES

Artinah., Z. Ahmad Nizan, Z, and Kassim, M.N. (2010). Malaysian gastronomy routes as a tourist destination. 15–24.

Bartlett, J.E., Kotrlik, K.W., and Hinggins, C. (2001). Organizational research: Determining appropriate sample size in survey research. *Information Technology, Learning and Performance Journal,* 19(1), 43–50.

Chandon, R.Y., Pierre-Yves, J. and Philippe, J. (1996). Service encounter dimension-dyadic perspective: Measuring the dimension of service encounter as perceived by customers and personnel. *International Journal of Service Industry Management,* 8(3), 65–86.

Chang, R.C.Y., Kivela, J., and Mak, A.H.N. (2010). Food preferences of Chinese tourists. *Annals of Tourism Research,* 37(4), 989–1011.

Cohen, E., and Avieli, N. (2004). Food in tourism: Attraction and impediment. *Annals of Tourism Research,* 31(4), 755–778.

Field, K. (2002). Demand for the gastronomy tourism product: Motivational factors. In: A.M. Hjalager, & G. Richards (Eds.), *Tourism and Gastronomy*. (pp. 37–50). London: Routledge.

Goeldner, C.R. and Ritchies, B.J. (2006). *Tourism: principles, practices and philosophies (10th Ed.)*. New Jersey: John Wiley and Son, Inc.

Hall, M., and Mitchell, R. (2000). Wine and food tourism. In N. Douglas and R. Derrett (Eds). *Special Interest Tourism.* Australia: John Wiley. 307–325.

Harrill, R. (2004). Residents' Attitudes toward Tourism Development: A Literature Review with Implications for Tourism Planning. *Journal of Planning Literature,* 18(1), 1–2.

Hjalager, A.M., and Richard, G. (Eds.). (2002). *Tourism and gastronomy.* London: Routledge.

Hjalager, A.M. (2004). What do tourist eat and why? Towards a sociology of gastronomy and tourism. *Tourism,* 52, 195–201.

Jang, S., and Wu, C. (2006). Seniors travel motivation and the influential factors: An examination of Taiwanese senior. *Tourism Management,* 27(2), 306–316.

Kim, Y.G., Eves, A., and Scarles, C. (2009). Building a model of local food consumption on trips and holidays: A grounded theory approach. *International Journal of Hospitality Management,* 28, 423–431.

Kivela, J and Crotts, R. (2006). Tourism and gastronomy: gastronomy's influence on how tourist experience a destination. *Journal of Hospitality and Tourism,* 3(30), 354–377.

Koster, E.P. (2009). Diversity in the determinants of food choice: a phychological perspective. *Food Quality and Preferences.* 20, 70–82.

Mak, A.H.N., Lumbers, M., Eves, A. and Chang, R.C.Y. (2012). Factors influencing tourist food consumption. *International Journal of Hospitality Management*. 31, 928–936.

Malaysia Planning Economic Unit (2011). *Tourism programs*. Retrieved July 15, 2011, from http://www.epu.gov.my/tourism.

McFarlane, T., and Pliner, P. (1997). Increasing the willingness to taste unusual food. *Phychological Reports*, 54, 739–745.

Mishra, A., and Mishra, H. (2010). We are what we consume: The influence of food consumption on impulsive choice. *Journal of Marketing Research*, 11(7), 1129–1137.

Mitchell, R. and Hall, M. (2003). Consuming tourists: food tourism consumer behavior. In M. Hall, L. Sharples, R. Mitchell, N. Macionis and B. Crambourne (Eds.). *Food Tourism Around the World*. Oxford: Butterworth Heinemann.

Nummedal, M., and Hall, M. (2006). Local food and tourism: An investigation of the New Zealand South Island's bed and breakfast section's use and perception of local food. *Tourism Review International*, 9, 365–378.

Okumus, B., Okumus, F., and McKercher, B. (2007). Incorporating local and international cuisine in the marketing of tourism destination: The case of Hong Kong and Turkey. *Tourism Management*, 28, 253–261.

Pelchat, M., and Pliner, P. (1995). Try it: you'll like it; effectd of information on willingness to try novel foods. *Appetite*. 24, 153–531.

Pliner, P., and Hobden, K. (1992). Development of a scale to measure the trait of food neophobia in human. *Prentice Hall Appetite*, 19, 105–120.

Quan, N., and Wang, N. (2004). Towards a structural model of the tourist experience: an illustration from food experiences in tourism. *Tourism Management*, 297–305.

Ritchey, P.N., Frank, R.A., Hursti, U., and Tuorila, H. (2003). Validation and cross-national comparison of the food neophobia scale (FNS) using confirmatory factor analysis. *Appetite 40*, 163–173.

Ryu, K., and Jang, S. (2006). Intention to experience local cuisine in a travel destination: The modified theory of reasonable action. *Journal of Hospitality and Tourism Research, 30*(4), 507–516.

Sajna, S.S. (2005). Food tourism and the culinary tourist. PhD Thesis. Clemson University. Retrieved April, 23, 2011, from Clemson University Digital Th.

Sobal, J., Bisogni, C.A., Devine, C.M. and Jastran, M. (2006). A conceptual model of the food choice process after the life course. In Shepherd, R., Raats, M. (Eds), The psychology of food choice. *CAB International, Oxfordshire*, 1–18.

Tabachnick, B.G., and Fidell, L.S. (2001). *Using econometrics: A practical guide*. Boston: Little Brown.

Tourism Malaysia. (2012). Tourists arrival to Malaysia for Januari–December. Retrieved on May 9, 2012, from http://www.tourismmalaysia.com/facts&figures/02/.html.

Tuorila, H., Lahteenmaki, L., Pohjalainen, L., and Lotti, L. (2001). Food neophobia among the finns and related responses to familiar and unfamiliar food. *Food Quality and Preference*, 12(1), 29–37.

Van Trijjp. H., & Steenkamp, J.B. (1992). Consumers' variety-seeking tendency with respect to foods: Measurement and managerial implications. *European Review of Agriculture Economics*, 19, 181–195.

Veeck, A. (2004). Extreme Foods: Expanding the Boundaries of Taste. Advances in Consumer Research, 31, 554.

World Tourism Organization UNWTO (2012). International tourism receipts surpass US$ 1trillion in 2011. Retrieved Mei 5, 2012, from http://www.media.unwto.org.

Hospitality and Tourism – Sumarjan et al. (eds)
© 2014 Taylor & Francis Group, London, ISBN 978-1-138-00151-0

Examining perceived value and tourists' satisfaction on Penang delicacies

Y. Atikahambar, A. Zainal, Z. Ahmad Nizan, A. Nor'Aini & M.K. Zain
Universiti Teknologi MARA, Malaysia

ABSTRACT: This present study focuses on examining perceived value and food tourists' satisfaction of Penang delicacies. The study utilized the self-administered questionnaire technique on international tourists who are having their vacation trips in the Island of Penang. Data was successfully gathered from a sample of 328 respondents, yielding almost 82% response rate. The data were analyzed using inferential statistics such as, reliability test and regressions analysis to confirm proposed hypothesis, simultaneously answering the posited research questions. The findings of this study revealed that perceived value is related to tourists' satisfaction. It indicates that 62% of the variance of the correlation between perceived value and food tourists' satisfaction. It is therefore suffice to say that the findings of this study enables greater comprehension on examining perceived value and tourists' satisfaction of Penang delicacies and give positive impact on tourism industry in order for targeting international tourists' arrival in Penang.

Keywords: perceived value, satisfaction, food

1 INTRODUCTION

Penang is an international tourist destination, famous for its many attractions which are steeped inhistory as well as for its scenic beauty. According to Socio-Economic and Environmental Research Institute (SERI, 2011) the number of tourists (Domestic and International) arrivals in Penang from Jan–Dec 2011 is 4559, 221.

Since Malaysian food is not one particular kind of food but a unique culinary spectrum originating from Malaysia's multi-ethnic population, that has blend together sampling tastes, ingredients and cooking methods from each other, the gastronomic routes in the country could be promoted robustly into local delicacies (Zainal *et al.*, 2010).

Penang as been proclaimed as outstanding by various organizations and this food heaven was recognized as having the Best Street Food in Asia by Time magazine in 2004 (Penang State Tourism and Development, 2010). Hence, among others this state can be highlighted as "the state for gastronomy routes" (Zainal *et al.,* 2010). Du Rand *et al.* (2003) stated, "the local food holds much potential to enhance sustainability in tourism; contribute to the authenticity of the destination; strengthen the local economy; and provide for the environmentally friendly infrastructure". It is noted that, New York Times readers have chosen Penang as second must-visit destination for 2009 and the top 12 places in the frugal and food categories (both readers' and the paper's recommendations) (Bernama, 2010).

The objective of this study is to analyze the effects of perceived value on food tourists' satisfaction. The hypothesis is formulated as follows:

H1: Perceived value affects food tourists' satisfaction.

2 LITERATURE REVIEW

Accordingly, a number of studies have been conducted to examine tourists' behavior using significant factors such as perceived value and satisfaction (Petrick 2002, 2004a, 2004b; Yoon & Uysal, 2005). Kim *et al.* (2011) stated that food tourism is fast becoming one of the most popular and interesting areas in the tourism industry today. Thus, many studies on perceived value and satisfaction revealed that they have a strong effect on future intention to revisit or repurchase products or services (Choi & Chu 2001; Tam, 2000; Yoon & Uysal, 2005). Besides that, studies suggest that gastronomy plays a significant role in creating a quality visiting experience for tourists and, therefore, influences their revisit intentions (Kivela & Crotts, 2006).

2.1 *Perceived value*

Perceived value, as defined by Zeithaml (1988) is "the consumer's overall assessment of the utility of a product based on perceptions of what is received and what is given". Kim *et al.* (2011) pointed out that perceived value is what consumers get for what they give, however value is not an apparent concept that can be described or identified easily. Zeithaml (1998) identified four diverse meaning of value:

i. *Value is low price.*
ii. *Value is whatever consumer want in a product.*
iii. *Value is the quality consumer get for the price paid.*
iv. *Value is what consumer get for what they give.*

2.2 *Quality*

Baker & Crompton (2000) stated that quality and perceived value are cognitive responses to a service experience. Meanwhile Sweeney *et al.* (1998), highlighted that quality is referred to how well the product was made, and emotional response to how a product made the consumer feel. Cronin & Taylor (1992) pointed out that SERVQUAL is one of the most extensively used measures of service quality.

Kivela & Crotts (2006) acknowledged that gastronomy can add value to the tourist experience and is associated with quality tourism for travelers in search of new products and experiences that yield a high level of satisfaction, wine tourism being a good example. Ruyter *et al.* (1998) noted that "quality has an impact on customer preference and the willingness to recommend the service to other consumers". It is highlighted that tourists request for the higher standard in quality services such as food quality. Thus, understanding tourists' eating habits will help in enhancing tourist satisfaction while travelling (Quan & Wang, 2004).

2.3 *Emotional response*

Emotional response was defined as a descriptive judgment regarding the pleasure that a product or service gives the purchaser (Sweeney *et al.* 1998). Past research has identified emotional response, or the joy received from purchased (Parasuraman & Grewal, 2000) and quality as dimensions of perceived value of a product/service.

2.4 *Monetary price*

The definition utilized for monetary price was the price of service as encoded by the consumer (Jacoby & Olson, 1977). Monetary costs include both direct and indirect costs (O'Sullivan & Spangler, 1998). The direct cost of the golf experience is the fee associated with playing the golf course. The indirect costs are the expenditures for equipment, clothing, transportation, and meals, if necessary.

2.5 *Behavioral price*

Behavioral price was defined as the price (non-monetary) of obtaining a service, which included the time and effort, used to search for the service (Zeithaml, 1988). Non-monetary costs "consist of opportunity and intrinsic prices paid for participating in an experience" (O'Sullivan & Spangler, 1998, p. 113). Opportunity costs refer to time and effort required by the experience. While intrinsic costs refer to association, social, psychological and sensory costs of the experience (Zeithaml, 1988).

2.6 *Reputation*

Reputation was defined as the prestige or status of a product or service, as perceived by the purchaser, based on image of the supplier (Dodds *et al.* 1991). Reputation has been interpreted as a competitive advantage (Balmer & Greyser, 2003) and represents an important intangible asset of the firm (Hall, 1993). From a corporate perspective, building a favorable reputation is deemed an effective way to gain market access and acceptance because reputation has been shown to be a determinant in consumers' purchase decision making (Carmeli & Tishler, 2005).

2.7 *Tourists' satisfaction*

Tourist satisfaction is important to successful destination marketing. According to Oliver (1997), "satisfaction is the consumer's fulfillment response. Researchers (Ziethaml, 1988; Oliver, 1997) noted that satisfaction can be defined as an effective response to experiencing a service.

Satisfaction refers to the perceived discrepancy between prior expectation and perceived performance after consumption (Oliver, 1980). Tourists are satisfied due to experiences compared to expectations results in feelings of gratification (Reisinger & Turner, 2003). While dissatisfied occur to tourists due to feelings of displeasure. Reisinger & Turner (2003) highlighted that the cultural differences between international tourist and the service providers may effects the social experience and finally effect the satisfaction of the tourists.

3 METHODOLOGY

3.1 *Participants and procedures*

The unit of analysis of this study involved international tourists' who visited or having their vacation trips in the Island of Penang from the month of February 2012 to April 2012. A total of 400 survey questionnaires were distributed, and only 328 (82%) were completed and met the screening requirement.

The instruments for the independent variables in this study were adopted from Petrick's multi-dimensional scale (Petrick, 2002) and modified to suit the cuurent study. The items are quality (4 items), emotional response (5 items), monetary price (6 items), behavioral price (5 items) and reputation (5 items) with scale measurement from "totally satisfied" (5) to "totally dissatisfied" (1).

The instruments for the dependent variables were aimed to measure satisfaction with 3 attributes created. It was modified and adapted from Oliver's (1997), the items were rate on a 5-point Likert scale, ranging from "totally satisfied" (5) to "totally dissatisfied" (1).

4 RESULTS

The descriptive statistic is used to analyze the respondents' demographic profiles focusing on the tourists' personal information including the purpose visiting Penang Island.

4.1 *Respondents profile*

The respondents were 55.5% male and 44.5% females. Majority of them was in the age group between 18 to 27 years olds (41.5%). Approximately, 35.4% of respondents indicated their occupation as professional level. 51.2% of respondents were single and 42.7% of the respondents are degree holders. Besides Australia, Middle East, United States, Singaporeans and Japan, approximately 41.7% of respondents originated from Switzerland, Russia, Indonesia and Thailand. Almost 41.5% respondents had visited Penang Island more than 7 times and 40.2% of the respondents visited Penang Island for the purpose of leisure.

4.2 *Hypothesis testing*

Table 1 provides the result of regression analysis for the relationship between perceived value and food tourists' satisfaction. The regression models shows that the perceived value is proved as predictor to food tourists' satisfaction ($\beta = 0.10$, $p < 0.01$**) explaining 62% of the variance of the correlation between perceived value and food tourists' satisfaction. The analysis indicated that three dimensions which are quality, emotional response and reputation is statistically significant predictor of tourists' satisfaction. The findings revealed that monetary price and behavioral price is not supported.

Table 1. Regression analysis of perceived value and food tourists' satisfaction.

Independent	Std. β	Sig. level
Perceive value	0.10**	0.000
Quality	0.17**	0.000
Emotional response	0.23**	0.000
Monetary price	−0.02	0.190
Behavioral price	0.04	0.193
Reputation	0.22**	0.000
R^2	0.62	
Adj. R^2	0.39	
R^2 Change	0.39	
F Change	212.92**	
Sig. F Change	0.000	

Scale: $p < 0.005$*, $p < 0.01$**.

5 DISCUSSIONS

Based on the findings presented, it was revealed that perceived value is related to tourists' satisfaction.

As elaborated in literature review, a number of findings showed that perceived value, which is a cognitive response, lead to satisfaction (Cronin *et al.* 2000). In addition, Baker & Crompton (2000) acknowledged that higher quality of performance and levels of satisfaction are perceived to result in increased loyalty and future visitation, greater tolerance of price increases, and an enhanced reputation.

6 CONCLUSION

In order for Penang to sustain in the global industry as a food tourism destination preference it is important for tourism bodies, government and related sectors to understand the international tourists' preferences and wants. Besides that, this study will give a clear view to local food provider to improve quality and taste of Penang food.

From the study, it can be concluded that tourism bodies, government and related sectors has to cooperate and support each other in revealing the significant influences on perceived value and satisfaction of Penang delicacies. Thus this will embarks and promote Penang Island as food destination to tourists'. One of the best suggestions is by including Penang food in the tour or promotion package.

7 LIMITATIONS & RECOMMENDATION

The empirical analyses are conducted based on data gauged at a specific time due to constrain in resources such as time and money. Another limitation that the present study faced was the lack of participation and cooperation from the respondents (international tourists). In addition, other constraint included time limitation and language barrier among the tour guides and respondents. Lastly, the constraint found in this study was on obtaining correct data and statistics regarding the international tourist's arrival in Penang. This is due to some information were obsolete and not being updated according to the current time interval when the research is being conducted.

There are several opportunities to extend this study. Findings of this study also contribute to more extensive understandings of theories that can be applied to hospitality and tourism industry. The changes of tourists' tastes are a factor due to the changes of tourist's character, attitude and lifestyle in a tourism destination. This condition has changed due to the influenced of activity, product and space in tourism destination, which is considered as out moded, created a feeling of boredom thus giving a negative perception. Therefore, these factors will influence a formation of tourist friendly destination concept, which will enable tourist to enjoy, and obtain in new knowledge and experience about certain tourism product such as food tourism.

REFERENCES

Baker, D.A., & Crompton, J.L. (2000). Quality, satisfaction and behavioral intentions. *Annals of Tourism Research*, 27(3), 785–804.

Balmer, J.M.T., & Greyser, S.A., 2003. Revealing the Corporation. Routledge, London and New York.

Bernama. (2010). Penang records 40 pct increase in number of international tourist arrivals. Retrieved at http://www.mysarawak.org (2010/09/10).

Carmeli, A. & Tishler, A., 2005. Perceived organizational reputation and organizational performance: an empirical investigation of industrial enterprises. *Corporate Reputation Review*, 8(1), 13–30.

Choi, T.Y. & Chu, R. (2001). Determinants of hotel guests' satisfaction and repeat patronage in the Hong Kong hotel industry. *International Journal of Hospitality Management*, 20(3), 277–297.

Cronin, J.J., Jr., & Taylor, S.A. (1992). Measuring service quality: a reexamination and extension. *Journal of Marketing*, 56(3), 55–68.

Cronin, J.J., Brady, M.K., & Hult, G.T.M. (2000). Assessing the effects of quality, value and customer satisfaction on consumer behavioral intentions in service environments. *Journal of Retailing*, 76(2), 193–218.

Dodds, W.B., Monroe, K.B., & Grewal, D. (1991). The effects of price brand and store information on buyers' product evaluations. *Journal of Marketing Reaseach*, 28, 307–319.

Du Rand, G.E., Heath, E., & Alberts, N. (2003). The role of local and regional food in destination marketing: a South African situation analysis. *Journal of Travel and Tourism Marketing*, 14(3/4), 97–112.

Hall, R., 1993. A framework linking intangible resources and capabilities to sustainable competitive advantage. *Strategic Management Journal*, 14(8), 607–618.

Jacoby, J., & Olson, J.C. (1977). Consumer response to price: an attitudinal, Information processing perspective, in Y. Wind and P. Greenberg (Eds), Moving Ahead With Attitude Research, 73–86. Chicago: American Marketing Association.

Kim, Y.H., Kim, M., & Goh, B.K, (2011). An examination of food tourist's behavior: Using the modified theory of reasoned action. *International Journal of Tourism Management*, 32, 1159–1165.

Kivela, J., & Crotts, J., (2006). Tourism and gastronomy: gastronomy's influence on how tourists experience a destination. *Journal of Hospitality and Tourism Research*, 30(3), 354–377.

Oliver, R. (1980). A cognitive model of the antecedents and consequences of satisfaction decisions. *Journal of Marketing Research*, 17, 460–469.

Oliver, R.L. (1997). Effect of expectation and disconfirmation on post exposure evaluation: an alternative interpretation. *Journal of Applied Psychology*, 62, 480–486.

O'Sullivan, E.L., & Spangler, K.J. (1998). *Experience marketing*. State College: Venture Publishing, Inc.

Parasuraman, A., & D. Grewal (2000). The impact of technology on the quality–value–loyalty chain: a research agenda. *Journal of the Academy of Marketing Science*, 28(1), 168–174.

Penang State Tourism and Development. (2010). *Practical Travel Guide Visit Penang 2010–2012*. Retrieved March 14, 2012, from *http://www.visitpenang.gov.my.*

Petrick, J.F. (2002). Development of a multi-dimensional scale for measuring the perceived value of a service. *Journal of Leisure Research,* 34(2), 119–134.

Petrick, J.F. (2004a). First timers' and repeaters' perceived value. *Journal of Travel Research*, 43, 29–38.

Petrick, J.F. (2004b). The roles of quality, value, and satisfaction in predicting cruise passenger' behavioral intention. *Journal of Travel Research*, 42, 397–407.

Quan, S. & Wang, N. (2004). Towards a structural model of the tourist experience: an illustration from food experience in tourism. *Tourism Management*, 25(3), 297–305.

Reisinger, Y., & Turner, L.W. (2003). Cross-cultural behaviour in tourism: Concepts and analysis. Oxford: Butterworth-Heinemann.

Ruyter, K.D., M. Wetzels, & J. Bloemer (1998). "On the Relationship Between Perceived Service Quality, Service Loyalty and Switching Costs." *International Journal of Service Industry Management*, 9(5):436–53.

Socio-Economic and Environmental Research Institute (SERI), "Penang statistics quarter 4, 2011". Available from http://www.seri.com.my/. Retrieved on 27 October 2011.

Sweeney, J.C., Soutar, G.N., & Johnson, L.W. (1988). Consumer perceived value: Development of a multiple item scale. American Marketing Association Conference Proceedings, 9, 138.

Tam, J.L.M. (2000). The effects of service quality, perceived value and customer satisfaction on behavioral intentions. *Journal of Hospitality and Leisure Marketing*, 6(4), 31–43.

Yoon, Y., & Uysal, M. (2005). An examination of the effects of motivation and satisfaction on destination loyalty: a structural model. *Tourism Management,* 26(1), 45–56.

Zainal, A., Zali, A.N., & Kassim, M.N. (2010). Malaysian gastronomy routes as a tourist destination. *Journal of Tourism, Hospitality & Culinary Art,* 15–24.

Zeithaml, (1998). Consumer perceptions of price, quality, and value: a means end model and synthesis of evidence. *Journal of Marketing*, 52, 2–22.

Hospitality and Tourism – Sumarjan et al. (eds)
© 2014 Taylor & Francis Group, London, ISBN 978-1-138-00151-0

An investigation of customers' satisfaction as a mediating variable between online discount programs and customers' behavioral intentions

A.H. Abdullah & N. Sumarjan
Universiti Teknologi MARA, Malaysia

ABSTRACT: This paper fills the gap in empirical research of online discount programs in Malaysia. The aim for this study is to investigate the customers' satisfaction, specifically as the mediating variable, between online discount program and customers' behavioral intentions. There are some factors that will influence customers' satisfaction and consumers' behavioral intentions regarding online discount programs. For this study, the researcher select three factors which are offer deals, site features and site security to test customers' satisfaction and consumers' behavioral intentions. Customers' behavioral intentions for this study consist of word-of-mouth, site revisit, and repetition of purchase that occurred after the customers satisfied with the online discount programs. Since the internet has been widely used as a medium of marketing and others, the booming of online discount program creates competitive advantage among businesses. Online discount programs can attract new customers who are not aware of the products or services that the businesses offer before by offer them an attractive discounted price. After the customers experience and satisfied with the products or services, they tend to repeat their purchase with or without to purchase online. Thus, it can be concluded that consumers are interested with the choices of discounted price that they will receive from the online discount programs.

Keywords: customers' behavioral intentions, customers' satisfaction, online discount programs

1 INTRODUCTION

The number of internet users worldwide has increased due to the growth of information technology (Internet World Stats, 2012). However, according to Paynter and Lim (2001), Malaysia can be considered as a late beginner in the internet growth since it was started on 1995 and the growth of internet users in Malaysia increased in 1996. It became popular as it promotes a new approach of doing business because through medium of internet is where people can share ideas, form communities, and communicate with each other in virtual place (Paynter & Lim, 2001).

As the internet has been widely used as a medium of marketing and others, the booming of online discount program creates competitive advantage among businesses (Edelman, Jaffe, & Kominers, 2011). A lot of companies make great deals to offer to potential customers such as handbags, restaurants, hotels, spa, vacations and others companies that create discount programs for their products and services. Customers can enjoy saving usually up to 50% or more through the discount programs. Occasionally the special discount can be enjoyed only by the redemption through the web site compare to walk in customers who have to pay normal price (Edelman et al., 2011).

Groupon is a good example of online discount program, which sometimes offer up to 70 or 80 percent discounts off the listed price. With the great discount, potential customers only need to pay minimal amount to the products or services and this will lead low risk for customers, who would like to try the new product or service offered by companies that are new in the market, or companies that the customers never experienced before (Dholakia, 2010).

The research objectives are as follows:

i. *To determine the relationship between online discount programs and customers' behavioral intention*
ii. *To investigate which dimension of online discount programs that most agreeable*
iii. *To examine the mediating effect of customers' satisfaction between online discount programs and customers' behavioral intentions*
iv. *To investigate the most agreeable dimension of customers' behavioral intention.*

2 LITERATURE

2.1 *Online discount program*

Previous studies that explore about online purchase (Coulter & Roggeveen, 2012; Dholakia, 2010;

Edelman et al., 2011; Harris & Goode, 2010; Blundo, Cimato & Debonis, 2005; Fortin, 2000) noted that online discount programs can benefit both customers and also the firms that offer discounted price and the customer can enjoy the offer deals and the firm can create brand awareness about its products and services. Although there are study explore about online purchase, there is still lack of study on how offer deals, site features and site security influence customer satisfaction and consumers' behavioral intentions in the context of Malaysia's retail industry. Thus, this study emphasis on these three dimensions in the context of Malaysia's retail industry in order to investigate their influence on customers' satisfaction and behavioral intentions.

2.2 *Offer deals*

According to Edelman et al. (2011), online customers in Malaysia believe that purchase online provides value for money since online products or services offer a better deal at lower prices as compared to the traditional way of purchasing products and services. Online discount programs will concentrate for promising great offers to the customers by giving price reductions. Coulter and Roggeveen (2012) mentioned that the offer deals also have time-to-expired whereby the deals can only be purchased in certain period of time depending on the availability of products and services. This strategy was used because when customers make decision making within a restricted time, they are forced to make the decision quickly.

2.3 *Site features*

Site features include the information about products or services and also the graphics used in the site that attracts the customers to visit the site. According to Mandel and Johnson (2002), consumers' appraisals of web sites performance is through information in the web pages which acts as a powerful influence to online customers. The authors further added, online layout focuses on the functionality of the site whereby online visual appeal concentrates on the graphic used in the site. Studies done by Donnelly (2001); Zeithaml, Parasuraman and Malhotra, (2002) noted that the functionality and usability of web sites are crucial criteria used by consumers to appraise the online environments.

2.4 *Site security*

Security is a main concern to internet user especially for online domestic or international purchase (Paynter & Lim, 2001). They are concerned about the safety of submitting personal billing information through the internet (Paynter & Lim, 2001).

Harris and Goode (2004) discovered that there is a relationship between consumers' trust and their purchase. Harris and Goode (2004) also noted that after the customers have the experience of using the secured website for their online purchase, then only they will trust the site and will lead to their purchase intentions.

2.5 *Customer satisfaction*

Study conducted by Horppu, Kuivalainen, Tarkiainen and Ellonen (2008) found out that, for internet customers, customers' satisfaction occur when they experienced web satisfaction. Muylle, Moenaert and Despontin (2003) suggest that satisfaction in the web site context as customers' satisfaction and positive experiences with the online navigation, performance and content in the web site. Shankar, Smith and Rangaswamy (2003); Anderson and Srinivasan (2003) noted that the consequences of customers' satisfaction on the internet such as that leads to customers' behavioral intentions.

2.6 *Relationship between online discount programs and customers' satisfaction*

Customers' satisfaction happen between service encounters (Shankar, Urban & Sultan, 2002) and service processes (Dube & Morgan, 1998). Service encounters are the connections between company and customer while service processes are all stages of buying process decision that the customers experience. Therefore, online discount programs that serve the best deal to offer will increase customers' satisfaction and influence their buying decision due to the discounted price that they received (Gounaris, Dimitriadis, & Stathakopoulos, 2010).

Based on the above literatures, the following hypotheses are thus suggested:

H1: Online discount programs positively influence customers' satisfaction.

- *H1a: Offer deals positively influence customers' satisfaction*
- *H1b: Site features positively influence customers' satisfaction*
- *H1c: Site security positively influences customers' satisfaction.*

2.7 *Relationship between customers' satisfaction and customers' behavioral intentions*

According to Nimoko (2012), customers' satisfaction is positively correlated with repurchase intention, possibility to promote, loyalty customer, positive word-of-mouth and customer retention. Behavioral intentions can be seen as the amount of effort that an individual will use to accomplish behavior (Yun & Good, 2007). Gounaris et al. (2010) mentioned that for customer to retain, they need to satisfy first. Therefore, customers' satisfaction is important to create customer behavioral intentions. Specifically, three types of customers' behavioral intention are studied in this research: word-of-mouth communication, revisit of the site, and purchase intention.

2.8 *Customers' behavioral intentions*

Gounaris et al. (2010) discovered that, customers' behavioral intentions in e-shopping are influenced by service quality and also satisfaction of product or service delivered. Paynter and Lim (2001) highlights decision making in online purchase received a great impact from customers' behavior and attitude. Customers' behavioral intentions for this study comprise of word-of-mouth, site revisit, and purchase intention that occurred after the customers are satisfied with online discount programs.

2.8.1 *Word-of-mouth*

Saha and Theingi (2009, p. 354) indicated that "word-of-mouth refers to a flow of information about products, services, or companies from one customer to another". A good reputation can attract new customers (Zeithaml, 2000). Poleretsky (1999) mentioned that in e-setting, word-of-mouth is much more crucial than physical setting because unhappy customers will tell their negative experiences to five friends, while on the internet they will spread to 5,000 people. This showed that the word-of-mouth can be a threat for the online service provider since the negative word-of-mouth can be easily spread to too many people through online and this can kill any online business if the negative word-of-mouth continued.

2.8.2 *Site revisit*

The decision to revisit a site shows that customer service converting behavior (Keaveney, 1995) which explained that a customer tend to use the same online service but will convert from one service provider to others. Generally, customers will use their past purchase experience as decision making in order for repetition of purchase. Smith and Sivakumar (2002) supported that if the customer is positively feel about the site after the interaction with the site, the more possibility the customer will revisit the site.

2.8.3 *Purchase intention*

Blackwell, Miniard and Engel (2001) defined intentions as subjective judgments about how we will act in the future. Purchase intentions represent what we think we will buy. Purchase intention eventually leads to purchase behavior as people attitude is measured through purchase intention. Therefore, it has been suggested as one of the way to predict customer purchase behavior (Follows & Jobber, 2000). In online context, purchase intentions occur when the customers are willing to purchase more via the internet and in the future, the customers tend to repeat their purchase (Gounaris et al, 2010). Based on the above literatures, the following hypotheses are thus suggested:

H2: Customers' satisfaction positively influences customers' behavioral intention.
- *H2a: Customers' satisfaction positively influences word-of-mouth*
- *H2b: Customers' satisfaction positively influences site revisit*
- *H2c: Customers' satisfaction positively influences purchase intention.*

2.9 *Relationship between online discount program and customers' behavioral intention*

Fortin (2000) suggest that, consumers see online discount program as a benefit to them since they can get discounted price to selected product or services. When customers purchase online, they can save time as time is value to people in busy life. According to Ahmad, Omar and Ramayah (2010), behavioural intentions can be considered as positive or negative. Positive behavioural intentions consist of positive word of mouth, spending more money with the company, and create customer loyalty while negative behavioural intentions are the other way around. Understanding consumer behaviour is an important part in marketing and is an essential necessity to organizations being marketing orientated and profitable because it supports all marketing activities pursued by them (Ahmad et al., 2010). Based on the above literatures, the following hypotheses are thus suggested:

H3: Online discount programs positively influence customers' behavioral intention.
- *H3a: Offer deals positively influence customers' behavioral intention.*
  - *Offer deals positively influence word-of-mouth*
  - *Offer deals positively influence site revisit*
  - *Offer deals positively influence purchase intention*
- *H3b: Site features positively influence customers' behavioral intention.*
  - *Site features positively influence word-of-mouth*
  - *Site features positively influence site revisit*
  - *Site features positively influence purchase intention*

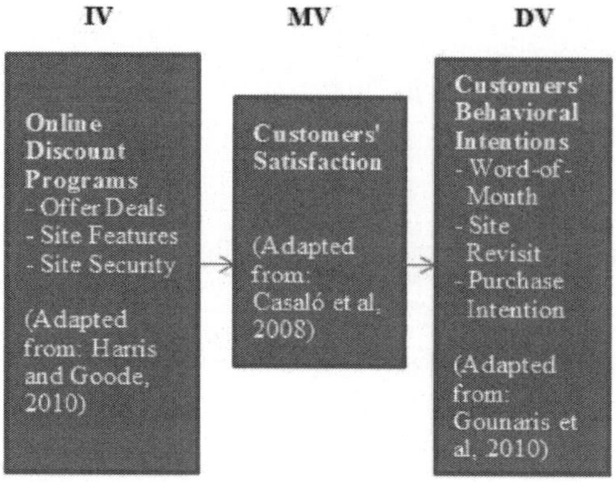

Figure 1. The proposed study framework.

H3c: Site security positively influences customers' behavioral intention.

- *Site security positively influences word-of-mouth*
- *Site security positively influences site revisit*
- *Site security positively influences purchase intention.*

2.10 *Study framework*

All directions of the research questions are shown in Figure 1. The proposed theoretical framework of the study is as below.

3 ISSUE

Online discount programs are important for businesses because this is one of the medium for businesses to attract new customers who are not aware of the products or services that the businesses offer before (Edelman et al., 2011). Customers are interested with the discounted price from online discount programs and will continue to purchase online since it benefitted them (Dholakia, 2010). However, poor web sites performance resulting to customers' dissatisfaction and will leave the site although the customers are interested with the discounted price (Akamai, 2006). The authors further added, quick page loading and easy-to-use web sites lead to online shopper satisfaction and loyalty.

Numerous studies has been conducted that explore about internet technology adoption such as study on purchase online (Jamaludin, 2013), online shopping (Gounaris et al., 2010; Smith & Sivakumar, 2002), e-coupons (Blundo, Cimato, & Debonis, 2005; Fortin, 2000), group buying websites (Coulter & Roggeveen, 2012; Dholakia, 2010; Edelman et al., 2011), and many more.

Most of the studies were not holistically investigated area of consumers' behavioral intention especially in the online discount programs. In Malaysia, the popularity of online discount programs is promising and due to little academic literature, this situation is opportunities for this study to be conducted. The information is vital to maintain the popularity and increase revenue. Hopefully, the outcome of this study will contribute to marketers and provide them with a better understanding regarding customers' satisfaction and behavioral intention toward online discount programs and provides some inputs for companies in improving their promotional effort and future plan.

This study also contributes to customers because through online discount programs, people will get benefits through the deals. Besides that, it is worth to conduct this study because when people spend, indirectly they can help the growth of nation's economy. The results from this study can also contribute the academia with crucial information and perceptions on the online discount programs, thus by discovery the most agreeable of online discount programs' dimensions and mediating influence of customers' satisfaction in determining customers' behavioural intentions by Malaysian perspectives can contribute the academic literature.

REFERENCES

Ahmad N., Omar A. & Ramayah T. (2010). Consumer lifestyles and online shopping continuance intention. *Business Strategy Series, 11(4), 227–243.*

Akamai (2006). Retail web site performance: Consumer reaction to a poor online shopping experience. *Vendor Research.* Retrieved from http://www.akamai.com/4seconds.

Anderson, R.E. & Srinivasan, S. (2003). E-satisfaction and e-loyalty: A contingency framework. *Psychology & Marketing, 20(2), 123–38.*

Blackwell, R.D., Miniard, P.W. & Engel, J.F. (2001). *Consumer behaviour, 9th ed.* Mike Roche, Grove City, OH.

Blundo, C., Cimato, S. & Debonis, A. (2005). Secure e-coupons. *Electronic Commerce Research, 5(1), 117–39.*

Coulter K.S. & Roggeveen A. (2012). Deal or no deal: How number of buyers, purchase limit, and time-to-expiration impact purchase decisions on group buying websites. *Journal of Research in Interactive Marketing, 6(2), 78–95.*

Dholakia, U.M. (2010). How effective are Groupon promotions for businesses? *Social Science Research Network.* Retrieved from http://ssrn.com/abstract¼1696327.

Donnelly, V. (2001). *Designing easy-to-use websites.* Addison-Wesley, Guildford.

Dube, L. & Morgan, M. (1998). Capturing the dynamics of in-process consumption emotions and satisfaction in extended service transactions. *Journal of Research in Marketing, 15, 309–20.*

Edelman, B., Jaffe, S. & Kominers, S.D. (2011). To Groupon or not to Groupon: The profitability of deep discounts. *Harvard Business School Working Paper, 11(63).*

Follows, S.B. & Jobber, D. (2000). Environmentally responsible purchase behaviour: A test of a consumer model. *European Journal of Marketing, 34(5/6), 723–46.*

Fortin, D.R. (2000). Clipping coupons in cyberspace: A proposed model of behavior for deal-prone consumers. *Psychology & Marketing, 17(6), 515–34.*

Gounaris, S., Dimitriadis, S. & Stathakopoulos, V. (2010). An examination of the effects of service quality and satisfaction on customers' behavioral intentions in e-shopping. *Journal of Services Marketing, 24(2), 142–156.*

Harris, L.C. & Goode, M. (2004). The four levels loyalty and the pivotal role of trust: A study of online loyalty, trust satisfaction, value, and service quality. *Journal of Retailing, 80(2), 139–58.*

Harris, L.C. & Goode, M. (2010). Online servicescapes, trust, and purchase intentions. *Journal of Services Marketing, 24(3), 230–243.*

Horppu M., Kuivalainen O., Tarkiainen A. & Ellonen H.K. (2008). Online satisfaction, trust and loyalty, and the impact of the offline parent brand. *Journal of Product & Brand Management, 17(6), 403–413.*

Internet World Stats (2012). *Internet growth statistics.* Retrieved November 8, 2012, from http://www.internetworldstats.com/emarketing.htm.

Jamaludin, A. (2013). The influence of website trust and loyalty on customer intention to purchase online. *4th International Conference on Business and Economic Research (4th Icber 2013) Proceeding.*

Keaveney, S.M. (1995). Customer switching behavior in service industries: An exploratory study. *Journal of Marketing, Vol. 59, April, pp. 71–82.*

Mandel, N. & Johnson, E.J. (2002). When web pages influence choice: Effects of visual primes on experts and novices. *Journal of Consumer Research, 29(2), 235–45.*

Muylle, S., Moenaert, R. & Despontin, M. (2003). The conceptualization and empirical validation of web site user satisfaction. *Information and Management, 41(5), 543–60.*

Nimako S.G. (2012). Linking quality, satisfaction and behaviour intentions in Ghana's mobile telecommunication industry. *European Journal of Business and Management, 4(7).*

Paynter J. & Lim J. (2001). Drivers and impediments to e-commerce in Malaysia. *Malaysian Journal of Library & Information Science, 6(2), 1–19.*

Poleretzky, Z. (1999). The call center and e-commerce convergence. *Call Center Solutions, 7, 76.*

Saha G.C. & Theingi. (2009). Service quality, satisfaction, and behavioural intentions: A study of low-cost airline carriers in Thailand. *Managing Service Quality, 19(3), 350–372.*

Shankar, V., Smith, A.K. & Rangaswamy, A. (2003). Customer satisfaction and loyalty in online and offline environments. *International Journal of Research in Marketing, 20(2), 153–75.*

Shankar, V., Urban, G. & Sultan, F. (2002). Online trust: A stakeholder perspective, concepts, implications, and future directions. *Journal of Strategic Information Systems, 11, 325–44.*

Smith, D.N. & Sivakumar, K. (2002). Flow and internet shopping behavior: A conceptual model and research proposition. *Journal of Business Research, 1, 1–12.*

Yun, Z.S. & Good, L.K. (2007). Developing customer loyalty from e-tail store image attributes. *Managing Service Quality, 17(1), 4–22.*

Zeithaml, V.A. (2000). Service quality, profitability, and the economic worth of customers: What we know and what we need to learn. *Journal of the Academy of Marketing Sciences, 28(1), 67–85.*

Zeithaml, V.A., Parasuraman, A. & Malhotra, A. (2002). Service quality delivery through web sites: A critical review of extant knowledge. *Journal of the Academy of Marketing Science, 30(4), 362–75.*

Hospitality and Tourism – Sumarjan et al. (eds)
© 2014 Taylor & Francis Group, London, ISBN 978-1-138-00151-0

Hotel customer-based experience, brand image and loyalty in Iran

H. Robabeh Sadat & A. Zainal
Universiti Teknologi MARA, Malaysia

ABSTRACT: Brand loyalty management is the essential point for the hotel industry in determining the success or failure of the organization. Being a service-dominant environment, the hospitality and tourism industry found that brand management is critical for managers to endeavor to achieve long-term success. Being aware of the importance and significance of the overall perceived quality of experience and the extent to which experiences of hotels guests are manifested in their motivation to stay in the hotels are crucial. Additionally, it is suggested that hotels managers develop cost-effective advertising, referral marketing and service programmers to improve hotel experience which in turn creates brand image for the development of better loyalty and satisfaction in the hospitality industry in Iran.

Keywords: hotel, customer-based experience, brand image, loyalty, Iran

1 INTRODUCTION

Tourism has gone and will continue to experience metamorphosis in reaction to the changes around us. The evolutionary process of tourism phenomenon can be witnessed through the changes that are experienced in the supply side (destinations), as well as, in the demand side (the market) (Inskeep, 1999; Gunn with Var, 2002; Hall, 2009; Edgell et al., 2008; Walton, 2009; Papatheodorou, 2004; Cooper and Hall, 2008). According to the United Nations Educational, Scientific and Cultural Organization (UNESCO), Iran is ranked as one of the world's top countries in terms of the number of ancient and historical sites. To date, 13 historical sites have been listed under the World Heritage list and fifty-nine sites are tentatively listed (UNESCO, 2011a). The seven world Intangible Cultural Heritages here include NowRuz, the Radif of Iranian music, the music of the Bakhshis of Khorasan, the Pahlevani and Zoorkhanei rituals, the ritual dramatic art of Ta'z_ıye and traditional skills of carpet weaving in Fars and Kashan regions (Robabeh Sadat and Artinah, 2013). Despite these attractions, local politics may have stunted the country's tourism growth. Furthermore, the lack of information and advertising aggravates the country's exposure to the outside world (Zamani Farahani, 2012).

Tourism, food and accommodation are closely related. Thus, being a service-dominant environment, the hospitality and tourism industry found that brand management is critical for managers striving to achieve long-term success (Berry, 2000; Berry, Lefkowith, & Clark, 1988; Laroche & Parsa, 2000). The hotel industry is also a typical ground where brand equity management becomes the focal point in determining success or failure of companies. pointed out that hotel managers realize the contribution of brand affiliation to their properties and market value. The value of a hotel brand is also manifested when it is able to sustain a competitive position for a company in the long run (Cai & Hobson, 2004).

Hotels are associated with branding and brand management as they are considered as top management priorities and strategic focuses of many organizations. In the service industry, brand management requires brand managers to take a holistic view of the brand that transcends marketing and service functions and transforms the brand into a rallying point for all components of the firm (Smith, 2004). Successful brand management can create added value for the consumer, thereby increasing brand franchise and brand equity.

Notably, elevated brand equity brings about opportunities for successful extensions, resilience against competitors' promotional pressures, and creation of barriers to new competitors from entering the market (Fayren & Lee 2011). Along with this, customer satisfaction is important to long-term business success and marketing (Jones & Suh, 2000; Pappu & Quester, 2006), and since customer satisfaction determines long-term business success, a majority of the research on customer satisfaction investigates its impact on customers' post consumption evaluations such as behavioral and attitudinal loyalty (Cooil, Keiningham, Aksoy, & Hsu, 2007).

2 IRAN AND HOTEL BRANDING

It is noted that hospitality sector is a key point in the development of tourism industry and service quality is an elusive and abstract construct that is difficult to define and measure. The hotel organizations are always determined to achieve sound service performance and delivery so that they can create tangible, symbolic, and original meanings in the process of developing their brand. As stakeholders, customers look forward to the promises that a hotel pledge about quality services and experiences.

Hence, tourism managers make considerable investment to improve hospitality industry, to reduce causes of declining service quality provided by hotels and as a result, lead to increase customers' satisfaction. (Akaba, 2006; Kao, Wub, & Deng, 2009).

Table 1. International tourist arrivals to Iran.

Year	By	Number	Total
2009	Aerial	634522	2,272,575
	Earthly	1619035	
	Nautical	16742	
	Rail	2276	
2010	Aerial	628624	3,121,283
	Earthly	2471744	
	Nautical	18524	
	Rail	2391	
2011	Aerial	638557	3,294,126
	Earthly	2626070	
	Nautical	27896	
	Rail	1603	
2012	Aerial	787420	3,130,126
	Earthly	2285544	
	Nautical	52927	
	Rail	4235	

Source: Khorasan's Cultural Heritage, *Handicrafts* and *Tourism Organization* (2012).

Customers, in turn, are most often concerned about their familiarity with a brand and the distinguished position of the brand that has been fostered in their mind. Therefore, refining consumer-based brand loyalty is a pertinent key driver of success in this industry. This is especially important for Iran since there are many international tourist arrivals yearly (Table 1).

One of the problems faced by the Iranian hospitality industry is the lag of Iran's hospitality brand. Outdated government policies which were adopted for such branding still remain in force and the market economy is not dynamic and does not allow Iranian investor to go for international branding, as one of the most important problems is the high cost of brand-building. It is noteworthy to say brand identity is slow in gaining ground for the hospitality and tourism industry in Iran. Some of the laws are a barrier to branding in Iran.

It is not erroneous to say that the hotels in Iran are lagging in identity, as such there is a necessity to create identity; and the remedy is branding. Since branding process is time consuming and requires technical knowledge, branding requires having a dynamic economy and support by the government. Although Iran does not have international brand hotels, but there is a possibility of building brand on their products to include self-identity and elements which consequently will increase the average stay of tourists to more than a week. A few examples of the local branded 4 and 5 star hotels found in major cities of Iran are chain hotels such as Homa and Pars, other than the stand alone hotels like Pardisan and Shahriyar.

Hotel brands are closely aligned with the service brands delivered and thus customers tend to build real trust through direct consumption experiences, which may confirm the indirect experiences gained. Direct experience with staff and the service environment reflect the brand image and brand equity for travelers. Likewise advertising and word-of-mouth both generated significant effects on brand associations.

Table 2. List of hotels in major cities of Iran.

City	5 Star hotel	4 Star hotel	3 Star hotel	1 & 2 Star hotels	Total
Tehran	5	10	22	61	98
Mashhad	4	9	23	213	249
Esfahan	2	4	9	33	48
Tabriz	2	1	2	9	14
Shiraz	1	5	4	25	35
Total	14	29	60	331	444

Source: http://hoteldari.net/listh/listh.php2012.

3 CONCLUSIONS

Since Iran is relatively new to the tourism industry it is necessary to measure the effectiveness of the hospitality industry and understand whether Iran is going in the right direction. To be successful, the product, price, promotion and distribution systems have to be favorable and competitive with other producers. Since branding process is time consuming and requires technical knowledge, branding requires having a dynamic economy and support by the government.

In view of the fact that branding is a long process of growth and reaching maturity, hence Iranian investors are working hard to produce a standard product since branding plays an important part in customer satisfaction.

Thus far in the absence of comprehensive discussions on the customer-based experience and motivational factors that leads to brand loyalty, with the suggested points above, this paper is expected shed some light to fill the gap and add to the volume of knowledge in customer hotel experience and branding.

REFERENCES

Akaba, A. (2006). Measuring service quality in the hotel industry: A study in a business hotel in turkey. *Hospitality management,* 25, 170–192.

Berry, L.L. (2000). Cultivating service brand equity. *Journal of the Academy of Marketing Science,* 28(1), 128–137.

Berry, L.L., Lefkowith, E.F., & Clark, T. (1988). In services, what's in a name? *Harvard Business Review,* 66(5), 28–30.

Cai, L.A. (2002). Cooperative branding for rural destinations. *Annals of Tourism Research,* 29(3), 720–742.

Cooil, B., Keiningham, T.L., Aksoy, L., & Hsu, M. (2007). A longitudinal analysis of customer satisfaction and share of wallet: Investigating the moderating effect of customer characteristics. *Journal of Marketing,* 71(1), 67–83.

Cooper, C. and Hall, C.M. (2008). Contemporary Tourism: An International Approach. Oxford: Butterworth- Heinemann.

Fayrene, Ch. Y.L., & Chai G.L. (2011). Customer-Based Brand Equity: A Literature Review. *Journal of Arts Science & Commerce.*

Gunn, C.A., and Var, T. (2002). Tourism Planning: Basics, Concepts, Cases. (4th. Ed). New York: Routledge.

Hall, C.M. (2009). Degrowing Tourism: Décroissance, Sustainable Consumption and—State Tourism. Anatolia. *An International Journal of Tourism and Hospitality Research,* 20(1), 46–61.

Inskeep, E. (1999). Tourism Planning: An Integrated and Sustainable Development Approach. New York: John Wiley & Sons. Inc.

Jones, M.A., & Suh, J. (2000). Transaction-specific satisfaction and overall satisfaction: An empirical analysis. *Journal of Services Marketing,* 14(2), 147–159.

Kao, Y.F., Wub, C.M., & Deng, W.J. (2009). The relationships among service quality, perceived value, customer satisfaction, and post-Khorasan's Cultural Heritage, Handicrafts and Tourism Organization (2012).

Laroche, M., & Parsa, H.G. (2000). Brand management in hospitality: An empirical test of the Brisoux-Laroche model. *Journal of Hospitality & Tourism Research,* 24(2), 199–222.

Papatheodorou, A. (2004). Exploring the Evolution of Tourism Resorts. *Annals of Tourism Research,* 31(1), 219–237.

Pappu, R., & Quester, P. (2006). Does customer satisfaction lead to improved brand equity? An empirical examination of two categories of retail brands. *Journal of Product and Brand Management,* 15(1), 4–14.

Robabeh Sadat, H & Artinah, Z., (2013). Place of Religious Tourism and How to Use Them to Improve On Existing Capacity and Potential: A Case Study of Neyshabour, Iran. International Conference On Religios Tourism And Tolerance 9–12 May, 2013 Konya, Turkey.

Smith, S.L.J. (2004). Brand experience. In R. Clifton & J. Simmons (Eds.), *Brands And branding.* Princeton, NJ: Bloomber Press.

UNESCO. (2011a). United Nations Educational, Scientific and Cultural Organization. World heritage URL http://whc.unesco.org/en/list. Accessed 29.8.11.

Zamani Farahani, H. (2012). The relationship between Islamic religiosity and residents' perceptions of socio-cultural impacts of tourism in Iran: Case studies of Sare'in and Masooleh. *Tourism Management,* 33 (2012), 802–814.

Hospitality and Tourism – Sumarjan et al. (eds)
© 2014 Taylor & Francis Group, London, ISBN 978-1-138-00151-0

A preliminary study on the benefits of loyalty programs in Malaysian city hotels

J. Anuar, N. Sumarjan & S.M. Radzi
Universiti Teknologi MARA, Terengganu, Malaysia

ABSTRACT: Hotels spend millions of dollars annually to grant benefits to their loyal members. However, despite having a huge portion of investment on this loyalty program benefits, the effectiveness of such programs in delivering benefits to members is still questionable. In-depth interviews of 14 hotel executives from either Front Office or Marketing department were primarily conducted to explore hotel's benefit structures. The taxonomical benefits offered by six (6) city hotel loyalty programs were compared and summarized. The transcripts and tapes derived from the interviews were analyzed, coded and summarized into categories. The findings were structurally compared. Findings revealed that although these six (6) city hotels sharing the same thirteen core benefits, but each hotel is trying very hard to differentiate themselves from their competitors. It is beneficial for hoteliers in improving their current loyalty programs or designing new loyalty programs for their hotel. Uniquely, this study provides benefit taxonomy summary of the six (6) city hotel loyalty programs in Malaysia and compares the benefits categorized into three (3) levels: core benefits, partially common benefits and individual benefits.

Keywords: benefits, hotel executives, loyalty programs, Malaysian city hotels

1 INTRODUCTION

The Loyalty programs were first introduced by American Airlines' through theirs (AAdvantage program) in the year 1981. Hotels and car rental agencies followed the suit and introduced their own loyalty programs to their customers. Then, Holiday Inn and Marriott navigates their Priority Club and Honored Guest, in year 1983 and was the first loyalty program in the hotel industry worldwide. Because of their ubiquity in the marketplace, worldwide complementary services companies including airlines, hotels, retailers and car manufacturers, have increasingly used loyalty programs to conduct their customer database (Mimouni-Chaabane, Volle, 2010). They believed that loyalty programs could act as the most important marketing tactic in driving customer loyalty.

Loyalty programs, which are in reality repeat-purchase rewards programs, were employed by various hospitality firms on the basis of mutual bonding between customer and firm. Management scholars and practitioners generally agree that loyal customers are vital to the long-term success of hospitality organizations (Skogland & Siguaw, 2004). As a marketing strategy, the program offers incentives and reward to its members with the objective of fostering more customers' loyalty to the company (Lacey, 2003; Yi & Jeon, 2003; Leenheer et al, 2007; Omar et al, 2011a). Loyal customers bring benefits such as more frequent purchasing cycles, greater profit margins and advertising effects for prospective customers (Han & Back, 2006, 2008; McCall & Voorhees, 2010). Businesses can also increase revenue from loyalty as reported by Omar, Abdul Aziz and Nazri (2011).

Recently, loyalty programs have become prevalent among luxury city hotels in Malaysia. A study by Steyn, Pitt, Strasheim, Boshoff and Abratt (2010) indicates that loyalty programs are experiencing a new wave in Asian Pacific country and Malaysia, whereas it has reached its maturity stage in the United States and Europe. Ganesan (2006) highlighted that loyalty programs in Malaysia in year 2006 is only RM2.4 billion while growing at RM3.5 billion in 2010. Probably this is due to the tremendous changes from the early days until today with the emergence of various types of hotel format, offers a wide assortment of products ranging from basic to prestigious product. The mature and saturated market has made it necessary for hoteliers to design their own strategies in fostering customer loyalty. A search for 3–5 star Malaysian hotels that adopt loyalty programs through the search engine Yahoo (http://www.yahoo.com, accessed in October, 2012) produced a list of 39 hotels all over the country. The continuous and marked change for local tourism and hospitality environment over the past decade, stable economy as well as large foreign investors to set up operations in Malaysia had contributed to the stiff competition among hotels in capturing customers.

Loyalty program membership benefits remain a vital factor in developing an interesting benefits program that can foster hotel guest loyalty. The impact is significant in addressing and maintaining hotel guests (Shanshan, Wilco & Eric, 2011). However, the effectiveness issue of the benefits of loyalty programs remains inconclusive (Kim, Cho, Han, 2012). Although each hotel believed that they are providing the best and valuable benefits to their guests, the type of benefits that customer really value remains unknown (Shanshan et al. 2011). In 1988, a table list of benefits accrued from 10 major hotel loyalty programs were established by Toh, Rivers and Withiam (1991).

Further, Omar, Musa, Che Wel and Abd Aziz (2012) develop a baseline model to examine the role of loyalty program benefits offered by a leading superstore in Malaysia on its members' satisfaction and ultimately loyalty towards the store. However, at present, there is a none empirical study that has been conducted on benefits of loyalty programs comprehensively in the Malaysian hotel scenario. This study adapted Shanshan et al. (2011) to explore hotel benefits structures. The taxonomical benefits offered by six (6) luxury hotel loyalty programs were compared and analyzed.

2 LITERATURE REVIEWS

Sharp and Sharp (1997) defined loyalty programs as a marketing effort providing rewards based on the cumulative purchases from a company. Dowling and Uncle (1997) interpreted it as 'the underlying belief of its popularity on the 80/20 Law', where revenue of the company (80%) comes solely from their customers (20%). Hence, it is vital to focus on 20% customers in retaining them. The current systems of loyalty programs allow members to collect points for a hotel stay and point's redemption through free night stays or upgrades room category. These programs normally offer an online membership account for guest reservations and personalize their bookings accordingly (Sigala, 2006).

A study by Hlavinka and Sullivan (2011) indicates that from the COLLOQUY's survey conducted in 2011, there were more than two billion loyalty program memberships in the United States. This is supported by another study that illustrated the development of loyalty programs worldwide (Van Heerde & Bijmolt, 2005). Retailers such as E. Leclerc in France invested ten millions of dollars per year to implement these loyalty programs (Saili et al, 2012). In the developing countries, the growth of loyalty programs is twice faster than that in developed countries (Saili et al. 2012). Hotels, airlines, casinos, car rentals and direct marketing companies established direct marketing services focusing on the mass market segment with the purpose of encouraging customer loyalty through program benefits (Vogt, 2010).

Loyalty programs on the other hand can be seen as effective tools in maintaining a high-end business travel market (Kwan, 2001). A well-established loyalty program benefits design can lead to a decisive competitive edge. This is due to customer information systems through loyalty program benefits that help identify individual guests' preferences, demographics and other forms of customer behavior (Hendler & Hendler, 2004; Lal & Carrolo, 2004). The customer information derived from the database can also produce valuable customized services that matched individual requests (Sigala, 2005; Xu & Walton, 2005).

3 METHODOLOGY

In-depth interviews of 14 hotel executives from either Front Office or Marketing department were performed to explore hotel's benefit structures. A total of six (6) loyalty programs in Malaysian city hotels was studied, including Sheraton Imperial Hotel, Renaissance, Park Royal, Double Tree by Hilton, InterContinental Hotel and Hotel A. The term 'Hotel A' will be used throughout this paper to respect the anonymity policy of one hotel brand. However, Hotel A has no reservation on sharing information pertaining to their loyalty programs. The Five-star hotels are specifically located in Kuala Lumpur area and listed among the 45 city/business hotels associated and registered with the Malaysian Association of Hotels (MAH, 2012). A search for MAH member listing (www.hotels.org.my, accessed in October 2012) shows that 14 Five-star hotels have adopted loyalty programs. All respondents were approached through email and phone calls to solicit their participation in this study. For those who replied the invitation and agreed to participate in this study, appointments were arranged. Six hotels turned down the invitation due to high occupancy and engagement with big events at their hotels. Another two hotels share the same loyalty programs with two of the six hotels studied (e.g. Hilton Hotel shared Hilton Honors with Double Tree by Hilton, Marriott Hotel shared Marriott Rewards with Renaissance Hotel). Hence, the two hotels mentioned above will not be included in this research.

This study chose in-depth interview since it offers 'rich' and 'complete' data from interviewees (Easter-Smith et al. 2002). Purposive sampling was

Table 1. The profile of interviewees.

Position	Number
Front office supervisor	5
Front office manager	3
Front office assistant manager	2
Marketing assistant manager	2
Marketing executive	2

*Self compiled by researcher.

employed for participants selected in the interview. Front Office or Marketing department hotel executives (holding at least supervisory positions or above) were selected mainly because they are the person in charge of handling loyalty program procedures as well as their in-depth knowledge of the associated mechanism and administrative logistics pertaining to the programs. Table 1 lists the detailed profile of the interviewees. The interviewees were 55% male and 45% female.

A set of questions was profoundly reviewed and used as a guideline for discussion. Each interview took an hour. The interview session was recorded, transcribed verbatim and summarized, from which themes, categories and concepts were identified. Studies conducted in qualitative research literatures suggested that the interview process should be 'audio-recorded if permitted, transcribed verbatim and analyzed in an iterative manner' (Fern, 2001; Jennings, 2001; Miles & Huberman, 1994). The transcripts derived from the interviews were repeatedly reviewed, coded and analyzed into categories.

4 FINDINGS AND DISCUSSION

The taxonomical benefits of the loyalty programs were developed after the in-depth interview process took place. Table 2 provides the taxonomy of loyalty program benefits among six (6) city hotels in Malaysia. While Table 3 details out the interviewed hotels with their summarized benefits. This study has managed to identify the core benefits defined as same benefits offered by all hotels, partially common benefits defined as part of benefits that are same among hotels studied and lastly is individual benefits defined as the unique benefit provided by the hotels.

The researcher has managed to identify 13 items (no. 1–13) for core benefits. All interviewed hotels provide free enrollment to their loyalty programs. It can be done online or upon checking in at the hotel. A pre-registration process also takes place to ease member's check-in. Guest preferences on the room as well as guests' information such as home address is already available during the pre-registration process. On top of that, as members of loyalty programs, check-in & checkout at exclusively designated area awaits them. Additionally waitlist for room reservations will be prioritized to members if they call for reservations during fully booked period. If the booking is made 72 hours prior to arrival, a room reservation is then guaranteed, together with the preferred bed type. For member's convenience, early check-in and late checkout are also available. Additionally, members enjoy free local and long distance calls as well as faxes at cost price. All six hotel loyalty program allows members to convert membership points to airline miles with participating partners. However, airline partners will vary among hotels. In line with the membership points, members can also enjoy offers such as weekend gateway as well as bonuses on reward points. The mechanics of membership levels among all hotels are similar. More benefits will be given to higher level of membership accrued. Membership levels are subject to the number of nights stayed.

For partially common benefits, ten (10) items were identified (no. 14–23). It was observed that online newsletter has become very popular in recent years. This might be due to the power of the internet and information technology, which influenced hoteliers to grab the opportunity and provide guests with the service. Members will receive latest promotion and program news concerning travel inspiration, insider knowledge and inspiring destination through their emails. Five out of the six programs offer lifetime membership while the remaining one program based its membership validity period on the active status of the member. The membership will expire if there are no points accrued during a 24-month period. Other than that, four programs offer exclusive customer support centre. All issues related to loyalty programs such as change of personal information and point's redemption will be entertained by the support staffs. Sheraton Imperial terms it differently, that is Special Elite customer service phone line. The majority of the hotels offers tangible or hard benefits such as high speed in-room internet access, complimentary continental breakfast and choices of welcome amenities.

Three programs however emphasized on intangible or soft benefits as well such as room upgrades and complimentary access to Club/Executive floor privileges. Interestingly, findings discovered that two programs promote no black-out dates. Members are free to redeem their stay at the hotel throughout the year with no restrictions/limitations on certain dates. Additionally, two loyalty programs have 'Moments/Discovering' program; which is an invitation to members to join hotels events and experiences. All pictures and information on the

Table 2. The taxonomy of loyalty program's benefits.

Core benefits	Partially common benefits	Individual benefits
1. Free enrolment	14. Online newsletter	24. Partner stays free
2. Pre-registration for easy check-in	15. Lifetime membership	25. Complimentary daily buffet breakfast
	16. Customer support centre	
3. Priority check-in & check-out	17. Complimentary high speed in-room internet access	26. Concierge services for non-staying guests
4. Priority waitlist for reservations		
5. Guaranteed room availability	18. Complimentary continental breakfast	27. Complimentary pressing
6. Guaranteed bed type		28. Complimentary use of selected resort facilities
7. Early check-in, late check-out	19. Choice of welcome amenities	
8. Free local calls	20. Room upgrade	29. Cash advance of upto USD 300 per stay
9. Long distance calls & faxes	21. Complimentary access to club-executive floors	
10. Convert membership points to airline miles		30. Complimentary international paper
	22. moments/discovering programmes	
11. Member offers		31. Online forum
12. Bonuses on rewards program points	23. No blackout dates	32. 5th night free
		33. Crossover rewards with airlines
13. Levels of membership		
		34 Marriott gift shop
		35. Complimentary bev.
		36. Membership level
		37. Points earn partners

*Source: Self-compiled by researcher.

Table 3. Hotels with summarized benefits.

Hotels	Core benefits	Partially common benefits	Individual benefits
Hotel A	All hotels shared the same benefits (items no. 1–13)	14–16, 19–21	24–31
Sheraton		14–19, 22–23	32–33
Imperial		14, 17–18	34
Renaissance		15, 17, 19, 22	35
Park royal		14–19, 20–21	36
Double tree by Hilton		14–17, 20, 23	37
InterContinental			

*Source: Self-compiled by researcher.

event will be uploaded in the hotel website after the event finish. For Park Royal, members do not just enjoy exclusive access to special activities, but also will be granted a chance to live like a local community in each city that events conducted.

The study found 14 items that were identified as individual benefits (no. 24–37). The benefits are offered by only specific hotels, recommending uniqueness of one loyalty program from the others (Shanshan et al. 2011). In an effort to create uniqueness, Hotel A has managed to provide 8 items differently such as complimentary pressing for one suit upon arrival, use of selected resort facilities, international newspaper on top of the free local newspaper, concierge services for non-staying guests and cash advance of up to USD 300 per stay. Advancement of technology is also utilized by Hotel A through the establishment of an online forum, which encourages members to share tips and recommendations based on their past experiences staying at the hotel. On top of that, the benefits are also extended to spouses; they are entitled for 'spouse stays free' offer which comes together with breakfast. As for Sheraton Imperial Hotel, it promotes benefits such as crossover rewards with Airlines Programs and the redemption of 5th night stays free. Renaissance and Park Royal offer only one individual benefit: a 10% Gift Shop discount on retail merchandise and souvenirs for Renaissance and complimentary beverages for Park Royal Hotel.

This study also found that Double Tree by Hilton provided four membership levels instead of three levels as practiced by the other five hotels. Their focus is to encourage members to stay at their hotel and use their services. Additionally, Inter Continental encourages points earning through participating partners. Members can easily earn points through everyday living such as renting cars, buying flowers and more. This study summarized that even though all hotels share the same 13 core benefits, each of them provides several individual benefits in attempts to differentiate themselves from their competitors. For example, considering the power of social media nowadays, the online forum approach by Hotel A seems to be taken very well by customers. However, this benefit may not remain unique forever, as the benefits of loyalty programs are easy to copy. More efforts need to be exerted in discovering ways to stay unique and different from competitors (Shanshan et al. 2011).

5 CONCLUSION AND FUTURE RESEARCH

Marketing and front office personnel that handle loyalty programs can gain valuable insights from this taxonomy summary. Hoteliers need to establish more competitive individual benefit elements that cannot be easily copied by their rivals. This study contributes to the loyalty program literature from the perspective of management or hotel employee. Future research could explore the perspective of the customers. Further study also may involve gap comparison between loyalty program members' preferences benefits and management/hoteliers perceptions of the benefits that their hotel offered. Additionally, future research also may focus on the lower scale hotel operations to give a different impact and scenarios. Lastly, the new secretary loyalty program conducted by Maya Hotel; Maya Pink Club may prove to be a very attractive case as the hotel is recognizing corporate secretaries and personal assistants who handle company bookings. It is interesting to find out from the hotel is rewarding specific customers instead of the common guest approach typically adopted by other loyalty programs.

REFERENCES

Dowling, G.R. & Uncles, M. (1997), Do customer loyalty programs really work?, *Sloan Management Review,* 38(4), 71–82.

Easterby-Smith, M., Thorpe, R. & Lowe, A. (2002), *Management Research: An Introduction,* London: SAGE.

Fern, E. (2001), *Advanced Focus Group Research,* London: SAGE.

Hendler, R. & Hendler, F. (2004), Revenue management in fabulous Las Vegas: combining customer relationship management and revenue management to maximize profitability, *Journal of Revenue and Pricing Management,* 3(1), 73–79.

Hlavinka, K. & Sullivan, J. (2011), The billion member march: The 2011 COLLOQUY loyalty census (Vol April 2011, pp 1–14), Cincinnati, OH.

Ganesan, V. (2006, May). Rich pickings with loyalty cards. *New Straits Times,* 40.

Han, H. & Back, K. (2006), Investigating the effects of consumption emotions on customer satisfaction and repeat visit intentions in the lodging industry *Journal of Hospitality and Leisure Marketing,* 15 (3), 5–30.

Han, H. & Back, K. (2008), Relationships among image congruence, consumption emotions and customer loyalty in the lodging industry, *Journal of Hospitality & Tourism Research,* 32 (4), 467–490.

Jennings, G. (2001), *Tourism Research,* Sydney: John Wiley & Sons.

Kim, Y., Cho, M-H., Han, H. (2012), Testing the model of hotel chain frequency program members' loyalty intentions, *Asia Pacific Journal of Tourism Research,* iFirst Article, 2012, 1–26.

Kwan, C. (2001), A study of alliance formation in the Asian hotel industry, PhD Thesis, The Hong Kong Polytechnic University.

Lacey R.W. (2003), Customer loyalty program: strategic value to relationship marketing, PhD Thesis, University of Alabama.

Lal, R. & Carrolo, P. (2004), *Harrah's Entertainment Inc.,* Harvard University, Harvard Business School Cases.

Leenheer J., Van Heerde HJ., Bijmolt THA, Smidtsd A. (2007), Do loyalty programs really enhance behavioral loyalty? An empirical analysis accounting for self-selecting members, *International Journal Research Marketing,* 24 (1), 31–47.

Malaysian Association of Hotels (MAH). Available at http://www.hotels/org.

Mattila, A.S. (2006), How affective commitment boosts guest loyalty and promoted frequent-guest programs, *Cornell Hotel & Restaurant Administration Quarterly,* 47 (2), 174–181.

McCall, M. & Voorhees, C. (2010), The drivers of loyalty program success, *Cornell Hospitality Quarterly,* 51(1), 35–52.

Miles, M. & Huberman, M. (1994), *Qualitative Data Analysis* (2nd Edition), Thousand Oaks, CA: SAGE.

Mimouni-Chaabane, A. & Volle, P. (2010), Perceived benefits of loyalty programs: Scale development and implications for relational strategies, *Journal of Business Research,* 63, 32–37. Doi:10.1016/j/jbusres.2009.01.008.

Omar, N.A., Abd. Aziz, N. & Nazri, M.A. (2011). Understanding the relationships of program satisfaction, program loyalty and store loyalty among cardholders of loyalty programs. *Asian Academy of Management Journal.* 16(1), 21–41.

Omar, N.A., Musa, R., Che Wel, C.A., Abd Aziz, N. (2012), Examining the moderating effects of programme membership duration in the retail loyalty programme: A multi group causal analysis approach, *World Applied Sciences Journal,* 19(3), 314–323.

Saili, T., Mingli, Z. & Zhichao, C. (2012), The effects of loyalty programs on customer loyalty: the mediating role of customer value and the moderating role of relationship benefits, *African Journal of Business Management,* 6(11), 4295–4309.

Shanshan, N., Wilco, C. & Eric, S. (2011), A study of hotel frequent-guest programs: Benefits and costs, *Journal of Vacation Marketing,* 17(4), 315–327.
Sharp, B. & Sharp, A. (1997), Loyalty programs and their impact on repeat-purchase loyalty patterns, *International Journal of Research in Marketing,* 14, 73–86.
Sigala, M. (2005), Integrating customer relationship management in hotel operations: managerial and operational implications, *International Journal of Hospitality Management,* 24, 391–413.
Sigala, M. (2006), e-customer relationship management in the hotel sector: guests' perceptions of perceived e-service quality levels, *Tourism: An International Interdisciplinary Journal,* 54(4), 333–344.
Skogland, I. & Siguaw, J.A. (2004), Are your satisfied customers loyal? *Cornell Hospitality Quarterly,* 45(3), 221–234.
Steyn, P., Pitt, L., Strasheim, A., Boshoff, C. & Abratt, R. (2010). A cross cultural study of the perceived benefits of a retailer loyalty scheme in Asia. *Journal of Retailing and Consumer Services,* 17(5), 355–373.
Toh, R.S, Rivers, M. and Withiam, G. (1991), Frequent-guestprograms: do they fly? *The Cornell hotel and Restaurant Quarterly,* 32(1), 46–52.
Van Heerde, H.J. & Bijmolt, T.H.A. (2005), Decomposing the promotional revenue bump for loyalty program members versus nonmembers, *Journal of Marketing Research,* 42 (4), 443–457.
Vogt, C.A. (2011), Customer relationship management in tourism: Management needs and research applications, *Journal of Travel Research,* 50(4), 356–364.
Xu, M. & Walton, J. (2005), Gaining customer knowledge through analytical CRM, *Industrial Management & Data Systems,* 105(7), 955–971.
Yi, Y., Jeon H. (2003), Effects of loyalty programs on value perception, program loyalty and brand loyalty, *Journal Academy Marketing Science,* 31(3), 229–240.

Hospitality and Tourism – Sumarjan et al. (eds)
 ISBN 978-1-138-00151-0

Psychological determinants of food neophobia and young culinarians

M.A. Ibrahim, R. Ahmad, M.S. Mohd Zahari & F.A. Hanan
Universiti Teknologi MARA, Malaysia

ABSTRACT: No doubt food neophobia is contributing to the diversity of human food preferences as it relates to the tendency of avoiding a novel and unfamiliar foods. Scholars suggest that understanding food neophobia from a psychological perspective, which relate to motivation, belief and knowledge especially among those individuals dealing with food is considered pivotal as it; only limit the food preferences, habits, exposure and knowledge of food but creativity or even career progression. This paper conceptually discusses the psychological determinants of food neophobia by reviewing the literatures.

Keywords: food neophobia, food preferences, novel foods, psychological factor

1 INTRODUCTION

In the early years, foods are only deemed and perceived as the basic necessity in life. Those perceptions however, gradually been altered and food are looking from a broader perspective in line with education, modernization and technological advancement lead to innovative of food products. Asperin, Phillips, and Wolfe (2011) stated that with dynamic and modern food technologies, more opportunities for people to opt for traditional, modification, new or the unique food. In this sense, new and unique food are considered as the novel foods and Tuorila, Lahteenmaki, Pohjalainen, and Lotti (2001) classified novel foods into five categories. (1) *functional foods or medicinal food* is the foods that claimed to promote health and immunity; (2) *genetically modified foods*, foods that are produced using gene technology; (3) *nutritionally modified foods*, foods that have better nutritional benefits than conventional foods; (4) *organic foods* refers to foods that are produced or farmed in traditional conditions and (5) *ethnic foods* that are specifically familiar to one culture but unfamiliar to others.

Novel foods in today's modern societies are aggressively introduced to suit the various changing demand of customers that affected by numerous dynamic factors ranging from psychological, biological, educational, cultural and religion perspectives. Nevertheless, the acceptance of novel foods may be clogged with the existence of food neophobia. Food neophobia is recognized as one of the factors contributes to the diversity of human food preferences as it relates to the tendency of avoiding a novel and unfamiliar foods. Food neophobia is referring to one's reluctance to eat and/or avoid novel foods (Birch & Fisher, 1998; Pliner & Hobden, 1992). Scholars suggest that understanding food neophobia from psychological perspective which relate to attitude, motivation and knowledge especially among those individuals dealing with food is considered pivotal as it can limit their food preferences, exposure and knowledge of food, creativity or even career progression. This research paper conceptually discusses the psychological determinants of food neophobia by reviewing the previous studies and relates the issue to the researchers proposed study.

2 LITERATURE REVIEW

2.1 *Food neophobia*

As previously mentioned, food neophobia is defined as one's reluctance to eat and/or avoid novel foods (Birch & Fisher, 1998; Pliner & Hobden, 1992). Asperin et al. (2011) refer food neophobia is the personality traits that subsist in human, which influences one's willingness to try and consume new and unusable foods and avoid any unpleasant taste that a person expects.

By far each individual in this world is experiencing food neophobia. However, what makes the differences is that the level of food neophobia is dissimilar in each human being. Based on novel food rejection according to Raudenbush and Capiola (2012), there are three types of food neophobia namely neophilic, average and neophobic. People with neophilic are more likely to accept new food while neophobic are those likely to reject new food and some may be in the middle of a continuum, which is being natural or average.

As food neophobia are extensively relate to the rejections of novel foods, the understanding on it categories is considering integral. Tourila et al. (2001) categorized the novel foods into five categories which are the functional foods, genetically modified foods, nutritionally modified foods, organic foods and the ethnic foods. Despite this, many available studies on neophobia have centrally focused on 'ethnic' foods (Pliner & Salvy, 2006) as the ethnic foods in the context of novel foods are the foods that are specifically familiar to one culture but unfamiliar to others (Tourila et al., 2001). Utami (2004) noted that ethnic food can be described the cuisine of the minority in multicultural societies which can be differentiate between the types of foods, method of cooking, the ingredients used and eating decorum based on ethics or states. Ethnic food can explicitly be associated with the Malaysia context owing to the multi-racial and multicultural characteristics. In this sense, *Tempoyak, Cencaluk, Budu, Umai* to name few could be associated with neophobia.

2.2 *Determinants of food neophobia*

Psychological factors are often associated with the human or consumer behaviour, an individual decision making in term of purchasing of certain items and food choices. Decision making is made by the consumers to achieve certain goals or fit a specific segment of lifestyle choices. Psychological factors such as motivation, personality and attitudes towards food and nutrition have been reviewed, and shown to essentially influence nutritional behaviour including food neophobia (Zielinska, 2006).

2.2.1 *Motivation*

Motivation is the driving force within the individual that impels them to action (Schiffman & Kanuk, 2010). In the perspectives of consumer behaviour, it is a driving force that impels human to action or not to action and these derived from positive and negative motivational forces. Need, desire and wants are the examples or the positive drives while fear or aversions are the examples of the negative drives (Schiffman & Kanuk, 2010). Both of these motivational forces lead a different intensity of the physical activities. Zielińska (2006) argued that every individual might have a diverse and dissimilar of need, desire and wants towards food preferences. For instance, to be healthy, slim or conserve some people act in the proper or improper toward food behaviour and this context health factor clearly are the motivational factors that contribute to levels of food neophobia (Pliner & Pelchat, 1993).

The motivation factor of "familiarity" of the foods also contributes to the level of food neophobia. For instance, if the individual is familiar with the food, the reluctance to try and consume may decrease and on the other hand, if the individual is not familiar with the food, the reluctance to try and consume the foods may increase. In two studies of Swedish families of children aged from 2–17 year-old, Koivisto Hursti and Sjöden (1996) and Koivisto and Sjöden (1997) revealed higher neophobia was associated with fewer uncommon foods being served and fewer foods being tried. Tourila et al. (2001) asserted that whether the foods were familiar or unfamiliar, less neophobic people had more frequently tasted and eaten them than to more neophobic people.

2.2.2 *Beliefs*

Belief is the other factors that contribute to the level of food neophobia. An individual might belief food that they are unfamiliar will cause harm and bad consequences towards their health thus tend to avoid and reluctant to consume. Martins and Pliner (2005) denoted that human may reject the food if they are believed or known to possess negative sensory properties (i.e., a bad taste, smell, texture or appearance; commonly referred as rejections based on "distaste") or if they are believed or known to promote harmful consequences, in either the short or long term (i.e. they are bad for you; commonly referred to as rejections based on 'danger').

Martins and Pliner (2005) used the items developed by Rozin and Fallon (1980) to assess the respondents' beliefs about various properties of foods. The items were assigned to subscales based on the three motivational dimensions food rejections which are the sensory affective, anticipated consequences and disgust. Rozin and Fallon (1980) defined the sensory affective as the reactions based on perceptions of the food taste, smell, texture, and appearance; anticipated consequences are the reactions based on perceptions of the consequences of ingesting the foods in both short and long term; while disgust is the reactions based on what the foods are, where they come from, or their social history.

2.2.3 *Knowledge*

Knowledge is also recognised as the determinants that may affect human behaviour towards the food they consume. Acquisition of knowledge can be obtained from various processes such as past experiences and the learning process. Socio economic and cultural diversity might also contribute to the exposure of knowledge. Flight, Leppard and Cox (2003), explained that exposure to diverse cultures and higher-economic status (SES) may increase knowledge of a wide variety of stimuli, including food which also leads to neophobia.

3 FOOD NEOPHOBIA AND YOUNG CULINARIANS

Tourila et al. (2001) stressed that study of food neophobia is crucial as it is influenced food preferences and changing of food habits among the individual and society. Food neophobia will also lead to the tendency in consuming the same types of food over and over again and contribute to the restriction of tasting other foods (Olabi, Najm, Baghdadi and Morton, 2009). Extensive available studies on food neophobia have been focusing among the children (Falciglia, Couch, Gribble, Pabst & Frank, 2000; Galloway, Lee & Birch, 2003). Using Health Eating Index, Falciglia et al. (2000) found that neophobic children are consuming less variety of food and had a higher intake of saturated fat. Galloway et al. (2003) stated that seven years old girls with food neophobia consumed fewer vegetables and less preference on variety of food intake. Looking at the level of food neophobia towards the children' with parent educational background, Mustonen, Oerlemans & Tuorila (2011) revealed children of well-educated parents had tasted a larger number of foods thus displaying lower behavioural neophobia compared to children with less educated parents.

On food neophobia among the young adults, Pliner and Hobden (1992) using the Food Neophobia Scale (FNS) by comparing the level of food neophobia between international college students of Lebanese and American. They found that the Lebanese students are higher in the FNS score in adapting the novel foods. The reason behind the higher score is owing to their current diets and their own culture that cause them reluctance to consume the novel foods (Rozin, 1997; Stallberg-white & Pliner, 1999). Choe and Cho (2012) revealed that Korean college students FNS score were slightly higher than American due to the wellness and growing preferences for higher diet thus avoiding and rejecting the novel food that they perceived to be high in salt and fat (Park, Park, & Choi, 2008).

Despite this, there is still a lack of studies at food neophobia among the young culinarians. Gisslen (2007) argued that to be knowledgeable, highly diversified in their skill the young culinarians are required to learn, prepare and taste enormous types of foods. They are also trained to be the future sensory panels that clearly involved with sensory of taste in evaluating the novel food product for new market entrance (McWilliams, 2010). In this sense, having food neophobia not only might limit young culinarians food preferences, habits, exposure and knowledge of food but creativity or even career progression. Therefore, food neophobia among these individuals is considered as interesting issue to be investigated.

ACKNOWLEDGEMENT

The research is funded by Universiti Teknologi MARA through Research Acculturation Grant.

REFERENCES

Asperin, A.E., Philips, W.J., Wolfe, K. (2011). Exploring Food Neophobia and Perceptions of Ethnic Foods: The Case of Chinese and Thai Cuisines, International CHRIE Conference.

Birch, L.L., & Fisher, J.O. (1998). Development of eating behaviors among children and adolescents, Pediatrics. 101, 539–549.

Choe, J.Y. & Cho, M.S. (2012). Food neophobia and willingness to try non-traditional foods for Koreans, Food Quality and Preference. 22, 671–677.

Falciglia, G.A., Couch, S.C., Gribble, L.S., Pabst, S.M., & Frank, R. (2000). Food neophobia in childhood affects dietary variety, Journal of the American Dietetic Association. 100, 1474–1478.

Flight, I., Leppard, P., & Cox, D.N. (2003). Food neophobia and associations with cultural diversity and socio-economic status amongst rural and urban Australian adolescents. Appetite. 41, 51–59.

Galloway, A.T., Lee, Y., & Birch, L.L. (2003). Predictors and consequences of food neophobia and pickiness in young girls, Journal of the American Dietetic Association.103(6), 692–698.

Gisslen, W. (2007). Professional Cooking. 10th edition. New Jersey: John Wiley & Sons, Inc.

Johnson, L.J., Raab, C., Champaner, E., & Leontas, C. (2002). Chefs' Perception of the Importance of Nutrition in Menu Planning. Pakistan Journal of Nutrition. 1(2), 85–88.

Knaapila, A., Tuorila, H., Silventoinen, K., Keskitalo, K., Kallela, M., Wessman, M., Peltonen, L., Cherkas, L.F., Spector, T.D., & Perola, M. (2007). Food Neophobia shows Heritable Variation in Humans, Physiology & Behavior. 91, 573–578.

Koivisto, U.K., Hursti, & Sjoden, P.O. (1997). Food and general neophobia andtheir relationship with self-reported food choice: familial resemblance in Swedish families with children of ages 7–17 years, Appetite. 29, 89–103.

Koivisto, U.K., & Sjoden, P.O. (1996). Food and general neophobia in Swedish families: parent-child comparisons and relationships with serving specific foods, Appetite. 26, 107–118.

Martins, Y., Pelchat, P.L., & Pliner, P. (1997). 'Try it; it's good and it's good for you': Effects of taste and nutrition information on willingness to try novel food. Appetite. 28, 89–102.

Martins, Y., & Pliner, P. (2005). Human food choices: An examination of the factors underlying acceptance/rejection of novel and familiar animal and nonanimal foods. Appetite.

McWilliams, M. (2010). Foods Experimental Perspectives, 7th Edition. New Jersey: Prentice Hall.

Mustonen, Sari., Oerlemans, P., & Tourila, H. (2012) Familiarity with and affective responses to foods in 8–11-year-old children. The role of food neophobia and parental education. Appetite. 58, 777–780.

Olabi, A., Najm, N.E.O., Baghdadi, O.K., & Morton, J.M. (2009). Food Neophobia levels of Lebanese and American college students, Food Quality and Preferences. 20(5), 353–362.

Park, K., Park, O., & Choi, S. (2008). A survey on the realities of consumers' recognition of Chinese cuisine, Food Service Industry Journal. 4(1), 9–34.

Pliner, P., & Hobden, K. (1992). Development of a scale to measure the trait of food neophobia in humans. Appetite. 19, 105–120.

Pliner, P., & Pelchat, M. (1993). Reduction of Neophobia in Humans by Exposure to Novel Foods, Appetite. 20(2), 111–123.

Pliner, P., & Salvy, S.J. (2006). Food neophobia in humans. In: Shepherd R, Raats M, editors. The psychology of food choice. Oxfordshire: CABI Publishing. 75–92.

Raudenbush, B., & Capiola, A. (2012). Physiological Responses of Food Neophobic and Food Neophilics to Food and Non-Food Stimuli, Appetite. 58(3), 1106–1108.

Rozin, P. (1997). The Use of Characteristic Flavoring in Human Culinary Practice. In C.M. Apt (Ed.) Flavor: Its Chemical, Behavioral and Commercial Aspects. Boulder, CO: Westview Press.

Rozin, P., & Fallon, A.E. (1980). The psychological categorization of foods and non-foods: A preliminary taxonomy of food rejections. Appetite. 1, 193–201.

Schiffman & Kanuk (2010). Consumer Behavior. 10th edition. New Jersey: Prentice Hall.

Stallberg-white, C., & Pliner, P. (1999). The effect of flavor principles on willingness to taste novel foods, Appetite. 33(2), 209–221.

Tourila, H., Lähteenmäki, L., Pohjalainen, L., & Lotti, L. (2001). Food neophobia among the Finns and related responses to familiar and unfamiliar foods, Food Quality and Preference. 12, 29–37.

Utami, H.D. (2004). Consumer behaviour and food processor response towards ethnic food in East Java, Indonesia. Unpublished doctoral thesis, Massey University, Palmerston North, New Zealand.

Zakaria, M.A., Wahab, S.H.A., & Jasmi, M.A. (2010). Culinary Art Education: A Demanding Profession in Culinary Tourism for Malaysian's New Economy Niche. Proceedings of the 1ST UPI International Conference on Technical and Vocational Education and Training Bandung, Indonesia, 10–11 November 2010.

Zielinska, F. B. (2006). The Role of Psychological Factors in Food Choice. Polish Journal of Food and Nutrition Sciences. 15/56(4), 379–384.

Hospitality and Tourism – Sumarjan et al. (eds)
 ISBN 978-1-138-00151-0

Factors attracting Malaysian Heritage Food (HF) consumption amongst international tourists in Malaysia

S.R. Omar & M.S. Ab Karim
Universiti Putra, Malaysia

ABSTRACT: This paper aimed to evaluate Malaysian Heritage Food (HF) as a powerful cultural and heritage component in attracting international tourism to Malaysia. As this paper goes on, the factors lie behind the issues of HF consumption in tourists' perspective, analyzing and discussing the underlying dimensions specifically focused on (a) socio-demographic and (b) motivational factors and (c) culture and religions factor. Predominantly, this conceptual paper attempts to elucidate the arguable findings from previous literatures with the response to the proposed objectives. Also, the factors influencing international tourists' food consumption on HF will be revealed. Conclusion from this study states that HF could play a significant contribution in affecting their food consumption and behavioral intention towards visiting Malaysia. Recommendation on further research for the establishment of Malaysian HF as crucial elements in food tourism industry need to be further investigated.

Keywords: Heritage Food (HF), tourists' behavior, food consumption, tourism destination

1 INTRODUCTION

Tourism is one of the utmost key sectors that have a significant impact towards economy growth for a country (Jalis, Zahari, Zulkifly, & Othman, 2009; Mohd Hairi, Mohd Salehuddin, Muhamad Izzat, & Zulhan, 2007). Among the sub-element in the tourism sector itself, food has becoming an utmost essential component and gaining a fruitful attention in the tourism industry (Hall & Sharples 2003) and has influence the tourists when choosing their destination to visit. Food acts as a significant instrument in promoting a tourism destination.

Gastronomy and tourism have mutual characteristic. Consumption of food among tourists is a crucial component of tourism by contributing significantly in tourism as it is a platform to expose one's cultures, heritage and traditions globally as well as promoting our country as a food image destination worldwide which in turn, could strengthen our economic development. In addition, Cusack (2000) suggested that food could become a great tool in portraying one's identity since heritage and culture elements embedded inside the food itself, supported by Hall and Macionis (1998) whereby wine and food can be expressive of a regional culture as well as a regional environment.

Indeed, there are possibilities to bring Malaysia as a food destination by exposing Malaysian HF to international tourists since there are strong tendency that HF could be experienced by all visitors upon their visit in Malaysia. This phenomenon has raised attention in the government sector. The government sector trying to highlight and to brand the local food as a powerful intangible culture for the purpose of marketing destination. Moreover, previous studies showed that the role of food in expressing Malaysian culture and heritage has always been neglected since they are more focused on tangible elements such as monuments, buildings and the picturesque natural geographical areas. Westering (1999) has coined that the heritage aspects of gastronomy almost been neglected since the cultural aspect outshine more and almost practically observed. Also, research on food consumption, food as a tourism attraction, marketing and promotional strategies regarding HF still incomplete and must be done continuously in ensuring Malaysia could be one of the most popular, established food image destinations worldwide.

Therefore, this paper aims to illuminate an understanding on the factors that could significantly affect the acceptance of HF consumption among international tourists in Malaysia by outlining 3 main objectives: (1) to determine the most significant contributing factor towards HF consumption in international tourism; (2) to examine the relationship of socio-demographic, culture and religions on their HF consumption (3) To evaluate tourists' overall acceptance towards HF consumption.

2 BACKGROUND OF STUDY

2.1 *Heritage food, taste and culture in Malaysia*

Malaysia is a remarkable country, with unique mishmash of rich, cultures make up of multiracial nations. When it comes to food, a variety of cuisines comprises of Malay, Indian, Chinese and Peranakan offers a wide range of colourful, flavourful gastronomical experiences to all local and international tourists. According to Francisco (2012), Malaysian heritage traditional cuisine is a blended of various culture and well-known for its rice, agro-food, seafood, leaves wrapping culture and synonym with the hot and spicy taste. The taste of food, ingredients, appearance and also preference are as a resemblance of multi-cultural Malaysians' people which has their own, unique, authentic value which could portray each culture, and ethnic that lives harmoniously in Malaysia. Jalis et al. (2009) mentioned that Malaysian cooking style is diverse, unique and authentic, mixed up of varieties of spices, herbs and local ingredient which bestow a flavorful spicy, sweet and salty savor.

Apart from that, each culture and has a different manner, beliefs and eating practices which inherited by the younger generation from time to time. Almost all Malays eat the food with their right hand which is similar with the Indians. Apparently, Chinese usually love to use chopstick upon eating their meals (Jalis et al. 2009). In addition, religiosity also reflects Malaysian's food cultural identity. Malays in Malaysia are Muslims, and prohibited to eat pork due to the Islamic rules and regulation (David & Chee-Beng 2001). Meanwhile, Chinese has a balanced eating habit which influenced by Yin-Yang concept; consuming both hot and cold meal creature symbolize their god (D'Sylva & Began, 2011).

The availability of HF is almost everywhere and every times in Malaysia. Recently, National Heritage Department of Malaysia has listed 100 multi ethnic HF that could be found in various places such as at the night market, stalls, Malaysian restaurants in rural areas or urban areas. Malaysian HF could be defined as food which embedded with cultural and historical value which consume daily, but nearly meet the extinction (Elis, 2009) in term of originality and authenticity of the recipes. Among the most popular HF in Malaysia: for example, breakfast time, *Nasi Lemak*, *Roti Canai*, *KuihKaripap* and *TehTarik*. However, some of them are still unsung and nearly meet extinction like *Manokpansoh, Kuihcincin, Nasi tumpang* and *Pulutkukusperiukkera.*

Although food conceded as a central attraction for tourist (Renko 2010), the awareness HF food among international tourists still remain puzzling and confusing. Therefore, marketing and promotional activities of HF should be boost up in the tourism sector in order to safeguard to Malaysian heritage value, and emphasizing the role of HF as a main tourism attraction among international tourists to Malaysia by enhancing the enforcement of HF as one of the outstanding tourist consumption products must be urgently accomplished as suggested by Renko (2010).

2.2 *Heritage food consumption in tourism*

Food consumption could be recognized as a social practice among tourists apart from fulfilling their basic necessity as for supplying energy, satiety and sustenance. It is also portraying the food as a cultural component (Wright et al., 2001) as a symbol, exclusive emerging style of experiencing other peoples' culture, and as a mean for socialization in a different milieu (Mak et al., 2012). Each individual tourist food consumption behavior is kept on changing which they will behave differently, driven by both internal and external forces. Mitchell and Hall (2003) clarified that tourists' decision making towards food consumption and purchasing driven by their attitude, beliefs (internal forces), and also strongly influenced by demographic, origin and their own culture (external forces). In addition, consumer behavior could be seen as a cyclical system, where both individual and collective forces have a mutualistic effect in a particular setting.

Undoubtedly, numerous studies have done by previous scholars regarding food and tourism; local food consumption experience in tourism destination (Mak et al., 2012; Chang et al., 2010; Kim et al., 2009; Okumus et al., 2007; Wee et al., 2011) and food preferences among tourist in tourism destination (Mak et al., 2012, Torres 2002). Robinson et al. (2012) has proven the interrelationship of HF consumption in British's tourism sector. The researcher pervades the tourists' gastronomic experience and motivation in heritage tourism and found that HF could play pragmatic role in promoting a destination. Tirca et al. (2009) claimed that HF should have at least eight rudiments: (1) perceived quality of the product; (2) awareness; (3) customer service attitude; (4) sustainability; (5) extent to which product perceived to be unique or special; (6) convenience; (7) community support and involvement; (8) management commitment and capability in order to be perceived as a potential tourism product. Apart from that, it found that by understanding the tourists' needs and wants, the heritage value could be preserved by injecting some innovation and maintaining the authenticity, historical elements in conjunction with their marketing initiatives. Undeniably, a paradigm in

Scottish Highland has reveal that Scotch whisky heritage value preservation lucratively marketed successfully and managed to come out with a very finest and exclusive and flavorful taste (Tirca et al. 2009).

Whilst there is insufficient literature specifically addressing the issue of HF in Malaysian gastronomy (Zainal et al., 2012, Jalis et al., 2009, Leong et al., 2012, Ab. Karim et al., 2012, Omar et al., 2011, Wee et al., 2011). Findings from previous scholars really aided very much as foundation for establishing the Malaysian HF consumption among international tourists in Malaysia. Due to globalization era, tourists' experiences are valuable as core influencing factor for their next repeat visitation (Arikan & Aksoz, 2008) upon immersing themselves with the authentic, heritage and cultural gastronomy. The globalization impact has uncovered the importance of food consumption experience among travelers (Shenoy, 2005). Hence, tourism industry must grab as much chance to maximize the attractive potential of HF in Malaysia.

2.3 *Factor influencing tourist' food consumption*

Food consumption plays a great role for tourists' experience in a particular destination (Long, 2004). Food is among one of the most promising intangible heritage which easily accessible nowadays (Okumus et al., 2007) and benefiting the economical aspect of a country (Quan and Wang 2004). The tourists' attraction towards food consumption in a destination perceived to be influenced by the reciprocal, warm integrity between host country and incoming tourists, the habit of food consumption and also the adapting process of tourists with the aesthetic eating customs and local mannerism in the tourism destination as reported by Ardabili et al. (2001). Overall, there are three broad perspectives related to food consumption as denoted by Randall and Sanjur (1981) in Mak et al. (2012) which act as intervening variables (Rozin et al. 2006) such as are tourist, food and environment perspectives. According to Mak et al. (2012, there are five potential factors which have a great influence towards food consumption in tourism: (1) Socio-demographic (2) Cultural and religious (3) Exposure effect and experience (i.e. Mere exposure effect, experience, past visitation and globalization; (4) Motivational Symbolic and (5) Food related personality traits as presented in Table 1. This conceptual paper attempted to investigate the factors that impress food consumption issue from tourists' milieu since "every tourist is a voyeuring gourmand" said Lacy and Doughlass (2002) as cited in Cohen and Avieli (2004).

Table 1. Factor influencing food consumption in tourism from tourists' perspective.

Factor	Dimensions
Socio-demographic	Socio-economic status Demographic status Social class Food as social marker Cultural capital
Cultural and religious	National culture Tourist's own food culture Cultural distance Flavor principles Core and peripheral foods Religious beliefs
Exposure effect and past experience	Mere exposure effect experience Past visitation Globalization
Motivational	Symbolic Obligatory Contrast Extension Pleasure
Food related personality traits	Food neophobia Food neophilia Variety-seeking

Source: Adapted from Mak et al., 2012.

2.4 *Proposed framework*

This study is attempt to investigate and compare international tourists' consumption factor specifically only focusing on (a) motivational factor and (b) socio-demographic factor and (c) culture and religious dimensions by Incorporating a theoretical framework introduced by Mak et al. (2012) as shown in Figure 1 due to space limitation.

Socio-demographic factor is very much related to tourists' food consumption pattern. Cultural, socio-demographic, economic background among tourists' have affected the sensory acceptance and buying decision among them in a holiday destination (Mäkelä 2000; prescott et al., 2002). Also, Kim et al. (2009) has contended that age, gender and also education are among the most influential factor in food consumption selection based on several research outcomes related to food consumption in tourism. In term of cultural and religious influence, both exert significant effect on food consumption (Torres, 2002; Tse & Crotts, 2005). For instance, recent study by Chang et al. (2011) found that tourists' awareness and degree of liking on strange food driven by their similarities in the food culture and flavor with their own place of origin. Various studies also has revealed the importance of tourist's travel motivations with regards to food based on their great experience moment

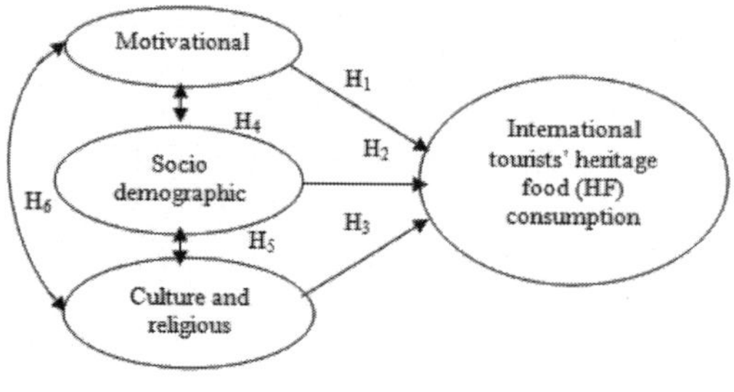

Figure 1. Proposed conceptual framework (Adapted from Mak et al., 2012).

as mentioned by Kivela and Crotts (2006). Satisfaction of local food consumption rely on physiological and psychological factors and exert a significant effect on increasing their motivation to savor local food in a particular tourism' destination areas (Mak et al., 2012). The interrelationship of motivation and culture facets proven by Kim et al. (2009). However, how strong the relationship the three variables towards HF consumption among tourist still unrevealed. Therefore, the hypotheses H1–H6 proposed.

3 PROPOSED METHODOLOGY

In this exploratory study, quantitative approach will be employed by developing survey instruments, operationalization of variables and pre testing the survey. The questionnaire is developed as a practical research instrument by adapting the research framework by Mak et al. (2012) which is to scrutinize the current international tourists towards HF food consumption based on their socio-cultural and psychological factors. A total of 200 self-administered questionnaire prepared in English language will be distributed to international tourists by convenience sampling, who had experienced HF upon their visit to Malaysia. Later on, the data collection will be conducted in Kuala Lumpur International Airport (KLIA) departure hall in October and November 2013. The data will be analyzed through correlation and multiple regression analysis with p-value (0.05) using Statistical Package for the Social Sciences, (SPSS) and *Analysis of Moment Structures* (AMOS).

4 CONCLUSIONS

To put into a nutshell, relevant perspectives on HF consumption behavior among international tourists in Malaysia are paramount important towards establishing HF as a tourism attraction. This study might help food service providers such as restaurateurs, caterers, tourism industry to understand and aware on the influencing factors that affect tourists' consumption behavior towards HF based on their motivation, different socio-demographic, culture and religious background. Moreover, it also very much assist in the economical aspect and help to improve the Gross Domestic Product (GDP) of our country. The research outcome also will manage to fill the gap in food tourism academic field of study.

Apart from that, valorization of HF also should be considered seriously in order to safeguard Malaysian heritage and culture value, and maintain its continuity for the sake of future generation. Further research need to be done to reinforce the role of HF as integral aspects of tourism attraction as a powerful catalyst in promoting Malaysian's tourism worldwide.

REFERENCES

Ab. Karim, M.S., Chua, B.L. & Salleh, H. (2012). Malaysia as a culinary tourism destination: International tourists' perspective. *Journal of Tourism, Hospitality and Culinary Arts*, 4, 65–78.

Ardabili, F.S., Rasouli, E., Daryani, S.M., Molaie, M., & Sharegi, B. (2011). The role of food and culinary condition in tourism industry. *Middle-East Journal of Scientific Research*. 9(6), 826–833.

Arıkan, İ. & Aksöz, O. 2008. An approach for heritage tourism destination development at Alibey (Cunda) Adası, Ayvalık-Turkey, an enterprise Odyssey: Tourism-Governance and Entrepreneurship, *4th International Conference, University of Zagreb*, 1557–1567.

Bessière, J. (1998). Local development and heritage: Traditional food and cuisine as tourist attractions in rural areas. *Sociologia Ruralis*. 38(1), 21–34.

Chang, R.C.Y., Kivela, J. & Mak, A.H.N. (2010). Food preferences of Chinese tourists. *Annals of Tourism Research*, 37(4), 989–1011.

Chang, R.C.Y., Kivela, J. & Mak, A.H.N. (2011). Attributes that influence the evaluation of travel dining experience: When east meets west. *Tourism Management*, 32(2), 307–316.

Cohen, E. & Avieli, N. (2004). Food in tourism. Attraction and impediment. *Annals of Tourism Research*, 31(4), 755–758.

Cusak, I. 2000. African cuisines: Recipes for nation building? *Journal of African Cultural Studies*, 13(2), 207–225.

David Y.H.W. & Chee-Beng, T. (2001). *Changing Chinese Foodways in Asia*. (Eds.). Hong Kong: The Chinese University Press.

D'Sylva, A. & Beagan, B.L. (2011). Food is culture, but it's also power: The role of food in ethnic and gender identity construction among Goan Canadian women. *Journal of Gender Studies*.

Elis, S. (2009). Foreign Tourist not so into nasi lemak. New Straits Times, p. 24.

Francisco, C.B. (2012). Traditional Delicacies in Malaysia You'd Want to Taste. Http://www.brighthubeducation.com/social-studieshelp/123358-traditional-delicasies-of-Malaysia/.

Hall, C.M. & Sharples, L. (2003). The consumption of experiences or the experiences of consumption? An introduction to the tourism of taste. In C.M. Hall, E. Sharples, R. Mitchell, N. Macionis & B. Cambourne (Eds.), *Food Tourism Around the World: development, management and markets* (pp. 1–24). Oxford: Butterworth-Heinemann.

Jalis, M.H., Zahari, M.S., Zulkifly, M.I. & Othman, Z. (2009). Malaysian gastronomic tourism products: Assessing the level of their acceptance among the western tourists. *South Asian Journal of Tourism and Heritage*, 2, 31–44.

Kim, Y.G., Eves, A. & Scarles, C. (2009). Building a model of local food consumption on trips and holidays: A grounded theory approach. *International Journal of Hospitality Management,* 28(3): 423–431.

Kivela, J. & Crotts, J.C. (2006). Tourism and Gastronomy: Gastronomy's Influence on How Tourists Experience a Destination. *Journal of Hospitality and Tourism Research,* 30(3) 354–377.

Leong, Q.L., Othman, M., Mohd Adzahan, N., & Ab. Karim, M.S. (2012). A model of Malaysian food image components: Towards building a sustainable tourism product. *Pertanika Journal of Social Sciences & Humanities,* 20(2), 299–315.

Mak, A.H.N., Lumbersa, M., Evesa, A., & Chang, R.C.Y. (2012). Factors influencing tourist food consumption. *International Journal of Hospitality Management,* 31(3), 928–936.

Mäkelä, J. 2000. Cultural definitions of the meal. In H.L. Meiselman (Ed.), Dimensions of the meal: The Science, Culture, Business, and Art of Eating, 7–18. Gaithersburg, M.L.: Aspen Publication.

Mitchell, R. & Hall, M. 2003. Consuming tourists: food tourism consumer behavior. In M. Hall, L. Sharples, R. Mitchell, N. Macionis, & B. Crambourne. (Eds.), *Food Tourism Around the World* Oxford: Butterworth Heinemann.

Mohd Hairi J., Mohd Salehuddin, M.Z., Muhamad Izzat, Z., & Zulhan, O. (2007). Gastronomic products as tourist motivational factors: An opportunity to Malaysia. *In proceeding of the 2007 Tourism Educators of Malaysia Conference.*

Okumus, B., Okumus, F., & McKercher, B. (2007). Incorporating local and international cuisines in the marketing of tourism destinations: The cases of Hong Kong and Turkey. *Tourism Management,* 28, 253–261.

Omar, M., Mohd Adzahan, N., Mohd Ghazali, H., Karim, R., Abdul Halim, N.M., & Ab. Karim, M.S. (2011). Sustaining traditional food: Consumers' perceptions on physical characteristics of KeropokLekor or Fish Snack. *International Food Research Journal,* 18, 117–124.

Prescott, J., Young, O., O'Neill, L., Yau, N.J.N., & Stevens, R. (2002). Motives for food choice: A comparison of consumers from Japan, Taiwan, Malaysia and New Zealand. *Food Quality and Preference,* 13: 489–495.

Quan, S. & Wang, N. 2004. Towards a structural model of tourist experience: an illustration from food experiences in tourism. *Tourism Management,* 25, 297–305.

Renko, S. 2010. Food in the function of rural development in the context of tourism industry (ICES 2010). Conference Proceedings: *Fifth International Conference of the School of Economics and Business in Sarajevo.*

Robinson, P. (2012) (Ed). Tourism: The Key Concepts. Oxford: Routledge.

Rozin, P. 2006. The integration of biological, social, cultural and psychological influences on food choice. In R. Shepherd & M. Raats (Eds.). *The Psychology of Food Choice:* 19–39. Oxfordshire: CAB International.

Shenoy, S.S. 2005. Food tourism and the culinary tourist. Doctoral Thesis, Clemson University, United States of America.

Tirca, A.M., Stănciulescu, G.C., & A. Chiş. (2009). Heritage tourism—A marketing focused approach. *International Conference "Marketing—From Information to Decision" 2nd Edition* 487.

Torres, R. (2002). Toward a better understanding of tourism and agriculture linkages in the Yucatan: Tourist food consumption and preferences. *Tourism Geographies*, 4(3), 282–306.

Tse, P. & Crotts, J.C. (2005). Antecedents of novelty seeking: international visitors' propensity to experiment across Hong Kong's culinary traditions. *Tourism Management,* 26, 965–968.

Wee, H, Wahid, N.A., Noordin, F., & Mohd Roslan, N.S. (2011). Food tourism in Malaysia, An exploratory Study on Type of factors influencing the international tourists toward local food consumption between two states, *Asia Pacific CHRIE Conference, Hong Kong.*

Wright, L.T., Clive, N., & Kwok P.M.H. (2001). Case study: Food taste preferences and cultural influences on consumption. *British Food Journal,* 103(5), 348–357.

Zainal, A., Zali, A.N., & Kasim, M.N. (2012). Malaysian gastronomy routes as a tourist destination. *Journal of Tourism, Hospitality & Culinary Arts,* 2(16).

Hospitality and Tourism – Sumarjan et al. (eds)
© 2014 Taylor & Francis Group, London, ISBN 978-1-138-00151-0

Guest purchase intention on hotel Room Internet Services

R. Salleh, N.H.C. Ahmat, S.M. Radzi & W.N.B. Wan Kamarudin
Universiti Teknologi MARA, Malaysia

ABSTRACT: Room Internet Services (RIS) have already become one of the hotel in-room facilities. This modern facility has become man's best friend since it helps in most of our work, social relationship and much more. Consequently, many hotels interested to deploy this technology in their hotel. Alas, this technology was too expensive to be set up for most of the hotel. Guest purchase intention has been studied in order to provide a data for the management in solving this crisis. Theory of Planned Behaviour (TPB) was used to identify variables which lead to the guest purchase intention. The result shows that most of the guests agree that RIS is a must for all hotels based on the guest attitude, behaviour control and subjective norms.

Keywords: Room Internet Services (RIS), Theory of Planned Behaviour (TPB)

1 INTRODUCTION

Information technology has become one of the crucial ingredients in assisting management decision making for the hospitality industry (Winata & Mia, 2005). This technology backbone was the internet. The internet nowadays becomes a necessity for most of the people in the world as stated by Suh & Han (2002). People use the internet to help in their work, to communicate with other peoples and to search for information. The importance of the internet has risen the attention for hotel operator as they also endorse it as one of their major departments in their organization structure. The uses of internet in the hotel were vast started from the front to back of the house. The system such as Point of Sales (P.O.S) and Property Management System was an example of technology that harvests the benefit of internet.

Not only that, hotel operator also aware that the internet was also a facility that fulfils the guest need and want. Installing a free WI-Fi in the lobby or public area becomes a common step done by the hotel operator to accommodate guest necessities. Nevertheless, guests always want it more than that. They want it internet as their friends in the rooms as what they have in their own house. This statement further been supported as what been found by Beldona & Cobanoglu (2007) which internet was highly important and highest satisfaction of their respondents. Brewer, Kim, Schrier & Farrish (2008) added that hotel operator may increase their revenue and enhance the guest experience with the application of this facility as it becomes the focus area in the future.

Despite the fact that the internet was important, the study done by the Colliers PKF Survey in 2010 show a backward trend in internet, fax and call revenue. The revenue drop around 86 percent from RM3, 865 (USD1,274) in year 2000 to only RM540 (USD178) in year 2009. To be worse, this figure is not from the USA alone but it was taken from all countries in the world. It becomes bitterer to hotel operator when they have to think of the cost of installing this facility in the room which was not a cheap deal. They major supplier of internet services in Malaysia, Telekom Malaysia revealed the cost of installing this facility were between RM50,000 to RM100,000 depend on number of room, internet speed and internet data usage.

The declining revenue of course will jeopardize the hotel strategic planning in short or even long term. Not only that, guest satisfaction and need also should be taken into consideration if the hotel planning to shut the facility down. To embark upon the crisis, this study was developed to look into guest attitude, behaviour and subjective norms that lead to the purchase intention of this facility. By study the gaps, the data collected hopefully can assist the hotel operator to make a right decision on this important yet inflict room facility.

2 LITERATURE REVIEW

Beldona & Cobanoglu (2007) indicated that home-based technology will change to guest-room technologies in the future. In Leung & Law (2012) study, they agree that information technology has become major attributes on guest selection of hotel. However, Siguaw, Enz & Namasivayam (2000) founded that hotel industry were more attentive in providing technology to employee for their work rather than enhancing the guest experience in the room. These also supported a study done by Ham, Kim & Jeong (2005) who stated that hotel managers believed investing in guest technology only give a minor effect on hotel performance. Furthermore, Singh and Kasavana (2005) noted that guest expect to find technologies in their hotel rooms that mirror those that they use in their daily lives. In summation, the findings reported in the academic literatures suggested that the industry slow to adopt guest-room technologies, because hotel operators and managers perceive that expenditures in the in-room entertainment technology amenities do not have a positive return on the investment (Siguaw et al., 2000; Ham et al., 2005; Singh & Kasavana, 2005).

In line with that, it is a challenge for lodging industry to determine when a potential new in-room technology is sufficiently accepted by the hotel guest. According to Beldona & Cobanoglu (2007), the technology adoption life cycle functions as an important framework in determining the feasibility of the implementation of a technology decision. For instance, high-speed Internet has been a part of the augmented lodging product for some time. However, lately, high-speed Internet access has become a mainstream technology. Hence, hotel operators should consider whether the offering of internet access has in fact crossed the trend in their industry and should therefore become part of their actual product. Given this situation, the way hotels package high-speed Internet access and how they charge for it, will change. As noted in the earlier examples, the cost of televisions in hotel guest rooms has been integrated into room rates, while the status of high-speed Internet is not fully known, it seems likely that soon cost of this service will also become included as part of the room rate (Beldona & Cobanoglu, 2007). This prediction supported the findings by the American Hotel & Lodging Association (2010) Lodging Survey indicates that the number of hotels charging for in-room Internet service has been continually decreasing.

In order to investigate the human behaviour, there was a lot of model been used by researchers but the most basic and famous was the theory of planned behaviour by Azjen. Theory of Planned Behaviour (TPB) is an expansion of Theory of Reason Action (TRA) by Ajzen and Fishbein. The key objective is to identify and forecast the human behaviour. In predicting the human behaviour, TPB looks into these three elements which are Attitude To Purchase (ATP), Subjective Norm (SN) and Behavioural Control (BC). Ajzen (1985) defined ATP as self-estimation on favourable or unfavourable to perform the actions. Whereby for subjective norm, Ajzen describe as the extent to which a person react to social pressure to carry out the particular behaviour. Chen (2007) referred BC as the power to manipulate the individual behaviour. Each element is generated by a number of beliefs and evaluations. They added that purchase intention is resolute by three factors which are ATP, SN and BC and each attributes is in turn produce by amount of values and appraisal. Study by Hyang, Teryy & Rob (2009) founded that all the independent variables (ATP, SN and BC) positively significant with the behavioral intention (dependent variable). Other studies also supported that there are positive relationship between ATP, SN and BC and behaviour intention (Suh & Han, 2002; Celik, 2008; Nor & Pearson, 2008).

3 RESEARCH METHODOLOGIES AND DATA COLLECTION PROCESS

This quantitative study was conducted in Pulau Pinang area which was a third popular tourist destination in Malaysia using self-administrated questioner. Using a non-probability sampling, 377 respondents from 5, 990, 864 total population were sufficient to be analysed based on Krejcie & Morgan (1970). The instrument has gone through a pilot study in UiTM hotel Pulau Pinang before distribute to the sample. The instrument has been separated into three sections which in first section it's inquire about sample demographic profile, followed by measuring dependent variable and in the last part it's developed to measure the guest's intention to purchase RIS. For demographic profile the instrument used closed ended questioner and the rest the instrument used five-point likert scale to measure the dependent and independent variables. The Likert scale ranging from 1-"Strongly Disagree" to 5-"Strongly Agree". The response to the questionnaire was around 65% which were adequate to be used for analysis (Pallant, 2007). The data were coded and keyed in Social Statistical Package (SPSS). The reliability test was then undertaken for every section separately.

4 RESULT & DATA ANALYSIS

Pilot test result shows that the instrument were significant to be used for the study. Subjective norms cronbach alpha value is 0.878, meanwhile attitude to purchase scores is 0.892 and behavioural control score 0.737. The dependent variable in this research, intention to purchase, scores a cronbach alpha value of 0.877. According to Pallant (2007), the scores above 0.7 is considered a reliable instrument with the current sample.

4.1 *Respondent profile*

Based on the percentage, the questionnaire is almost fairly distributed between both genders where 127 of the respondents were male and the balance of 121 respondents was female. From Table 1, it also shows that the majority of the respondents were between 18 to 40 years old with 44 percent were age 18–30 and 45 percent at age 31–40. 9 percent of the respondents at age 41–50 and the balance 2 percent were age 51 years old and above. Age have an impact on the study due to the independent variables which lead to decision on purchasing the RIS. Almost half of the respondents have their Master and PHD which were 118 persons. 93 respondents with a Bachelor Degree represent the second largest sample of the respondents, followed by the Diploma holder with 14 percent and another 1 percent goes to SPM holder. These reflected the level of knowledge among respondents with IT literate which they did know how to used internet. More than half worked in the service industry (55%); followed by 27 percent work in the hospitality industry and retail as at 9 percent. 7 percent in manufacturing industry and 2 percent were not working.

4.2 *Room internet services attributes*

Table 2 shows the result of descriptive analysis on all variables. Subjective norms have a mean score of 3.72 and standard deviation of 0.59. Table 4.6 below show the all the items in subjective norms attributes of RIS. The results show that lowest mean score for SN is $M = 3.05$ with $SD = 1.211$ which is SN4 (I feel uncomfortable to use room internet services provided by the hotel). The highest mean score for SN is $M = 4.17$ with $SD = 0.711$ which is SN2 (I think society is ready to accept the room internet services as a room facility in the hotel). With the mean of 3.05 for SN4, it shows that a large amount of the respondents neither agree nor disagree with the question ask. This happened due to the respondent misunderstanding or overlook on the word of "uncomfortable" in SN4 statement but the majority of the respondent agree that society nowadays already accept and look toward to use internet in their daily life. This can be explained from several researches that support internet will become a basic necessity for most of the hotel guests (Bilgihan, 2009; Beldona & Cobanoglu, 2007).

Table 1. Respondents' demographic profile.

	All	
Variables	n	%
Gender		
Male	127	51
Female	121	49
Age		
18–30 years	109	44
31–40 years	111	45
41–50 years	21	9
51 years and above	7	2
Level of education		
SPM	2	1
Diploma	35	14
Bachelor degree	93	37
Master & PhD	118	48
Working industry		
Manufacturing	17	7
Service	136	55
Retail	23	9
Hospitality	67	27
Others	5	2

Table 2. Descriptive analysis result for all variables.

	Mean	Std. deviation
Subjective norms	3.7195	0.5903
Attitude to purchase	4.0121	0.6604
Behavioural control	4.0524	0.5634
Intention to purchase RIS	3.5066	0.6318

Table 3. Relationship analysis between RIS attributes and guest purchase intention.

	Beta coefficients	Sig. value
Subjective norms	0.010	0.984
Attitude to purchase	0.498	0.000
Behavioural control	0.289	0.000
R^2	0.521	
Adjusted R^2	0.515	
F	88.430	

4.3 *Variables relationship*

From Table 3, all the independent variables have a relationship with the dependent variable. Result of subjective norms was $\beta = 0.001$, $p > 0.984$, attention to purchase was $\beta = 0.498$, $p < 0.0005$ and for behavioural attention was $\beta = 0.289$, $p < 0.0005$. The result shows the attitude that has greater influence on intention to purchase RIS. This finding has the support of other researchers which there is a close relationship between attitude to purchase and behavioural intention to purchase RIS (Tan & Teo, 2000; Brewer et al., 2008). This conclusion we rightly due to the most of the Malaysian people usually have a positive mind-set which eventually will agree to most things that have benefits for themselves.

5 DISCUSSION, RECOMMENDATION & CONCLUSION

Data analysis shown that guest purchase intention has a correlation with RIS attributes which in this case were subjective norms, attitude to purchase and behavioural control. The data show those guests need this facility to be at their hotel room. This RIS were important for hotel guest either for their work or for social communication. The rate of usage, the speed of the internet and the easiness of use become the main factor to evaluate the satisfaction level of the guest when the purchase the RIS.

This finding will have an implication to two both lodging industry and also to the academic world. The first implication will directly to the hotel itself. From the study, most of the respondents did agree and want this type of facility to be provided in the hotel room for their basic necessities. This actually follows the current trends whereby most of the people use internet in daily basis. A literature also adds that almost every person in the world has at least one electronic devices that use the internet. Hence, this finding will help the managerial level to make a decision regarding this in-room internet service. To be added, the study also uncovers that guest would like to have it free and they love it if the hotel operator provides a high-speed internet facility. To those hotel operators which already set up this in-room internet services, it is the best time for them to check again their internet service in term of the speed and charge either it's acceptable or not. By looking into this facility, the hotel operator actually will help them to compete with each other. Those who satisfied their guest will gain a great benefit especially in term of financial matter. The second implications were to the academic world. The study is a rare topic that's being looked by Malaysian researchers which will help the others to come out with the help of this study. Furthermore, it gives a glance to those researchers in Malaysia especially to look into technology in hotel that should be taken into a deep review for satisfaction of the guest.

There are several areas of possibility for future research regarding this topic. This study adopts the Theory of Planned Behaviour (TPB) and its background was to understand the behavioural attributes that affect the intention of guest to do something which in this case were to purchase the RIS. From this study, the data gather show there were slightly little different preferences to purchase in-room internet services for business and leisure traveller. Future research could therefore focus on identifying why such differences are present in the study.

Besides that, future study also can adopt or adapt this study to look into other in-room facility provided by the hotel such as in-room entertainment. This could further the understanding for hotel operator to develop a plan to alter their facility accordance to guest preference. Since the study focus in-room internet, other researcher could look into other area in hotel that usually set-up an internet such as lobby area and dining area. For lobby area which commonly been set-up with the free internet Wi-Fi, other researcher can look the guest level of satisfaction for the internet provided. In dining area which usually hotels are rarely installed the internet; another researcher could use the model used in this study to investigate either the guest were needed this facility or not.

Finally, the result of behavioural intention to purchase the room internet intention is not the only factor to be considered to implement the room internet services. Another factor such as budget, location, availability also needs to take into consideration because this investment needs big amount of money. Besides that, the prediction of future technology trends should be put into consideration because as we all know the technology advancement is very rapid and uncertain which it also can jeopardize the hotel plan for the internet facility.

To wrap up the study, hotels operators should offer this internet service as their basic facility in the room in order to fulfil the guest need and wants since the internet has become one of the basic necessities for most of the people. This technology has helped many of us, whether for human relations or to help in our daily work.

REFERENCES

Ajzen, I. (1985). From intentions to actions: a theory of planned behaviour. Springer, New York, pp. 11–39.

American Hotel & Lodging Association (2010). *2010 Lodging industry survey.* Retrieved Jan 12, 2011, from http://www.ahla.com/.

Beldona, S. & Cobanoglu, C. (2007). Importance-performance analysis of guest technologies in the lodging industry. *Cornell Hotel and Restaurant Administration Quarterly,* 48(3), 299–312.

Bilgihan, F.A. (2009). An analysis of in-room entertainment technologies in hotels. *Journal of Hospitality and Tourism Technology,* 3(1), 24–31.

Brewer, P., Kim, J., Schrier, T.R., & Farish, J. (2008, May). Current and future technology use in the hospitality industry. *American Hotel and Lodging Association & University of Nevada, Las Vegas.*

Celik, H. (2008). What determines Turkish customers' acceptance of internet banking? *Journal of Marketing*, 26(5), 353–370.
Chen, M. (2007). Consumer attitudes and purchase intentions in relation to organic foods in Taiwan: moderating effects of food-related personality traits. *Food Quality and Preference*, 18(7), 1008–21.
Ham, S., Kim, W.G. & Jeong, S. (2005). Effect of information technology on performance in upscale hotels. *Journal of Hospitality Management*, 281–294.
Hyang, J.H., Teryy T.K., & Rob, L. (2009). A comparison of theoretical models for understanding acceptant behaviour of information system in upscale hotels. *International Journal of Hospitality Management,* 2(2), 11–34.
Krejcie, R.V., & Morgan, D.W. (1970). Determining sample size for research activities. *Educational and Psychological Measurement*, 30, 607–610.
Leung, R. & Law, R. 2012. Evaluation of Hotel Technologies and EDI Adoption: The Perspective of Hotel IT Managers in Hong Kong. *Cornell Hospitality Quarterly,* 54(1), 25–37.
Nor, K. & Pearson, J. (2008). An exploratory study into the adoption of internet banking in a developing country: Malaysia. *Journal of Internet Commerce*, 7(1), 29–73.
Pallant, J. (2007). *SPSS Surviving Manual* (3th ed.). Allan & Unwin.
Siguaw, J.A., Enz, C.A., & Namasivay, K. (2000). Adoption of Information Technology in U.S. Hotels: Strategically Driven Objectives. *Journal of Travel Research*, 192–2001.
Singh, A. & Kasavana, M. (2005). The impact of information technology on future management of lodging operations: A Delphi study to predict key technological events in 2007 and 2027. *Tourism & Hospitality Research*, 6(1), 24–37.
Suh, B. & Han, I. (2002). Effect of trust on customer acceptance of internet banking. *Electronic Commerce Research and Applications*, 1(3), 247–263.
Tan, M. & Teo, T.S.H. (2000). Factors influencing the adoption of internet banking. *Journal of AIS*, 1(5), 1–44.
Winata, L., & L. Mia. 2005. Information Technology and the Performance Effect of Managers' Participation in Budgeting: Evidence from the Hotel Industry. *International Journal of Hospitality Management,* 24(1), 21–39.

Hospitality and Tourism – Sumarjan et al. (eds)

Hotel customer loyalty antecedents and behavioral loyalty

A.A. Azdel, N.N. Azizan, M.F.S. Bakhtiar & N.A. Ahmad
Universiti Teknologi MARA, Malaysia

ABSTRACT: In order to ensure the customer will be a loyal customer is not simple and need much effort as different customer may have their own reason on why they become loyal to a certain service provider and perhaps their experience would play a role in transforming them to be a loyal customer. This paper study the perspective of hotel's customer in determining their loyalty as the loyalty behavior of customers will bring enormous profit and prosperity to the hotel's organization. The customers that had stayed and experienced the services at five star hotels in Kuala Lumpur were the target population. Pearson Correlation analysis used to test the relationship, and it showed the relationships between customer loyalty antecedents and behavioral loyalty. All of the findings were statistically significant. This study had shown the consequence of understanding the customer loyalty antecedents as it can give an impact towards retaining the loyal customer.

Keywords: customer loyalty antecedents, behavioural loyalty

1 INTRODUCTION

Hotel industry falls into the category that is intensely competitive as the number of hotels is growing rapidly, and its offer the commodities product and services. This situation happens to the hotel industry all over the world including Malaysia. In order to compete with the other competitors, the hotel itself needs to think wisely on how to make the customer choose their properties and retain them. One of the ways is to create loyalty among the customer to ensure that they will visit the hotel that they are loyal with every time they want to spend their vacation.

Previous study done by Kandampully and Suhartanto (2000) noted that one of the necessary requirements for the future survival in hotel organization is by focusing on customer loyalty. By creating loyalty among the customer, it will strengthen the relationship of the service provider with the customer. According to Mechinda, Serirat and Gulid (2009), the cost to retain the customer is lower than to attract the new customer. Hence, it can help the hotel to reduce their cost in seeking for new customer by just figure out the way to make the customer loyal with them and perhaps delivering memorable experience to their customer. In addition, the same authors also noted that loyal customers can be more profitable than the new customer as they will spend more in the hotel that they are familiar with and less price sensitive. Thus, it can lead to the increasing in the hotel's revenue.

The previous research done by Lau, Akbar and Yong (2005) noted that intense competition in the Malaysia hotel environment had brought an attention among researchers to conduct a research regarding this industry. As defined by Tepeci (1999), understanding the aspect of loyalty is crucial as loyalty will assist hotels to increase their market share and profit. Thus, this study will help the researchers to identify the customer loyalty antecedents that will trigger the customer to be loyal and helps to find ways to tackle this type of customer that will benefit the hotel itself. Therefore, it raised an interest for the researcher to investigate the antecedents from the perspective of the hotel's customer in Malaysia and see whether it will lead to behavioural loyalty. Furthermore, this study will help the hotel industry to understand the customer loyalty antecedents and make use of it to tackle this target customer. Hence, this study will be using different settings from the previous research.

2 LITERATURE REVIEW

2.1 *Customer loyalty antecedents*

All industry is very concern about how to retain their customer or perhaps their supplier and turn them to be a loyal customer. Same guest re-visit the hotel as retaining customer. Customer loyalty is crucial nowadays as the hotel industry is very competitive and offer the same commodities of product and service (Lewis & Chambers, 2000).

Recent study by Huei and Easvaralingam (2011) defined customer loyalty in their study as the willingness of the customer to carry on engaging in the services of a hotel in spite of differences in prices and other factors.

However, the hoteliers did not have specific agreement on what makes the customer being loyal to the hotel, how to raise their loyalty and how to characterize it. Hence, this study had identified customer loyalty antecedents that will help the hoteliers or academia to view that those antecedents will give an input to them in executing their strategies. Previous study done by Blackwell et al. (1999), Mechinda et al. (2009) and Velazque et al. (2011) had stated different dimension of customer loyalty antecedents in their study. The antecedents that had been review and will be discussed in this study are perceived value, familiarity and attachment.

2.1.1 *Perceived value*

According to Bolton and Drew (1991), perceived value can be defined as the difference between quality or benefits that the customer gain and the cost or price that they need to pay. The same authors also stated that the differences can be in term of monetary costs, nonmonetary cost, customer tastes and customer characteristics. While Clemes, Gan and Ren (2011) identified that the value customer perceived will influence the satisfaction level of the customer.

Mechinda et al. (2009) had defined value in four ways which are value is low price, value is anything that the customer wants in a product, value is the quality that the customer will receive when they paid for a certain price and value is what the customer will acquire which means the quality with the price that they gave. Parasuraman and Grewal (2000) stated that perceived value is a function of a 'get' component which means the benefits that a buyer derives when purchase a product or service and a 'give' component that is the buyer's monetary and non-monetary costs of acquiring the product and service.

Apart from that, Cengiz and Kirkbir (2007) noted that customer perceived value is believed to be as one of the most important management practices as it can help to catch the attention of the customer after quality, improve their satisfaction and loyalty. However, Woodruff and Gardial (1996) argued that the un-dimensional is lack of validity and Petrick (2004) add up that from the perceived value, the customer might have a shared meaning of value. So, it had led to the various perceptions from different authors regarding the perceived value of a customer.

2.1.2 *Attachment*

Attachment of the hotel and customer makes the customer feel that they belong to the hotel. As being defined by Williams, Patterson, Roggenbuck and Watson (1992), the authors had divided the place attachment to the two dimensions which are place identity and place dependence. The authors stated that place identity is the attachment of a place that will create emotional bond between customer and particular setting.

Conversely, place dependence is more likely to be defined as a functional meaning it refers to the specific functions and conditions of a place that compulsory to satisfy the needs and goals of a customer. So, for this study the particular setting will be the hotel itself as the customer will perceived whether the hotel that has stayed had given a feeling of attachment for them or not. Hou, Lin and Morais (2005) mentioned that the customer self-identity and bonding with a particular place will lead to a feeling of attachment. In addition, the attachment is also able to link psychologically of customer with the destination and from that the customer will feel the sense of secure, trust, confidence and attractiveness (Alexandris, Kouthouris & Meligdis, 2006).

2.1.3 *Familiarity*

Familiarity is being defined by Moorthy, Ratchford and Talukdar (1997) as the customer perception on how much he/she knows about the place that means the hotel. Hu and Ritchie (1993) noted that familiarity can manipulate customer's perception and attractiveness of the particular place. Apart from that, Bettman and Park (1980) identified that when the customer familiar with a place, it will help in the decision making process. Consequently, it will ease the customer to select a hotel and being loyal to them as they feel that if they well known about a hotel, they will perceived less risk to being dissatisfy with the hotel.

Black, Lockett, Ennew, Winklhofer and Mckechnie (2002) had make a suggestion in the research that lack of familiarity might lead to the high perception of risk and it might head to the possibility that the customer will not purchase the product or service. Besides that, less familiarized customer might give an impact to the reputation of the product as they will use external information resources to hear about the product whereby the more familiarized customer will decide according to their satisfaction or usability of a product and service (Abu-Ghazzeh, 1996).

2.1.4 *Relationship between customer loyalty antecedents and behavioural loyalty*

Latest study by Khan (2012) also noted a truly loyal customer must not return to the firm or had a sense of behavioural loyalty but at the same time, the customer will feel emotionally attached that he/she felt no other company can fulfil his/her needs and consequently resist them to switch to another providers. Previous study done by Reichheld and Sasser (1990) pointed out the loyal customers had acted as marketing agents for the company by recommending the company to others.

On the other hand, Soderlund (2002) found the high familiarity customer will tend to repurchase and create word of mouth to be compared with low familiarity customer. It was because the customer that already familiar with the service provider knows well about the service offered and will not hesitate to come again and recommend to others. The above statement had been supported by the study done by Chi (2012).

2.2 *Methodology*

This research adopts the descriptive design research, and conducted in quantitative manner because it targeted to examine and measure the relationship of hotel customer antecedents and behavioral loyalty of hotel customers. A quantitative approach was used to gather the data by distribution of questionnaire to the sample. It involves gathering numerical data using structured questionnaires to collect primary data from individuals. The research took place at the 5 stars hotels around Kuala Lumpur. The customers that had stayed and experienced the services at five star hotels in Kuala Lumpur were the target population. From the recent record, there are 25 hotels in Kuala Lumpur are 5-stars rating (Tourism Malaysia, 2012). The researchers choose the convenience sampling and distributed 400 questionnaires to 25 hotels that involved in this study.

3 FINDINGS

3.1 *Response rate*

The response rate for this study was only 42%. In spite of that, there was no missing value as the researcher approached the respondent directly and waited for each respondent answering their questionnaire. On the other hand, since the number of respondents that had been obtained for this study were not achieved the required sample size, the researcher had run a test to check the power of test for this study and see whether the data was sufficient to run for further analysis or not. G*Power 3 develop by Faul, Erdfelder, Lang and Buchner (2007) as a test of statistical power analysis for the social, behavioural, and biomedical sciences. The rule of thumb for power of test is more than 0.8 and then the study can proceed for future analysis. After the researcher had tested the analysis using G*Power, the result for power of test was 1.0. Hence, further analysis can be done from this point onward.

3.2 *Respondent's background profile*

Among the 169 respondents who answered the questionnaires, it appears that majority (57.4%) of them were female while almost half (42.6%) were men. Meanwhile, most of the of them (38.5%) were in the age of 20 to 30 then followed by the respondents that were in age of 31–40 that contributed 30.8%. The respondents in the age of 41 to 50 and 51 to 60 contributed to the same percentages which were 13% respectively. The rest of the respondents were above the age of 60 or below 20 (4.1% and 0.6% accordingly). For the annual income, majority of the respondents (33.7%) received more than RM60001 for their annual income followed by 14.8% of them had the annual income at the level of RM40001 to RM50000. Subsequently, the respondents who received the annual income of RM20001 to RM30000 and RM30001 to RM40000 were holding the same percentage which was 14.2% while 12.4% of the respondents' annual income was less than RM20000. Minority of the respondents had the annual income of RM50001 to RM60000. Almost half (47.9%) of them stayed at the five stars hotel for holiday purposes. Other than that, the respondents who stayed for the purpose of business contributed 45.0% followed by the respondents stayed for visiting friends and relatives which was 7.7%. This study also showed that minority (5.3%) of the respondents stayed at five stars hotel in Kuala Lumpur for honeymoon.

3.3 *Pearson correlation of hotel customer antecedents and behavioral loyalty*

According to Table 1, there was a significant correlation between perceived value and

Table 1. Result of pearson correlation for customer loyalty antecedents (perceived value, attachment, familiarity) and behavioural loyalty.

	Behavioural loyalty
Perceived value	
Pearson correlation	0.760*
Sig. (2-tailed)	0.001
Attachment	
Pearson correlation	0.779*
Sig. (2-tailed)	0.001
Familiarity	
Pearson correlation	0.560*
Sig. (2-tailed)	0.001
Customer loyalty antecedents	
Pearson correlation	0.824*
Sig. (2-tailed)	0.001

*Correlation is significant at the 0.05 level (2-tailed).

behavioural loyalty, $r = 0.760$; $p < 0.001$. Moreover, the result also showed that there was a positive relationship too between attachment and behavioural loyalty, $r = 0.779$; $p < 0.001$. The two variables were indicating very strong relationship with behavioural loyalty. Meanwhile, the correlation between familiarity and behavioural loyalty presents that there was a strong relationship, $r = 0.560$; $p < 0.001$ for those variables.

For the combination of the variables of customer loyalty antecedents (perceived value, attachment and familiarity), the results indicate that there was a significant correlation between customer loyalty antecedents and behavioural loyalty (sig. value = 0.001 < 0.05). This was supported by the very strong positive coefficient of correlation value, $r = 0.824$.

4 CONCLUSION

4.1 *Summary of the findings*

The purpose of the first hypothesis is to find the relationship between customer loyalty antecedents which are perceived value, attachment and familiarity and behavioural loyalty. The finding showed the positive relationship between customer loyalty antecedents as a whole and also for each dimensions towards behavioural loyalty.

The finding was in line with the previous research done by Veloutsou, Gilbert, Moutinho and Good (2005) as the authors noted in their findings that loyal customers had make a contribution to company prosperity via repeat purchasing and recommending about the organization to other consumers and it showed the similarity of the results. Therefore, the antecedents from the customer loyalty were positively lead to the behavioural loyalty as when the customer perceived that it was worth to stay at the hotel with the value that they offered, felt attached and be familiar at the same time, consequently the customer will not be indecisive to come again to the hotel. What is more, the customer will also recommend to their friend and relative and spread positive word of mouth about the hotel.

Meanwhile, perceived value also significance towards behavioural loyalty and the result were aligned with previous study done by Hutchinson et al. (2009) and Meng et al. (2011). Attachment also had relationship towards behavioural loyalty and the result was supported by current study conducted by Lee et al. (2012) as the authors indicated in their research regarding festival satisfaction and destination loyalty that attachment can lead to the loyalty. In addition, the result of this study also parallel with study done by Yuksel, Yukseland Bilim (2010).

Customer's familiarity also showed the positive relationship with behavioural loyalty. The finding was parallel with the study done by Chi (2012) as when the customer are familiar with a place, they will feel secure, comfortable and would likely to come again.

However, for this study the most important attribute or the strongest one was attachment followed by perceived value and the least important attribute was familiarity. The distinction might be because the previous study was confined to tourist whereby this study was focused towards hotel's customer. For tourist, it was important to be familiar with the destination and felt attached as it will ensure that they will revisit the destination again.

Meanwhile, from the point of view of hotel's customer, familiarity was a small issue and they regard the attachment towards hotel was essential as they were likely to have emotional attachment with the hotel and means a lot for them. The different setting and situation had caused the result to be differed from previous study.

4.2 *Implication*

This study had shown the consequence of understanding the customer loyalty antecedents as it can give an impact towards retaining the loyal customer. It can be doubted that there are other antecedents besides these three antecedents that had been discussed in this study that also gave an impact towards loyalty. However, the contribution of the results was very essential as it provides the useful information to the hoteliers on how they can make use of these antecedents in retaining their customer. Among these three antecedents, attachment seem to be the most valuable antecedents that been selected by the respondents. Hence, the managerial can create the right strategy to boost the attachment value to their customer.

The hoteliers must keep in their minds that they cannot treat customer the same and anticipate to receive the same evaluation of satisfaction from them as different customer have their own expectation and level of satisfaction.

On the other sides, for the academician, this study had helped to identify the antecedents of customer loyalty and see whether it really contribute to the behavioural loyalty. The findings of this study might vary for other studies. Thus it can help the academicians to improve this model and add up any variables that are appropriate to get the desirable result.

4.3 *Limitations and recommendations for future research*

This study is also confined to local customer perspective. In spite of that, the future research should study the view of foreign customer and distinguish the distinction between local and foreign customer perspective. Thus, it will help the hoteliers and academicians to acknowledge that there might be a distinction in term of local and foreign customer point of view. In addition, there were only three antecedents of customer loyalty discussed throughout this study. So, the future research can add up any variable that is suits with the model to get better result.

ACKNOWLEDGEMENT

This research is funded by Universiti Teknologi MARA, under the Research Acculturation Grant Scheme (RAGS).

REFERENCES

Abu-Ghazzeh, T.M. (1996). Movement and way finding in the King Saud University built environment: A look at freshman orientation and environmental information. *Journal of Environmental Psychology, 16*, 303–318.

Alexandris, K., Kouthouris, C., & Meligdis, A. (2006). Increasing customers' loyalty in a skiing resort: the contribution of place attachment and service quality. *International Journal Contemporary Hospitality Management, 18*(5), 414–425.

Bettman, J.R., & Park, C.W. (1980). Effects of Prior Knowledge and Experience and Phase of the Choice Process on Consumer Decision Processes: A Protocol Analysis. *Consumer Research, 7*(3), 234–248.

Black, N., Lockett, A., Ennew, C., Winklhofer, H., & McKechnie, S. (2002). Modelling consumer choice of distribution channels: An illustration from financial services. *The International Journal of Bank Marketing, 20*(4), 161–173.

Blackwell, S.A., Szeinbach, S.L., Barnes, J.H., Garner, D.W., & Bush, V. (1999). The Antecedents of Customer Loyalty: An Empirial Investigation of the Role of Personal and Situational Aspects on Repurchase Decisions. *Journal of service research, 1*(4), 362–375.

Bolton, R.N., & Drew, J.H. (1991). A Multistage Model of Customers' Assessments of Service Quality and Value. *Consumer Research, 17*, 275–284.

Cengiz, E., & Kirkbir, F. (2007). Customer perceived value: the development of a multiple item scale in hospitals. *Problems and Perspectives in Management, 5*(3), 252–286.

Chi, C.G.Q. (2012). An examination of destination loyalty differences between first-time and repeat visitors. *Journal of Hospitality & Tourism Research, 36*(1), 3–24.

Clemes, M.D., Gan, C., & Ren, M. (2011). Synthesizing the effects of service quality, value and customer satisfaction on behavioral intentions in the motel industry: An empirical analysis. *Journal of Hospitality and Tourism Research, 35*(4), 530–568.

Faul, F., Erdfelder, E., Lang, A.-G., & Buchner, A. (2007). G*Power 3: A flexible statistical power analysis program for the social, behavioral, and biomedical sciences. *Behavior Research Methods, 39*, 175–191.

Hou, J., Lin, C., & Morais, D.B. (2005). Antecedents of Attachment to a Cultural Tourism Destination: The Case of Hakka and Non-Hakka Taiwanese Visitors to Pei-Pu, Taiwan. *Journal of Travel Research, 44*(4), 221–233.

Hu, Y., & Ritchie, J.R.B. (1993). Measuring Destination Attractiveness: A Contextual Approach. *Travel Research, 32*(2), 25–34.

Huei, C.T., & Easvaralingam, Y. (2011). Perceptions of Service Quality, Corporate Image and Customer Loyality in the Hotel Industry of Malaysia. *International Research Symposium in Service Management*, 126–135.

Hutchinson, J., Lai, F., & Wang, Y. (2009). Understanding the relationships of quality, value, equity, satisfaction, and behavioral intentions among golf travelers. *Tourism Management, 30*(2), 298–308.

Kandampully, J., & Suhartanto, D. (2000). Customer loyalty in the hotel industry: the role of customer satisfaction and image. *International journal of contemporary hospitality management, 12*(6), 346–351.

Khan, O. (2012). Towards understanding customer loyalty: An empirical study on emotional attachment. *International Journal of Innovations in Business, 1*(3), 241–267.

Lau, P.M., Akbar, A.K., & Yong, G.F.D. (2005). Service Quality: A study of the luxury hotels in Malaysia. *The Journal of American Academy of Business, Cambridge, 7*(2), 46–55.

Lee, J., Kyle, G., & Scott, D. Scott. (2012). The Mediating Effect of Place Attachment on the Relationship between Festival Satisfaction and Loyalty to the Festival Hosting Destination. *Journal of Travel Research, 51*(6), 754–767.

Lewis, R.C., & Chambers, R.E. (2000). *Marketing leadership in hospitality*. New York: John Wiley & Sons.

Mechinda, P., Serirat, S., & Gulid, N. (2009). An examination of tourists' attitudinal and behavioral loyalty: comparison between domestic and international tourists. *Journal of Vacation Marketing, 15*(2), 129–148.

Meng, S.M., Liang, G.S., & Yang, S.H. (2011). The relationships of cruise image, perceived value, satisfaction, and post-purchase behavioral intention on Taiwanese tourists. *African Journal of Business Management, 5*(1), 19–29.

Moorthy, S., Ratchford, B.T., & Talukdar, D. (1997). Consumer information search revisited: Theory and empirical analysis. *Consumer Research, 23*(4), 263–277.

Parasuraman, A., & Grewal, D. (2000). The impact of technology on the quality–value–loyalty chain: A research agenda. *Journal of the Academy of Marketing Science, 28*(1), 168–174.

Petrick, F. (2004). First timers and repeaters perceived value. *Journal of travel research, 43*, 29–38.

Reichheld, F.F., & Sasser, W.E. (1990). Zero defections: Quality comes to services. *Harvard Business Review, 68*(5), 2–9.

Soderlund, M. (2002). Customer familiarity and its effect on satisfaction and behavioral intentions. *Psychology & Marketing, 19*(10), 861–880.

Tepeci, M. (1999). Increasing brand loyalty in the hospitality industry. *International Journal of Contemporary Hospitality Management, 11*(5), 223–229.

Tourism Malaysia (a). (2012). Accomodation List. Retrieved 16th May 2012 from http://www.tourism.gov.my/accommodations/list.php?start=10&keyword=&acc_type=&star_rating=5&price_range=&state=kl&sort_by=&current_sort.

Velázquez, B.M., Saura, I.G., & Molina, M.E.R. (2011). Conceptualizing and measuring loyalty: Towards a conceptual model of tourist loyalty antecedents. *Journal of vacation marketing, 17*(1), 65–81.

Veloutsou, C., Gilbert, R.G., Moutinho, L.A., & Good, M.M. (2005). Measuring transaction specific satisfaction in services. *European Journal of Marketing, 39*(5–6), 606–628.

Williams, D., Patterson, M., Roggenbuck, J., & Watson, A. (1992). Beyond the commodity metaphor: Examining emotional and symbolic attachment to place. *Leisure Sciences, 14*, 29–46.

Woodruff, R.B., & Gardial, S.F. (1996). *Know your customer: New approaches to understanding customer value and satisfaction*. Cambridge: Blackwell Publishers.

Yüksel, A., Yüksel, F., & Bilim, Y. (2010). Destination Attachment: Effects on Consumer Satisfaction and Cognitive, Affective and Conative Loyalty. *Tourism Management, 31*(2), 274–284.

Hospitality and Tourism – Sumarjan et al. (eds)
© 2014 Taylor & Francis Group, London, ISBN 978-1-138-00151-0

The influence of reward criteria towards the effectiveness of loyalty program in Kopitiam restaurant

Z.N. Nadiah, S.A. Talib & M.F. Mohd Faiz
Universiti Teknologi MARA, Malaysia

ABSTRACT: As the supply of food and beverage increased, the need to attract new customer and to retain of them are needed in order for food establishment to sustain and succeed. In line with that, a successful customer loyalty program can further strengthen the relationship between existing customers, lengthening customer life cycle, and increasing its share of customer expenditure. This study therefore tries to fill the gap in the literature by looking at the relationship between type and timing of reward towards the effectiveness of loyalty program in a restaurant. The result from this study can be considered to be significant as they were intended to have an understanding of which dimensions under the reward program that can improve the business performance.

Keywords: loyalty program, type of reward, timing of reward

1 INTRODUCTION

Loyal customers offer businesses a steady customer base, more frequent purchase cycles, higher profit margins, and a group of advocates who voluntarily market the firm to potential customers (Buchanan and Gilles 1990; Reichheld and Sasser 1990; Rigby, Reichheld, & Schefter 2002). For these reasons, academics have spent decades attempting to explain the mechanisms that affect consumers' loyalty, while practitioners have developed and applied numerous strategies focused on developing a loyal customer base. Among these efforts, one of the most significant and innovative breakthroughs in loyalty management occurred in 1981, when American Airlines launched AAdvantage, the first contemporary customer reward program, with the goal of increasing repeat purchases.

In the global marketplace, competitions for customers are stiff. Company need to look for new ways to attract customers. Meeting the diverse customer demand to achieve customer loyalty is an important issue for foodservice management today (Byron & Sharp, 1997). Customer loyalty becomes possible through the development of long term, mutually beneficial relationships with customers (Keh & Lee, 2006). One of the most valuable marketing strategies that executives can apply is a customer loyalty program, which is a promotional plan designed to increase loyalty by providing incentives to customers through added benefits (García Gómez, Gutiérrez Arranz, & Gutiérrez Cillán, 2012).

Customer loyalty programs are a widely used marketing tool to increase sales and maintain a stable market share (Henderson, Beck, & Palmatier, 2011). Previous studies regarding loyalty program focused more on the other field of service such as airlines and retail establishment (Kwong, Soman, & Ho, 2011). An increasing number of foodservice establishments are embracing the concept of loyalty program and becoming a main idea in attracts customers. Past studies mainly focus on the perspective of retail establishment.

In relation with the above issue, consumers now more attracted to dine in the Kopitiam restaurant. Thus, the concept of loyalty program can lead to potentially attract new customer and to retain them order for food establishment to sustain and succeed. The idea hit it off immediately and now customer can see these new age Kopitiam spreading through the power of franchising. The Kopitiam whirl windswept Malaysia since the middle of the last decade and currently captions has ventured overseas in China and Indonesia (Cheng, 2010).

This is how the evolution of the coffee culture in Malaysia is, from the long standing years of Kopitiam that our grandparents and their parents used to enjoy, though now declining from its previous glories, to the coffee giants invasion to Malaysia where loads of youth couldn't resist. Due to this notion, Kopitiam restaurant has potential in implementing the concept of loyalty program to attract the consumer.

2 LITERATURE REVIEW

2.1 *Kopitiam restaurant*

A Kopitiam or kopi tiam is a traditional coffee shop found in Southeast Asia, patronised for meals and beverages. The word kopi is Malay for coffee (as borrowed and altered from English) and tiam is the Hokkien/Hakka dialect word for shop. Menus typically feature simple offerings: a variety of foods based on egg, toast, and kaya, plus coffee, tea, and Milo, a malted chocolate drink which is extremely popular in Southeast Asia, particularly Singapore, Malaysia and Riau Islands (Cheng, 2010).

Each shop has its own unique flavour and personality, but there are certain similarities shared by most. The atmosphere in a kopitiam is uniquely Malaysian, meaning that it blends several different cultures. One similarity is the kopitiam table and chairs. In general, the tables are marble topped and the chairs are made of wood with round backs. There are other common features of the decor that identify this particular type of coffee shop, even to someone who has never been inside one (Cheng, 2010). The development of kopitiams is in many ways closely related to the Hainanese migrants. The early Hainanese migrants who came to South East Asia in 1850's and as late migrants, the Hainanese was forced to find employment in less profitable trades. By the time the Hainanese arrived, the Hokkiens and Teochews, for instance, were already well established in agriculture, commerce and trade (Food Blog Malaysia, 2010).

As fate has it, the Hainanese crafted a place for them in the service sector, working as cook boys, waiters or servants in the local hotels, restaurants, bakeries and bars and as cooks or domestic servants for wealthy European and Peranakan households. At one time, some also worked as canteen operators in the military bases. Some of them became seamen and sailors and worked as cooks on board the ships. Later, many Hainanese became involved in the kopitiam or "coffee shop" business as stall holders or assistants (Food Blog Malaysia, 2010). In fact, the Hainanese community has been credited with introducing the kopitiam culture to Malaysia and Singapore. Many old Hainanese Kopitiam operators recalled that in its hey days in the 1950s to 1970's captions were very popular and caters to almost The popularity of the old-fashioned outlets along with society's obsession with nostalgia and increasing affluence has led to the revival of these captions. The revival of the traditional culture started in the early 2000's with the introduction of new age kopitiams (Food Blog Malaysia, 2010).

The revival of the old Kopitiam culture started in the early 2000s with the introduction of new age kopitiams. The new kopitiams are fast-food outlets which are reminiscent of the old kopitiams in terms of decor, but are usually built in a more modern, hygienic setting such as a shopping mall rather than in the traditional shop house, catering mainly for young adults. Modern kopitiams have retained the old-fashioned marble-topped tables, wooden chairs and chunky crockery, but now they are to be found in the cavernous air-conditioned shopping malls (Cheng, 2010).

2.2 *Loyalty program*

In the 1970s European researchers studying business-to-business marketing discovered that suppliers who formed close working relationships with their customers tended to have better customers. That is, they were more loyal to their suppliers and often gave these suppliers a greater share of their business. The customers also reported having better suppliers. In short, it was a win-win arrangement.

For a company to practice loyalty marketing however, it is first necessary to know who the loyal customers are. This is a lot easier for smaller than larger businesses. As the number of customers increases, database marketing and market research have to be used in the absence of personal knowledge. Each year, as computerized database technology becomes more sophisticated, so does a firm's ability to monitor its customers' behavior. These techniques are being used by retailers and packaged-goods manufacturers to help efficiently target products and services, and allocate marketing resources to achieve the maximum return (Dowling and Uncles, 1997).

In times of severe competition, a loyalty program, which is often called a reward program, is usually introduced to build customer loyalty through the planned reward scheme based on a customer's purchase history. The goal of a loyalty program is to establish a higher level of customer retention in profitable segments by providing more satisfaction and value to certain customers (Bolton, Kannan, and Bramlett 2000). Thus, it becomes a part of the value chain or points of product differentiation. Although loyalty programs are widely in use, little empirical research has investigated whether the loyalty program is indeed perceived as valuable to the customer and whether it actually contributes to building brand loyalty.

In line with previous research (Sharp & Sharp, 1997; Yi & Jeon, 2003), loyalty program defined as an integrated system of marketing actions that aims to make member customers more loyal. In line with that, Yi and Jeon stated that loyalty program is a marketing program that is designed to build customer loyalty by providing incentives to potential customers.

2.3 *Reward criteria*

Dowling and Uncles (1997) classified reward programs into either direct rewards or indirect rewards. Direct rewards are rewards that directly support the value proposition of a given product whereas indirect rewards refer to incentives that are not relevant to a given product (other types of reward that have no linkage with the product/service) (Mattila, 2001). The timing of reward is also classified into two categories: immediate versus delayed rewards. Immediate rewards can be seen as rewards given for every visit, whereas delayed rewards are rewards provided for every visit (Yi & Jeon, 2003).

The potential of a loyalty program to attract members depends not only on the value of its rewards but also on when the rewards are available. (Sunny Hu, 2010) illustrated immediate and delayed rewards are used for timing of redemption. The immediate rewards are psychological, namely, a feeling of participation, the anticipation of future rewards, and a sense of belonging reward (Uncles, Dowling, & Hammond, 2003). The reward with a discount feature gives price discounts on certain items of the assortment for loyalty program members only.

Discount feature supplies member customers with immediate rewards for their purchases (Sunny Hu, 2010). The delayed rewards are a bimonthly summary of accumulated points and sometimes the qualification for a reward (Uncles et al., 2003). The reward with a saving feature gives loyalty program members saving points, dependent on the monetary amount spent in the company (Mägi, 2003).

3 ISSUES

Previous studies regarding loyalty program focused more on the other field of service such as airlines and retail establishment (Kwong, Soman, & Ho, 2011). An increasing number of foodservice establishments are embracing the concept of loyalty program and becoming a main idea in attracts customers.

However, all of the above studies have been conducted in the Western countries. Limited studies have been conducted in Asian countries especially in Malaysia (Leenheer, Heerde, Bijmolt & Smidts, 2007). Further, this study will address the rarity of the research on the loyalty program looking at the dimension of type of reward and timing of reward. Past studies mainly focus on the perspective of retail establishment.

A lot of the study regarding loyalty program has been conducted in the retail field management. The food service industry should also conduct a similar study to improve customer loyalty. Thus, this study is designed to engage the loyalty program through the approach of type of reward and timing of reward in Kopitiam. It is important for us to know in what degree type and timing of reward hold the influence over the overall loyalty program in Kopitiam due to the booming of this type of food service establishment. This study holds the potential to enable the foodservice operators to adapt the right measure in further improving their loyalty program.

REFERENCES

Buchanan, R, & C. Gilles. (1990). Value managed relationship: The key to customer retention and profitability. *European Management Journal, 8*(4), 523–526.

Byron Sharp, A.S. (1997). Loyalty programs and their impact on repeat-purchase loyalty. *International Journal of Research in Marketing, 14*, 473–486.

Food Blog Malaysia (2010). Kopi Tiam, The Original Starbucks? Retrieved from http://www.feedmelah.com/.

García Gómez, B., Gutiérrez Arranz, A.M., & Gutiérrez Cillán, J. (2012). Drivers of customer likelihood to join grocery retail loyalty programs. An analysis of reward programs and loyalty cards. *Journal of Retailing and Consumer Services, 19*(5), 492–500.

Henderson, C.M., Beck, J.T., & Palmatier, R.W. (2011). Review of the theoretical underpinnings of loyalty programs. *Journal of Consumer Psychology, 21*(3), 256–276.

Keh, H., & Lee, Y. (2006). Do reward programs build loyalty for services? The moderating effect of satisfaction on type and timing of rewards. *Journal of Retailing, 82*(2), 127–136.

Kwong, J.Y.Y., Soman, D., & Ho, C.K.Y. (2011). The role of computational ease on the decision to spend loyalty program points. *Journal of Consumer Psychology, 21*(2), 146–156.

Lai Ah Eng. (2010). The Kopitiam in Singapore: An Evolving Story about Migration and Cultural Diversity. *Asia Research Institute.* Working paper series no. 132. Retrieved from http://www.ari.nus.edu.sgdocswpswps10_132.pdf.

Leenheer, J., van Heerde, H.J., Bijmolt, T.H.A., & Smidts, A. (2007). Do loyalty programs really enhance behavioral loyalty? An empirical analysis accounting for self-selecting members. *International Journal of Research in Marketing, 24*(1), 31–47.

Mägi, A.W. (2003). Share of wallet in retailing: the effects of customer satisfaction, loyalty cards and shopper characteristics. *Journal of Retailing, 79*(2), 97–106.

Mattila, A.S. (2001). Emotional bonding and restaurant loyalty. *Cornell Hotel and Restaurant Administration Quarterly, 42*(6), 73–79.

Ou, W.-M., Shih, C.-M., Chen, C.-Y., & Wang, K.-C. (2011). Relationships among customer loyalty programs, service quality, relationship quality and loyalty: An empirical study. *Chinese Management Studies, 5*(2), 194–206.

Reichheld, F., and W. Sasser. (1990). Zero defects: Quality comes to services. *Harvard Business Review,* 70, 105–111.
Roehm, M.L., E.B. Pullins, and H.A. Roehm Jr. 2002. Designing loyalty building programs for packaged goods brands. *Journal of Marketing Research,* 39, 202–213.
Rosenbaum, M.S., A.L. Ostrom, and R. Kuntze. (2005). Loyalty programs and a sense of community. *Journal of Services Marketing,* 19, 222–233.
Sunny Hu, H.-H.H., Chun-Te Chen, Po Tsang. (2010). Do reward programs truly build loyalty for lodging industry? *International Journal of Hospitality Management, 29*(1), 128–135.
Uncles, M.D., Dowling, G.R., & Hammond, K. (2003). Customer loyalty and customer loyalty programs. *Journal of Consumer Marketing, 20*(4), 294–316.
Yi, Y., & Jeon, H. (2003). Effects of Loyalty Programs on Value Perception, Program Loyalty, and Brand Loyalty. *Journal of the Academy of Marketing Science, 31*(3), 229–240.

Hospitality and Tourism – Sumarjan et al. (eds)
© 2014 Taylor & Francis Group, London, ISBN 978-1-138-00151-0

Customer purchase intention towards organic foods

M.S.M. Nur Hidayat & M.S. Fadzilah
Universiti Technologi MARA, Malaysia

ABSTRACT: Organic foods are products that were produced without the aid of fertilisers or pesticides and with minimal process in order to maintain the integrity of the foods. Recently, knowledge about organic foods starts to bloom in the market and among customer, but many have the perceptions that organic foods are too expensive, limited availability, unsatisfactory quality, limited choice and lack of perceived value. Customer purchase intention will be quantitatively explored through self-administrated questionnaire and will be distributed to the organic food consumers in the supermarkets located in Bangsar Selangor. The data collected will be analysed using Statistical Package for Social Science (SPSS) software. The study discusses on the customer purchase intention towards organic foods by reviewing the past studies and it attributes and relates the issues to the contextual setting where the research is undertaking.

Keywords: health conscious, organic foods, price, purchase intention, quality

1 INTRODUCTION

The story about health conscious is not new, as indeed, it always becomes the importance to human, as foods are the central axis for main supply. As stated by Teng, Rezai, Mohamed and Shamsudin (2012), food is the basic needs for human being to survive therefore the consumers have the right to pick good and safe foods for their consumption. Selecting food products are easy task however there are several aspects must be concern, such as the safety of the food, hygiene, food quality and food storing area.

According to Teng et al. (2012), green foods concept refer to foods that are safe to be consumed, fine quality and nutritious food produces under the principle of sustainable development. It can be divided into two categories. First category allows the use of certain amount of chemicals and in second category which is also known as organic foods. The term "organic" is rooted in "bio" from Greek "*bios*" meaning life or way of living, and no clear standard and definition found for "natural food" (Essoussi and Zahaf, 2008). 'Organic' refers to products that are cultivated without the aid of fertilisers or pesticides, however not all natural foods can be classified as organic (Smith and Paladino, 2010).

Nowadays, the sales of organic foods are growing due to customer's health and food safety concern. Consumer also found that organic foods are healthier than conventional food products. This has been redirected in increasing demand for organic product, which is perceived as less damaging to the environment and healthier than conventionally grown foods (Saba and Messina, 2003). This is strengthened in the study of Zhen and Mansori (2012), where in Malaysia food industry, health consciousness and organic foods become new important trend in the past decades.

The study of Australians' organic food beliefs, demographics and values by Lea and Worsley (2005), shown that, sales of organic foods have increased considerably over recent years. Global sales of organic foods and drinks reached around US$19 billion, representing a US$3 billion increase from the previous year. In Malaysia the total of land area planted with organic crops has increased from 130 hectare in 2001 to more than 2400 hectare in 2007 (Dardak, Zainol Abidin and Ali, 2009). It is estimated that between 60–90% of Malaysian organic products are imported, with the main imported was fresh organic products are vegetables and fruits.

2 LITERATURE REVIEW

In order to discover why customers purchase organic foods, it is important to understand the purchasing trends in purchasing and the reason on what motivate them to purchase organic foods. It will fulfil the need and demand of the customer. As stated in Yakup and Jablonsk (2012) in order to fulfil the needs, there are five different levels in Maslow's hierarchy of needs; physiological needs, security needs, social needs, esteem needs and self-actualizing needs. The writer further stated that, the customer are not aware of their buying behaviour are mostly shape by the psychological needs.

According to Magistris and Gracia (2008), consumers decide to purchase organic foods under asymmetric information due to the deficiency of information about organic food production. Thus, they cannot clearly differentiate the unique qualities of organic from conventionally grown substitutes. As shown in the study of Essoussi and Zahaf (2009), consumers purchase organic foods because it is seen as healthier with no chemicals substance are used, tastes better than conventional food, and is environmental friendly. Kriwy and Mecking (2012) stated that, the organic foods have a relation with ecological and health motives and it is the main factors discussed as predictors of attitudes and consumption.

2.1 *Product knowledge*

There also some reasons why people or consumer not motivated to purchase organic foods. Lack of knowledge about the benefits of the products will not lead them to consume the organic foods. Aertsens, Mondelaers, Verbeke, Buysse and Huylenbroeck (2011) strengthen that, lack of knowledge and awareness is considered to be the main factor for consumers not purchases organic food in the USA—59 per cent of respondents point out that they have never considered organic products because they did not know about it.

Consumer value is the difference between what a customer obtains from a product, and what he/she has to provide in order to acquire it. Hence, consumer value helps in motivating the consumer whether to purchase organic food or not. Consumer value supports marketers to ascertain the elements on acceptability in order to motivate consumer to purchase organic food (Zhen and Mansori, 2012).

2.2 *Attitude of consumer towards organic foods*

Customer will give a positive attitude when asking about the organic food products. Radmans (2005) stated that, in the past study consumers had a positive attitude towards organic foods. This is strengthened by Aertsens et al. (2011) said that, amongst adolescents of 15–16 years that more knowledge results a positive attitude towards organic food, just making them read a definition of the organic foods. The reasons for their purchase, because of the organic foods are more nutritious than conventional foods.

The reasons for customer purchases, because of the organic foods are more nutritious than conventional foods. Other researchers also stated that attitude towards the customer purchase intention (consumer behaviour) related with the customer positive or negative perception about the specific products. Massive intentions to purchase organic foods are owing to the high level of positive attitude towards purchase behaviour (Zhen and Mansori, 2012).

According to Babicz-Zielinska (2006), the concern about healthy life seems to be the main reason for purchase of organic foods. The author convinced that, the purchase reasons are because of the higher value of health, no colouring substances, pesticides, artificial flavours or preservatives, environment-friendly and better taste. This also stated in Organic Trade Association (2009) nearly three-quarters (73 per cent) of U.S. families buy organic products at least occasionally, chiefly for health concern.

The impact of customer purchase intention is also owing to the higher educational level that leads to the higher percentage of organic foods purchases. Magnusson et al. (2001) and Wier and Calverley (2002), stated that, people with higher educational level seems to be more willing to pay more for organic foods. Tsakiridou, Boutsouki, Zotos and Mattas (2008) strengthened that, people with high education level tend to express positive attitudes towards organics food products; require lots of information about the production and process methods of organics have the confidence to negotiate conflicting claims in relation to organic food and willing to purchase for a premium price of the organic foods.

Mutlu (2007) in a study on consumer attitude and behaviour towards organic food found that regular organic consumer seems to have positive product knowledge and buying behaviour. On the other hand education was found to be a significant factor of attitudes towards organic food and consumer with high education level (Tsakiridou et al., 2008). Organic food had a positive influence on the customer attitude towards and consumption of organic if the customer has high awareness and knowledge about the products (Pieniak, Asetsens, and Verbeke, 2010). Kriwy and Mecking (2012) assert that, higher educational level of a customer increase the level of health investment and them also more efficient in health production. They also stressed that, many organic consumer likely have a higher education background.

3 ISSUE

Organic foods start to burgeoning in markets and among customer, yet some of them are still suspicious about the nutritious content and unconvinced about the quality of the organic foods. This is strengthened when Essoussi and Zahaf (2008), stressed that, organic foods are too expensive, limited availability, unsatisfactory quality, satisfaction with current purchases, lack of trust, limited choice, lack of perceived value and lack of understanding of organic ways of productions. According to Zhen and Mansori (2012), maintaining the organic foods quality can emphasizing the consumer opinion to purchase and at the same times counter the negative perception such as high price of organic foods.

In line with that people are sceptical whether organic products really give a positive effect on their health and quality of life (Dardak et al., 2009). In 2006, Saher, Lindeman and Hursti asserted that there are lacks of scientific support for the common beliefs that organic foods would be healthier or nutritious rather than the regular foods. Other researcher stated that healthiness has become an important criterion for food purchases and now consumers have becoming more conscious about the nutrition, health, and quality of the food they eat (Magnusson, Arvolla, Hursti, Aberg and Sjoden, 2001; Wier and Calverly, 2002; Chen, 2009; Kriwy and Mecking, 2012). The authors further stated that, health concern is the commonly stated motive that lead customer to purchase organic food products.

Owing to the certain circumstances, some of the customers are not going to purchase the organic foods because of the higher price. Lili and Tong (2007) assert that, the customer willing to pay more for organic food products if the price is cheap. Most of the consumers concern when they spend their money on organic foods because they do not know whether it is worth spending (Zhen and Mansori, 2012). In addition they also stated that, suppliers of organic foods are not certified by local or international organic foods certificate because they are complying with high price of standard of qualification and they are protected by organic food certification. With all the resources support the quality, health and prices are the obstacles on the customer purchase towards organic foods. The underlying reasons of the highlighting issues are still under investigation.

REFERENCES

Aertsens, J., Mondelaers, K., Verbeke, W., Buysse, J., & Huylenbroeck, G.V. (2011). The influence of subjective and objective knowledge on attitude, motivations and consumption of organic food. *British Food Journal, 113*(11), 1353–1378.

Ahmad, S.N.B., & Juhdi, N. (2010). Consumer's perception and purchase intentions towards organic food products: exploring the attitude among Malaysian consumers. Unpublished doctoral dissertation, Tun Abdul Razak University.

Babicz-Zielinska, E. (2006). Role of psychological factors in food choice—a review. *Polish Journal of Food and Nutrition Sciences, 15/56*(4), 379–384.

Chen, M.F. (2009). Attitude toward organic foods among Taiwanese as related to health consciousness, environmental attitudes, and mediating effects of a healthy lifestyle. *British Food Journal, 111*(2), 165–178.

Dardak, R.A., Zainol-Abidin, A.Z., & Ali, A.K. (2009). Customer' perceptions, consumption and preference on organic product: Malaysian perspective. *Economic and Technology Management Review, 4,* 95–107.

Essoussi, L.H., & Zahaf, M. (2008). Decision making process of community organic food consumers: An exploratory study. *Journal of Consumer Marketing, 25*(2), 95–104.

Essoussi, L.H., & Zahaf, M. (2009). Exploring the decision-making process of Canadian organic food consumers: Motivations and trust issues. *Qualitative Market Research: An International Journal, 12*(4), 443–459.

Kriwy, P., & Mecking, R.A. (2012). Health and environmental consciousness, costs of behaviour and the purchase of organic food. *International Journal of Consumer Studies, 36*, 30–37.

Lea, E.,& Worsley, T. (2005). Australians' organic food beliefs, demographics and values. *British Food Journal, 107*(11), 855–869.

Lili, Z. & Tong, C. (2007, March 8–10). Consumer perception of organic food in Urumqi. Contributed paper prepared for presentation at the 105th EAAE Seminar 'International Marketing and International Trade of Quality Food Products', Bologna, Italy.

Magistris, T., & Gracia, A. (2008). The decision to buy organic food products in Southern Italy. *British Food Journal, 110*(9), 929–947.

Magnusson, M.K., Arvola, A., Hursti, U.K.K., Aberg, L., & S, P.O. (2001). Attitudes towards organic foods among Swedish consumers. *British Food Journal, 103*(3), 209–227.

Mutlu, N. (2007). Consumer attitude and behaviour towards organic food: Cross cultural study of Turkey and Germany. Unpublished doctoral dissertation, Stuttgart-Hohenheim University.

Pieniak, Z., Aertsens, J., & Verbeke, W. (2010). Subjective and objective knowledge as determinants of organic vegetables consumption. Food Quality and Preference 21, 581–588.

Radman, M. (2005). Consumer consumption and perception of organic products in Croatia. *British Food Journal, 107*(4), 263–273.

Saba, Anna., & Messina, F. (2003). Attitudes towards organic foods and risk/benefit perception associated with pesticides. *Food Quality and Preference, 14*, 637–645.

Saher, M., Lindeman, M., & Hursti, U.K. (2006). Attitudes Towards Genetically Modified and Organic Foods. *Research Report, 46*, 324–331.

Smith, S., & Paladino, A. (2010). Eating clean and green? Investigating consumer motivations towards the purchase of organic food. *Australasian Marketing Journal, 18*, 93–104.

Teng, P.K., Rezai, G., Mohamed, Z., & Shamsudin, M.N. (2012). Factors influencing public intention towards purchasing green food in Malaysia. *International Journal of Sustainable Development, 4*(2), 51–60.

Tsakiridou, E., Boutsouki, C., Zotos, Y., & Mattas, K. (2008). Attitudes and behaviour towards organic products: an exploratory study. *International Journal of Retail and Distribution Management, 36*(2), 158–175.

Wier, M., & Calverley, C. (2002). Market potential for organic foods in Europe. *British Food Journal, 104*(1), 45–62.

Yakup, D., & Jablonsk, S. (2012). Integrated approach to factors affecting consumers purchase behavior in Poland and an empirical study. *Global Journal of Management and Business Research, 12*(15), 61–87.

Zhen, J.S.S., & Mansori, S. (2012). Young female motivations for purchase of organic food in Malaysia. *International Journal of Contemporary Business Studies, 3*(5), 61–72.

Hospitality and Tourism – Sumarjan et al. (eds)
 ISBN 978-1-138-00151-0

The effect of brand image on purchase intention mediates by use experience towards Furama hotel

P.S. Jalal, N.A. Ishak, Z. Mohi & N.S. Shariffuddin
Universiti Teknologi MARA, Malaysia

ABSTRACT: This research attempts to study the effect of brand image on purchase intention mediates by use experience towards Furama brand. The purpose of this study is to incorporate the core brand image, brand awareness and brand preferences with brand experience to testify the effect on the consumer purchase intentions.

Keywords: brand image, brand awareness, brand preference, use experience, purchase intention

1 INTRODUCTION

Marketing is one of the most important tools to boost up sales in an organization and without good marketing strategy there is nothing more rather than wish and hope for miracle to come. Hotel without advertising is like breathe in the mud. The first step to build advertising objectives is by creating brand awareness towards customer hence before creating a great impression for guest to purchase they have to be aware of the brand and the meaning. The common used of marketing message that delivered through media are to communicate the brand name and to highlight the important messages tied to hotel product. Wake people up helps drive traffic to business thus create a buzz in the market feedback. Thus enable them to resolve problems quickly and work to improve future brand performance.

Same goes to Furama Hotel which located nearby the golden triangle which known as Kuala Lumpur commercial, shopping and entertainment hub and actually a large area that encompasses a number of major streets which are Jalan Imbi, Jalan Sultan Ismail and Jalan Raja Chulan. There are a lot of competitor surrounding and hotel is growing everywhere doesn't matter in what rating they are as long they can offer good price and of course with a good marketing strategy they can survive. Furama Hotel tries to compete the market of the Four-star hotels. But, Furama hotel is still in the beginning path to local people which they never heard about this brand, more over the brand awareness still not in their mind. Thus, brand awareness towards local people can affect the hotel performance in term of their sales as not many people noticed about Furama Hotel. Therefore, marketing team should play a big role in promoting Furama to public.

2 ISSUE

Furama Hotel is under rating of Four-star that have been given by Ministry of Tourism Malaysia in 2012. The minimum requirements for Four-star rating of hotel to have a business centre, room for the disabled, banquet/conference hall with adequate conference facilities and services and so forth. Furama hotel has set their target market by putting the corporate guest as their target but unfortunately until this tense they could not reach their corporate guest as a target market of the hotel due to some circumstances that lead to the matter. Currently Furama hotel just cater guests from group which this is out of their target and aims. It shows that they are lacking in promoting their brand image and marketing strategies of the hotel.

In line with the previous notion, marketing and sales department should revise their tactics in attracting and convinced guest to stay and experience the services because good tactics will create the repeated guest and with more guest hotel will get more recommendation based on word of mouth. In addition, with the sophisticated technology nowadays potential target just need to use their fingertips to search for information but it depends on how good the marketing team promotes and attract people to come. Marketing and sales department should aggressively improves and encounter the problem of their team if they want to survive in hotel industry and also to earn maximum hotel revenue.

Due to the lack of marketing strategy, some of the reason has been identify which contribute to this problem which is the brand awareness where guest/individual might not recognize this hotel brand due to the new market entrant also the hotel weaknesses in advertise the hotel, even though the hotel has been operated almost two years but they still need an improvement. Furthermore, unattractive marketing strategy could lead to failure of an organization. In addition, inconsistency of brochure regarding the hotel information keep changing all the time it's like a wasting of money because there are like promoting something without confident. As a Four-star hotel they should compete with other Four-star hotel to grab an opportunity of getting as much guest as can or else they should change their target market.

Consumers are able to recognize a product, evaluate the quality, lower purchase risks, and obtain certain experience and satisfaction out of product differentiation through brand image. When it comes to experiential product evaluation, a positive brand image may make up for an inferior image of the origin country and raise the possibility of the product being selected (Thakor & Katsanis, 1997). Besides, a brand that has some level of brand awareness is far more likely to be considered, and therefore chosen, than brands which the consumer is unaware of. Additionally, the strength of awareness of the brands within the consideration set can also be significant. Wilson (1981 cited in Woodside & Wilson, 1985) confirmed the importance of top-of-mind awareness in a study which found that the higher the position of the brand in the consumer's mind measured by unaided recall, the higher the purchase intention and the higher the relative purchase of the brand. Furthermore, brand preference is closely related to brand choice that can simplify consumer decision making and stimulate brand purchase intention. However, forecasting consumer's preferences between brands is not an easy task.

Therefore, it is worth to investigate and revealed the causation. This study will be carried out in order to strengthen the theoretical and empirical the role of brand awareness and brand preference towards purchase intention.

3 LITERATURE REVIEW

3.1 *Brand image*

According to Hsieh, Pan and Setiono (2004, p. 252), "a successful brand image enables consumers to identify the needs that the brand satisfies and to differentiate the brand from its competitors, and consequently increases the likelihood that consumers will purchase the brand". A company or its product/services which constantly holds a favorable image by the public, would definitely gain a better position in the market, sustainable competitive advantage, and increase market share or performance (Park, Jaworski, & MacInnis, 1986). Keller (1993, p. 3) defined brand image as "the perceptions about a brand as reflected by the brand associations held in consumer memory". In other words, brand image is what comes to the mind of the consumer when a brand name is mentioned (Arslan, 2010). A similar definition to Keller's (1993) was proposed by Aaker (1991, p. 109), whereby brand image is referred to as "a set of associations, usually organized in some meaningful way".

Wu and Lo (2009) portrayed researches on a larger scale suggesting the major components of core-brand image are awareness and preference for the brand. The two components of core brand image; awareness for the brand and preference for the brand has an impact on core-brand image and is positively associated with the attitude of the core-brand (Wu & Lo, 2009). Moreover for the effective branding, relevance and awareness both are vital (Sevier, 2001). Bogart and Lehman (1973) explained that brand awareness came into being from the big consists of dissimilar parts and dependent upon extension to different fields of advertised messages. Furthermore, attributes (the thing which exists and can be distinguished from each other) that are rational might be in true sense viewed as "causes" to the image of the brand rather saying than the image itself (Da Silva & Alwi, 2006).

3.1.1 *Brand awareness*

Rossiter and Percy (1987) defined brand awareness as being vital for the communications process to occur as it precedes all other steps in the process. No other communication effects can happen without brand awareness. For a potential target to buy a brand they must first be aware of the brand. Brand attitude cannot be formed, and intention to buy cannot occur unless brand awareness has occurred (Rossiter & Percy, 1987; Rossiter et al., 1991). In memory theory, brand awareness is positioned as an important first step in building the "bundle" of associations which are attached to the brand in memory (Stokes, 1985). The brand is conceptualized as a node in memory which allows other information about the brand to be "anchored" to it (Aaker, 1991b). The conceptualization of a network of brand associations in memory with the brand as a central core has been put forward by many others (e.g. Holden, 1993; Holden & Lutz, 1992; Keller, 1993).

In addition, brand awareness refers to the consumers' ability to recall and recognize the brand.

Brand awareness is more than just consumers knowing the brand name and having previously seen the brand, perhaps even many times. Brand awareness also involves concerning the brand name, logo, and symbol to certain associations in memory. Brand awareness can be classified in two key dimensions: Depth of brand awareness refers to how easily customers can recall or recognize the brand; breadth of brand awareness refers to the range of purchase and consumption situations in which the brand comes to mind. Preferably, a brand would have both depth and breadth of brand awareness.

3.1.2 *Brand preference*

Several brands may have the visibility and credibility to be considered, but to be selected the brand has to be preferred to others. It usually means beating the competition in at least one of the dimensions that define the product category and by being at least as good as the other compared brands in the rest of the dimensions (Aaker, 2011). For this reason, marketers develop strategies to increase the probability that a brand will be activated from consumers' memories and also be included in the consumer's consideration set. One mean to achieve this can be heavy expenditures on advertising campaigns in order to create "top-of-mind-awareness" and thus, increase the probability of a purchase (Peter & Olsson, 2008).

3.2 *Use experience*

Use experience is associated but different from some of the popular concepts in branding literature, such as brand awareness, image, attitude, credibility and consumer based brand equity. In fact it is the acceptance of its distinct conceptualization by both the academicians and practitioners that has facilitated its practical implementation and raised its status as one of the most essential aspects of a firm's marketing mix strategy (Brakus et al., 2009; Chang & Chieng, 2006; Choi et al., 2011). Hence, it is not startling to note that crafting and providing best consumer experiences has become one of the strategic objectives of most renowned brands in the world.

Use experience is described as "subjective, internal (sensations, feelings, and cognitions) and behavioral responses evoked by brand related stimuli that are part of a brand's design and identity, packaging, communications, and environments" (Brakus et al., 2009, p. 53). Although the term use experience encompasses all types of consumer experiences, emotional, contextual, symbolic and non-utilitarian are the aspects that are receiving greater attention in recent times (Arnould & Thompson, 2005; Brakus et al., 2009; Hultén, 2009; Zarantonello & Schmitt, 2010).

If a consumer considers his past purchases and experiences for deciding his future purchases, it can better predict his purchase intentions for future purchases (Conner & Armitage, 1998). This phenomenon is based on behavioral learning theories (Bentler & Speckar, 1979). If consumer gets benefits by purchasing a product or service then it is sure that consumer will purchase again and also tell other about their positive experience to others (D'Souza et al., 2006). Consumers' use experience towards brand mediates the impact of purchase intention in brand on customer.

3.3 *Brand image and use experience*

There are numerous studies indicating that the accumulation of brand image and use experience is key determinant of attitude towards core-brand attitude (Alpert & Kamins, 1995; Carpenter & Nakamoto, 1989; Ghen & Liu, 2004; Kardes & Kalynaram, 1992; Martinez & Chernatony, 2004). This indicates that core-brand attitude can be influenced by core-brand image and use experience of core-brand. Brand image that consists of brand awareness and brand preference play an important role in the success of hotel organization. How the potential target would like to create the brand awareness and brand preference if they did not experience the products and services. Hence, the experience staying at the hotel will activate from consumers' memories and make them to recall and recognized the brand.

3.4 *Purchase intention*

Purchase intention can be describe as individual's intention to buy a specific brand individuals who want to buy a specific brand which they has chosen for themselves after certain evaluation; there are variables by which we can measure purchase intention for instance consider the brand for purchasing and expecting to purchase the brand in the future (Laroche & Sadokierski, 1994; Laroche & Zhou, 1996; MacKenzie & Belch, 1986).

Intent to purchase is a kind of decision in which studied why a customer purchases a brand in particular. Constructs like considering something purchasing a brand and anticipating to purchase a brand aids to scope the intentions of purchasing (Porter, 1974). Porter (1974) also elaborated customers' intention to purchase a focused brand is not merely by his same brand attitude, but also by his attitudes leading to other brands in choice of set considered.

3.5 *Use experience and purchase intention*

The influence of marketing strategy in attract guest will lead to use experience of core-brand. Smith and Park (1992) stated that products can be divided into (a) product use experience and (b) product availability for visual evaluation. They found that when products were categorized under product experience, consumers were inclined to utilize their experience of the core-brand during product quality evaluation on frequency, actual purchase behavior and satisfaction after use.

Furthermore, when guest attracted to thing for example the service, environment and the lovely room for sure the guest will have intent to purchase the product. Thus, when it comes to the actual situation where guest already check in and attempt to the products and services, experience it, guest will have their own impression. Hence, the use experience will lead either the consumers have intent to purchase the product or not.

3.6 *Brand image and purchase intention*

Brand image can provide a sustainable competitive advantage that can moderate the impact of price and other product attributes on consumers' purchase decisions. Brand image is essential in products and services purchase behavior because it impacts consumers' preferences and purchase intentions as well as their willingness to pay a premium price and recommend the brand to others. Besides, brand awareness is defined as the extent to which how consumer can easily memorize the brand when he is going to buy a product. For the formation of brand awareness, brand image is an essential condition (Keller, 1993). To evaluate brand awareness brand recall is a tool by which we can easily measure and predict brand awareness (Esch et al., 2006).

When consumers strongly recommend the brand, and have unique relations with a brand, that generates positive attitude toward the brand (Keller, 2003). We consider brand image as an important factor because it adds to the consumer's decision process either he should purchase the brand or not (Dolich, 1969). Brand image has a great effect on consumers' next behavior (Fishbein, 1967; Johnson & Puto, 1987), and have great contribution in the brand equity (Biel, 1992). Furthermore, consumer's core-brand attitude will affect their intention to purchase extended products (Aaker & Keller, 1990; Faircloth et al., 2001; Miller et al., 1971; Nan, 2006). For this reason, the current study seen that brand image that consists of brand awareness and preference will play a vital role and will affect the purchase intention of potential target.

4 CONCLUSION

The importance of this study is to identify the effect of core-brand image that contribute to the purchase intention towards Furama hotel brand mediates by use experience. Besides, to find out the most importance caused that lead to the unsuccessful of Furama in promoting their brand in to public. Furthermore, this study will help to increase the knowledge of marketers on how to craft and execute the strategy in order to ensure that consumers/individual aware about the brand that will lead to the purchase intention. In addition, the manager will know what the consumers' preference is. Thus, marketing team will get a good number of profits to sustain the business and compete with other hotel as well. However, the highlighting issue is still under investigation.

REFERENCES

Aaker, D.A. (1991). Managing Brand Equity: Capitalizing on the Value of a Brand Name. *Free Press*, New York, NY.

Aaker, D.A. (1991b). Managing Brand Equity: Capitalizing On the Value of a Brand Name. New York: *The Free Press*. Aaker (1991). 109.

Aaker, D.A. & Keller, K.L. (1990). Consumer Evaluations of Brand Extensions. *Journal of Marketing,* 54. 27–41.

Aaker, D.A. (2011). Preferences vs. Relevance. *Marketing News*, February 28.

Alpert, F.H. & Kamins, M.A. (1995). An Empirical Investigation of Consumer Memory, Attitude, and Perceptions toward Pioneer and Follower Brands. *Journal of Marketing*, 59. 34–45.

Arnould, E.J. & Thompson, C.J. (2005). Consumer Culture Theory (CCT): Twenty Years of Research. *Journal of Consumer Research*, 31. 868–82.

Arslan, F.M. (2010). The Effect of Brand Extensions on Product Brand Image. *Journal of Product & Brand Management*, 19/3. 170–180.

Bentler, P. & Speckar, G. (1979). Models of Attitude-Behavior Relations. *Psychological Review*, 86(5). 452–64.

Biel, A. (1992). How Brand Image Drives Brand Equity. *Journal of Advertising Research*, 32(6). November/December. RC6–RC12.

Bogart, L. & C. Lehman, 1973. What Makes a Brand Name Familiar? *J. Market. Res.* 10(1). 17–22.

Brakus, J.J., Bernd H., & Zarantonello, L. (2009). Brand Experience: What Is It? How Is It Measured? Does It Affect Loyalty? *Journal of Marketing*, 73(3). 52–68.

Carpenter, G.S. & Nakamoto, K. (1989). Consumer Preference Formation and Pioneer Advantage. *Journal of Marketing Research*, 26. 285–98.

Chang, P., & Chieng, M. (2006). Building Consumer-Brand Relationship: A Cross-Cultural Experience View. *Psychology & Marketing,* 23(11). 927–59.

Choi, Y.G., Ok, C. & Hyun, S.S. (2011). Evaluating Relationships among Brand Experience, Brand Personality, Brand Prestige, Brand Relationship Quality, and Brand Loyalty: An Empirical Study of Coffee House Brands. *Available At: Scholarworks.umass.edu* (accessed 14 July).

Conner, M. & Armitage, C. (1998). Extending the Theory of Planned Behavior: A Review and Avenues for Further Research. *Journal of Applied Social Psychology*, 28(15). 1429–64.

Da Silva, R.V. & Alwi S.F.A. (2006). Cognitive, Affective Attributes and Cognitive Behavioral Responses in Retail Corporate Branding. *J. Prod. Brand Manage*, 15(5). 293–305.

Dolich, I.J. (1969). Congruence Relationships between Self Images and Product Brands. *Journal of Marketing Research*, Vol. 6 No. 1, February, pp. 80–4.

D'Souza, C., Taghian, M., Lamb, P. & Peretiatkos, R. (2006). Green Products and Corporate Strategy: An Empirical Investigation. *Society and Business Review*, 1(2). 144–57.

Esch, F. & Langner T., Bernd H. Schmitt, Patrick Geus. (2006). Are Brands Forever? How Brand Knowledge and Relationships Affect Current and Future Purchases. *Journal of Product & Brand Management*, 15(2). 98–105.

Faircloth, J.B., Capella, L.M. & Alford, B.L. (2001). The Effect of Brand Attitude and Brand Image on Brand Equity. *Journal of Marketing Theory and Practice*, 9(3). 61–75.

Fishbein, M. (1967). Attitude and the Prediction of Behavior. *In Fishbein, M. (Ed.), Attitude Theory and Measurement*. John Wiley & Sons, *New York*, NY. 477–92.

Ghen, K.J. & Liu, G.M. (2004). Positive Brand Extension Trial and Choice of Parent Brand. *Journal of Product and Brand Management*, 13(1). 25–36.

Holden, S.J.S. (1993). Understanding Brand Awareness: Let Me Give You A Clue!. *Advances in Consumer Research*, 20. 383–388.

Holden, S.J.S. & Lutz, R.J. (1992). Ask Not What The Brand Can Evoke; Ask What Can Evoke the Brand. *Advances in Consumer Research*, 19. 101–7.

Hsieh, M.H., Pan, S.L., & Setiono, R. (2004). Product, Corporate, and Country-Image Dimensions and Purchase Behavior: A Multi-country Analysis. *Journal of the Academy of Marketing Science*, 32(3). 251–70.

Hulten, B. (2009). Sensory Marketing: The Multi-Sensory Brand-Experience Concept. European Business Review, 23(3). 256–73.

Johnson, M.D. & Puto, C.P. (1987). A Review of Consumer Judgement and Choice. In Houston, M. (Ed.), *Review of Marketing, American Marketing Association*, Chicago, IL. 236–92.

Kardes, F.R. & Kalyanaram, G. (1992). Order-Of-Entry Effects on Consumer Memory and Judgment: An Information Integration Perspective. *Journal of Marketing Research*, 29. 343–57.

Keller, K.L. (1993). Conceptualizing, Measuring, and Managing Customer-Based Brand Equity. *Journal of Marketing*, 57. 1–22.

Keller, K.L. (2003), Strategic Brand Management: Building, Measuring and Managing Brand Equity, 2nd Edition. *Prentice-Hall*, Englewood Cliffs, NJ.

Laroche, M. & Sadokierski, R.W. (1994). Role of Confidence in A Multi-Brand Model of Intentions for A High Involvement Service. *Journal of Business Research*, 29(1). 1–12.

Laroche, N., Kim, C. & Zhou, L. (1996). Brand Familiarity and Confidence as Determinants of Purchase Intention: An Empirical Test in a Multiple Brand Context. *Journal of Business Research*, 37(10). 115–20.

MacKenzie, S.B., Lutz, & Belch, G.E. (1986). The Role of Attitude toward the Ad as a Mediator of Advertising Effectiveness: A Test of Competing Explanations. *Journal of Marketing Research*, 23(2). 130–43.

Martinez, E. & de Chernatony, L. (2004). The Effect of Brand Extension Strategies upon Brand Image. *Journal of Consumer Marketing*, 21(1). 39–50.

Park, C.W., Jaworski, B.J., & Maclnnis, D.J. 1986). Strategic Brand Concept-Image Management. *Journal of Marketing*, 50(4). 135–45.

Peter, J.P. & Olsson, J.C. (2008). Consumer Behavior and Marketing Strategy. *New York*: McGraw-Hill Book Company.

Porter, M.E., 1974. Consumer Behavior, Retailer Power and Market Performance in Consumer Goods Industries. *Rev. Econ. Stat.*, 56(4). 419–36.

Rossiter J.R. & Percy L. (1987). Advertising and Promotion Management. *Singapore*: McGraw-Hill.

Rossiter J.R., Percy L. & Donovan R.J. (1991). A Better Advertising Planning Grid. *Journal of Advertising Research*. 11–21.

Sevier, R.A., 2001. Brand as Relevance. *J. market. Higher Educ.*, 10(3). 77–97.

Smith, D.C. & Park, C.W. (1992). The Effects Of Brand Extensions On Market Share And Advertising Efficiency. *Journal of Marketing Research*, 29. 296–313.

Stokes R.C. (1985). The Effects of Price, Package Design, and Brand Familiarity on Perceived Quality. *In J Jacoby & J Olson (eds.). Perceived Quality. Lexington*: Lexington Books.

Thakor, M.V. & Katsanis, L.P. (1997). A Model of Brand and Country Effects on Quality Dimensions: Issues and Implications. *Journal of International Consumer Marketing* 9(3). 79–100.

Wilson, C.E. (1981). A Procedure for the Analysis of Consumer Decision Making. *Journal of Advertising Research*, 21. 31–8.

Woodside, A.G. & Wilson, E.J. (1985). Effects of Consumer Awareness of Advertising on Preference. *Journal of Advertising Research*, 25. 41–8.

Wu S. & Lo C. (2009). The Influence of Core Brand Attitude and Core-Brand Perception on Purchase Intention towards Extended Product. *J. Mark. Logistics*, 21(1). 174–94.

Zarantonello, L. & Schmitt, B.H. (2010). Using the Brand Experience Scale to Profile Consumers and Predict Consumer Behavior. *Journal of Brand Management*, 17(7). 532–40.

Hospitality and Tourism – Sumarjan et al. (eds)
© 2014 Taylor & Francis Group, London, ISBN 978-1-138-00151-0

The attractiveness of loyalty program attributes towards customers' intention to join hotel loyalty program membership

H. Nurhaizura & R.B. Ningkan
Universiti Teknologi MARA, Malaysia

ABSTRACT: Loyalty program is a marketing tool used to rewards customers for their purchase and encourage loyalty behavior. The attractiveness of a loyalty program plays a significant role in determining potential membership. This research note discusses on the attractiveness of loyalty program attributes by reviewing the components of hard benefits and soft benefits of hotel loyalty program, and its relation to customers' intention to join the loyalty program membership.

Keywords: loyalty program, attributes, membership

1 INTRODUCTION

The past decade has seen many companies shifting their marketing practices from commodity orientation to customer orientation. Customer orientation is a modern marketing practice which focuses effort on the customer and building relationship with the customer (Kotler, Bowen, & Makens, 2003), known as Customer Relationship Management (CRM). The main purpose of customer relationship management is to learn more about customers' needs and behaviors in order to develop stronger relationships with them. Once successfully implemented, companies will be able to reduce costs, increase profitability by solidifying customers' satisfaction, loyalty and advocacy (Rowley, 2007).

Latest advances on information technology have taken CRM to a new level. The most successful component of CRM is the development of loyalty program (Uncles, Dowling, Hammond, 2003). Loyalty program has become increasingly sophisticated marketing tools in the recent years particularly to service industry. Generally, the strategy based on attainment of information connecting to customers' spending habits, and offering incentives such as rewards, gifts, discounts and rebates intended to capture customers' loyalty to the company. Alternatively, loyalty programs also known as frequent purchase programs (Long & Schiffman, 2000) or reward programs (Kim, Shi, & Srinivasan, 2001). Previous studies (e.g., Casado & Lara, 2009; Gomez, Arranz, & Cillan, 2006; Palmer, McMahon-Beattie, & Beggs, 2000) found that many companies are employing loyalty programs to attract potential customers as well as retain existing ones (Bolton, Kannan and Bramlet, 2000; Uncles et al., 2003; Rowley, 2007).

The main objective ofloyalty program is to reward customers fortheir repeat purchase and create loyalty to the company (Lacey & Sneath, 2006). Many companies implementing loyalty programs, also believed that they will be able to gain positive effect on customers' evaluation and behavior towards the company.

A study by Reichheld and Sasser, 1990 (as cited in O'Malley, 1998) highlighted that many companies are rigorously developing and offering reward scheme especially to a new customer to cover up losses of existing ones. This is due to lack of proper development strategy and poor retention effort. It was not until the early 1990's that this approach strike debates between experts in the marketing field. The study noted that an organization can achieve profit of near 100% by retaining only 5% of their customers.

In order to retain them, organization should design multiple rewards scheme and features targeted at a different stage of involvement and one of them is by offering attractive rewards to non-members. Attaining guests' membership from non-members is of all important. Primarily this can be done by offering attractive loyalty programs attributes to non-members in order to capture or attract them to join the loyalty program.

Before a loyalty program implemented, it is worth to understand how to develop attractive loyalty program and what are the key attributes of a successful loyalty program or else no one interested in partake in such a program.

2 LITERATURE

2.1 *Loyalty programs*

Customers are becoming more cautious and selective in their purchases. Kotler et al. (2003) stated that with the present economic downturn and fluctuation of the market, companies are competing with each other to earn bigger pieces of the market share by shifting their interest to the customer. In line, this, Customer Relationship Management (CRM) allows companies to formulate a mechanism which can be referred as a loyalty program. Loyalty program designed to recognize the customers, attract them to participate and reward them for their membership. To date, there are numbers of loyalty programs developed by companies including retailer loyalty schemes, financial loyalty schemes, online rewards schemes, frequent flyer, travel reward program and multiple partners loyalty scheme (Rowley, 2007).

The development of modern loyalty programs originated from the tourism sector (Casado et al., 2009). It was in the early 1980s when American Airlines first introduced its frequent flyer program aim at rewarding travellers for their frequent patronage. The program allows frequent travellers to accumulate points based on miles travelled and redeem the points for free flights or free upgrades. Soon, frequent travellers accumulate points program adopted by various airlines worldwide from distinct premium, airlines to low cost carriers airlines, which proven successful with the increasing number of membership and travellers commitment towards the company (O'Malley, 1998). It was copied by the broad spectrum of the consumers market including retails, car rentals, credit cards issuers, entertainment, foodservice and hotels industry.

2.2 *Definition of loyalty programs*

According to Meyer-Waarden and Benavent (2009), loyalty programs are an integrated system of individualized marketing action and personalized customer relationship that seeks to increase customers' loyalty for their repeat purchase. Similarly, Hansen, Deitz and Morgan (2010) described loyalty programs as structured marketing effort that rewards and compensate customers for their loyal buying behavior. This behavior is potentially beneficial to the company as it encouraged customers to make a return visit and purchase in this context, hotels membership. This ties the customer closely to the hotel thus creating customer loyalty (Palmer et al., 2000).

2.3 *Role and importance of loyalty programs*

The main role of loyalty programs is to offer rewards to potential customers as well as retaining existing ones. Much emphasized on the strategic role of loyalty programs towards sustaining a high level of customers' retention (Bolton et al., 2000). Repeat customers' not only result to repeat purchase but also customers' commitment to the brand and the hotel.

Having highlighted the role of loyalty programs, it is vital for hotels operators to know that loyalty programs affect hotels in a positive way if appropriately managed. As stressed by Casado et al. (2009), consideration allows for structured and effective actions to manage, select, relate and control customers' buying behavior. One of the ways this can be done is through the implementation of loyalty program. By offering highly sought rewards, it will encourage and reinforce customers' commitment to the hotel and the brand. In other word, loyalty programs could become the center for effective marketing solution in both attaining and retaining customers.

2.4 *Loyalty program attributes*

Loyalty program attributes are what customers get in return for their participation in the loyalty program. Nearly all loyalty programs attribute provide customers with a broad range of hard and soft benefits as a way to attract customers to join the programs increase their purchases and act as advocates to the hotel (Gable, Fioroti, & Topol, 2008). Ho, Huang, Huang, Lee, Rosten and Tang (2009) commented that both loyalty program attributes' of hard and soft benefits are the driving forces for customers to participate in the program. In fact, numerous hotels have introduced multiple rewards scheme offering differential benefits to customers depending on their level of involvement in the membership.

Hard benefits are tangible benefits such as coupons, special discounts or rebates for past purchases or savings on future purchases. Gomez et al. (2006) argued that while immediate consumption of rewards affect positively on customers' intention to join hotel' membership program delaying rewards may restrict customers' perceived value of these rewards. This may result to their postponement to participate in the program.

In contrast, soft features are intangible benefits that can be highly valued by customers (Gable et al., 2008). These benefits may not be highly sought after, but they provide exclusiveness beyond financial element. Some examples of soft benefits hotels can make use of are members only, exclusive invitations and special privileges. Soft benefits focus on building a relationship with customers and facilitating information. Though difficult to implement and copied by competitor, soft benefits serve as a vehicle that ties the customer closely to the hotel.

2.5 *Hotel loyalty program membership*

O'Malley (1998) noted that customers develop feelings towards loyalty programs in different ways. Satisfied customer will become loyal to the hotel while dissatisfied customer will defect. Customers' loyalty is a bundle of mixed view of affection, faith or commitment towards a program. This has led to the use of customers' satisfaction as a reliable measure of loyalty because it presumed that satisfaction affects buying intentions in a positive way. Rowley (2007) supported that customers' membership is likely to be affected by the satisfaction customers' feel towards the loyalty programs and the degree of returns they are receiving.

Loyalty program enrollment may reflect on the effectiveness of the loyalty program itself in promoting the rewards and benefits that trigger customers' behavior intention to become a member in order to enjoy the rewards. Meyer-Waarden and Benavent (2009), pointed that psychological, sociological and relational drivers may enhance customers trust and commitment to the hotel. Meyer-Waarden and Benavent (2009) also added that the loyalty programs stimulate the feeling of pride, associated with such a program and earned something as rewards without having to pay the regular price.

Based on the rewards and benefits discussed, Palmer et al. (2000) claimed that loyalty program benefits that most suitable for different category of hotels. Budget hotels are least likely to benefit from complex loyalty programs. This is due to their limited product and service offerings. As such, they should consider designing rewards based on hard features such as reduced room rates to attract customers' participation. For medium to large hotels, an extended loyalty programs offering customized rewards is most appropriate. For example, point accumulation system that would able customer to accumulate points in exchange for free nights or room upgrades. On the other hand, luxury hotels deemed for sophisticated and highly personalized loyalty programs. Customers' profiles and status are the base for designing rewards and require creativity to differentiate their loyalty programs from others. Reward designs heavily revolve on soft features tailored made to fit the customers' lifestyle and preference.

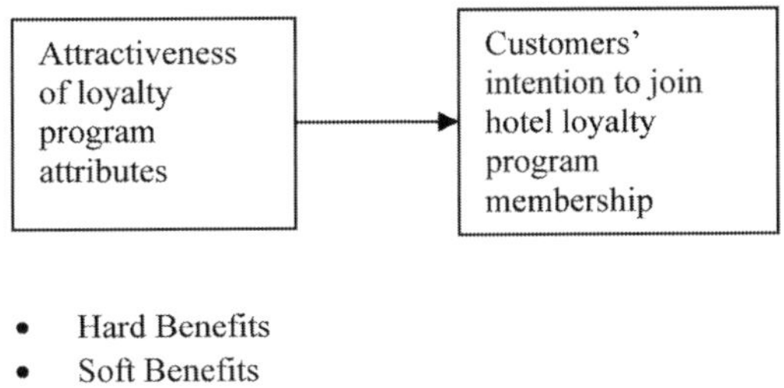

Figure 1. Proposed study framework.

3 STUDY FRAMEWORK

Based on the literature, Figure 1 depicts framework for the study adapted from Gable, Fioroti and Topol (2008).

REFERENCES

Bolton, R.N., Kannan, P.K., & Bramlett, M.D. (2000). Implications of loyalty program membership and service experiences for customer retention and value. *Journal of the Academy of Marketing Science*, *2*(1), 95–108.

Casado, J.M.P. & Lara, P.R. (2009). Loyalty programmes as a tourism service sales channel: a Spanish multi-sponsor programme case study. *Tourism Review*, *65*(3), 35–45.

Divett, M. Crittenden, N. & Henderson, R. (2003). Actively influencing consumer loyalty. *Journal of Consumer Marketing*, *20*(2), 109–126.

Gable, M., Fiorito, S.S. & Topol, M.T. (2008). An empirical analysis of the components of retailer customer loyalty programs. *International Journal of Retail & Distribution Management*, *36*(1), 32–49.

Gomez, B.G, Arranz, A.G. & Cillan, J.G. (2006). The role of loyalty programs in behavioral and effective loyalty. *Journal of Consumer Marketing*, *23*(7), 387–396.

Hansen, J.D, Deitz, G.D. & Morgan, R.M. (2010). Taxonomy of service-based loyalty program members. *Journal of Services Marketing*, *24*(4), 271–282.

Ho, R., Huang, L., Huang, S., Lee, T., Rosten, A. & Tang, C.S. (2009). An approach to develop effective customer loyalty programs: The VIP program at T & T Supermarkets Inc. *Managing Service Quality*, *19*(6), 702–720.

Jackson, S.E. (2009). Membership has its shareholder rewards. *Journal of Business Strategy*, *30*(1), 53–55.

Kim, B., Shi, M. & Srinivasan, K. (2001). Rewards programs and tacit collusion. *Marketing Science*, *20*(2), 99–120.

Kotler, P., Bowen, J. & Makens, J. (2003). *Marketing for Hospitality and Tourism*. Upper Saddle River, New Jersey: Prentice Hall.

Lacey, R. (2006). Customer loyalty programs: are they fair to consumers?. *Journal of Consumer Marketing*, *23*(7), 458–464.

Long, M. & Schiffman, L. (2000). Consumption values and relationships segmenting the market for frequency programs. *Journal of consumer Marketing, 17*(3), 214–243.

Meyer-Waarden, L. & Benavent, C. (2009). Grocery retail loyalty program effects: self selection or purchase behavior change?. *Journal of Academy of Marketing Science*, *37*, 345–358.

Nor Asiah, O., Norzalita, A.A. & Muhamad, A.N. (2011). Understanding the relationship of program satisfaction, program loyalty and store loyalty among cardholders of loyalty programs. *Asian Academy of Management Journal*, *16*(1), 21–41.

O'Malley, L. (1998). Can loyalty schemes really build loyalty?. *Marketing Intelligence & Planning*, *16*(1), 47–55.

Palmer, A., McMahon-Beattie, U. & Beggs, R. (2000). A structural analysis of the hotel sector loyalty programmes. *International Journal of Contemporary Hospitality Management*, *12*(1), 54–60.

Papadatos, C. (2006). The art of storytelling: how loyalty marketers can build emotional connections to their brands. *Journal of Consumer Marketing*, *23*(7), 382–384.

Rowley, J. (2005). Building brand webs: Customer relationship management through the Tesco Clubcard loyalty scheme. *International Journal of Retail & Distribution Management*, *33*(3), 194–206.

Rowley, J. (2007). Reconceptualising the strategic role of loyalty schemes. *Journal of Consumer Marketing*, *24*(6), 366–374.

Tepeci, M. (1999). Increasing brand loyalty in the hospitality industry. *International Journal of Contemporary Hospitality Management*, *11*(5), 223–229.

Uncles, M.D, Dowling, G.R. & Hammod, K. (2003). Customer loyalty and customer loyalty programs. *Journal of Consumer Marketing*, *20*(4), 294–316.

Hospitality and Tourism – Sumarjan et al. (eds)
 ISBN 978-1-138-00151-0

Analysis on the effect of servicescape toward customer satisfaction: A case study at Dataran Cendekia café UiTM Shah Alam

N.A. Abdul, N.S. Karjanto, R.P.S. Raja Abdullah & N.A. Baba
Universiti Teknologi MARA, Malaysia

ABSTRACT: The service environment which is closely related to 'servicescapes' can be defined as the physical environment surrounding the service that can be controlled by the service organization to enhance or hinder employees and customers' activities. This conceptual paper is specifically meant for assessing the students' satisfaction level towards servicescapes offered by the Dataran Cendekia Café Universiti Teknologi MARA (UiTM) Shah Alam. This place has been selected since numbers of negative comments were made by the students especially regarding the servicescape elements of the facility. Students who had the experience eating at Dataran Cendekia Café, UiTM Shah Alam will be selected as samples for the study. Results will be useful in validating and complementing the students' comments and consequently assist the university to come out with strategies for improvisation and corrective action towards the facility.

Keywords: customer satisfaction, Dataran Cendekia café, servicescapes, service quality

1 INTRODUCTION

Malaysia is facing the phenomenon as statistics from Higher Education Indicators 2009 to 2010 showed that the number of university students including public institutions of higher learning and private institutions of higher learning is increasing, from 628,479 students in 2001 to 1,134,134 students in 2010 (Ministry of Higher Education Malaysia, 2013) which is an increment of 80.5 percent after ten years.

The numbers of higher education students at university campuses are growing and this would spark demand for foodservice availability and quality on campus food (Shanka & Taylor, 2005). Students besides others definitely need to receive good food service facilities towards comfortable life (Noraishah & Abdullah, 2012). In line with this, there is increasing number of foodservice operations in the universities that are providing varieties of food and delivering fresh, healthy and tasty ingredients for the students (Kim & Moon, 2009).

With advance education today's university students are leaning more towards good quality and nutritional value of the food (Gramling *et al.*, 2005). In order to maintain and continuously reach the students' expectation, the university food service operation including canteen and cafeteria must complied or monitoring the overall service environment or popularly known as servicescape. This service servicescape include the physical environment that surrounds with overall services through operational activities, employees and customers' activities (Sheriff, 2007).

Having highlights the servicescape, research proposal is to look on students' satisfaction towards the servicescape at the Dataran Cendekia cafés located in the Universiti Teknologi MARA, Shah Alam, Selangor. The main objective is further supported with immediate objectives.

i. *To examine customer level of satisfaction toward ambiance of Dataran Cendekia café.*
ii. *To examine customer level of satisfaction toward space and function of Dataran Cendekia café.*
iii. *To examine customer level of satisfaction toward layout and accessibility of Dataran Cendekia café.*
iv. *To examine customer level of satisfaction toward equipments of Dataran Cendekia café.*

2 LITERATURE REVIEW

Servicescape refers to an environment which has physical surrounding of the foodservice operation (Bitner, 1992). Han and Ryu (2009) state that servicescape can be either facility exteriors or interiors elements, whereby interiors contain elements such as design of the building, equipment such as furniture or fixtures, decoration, temperature, smells and so on. While facility exteriors contain elements such as the signage of the building, building's architecture, parking and also include landscaping. Studies found that customers behavioral and intention effects with color, light, background music and also odors (Yeoh & North, 2010; Gueguen & Petr, 2006; Chebat & Michon, 2003).

In the foodservice operation environment, customer satisfaction can be determined as "a cognitive or affective reaction that emerges in response to a single or prolonged set of service encounters" (McDougall & Levesque, 2000). The tangible physical environment development is one of the important factor but often been neglected in the food service operation (Barber & Scarcelli, 2010). The tangible physical environments are composed of ambient conditions, layout, space and function and also equipment, all of which are important predictors of service quality.

According to Wall & Berry (2007), ambiance refers to mechanic clues in the service environment that contain non-human elements in an equipment, facility layout, lightning, and color. Some authors stated that ambiance conditions refer to the background features of the environment such as temperature, lightning, noise, music, air quality, visual attractiveness and scent and in the same time, ambient affects all the five senses in human as a general rule (Simpehand Simpeh, 2011; Bitner, 1992; Lee, 2011; Johnson, Mayer, & Champaner, 2004).

Bitner (1992) defined spatial layout as a condition where equipment, machines, furniture, and any other physical objects are arranged and coordinated in the foodservice facility. In a simple word, it is all about the arrangement of seats, corridors, hallways and sidewalks, food service lines, restrooms, and entrances and exits of a particular restaurant. This Spatial layout is said to have a positive relationship on customer satisfaction, according to Han & Ryu (2009) in their studies.

Functionality on the other hand refers to the objects' physical ability to improve the service performance. Kincaid & Baloglu (2010) stated that the arrangement of furnishings in spatial layout and functionality is highly related to customers' and workers' needs as they would spend extra cash in the facilities where the food and drinks is served. Customers' satisfaction and behavior in this case is heavily influenced by the on-service operations and settings provided by the facilities, as it is the place where customers 'interact' with the organization (Wall & Berry, 2007).

Equipment, according to Wall & Berry (2007) can be described as one of the non-human element in the mechanic clue in the service environment. The authors also found that successful dining places should have good mechanic clues to meet customers' expectations. This is particularly true as dining equipment's are expected to be one of the most crucial elements in upscale restaurants, as it will influence the customers' quality perception and also the customers' emotional states in one way or another.

Kivela, Inbakaran, & Reece (2000) had embarked a research in Hong Kong which analyzed the customers' satisfaction factors based on service excellence, ambience excellence, food excellence and feeling comfortable on eating, reservations and parking. The study suggested that customer satisfaction is truly related with the engagement with physical environment or 'servicescape'.

3 ISSUES

There were hundreds of researches have been conducted on service quality and customer satisfaction. The importance of customers' satisfaction is also stressed by Syed and Conway (2006), where they claimed that it provides an important indicator of their return intention. The dissatisfaction of customers through service quality to element like the cleanliness of the table and cutlery cannot be overlooked. Aksoydan (2007) found that the foodservice establishment which failed to meet the standards of food hygiene and cleanliness expected by the customers who would assess the establishment as offering poor or low service quality (Sheriff, 2007). Beside the ambient of the cafeteria, space and layout might also affect the social interaction among the customers as the spaces between seats influence affective perception of the customers (Lam, Chan, & Lo, 2011). If the seats were too close between one and another, this could give dissatisfaction to the customers who came to dine at the facility.

In relation with the above notion, Universiti Teknologi MARA (UiTM) Shah Alam has exposed the same displeased condition on servicescape in it Dataran Cendekia (DC) cafe. With all the resources support, complaints, observation and interview about the facilities, servicescape in DC were arguably very low and poor in the last couple of years. There were discontentment on the number and the condition of tables, sewerage sytem, water piping and availability of sinks or washing basin. The positioning of disposal area was one of the greatest concerns where the smell from is intolerable and 'agonizing' for most of the customers.

4 METHODOLOGY

As this study seeks to identify customer satisfaction from servicescape perspective by customers in Dataran Cendekia café at UiTM, Shah Alam, a quantitative approach through a set of structured questionnaire is considered to be the most appropriate method as research instrument for the data collection method. A descriptive research design will be used for this purpose. The study will be cross-sectional in nature where the data will be collected at same point of time (Sing, Fook, & Sidhu, 2006).

As for the study sample, this study will probe the customers who dine at Dataran Cendekia café, University Teknologi Mara (UiTM), Shah Alam as the respondents.

5 CONCLUSION

This study is very much important to validate the comments or feedbacks made by the students regarding the Dataran Cendekia café, University Teknologi Mara (UiTM), Shah Alam. The students' satisfaction could be vital in indicating the service quality or servicescape level of the facility hence warrants improvisation and corrective action to be made by the university in the latter stage after the study completion. Furthermore, the importance in developing servicescape would not only contribute a good image to the cafeteria but it will also give the same impact to the university. UiTM as education providers need to have a systematic procedure to combat these issues for the benefit of students as customers and at the same time increase the confidence of parents or students.

In conclusion, the result of the study will assist UiTM management as an educational provider to come out with a more systematic system in encountering the above mentioned issues for the sake of students and other stakeholders. Most importantly, the students are not only getting the best education in this university but also provided with the best satisfaction when dining in the university's cafes.

REFERENCES

Aksoydan, E. (2007). Hygiene factors influencing customers' choice of dining-out units: findings from a study of university academic staff. *Journal of Food Safety*, 27(3), 300–316.

Barber, N., & Scarcelli, J.M. (2010). Enhancing the assessment of tangible service quality through the creation of a cleanliness measurement scale. *Managing Service Quality*, 20(1), 70–88.

Bitner, M.J. (1992). Servicescapes: the impact of physical surroundings on consumers and employees. *Journal of Marketing*, 56, 57–71.

Chebat, J.C., & Michon, R. (2003). Impact of ambient odors on mall shoppers' emotions, cognition, and spending. *Journal of Business Research*, 529–539.

Gramling, L., Byrd, R., Epps, L., Keith, D., Lick, R., & Tian, R. (2005). Foodservice management and its impact on college operations: a business anthropological case study. *Foodservice Research International*, 16(1/2), 15–43.

Guéguen, N., & Petr, C. (2006). Odors and consumer behavior in a restaurant. *International Journal of Hospitality Management*, 25(2), 335–339.

Han, H., & Ryu, K. (2009). The Roles of the Physical Environment, Price Perception, and Customer Satisfaction in Determining Customer Loyalty in the Restaurant Industry. *Journal of Hospitality & Tourism Research*, 33, 487–510.

Johnson, L., Mayer, K.J., & Champaner, E. (2004). Casino atmospherics from a customer's perspective: A re-examination. *UNLV Gaming Research & Review Journal*, 8(2), 1–10.

Kim, W.G., & Moon, Y.J. (2009). Customers' cognitive, emotional, and actionable response to the servicescape: A test of the moderating effect of the restaurant type. *International Journal of Hospitality Management*, 28, 144–156.

Kincaid, C., & Baloglu, S. (2010). What really brings them back? The impact of tangible quality on affect and intention for casual dining restaurant patrons. *Journal of Contemporary Hospitality Management*, 22(2), 209–220.

Kivela, J., Inbakaran, R., & Reece, J. (2000). Consumer research in the restaurant environment. Part 3: analysis, findings and conclusions. *International Journal of Contemporary Hospitality Management*, 12(1), 13–30.

Lam, L.W., Chan, K.W., & Lo, F. (2011). Does the look matter? The impact of casino servicescape on gaming customer satisfaction, intention to revisit, and desire to stay. *International Journal of Hospitality Management*, 30, 558–567.

Lee, S. (2011). Evaluating serviceability of healthcare servicescapes: Service design perspective. *International Journal of Design*, 5(2), 61–71.

McDougall, G.H.G., & Levesque, T. (2000). Customer satisfaction with services: putting perceived value into the equation. *Journal of Service Marketing*, 14(5), 392–410.

Ministry of Higher Education Malaysia (2013). *Statistics of students enrollment in higher education* from 2001 to 2010. Retrieved Jun 25, 2013, from http://www.mohe.gov.my/web_statistik/indikator_pengajian_tinggi_2009–2010.pdf.

Noraishah, S., & Abdullah, D. (2012). Exploring the Major Determinants of Student Satisfaction on University Cafeteria Food Services: A Malaysian Case, 2(2000), 62–73.

Qin, H., & Prybutok, V.R. (2009). Service quality, customer satisfaction, and behavioral intentions in fast-food restaurants. *International Journal of Quality and Service Sciences*, 1(1), 78–95.

Ryu, K., & Jang, S.S. (2007). The Effect of Environmental Perceptions on Behavioral Intentions Through Emotions: The Case of Upscale Restaurants. *Journal of Hospitality & Tourism Research*, 31(1), 56–72.

Simpeh, K.N., & Simpeh, M. (2011). Servicescape and customer patronage of three star hotels in Ghana's metropolitan city of Accra. *European Journal of Business and Management*, 3(4), 119–130.

Singh, P., Fook, C.Y., & Sidhu, G.K. (2006). *A comprehensive guide to writing a research proposal.* Batu Caves: Venton Publishing.

Shanka, T., & Taylor, R. (2005). Assessment of university campus café service: The students' perceptions'. Asia Pacific *Journal of Tourism Research*, 10(3), 329–340.

Sheriff, N.M. (2007). Service environment of private colleges: Revisiting the basics, (July), 62–64.

Syed, S.A., & Conway, C. (2006). Customer Satisfaction in the Restaurant Industry: An Examination of the Transaction-specific Model. *Journal of Service Marketing*, 20(1), 3–11.

Wakefield, K.L., & Blodgett, J.G. (1996). "The effect of the servicescape on customers' behavioural intentions in leisure service settings", *The Journal of Services Marketing*, 10(6), 45–61.

Wall, E.A., & Berry, L.L. (2007). The combined effects of the physical environment and employee behavior on customer perception on restaurant service quality. *Cornell Hotel and Restaurant Administration Quarterly*, 48(1), 56–69.

Yeoh, J.P.S., & North, A.C. (2010). The effects of musical fit on choice between two competing foods. *Musicae Scientiae*, 14(1), 165–180.

Hospitality and Tourism – Sumarjan et al. (eds)
© 2014 Taylor & Francis Group, London, ISBN 978-1-138-00151-0

Customer satisfaction and repurchase intention towards Darul Rahmat Chicken in Shah Alam

R. Muhammad & S. Amirrah
Universiti Teknologi MARA, Malaysia

ABSTRACT: The fast food industry in Malaysia is facing increasingly competitive challenges as with other industries around the world. There are over 2,500 fast food outlets in the country representing more than 240 products (Mohamed et al., 2012). Darul Rahmat Chicken is one of the local fast food that has been rebranding their name from the Radix Fried Chicken since May 2012. The main objective of the creation of the Darul Rahmat Chicken or D'RC to give consumers the opportunity to choose foods that are prepared to meet halal and organic methods ranging from chicken farming system that will be used in the D'RC. The purpose of this study: (1) to determine the attributes that influence the level of customer satisfaction towards D'RC; (2) to examine the level of customer's satisfaction towards D'RC; and (3) to identify whether customer intend to repurchase D'RC products. This study will be conduct in D'RC Shah Alam and aimed to identify the significant relationships between customer satisfaction and customers repurchase intention. Understanding the impacts from the previous study will lead to the empirical investigation in D'RC issues.

Keywords: fast food industry, Darul Rahmat Chicken, halal

1 INTRODUCTION

In Malaysia, fast food outlets such as McDonald's, KFC, and Pizza Hut are undergoing dramatic transformations and are experiencing heightened competition (Yap & Kew, 2007). A fast-food restaurant characterized as quick services of franchised restaurant chain, which supplied the food quickly after ordering and minimal service offering for dining and takeout facilities (Elliot & Reed, 1999). Nowadays, the Fast Food Restaurants (FFRs) focus on several ways to increase their service quality in order to increase the level of satisfaction among their customers and thus increase their purchase intentions, as well as loyalty (Qin & Prybutok, 2008; Gillbert, Veloutsou, Goode, & Moutinho, 2004; Kara, Kaynak, & Kucukemiroglu, 1995). When FFRs are able to achieve or exceed the expectations of customers, the customer will satisfy with the service. The fast food industry is especially attractive for such research on customer satisfaction because it is a worldwide phenomenon and growing rapidly (Ball, 1992).

1.1 *Problem statement*

There are many fast food restaurants in Malaysia. To name a few, McDonalds, Pizza Hut, KFC, Subway, A & W are some of the examples. Radix Fried Chicken considered a new fast food restaurant that founded by Mr. Haji Ismail Bin Haji Ahmad, the Managing Director of the Herba Penawar Al-Wahida or HPA (www.hpa.net.my, 2012). Radix Fried Chicken began its operations on 1 June 2008. Recently, Radix Fried Chicken or RFC already has branches in other country such as Saudi Arabia. In addition, Radix Fried Chicken is also famous for its coffee known as Radix coffee. RFC is a local fast food, represent as an alternative to the restaurant introduced by foreign companies. The main objective of the creation of the RFC to give consumers the opportunity to choose foods that prepared to meet halal and organic methods, ranging from chicken farming system serving in all RFC restaurant chain. The beginning of an idea occurs after exposure in the Malaysian media on the level of doubt broiler production.

On May 2013, Darul Rahmat Chicken (D'RC) was rebrand from the RFC due to the management issues. D'RC managed by the Darul Rahmat Sdn and has four franchises in Klang Valley area: Shah Alam, Sungai Besi, Bandar Baru Bangi and Setiawangsa. D'RC still maintained their fast food service with the organic chicken in the menu. Nevertheless, most food sellers in Malaysia are lack in creativity and new ideas, therefore, the customers are being served with ordinary food. This is the reason why the customers saw a little impact in the innovation towards their decision to purchase fast food restaurants. The innovation

DR'C restaurant to customers is at no difference between the competing fast food providers. Radix Fried Chicken is offering almost similar business settings compared to KFC for fried chicken, Subway Restaurant and O'Briens Irish Sandwich Bars for a sandwich, and McDonald's and Burger King for a product of a kind of burger.

In Shah Alam, the competition is high between McDonald's, KFC, Subway and A & W. Based on a forum in (http://www.carigold.com/portal/forums/archive/index.php/t-375680.html), 70% consumer complaints about the standard quality of food and the service of D'RC in Shah Alam. Yap and Kew (2011) noted that quality and customer satisfaction have long been recognized as playing a crucial role for success and survival in today's competitive market. In addition, consumers are seeking a quality, value and desirable environment away from the daily life pressures (Soriano, 2002). With the growing affluent and educated society, consumers nowadays are demanding safe and better food quality to consume. Factors such as safety, freshness, storage, preparation, consumption, price, packaging, and place of purchase are the factors that may create the consumers' purchase intentions towards their food purchasing behaviour. Quality and safety are two key elements in consumer food perceptions and decision making associated with food choice (Grunert, 2005; Röhr et al., 2005). Price is not the prime factor and those sellers who are able to serve excellent quality food will always win the customer. Nevertheless, consumers always are seeking a better value for their money (Klara, 2001). Food quality and fresh ingredients are the most prominent reasons why customers return to the restaurant (Brumback, 1998).

In contrast, failure to deliver quality food can only threaten marketers' long term profitable relationship with the customer. The future of the food product will be short-lived, and the consumer acceptance will fall drastically. As the consumer perception is getting lower, it would seem reasonable to suppose that the overall chances to induce the consumer intention to purchase the food product will undoubtedly be an uphill task to the marketer. Significance to this, researcher would like to find out to what extent D'RC can satisfy their customers, and this issue are still questionable due the company is still new and did not well establish. Therefore, this study will explore this area of interest.

2 LITERATURE REVIEW

2.1 *Customer satisfaction*

Satisfied customers are central to the long term success of business, and their linkage between customer satisfaction and customer behavior has well documented in marketing literature (Gilbert et al., 2004). Customer satisfaction is an essential indicator of a company's past, current, and future performance and, therefore, has long been critical focus among marketing practitioners and scholars (Oliver, 1999). Customer satisfaction defined as a post-purchase evaluative judgment concerning a specific buying decision (Homburg & Giering, 2001). In modeling satisfaction, two general conceptualizations of satisfaction exist in the literature: transaction-specific satisfaction and cumulative satisfaction (Anderson & Fornell, 1993; Boulding, Kalra, Staelin, & Zeithaml, 1993). In the transaction-specific viewpoint, satisfaction is transient: How happy is a customer with a product or service at a certain point in time (Cronin & Taylor, 1992). Alternatively, cumulative satisfaction is a customer's evaluation of the total consumption experience with a product or service to date, which directly affects post purchase phenomena such as attitude change, repeat purchase, and brand loyalty (Johnson & Fornell, 1991).

For decades, customer satisfaction has been one of the key global constructs predicting customers' behavioral intentions and actual behavior (Oliver, 1999; Westbrook & Oliver, 1991; Yi, 1991). Customer satisfaction has a positive effect on attitude change in the post purchase period (Oliver, 1999; Westbrook & Oliver, 1991; Yi, 1991), and these positive attitudes do enhance repurchase intentions (Yi, 1990).

Furthermore, previous researchers suggest that it is necessary to include affective variables n modeling satisfaction evaluation along with cognitive variables, especially when consumers involved in a service setting (Pfaff, 1997). More recently, the focus of satisfaction research seems to have moved to emotional responses, instead of purely cognitive informing behavioral outcomes (Namkung & Jang, 2007). As an example, Oliver (1993) argued that consumer attributions about satisfaction with specific product characteristics lead to positive and negative responses, which influence overall satisfaction.

2.1.1 *Food quality*

Generally, the food quality viewed as a fundamental element of overall restaurant experience (Kivela et al., 1999, Rajpoot, 2002; Sulek & Hensley, 2004; Namkung & Jang, 2007). According to Peri (2006), food quality is a necessary condition to satisfy the needs and expectations of customers. Sulek and Hensley (2004) investigated the relative importance of food quality, physical settings, and service in a full-service restaurant and reported that food quality appeared to be the most significant predictor of customer satisfaction although food quality explained only 17% of repeat-patronage intentions. Susskind and Chan (2000) also found food quality was one significant determinant of customer assessments of restaurants in Toronto, Canada, and restaurant market.

Despite the importance of food quality in the restaurant business, there is no consensus on the individual attributes that constitute food quality. Sulek and Hensley (2004) grouped all food attributes into only one variable, food quality, whereas Kivela et al. (1999) divide the food quality attributes to presentation, tastiness, menu item variety, and temperature. Even so, little has done with the of food quality attributes in relation to customer satisfaction and repurchases intention.

2.1.2 *Service quality*

Ghobadian et al. (1994) posit that most of the service quality definitions fall within the "customer led" category. Juran (1999) elaborates the definition of customer led quality as "features of products which meet customers' needs and thereby provide customer satisfaction." As service quality relates to meeting customers' needs, we will be looking at "perceived service quality" in order to understand consumers (Arnauld et al., 2002). Grönroos (1984) and Parasuraman et al. (1985) looks at perceived quality of service as the difference between customers' expectation and their perceptions of the actual service received. Other researchers look at perceived service quality as an attitude. Arnauld et al. (2002) defined perceived quality "whether in reference to a product or service" as "the consumers' evaluative judgment about an entity's overall excellence or superiority in providing desired benefits" (p. 327). Hoffman & Bateson (2001) defines service quality as an attitude "formed by a long-term, overall evaluation of a performance". Attitude defines as "a consumer's overall, enduring evaluation of a concept or object, such as a person, a brand, or a service", (Arnauld et al., 2002). Service quality as "an attitude" is consistent with the views of Parasuraman et al. (1988), Cronin and Taylor (1992) and Sureshchandar et al. (2002). Basis of view elaborated by the latter:

As perceived service quality portrays a general, overall appraisal of service i.e. a global value judgment on the superiority of the overall service, it is viewed as similar to attitude" (p. 364) Khalidah Abu/Proceeding of IBBC 2004 635. Feinburg and de Ruyter (1995) pointed the importance of adapting the definition of service quality in different cultures. Ueltschy and Krampf (2001) contended that differences in culture affect measure of quality in a service sector. They encapsulated service quality measures as "culturally sensitive" and "may not perform properly or comparatively in a culturally diverse group domestically or abroad".

Moreover, since perceived levels of service in restaurants based on the relationship between customers and service providers, customer's perceptions and evaluation of service quality may be highly dependent upon the service provider's performance during service delivery (Nikolich & Sparks, 1995). Reliable, prompt, and assured service can be considered intangible cues that influence satisfaction and post-consumption behaviors (Brady & Robertson, 2001). Thus, in this study service quality refers to the level of service provided by restaurant employees.

2.1.3 *Atmosphere and novelty*

The atmosphere of a service setting has deemed essential in determining customer satisfaction (Bitner, 1990, 1992). It has also been established that atmosphere quality can be influential in determining repeat patronage (Wakefied & Blodget, 1994). Novelty seeking behavior relates to a person's desire to seek out novel experiences, or alternative sources of stimulation and willingness to try new things (Hirschman, 1984). It is thought that consumers' novelty seeking drive can have implications in buyer choice and brand loyalty within the service industries and this phenomenon influences purchase behavior. It has been found that novelty seeking in a motivator in desiring to travel (Lee & Crompton, 1992). In addition, it has been found that people who visit novel travel locations are less likely to return to the same location again (Bello & Etzel, 1985).

2.1.4 *Brand preference*

Customer satisfaction can influence attitudinal change (e.g., service and supplier preference) which in turn affects repurchase intention (Innis, 1991; Oliver, 1980; Oliver & Bearden, 1985; Stauss & Neuhaus, 1997; Phillip et al., 2002). A high level of satisfaction is likely to increase the probability that the brand in question will retain in the customer's consideration set and will increase the customer's preference for the brand (Westbrook & Oliver, 1981). The effect of brand preference on willingness to buy has rarely been examined (Dodds et al., 1991). Encouraging approaches to be more detailed description of customer choice behavior are provided by developments in the consideration set theory proposed by several researchers (Kardes et al., 1993; Roberts & Lattin, 1991, 1997; Shocker et al., 1991).

2.2 *Repurchase intention*

Satisfaction may be merely a judgment with cognitive and affective response to the target whereas the repurchase intention also involves a behavioral component (Mittal, Ross, & Baldasare, 1998; Kim & Ok,

2009). Thus, to predict repurchase intentions, other variables than satisfaction should be considered such as brand preference and attitude. Improving current customers' repurchase intentions is economically more advantageous than constantly seeking new customers (Kim & Ok, 2009; Reichheld & Sassers, 1990; Sheth & Partvitiyar, 1995).

REFERENCES

Acebron, L.B., & Dopico, D.C., (2000). The importance of intrinsic and extrinsic cues to expected and experienced quality: An empirical application for beef. *Food Quality and Preference,* 11(3), 229–238.

Anderson, E.W., & Fornell, C. (1993). A customer satisfaction research prospectus. In R.L. Oliver & T.R. Rust (Eds.), *Service Quality: New directions in theory and practice.* Newbury Park, CA: Sage.

Arnauld, E.J., Price, L.L., & Zinkhan, G.M., (2002). *Consumers, New York, NY: Mc-Graw-Hill Higher Education.*

Ashton, S.A., Scott, N., Solnet, D., & Breakey, N. (2010). Hotel restaurant dining: The relationship between perceived value and intention to purchase. *Tourism and Hospitality Research,* 10 (206).

Ball, S. (Ed.) (1992). Fast food operation and their management, *Stanley Theories Ltd, London.*

Bello, D.C., & Etzel, M.J., (1985). "The Role of Novelty in the Pleasure Travel Experience," *Journal of Travel-Research,* Summer, 20–26.

Bhattacharya, S., Sengupta, P.P., Mishra, R., (2011). Exploring customer loyalty: A study on fast food outlet in the City Of Chennai retrieved at www.zenithresearch.org.in on 27 May 2013.

Bitner, M., (1990). Evaluating service encounters: The effects of physical surroundings and employee responses. *The Journal of Marketing,* 54(2), 69–82.

Bitner, M.J., (1992), Servicescapes: The impact of physical surroundings on customers and employees. *Journal of Marketing*, 56, 57–71.

Bolton, R.N., & Drew, J.H., (1991). A multistage model of customers' assessments of service quality and value. *Journal of Consumer Research*, 17(4), 375–385.

Boulding, W., Kalra, A., Staelin, R., & Zeithaml, V.A. (1993). A dynamic process model of service quality: From expectations to behavioral intentions. *Journal of Marketing Research,* 30(1), 7–27.

Brady, M.K., & Cronin Jr., J.J., (2001a). Customer orientation: effects on customer service perceptions and outcome behaviors. *Journal of Service Research,* 3 (3), February, 241–251.

Brady, M.K., & Robertson, C.J., (2001). Searching for concensus on the antecedent role of service quality and satisfaction: An exploratory cross-national study. *Journal of Marketing Research,* 30(1), 7–27.

Breffni M. Noone, Sheryl E. Kimes, Anna S. Mattila, & Jochen Wirtz (2007). The effect of meal pace on customer satisfaction. *Cornell Hotel and Restaurant Administration Quarterly 2007,* 48:231.

Cortese, A., (2003). Fine dining? Just across the lobby: Some of the best new restaurants are popping up in hotels. *Business Week*, pp. 130–131.

Creswell, J.W. (2003). Research design: Qualitative, quantitative, and mixed methods approaches (2nd ed.). Thousand Oaks, CA: Sage.

Cronin, J., and Taylor, S. (1992). Measuring Service Quality: A Reexamination and Extension. *Journal of Marketing, 56, July,* 55–68.

Dodds, W.B., Monroe, K.B., & Grewal, D. (1991), Effect of price, brand and store information on buyers' product evaluations. *Journal of Marketing Research,* 28(3), 307–319.

Elliot, P., Reed, R., (1999). The Valuation of fast-food outlets: analysis, methodology, and reliability. *The Appraisal Journal,* 67, 359–369.

Feinburg, R.A. & de Ruyter, K. (1995). Consumer-defined service quality in international retailing. *Total Quality Management*, 6 (1), March.

Ghobadian, A., Speller, S., & Jones, M., (1994). Service quality: Concepts and Models, *International Journal of Quality & Reliability Management*, 11(9), 43–66.

Gillbert, G.R., Veloutsou, C., Goode, M.M.H., & Moutinho, L., (2004), Measuring customer satisfaction in the fast food industry: a Cross-national approach. *Journal of Services Marketing,* Vol 18. No. 5, 2004. pp. 371–383.

Goncalves. G.H., & Sampaio, P. (2012). The customer satisfaction—customer loyalty relationship: Reassessing customer and relational characteristics moderating effects. *Management Decision; Emerald Group Publishing Limited.* Vol. 50 No. 9, 2012.

Grönroos, C., (1984). A service quality model and its marketing implication. *European Journal of Marketing*, 18(4), 36–44.

Ha, J., & Jang, S.C.S, (2010). Effects of service quality and food quality: The moderating role of atmospherics in an ethnic restaurant segment. *International Journal of Hospitality Management.* 29(3), 520–529.

Halim, W.W.Z, & Hamed, A.B., (2005). Consumer purchase intention at a traditional restaurant and fast food restaurant, ANZMAC *2005 Conference: Consumer Behavior.*

Hirschman, A., (1984). A dissenter's confession: The strategy of economic development revisited, in G. Meier and D. Seers (Eds.). *Pioneers in Development*, New York, World Bank.

Hoffman, K.D. & Bateson, J.E. (2001). Essentials of service marketing: Concepts, strategies, and cases. *(2nd. Ed.). Australia: South Western Thompson Learning.*

Homburg, C., & Giering, A., (2001). Personal characteristics as moderators of the relationship between customer satisfaction and loyalty—an empirical analysis. *Psychology & Marketing*, 18(1), 43–66.

Hui, M., Dubé, L., & Chebat, J.C., (1997), The impact of music on consumers' reaction to waiting for services, *Journal of Retailing*, 73(1): 87–104.

Innis, H.A. (1991). *The bias of communication. Toronto: University of Toronto Press.*

Johns, N., & Tyas, P., (1996). Investigating of the perceived components of the meal experience using perceptual gap methodology. *Progress in Tourism and Hospitality Research*, 2(1), 15–26.

Johnson, M.D., & Fornell, C., (1991). A Framework for comparing customer satisfaction across individuals and product categories. *Journal of Economic Psychology,* 12(2), 267–286.

Kahkonen, P., Tuorila, H., & Hyvonen, L., (1995). Dairy fat content and serving temperature as determinants of sensory and hedonic characteristics in cheese soup. *Food Quality and Preference*, 6(2), 127–133.

Kara, A., Kaynak, E., & Kucukemiroglu, O. (1995). Marketing strategies for fast-food restaurants: a customer view. *International Journal of Contemporary Hospitality Management*, 9(3), 248–256.

Kardes, F.R., Kalyanaram, G., Chandrashekaran, M. and Dornoff, R.J. (1993) Brand Retrieval, Consideration Set Composition, Consumer Choice, and The Pioneering Advantage, *Journal of Consumer Research,* 20(1), 62–75.

Kim, W., & Ok, C., (2009). The effects of relational benefits on customer's perception of favorable inequity, affective commitment, and repurchase intention in full-service restaurants. *Journal of Hospitality and Tourism Research* 2009, 33:227.

Kivela, J., Inbakaran, R., & Reece, J., (1999). Consumer research in the restaurant environment, Part 1: A Conceptual model of dining satisfaction and return patronage. International Journal of Contemporary Hospitality Management, 11(5), 205–222.

Kotler, P., (1973). Atmospherics as a marketing tool, *Journal of Retailing*, 49(4), 48–64.

Lee, Tae-Hee and John L. Crompton (1992), "Measuring Novelty Seeking in Tourism," *Annals of Tourism Research,* 19, 732–751.

Leedy, P., & Ormrod, J., (2001). Practical research: Planning and design (7th ed.). Upper Saddle River, NJ: Merrill Prentice Hall. Thousand Oaks: SAGE Publications

Milliman, R.E., (1986). The influence of background music on the behavior of restaurant patrons. *Journal of Consumer Research*, 13, 286–289.

Mittal, V., Ross, W.T, Jr., & Baldsare, P.M., (1998). The Asymmetric impact of negative and positive attribute-level performance on overall satisfaction and repurchase intentions. *Journal of Marketing*, 62(1), 33–47.

Myungsook, A., & Noh, Y. (2009). Airline customer satisfaction and loyalty: impact of in-flight service quality. *Serv Bus (2009) 3:293–307*. DOI 10.1007/s11628-009-0068-4.

Namkung, Y., & Jang, S.S., (2007), Does Food Quality Really Matter in Restaurants? Its Impact on Customer Satisfaction and Behavioral Intentions. *Journal of Hospitality and Tourism Research,* 31(3), 387–410.

Ndubisi, N.O. (2003). Service Marketing: Are perceptions of Service Quality Predictors of Behavioral Intentions? The Malaysian Borneo Banking Industry Perspective, *Proceedings of the Academy of Marketing Studies*, 8 (1), 37–44.

Nikolich, M.A, & Sparks, B., (1995). "The hospitality service encounter: the role of communication", *Hospitality Research Journal*, 19(2), 43–56.

Oliver, R.L., (1980). A Cognitive model of the antecedents and consequences of satisfaction decisions, *Journal of Marketing Research*, 460–469.

Oliver, R.L., (1993). Cognitive affective & attribute basis of the satisfaction response. *Journal of Consumer Research*, 20, 418–430.

Oliver, R.L, (1999). Whence consumer loyalty? *Journal of Marketing*, 63(4), 33–34.

Oliver, R.L., (1999). The red queen rules. *Journal of Business Strategy* 20(3), 8.

Oliver, R.L., & Bearden, W., (1985). Disconfirmation Processes and Consumer Evaluations in Product Usage. *Journal of Business Research*, 31, 235–246.

Parasuraman, A., Zeithaml, V.A., & Berry, L.L. (1985). A Conceptual Model of Service Quality and Its Implications for Future Research. *Journal of Marketing,* 49, Fall, 41–50.

Parasuraman, A., Zeithaml, V.A., & Berry, L.L. (1988). SERVQUAL: A Multiple-item Scale for Measuring Consumer Perceptions of Service Quality. *Journal of Retailing,* 64(1), Spring, 12–40.

Peneau, S., Hoehn, E., Roth, H.-R., Escher, F., & Nuessli, J., (2006). Importance and consumer perception of freshness of apples. *Food Quality and Preference*, 17(1/2)9–19.

Peri, C., (2006). The universe of food quality, *Food Quality and Preference*, 17(2), 3–8.

Pfaff, M. (1997). The index of consumer satisfaction: Measurement problems and opportunities. In H.K Hunt (Ed.), *Conceptualization and Measurement of Consumer Satisfaction and Dissatisfaction* (36–71). Cambridge, MA: Marketing Science Institute.

Phillip K. Hellier, Gus M. Geursen, Rodney A. Carr and John A. Rickard (2002), Customer Repurchase Intention: A General Structural Equation Model. *European Journal of Marketing,* Vol. 37 No. 11/12, 2003 pp. 1762–1800.

Qin, H., & Prybutok, V.R. (2008). Determinants of customer-perceived service quality in fast food restaurants and their relationship to customer satisfaction and behavioral intentions. *Quality Management Journal*, 15(2), 35–50.

Rajpoot, N.A., (2002). TANGSERV: A multiple item scale for measuring tangible quality in foodservice industry, *Journal of Foodservice Business Research*, 5(2), 109–127.

Raza, M.A., Siddiquei, A.N., Awan, H.M., & Bukhari, K., (2012). Relationship Between Service Quality, Perceived Value, Satisfaction and Revisit Intention in Hotel Industry. *Interdisciplinary Journal of Contemporary Research In Business.* 4(8).

Reichheld, F.F., & Sasser, W.E., Jr., (1990). Zero defections: Quality comes to service. *Harvard Business Review*, 68(5), 105–111.

Roberts, J.H. & Lattin, J.M., (1991). Development and Testing of a Model of Consideration Set Composition, *Journal of Marketing Research,* 28 (4), pp. 429–440.

Roberts, J.H., & Lattin, J.M., (1997). Consideration: Review of research and prospects for future insights, *Journal of Marketing Research,* 34 (3), pp. 406–410.

Robson, S.A. (1999). Turning the tables: The psychology of design for high-volume restaurants. *Cornell Hotel and Restaurant Administration Quarterly*, 56–63.

Ryu, K., & Han, H., (2007). Influence of the Quality of Food, Service, and Physical Environment on Customer Satisfaction and Behavioral Intention in Quick-Casual Restaurants: Moderating Role of Perceived Price. *Journal of Hospitality & Tourism Research* 2010, 34: 310.

Ryu, K., & Jang, S.S., (2007). The Effect of Environmental Perceptions on Behavioral Intentions Through Emotions: The Case of Upscale Restaurants, *Journal of Hospitality & Tourism Research*, 31, 56–72.

Salami, C.G.E., & Ajobo, R.T., (2012), Consumer Perceptions About Fast Food Restaurants in Asaba. *Global Journal of Management and Business Research.* 12(1), 75–81.

Sayuti, S.A. (2011), Customers Satisfaction Toward Service Quality In RFC Restaurant: A Case Study In Kedah. Retrieved at: http://lintas.uum.edu.my:8080/elmu/index.jsp?modulon 15 April 2013.

Shaharudin, M.R, Pani, J.J., Mansor, W.S., & Elias, J.S., (2010). Factors Affecting Purchase Intention of Organic Food in Malaysia's Kedah State, *Cross Cultural Communication,* 6(2), 105–116.

Shocker, A.D., Ben-Akiva, M., Boccara, B. & Nedungadi, P. (1991). Consideration set influences on consumer decision-making and choice: issues, models, and suggestions. *Marketing Letters*, 2(3), 181–197.

Skogland. I., and Siguaw, A.J., (2004). Are Your Satisfied Customer Loyal? *Cornell Hotel and Restaurant Administration Quarterly* 2004, 45: 221.

Stauss, B., & Neuhaus, P., (1997), "The qualitative satisfaction model", *International Journal of Service Industry Management*, Vol. 8 No. 3, pp. 236–249.

Sulek, J.M., & Hensley, R.L., (2004). The relative importance of food, atmosphere, and fairness of wait. *Cornell Hotel and Restaurant Administration Quarterly*, 45(3), 235–247.

Suprihanti, A. The effect of services to customer's loyalty: case study of kentucky fried chicken restaurant at Galleria Mall Yogyakarta. *The 2nd International Research Symposium in Service Management 2011* retrieved at http://irssm.upnyk.ac.id/userfiles/file/papers/050.pdf on 14 April 2013.

Sureshchandar, G.S., Rajendran, C., & Anantharaman, R.N., (2002). "The relationship between service quality and customer satisfaction—a factor specific approach", *Journal of Services Marketing,* 16(1), 9–34.

Susskind, A.M., & Chan, E.K., (2000). How restaurant features affect check averages. *Cornell Hotel and Restaurant Administration Quarterly*, 41(6), 56–63.

Tat, H.H., Min, S.S., Chin, T.A., Rasli, A., & Hamid, A.A.B. (2011). Consumer's Purchase Intentions in Fast Food Restaurants: An Empirical Study on Undergraduate Students. *International Journal of Business and Social Science,* 2(5).

Taylor, S.A., & Baker, T.L., (1994). An assessment of the relationship between service quality and customer satisfaction in the formation of consumers' purchase intentions. *Journal of Retailing*, 70, 163–178.

Tony Wilson And Khor Yoke Lim (2008). Corporate And Consumer Conceptions Of "Fast Food" Branding In Malaysia; *17th Biennial Conference Of The Asian Studies Association Of Australia In Melbourne 1–3 July 2008.*

Ueltcschy, L., & Krampf, R., (2001). Cultural sensitivity to satisfaction and service quality measures, *Journal of Marketing Theory and Practice*, 9(3), 14.

Wakefield, K.L. & Blodgett, J.G. (1996). The effect of the servicescape on customers' behavioral intentions in leisure service settings, *Journal of Services Marketing*, 10 (6): 45–61.

Weiss, R., Feinstein, A.H., & Michael Dalbor (2004). Customer Satisfaction of Theme Restaurant Attributes and Their Influence to Return Intent. *The Harwoth Press Inc.*

Westbrook, R.A., & Oliver, R.L., (1981), Developing better measures of consumer satisfaction: Some preliminary results, *Advances in Consumer Research* 8(1), 94–99.

Wood, R.C. (1994). Dining Out on Sociological neglect. *British Food Journal,* 96(10), 10–14.

Yap Sheau Fen, Kew Mei Lian (2007), Service Quality and Customer Satisfaction: Antecedents of Customer Re-Patronage Intentions; *Sunway Academic Journal,* Vol 4. pp. 60–72.

Yi, Y., (1991). A critical review of consumer satisfaction. In V.A Zeithaml (Ed.), Review of marketing (pp. 68–123). *Chicago: American Marketing Association.*

Zeithaml, V.A., (1988). Consumer perceptions of price, quality, and value: A means-end model and synthesis of evidenced. *Journal of Marketing*, 52(3), 2–22.

Hospitality and Tourism – Sumarjan et al. (eds)
© 2014 Taylor & Francis Group, London, ISBN 978-1-138-00151-0

Ready to eat food attributes and consumer purchase intention towards 7-Eleven convenience foods

K.N. Sabri, N.N. Johari, N. Baba & R.P.S. Raja Abdullah
Universiti Teknologi MARA, Malaysia

ABSTRACT: Development of the metropolitan cities due to increase in population, emergence of industries, evolution of various new factors, time factor and many others created the need for Ready-To-Eat (RTE) foods in the market. This study is generally to understand the intention towards buying ready to eat food. Individual dietary choices are primarily influenced by such considerations as taste, cost, convenience and nutritional value of foods. Food safety and quality knowledge are also a crucial factor of considering the level of student's purchase behavior. Price incentives can be an effective intervention strategy to influence individual food purchases. Price reductions had consistent and strong effects on purchasing patterns. These consumers are among those who have made their purchases along with RTE food at 7-Eleven Kuala Lumpur Sentral and who are in longer time dedicated themselves as a RTE food buyers and consumers.

Keywords: ready to eat, convenience store, purchase intention, food attributes

1 INTRODUCTION

Ready to eat food is food offered or exposed for sale without additional cooking or preparation, which is packed on the premises where sold and is ready for consumption (De Boer & McCarthy, 2004). There are varieties of Ready To Eat (RTE) food available in the market to choose from and now a day they have become a part of everyday life. RTE foods at home are those purchased at a supermarket or a convenience store for consumption (Candel, 2010).

Convenience stores may provide greater access to foods and beverages than supermarkets or small grocery stores and thus influence food choice and consumption, especially in underserved areas and among children and adolescents (Sharkey, Dean, & Nalthy, 2012). Food attributes has been regarded as an important factor in predicting consumers' perceptions in RTE food choices decision complexity. According to Lewis (2008), food attributes are referred as the product features in a distinction as compared to competitors' products while food attributes in the research includes food safety concern, price, convenience, quality and portion size.

2 LITERATURE

2.1 *Ready to eat food*

Ready-to-eat food is food offered or exposed for sale without additional cooking or preparation, which is packaged on the premises where they are being sold and are ready for consumption (Hirekenchanagoudar, 2008). Moreover, the other researcher also defines ready to eat food as the food which can be directly consumed from the package with or without warming or thawing, and preparations (Suneetha, 2012). According to Sarathy and Gopal (2011), they stated that rapid demographic and socioeconomic changes, such as the massive entrance of women into the workforce and increasing multi-ethnicity, are fundamental drivers of food-buying and dietary patterns and this has led to the entry of ready-to-eat category of food products.

2.2 *Convenience store*

Convenience stores are still a fairly new retail store concept in Malaysia (Shamsudin & Selamat, 2009). They have around 11% of the total retail sales. These stores are mainly located in major urban centers and along the North-South Highway to capture busy customers who seek convenience. They concentrate on selling a small range of convenience foods in medium to small sized packaging, usually single serve sizes. The main convenience store is 7-Eleven although, the other smaller local convenience stores exist (Shamsudin & Selamat, 2009). Fraser, Poole and Doishita (2009) also stated that convenience stores tend to attract purchasers of hot ready to eat meals.

2.3 *Factor affecting consumer intention towards RTE foods*

Food selection and consumption are complex phenomena influenced by various factors, which can be classified as marketing-related, psychological and sensory related. Strategies used by consumers to choose a food product may influence both the initial product evaluation and satisfaction after use (Kim, 2009). In this study, there are five ready to eat food attributes that might influence consumer intention towards RTE foods such as food safety concern, price and convenience, quality, and portion size.

2.3.1 *Food safety concern*

Food safety has been defined as the conditions and measures that are necessary during the production, processing, storage, distribution and preparation of food to ensure that it is safe, sound wholesome and fit for human consumption (AbdPatah, Mat Issa, & Mohammad Nor, 2009). Food safety issue is not new in Malaysia as problems such as unsafe food handling, doubtful food preparation, food poisoning and spreading of infectious food borne illness have been discussed by the public more often than before (AbdPatah et al., 2009).

2.3.2 *Price*

According to Shamsudinand Selamat (2009), Malaysian consumers, particularly those in the 20–40 age groups, who can afford and are willing to pay for the price of convenience, have a preference for semi-prepared foods and/or takeaway meal. Price incentives can be an effective intervention strategy to influence individual food purchases, and price reductions had consistent and strong effects on purchasing patterns of targeted foods in work site or school cafeteria settings (Simone, 2009).

2.3.3 *Convenience*

The strong economic growth in the late 80's and early 90's contributed to major changes in consumer purchases and consumption patterns. As stated by (Boyce, Broz, & Binkley, 2009) there is a growing demand for the convenience of take-out food and foodservice establishments are seeking to satisfy this demand. According to Simelane (2008), convenience in food is generally defined as anything, that saves or simplifies work, adding to a person's ease or comfort. Other than that, consumers buying and consuming convenience food are also associated with a perception of the quality of these products, as well as with conflict between saving time or saving money (Olsen, Prebensen, & Larsen, 2009).

2.3.4 *Quality*

Quality control of processed foods in Malaysia is in the transformation stage from the conventional method where product quality is only inspected at the end of the processing line to the quality assurance system where HACCP concept and plan are used. Nowadays, quality assurance standards and guidelines are widely applied in the food industry because of the increasing demand by consumer (Verlegh & Candel, 1999).

2.3.5 *Portion size*

Food comes in many shapes, types, sizes, colors and textures. Hence, a portion of food can be presented in different ways (Marchiori, 2012). Generally, a portion of food is served on a plate, bowl, glass, and container or simply in a package. The portion of food suggests a situational consumption norm, an authoritative judgment as to how much is appropriate, typical, normal and reasonable to consume.

2.4 *Consumer's purchase intention*

According to Ajzen (2007) intentions have four dimensions, behavior which is intended to perform, the target object at which the behavior is directed, the situation in which the behavior is to be performed and the time at which the behavior is to be performed. Moreover, specific intention in contrary to the general intention covers all the four elements specified for a given behavior. Literature has identified that consumers' behavioral intentions are dependent upon certain factors like perceived value (McEachern & Warnaby, 2010) positive attitudes, subjective norms and behavioral control (Shaw & Clarke, 2007). Sparks and Shepherd (2002) explained significant role of consumers' self-identity, attitude, social norm and perceived behavioral control on their purchase intentions for RTE foods.

3 CONCLUSION

Practically, the study of ready to eat food attributes and consumer purchase intention could help marketers to understand the buying behavior and preferences of different consumers. Consumer's taste and preference were found to change rapidly especially in a dynamic environment. This study will help in understanding the consumer behavior and would help the firms in formulating strategies to cater to the needs of the consumer and thereby increase their market share.

REFERENCES

Abd Patah, M.O.R., Mat Issa, Z., & Mohammad Nor, K. (2009). Food Safety Attitude of Culinary Arts Based Students in Public and Private Higher Learning Institutions. Internation Education Studies, 2, 4.

Ajzen, I. (2007). The theory of planned behaviour. Organizational Behavior and Human Decision Process, 50, 179–211.

Boyce, J., Broz, C.C., & Binkley, M. (2009). Consumer Perspectives: Take-out Packaging and Food Safety. British Food Journal, 110, 8, 819–828.

Candel, M. (2010). Consumer's convenience orientation towards meal preparation. Conceptualization and measurement. Appetite, 36, 15–28.

De Boer, M., & McCarthy, M., (2004). The influence of lifestyle characteristics and beliefs about convenience food on the demand for convenience food in the Irish market. Food Quality and Preference, 15, 155–165.

Fraser, J.H., Poole, N., & Doishita, M. (2009). The failure of Multinational Food Retailers in Japan: A Matter of Convenience. British Food Journal, 111, 4, 327–348.

Hirekenchanagoudar, R. (2008). Consumer Behaviour Towards Ready To Eat Food Products.

Kim, C. (2009). Working wives' time-saving tendencies. Durable ownership, Convenience food consumption and meal purchases. Journal of Economic Psychology, 10, 391–409.

Lewis, G., & Salsbury, P.A. (2008), "Safe food at retail establishments", Food Safety Magazine, Vol. 7 No. 4, pp. 13–17.

Marchiori, D. (2012). Size Matters! The Joint Influence of the Size of Portion, Food Intake and Container on Food Intake.

Market Watch 2012. (2011). The Malaysian Food Industry.

McEachern, M.G., & Warnaby, G. (2010). Food shopping behaviour in Scotland: the influence of relative rurality. International Journal of Consumer Studies, 30 (2), 189–201.

Meenambekai, R., & Selvarajan, P. (2012). Consumer Attitudes towards Ready To Eat Packed Food Items. The Seventh International Research Conference on Management and Finance.

Olsen, S.O., Prebensen, N., & Larsen, T.A. (2009). Including Ambivalence as a Basis for Benefit Segmentation: A Study of Convenience Food in Norway. European Journal of Marketing, 43, 5/6, 762–783.

Sarathy, T., & Gopal, S. (2011). Managing the Diffusion of Innovation in Ready To eat Food Products in India, 25.

Shamsudin, M.N., & Selamat, J. (2009). Changing Retail Food Sector in Malaysia.

Sharkey, J.R., Dean, W.R. & Nalty, N. (2012). Convenience Stores and the Marketing of Foods and Beverages through Product Assortment. Journal of Preventive Medicine, 43, 109–115.

Shaw, D.S., & Clarke, I. (2007). Belief formation in ethical consumer groups: an exploratory study. Marketing Intelligence and Planning, 17 (2), 109–119.

Simone, A. (2009). Pricing Effects on Food Choices, FIU Hospitality Review, Vol. 18 No. 2, pp. 89–97.

Sparks, P., & Shepherd, R. (2002). Self-identity and the theory of planned behaviour: assessing the role of identification with green consumerism. Social Psychology Quarterly, 55 (4), 388–399.

Verlegh, P.W.J., & Candel, M.J.J.M. (1999). The consumption of convenience foods: reference groups and eating situations. Food Quality and Preference, 10, 457–464.

Hospitality and Tourism – Sumarjan et al. (eds)
© 2014 Taylor & Francis Group, London, ISBN 978-1-138-00151-0

Factors influencing international tourists' food preference towards Malaysian local food

H. Selamat & F.A. Hanan
Universiti Teknologi MARA, Malaysia

ABSTRACT: In the age of post modernity, the experiences of tourist play an important role in economic and social life. As tourist demand a higher standard in quality services, quality food and hospitality experiences particularly have caught the full attention in the tourism industry on one hand and the literature on hospitality on the other. Food has been regarded as not only being a basic necessity for tourist consumption but also as an essential element of regional culture. Local food and beverages have urbanized from being crucial necessity for tourists being recognized as an imperative part of the local culture that tourists consume. Food can at least provide superfluous opportunities for tourist to be in a more memorable and enjoyable holiday atmosphere than they expected, no matter whether trying different kinds of food is the main purpose for tourist to travel. Gastronomy, thus, seen as an important source of marketable images and experiences for the tourist. The implications of the findings are discussed, and directions for future research are suggested.

Keywords: neophobia, neophilia, motivational factors, psychological factors, food preference

1 INTRODUCTION

Most international tourists' has a common question of what is that dish, when looking at the local dishes in Malaysia for the first time. Indeed, for some tourists who are otherwise fairly adventurous are frequently fastidious regarding local food and reluctant to eat the local fare. According to Cohen and Avieli (2004), the crucial trait of tourist culinary situations appears to be that they exacerbate the tension between neophobic and novelist tendencies in most individuals. On the one hand, since eating involves actual bodily involvement with the unfamiliar environment of the destination, the intake of food, the neophobic tendencies will become more prominent. Thus, tourists will be generally reluctant to taste or eat "strange" foods, such as traditional local food or unfamiliar to them (Cohen & Avieli, 2004).

Furthermore, the central role of dining behaviours in the holiday's experiences, the interface between food and tourism has received limited research attention. Referring to Cohen and Avieli (2004), underscore that eating and drinking remain unexplored in the sociological and anthropological study of tourism. Even though there is a growing interest in research into food experiences in tourism, Kim, Eves and Scarles (2009) indicate that research on tourist food consumption and preference is in its infancy, and is still establishing its basic tenets.

In line with the statement, what are the important factors that influenced international tourists' food preference towards Malaysian local food products? In understanding these questions, these papers investigate the factors that influencing international tourists' food preference towards Malaysian local food. This objective is further support with the following hypothesis.

H1: There is a relationship between motivational factors and psychological factors towards international tourists' food preference.

2 LITERATURE REVIEW

2.1 *Motivational factors*

There are four motivational factors criteria, which need to be focused in order to determine the factors influencing international tourists' food preference towards Malaysian local food products. The four motivators are physical, cultural, interpersonal and status and prestige. In the tourism context, motivation is refers to a set of internal psychological needs, that cause a person to act in a certain way or stimulate their interest in travel and participation in a tourist activity (Crompton & Mackay, 1997; Fodness, 1994). From the previous studies, exploring tourist motivation has also suggested that tourists tend to choose the type of holiday that can satisfy their desires or needs (Crompton & Mckay, 1997; Fondness, 1994; Mcintosh, Goeldner, & Ritchie, 1995). Indeed, according Mcintosh Goeldner, & Ritchie (1995), they emphasised that tourist motivation is fundamental to gaining an understanding of tourist behaviour, and they further indicated that improved tourist motivation theory should benefit research on both travel behaviour and travel choice.

There are nine motivational factors as stated by Kim, Eves and Scarles (2009); exciting experience, escape from routine, health concern, learning knowledge, authentic experiences, togetherness, prestige, sensory appeal and physical environment. However, in this study, only four motivational factors have been chosen, which are, physical, cultural, interpersonal and status and prestige. These four motivational factors, is seen as an important aspect to determine the factors influencing international tourists; food preference towards Malaysian local food.

Travel is one of the most common means of alleviating or escaping boredom. Scarpato (2002) pointed out that wealthy tourists sometimes travel long distances to sample the fresh and exotic fruits of poverty, regarded as an inversion of daily life. Hjalager (2002) also supported that, some tourists are interested in tasting local food as 'gastronomy tourist', seeking to escape from their everyday eating habits, such as preparing dish, food shopping and routine dining. Similarly, Kim, Eve and Scarles. (2009) found that, some tourists may taste local food as they want to escape from daily routines or experience something different.

2.2 *Psychological factors*

Psychological factors contain both food neophobia and food neophilia which is described as personality in relation to whether or not people enjoy trying new foods (Ritchey et al., 2003). Neophobia and neophilia among people are engendered by both physiological and cultural influences (Cohen and Avieli, 2004; Pliner and Hobden, 1992). Thus, Cohen and Avieli (2004) emphasized that these are a considerable issue because differences between cultures may have effects on tourist experiences of local food.

Food neophobia is generally characterized as a personality trait, a continuum along which people can be placed in terms of their tendency to accept or avoid new foods (Pliner & Salvy, 2006). Furthermore, food neophobia issued has been discussed as a form of behaviour, involving the avoidance of novel foods in a particular situation (Pliner & Salvy, 2006). Food neophilia are an individual overtly willing to try new foods (Raudenbush, 1999). Typically, tourist on a trip may be eager or more willing to engage in a novel or unusual food experienced.

3 METHODOLOGY

3.1 *Sampling and instrument*

The sample populations were among the international tourists at the Kuala Lumpur area. The respondents involved were international tourists' who were staying in Malaysia for more than three days and less than two weeks. The international tourists must consume and experienced Malaysian local food during their holiday in Malaysia (Table 1).

3.2 *Data collection process*

In order to determine the reliability of measures, pilot test was conducted before the final process of data collection and 50 questionnaires was distributed to international tourists at Kuala Lumpur City Centre (KLCC). In light of the positive feedback and cooperation from the international tourists, the instrument are processes, good response with a total of 398 completed questionnaires were collected (Table 2).

Table 1. Response rate.

Description	Number and percentage
Sample size	384
Questionnaires distributed	450
Questionnaires returned	450
Invalid/rejected	52
Useable	398
Percentage	88.4%

Table 2. Reliability analysis.

	Cronbach alpha	N
Physical motivator	0.892	5
Cultural	0.950	5
Interpersonal	0.877	3
Prestige	0.972	4
Psychological	0.828	5
Tourist food preference	0.746	5

4 RESULTS AND ANALYSIS

Based from Table 3, the most influential factor is the motivational factor ($\beta = 0.819$) as compared to the other components. This variable gave the strongest contribution to explain the dependent variable when the variance explained by all other variables in the model is controlled for. As shown in Table 3, Psychological factor ($\beta = -1.262$) contributed less in tourist preference.

From the previous studies, exploring tourist motivation has also suggested that tourists tend to choose the type of holiday that can satisfy their desires or needs (Fodness, 1994; Mcintosh, Goeldner, & Ritchie, 1995). Indeed, according Mcintosh Goeldner, & Ritchie (1995), they emphasized that tourist motivation is fundamental to gaining an understanding of tourist behavior, and they further indicated that improved tourist motivation theory should benefit research on both travel behavior and travel choice. Tourist can build their knowledge of the Malaysian local food by eating and exploring the new cuisines that they are not likely to encounter at home (Mak et al., 2009). Furthermore, he stressed that, the act of eating is involving sensory perceptions to appreciate the food or tourist need for substance.

As supported by Kivela and Crott (2006), motivation to travel for food gastronomy is a valid construct and that food plays an important role in affecting the overall tourist experience and intention to revisit a destination. Furthermore, according Crompton and McKay (1997) tourist motivation is recognised as an important construct in understanding tourist choice and behaviour.

Since tourist motivation exerts significant influence exerts significant influence over tourist choice and behaviour, it can be a significant force affecting tourist choice and behaviour, and it can be a significant force affecting tourist food consumption (Mak et al., 2009). Food can be a contemporary culture resource as it satisfies all the conventional requirements of cultural tourism product (Scarpato, 2002). Fields (2002) states that, when tourists are experiencing new local cuisines, they automatically experiencing a new culture.

Table 3. Relationship between motivational factors and psychological factors towards international tourists' food preference.

Model	Standardized coefficients Beta	Sig.
1		
(Constant)		0.000*
Motivational factor	0.819	0.000*
Psychological factor	−1.262	0.000*

a. Dependent Variable: International tourists' preference.

Since food reflects human cultures (Beardsworth & Keil, 1997), various aspects of the local food such as the way the local people eat the food preparation process and the taste of the local food provide clues that are critical for discerning the local culture. Accordingly, this explained one of the underlying reasons why many participants were enthusiastic in trying local Malaysian food.

5 LIMITATIONS OF THE STUDY

Although the result highlighted a range of interesting and meaningful findings, there were also limitations encountered during the research process. There is a need to expand this present study by looking at a broader scope especially on the minority ethnic groups' food products such as 'Babas and Nyonyas' and those tribes in the Borneo region (Sabah and Sarawak). Those works will give a thorough picture and understanding on tourists' perception of the Malaysian local food products. Furthermore, to examine whether this perception holds true, a study on the international tourists' level of consumption of the Malaysian local food needs to be conducted in the future.

6 CONCLUSION AND IMPLICATIONS

This study consolidated the relevant perspectives on tourism and food preference to identify the salient factors affecting tourists' food preference. The findings open up many new possibilities for research into food preference in tourism.

Majority of the international tourist perceived that besides other tourism activities and products, Malaysia also offers a great selection of food, beverage and food culture to attract the international tourist. With this result, it is clearly shown that, Malaysian local food can be a catalyst to boost the local economy. In line with this statement, those individuals (business operators), organizations (travel agencies and non-government organizations) and government authorities (Ministry of Tourism, Ministry of Culture, Arts) who are closely associated with the tourism industry therefore, should not ignore this good sign but take a proactive action in promoting this Malaysian local food to the tourists because these products can be the 'pulling-factors' of visiting Malaysia.

As for the food operators such as restaurants, hawker stalls and food courts particularly those who operate their business at frequently visited areas by the international tourists, therefore, take this opportunity to promote and provide more varieties of Malaysian and beverages which are of good quality clean and decently presented. Travel agencies and other tour organizers should offer more tour packages specifically focusing on food and food culture, indigenous cuisines, dining out, cooking and shopping for food at local markets. This is in line with Kivela and Crotts (2006) who noted that many international tour agencies and tour organizers around the world have included the gastronomic products as part of their tour itineraries, and consequently this has successfully attracted more international tourists to participate in those activities. This approach has been proven to be successful in marketing and promoting the exquisiteness of the local gastronomic products.

As a conclusion, the collaboration and cooperation of food operators, travel agencies and government related authorities and other stakeholders are therefore, crucial or should be further strengthened if the 'total tourism experience' among the international tourists in this country is to be achieved. The wonderful tourism products in this country would be left wasted if such commonality of approach were not developed.

REFERENCES

Athena, H.N., Mak, M.L., Anita, E., Richard C.Y., & Chang (2012). Factor influencing tourist food consumption. *International Journal of Hospitality Management*, 31(3) 928–936.

Au, N., & Law, R. (2002). Categorical classification of tourism dining. Annals of Tourism Research, 29(3), 819–833.

Babiez-Zielinska, E. (1998). Food preferences among the university students. Polish Journal of Food and Nutrition Science, 7/48(1), 136–139.

Basala, L.S. & Klenosky, D.B. (2001). Travel style preferences for visiting a novel destination: a conjoint investigation across the novelty familiarity continuum. *Journal of Travel Research,* 40(16), 172–182.

Bessiere, J., 1998. Local development and heritage: traditional food and cuisine as tourist attractions in rural areas. Sociological Ruralis 38(1), 21–34.

Boniface, P. (2003). Tasting tourism: Travelling for food and drink. Aldershot; Burlington, Vt.: Ashgate.

Botha, C., Crompton, J.L, & Kim, S. (1999). Developing a revised competitive position for Sun/Lost City, South Africa. Journal of Travel Research, 37, 341–352.

Creswell, J.W., 2008. Research design: qualitative, quantitative and mixed methods. Sage Publications, Thousand Oaks, CA.

Cohen, E., Avieli, N., 2004. Food in tourism: attraction and impediment. Annals of Tourism Research 31(4), 755–778.

Cornell, J. (2006). Medical Tourism: sea, sun, sand and ... surgery. Tourism Management, 27(6), 1093–1100.

Cox, D.N., & Evans, G. (2008). Construction and validation of a psychometric scale to measure consumers' fears of novel food technologies: The food technology neophobia scale. Food Quality and Preference, 19(8), 704–710.

Crompton, J.L., & McKay, S.L. (1997). Motives of visitors attending festival events. Annals of Tourism Research, 24, 425–439.

Dann, G.M., Jacobsen, J.K., 2002. Leading the tourist by the nose. In: Dann, G.M. (Ed.), The Tourist as a Metaphor of the Social World. CABI Publishing, New York, pp. 209–236.

Devine, C.M., Connors, M., Bisogni, C., & Sobal, J. (1998). Life course influences on fruit and vegetable trajectories: qualitative analysis of food choices. Journal of Nutrition Education, 30, 361–370.

Dovey, T.M., Staples, P.A., Gibson, E.L., & Halford, J.C.G. (2008). Food neophobia and "picky/fussy" eating in children: A review. Appetite, 50(2–3), 181–193.

Fields, K. (2002). Demand for the gastronomy tourism product: motivational factors. In A. Hjalager, & G. Richards (Eds.), Tourism and gastronomy (pp. 37–50). London: Routledge.

Fodness, D. (1994). Measuring tourist motivation. Annals of Tourism Research, 21(3), 555–581.

Flight, I., Leppard, P., & Cox, D.N. (2003). Food neophobia and associations with cultural diversity and socio-economic status amongst rural and urban Australian adolescents. Appetite, 41(1), 51–59.

Kim, Y., Eves, A., & Scarles, C. (2009). Building a model of local food consumption on trips and holidays: A grounded theory approach. *International Journal of Hospitality Management,* 28(3), 423–431.

Kivela, J., & Crotts, J. (2006). Tourism and gastronomy: gastronomy's influence on how tourists experience a destination. Journal of Hospitality and Tourism Research, 30(3), 354–377.

Mak, A.H.N., Wong, K.K.F., Chang, R.C.Y., (2009). Health or self indulgence? The motivations and characteristic of spa-goers. International Journal of Tourism research 11(2), 185–199.

McIntosh, R.W., Goeldner, C.R., & Ritchie, J.R.B. (1995). Tourism principles, practices, philosophies (7th ed.). New York: Wiley.

Pliner, P., & Hobden, K. (1992). Development of a scale to measure the trait of food neophobia in humans. Appetite, 19(2), 105–120.

Raudenbush, B., & Frank, R.A. (1999). Assessing food neophobia: The role of stimulus familiarity. Appetite, 32(2), 261–271.

Scarpato, R. (2002). Gastronomy as a tourist product: The perspective of gastronomy studies. In A.M. Hjalager & G. Richards (Eds.), Tourism and gastronomy (pp. 51–70). London: Routledge.

Shenoy, S.S. (2005). Food tourism and the culinary tourist. Published PhD thesis: Clemson University, December.

Shepherd, R., & Sparks, P. (1994) Modelling Food Choice. In H.J.H. Macfie, & D.M.H Thomson (Eds.), Measurements of food preferences (pp. 202–206). London: Blackie Academic & Professional.

Shuai Quan, N. Wang., Tourism Management 25(2004) 297–305.

Symons, M. (1994). Simmel's gastronomic sociology: an overlooked essay. Food and Foodways, 5(4), 333–351.
Sobal, J., Bisogni, C.A., Devine, C.M., Jastran, M., 2006. A conceptual model of then food choice process over the life course. In Shepherd, R., Raats, M. (Eds.), The Psychology of Food Choice. CAB International, Oxfordshire, pp. 1–18.
Sparks, B., Bowen, J., & Klag, S. (2003). Restaurant and the tourist market. International Journal of Contemporary Hospitality Management, 15(1), 6–13.
Telfer, D.J., & Wall, G. (2000). Strengthening backward economic linkages: Local food purchasing by three Indonesian hotels. Tourism Geographies, 2(4), 421–447.
Torres, R. (2002). Toward a better understanding of tourism and agriculture linkages in the Yucatan: Tourist food consumption and preferences. Tourism Geographies, 4(3), 282–306.
Tuorila, H., Huotilainen, A., Lahteenmak, L., Ollila, S., Tuomi-Nurmi, S., & Urala, N. (2008). Comparison of affective rating scales and their relationship to variables reflecting food consumption. Food Quality and Preference, 19, 51–61.
Tuorila, H., Lähteenmäki, L., Pohjalainen, L., & Lotti, L. (2001). Food neophobia among the Finns and related responses to familiar and unfamiliar foods. Food Quality and Preference, 12(1), 29–37.
Tuorila, H., Meiselman, H.L., Bell, R., Cardello, A.V., & Johnson, W. (1994). Role of sensory and cognitive information in the enhancement of certainty and linking for novel and familiar foods. Appetite, 23(3), 231–246.
Urry, J. (2002). The tourist gaze (2nded.). London: Sage.
Valli, C., Traill, B., 2005. Culture and food: a model of yoghurt consumption in the EU. Food Quality and Preference 16, 291–304.

Hospitality and Tourism – Sumarjan et al. (eds)
© 2014 Taylor & Francis Group, London, ISBN 978-1-138-00151-0

Investigating the market potential for processed banana snack product: Consumer acceptance level on banana chip in Malaysia

M.A. Tokijan & M.Z. Hasni
Universiti Teknologi MARA, Malaysia

ABSTRACT: As a result of the growing trend towards consumption of snacks, it has led to the flood of these products in the supermarket and hypermarket tremendously. Numerous brands from global and local product can easily be found on the shelves with different prices, packaging, flavour and health benefits. It simultaneously injected the intention to find the key features on what is the main reason that this product will be the contributing factors towards consumer acceptance. The consumer acceptance level is abstract whereby different consumer has different preference to accept a certain product. Due to the concern of healthy well-being and ethnic food product, the amount of consumption of the banana chip product in Malaysia seems to grow (Anne N.H., 2006). For that purpose, this study will be conducted mainly to measure consumer acceptance level of banana chip, based on four main attributes (i) hedonic: texture, appearance; (ii) perceived value: price; (iii) nutritional value and (iv) packaging. A better understanding of the product and consumer profiles may signify the potential to penetrate international snack's market.

Keywords: acceptance level, banana chip, banana snack, hedonic, nutritional value packaging, perceived value

1 INTRODUCTION

Malaysia's food industry is varied in terms of tropical and agricultural resources reflecting diverse cultures in Malaysian society such as Malay, Chinese and Indian which have led to captivating range of processed food with an Asian twist. In Malaysia, the food industry is dominated by small and medium scale companies. The major sub-sectors are the fish and fish products, livestock and livestock products, fruits, vegetables and cocoa (Market Watch, 2012). In line with the government's emphasis on the agriculture sector, the processed food and beverage industry had become an important element of the agro-based industry. In 2010, the food processing industry contributed about 10% of the Malaysian manufacturing output attracted a total of RM1.972 billion in 69 projects (Market Wach, 2011).

Malaysia also has a large and rising food retail market that is supplied by local and imported products. The current total retail sales of food and beverages are appraised at US$11 billion. The estimate for this sector is likely to grow by around 10% per annum over the next three to five years. Malaysian households spend an average 24% of their household income on retail purchases of foods. Due to rising affluence and higher education level, Malaysian consumers have become more sophisticated and demand higher quality for the goods that they purchased (Market Watch, 2012).

Living in this modern era, people seek for something unique and valuable. The preference toward food also changes due to the switching of lifestyle and other social, technological and economic factors. According to Euromonitor (2012), the overall packaged food industry in Malaysia reached US$5.5 billion in 2011 and is forecasted to grow to US$5.9 billion by 2016. The highest value segments of packaged foods in 2011 were estimated to be 'dried processed foods' at US$955.9 million, followed by 'dairy' at US$948 million, and 'bakery' at US$765.9 million.

These positive numbers were contributed by snack product produced locally and internationally. Chips represent the biggest market share of the snack industry with their sales figure amounting to RM102.1 million (FAMA, 2013). Locals' manufacturer has focused on traditional chips such as salty cassava chips, hot or chilly cassava chips, sweet potato chips, yam chip, and the most popular banana chips.

Banana chips consist of banana figs, sweet banana chips and savoury banana chip. The most common banana chip in Malaysia is savoury banana chips. The global volume sale of this particular product grew at 4.6% from 1,925,000 tonnes in 2010 to 2,010,000 tonnes in 2011. The figure shows that banana chip is performed well in the market (Anon, 2010). Therefore, a better understanding of the current status of banana chip can significantly help our local manufacturer in determining their market potential. Thus, this study will investigate on in what extent those consumers choose to consume this snack rather than other meals. Factor affecting the acceptance levels of consumer on processed banana snack product (banana chips), will be measured based on four independent variables (i) hedonic: texture, appearance; (ii) perceived value: price; (iii) nutritional value and (iv) packaging. Perhaps, this study can be an assist or guideline for the local authority and manufacturer to get a better understanding of the consumer perception and improvement of local snack products in the future.

2 LITERATURE REVIEW

2.1 *Snack trend*

An exact consumer definition of snack food according to US Market Trends and Forecast (2011) is illusive and commonly allocated to salty foods eaten between regular meals. Synonymous with snack foods are savoury snacks, potato chips, snack nuts, and popcorn (US Market Trends and Forecasts, 2011). Savoury snacks consist of pretzels, tortilla chips, cheese curls/puffs, and meat snacks. Other foods eaten as snacks may be sweet and generally referred to as candies. Chocolate, alone or in combination with other ingredients, is one of the most popular sweet snacks. Beckley & Moskowitz (2002) who studied on the expanded definition of snack foods of both salty and sweet foods, to assess a number of components of a single food as the consumer might see them. The value of snack food and how it catches the consumer eye outside-in seems to be a phenomenon.

According to a survey done by the American Snack Food Association (2007), potato chip was the most popular snack food exported to Canada, Western Europe, The Asia Pacific Rim and the Middle East. This was followed by tortilla chips, other corn-based products, pretzels, extruded snacks, and ready-to-eat popcorn. The number of consumers who snacked between meals at least once a day was highest in the United States (80%), followed by Germany (74%), united kingdom (69%), Japan (66%), Brazil (64%), Italy (62%), Spain (57%), and France (53%) (Euromonitor, 2010).

In 2009, Japan had the highest per capita consumption of snack food in term of value at US$70.6, but in term of quantity, the United States had the highest per capita consumption at 8.6 kg. However, in the year 2010, the United Kingdom overtook Japan as the main consumer of savoury snacks in term of value. In the year 2011, the consumption of savory snacks in Japan did not show a very encouraging trend because Japan was still suffering from the recession. The consumers were still very caution with their expenditures and there was also increased competition from fast foods. The recession and competition from fast food had also affected the consumption of snack foods in the other Asia Pacific region (Auni N.H., 2006). Australian Trade Commission (2005) in their reported shows that nutritional snacks such as fruit snack products have the opportunity to enter the Malaysian markets.

2.2 *Banana snack*

Local chips made of tapioca and bananas are not only having loyal followers within the country, but also from abroad. Under the concept of One Village One Industry, the snack industry has played a significant role in driving the rural economy and expands the rural people's income, where many of the chip makers work in groups (Auni N.H., 2006). Released report from FAMA showed that Malaysia's traditional snack industry is growing rapidly with the sales figure hitting RM200 million (66 Million USD) in 2012. While consumers have plenty of choices when comes to local snacks, crispy chips made of tapioca, banana, sweet potato and onion, are the favourite ones. Presently, the banana chips are exported to the Middle East and Arab countries. However, due to the shortage of raw ingredients, it was a serious impediment whereby this product cannot be supplied to other potential countries. In order to overcome the shortfall in the ingredients to process the chips, FAMA has established the contract farming system to purchase the yield which planted by farmers to fulfil the industry needs.

The chips represent the biggest market share of the snack industry with their sales figure amounting to RM102.1 million (34 Million USD). FAMA's Chairman, Datuk Badruddin Amiruldin said that the increase in the value in sales of the snack industry in Malaysia at present has been very encouraging especially looking at the high demand. Initially, this cottage product received an unenthusiastic reaction but things have changed as they stand at par with imported snacks with numerous flavours, in uniform shapes and sizes, impressive packaging and health benefits (BERNAMA, 2013).

2.3 *Key attributes*

The consumer's preferences and acceptance level become sophisticated. Imran (1999) in their previous study mentioned that consumer's perceptions of (and attitudes towards) food attributes can be assumed to be key determinants of the success of any particular brand within a product category, including the fast-growing one of healthy snacks. Otherwise, brand attitudes and loyalty come from memory of previous experience of sensory attributes such as taste and odour along with marketing sensory inputs such as point-of-purchase information for advertising, labelling/nutritional contents, display aspects, pricing, brand/product prestige, familiarity and health beliefs.

Raging with the competitive food market today, with many new products, and changing customer preferences, companies progressively use sales-shelf appearance of food products, including the colour of the packaging, appearance, terminology and brand name to gain trial through gaining an immediate perception of the foods' characteristic sensory qualities (Martin, 1990). Psychologically, consumers are provided with extensive choice attributes which impact on the consumer's ability to make informed and confident purchasing decisions. These include the provision of packaging which hopefully reflects the product's desired qualities.

2.4 *Hedonic*

Adapted from Stone & Sidel (1993 and Ramcharitar et al. (2005) model, product hedonic characteristic consists of appearance, colour, taste and texture. Taste-motivated preferences have long been shown to provide an enduring motivation for dietary change (Gladstone, 1941; Sweeney, 1942). The taste remains one of the most important motivations in choosing a certain type of food (Steptoe et al., 1995; Verbeke, 2006; Enneking et al., 2009). In sensory study, taste is signified in terms of five basic tastes, consist of sweet, sour, salty, bitter and umami, the latter a savoury meaty kind of taste. It has been suggested by Pangborn and Giovanni (1984) that a preference for sweet taste correlates with the consumption of sweet-tasting products.

2.5 *Pricing*

Psychological pricing has been used by marketers over the years to influence buying behaviour of consumers. Many businesses use psychological tactics in pricing their product or service; sometimes unknowingly. Psychological pricing is the practice of structuring and presenting prices to appeal to consumers' emotions and to influence their decision-making processes (Pride and Ferrell, 1997).

2.6 *Packaging*

Packaging seems to be one of the most important factors in purchase decisions made at the point of sale (Prendergast and Pitt, 1996), where it becomes an essential part of the selling process (Rettie and Brewer, 2000). Packaged food products are moving into ever larger supermarkets and hypermarkets, and there is a proliferation of products, offering consumers vast choice.

2.7 *Nutritional value*

Consumer demand for sustainability is not yet strong in Malaysia, even though there is a concern to food safety that is at times associated with organics. The general perception among consumers is that 'organic' means 'safer'. With a growing interest in sustainability in other sectors (for example green building, clean technology, and bio fuels), Malaysian consumers are likely to become more sensitive when making food and beverage purchases in the future. The three keys, broad consumer brand-selection elements of Lawer and Knox (2006) can be considered in healthy purchases relating to choice criteria (of health claims and product alternatives), involvement and knowledge (both of specific interests in areas of health, or product/brand claims and reputation, or both).

3 ISSUE

Consumption of snacks has become a trend in Malaysia. Snack food provides a feeling of convenience and quick health enhancer. The snack food includes the manufacture of potato chips, corn chips, popcorns, pretzels, extruded snacks, seeds and nuts (peanuts) and chocolates. The locally produced savoury snacks are chips and *Keropok*. In general, the demand for snack food in Malaysia, is an increase in the future, as there is an increasing number of working women from 4.7 million in 2010 to 5.3 million in 2011 (Market Watch, 2012). A recurrent issue in both science and business for food products is that absence of a significant part of data about what features are important to consumers. In some product categories such as snack food, there is a great deal of information, but much of that information concerns on the well-known product. The lack of information on the consumer's preferences on local snack product indirectly affects the local food industry environment. A better exposure of this product can signify the potential to penetrate international snack's market. Underlying of this issue, this survey will undertake to study the current status of the banana chip industry, to identify the key attributes of acceptable level of consumer towards this particular product and to define the specific needs to be promoted for the market potential of this product.

REFERENCES

Alreck, P.L. and Settle, R.B. (1999), Strategies for building consumer brand preference, *Journal of Product & Brand Management*, 8(2), 130–44.

American Snack Food Association. (2007). http://www.sfa.org/.

Anon. (n.d). Snack and sweets statistics. http://www.fiery-foods.com.

Austrade (2002). Australian Trade: Processed food to Malaysia. Australian Trade Commission. Retrieved from http://www.austrade.gov.au/print_templat.

DOA. (2000). TeknologiPisang. Department of Agriculture, Kuala Lumpur.

Euromonitor. (2003). Savoury Snack. Retrieved from http://www.euromonitor.com.

Euromonitor. (2010). Euromonitor International: Country Market Insight (November 2010), Packaged Food—Malaysia.

Euromonitor. (2011). Euromonitor International: Country Market Insight (November 2010), Packaged Food—Malaysia.

Euromonitor. (2012). Euromonitor International: Country Market Insight (November 2010), Packaged Food—Malaysia.

Gladstone, I. (1941), "Motivations in health education", Journal of the American Dietetic Association, 25, 745–51.

Imran, N. (1999), "The role of visual cues in consumer perception and acceptance of a food product", *Nutrition & Food Science*, 99(5), 224–8.

Lawer, C. and Knox, S. (2006), Customer advocacy and brand development, *Journal of Product & Brand Management*, 15(2), 121–9.

Market Watch (2011). Retrieved from http://ec.europa.eu/trade/creatingopportunities/bilateralrelations/countries/malaysia/. Accessed on 21st November 2011.

Martin, D. (1990), "The impact of branding and marketing on perception of sensory qualities", *Food Science and Technology*, 4, 44–9.

MIDA, (2012), http://www.mida.gov.my/env3/index.php?page=food-industries.

Moskowitz H., Rachel K., Beckley J., and Ashman H. (2004), Understanding Conjoint Analysis, Food Technology Magazine, 58(1).

Pangborn, R.M. and Giovanni, M.E. (1984), "Dietary intake of sweet foods and of dairy fats and resultant gustatory responses to sugar in lemonade and fat in milk", Appetite, Vol. 5 No. 4, pp. 1217–8.

Prendergast, G. and Pitt, L., (1996). Packaging, marketing, logistics and the environment: are there trade-offs? International Journal of Physical Distribution & Logistics Management 26(6), 60–72.

Pride, W.M. and Ferrell, O.C., (1997): Marketing Concepts and Strategies, (10th ed.), New York, HoughtonMiffl in Co.

Rettie, R and Brewer, C. (2000), The Verbal and Visual Components of Package Design, *Journal of Product and Brand Management,* 9(1), 56–70.

Schutz, H.G. (1965). A food action rating scale for measuring food acceptance. Journal of Food Science, 30, 365–74.

Shewfelt (1990). Quality of fruit and vegetables. *Food Technology,* 44(6), 99–106.

Steptoe, A., Pollard., T. & Wardle, J. 1995. Development of a measure of the motives underlying the selection of food: the food choice questionnaire. Appetite 25, 267–84.

Hospitality and Tourism – Sumarjan et al. (eds)

Guest perception on service quality among front office personnel

S.H.S. Mohamad Khairizam, M.A. Hafiy Azmi & N.H.C. Ahmat
Universiti Teknologi MARA, Malaysia

ABSTRACT: The problem with service quality may perhaps relate with the difference of perception about service quality between the front office personnel and the hotel guests. Using SERVQUAL dimensions, a closed-ended survey was created and 300 respondents were selected to participate in this study. The data was analysed by using Cronbach's Alpha for reliability test, together with the descriptive statistics of frequencies, mean scores, and standard deviation. Findings revealed that tangibility aspect is very important to influence the level of customer perception on service quality. Tangibility dimension include physical aspect such as, professionalism and neatness of personnel and also the physical appearance of the hotel (Parasuraman et al., 1991; Sriyam, 2010). Further study should investigate the other departments that have direct contact with the guest such as food and beverage and housekeeping departments.

Keywords: customer perception, SERVQUAL, service quality

1 INTRODUCTION

Customer satisfaction has been identified as a key performance indicator in hotel industry. According to Ching (2008), the demand for hotels in Malaysia increases as the quality standard increases parallel to the Visit Malaysia campaign in 1990 and also Commonwealth Games in 1998. For most customers, the quality perception of a service operation starts at the front entrance and in what the facilities appears to offer them. It is known that the front office is the nerve centre of the hotel which requires them to keep up with the latest information about the hotel operations. Service quality for hospitality industry is defined as "the global evaluation or attitude of overall excellence of services" or in other words of saying, it is about how well a delivered service conform the client expectation (Parasuraman, Berry and Zeithaml, 1985). It is important to realize that, there are guests who require only basic necessities while others need the full resources from the hotel. With that, understanding the guest needs is the number one priority.

Service quality is an important subject in service industry to evaluate the extent to which the service meets and exceeds customer needs and expectation (Zahari, Yusoff, & Ismail, 2008). It has become a major area of attention to managers and researchers because of its strong impact towards business performance. Nevertheless, the problem with service quality may perhaps be related with the difference of perception about service quality between the front office personnel and the hotel guests. Some factors that could affect the service quality are comfortable and welcome feeling, need appearance of staff, professionalism of staff, friendliness, special attention given by staff, quick check in and check out, and handling the complaint (Sriyam, 2010). Previous study by Kuveya (2001) had mentioned "front office staffs have to know that the Open Distance learner is not homogenous but the heterogeneity". Meaning that, delivering a service requires team chemistry and the initiative in order to perform better to exceed the expectation of hotel guest. Several scholars argued that to retain customer through loyalty is much more difficult compared to obtaining a new customer (Soteriou & Zenious, 2000) and this is another barrier for a hotel to shine brightly in the industry. The moment of truth is essential and begins at the front office department because this stage is where perceptions are made. Prior study on service quality in Mauritian hotel found out that by focusing on reliability and assurance, they can achieve high level of satisfaction and service quality (Juwahee & Ross, 2003).

In line with the above mentioned statement, this paper intends to examine the guest perception on service quality among front office personnel by looking into a different contextual setting. This study is vital for better understanding on what to prepare when delivering the service and to ensure the hotel establishment realize that they have to get back on track on what remarkable service is defined. Also, it will become practical guidelines in training the staff for the development of skills among employees in front office department so the expectation of guest will be exceeded. The research was carried out to examine the guest perception on service quality among front office personnel.

2 REVIEW OF LITERATURE

The key to success in hotel business is when the service initiates with the first service encounter between guest and front office personnel (Roldan & Amelia, 2004) because the interaction between the guest and the front office personnel occurs more than any other department (Beatriz, Nullar, Blesilda, & Reside, 2009). This is the stage where the first impressions are formed based on the level of quality of service provided. In front office department, there are sub-departments which are, front desk, reservation, operator, concierge, and bellman. Ankara (2009) stated that front office is a place where guest arriving to the accommodation business or welcome, accommodation record and obtain, reservations are taken, account and record are kept, consultancy services are provided and cooperation with other departments is ensured and thus it can be considered as the heart of facility and the responsibilities of the front office personnel.

Hospitality services are a harmonious mixture of three elements material products, behaviour and attitude of employees and the environment (Reuland, Coudrey, & Fagel, 1985). Previous study had shown most of the service breakdown caused by the hotel started with the employees during the service delivery. Not only the management and the recognition of quality among employers but also the quality provides by the service employees (Hang, 2008). Moreover, Hang (2008) stressed that hotel leaders' ignorance on service management will also directly influence each of the employees, hence will cause service personnel's weak consciousness of customer service and lower the entire service level of the hotel. In brief, it could be said that there was a causal link between a good organizational climate and the level of service quality in a hotel (Davidson, 2003) which means that, the service quality of a hotel business is also depending on the management performance thus influence the employee who is responsible for the service delivery system.

Previous study had shown SERVQUAL model developed by Parasuraman et al. (1985) is the most appropriate tools to measure the level of customer satisfaction on service quality. This instrument was created to measure the scale of quality in the service sector in the aspect of tangibility, reliability, responsiveness, assurance, and empathy. It is believed that these factors drives hotel guest to return back to the same hotel in the future. Tangibility means the level of physical appearance of product or service. Nevertheless, in hotel area, intangible products such as the ambiance and environment plays important role to be measured and evaluated. Albayrak, Caber and Aksoy (2010) affirmed that even both tangible and intangible products are dissimilar in characteristics, but they can be differentiated by their effects on customer satisfaction and for a better impact is to combine both for improving customer satisfaction (Xu, 2009).

Furthermore, reliability is the ability of a worker to perform his work as promised. It is also can be defined as competence which means the front office staff must be knowledgeable and confidence when handling customers' questions and request (Zeithaml, Parasuraman, & Berry, 1990). Meanwhile, responsiveness is related to the willingness of the employees to assist guests and provide prompt service. Hotel establishments are responsible to provide better experience to the guests along their stay in the hotel as the level of hospitality skills determine the guests will return to the particular hotel again (Arrifin & Magzhi, 2012).

In addition, assurance is the knowledge and courtesy of the employees and their ability to build trust and confidence to the customers (Parasuraman et al., 1988). Customers usually compare their expectation towards the service provided with the performance outcomes (Brunner-Sperdin, Peters, & Strobl, 2012) and they will always expect a high quality service in front office department as the personnel play an important role that effects customers' perception in service settings. Therefore, front office personnel should create a desirable atmosphere to optimize social interactions to meet customer expectations and later affects their satisfaction level (Brunner-Sperdin et al., 2012). On the other hand, empathy relates to the individualized attention or caring that the front office personnel need to provide to the guest. It is important to note that, many specific experiences involve a broad range of mixed emotions or ambivalence (Roster & Richins, 2009). There might be customers who have a high level of expectation when visiting a hotel that consist of negative and positive feelings. Negative feelings consist of sad, fury, anxiety, regretful, and in hurry while positive feelings consist of pleasure, and happiness. Hence, service with a more personalized touch will impact more tranquil, safe, and harmonious (Thuy & Hau, 2010) other than helping the customer to feel the value of their money whether it is reasonable compare to what he or she received from the service provided by the hotel (Ladhari & Morales, 2007).

In line with the above notion, Gronroos (1984) defined perception of quality as the result of an evaluation process, where the customer compares expectation and the perception of service quality received. Gronroos (1984) also placed the perception of service and service expected is different between each other. In accommodation service research, the study focuses on hotels, without particular reference to the perception comparison between employee and customer (Saleh & Ryan, 1991) while other said service quality is measured by evaluating perception and expectation together (Brown, Churchill, & Peter, 1993). So the criteria, requires a clear definitions on the perception of quality of service.

3 METHODOLOGY

Descriptive cross-sectional study was used to describe the characteristic of the variable of interest in this research. This research intended to examine the guest perception on service quality among front office personnel. According to Tourism Malaysia (2013), the local tourist arrival in Kedah for 2012 was 1,640,341 million. The study was conducted in Alor Setar as the capital of Kedah. It is presumed that local tourists transit in Alor Setar before travelling to Langkawi Island. Based on Malaysia Association Hotel (MAH, 2010), the number of hotels in Alor Setar is 26 and it is important to highlighted that the study will only focusing on the several 3 and 4 stars hotels in the city of Alor Setar which are Holiday Villa City Centre, Grand Crystal Hotel, Sentosa Regency Hotel, The Regency Hotel, and the Star City Hotel. Purposive sampling method was used as researchers believed that the respondents who had experienced staying in the hotels in Alor Setar are able to answer the survey accurately. Thus, a closed ended survey was used to screen out the level of service quality provided by the front office personnel in the 3 and 4-star hotel in Alor Setar, Kedah. The questionnaires were distributed evenly among all five hotels mentioned above to be answered by the hotels' guests. For adequate analysis, this study requires 300 respondents. By using 5 point Likert's scale, the items used to measure the level of satisfaction or dissatisfaction of the respondents from Strongly Disagree (1) to Strongly Agree (5). The questionnaire consisted of two sections, Section A (demographic profiles) and Section B (guest perception on service quality of front office personnel).

4 FINDINGS AND CONCLUSION

The internal consistency of the SERVQUAL item was assessed by computing the total reliability scale with the result of 0.926 and each of the dimensions were reliable and acceptable with result mostly more than 0.686. Though, out of 300 surveys, only 127 responded and analysed using SPSS version 20. Based from the frequencies statistics, it was found that, more female respondents (64.6%) participated in this study than male respondents (35.4%). Majority of the respondents was less than 25 years old (44.9%) or can be categorized as young travellers. It was found that more than 50 respondents working as government official and more interestingly, the frequency level of the respondent who stayed in hotels around Alor Setar is only one time (n = 88) or it was their first visit to the hotel. The frequency of visit can influence the perception of the guests towards the quality of service provided by the hotel. Furthermore, anecdotal evidences show that most probably majority of the respondents just transit to Alor Setar and they were intended to travel to Langkawi for the next day. Most importantly, a vast majority of the respondents (n = 69) confirmed that they have no intention to return to the same hotel in Alor Setar, Kedah anymore in the future. Based on the observation, it was hypothesized that among the reason why they refused to return back is because they had figured out how to travel to Langkawi without transit or staying in hotels in Alor Setar. The critical part is most of the hotels' customers were not satisfied with the quality of service provided by the front office personnel throughout their stay in the hotel. This requires serious attention from the management to do further service recovery. As predicted, most of the respondents visited Alor Setar for vacation purpose (66.1%).

Descriptive analysis in Table 1 provide an information of continues variable that influence the guest perception on service quality provided among front office personnel. The outcomes revealed that the highest mean scores is tangibility with mean score of 3.6811 (*SD* = 0.68623) which supported study done by Wong, Mei, Dean, and White (1999) and Sriyam (2010) who investigated that tangibility is the important key dimension of service quality. Tangibility dimension includes physical aspect such as, professionalism and neatness of personnel and the physical appearance of hotel. The result showed the guest agreed that the

Table 1. Mean scores and standard deviation for service quality dimension.

	(M)	*(SD)*
Tangibility	3.6811	0.68623
Reliability	3.6713	0.67532
Responsiveness	3.4751	0.75770
Assurance	3.6472	0.64609
Empathy	3.4724	0.78631

Valid n = 127.

front office personnel able to fulfil the tangible aspect. Moreover, empathy dimension showed the lowest mean scores of 3.472 (SD = 0.78631). Respondents slightly agreed that the front office personnel managed to show their empathy in terms of willingness, caring, effective communication, and understanding the customers (Parasuraman et al., 1991). Based on the result, it could be said that the front office personnel did not have a good communication skill and not too concern about the guest need.

From the mean scores tabulated in Table 2, respondents agreed that the staff uniform is clean (M = 3.88) and they also agreed that the front office personnel managed to inform them exactly when the service will be provided (M = 3.76). The result indicates, when the personnel is wearing a clean and neat uniform, the guest will feel welcome and comfortable to interact with the personnel. Meanwhile, it could be concluded that most of the respondents at least slightly agreed towards all the items proven through the mean scores of above 3.00. Nevertheless, under responsiveness aspect, the respondents neither agree nor disagree that the front office personnel gave them prompt service during their stay in the hotel. This mirror the quality of service provided and the management should seriously look into that since service quality will determine repeat patronage. Moreover, the personnel must have determination when helping a guest and always ready to provide the service.

Table 2. Mean scores and standard deviation for each item in the dimensions.

	(M)	(SD)
Tangibility		
1. The staff dresses appropriately.	3.73	0.791
2. The staff uniform is clean.	3.88	0.720
3. The staff provides the service with smiling.	3.56	0.851
4. The staff has attractive appearance.	3.55	0.888
Reliability		
1. The staff can provide you the service as promised.	3.65	0.761
2. The staff provides you accurate information.	3.65	0.858
3. The staff performs the service right at the first time.	3.73	0.830
4. The staff offer you some help.	3.65	0.903
Responsiveness		
1. The staff tells you exactly when service will be provided.	3.76	0.877
2. The staff give you prompt service.	3.00	1.113
3. The staff are willing help you.	3.66	0.893
Assurance		
1. The staff has product knowledge of hotel information.	3.65	0.802
2. The staff has required skill to perform service.	3.70	0.759
3. The staff speaks with you by using appropriately address forms.	3.72	0.879
4. The staff is trustworthy.	3.56	0.803
5. The staff makes you feel safe when staying at the hotel.	3.61	0.808
Empathy		
1. The staff is able to communicate with you in English.	3.35	0.954
2. The staff is able to communicate effectively with you.	3.68	0.853
3. The staff shows the personal attention to you.	3.45	1.006
4. The staff knows your specific need.	3.42	1.019

For assurance dimension, the personnel speck with the guest by using appropriately address form earned mean score of 3.72 (SD = 0.879). The personnel must have knowledge and confidence when addressing guest by using Dato', Datin, Prof, Mister, Miss, Sir, or Madam so that the guest will feel honoured when the staff acknowledged them. Also, Parasuraman et al. (1991) depicted empathy dimension represent the provision of individualized attention and caring towards customer. Customers want the service personnel to give attention towards their needs and wants. This finding aligned with Sriyam (2010) who identified that effective communication among front office personnel is the main factor that contributes to customer perception.

5 CONCLUSION AND RECOMMENDATIONS

It was found that most of the respondents are young travellers and vacation is the major purpose of trip to stay in hotel at Alor Setar areas. Nonetheless, most of the respondents were the first time travellers or first time staying in the hotels which somehow affect the result. Their expectation maybe higher than those who had experienced stayed in the same hotel more than one time. Furthermore, it can be concluded that the respondent probably stayed in Alor Setar hotels just for their transit before travelling to Langkawi. The study also found that the district of Alor Setar itself could be one of the factors that affect the guest perception toward service quality among front office personnel. In other words, it is important for the hotel management to train their service personnel to follow the attributes in the service quality to ensure that they provide a quality service required by the guest. The guest agreed that the service personnel fulfil the tangibility aspect while slightly disagreed that the front office personnel managed to show their empathy towards them.

Therefore, hotel management should arrange special courses to their service personnel to improve their communication. Among all dimensions, tangibility was found to be the strongest dimension of perception. Therefore, hotel management should maintain the attributes of tangible service quality at the hotel. For the next step, human resource management should arrange in-house training program to improve the main work of front office staff and to promote them in their careers. Other than that, the scope of the study is towards the front office personnel only hence further study should embark on few other departments which have direct interactions between customer and employees for example, Housekeeping, Food and Beverage, and Sport and Recreational facilities.

REFERENCES

Albayrak T., Caber M., & Aksoy S., (2010). Relationship between Tangible and Intangible Elements of Tourism Products with overall Customer Satisfaction. *International Journal of Trade, Economics and Finance. 1*, pp. 140–143.

Ankara, (2009). Directorate General of Research and Training. *Ministry of Culture and Tourism Publications.*

Arrifin, A.A.M. & Magzhi, A. (2012). A preliminary study on customer expectations of hotel hospitality: Influences of personal and hotel factors. *International Journal of Hospitality Management, 31*(1), 191–198.

Beatriz M., Nullar M.B., & Blesilda D., (2009). Front Office of Accredited Hotels in Baguio City: Profile, Problems and Strategies for Quality Service. *University of the Cordilleras Research Journal, 1.*

Brown, T.J., Churchill Jr, G.A., & Peter, J.P. (1993). Improving the measurement of service quality. *Journal of Retailing, 69*(1), 127–139.

Brunner-Sperdin, A. Peters, M. & Strobl, A. (2012). It is all about the emotional state: Managing tourists' experiences. *International Journal of Hospitality Management. 31*(1), 23–30.

Ching, P.W. (2008) *Malaysian Economy. Malaysia*: Prentice Hall, 2008 Davidson, M.C.G. (2003). Does organizational climate add to service quality in hotels? *International Journal of Contemporary Hospitality Management.* (15), 4, 206–213.

Chung-Wei Hang, (2009). *"From Quality to Utility: Adaptive Service Selection Framework."* Proceedings of the 8th International Conference on Service Oriented Computing (ICSOC), pp. 456–470. San Francisco, USA.

Gronroos, C., (1984). *Service Management and Marketing.* Lexington, MA: Lexington Books.

Juwaheer, T.D. & Ross, D.L. (2003). A study of hotel guest perceptions in Mauritius. International *Journal of Contemporary Hospitality Management, 15*(2), 105–115.

Kuveya, (2001). *Strategic Management.* Module MGT 204. Zimbabwe Open University, Harare.

Ladhari, R. & Morales, M. (2007). Perceived service quality, perceived value and recommendation: A study among Canadian public library users. *Library Management, 29*(4/5), 352–366.

Malaysian Association of Hotel MAH (2010) Retrieved on April 8, 2013 from http://www.hotels.org.my/index.php/member-listing?catid=25.

Ministry of Tourism Tourist Arrivals & Receipt to Malaysia (2011). Retrieved on April 8, 2013 from *http://corporate.tourism.gov.my/research.asp?page=facts_figures.*

Parasuraman. A., Zeithaml, V.A. & Berry, L.L., (1985 & 1988). A conceptual model of service quality and its implication. *Journal of Marketing, 49*, 41–50.

Reuland, R. Coudrey, J. & Fagel, A., (1985). Research in the field of hospitality. *International Journal of Hospitality Management, 14*(4), 141–146.

Roldan & Amelia, S. (2004). Front office procedures and guest relations. *Paranaque: AR Skills Development & Management Services.*

Roster, C.A. & Richins, M.L. (2009). Ambivalence and Attitudes in Consumer Replacement Decisions. *Journal of Consumer Psychology, 19*, 48–61.

Saleh, F. & Ryan, C. (1991). Analyzing service quality in the hospitality industry using the SERVQUAL model. *The Service Industries Journal, 11*(3), 324–343.

Sekaran U. & Bougie R. (2010). *Research Methods for Business* (5th ed.).

Soterious, A. & Zenious, S. (1999). Operations, Quality, and Profitability in the Provision of Banking Services. *Management Science, 45*(9), 1221–1238.

Sriyam, A. (2010). *Customer's Satisfaction towards Service Quality of Front Office Staff.* Paper Presented in Partial Fulfilment of the Requirements for the Master of Arts Degree in Business English for International Communication at Srinakharinwirot University, Thailand.

Thuy, P.N. & Hau, L.N. (2010). Service personal values and customer loyalty: A study of banking services in a transitional economy. *Journal of Bank Marketing, 28*(6), 465–478.

Wong Ooi Mei, A., A.M. Dean, & C.J. White. (1999). Analysing service quality in the hospitality industry. *Managing Service Quality, 9*(2), 136–143.

Xu, J.B. (2009) Perceptions of tourism products, article in press, Tourism Management 2009, doi: 10.1016/j.tourman.2009.06.011.

Zahari, W., Yusoff, W., & Ismail, M., (2008). FM-SERVQUAL: a new approach of service quality measurement framework in local authorities. *Journal of Corporate Real Estate, 10*(2), 130–144.

Zeithaml V.A., Parasuraman A., & Berry L.L., (1990). *Delivering quality service: Balancing customer perceptions and expectations.* The Free Press, New York. NY.

Hospitality and Tourism – Sumarjan et al. (eds)
© 2014 Taylor & Francis Group, London, ISBN 978-1-138-00151-0

Advertising effects toward the consumer purchase behaviour on the SMEs product

A.D. Aminuddin, M.S.Y. Kamaruddin & N.H. Zainul
Universiti Teknologi MARA, Malaysia

ABSTRACT: In the emerging economies, Small and Medium Enterprise (SMEs) is one of the sectors in Malaysia becoming increasingly significant in the process of exporting and industrialization. Despite the fact that SMEs industry is growing, there are some barriers for the SMEs need to face related to problem in entering the market and also heavy competition with others well establish company in order to keep them sustain in the industry. Therefore, the marketing strategies need to well developed and the focus is toward the advertising and promoting activities. The advertising is one of a process of communication that function to deliver information which also use as an attention grabber tool and at the same time to convince of consumers to purchase of certain products or services. This study will focus more into these two elements in order to identify the awareness at the same time lead to the purchase intention of consumers toward SMEs products especially in food and beverage production. The study will be conducted by using the AIDA model and more than 300 respondents in an area of Selangor will be selected by using the convenience sampling to accomplish the study.

Keywords: AIDA model, advertising and promotion, consumer purchase behavior, SMEs

1 INTRODUCTION

Nowadays, the successful companies usually strongly focused on customer and greatly dedicated to marketing strategies or decisions (Mohammad et al., 2012). As were mentioned in the previous study, marketing decisions are the most important that small business owners to focus, ahead of other business decisions in the matter of finance and employee compensation (Pineda et al., 1998). Related to this, the 4Ps framework which represented of product, price, promotion and place are the base of marketing decision making. The function of the 4Ps classification on the marketing mix is actually in order to understand the fundamental of marketing (Keh, Nguyen and Ng, 2007). The fundamental of marketing is able to assist the small business to survive and at the same time discover the needs and wants of the target markets more effectively and efficiently compared to the other competitors (Emmanuel, 2009).

Therefore, in order to keep effective and efficient, marketers need to develop strong marketing strategies. By that, it is not only significant to help the small business to survive, but it could also help them to attract potential customers to purchase the offered products or services at the same time enhance the sale and profitability of the business (Mohammad et al., 2012). From the same sources, the marketing strategies were dominated by large companies. Compare to the SMEs, it much differs which they are less preferred in the market due to their low marketing strategies in term of promotion and advertise their products or services. Hence, the SMEs actually should plan strategies of marketing that significantly help them to influence their target customer in making the decision to purchase their products (Marjanova and Stohanovski, 2012). As a result, this is where the advertising and brand management can play important role in SMEs to influence the purchase intention toward products offered (Muhammad et al., 2012). From the earlier 4ps models, this study is focusing on the aspect of promotion, of which the advertising is the subset matter. The challenges in SMEs are to create product awareness that could lead into customer purchase intention. Thus the purpose of this study is to investigate on how the advertising activities could create awareness and influence the purchase intention of the SMEs products.

2 LITERATURE REVIEW

2.1 *SMEs in Malaysia*

Small and Medium Enterprise (SMEs) refers to any manufacturing company or companies offering manufacturing related services with yearly turnover not above than RM25 million and occupy full-time employees less than 150 (Mohammad, Setapa and Rahman, 2012). But it is different when it comes to the international business, the SMEs defined based on the socioeconomic growth of every country. In the United States, SMEs is any companies with fewer than 500 employees (Basefsky and Sweeney, n.d.). Meanwhile, In Taiwan the SMEs defined as an establishment with 650 employees or less (Chelliah, Sulaiman and Yusoff, 2010). In a simple explanation, the term of SMEs can also be defined as an independent management, the person who takes risk are entrepreneurial, the competitive power in the current markets, the marketing power in the active market, close cooperation between business owner and staff, cost of business and composition of the expenditures of business owner, low profile of expert level (Kaya and Alpkan, 2012)

Compare to now and then, the SMEs is nowadays becoming increasingly significant in the process of exporting and industrialization in the emerging economies (Hashim, 2012). As has been explained by one of the group CEO of Malaysia SMEs, the number of SMEs companies were estimated to grow over to a million in 2012 (Wong, 2012) and will be estimated to increase towards a high-income economy by 2020 (Department of Statistics Malaysia, 2012). In addition, it has been recorded on an annual report 2010/2011, the SMEs represented about 99.2% of the entire business formations in Malaysia in 2010 (Moorthy et al., 2012). The performance of SMEs is important as they will change Malaysia into a high-income and knowledge-based economy through their role to the national GDP. In addition, as been stated by the Department of Statistics Malaysia (2012), the services sector was the largest in SMEs. The total in this sector is more than 580,000 establishments (98.2%).

In the census 2011 showed that the percentage of micro establishments stays the same as different to the census of 2005 (Department of Statistics Malaysia, 2012). Overall, the majority 77% of SMEs were micro-sized establishments, 20.0% of small establishments, and 3.0% of medium-sized establishments. The services sector with percentage of 79.6% is the highest percentage of micro-sized establishments, followed by manufacturing and agriculture 57.1% and 56.3% respectively. In addition, with the percentage of 42.1, the mining and quarrying sector were mainly small size of SMEs.

It also reported most of the SMEs in represented more than 50% were concentrated in Selangor, WP Kuala Lumpur, Johor and Perak, Meanwhile, in Sabah and Sarawak the total of SMEs represented up to 13.1% in Malaysia (Department of Statistics Malaysia (2012).

From all of the established SMEs in the country that has been stated, it has shown that the contribution to GDP has increased from 29.4% in 2005 to 32.5% in 2011 (The Department of Statistics Malaysia, 2012). Between the periods of 2006 to 2011, growth of SMEs is the most excellent in overall economic growth whereby in 2011, the growth registered by SMEs was 6.8% compared to the overall growth of 5.1%. Refer to the details given, it is proved the SMEs getting move forward to the economic growth and in Malaysia it becomes one of the important industry. Therefore, to be competitive in the global market, the strategy practices by SMEs in Malaysia are important and need to develop (Muhammad et al., 2010).

2.2 *General challenges of SMEs*

According to a study from the World Bank Productivity and Investment Climate Surveys has revealed that there are six factors will influence the performance of Malaysian SMEs (SME Cooperation Malaysia, 2012). The analysis found that innovation and technology adoption was the most important performance force, having the highest impact on total factor productivity and the growth of employment. This was followed by human capital development, access to financing, market access and to a lesser extent regulations and infrastructure. Meanwhile, based on the Report of ACCCIM 2012 SMEs Survey (2012), in the report stated that the challenges that with the hard competition in the global market, Malaysian SMEs have to plan for these changes related to innovation, human capital, market entrance, financing etc. in order to grow and be competitive.

Some of others challenges are included having a complexity in facing recession, obstacle from global sourcing, low productivity, lack of managerial skills and financing, difficulty in accessing management and technology, heavy regulatory burden and others (Muhammad et al., 2010). Nevertheless, other challenges found are the barriers to entrepreneurship namely lack of access to credit and lack of access to formal business and social networks. Besides, as been explained, SMEs is dealing with intensified global challenges, new emerging technologies in ICT and production process as well as increasing factor costs, which affect the export competitiveness.

2.3 *Marketing, promotion and advertising of SMEs toward the purchase behaviour*

Based on a book entitled "Advertising and Promotion" (Belch and Belch, 2003), the advertising and promotion are considered as the primary part of the social and economic systems.

In addition, the authors also mentioned that, based on our multipart society, the advertising has changed into a very important communications system for the business and also consumers. Advertising can be defined as a persuasive communication in paid form in order to catch the intention of the target audience at the same time to give information about the product either goods or service (Moriarty et al., 2009). Strategic communication that targets to achieve something whereby to build impact toward consumers in term of their understanding information or persuading a person to do something is known as a modern advertising (Wijaya, 2012).

To generate a well-known status in the market and at the same time to achieve the power of brand acceptance of certain products, the company needs to be done it through the advertisement and marketing promotions (Hemilä and Oinas, 2009). However, the promotion and market research were the most frequent problems of SMEs in the areas of marketing (Simpson and Taylor, 2002). Due to this issue, in order to make their position and competitiveness on the market, their skills and knowledge needs to improve (Mirchevska and Mirchevski, 2007) especially in the role of marketing. The organization can view the role of marketing within an organization as an inner focus on the use of marketing. Therefore, the marketing will be integrated into the business plans and at the same time use to generate strategies and the future of the company can be planned (Simpson and Taylor, 2002).

Logically, marketers cannot convey any information or message about their product or services to their audience or buyers if there is no advertising in today's environment. Because of that, it is very important the analysis of consumer needs to be conducted in the purpose to analyse the consumer buying behaviour through the advertisement (Niazi et al., 2012).

2.4 *AIDA model*

Vakratsas and Ambler (1999), in the matter of promoting the SMEs products, the usage of advertisers is one the priority medium should be concerned. However, the most important thing been highlighted by the authors are the manufacturer or owner of certain company need to understand with how advertising that could affect their main customers and at the same time how it works in the reason to prepare more effective advertising strategies. Therefore, attributed to the E. St. Elmo Lewis in 1898, the AIDA model which is included of attention, interest, desire and action is the first formal advertising model can be used in the activity of the marketing process.

The AIDA model is a progressive method of separating the information of purchasing into four stages, and advertising is an exterior way to guide the customer from one step to another (Sanayei et al., 2013). This model has also been called as a hierarchy of effects model. This model is a behavioural model that has as function to make sure that advertisement increase awareness, stimulate interest, and leads the customer to desire and eventually action (Aryal, 2005). With the AIDA model the attributes proposes that for an advertisement to be effective it has to be one that:

1. Commands Attention
2. Leads to Interest in the product
3. Desire to own or use the product.

3 PROBLEM STATEMENT

The Small and Medium Enterprises (SMEs) nowadays are increasingly significant in the process of exporting and industrialization of the emerging economies (Hashim, 2012). Accordingly, the number of SMEs companies were estimated to grow over to a million in 2012 (Wong, 2012) and estimated to increase towards a high-income economy by 2020 (Department of Statistics Malaysia, 2012). In addition, it has been recorded on an annual report 2010/2011 that, the SMEs represented about 99.2% of the entire business formations in Malaysia in 2010 (Moorthy et al., 2012). The performance of SMEs is important as they will change Malaysia into a high-income and knowledge-based economy through their role to the national GDP.

However, the SMEs is having a challenge whereby it is kind of difficult to enter the market (Muhammad et al., 2012) at the same time to differentiate their products from the competitors (Malaysian External Trade Development Corporation—MATRADE, 2008). Importantly, it has also been highlighted that the promotion and market research were the most frequent problems of SMEs in the areas of marketing (Simpson and Taylor, 2002). Usually, the problems occur where the marketers themselves cannot convey any information or message about their product or services to their audience or buyers if there are no advertising in today's environment (Niazi et al., 2012).

Consequently, it will affect the performance of SMEs companies if there is no action taken. Poor marketing strategies reflected back to the wrong target market or the message are not being well interpreted by the consumer thus create a wrong impression towards the product. This will lead to less people buying and it will put the company performances at stake. Although the SMEs often gives a more competitive advantage than their competitors, in terms of cost and product differentiation, Moorthy, Tan, Choo, Wei, Ping and Leong (2012), stated that the SMEs' product was not being fairly treated among the researcher as they are more intended to study about the larger enterprise in Malaysia. The unfair treatment was due to the strong brand that have a high tendency in getting the attention from various group. In line with the above notion, although the SMEs product was available in the market, it was being neglected due to the existing brand that offered the same goods.

Conversely, the SMEs product need to work hard in order to get the recognition and market share among the international brand product to ensure that they will be sustained thus created the demand from the consumers. For that reason, the SMEs really need to plan their own marketing strategies in order to enhance their future performance (Taiwo, 2010). To achieve that particular accomplishment, awareness needs to be created as it will alert the consumer about the existing product hence influenced them in decision making.

For that reason, in order to ensure the consumer awareness and purchase the products offered, the advertising plays an important role whereby it becomes the communications system for the business and also consumers (Belch and Belch, 2003). The advertising will be used in helping the any of SMEs companies to realize and retain their position. In that case, this study is mainly to examine the influence of advertising on the products offered by the SMEs company in Malaysia which is will lead to the consumer purchase intention.

REFERENCES

Aryal, B.R. (2005). AIDA Model Application for TV Commercials in Nepal. *The Journal of Nepalese Business Studies.* 2(1), 65–71.

Basefsky, S., & Sweeney, S. (n.d.). Employment Relations in SMEs: The United States. Retrieved 2013, Mei 20th from http://www.ilr.cornell.edu/international/news/upload/BERLIN%20SME%20report.pdf.

Belch, G.E. & Belch, M.A. (2003). *Advertising and Promotion. An Integrated Marketing Communications Perspective,* (6th Ed). New York: The McGraw-Hill.

Chartered Institute of Marketing (2009). Marketing and the 7 Ps: A Brief Summary on Marketing and how it Works. Retrieved 2013, June 3rd from http://www.cim.co.uk/files/7ps.pdf.

Chelliah, S., Sulaiman, E.M., and Yusoff, Y.M. (2010, June). Internationalization and Performance: Small and Medium Enterprises (SMEs) in Malaysia. *International Journal of Business and Management.* 5(6), 27–37.

Department of Statistics Malaysia (2012, September 18th). Media Conference, Census Report on SMEs 2011. Retrieved 2013, April 29 from http://www.statistics.gov.my/portal/images/stories/files/LatestReleases/BE/BI/BE2011_SMEsBI.pdf.

Emmanuel, J. (2009, December 15th). The Importance of Marketing Management to SMEs. Retrieved 2013, Mei 2nd from http://www. commerce.gov.lc/articles/view/18.

Hamid, H. (2009, March 2nd). Target to build 100 SME brands by 2020. *Business Times*. Retrieved 2013, May 15th from http://www.btimes.com.my /Current_News/BTIMES/articles/28BAM/28BAM.pdf.

Hashim, F. (2012). Challenges for the Internationalization of SMEs and the Role of Government: The Case of Malaysia. *Journal of International Business and Economy*. 13(1), 97–122.

Hemilä, J. & Oinas, M. (2009, May). How Marketing Affects an SME Firm's Profitability, Growth and Success. *POMS 20th Annual Conference*. Retrieved 2013, June 3rd from http://www.pomsmeetings.org/ConfPapers/011/011-0071.pdf.

Kaya, S. & Alpkan, L. (2012). Problems and Solution Proposals for SMEs in Turkey. *Emerging Markets Journal*, 2, 31–45.

Keh, H.T., Nguyen, T.T.M. & Ng, H.P. (2007). The Effects of Entrepreneurial Orientation and Marketing Information on the Performance of SMEs. *Journal of Business Venturing*, 22, 592–611.

Malaysian External Trade Development Corporation—MATRADE (2008, March 6th). Branding, Advertising and Marketing for SMIs/SMEs. Retrieved 2013, June 1st from http://www.Smibusinessdirectory.com.my/smismeeditoral/more/more-business-tips/396-branding-advertising-a-marketing-for-smissmes.html.

Marjanova, T.J. & Stojanovski, M. (2012). *Marketing Knowledge and Strategy For SMEs: Can They Live Without It?* Thematic Collection of papers of international significance: "Reengineering and Entrepreneurship under the Contemporary Conditions of Enterprise Business". pp 131–143.

Mirchevska, T. and Mirchevski, V. (2007, February). The Role of Marketing in SME's Development in Republic Macedonia. Retrieved 2013, June 1st from http://www.Amfiteatrueconomic.ase.ro/arhiva/pdf/no21/articol_fulltext_pag 40.pdf.

Mohammad, N., Setapa, M., & Rahman, K. (2012). Factors Affecting the Purchase of Local Beverage Product. *International Journal of Academic Research in Business and Social Sciences.* 2(4), 2994–3009.

Moriarty, S., Mitchell, N., and Wells, W.D. (2009). Advertising: Principles & Practice. New Jersey: Pearson.

Muhammad, M.Z., Char, A.K., Yasoa', M.R., & Hassan, Z. (2010) Small and Medium Enterprises (SMEs) Competing in the Global Business Environment: A Case of Malaysia. *International Business Research.* 3(1), 66–75.

Niazi, G.S.K., Siddiqui, J., Shah, B.A. & Hunjra, A.I. (2012). Effective Advertising and its Influence on Consumer Buying Behavior. *Information Management and Business Review*. 4(3), 114–119.

Pineda, R.C., Lerner, L.D., Miller, C.M., & Phillips, S.J. (1998). An Investigation of Factors Affecting the Information Search Activities of Small Business Managers. *Journal of Small Business Management.* 36, 60–71.

Pineda, R.C., Lerner, L.D., Miller, C.M., & Phillips, S.J. (1998). An Investigation of Factors Affecting the Information Search Activities of Small Business Managers. *Journal of Small Business Management.* 36, pp 60–71.

Report of ACCCIM 2012 SMEs Survey (2012, March 15th). Retrieved 2013, June 1st from http://www.acccim.org.my/file/2012%20SME_EN.pdf.

Sanayei, A., Shahin, A., and Amirosadt, S.N. (2013). Evaluating the Effectiveness of TV Advertisement and Analyzing its Influence on Attraction of Saving Deposit Accounts of Ansar Bank in the City of Isfahan. *Journal of Business Administration and Management Sciences Research.* 2(2), 53–58.

Simpson, M. & Taylor, N. (2002). The Role and Relevance of Marketing in SMEs: Toward a New Model. *Journal of Small Business and Enterprise Development.* 9(4), 370–382.

SME Cooperation Malaysia (2012). Summary SME Masterplan 2012–2020: Catalyzing Growth and Income. Retrieved 2013, Mei 26th from http://www.smecorp.gov.my/vn2/node/190.

Smith, M. (n.d.). Effects of Bad Marketing. Retrieved 2013, June 15th from http://www.selfgrowth.com/articles/effects-of-bad-marketing.

Vakratsas, D., & Ambler, T. (1999). How Advertising Works: What Do We Really Know? *Journal of Marketing.* 63, 26–43.

Wijaya, B.S. (2012). The Development of Hierarchy of Effects Model in Advertising. *International Research Journal of Business Study.* 5(1), 73–85.

Zainudin, A. (2011). Elak Hutang Kunci Kejayaan Hasim. *MyMetro.* Retrieved 2012, January 12 from http://www.hmetro.com.my/myMetro/articles/ElakhutangkuncikejayaanHasim/Article/index_html.

Hospitality and Tourism – Sumarjan et al. (eds)
© 2014 Taylor & Francis Group, London, ISBN 978-1-138-00151-0

Consumer perceptions of quality of food at night market

N.F. Ramlan
Universiti Teknologi MARA, Malaysia

ABSTRACT: A point located with respect to surface features of some region, the night market is a very unique place to be visit. In the multi society, and night market had their owned culture, often offer something unique and special. In regard, because the potential in making profit, many local people trying to do the business. Without considering the safety and quality factors, yet people still wanted to run the business. Thus, this will lead to the unsatisfied consumer due to the perceived value that they get. Due to that, there is an increasing consumer's concern regarding on food safety and quality issues. Currently, the numbers of food consumption in food stall are increasing due to the today's lifestyle. This may expose the consumer to the risk of the foodborne disease. The needs to create the awareness and change consumer perception are very crucial. This research also presented here to aim the consumer perception of the concept of food quality that give greater impact on how consumer perceived food and used in making the decision as well as purchase decision. Overall, this study setting focuses on the attributes of good quality at night market.

Keywords: consumer perception, food quality, night market

1 INTRODUCTION

In the multi society, night market had their owned culture, often offer something unique and special (Hsieh and Chang, 2006). It portrays the local culture, as a leisure activity and provides the local people as well tourist to enjoy the evening shopping experience (Huang, 2009). Most of the night markets provide local foods. According to Choy (2012), food acts as one of the medium and easy ways to represent the culture. Thus, foods at night markets are cheaper, convenience, delicious yet represent the local cultures (Halvorson and Raundi, 1992).

At present, night market activity becomes a trend among Malaysian (Virtual Malaysia, 2013). Most of the local people were enjoying shopping at the night market. The uniqueness of night markets makes it must visit for both local people and tourist. It offers suitable places to shop and spend leisure time with families and friends (Khursiah and Avini, 2012). Due to that, it attracts an entrepreneur to run a business in this study setting (Khursiah, 2013).

Next, there are increasingly in sales of the commercial food sector which from 28 percent to 45 percent between 1970 and 1990 (Toursim Bureau, 2010). This also includes the fast-food segment and also night markets. In other words, in every one-fifth of the consumer money contribute to the food expenditure in food sectors. Therefore, this offers the growth opportunity of the night market industry. It is also considered as a good potential business option for local peoples. In 1989, Grice stated that night market setting offers huge prospect for becoming next entrepreneurs, since it require small capital to start up the business.

The wide opportunity to open setting up the stalls at night markets especially in the food business are mushrooming. In fact, without considering the safety and quality factors, yet people still wanted to run the business. Due to that, there is an increasing consumer's concern regarding on food safety and quality issues. Thus, at the same time, there has been a significant consumer's augmentation in differentiated or high value product consumption (Rodriguez et al. 2006).

This study focuses on the attributes of food quality at night market in Shah Alam, Selangor. This study is basically being done due to the increasing in food poisoning and any other foodborne disease cases. There were 11,226 cases from January till September in 2007, a 100 percent increase compared to the same period in 2006 (Ministry of Health, 2006). All these cases had been declared as major public health concern (Ministry of Health, 2006). The night market or street foods are the major causes that associated with the microbiological contamination due to the poor quality of food and low hygienic practice by the food operators (WHO, 2006). Thus, this will lead to the unsatisfied consumer due to the perceived value that they get (Grunert, 2005).

As a result, this paper aim to give an overview of research that carried out on consumer perception on the quality of food at night market. It gives an input on the quality perception from a consumer point of view. In fact, the quality is not the aim in itself, however is desired because it helps to deliver good safety and quality of food product, consumer satisfy with the purchase value and intention to re-visit for next time (Namkung and Jang, 2003; Grunert, 2005).

2 LITERATURE REVIEW

In Malaysia, night markets are known as bazaars or pasar malam (Khalilah, 2010). It is clearly popular among local people where it's easily found in nearly every small town and city in every state (Virtual Malaysia, 2013). Generally, the suburban night markets commonly operate near residential areas, and are usually located near familiar and visible zones such as the town centre or within the vicinity of other business operations such as shops and eateries (Khursiah et al. 2010).

By definition, night markets were defined as a trading place during the evening for the small traders and often offered a variety of products include foods at affordable prices, (Huang et al. 2009). In accordance with truth or fact, this statement supported by Chang (2010), the author stated that there are a lot of stalls that gather in a place and offer a non-casual place for shopping, leisure, social gathering and relaxation. Thereof, night market also well-known with the local activity and place that sparks during the night time compared with the business such day markets.

On the other way, according to the Food safety of Ireland (2006), defined that street food, hawkers or night market also were defined as any vehicle or portable stalls or structure which not included in a food premises in or on which is offered for sale or from which food is sold. Therefore, the goods are sold usually were prepared in somewhere else. In many developing countries like Malaysia and many others, street food vendors such in night market are an important component of the food supply chain. Being reasonably priced and conveniently available, night market still can satisfy the population needs and demands.

Nowadays, consumers have become more demanding when choosing their foods. Consumers were ensuring that every dollar that they spend will give them maximum value from each food that they eat (Tristano, 2011). Usually, consumers will spend the amount of money for having a good quality of foods. However, in the night market, the average of food prices quite cheaper and the quality of foods often worst and high risk has been contaminated (Andrew et al. 2007). For people that come from high-risk groups such as infant, elderly, children and others will highly expose to the diseases.

For a few decades, the issues regarding food quality and safety had a major concern among the consumers (Boyce et al. 2008). Thus, as a result very low confident and scared among consumers. For food quality it is known as quality is the totality of characteristics of an entity (product, service, process, activity, system, organization and person) that bear on its ability to satisfy and implied needs. Yet food safety and food quality are associated with all the activities that related to the standardization, quality control, certification and accreditation (Bettman et al. 2003).

Quality uncertainty played the significant role in term of the safety and the product liability (Rodríguez et al. 2006). According to the same author, it is due to the aim of the food consumption, which is not only to nourish human body, yet it is for the health enhancement. In addition, if the food that's available in night market were not safe and does not meet the qualifications, it will not lead to the food security. From that statement, Kinsey (2003) pointed out that food quality does not threaten food security, but there are interrelated in improving the human health.

Currently, in order to get the good and quality of food, consumers willing to pay for the higher price. However, for medium to high income people, it should be no problem because they have choices. This statement was supported by the Tillotson (2003), the author revealed that generally most of the customers in the restaurant segment were from medium and high end income. Vice versa, generally for the low end income people, there are no various choices for them. The low end income people may typically prefer to the economical foods which means the lower quality items (Kaufman et al. 1997).

As mentioned in above, it demonstrates that different type of consumer had different perceptions about the food quality regarding the background (Lazarova and Krystallis, 2010). In defining the perception, back in the year 1988, Zeithaml claimed that perception is all about the consumer shrewdness about the overall quality. In others study, consumer perception is defined as the human sense that used to assemble information about their surroundings (Wade and Tavris, 1987). Through this statement it explains that perception indirectly influences the consumer behaviour and intention (Walter et al. 1989).

2.1 *Consumer perception on safety attributes*

At first place before consumer choosing healthier food, most of the consumer will choose the safety food. In night market there are several factors that make the safety attributes in the dangerous place that give significant impact toward food safety. These include the store the food product in improper temperature, cross contamination with the equipment and utensils, poor personal hygiene among food handlers in the night market and others (Medeiros et al. 2001).

2.2 *Consumer perception on healthy attributes*

Along with the food quality attributes, healthy food is one of the factors that influence consumer perception. According to the Rodriguez et al. (2006), healthy is included in the intrinsic factor Similarly, the perception towards food quality is one of the psychological interpretation which influence the consumer perceptions, attitudes and behaviour in making the decision to perceive the food as quality and meet the expectation. Usually, consumers perceived the quality usually by the attributes of the food.

Health attributes nowadays become a trend for consumer in making their choice. This quality dimension gives significant effect to many consumers. There are a number of studies that indicate health is as important as taste and consequent from that, this dimension perceived as a motivation for a longer and higher quality life (Roininen et al. 2001).

2.3 *Consumer perception on environment attributes*

Back in a 1973 study by the Kotler, claimed that the environment attributes aids to influence consumer specific emotional and their purchase intention. There is numerous empirical study by the researcher regarding the relationship between the environment with the consumer perception and behaviours (Bitner, 1992; Donovan et al. 1994; Mattila and Wirtz, 2001; Wakefield and Blodgett, 1994; Ryu and Jang, 2007, 2008).

For instance, on 1992, Bitner came out with the framework that explaining the dimension of environment in service settings. Meanwhile for the Donovan et al. (1994) stated that in the place for shopping have a pleasant environment, usually consumer will spend more time in that place. The statement in the above was supported by Wakefield and Blodgett (1994) in leisure and relaxing environment in night market it reflect to the consumer perception to become satisfied and will re-visit again. Furthermore, Ryu and Jang (2007) said that the positive perception toward environment will lead to the more positive emotions and good product and services from the food vendors.

2.4 *Issue and problem in context*

Generally, there are several studies had been done in developing countries and mostly there are about 20 until 25 percent of the household expenditure goes to the food expenditure (FAO, 2013). In which made up from the outside food which include hawkers, street foods as well night markets. Resulted from the rapid urbanization, millions of working people having no enough time and access to the kitchen and deeply depend on the street foods, hawkers and/or night market.

Recently, the cases of the poor quality of food product cause of the food poisoning and decease the consumer trust (Rijswijk and Frewer, 2008). Thus, this resulted in a growing awareness of and concern for the foods in Malaysia (Deloitte, 2010). Food operators as well as MOH need to think how to solve these problems that each day the numbers of cases were increased. Below are the certain cases that were highlighted by the newspapers. In the year 2003, a total of 6624 food poisoning cases had been reported to the Department Ministry of Health Malaysia.

According to The Star, in 2007, food poisoning in Malaysia has increased by 100 per cent from 2006 for the period of January until September, with the 67 percent of such as cases were confined to primary and secondary students. These statistics also include in the night market and hawkers too. Most of the cases happened in the school canteen and in the night market and street food settings.

Next, according to the MOH statistic, there were 11,226 cases from January till September in 2007, a 100 percent increase compared to the same period in 2006. Despite a perceived higher awareness about food safety and quality, Malaysia continues to be overwhelmed with food poisoning cases in 2010 which about 50 percent involved night market and school canteens. According to the MOH (2002), food poisoning in Malaysia was increasing every year. For instance in 1995, the numbers of cases were about 1438 compared to 8640 cases in 1999.

In addition, Health Ministry Datuk Seri Liow Tiong Lai revealed that there are 311 cases was an increase of 29 cases from previous year which in 2009 and mostly due to poor quality of food during preparation, distribution as well as serving and/or consuming (Bernama, 2010). Meanwhile, in 2000, World Health Organization (WHO) claimed that, many developing countries do not have a comprehensive and proper food safety and quality programme.

In light of these, since most of the Malaysian enjoys the food from the night market, or any other street food, food quality is crucial (Virtual Malaysia, 2013). From the above statement, it revealed that there are important to note that, since it gives an impact to the food safety and quality cases (Grunert, 2005). Indeed, through increasing the food quality at night market, it may help to increase the food quality, reduce the number of food poisoning cases or other disease that related to foodborne disease and food security (Lakni and Mudalige, 2009).

ACKNOWLEDGEMENT

The author are grateful to the University Teknologi MARA Malaysia.

REFERENCES

Andrew, J.M., Worosz, M.R. & Todd, E.C.D. (2007). Serving food safety: Consumer perceptions of food safety at restaurant. *International Journal of Contemporary Hospitality Management,* 19(6), 476–484.

Bitner, M.J. (1992). Servicescapes: the impact of physical surroundings on customers and employees. *Journal of Marketing*, 56, 57–71.

Bernama. (2011). Half of Food Poisoning Cases Involve School Canteens. Retrieved on May 23th 2012 from http://www.bernama.com.my/bernama/v6/newsgeneral.php?id=574886.

Bettman, J.R., Luce, M.F. and Payne, J.W. (2003). Constructive consumer choice process. Journal of Consumer Research. Volume 25, page 187–217.

Cogan, T., Slader, J., Bloomfield, S. & Humphrey, T. (2002). Achieving hygiene in the domestic kitchen: The effectiveness of commonly used cleaning procedures. *Journal of Applied Mircobiology,* 92(5).

Choy, A. (2012). The Daily: TSA and TOSA collaborate on night market for the first time. Volume 136, Issue 40.

Deloitte. (2010). Food and Product Safety. Retrieved on May 26th, 2012 from http://www.deloitte.com/view/en_US/us/Industries/ consumerproducts/e11.

Donovan, R.J., Rossiter, J.R., Marcoolyn, G., & Nesdale, A. (1994). Store atmosphere and purchasing behavior. *Journal of Retailing,* 70(3), 283–294.

Fauzilah, S., Noryati, Y., Kamariah, Y., Mazuri, A.G. and Kuntom, W.W.S. (2012). Factors Influencing the Night Market Traders' Performance in Malaysia. International Journal of Business and Management. 7(14).

FAO. (2013). Strategy for a Safe and Nutritious Food Supply.

Food Safety of Authority Ireland. (2006). Hygiene of Foodstuff.

Grice, K. (1989). The Institutionalization of Informal Sector Activities: A Case Study of Cooked Food Hawkers in Singapore. Unpublished PhD Theses, Keele: University of Keele.

Grunert, G.K. (2005). Food quality and safety: consumer perception and demand. *European Review of Agricultural Economics* 32(3), 369–391.

Halvorson, Q.R. (1992). The growing Potential of Micro-Enterprises Organisation for Economic Cooperation and Development.

Huang, C.H. (2012). Service Quality at Night Markets in Taiwan. *International Journal of Marketing Studies.*

Huang, S.O., Liou, Y.H. and Tzeng, G.H. (2009). Development strategies for improving the service of tourist night market through hybrid MCDM technique.

Hsieh, A.T. and Chang, J. (2006). Leisure motives of eating out in night market. *Journal of Business Research.* 59(12), 1276–1278.

Jang, S., Namkung, Y. (2009). Perceived quality, emotions, and behavioral intentions: application of an extended Mehrabian–Russell model to restaurants. *Journal of Business Research* 62(4), 451–460.

Kinsey, J. (2003). Will food safety jeopardize food security?. Proceeding 25th International Conference of Agricultural Economists, Durban, South Africa, August 2003.

Khalilah Zakariya. (2010). Walking through night markets: a study on experiencing everyday urban culture. The 11th International Joint World Cultural Tourism Conference, 2010, Hangzhou, China.

Khomar, N.I., Khursiah, A.A. & Amri. A. (2013). Dynamism of Night Market. *Journal of Case Research in Business and Economics.*

Khursiah, A.A & Avini, S. (2012). Consumer Buying Behavior at Night Market. Interdisciplinary Journal of Contemporary Research in Business.

Lakni, W.A.K.I., & Jayasinghe-Mudalige, U.K. (2009). Customer-defined quality of food: An empirical investigation based on KANO Analytical methods for milk powder. Journal of Agricultural Sciences, 4, 45–58.

Mattila, A.S., & Wirtz, J. (2001). Congruency of scent and music as a driver of in-store evaluations and behavior. *Journal of Retailing,* 77, 273–289.

Medeiros. L., Hillers, V., Chen, G., Bergmann, V., Kendall, P. and Schroeder, M. (2009). Design and development of food safety knowledge and attiudc scales for consumer food safety education. *Journal of the American Dietetic Association,* 104(11), 1671–1677.

Ministry of Health. (2006). National plan of action for nutrition of Malaysia 2006–2015. National Coordinating Committee on Food and Nutrition.

Namkung, Y., Jang, S. (2007). Does food quality really matter in restaurant: its impact of customer satisfaction and behavioral intentions? *Journal of Hospitality and Tourism Research,* 31(3), 387–410.

Nasir, M.S. and Jinab, S. (2005). Changing Retail food sector in Malaysia.

Rijswijk, W.V. and Frewer, L.J. (2008). Consumer perceptions of food quality and safety and their relation to traceability. *British Food Journal,* 110(10), 1034–1046.

Rodriquez, E., Lupin, B. and Lacaze, M.V. (2006). Consumer perceptions about food quality attributes and their incidence in Argentinean organic choices. *Agricultrual Economics Research Group.*

Roininen, K., Tuorila, H., Zandstra, E.H., de Graaf, C. and Vehkalahti, K. (2001). Differences in health and taste attitudes and reported behaviour among Finnish, Dutch, and British consumers: A crossnational validation of the Health and Taste Attitude Scales (HTAS). *Appetite*, 35(1), 33–45.

Ryu, K., Jang, S. (2007). The effect of environmental perceptions on behavioral intentions through emotions: the case of upscale restaurants. *Journal of Hospitality and Tourism Research* 31(1), 56–72.

Tourism Bureau. (2010). Annual Report on tourism. Taipei, ROC: Tourism Bureau, Ministry of Transportation and Communication.
Tristano, D. (2011). Restaurant Startup and Growth. For Casual Dining Restaurant Value is Key. Retrieved June 1st from http://www.rsgmag.com.
Virtual Malaysia. (2013). Malaysian Night Market. Retrieved on May, 21st, 2013 from www.virtualmalaysia.com.
Wakefield, K.L., Blodgett, J.G. (1994). The importance of servicescapes in leisure service settings. Journal of Services Marketing 8(3), 66–76.
Wandel, M. and Bugge, A. (1994). Consumers, Food and the Market; Consumer Valuations and Priorities in the.hfineties. SIFO-report no. 2–94. National Institute for Consumer Research, Lysaker, Norway.
WHO. (2002). WHO Global Strategy for Food Safety: Safer food for better health, World Health Organization, Geneva.
WHO. (2006). A Model for Establishing Upper Levels of Intake for Nutrienta and Realted substances. Report of a Joint FAO/WHO Technical Workshop on Nutrient Risk Assessment. Retrieved on May 20th, 2013 from http://www.who.int/ipcs/methods/nra/en/index.html.
Zeithaml, V.A. (1988). Consumer perceptions of price, quality, and value: A means-end model and synthesis of evidence. *Journal of Marketing*, 52(3), 2–22.

Hospitality and Tourism – Sumarjan et al. (eds)
© 2014 Taylor & Francis Group, London, ISBN 978-1-138-00151-0

Tangible benefits of green branding

M.R. Jamaluddin, H. Hamzah & N.H. Maskor
Universiti Teknologi MARA, Malaysia

ABSTRACT: Building a strong brand in the market is not only become the mission of organizations but has become a never ending process. The fact that marketers stressed the outcome of the strong brand had resulted in extensive studies done to prove the relationship of strong brand towards the financial and non-financial benefits. However most of the studies done previously answered the question of what the strong brand is but not how to build stronger brands. This study plans to stress the importance of establishing tangible benefits that the brand may offer in particular for the green branding effort. This in turn will create a new basis for an excellent branding strategy that will be able to relate the relationship between firm's profit, human benefit and environmental sustainability. In order to achieve this objective, this study will diagnose the current practice and awareness, and later unveil the new strategy that will encourage all stakeholders in the hotel industry to appreciate green practice as a critical measure for sustaining business, environment and resources.

Keywords: branding, brand loyalty, green, tangible benefit

1 INTRODUCTION

The study of green branding has become one of the major current research topics as environmentalism has emerged as the hot issue of global concern. More consumers nowadays believe their purchase behaviour will have a direct impact on ecological problems (Mostafa, 2007). As stated by Jamaluddin, Hashim and Hanafiah (2011), the concept of "green consumerism" has taken place with the emergence of environmental awareness and sensitivity consumers that willing to pay extra for 'green' products. In the tourism products, environmentally responsible behaviour has grown significantly from hardly being talked to have been a major concern for industry players (Gustin and Weaver, 1996). Even though environmentalism has caught up in many Asia countries including Malaysia, the research involving green branding and the benefit it provides remained noticeably low.

Green practices and its image are no longer a jargon to consumers. The reason is consumers aware about the consequences avoiding green practices towards global warming. In Malaysia, the awareness of environmentally friendly has resulted in various campaigns like cleaning the beach and river, recycling waste, and the most popular one "No Plastic Bag Day" campaign. This campaign received overwhelming support not only to the company but also from individual. Hotel industry effort in this campaign can be seen from the way it promotes conservation in energy and water, and the concept of reuse and recycle of the room amenities. However, not all consumers participate in the green practice effort because of the differences in motivational in consuming hotel products. Since the campaign requires guests to sacrifice in terms of energy, water, and amenities consumption, several questionsarise: (i) Will they pay a slightly lower price for agreeing to adopt green practices (ii) Will they get something in return for agreeing to adapt green practice. To answer this question, this study will explore the consumer perception towards green branding and the effect of tangible benefits towards behavioural brand loyalty. The fact that industry players invested in expensive technologies without properly diagnosed the market readiness to accept green practice has triggered this issue to be explored.

For further explanation, hotel guests eager to know the value they get in return for collaborating in the effort of green practice. For example, if the price for green product is similar to non-green product, what will they get in return for sacrificing the full services offered by five-star hotel. In order to be green, guests must will to reuse the towel and linen, and comfortable with energy-save lighting in the room. Not all guests agree with this idea because, with the premium price they paid for staying in the five-stay hotel, it must be followed with the same premium service. Here comes the concept of tangible benefit in which the introduction of souvenir can act as a token of appreciation for agreeing to participate in the green effort.

The role of souvenir is important in the green branding because it will answer the question of "Will they get something in return for agreeing to adapt green practice".

The role of souvenir in complementing the travel satisfaction is proven as indicated by Hugh (2009), gift giving has become part of the rituals associated with vacations. Usually tourists look for souvenir everywhere they go to recall the sweet memories of visiting the place of interest. Cited in Godwin (2012), souvenirs act as tangible evidence to re-live the experience and retain the memory that fades over time. In other word, souvenir reminds to special occasion and location. To complement the service, souvenir is a key determinant to predict the outcome variable such as behavioural intention and brand loyalty. The word souvenir comes in two forms in which it can be bought from a service provider, or it can be given away by a service provider. To differentiate this, the word tangible benefit is used to act as a gift, a memento or evidence that will be given away as an appreciation to collaborate in the green practice. The items included as the tangible benefit are not available for purchase to ensure it meets the purpose of acting as mediating variable in the green branding process.

In offering green products, the dilemma faced by the service providers is the willingness to pay for the premium product but agree to sacrifice in return. So far, such popular programs that offered by five-star hotels encompass reuse of tower and linen, and energy saving rooms. Godwin (2012) argued that guests expecting more from green hotels to be social-environmentally responsible. Besides, Manaktola and Jauhari (2007) found that "11 percent of guest were willing to pay 25 percent of the cost, 40 percent felt the hotels should bear 50 to 100 percent of the cost, and 40 percent agree to pay 4 to 6 percent of the total cost, to stay in the green hotel. The self-responsibility in turning off the light while away from room is one of the action that fall in the green practice in line with Yesawich (2008) that states "Eight out of ten Americans turning off light to support green practice". This is not enough because to be green, guest should turn off electricity while away from room. Yesawich (2008) also proves that four out of ten travelers pick an environmental-friendly hotel as their requirement to select a place to stay. Here it indicates that guest pick the hotel based from other characteristics such as price, location and promotion but the green hotel accidentally fall within their searching result. This prompts another question, "Do they stay in the green hotel because to oblige with the hotel policy or, truly because of green concern". Currently, it is very difficult to find hotels that offer tangible benefit as a reward to their guest, and If any, it comes with premium price and act as normal souvenir that complement the stay.

2 RELATED WORKS

2.1 *Green practice*

Gilbert (2007) identifies green practice as an activity that is performed in a manner that has either limited negative ecological impact, or directly gives benefits to the natural environment in several ways. In the hotel business, the way the actors implemented green practice are by reducing the water conservation, energy consumption as well as reducing the solid waste. Using water as a way to improve the technologies is important so that the hotel can deliver better service and can avoid the water problem in the future. For energy consumption, Alexander (2002) explains the various options such designing the passive solar heating to something as simple as providing better insulation which can help reduce or eliminate the need for costly heating and cooling. The most common strategy is to automatically shut down electricity in the room except for refrigerators, alarm clocks and other important appliances. And lastly for solid waste management, Alexander (2002) adds that it is vital for the hotel to always keep informing employees to reduce solid waste by recycling paper, aluminium, and plastic.

2.2 *Tangible benefit*

Wilkins (2009) explains souvenir as a vital component to accompany the visitor's experience that represents good memories or valuable evidence. For the study context, tangible benefits act as an evidence to confirm the collaboration between two parties specifically guest and hotel. The tangible benefits play an influential role as a gift, evidence, and create a memory for involving in green products that is proved by Kim and Littrell (2001). The transformation of intangible experiences into tangible memories enables the guests freeze a passing moment in an extraordinary reality and to interweave it into their daily lives (Zins and Kreiner, 2011). There are many types of tangible benefits that can be offered by the hotel such as Keychain, mug, napkin towel, pin badge or stickers. However, all these souvenirs should symbolize the image green or, in other word "green branding" that will promote behavioural intention such as positive word-of-mouth, revisit intention or brand loyalty.

2.3 *Behavioural brand loyalty*

Behavioural brand loyalty can be defined as "the situation in which a consumer repeatedly purchase the same product or service over time rather than purchasing from multiple suppliers within the category" or "the degree to which a consumer consistently purchases the same brand within a product class" (Moisescu, 2006). Brand loyalty measures the level of consumer loyalty to a particular brand (which focused on green branding) and explained through their repeat purchases. It requires commitment and followed by the perceived value, satisfaction and brand trust, regardless of the competing brands. Kotler (2002) explains four types of brand loyalty which are (i) Complex loyalties—do research, develop beliefs and attitudes, and make thoughtful choices of the brand; (ii) Dissonance loyal—shop around first and seek information to support final choice; (iii) Habitual loyal—decide based on familiarity and keep buying the same brand; and (iv) Variety-seekers—seek different brand and evaluate during consumption.

3 CONCEPTUAL FRAMEWORK

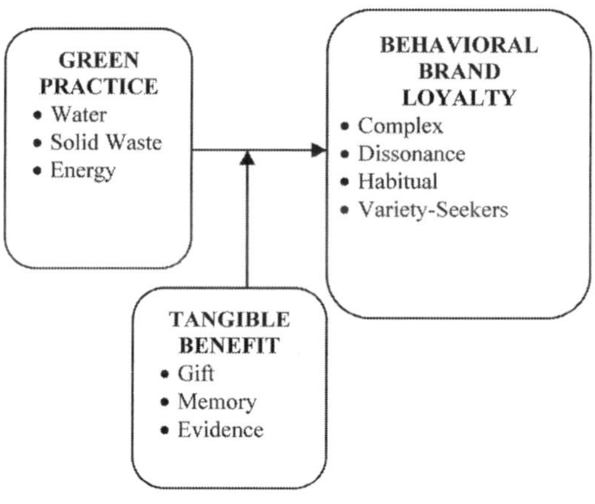

Figure 1. Conceptual framework.

4 RESEARCH VALUE

There are several questions highlighted in this conceptual paper that need immediate responds that could benefit industry players to craft better policy and guideline in green branding effort. Responses from guests in Malaysia will provide better understanding on how they perceive the effort implemented by the five-star hotel players in providing environment-friendly products and services. Since price plays an important role in promoting green practice, the effect of tangible benefits may eliminate the dilemma faced by the guests in opting for green products. Tangible benefits may create a competitive weapon to survive in the saturated market place. If the hotel is ready to provide the tangible benefits, this research will provide the type of tangible benefits demanded by the guest. Lastly, this research will pave the way for other research for the holistic green branding effort that will include other outcome variable like satisfaction, trust, brand image, and intention to practice in daily life which will analyze guest from diverse views of future motivation.

REFERENCES

Alexander, S. (2002). Green Hotels: Opportunities and Resources for Success (On-line) Available http://www.zerowaste.org/publications/GREEN_HO.PDF.

Gilbert, A.J. (2007). The value of green marketing education at the University of Wisconsin-La Crosse (Online) Available www.uwlax.edu/URC/JURonline/PDF/2007/gilbert.pdf.

Godwin, C.O. (2012). Perception of green hotels in the 21st century. *Journal of Tourism Insights*, 3(1).

Gustin, M. & Weaver, P. (1996). Are hotels prepared for the environmental consumer? *The Cornell Hotel and Restaurant Administration Quarterly*, 20(2), 1–14.

Hugh, W. (2009). Souvenirs: What and why we buy. Griffith University. Retrieved from http://www98.griffith.edu.au/dspace/bitstream/handle/10072/41195/71210_1.pdf?sequence=1.

Jamaluddin, M.R., Hashim, R., & Hanafiah, M.H. (2011). Preserving in five-star hotel segment. *The 12th International Research Symposium on Service Excellence in Management*, 607–616.

Kim, S., & Littrell, M.A. (2001). Souvenir buying intentions for self versus others. *Annals of Tourism Research*, 28(3), 638–657.

Kinley, T.R., Forney, J.A., & Kim, Y.K. (2012). Travel motivation as a determinant of shopping venue. *International Journal of Culture, Tourism and Hospitality Research*, 6(3), 266–278.

Kotler, P. (2002). Marketing Management, Millennium Edition, Prentice Hall Inc., Upper.

Manaktola, K. and Jauhari, V. (2007). Exploring Consumer Attitude and Behaviour toward Green Practices in the Lodging Industry in India. *International Journal of Contemporary Hospitality Management*, 19(5), 364–377.

Moisescu, O.I. (2006). A Conceptual Analysis of Brand Loyalty As Core Dimension of Brand Equity. Published in: Com petitiveness and Stability in the Knowledge-Based Economy No. International conference proceedings, Craiova, Romania.

Mostafa, M.M. (2007). Gender differences in Egyptian consummers green purchase behaviour: the effects of environmental knowledge, concern and attitude. *International Journal of Consumer Studies*, 31(3), 220–229.

Wilkins, H.C. (2011). Souvenirs: what and why we buy. *Journal of Travel Research*, 50(3), 239–247.

Yesawich, P.C. (2008). Turning Green Practices into More Green. *Hotel & Motel Management*, 223(16), 14–24.

Zins, Y., & Collins-Kreiner, N. (2011). Tourists and souvenirs: changes through time, space and meaning. *Journal of Heritage Tourism*, 6(1), 17–27.

Hospitality and Tourism – Sumarjan et al. (eds)
© 2014 Taylor & Francis Group, London, ISBN 978-1-138-00151-0

Iranian restaurant business in Malaysia: Diagnosing operators' views

A.H. Raeis, M.S. Mohd Zahari, M.A. Hemdi & M.Z. Suhaimi
Universiti Teknologi MARA, Malaysia

ABSTRACT: The international environment of the hospitality industry especially the restaurant industry reflected the global nature that leads the main players to expand their wings internationally. Fast food restaurants are one of the examples of the fastest growing types of restaurants globally. They are establishing themselves all over the world with a well-known brand and trademarks. Besides this, there are also remarkable growths of ethnic based restaurants especially from the Middle Eastern countries all over the world with Malaysia without an exception. Through that, this study investigated the phenomenon of Iranian restaurant in Malaysia as this ethnic based restaurant positively mushrooming and in fact, the development of Iranian restaurant is penetrating the other states beside Kuala Lumpur as well. Qualitative technique through a series of interviews with the restaurant operator's is conducted to discover their attitudes and experiences on this phenomenon.

Keywords: ethnic, internationalization, Iranian, restaurants

1 INTRODUCTION

Hospitality is by nature is an international industry especially with regard to the hotel. This is evident when many of the international hospitality companies are operated internationally. A cursory look at demand sources shows that part of the demand for a hospitality service is derived from international tourists and international business. The global economy makes the hospitality sub-divisions like hotels, tour operators and convention organizers including restaurant operators going internationally. Teare and Olsen (1992) argue hotel and restaurant is the fastest sub-divisions of hospitality penetrating the global market. The international environment of the hospitality industry reflects the global nature that leads the main players to expand their wings internationally. According to Johnson (2002) in the last two decades the multinational companies in the hospitality industry basically are among the American companies but nowadays the situation changed whereby many other international companies across the globe venturing into the international arena.

In Malaysia market, besides fast food restaurants like McDonald's, Burger King, subway, Wendy's, KFC, western and local ethnic restaurants, other ethnic country restaurant such as Italian, Moroccan, Thai, Indian, Pakistani, Turkish and Middle Eastern including Iranian restaurants are flourishing especially in the big cities like Kuala Lumpur, Penang, and Melaka (Euromonitor, 2006). As Iran is part of the Middle Eastern countries, the highlighting issues are specifically focusing on the development of this restaurant in Malaysia. In the last 10 year, there were not many Iranian and Middle Eastern restaurants operating in this country (Euromonitor, 2009). However, this scenario has changed. As reported by Euromonitor (2009), Middle Eastern restaurants are mushrooming in Malaysia which occupied in many places and states. In Langkawi for instance, there are three Middle Eastern restaurants operated in the Island. The number is even larger in Kuala Lumpur city particularly in the area of KLCC, Ampang and Bukit Bintang. The developments of this restaurant have encouraged many restaurant investors from Iran to venture in this called sub-sector of the hospitality in this country. This is evident when more than fifteen Iranian owned restaurants available in the Kuala Lumpur and its vicinity as compared to two restaurants in the last five years (Iranian Embassy, 2011). In Ampang alone there are more than seven Iranian restaurants such as Naeb, Tehran, Dolphin and Gol, Ali Baba and Roz in operation. Naab, Jam-e-jam, Kolbeh Sharghi and Shiraz are the other Iranian restaurants occupied in the Bukit Bintang and KLCC area. Similar to other areas like in Mont Kiara (Naab), Jalan Sultan Ismail (Palm in Soltan Ismaeel and Zeytoon) and in Subang Jaya (Hanni and Javan). In fact, as previously mentioned the development of this country restaurant is penetrating the other states as well. This positive development raises critical questions as to why the Iranian restaurants investors in particular venturing or opening restaurants in Malaysia? In another way of saying, what are the factors influencing their investing mood and what are characteristic of customers and level of performance of their restaurant business in this country?

To date, most of the available studies related to restaurant development overseas were undertaken in western counties setting (Hua & Upneja, 2006) with the lack of this study looking at those developments from Asian countries particularly with regard to the Iranian restaurant. With that, this study is attempt to look on the Iranian restaurant operators who currently running restaurant business in Malaysia.

2 LITERATURE REVIEW

2.1 *Internationalization*

Based on the popular belief, internationalization sometimes called as "I18 N" is the conversion of something in order to make it international or process of making a product suitable for other countries market or the principles and practices used in the designing and modification of a product to ensure it performs as expected when adapted for use in different locales (Madsen, 2005; Mockaitis et al., 2006). It is also referred to the process of planning, implementing and marketing the products and services and being accepted internationally. Internationalization process is sometimes called translation or localization enablement and Paul (2010) noted that, internationalization from computerizing perspective is the process of designing a software application so that it can be adapted to various languages and regions without engineering changes. In the business organization context, internationalization is measured to be the process through which a firm moves from operating only in the domestic market to the international marketplace (Buckley & Casson, 1998; O'Farrell et al., 1998) and the internationalization of services differs from domestic services in that they cross borders and embrace a foreign culture (Clark and Rajaratnam, 1999).

According to Machková (2006), beginning of the 21st century, business activities are really changing and connection with the world and penetration into the foreign markets and internationalization offers new opportunities to the enterprises. Entering the international market enables companies not only increased their income but at the same time it increase the volume of products and services internationally recognized. Machková (2006) also contend that the new international market offer new opportunities for firms and companies and the globalised competitive environment requires the firms to respond quickly to the consumers' needs and problems for which they ought to have enough knowledge (Hron, 2006). This is the main idea from the viewpoint of the adaptation of knowledge management in internationalization of business activities (Látečková, 2007). Bielik and Rajčániová (2004) argued that the competitiveness can be examined at various levels of the economy ranging from the product level, the enterprise level, the sector level and the level of the entire economy. The international competitiveness as argued by Podolák (2000) and Kapsdorferová (2004) is an ability of the country to take a place in the world competition which relate to the production and the tradeoff of its products and services to reach a new marketplace in the regional and global environment. As reported, at present decades, the regional, national and international companies are under the pressure of global competitiveness and companies must take into consideration that competitiveness issues from their resources while they must adapt to the new political, social and economic conditions (Hambalková, 2001) and (Blaas, 2008).

Teare and Olsen (1992), argued two different types of reasons for companies to enter new foreign markets and that stem from inside or outside of the company which are proactive and reactive. The proactive reasons consist of growth and profit motives, managerial urge, the opportunities that exist in the foreign markets, agents encouragement, extension of products and unique products. These unique products and extension of products and product range, bring with them some problems at the same as supply chain problems. Reactive reasons for internationalization includes, the desire to diversify, risks and saturation of domestic markets, as experienced by the fast food restaurant like McDonald's, KFC and many others restaurant companies in the United States. According to Technomic (1998), in 1997, the number of international units restaurants and sales of fast food restaurant chain has raised by 12.7 percent compared 4.8 percent of their domestic counterparts. This clearly shows the importance of foreign markets for the restaurant industry. Technomic (1998) asserted the exploitation of international market chance and opportunities for American branded food concept is accelerating as domestic market saturation which some brands being reached.

Internationalization also receives significant concern with regard to the operational issues, availability of capital, diligence associated with new expansion opportunities, management of financial risks, geopolitical risks, human resources issues (getting the right people on the ground in foreign locations), understanding and complying with foreign laws and regulations, export/import issues and manoeuvring cultural and language differences. O'Farrell (2006) believed there are so many locations that every new one has to be stolen from each other's market by competitors and this situation is also occurring in the restaurant industry especially among the multinational food companies.

3 METHODOLOGY

The qualitative approach was employed in this study as it provides in depth understanding on the issues investigated. In addition, qualitative are more subjective in nature and involve examining and reflecting on perception to gain an understanding of the social and human activities, that this research dealt with the Iranian restaurant operators. The informants for this study were among fifteen operators of Iranian restaurants' in Kuala Lumpur and Petaling Jaya.

In tapping the operators view, a pre-determined set of semi structured interview questions were developed in a consistent order to ensure the uniformity of information collected. The questions were designed in an interactive way to clearly bring out and develop a relaxing, friendly atmosphere to gain high quality information through the interview.

Before the actual data collection, all fifteen Iranian restaurant operators in Kuala Lumpur and the surrounding area were contacted to acquire their willingness to participate in the research study. Owing to business matters and barely available only 5 restaurants operators were willing to be part of the sample. With operators agreement on the dates and times the interviews were successfully undertaken. Persian language was used during the interview, tape recorded and all information was then translated into English.

4 RESULTS AND DISCUSSION

4.1 *Restaurant profiles*

Five Iranian restaurant operators in Kuala Lumpur and the surrounding were participated in this study. All of the restaurant operators except one have been in the operation more than two years and all of those operators were Iranian. Their profiles are simplified and presented in Table 1.

4.2 *Reasons of opening restaurant in Malaysia*

The first question relates to acquiring the information about Malaysian restaurant industry and factors that influencing interest of entering into the market. Most of the informants revealed about getting the information from friends who are living as well as students who studying and working in Malaysia. Stability, economy, good investment opportunities and the increased in number of Middle Eastern tourists' and students including Iranian creating interest among the operators.

Table 1. Restaurant details and demographic.

Restaurant	Types of food	Type of food	Number of employees
A	8 years	Persian and Middle Eastern Food	65
B	2 years	Persian and Middle Eastern Food	7
C	1 years	Persian and Middle Eastern Food	10
D	2 years	Persian and Middle Eastern Food	7
E	3 years	Persian and Middle Eastern Food	10

Table 2. Reasons of opening restaurant in Malaysia.

Operators	Answers
Restaurant A	*"Let me tell you, before involving myself in restaurant operation here, I actively involved in trading business … importing and exporting goods from Iran to Malaysia and from Malaysia to Iran. That's how I got information on the restaurant industry and we are the first Iranian restaurant in this country"*
Restaurant D	*"Let me tell you, if the country is stable in term of politic, economy and center of education, who do not want to go there inventing money. Malaysia is one of the most stable country in this region. Many international investors' feelings safe by investing money here and so are we the Iranian. How many Iranian students currently studying here and however Iranian businessmen doing business and many Iranian tourists visiting Malaysia? This in actual makes me be part of Iranian investors. Only that … I am more toward restaurant business"*

4.3 *Type of menus and customers*

All of restaurant operators disclosed that Persian dishes are the main items incorporated into

Table 3. Type of menus and customers.

Operators	Answers
Restaurant B	*"Yup ..., in this restaurant ... .our main customers are among Middle Eastern tourists. I should say, Arabs from Middle Eastern countries contribute around fifty percent to sixty percent customer of this restaurant especially during peak seasons. You know Iranian food is similar to Arabs food and they like the taste it. Another forty percents of customer are the Iranian tourist, students and ten percent are among the others including local"*
Restaurant E	*"This is Iranian restaurant and of course we serve Iranian foods. Mirza Ghasmi, Nesa Polo and Reshteh polo some of them. However, because we have lots of Middle-Eastern customers some their foods including appetizers are also offered in the menu"*

Table 4. Financial performance.

Operators	Answers
Restaurant B	*"... We are too young here ... therefore I can't tell on our financial performance. At this stage ... enough for me pay salary, overhead and rental ... That's good to me. But I am working hard for it ... I believed if my friend like NABB restaurant can do it ... why not us. So far, our customers increased day by day especially this coming June to December. Can't wait to see good money during that period. Whatever, Alhamdulillah we still survived"*
Restaurant C	*"Entering a new market is not easy as some people thought. One year in operation, lot things need to be done ... customers, service, connection and many others. Therefore, don't think about profit yet ... Survival is more important at this infant stage. I need to use lots of strategies to attract customers. More customers mean more money ... you know. I have also to promote my restaurant to the local people. For whatever reasons, a Persian restaurant besides Iranian, Arabs will have better position if we have more local customers. In short, I would say making profit or paid back investment is still a long way for us but we survive"*

Table 5. Non financial performance.

Operators	Answers
Restaurant C	*"Same as other restaurants, we did the same thing by advertising our restaurant in Persian Magazines or sites. Travel agents are also our focus ... I am happy and feel so proud when I see a lot of old and new customer dining here ... One more thing tourists, travel agents that we engaged honestly helping us"*
Restaurant D	*"Ok, you know there are Iranian magazines which are published in Malaysia ... I think this is one of important advertising tools for us. Normally I have advertisements there. I also promote my restaurant in any Iranian festivals to businessmen, university and others through flyers and brochures. From there, I my customers increased and some repeated and bring new friends here ... it's really works"*

the menu. However, few restaurants offer other food like Middle Eastern countries' food to make it more varieties as Arab peoples are one of the clients of their restaurants. That Iranian restaurant not only attract the Iranian students, businessmen and Iranian tourists through travel agents but received range of customers including Arabs, Malaysians, Indians, Chinese, Caucasians and others.

4.4 *Financial and non financial performance*

In the last part of the interview, the questions asked were dealing with financial and non financial performance of the restaurant. In this context, financial performance deal with the overall monetary performance while non financial performance relates to a marketing initiative undertaken by the restaurant operators, customer turnover and repeat patronization. However, it is worth mentioning that the interview is not trying to reveal the profit and loss of the restaurant but just an explanatory whether there are making money or otherwise.

On the financial performance, the question probed *"How is your restaurant financial performance?"* As this question a quite sensitive to any restaurant operators, therefore the answers given were more toward explanatory rather than revealing how much profits they make. The majority of the Iranian restaurant operators in this study revealed that despite some difficulties especially during first year in operation but overall their restaurant financial performances gradually picked up year by year as the cost of opening restaurant Malaysia slight lower than in Iran particularly Tehran.

On the non-financial performance, the first question asked *"how do you market your restaurant?"* On the overall answers obtained, Free Iranian Magazines which are published in Malaysia, Iranian travel agencies, websites, internet, flyers, pamphlets and word of mouth were the distribution channel used most Iranian restaurant operator in marketing themselves. Furthermore, most restaurants received a good customer turnover and repeated customers among the students and new customers among the tourists especially during peak season.

5 IMPLICATIONS AND CONCLUSIONS

Although a small research, this study has revealed a range of interesting and significant findings and giving some practical implications. Findings of previous studies clearly indicate political stability and economic, good foreign policy practices attract foreign investor to venture into a particular country. Similar results apparent in this study whereby significant numbers or Iranian restaurateurs opening their country restaurant in this country. In addition, open educational policy and the continuing growth of tourism industry also encourage the Iranian restaurant operators of running this risky business. The result of this also of shows that although Malaysia is considered a new destination in term business investment and education among the Iranian and middle eastern countries but significantly provides opportunity for businessmen from this region to expand their businesses. This is evident when few restaurants even just spreading their wings but showing some positive indicator in term of non-financial and financial performances.

The above good indicator should not be looking coolly by the existing Iranian restaurateurs but should be looking on the positive side. The existing Iranian operators should maintain or increasingly improve on what they are doing now in attracting customers not only among Iranian, Middle East but also to the local or Malaysian or other foreigners through a good food, services and other attributes. For the new investors and those who intent to venture into restaurant businesses that this indicator denote that Malaysia is one of the fast prospering countries in this region that creating vast opportunities for foreign investors. As a conclusion, it is hoped that the recommendations and information flow from this study will facilitate the not only the Iranian restaurant operators and other international restaurant investors in understanding the overseas restaurant opportunity subsequently position themselves well and be more competitive in the fast global growing restaurant businesses.

REFERENCES

Bielik, P., & Rajčániová, M. (2004). Competitiveness analysis of agricultural enterprises in Slovakia. Agricultural Economics—Czech, 50, 556–560.

Blaas, G. (2008). Agriculture in EU 15 and in Slovakia: Comparison of trends. In Konkurencieschopnosť a ekonomický rast (pp. 92–99). SPU: Nitra.

Buckley, P.J., & Casson, M. (1998). The Internationalization of the Firm, London, U.K.: The Internationalization of the Firm, Academic Press.

Clark, T., & Rajaratnam, D. (1999). International services: perspectives at century's end. Journal of Services. 13(4), 298–310.

Euromonitor. (2006, July). Ethnic and International Restaurant in Malaysia. Euromonitol. Retrieved from www.euromonitor.com.

Euromonitor. (2009, January). Hospitality Industry in Malaysia. Retrieved from www.euromonitor.com/.

Hambalková, M. (2001). Competitiveness of the selected agrarian commodities in SR, CRCR. In Poland and Hungary before entrance into EU (pp. 492–496). Nitra.

Hron, J. (2006). Knowledge and strategic management. Agricultural Economics—Czech, 101–106.

Hua, N., & Upneja, A. (2006). The degree of internationalization's effects on the market capitalization of publicly traded US restaurant firms. Advances in Hospitality and Leisure, 159–178.

Iranian-Embassy. (2011, January). Iranian Restuarants In Malaysia. (M.H. Raies, Interviewer).

Johnson. (2002). Locational Strategies Of International In Eastern Central Europe. Switzerland: University Of Fribourg.

Kapsdorferová, Z. (2004). Current situation on dairy market in time of entering Common European market (Vols. 80-8096-356-0). Európska integrácia: Nitra.

Látečková, A. (2007). Uplatnenie znalostí v podnikovom manažmente (Applying of knowledge in enterprise management). Z Studio, spol. s r.o.

Machková, h. (2006). Mezinárodní marketing (Vols. 80-247-1678-X). Praha: grada Publishing.

Madsen, T. (2005). Internationalization research: the impact of the Carnegie school. Scand. J. Manage, 21, 273–284.

Mockaitis, A., Vaiginiene, E., & Giedraitis, V. (2006). The internationalization efforts of lituanian manufacturing firms—strategy or luck. Res. Int. Bus. Financ, 20(1), 111–126.

O'Farrell, P.N. (1998). Internationalisation by Business Service SMEs: An Inter-Industry Analysis. International Small Business Journal, 16(2), 13–33.

O'Farrell, P., Wood, P., & Zheng, J. (1998). Internationalisation by business service firms: towards a new regionally based conceptual framework. Environment and Planning, 30, 109–128.

Paul. (2010). Internationalization and localization. Retrieved 13, 2011, from Wikipedia: http://en.wikipedia.org/wiki/Internationalization_and_localization.

Podolák, A. (2000). Methodological access to evaluation of commodity agrarian. Aktuálne problémy agrárneho trhu, 32(2), 31–34.
Teare, R., & Olsen, M. (1992). International Hospitality Management: Corporate strategy in practice. London: John Wiley.
Technomic, Inc. (1998). Update of Analysis of the 100 Largest US Chain Restaurant Companies Technomic Information Services. Technomic Information Services, Chicago, IL.

Current trends in hospitality & tourism

Hospitality and Tourism – Sumarjan et al. (eds)
© 2014 Taylor & Francis Group, London, ISBN 978-1-138-00151-0

Perceived importance: Last chance tourism in Malaysia

N.A. Ahmad, I.M. Azman, M.D. Darson & N.S. Shamsudin
Universiti Teknologi MARA, Malaysia

ABSTRACT: Climate change has affected tourist decision on their travel plan. Nowadays, there is the new travel trend whereby tourists increasingly seek to experience most endangered sites before they vanish. The term "last chance" tourism brings a desire for tourist to witness disappearing flora, fauna, landscapes and seascapes. Enriched with natural beauty, Malaysia has the potential to be promoted as a last chance tourism destination. This study focuses its scope to the local tour operator rather than tourist. Hence, how tour operator perceived the importance of last chance tourism product and what are their behavioral intentions on promotion were examined. Self-administered questionnaire were distributed to the local tour operators in Klang Valley. Through a sequence of analyses, some useful understandings on the issues of interest were obtained. The majority of the respondents agreed that it is important to sustain endangered flora and fauna as a last chance tourism product and it shows that they do promote the last chance tourism destination to the tourist. Most of the tour operator agreed that promoting and preserving endangered flora and fauna is important and valued the potential of this new market niche.

Keywords: endangered, fauna, flora, last chance tourism, perceived importance

1 INTRODUCTION

1.1 *Last chance tourism*

Tourism is composite of activities, service and industries that deliver a travel experience included transportation, accommodations, eating, drinking establishment and others available for individuals or group that are travelling away from home (Charles et al., 2008). As mentioned by Bowen and Clarke (2009), there is a relationship between tourism, leisure and recreation. It is related to one another but with different concepts. As cited by Niemela (2010), the subgroup of tourists around the world consist of educational tourism, cultural tourism, special interest tourism and heritage tourism, health tourism, winter tourism, mass tourism and niche tourism.

One of the new niche tourism is "doom tourism" also known as "last chance tourism" or "disappearing tourism". According to Dawson (2009), cited in Lemelina et al., 2010, last chance tourism is defined as a niche tourism market where tourists plainly seek vanishing landscapes or seascapes, and or disappearing natural and or social heritage. Shapiro (2007) stated that travellers are purposely seeking out imperilled destinations and try to experience their splendour before they vanish. In addition, the main attraction of doom tourism is visiting endangered species of flora and fauna and natural or social heritage that may disappear (Bobyleva, 2011). Awareness of these growing environmental threats has created a new trend in the travel industry, known variously as climate change travel, climate change sightseeing, global warming travel, even the tourism of doom (Schlichter, 2010).

According to Bobyleva (2011), last chance tourism is mainly related to negative impacts such as climate change. However, as a trend, it may appear to be a conflicting issue, since it can be relatively sustainable as well as unsustainable, depending on disappearing attraction, type of tourists and some more. Tourism is currently responsible for the largest, annual human migration in history. This great movement of people has significant positive and negative consequences on nature, societies, cultures and economies. Nevertheless, tour operators play the responsible for sustainable tourism rather that keep promoting it without any sustain activities (Adriana, 2003).

Malaysia is one of the world's twelve mega diversity areas however, it's facing a problem of endangered species of flora and fauna and need to be sustained and conserve. According to NRE (2006), the biological diversity of life on earth is under serious and continuing threat from the impacts of human activities and it reported that Malaysia's mammals face extinction in the next 30 years. It also stated that, studies estimate that 22% to 40% of plants will be extinct by 2050.

Thus, the tour operator is responsible for taking care of all the necessary activities in organizing a tour so that the overall trip of the client goes smooth and pleasant from the beginning till the end. Other roles of tour operator such as planning a tour, arranging a tour, advice on sightseeing, promoting, sales and marketing and others (Mastbusiness, 2012). This growth has led to an increase in the negative environmental and social impacts of tourism and tour operators are in part responsible for these negative impacts. Because of the factors stated before, this study will investigate on tour operator perspectives what their perceived importance on last chance tourism products.

2 LITERATURE REVIEW

2.1 *Last chance tourism and perceived importance*

In recent years, global warming, pollution and greenhouse effect generate various kinds of threatening impacts on sustainability natural and artificial resource. Climate change is a warming planet alters weather patterns, water supplier, seasonal growth for plans and a sustainable way of life for us, and the world's wildlife. With the passage of time, climate change is changing the landscape features and world's atmosphere drastically.

Air pollution, increasing temperature, rising sea levels and deforestation bring away some of the most perfect places and are changing the landscape. In fact, climate change has bothered the decline process of some of the unique tourism destination in the world (Chandran, Bhaduri & Swamy, 2012). Furthermore, destinations or attraction is threatened from several scales including local interactions between tourists and the environment as well as non-linear global processes such as climate change (Maher, 2007).

UNWTO and UNEP (2008) stated that tourism is a very climate sensitive economic sector but as the impacts of climate change become more obvious, it is increasingly important to understand the relationship of climate to the tourism system. Moreover, climate change often described as a motivator for travel. One of a common motivation theory identifies is push and pull factors for travel (Crompton, 1979). The weather is considered a pull factor by several authors (Turnbull & Uysal, 1995). Warm weather motivates tourist to visit and climate is a common theme in tourism promotion.

Consequences from the climate change make traveller rush to endangered destinations before they disappear. For example, Valda (2002) stated, the opportunities to watch upon glacial landscapes, mega-fauna and unique original cultures have attracted visitors to the Arctic for centuries. Buhasz, (2007) on the other hand mentioned that, the potential loss of these unique glacial landscapes has attracted more tourism and has provided tour operator with an opportunity to market their products as last-chance tourism.

On the other hand, the word perceived was defined as an act of being aware of one's environment through physical feeling, which denotes an individual's ability to understand (Unumeri, 2009). However, Rao and Narayan (1998: 329–330) as cited by Unumera, highlight that perception ranks among the important cognitive factors of human behaviour or psychological mechanism that enable people to understand their environment. On the other hand, our attention, feelings and the way we act are influenced by our environment. According to Unumeri (2009), the word perceived has a relationship with perception word where it means perception is the processes whereby people select, organize, and interpret sensory stimulations into meaningful information about their work environment.

Perceived and perception in the tourism industry as stated by Talisman & Kosuta (1989) is insisting that a tourist's decision is determined not only by the destination's potential for performance, but also by the perception of its personality or image. The fact before supported by (Davidoff & Davidoff 1994), where when initial credibility differs from the public's perception of a product. The perception of the image will determine that product's successor failure. Um and Crompton (1990) argued that the image of a place as a pleasure travel destination is lacking in originality, to a greater or lesser extent, from attitudes towards the destination's perceived tourism attributes. Apart from that, Pizam (1978) mentioned on the perceived importance of tourism tended to focus on the economic and the positive effects of tourism.

3 METHODOLOGY

An exploratory research design using a quantitative approach is considered as the most appropriate method to be used to tackle the research objectives. The samples were the tour operators in Klang Valley and Kuala Lumpur. The self-administrated questionnaire was performed to accomplish the objective of the study. Sekaran (2003) argued that self-administered questionnaires are less time consuming and less expensive. In this study the questionnaire is divided into four stages covering the dimensions of three variables: section (A) demographic profile of the tour operator, section (B) last chance tourism attributes, section (C) perceived importance of last chance tourism product and the last section (D) behavioural intentions (intention to promote). A five point Likert scale ranging from 1 = strongly disagree and 5 = strongly agree to help the respondents express their perceived importance and behavioural intentions. A total 166 questionnaires were returned, usable and coded. A quantitative research design was selected in order to 1) To investigate the local tour operator perceived importance of last chance tourism attributes. 2) To determine the relationship between last chance attributes and perceived importance.

4 RESULTS AND ANALYSES

A series of analyses were undertaken to answer the research objectives of the study by looking at the mean scores rated by the respondents. The analyses divided into two separate sections. The first section looks on the descriptive statistic and overall analysis on the perceived importance of tour operator on last chance tourism.

4.1 *Last chance tourism attributes*

Most of the respondents are having the same perception of the importance of last chance tourism attributes toward all items in this dimension. The magnitude of mean from 3.5 to below 4.0 indicate respondents were having agreed and aware of the importance of last chance tourism.

4.2 *Perceived importance of last chance tourism*

From the result above, it clearly shows that the tour operator perceived the importance of last chance tourism. Obviously, most of the respondents agreed on the importance to include last chance tourism attraction in their tour packages.

Table 1. Mean score for last chance tourism attributes (n = 166).

No.	Item	Mean	S.D.
1	Promoting endangered flora such as Rafflesia, pitcher plant, and orchid gives knowledge to tourist about last chance tourism	4.78	0.415
2	It is important to promote endangered fauna as a last chance tourism product	4.94	0.243
3	It is important to preserve and protect endangered fauna as a unique attraction to tourist	4.88	0.325
4	Preserving and protecting endangered fauna as part of last chance tourism product are very important	4.99	0.111
5	Promoting protected endangered fauna such as orang utan, hornbills and leatherback turtles as a last chance tourism product are important to increase tourist visitation to Malaysia	3.80	0.642
6	Malaysia offer great choice of last chance tourism products	3.57	0.880
7	Last chance tourism attraction in Malaysia are internationally recognized	3.29	1.030
8	Last chance tourism contributes revenue to the country's economy	3.72	0.877
9	Malaysia has many last chance tourism attraction	3.50	0.991
10	Preserved last chance tourism attraction provide value to the tourist	3.50	0.793
11	It is important to promote endangered flora as a last chance tourism product	4.09	0.642
12	It is important to preserve and protect endangered flora as a unique attraction to tourist	3.88	0.815
13	Preserving and protecting endangered flora as part of last chance tourism product are very important	3.79	0.719
14	Promoting protected endangered flora as a last chance tourism product are important to increase tourist visitation to Malaysia	3.77	0.746

Table 2. The mean scores for perceived importance of last chance tourism (n = 166).

No.	Item	Mean	S.D.
1	It is important to include last chance tourism attraction in company tour packages	3.34	1.058
2	It is important for tour operator to promote endangered flora and fauna as main product	3.12	1.012
3	It is important to have knowledge about last chance tourism	3.19	1.065
4	Preserving and protecting endangered flora and fauna is part of tour operator responsibility	3.14	1.002
5	It is important to for company doing preservation activities	3.17	0.953

Table 3. Multiple regressions of last chance tourism and perceived importance.

Predictors	Model 1/std. β
Step 1: Model variables Last chance attributes	0.807***
R^2	0.651
Adj. R^2	0.649
R^2 change	0.651
F-change	294.601

Note: *p < 0.05, **p < 0.01, ***p < 0.001.

4.3 *Relationship between perceived importance and last chance tourism attributes*

Last chance tourism attributes were able to explain 65.1% of the overall dimension ($R^2 = 0.651$, F-change = 294.601, $p < 0.003$***) of the variance in last chance tourism attributes. The outcomes demonstrated that last chance tourism attributes significantly contributed to the prediction of the last chance tourism perceived important. It evidently shows that last chance attributes was found to significantly affect the perceived importance.

5 FINDINGS

The result of the study revealed that last chance tourism has a positive impact towards perceived importance. Based on the local tour operator's perceived importance of last chance tourism attributes, analysis of the result revealed that most of the tour operator knows and agreed on the importance of last chance tourism products. Analysis clear revealed that all the last chance tourism attributes such as endangered flora and fauna including rifles, pitcher plants, orchard, Orang Utan, Turtle, Sumatran rhinoceros and sun bear give a positive relation on the perceived importance of tour operator. Analysis revealed that the overall attributes of last chance tourism have a significant relationship with perceived importance. Most of the tour operator agreed that promoting and preserving endangered flora and fauna is important and agreed that the potential this new market niche. There is a strong relationship between these two variables.

Although this study highlighted a range of remarkable, significant and meaningful findings, there are few limitations occurred. The survey was conducted primarily in Klang Valley area instead of other areas or regions. Therefore, the results obtained represent merely within this particular area and might not be represented as a whole of Malaysia's tour operator perception or point of view. A larger sample will enhance the understanding of the issues. Moreover, the researcher only focus on two attributes which is endangered flora and endangered fauna. Based on previous research, last chance tourism is included endangered flora, fauna, landscapes, seascapes, culture and heritage. Due to time constraint, only two attributes have been discovered. For future research, other attributes specially culture and heritage should be included in the survey.

REFERENCES

Adriana, B. (2003). Impacts and responsibilities for sustainable tourism: A tour operator's perspective. International Institute for Environmental Economics.

Bobyleva, A. (2011). Sustainability of Last-Chance Tourism. Retrieved November 23, 2012, from World Wide Web: http://www.preservearticles.com.

Bowen, D. & Clarke, J. 2009. Contemporary Tourist Behaviour.

Chandran, A., Bhaduri, R., & Swamy, A. (2012). The role of doomsday tourism in setting new paradigms towards addressing climate change issues A qualitative study on the Indian perspective. *Tourism Management*, 6(3), 100–110.

Charles, R., Goeldner, J.R., & Brent, R. (2008). Tourism: principles, practise, philosophies.

Crompton, J. (1979). Motivations for pleasure travel. Annals of Tourism Research, 6(4), 408–424.

Davidoff, Philip, G., & Davidoff, D.S. (Eds.). (1994). Sales and Marketing for Travel and Tourism (2nd ed.). Engle-Durst, NJ: Prentice-Hall.

Lemelin, H., Dawson, J., Stewart, E., Maher, P., & Lueck, M. (2010). Last-chance tourism: the boom, doom, and gloom of visiting vanishing destinations. *Current Issues in Tourism*, 13(5), 477–493.

Maher, P.T. (2007). Footsteps on the ice: visitor ex periences in the ross sea region, antarctica. Unpublished Doctor of Philosophy thesis, Lincoln University, Christchurch, New Zealand.

Mast business directory. (n.d). Retrieved November 15, 2012, from World Wide Web: http://www.mastbusiness.com/bizguides/bg2/.

Niemela, T. (2010). Motivation Factors in Dark Tourism: Case House of Terror at Lahti. Nature and soft Adventure Tourism, 6. Other as Tourists: CABI.

NRE (2006). Biodiversity in Malaysia: Multilateral Environmental Agreements: Capacity Building and Implementation Project (MEA Project) Malaysia. Retrieved November 15, 2012, from World Wide Web: http://www.nre.gov.my.

Schlichter, S. (2010). Climate Change Travel: The World's Most Endangered Places. Retrieved November 30, 2012, from Independent Traveler.com World Wide Web: http://www.independenttraveler.com/travel-tips/none/climate-change-travel-the-worlds-most-endangered-places.

Sekaran, U. (2003). Research methods for business (4th ed.). Hoboken, NJ: John Wiley & Sons.

Shapiro, K. (2007). The tourism of doom Travel Age West. Retrieved November 30, 2012, from World Wide Web: http://www.travelagewest.com.

Telisman, & Kosuta (Eds.). (1989). Tourist Destination Image. In Tourism Marketing and Management Handbook, S.F. Witt and L. Moutinho, eds., pp 557–561. Prentice-Hall.
Turnbull, D., & Uysal, M. (1995). An exploratory study of german visitors to the Caribbean.
Um, S., & Crompton, J.L. (1990). Attitude Determinations in Tourism Destination Choice. Annals of Tourism Research, 17, 432–448.
Unumeri, G.O. (2009). Perception and Conflict. Social Scientist, 18–240.
UNWTO & UNEP (2008). Climate change and tourism responding to global challenges: UNWTO.
Valda, B. (2002). Aboriginal cultural tourism in Canada. In J. Nicks & J. Sloniowski (Eds.), Slippery pastimes: Reading the popular in Canadian culture 37–59. Waterloo, ON: Wilfrid Laurier University Press.

Hospitality and Tourism – Sumarjan et al. (eds)
 ISBN 978-1-138-00151-0

Perception of international tourists on sex tourism in Malaysia

N.E. Zainal, M.S.Y. Kamaruddin, M.F.S. Bakhtiar & N.A. Ahmad
Universiti Teknologi MARA, Malaysia

ABSTRACT: Sex tourism has being argued since 1980s. Some country accept it as a bad image contributor, but some taking the advantage and promoting it as a part of economic booster. Unlike other type of tourism, this activity has been perceived as a bad image to the certain countries. Malaysia therefore without doubt is no exception has this kind of activities. However, the extent of this activity takes place is still vague. Using a qualitative approach employing interview method a group of international tourists consist of four respondents were interviewed. The respondents selected as they have visited Malaysia. Findings showed tourists do not perceive Malaysia as a sex tourism destination but rather the appreciating the nature of this country.

Keywords: sex, trafficking, international tourists, perception, bad image

1 INTRODUCTION

In Asia, countries such Thailand, Philippines and Cambodia are places which offer sex tourism with no exception Malaysia. Meanwhile, the sex industry was changing rapidly as reported by World Health Organization (2001), and become increasingly complicated with highly differentiated sub sectors. Most studies from the blogs, website, online articles and newspaper report along with anecdotal evidence, suggested that the commercial sex is becoming more common which involving the large number of people in a large variety of sites. Up until now, the activity somehow is still vague. The Southeast Asian countries which consists of Thailand, Indonesia, Philippines and Malaysia, report that the industry estimated to account for between 2%–14% of Gross Domestic Product (GDP). Research regarding sex tourism in Malaysia is only quite a number. The size of the sector due to this activity is operating underground, especially in Malaysia as known for Islamic image. Running 'places of assignation' is illegal where the establishment can link clients to the sex workers (WHO, 2001). Therefore, it is hard to control and monitor the result of underground sex trading. The sex market has become mobile and innovative. Sex workers wish not to be identified because of the legal restraint. Depending on some part of the globe sex tourism is a extremely moneymaking industry (Gore, 2009). She added that the Philippines, Thailand, Mexico Central America including Malaysia are countries that have become prime destinations for tourists seeking for sex. It estimated that more than one million children worldwide drawn into the sex trade every year as reported by the non-profit organization End Child Prostitution, Child Pornography, and the Trafficking of Children (ECPAT) (Gore, 2009). Fitness clubs, hairdressing salons, karaoke bars, coffee shops, nightclubs and hotels are venues that provide access to sexual services, which operate underground and illegal. It is vital to look for the truth about sex tourism or illegal sex activity in Malaysia.

2 LITERATURE REVIEW

2.1 *Sex tourism*

In Asia, countries like Thailand, the Philippines and Cambodia are seen as the most sexpats in the region while Indonesia and Vietnam considered growing. South East Asia is the popular chosen tourists' destination which means that those countries supply local prostitutes seriously (Xtremetravelstories.com, 2012).

Burton (1995), suggested that the 'cheaper' sex available in South East Asia prostitution is an inadequate explanation for the persistence of sex tourism since the encouragement to blend travel with consumption of commercial sex can be distinguish from the employment of prostitutes within the ordinary environment of the individuals involved. It is undeniable that the growth of prostitution in South East Asia has a correlation between the location that caused by tourism. In combination with seductive marketing of South East Asia and cheap accessibility to travel in developed countries, it helped to promote the region as a specialist tourist destination where catering to the erotic fantasies of men. The massive expansion of prostitution in South East Asia, therefore, represents the outgrowth of the indigenous industry to cater for a mass market. The questioning of religious laws that prevent prostitution with the operation of red light districts and brothels as seen in such Islamic countries as Pakistan and Malaysia (Wall & Mathieson, 2006).

2.2 *Sex industry in Malaysia*

Lim (1998) reported that there are 43000 and 142000 sex workers in Malaysia. During the period 1986–1990, numbers of 2626 women and girls caught by police from brothels, bars, massage parlours and house of prostitution. 50% of them were under 18 years old, and the rest were between 18–21 years old. In Malaysia, The Coalition Against Trafficking in Women (CATW) estimated that there are around 142000 women in prostitution with between 8000–10000 located in Kuala Lumpur.

The Star Online reports (2013) by Lee saying that a study done by a sex search incriminates Malaysians prefer to watch home grown material for internet pornography. PornoMD.com saying that Malaysians top three searches, over a six month period, starting from August 2012 to January 2013 was Malay, Malaysia and Japanese. As been told by the chief information officer of Porn.MD.com, George Robinson, and these searches were some of the 4.04 million or around 22000 per day internet searches by Malaysians on the website. Malaysia is the 19th most famous country on the site. In addition, PornMD.com received over 270 million searches worldwide in total during the same period.

2.3 *Sex trade a lucrative business in Malaysia*

Article by Sagayam from The Star Online (2006) reports that the lucrative sex trade is attracting both local as well as foreign women whereby some as young as 14 years old with vice syndicates in the country raking in at least RM1 mil daily. This syndicates operated by local along with international connections proved by the police. Based on the police's statement, sex workers from all over the world line up as China, Indonesia, Thailand and the Philippines providing the largest numbers. Russia, Uzbekistan and India are the sources for sex workers, while syndicates started taking Vietnamese women following "current market demand".

2.4 *Over 600,000 seek for a "sex buddies"*

Metro Ahad newspaper (2013) reports that more than 600000 individuals in the country including school dropout and senior citizens, detected enrol in an international social networking site just to find new contacts to be 'sex buddies'.

2.5 *Online websites listed Malaysia in the internet for sex services*

There are several of websites that offer sex services in the internet such aslaksaboy.com, sammyboyforum.com, locanto.com, asianescorts.com and klescorts.com. Several websites offered sex workers from Malaysia. A website named asianescorts.com (2013) claimed that Malaysia is the second highest number that provides sex workers after Japan in Southeast Asia. These numbers are for registered members only, not including the unregistered sex workers.

3 METHODOLOGY

Using a qualitative approach employing interview method, four informants (international tourist) interviewed. Three questions asked during the data collection are as following:

i. *Have you heard about sex tourism in Malaysia? From where do you hear it?*
ii. *Do you think that Malaysia has this kind of activity?*
iii. *If yes, what made you think like that?*
iv. *Have you been offer/encounter by this service while staying here?*

Pilot study was conducted to ensure the questions were well understood. Data were collected among tourist in airport who waiting for departure recorded with media player (MP3) recorder.

4 RESULT AND ANALYSES

4.1 *Findings*

Based on guideline by Silverman (2001) and Denzin (2001), for qualitative, the pen portraits analysis on the interview. Discussions of the profile of informants are simplified in Table 1.

Table 1. Informants profile of international tourists.

Informant	Country	Duration of stay	Age
1	Singapore	2 weeks	30
2	United States	1 week	40
3	Netherlands	Not specify	34
8	France	12 days	32

1. Have you heard about sex tourism in Malaysia? According to the Table 2, only 25% stated that have heard about sex activity in Malaysia while the other 75% saying that they never heard about sex activity in Malaysia.

Table 2. Perception of international tourists on sex tourism in Malaysia.

Informant	Yes	No	Indifferent
1	√		
2		√	
3		√	
4		√	

2. Do you think Malaysia has this kind of activity?
   a) If yes, what made you think like that?

Informants	Answers
Informant 1	"They do offer (sex service) but they don't promote it. Malaysia is not well-known about that (sex service). It's just that being the oldest profession in the history of working life so definitely there is services like that, you cannot run away from that. That doesn't surprise me that also happen here in Malaysia".
Informant 2	"I don't think Malaysia is known for that (offering sex service) at all in my experience. I never saw anything overt about that business (sex industry) here"
Informant 3	"I don't portray Malaysia as a country that offers this kind of services. Like Thailand is well-known for it. Like Philippines and Jakarta is known for it but Malaysia is not known for it I think. I know that people can go to place like karaoke rooms, pubs, and hotels. I think here in Malaysia is not common as in other places like Thailand, Philippines and Indonesia maybe. If a person wants to make use of it is available then there a places where they can go for it like in Kuala Lumpur and even in Setiawan, Perak. You have to find, you have to be long enough in that area, you have to hear from people who knows about it. You can find it if you search for it".
Informant 4	"I don't know sex tourism in Malaysia but I know about the sex tourism in South East Asia in general. I know that something is happening. I don't know much but I know that most guys come to South East Asia to look for this kind of purposes".

3. Have you been offer/encounter by this service while staying here?

Informants	Answers
Informant 1	"Yes, I've been offered by this service. I have encountered a few times. Mostly is in Kuala Lumpur. She is a Malay woman at Beach club in Kuala Lumpur. She (the sex worker) asked me that you would like to bring me home?"
Informant 2	"I never had in Malaysia (offer by this activity)".
Informant 3	During my stays in Malaysia, I never had been offer for this kind of activity. Not offered like what you have like in Thailand, that people are approach you obviously.
Informant 4	I haven't had the experience (offers for sex service) yet.

5 DISCUSSION AND CONCLUSION

Based on the summary of previous findings has clear cut the phenomenon of this activity in Malaysia as overall. From this study clearly shows evidence that from the eyes of genuine international tourists, they only hear sex tourism as a whole in South East Asia region particularly in Thailand, Vietnam and Indonesia while they are hardly heard about sex tourism in Malaysia. However, it is undeniable that sex activity also may exist in Malaysia in vague scale. Furthermore, for some international tourists in their opinion saying that if the international tourists often come visiting Malaysia, they might know where to find this activity even in rural area. In addition, there is an international tourist respondent had an experience whereby he had been approach and get offered of sex services from a girl at a bar. However, they disagree if Malaysia promoting the sex activity as part of their tourism product. Malaysia has not reach to the level that sex services becoming sex tourism but only as sex industry. Unlike other countries such Thailand, Philippines and Singapore where sex industry is widely known by the world, Malaysia also have this activity in vague scale.

In addition, whenever people talking about the main reason to go to Malaysia, people do not simply portray as looking for sex services. This is due to the Islamic practises runs Malaysia and the majority of Malaysia's citizen are Muslim where at least control this situation from being well known worldwide. The sex trade is a thriving business and it is not limit to only one country or region. It happened globally. This business are highly profitable but only if the business are well arrange and runs by the responsible person to avoid from getting caught by the authority. For some countries, it is also undeniable that due to the existence of this kind of activity is attracting more tourists. However, the implication or the dark side of this trade is also fearsome. There are lots of negative impacts in sex tourism industry. Firstly, let's look at impact to the country. If this industry is widely operates in Malaysia, surely it will create the bad image as a country which practices Islamic tourism.

With the majority of Muslims population in Malaysia, of course the Muslims feel ashamed and frustrated if Malaysia is well-known with this kind of tourism. It also same goes to other races whereby they will feel the same too. To Malaysian, Malaysia has lots of other sources that can be promoted such heritage tourism, the cultures, foods, the buildings, beaches and islands. Besides, if this tourism successfully flourishes, it shows that there are irresponsible party whom dare to challenge the law and regulations of Malaysia and being underestimated. This scenario proof that the irresponsible party being disrespectful and tries to jeopardize the integrity of Malaysia's authority. With the listed of implications above, it proven that the negative impacts emerge more than positive side. The sex tourism industry might only boast up the economy of the particular country but it leads to nowhere. In Islamic religion, prostitution or sex activity is legal offence.

To conclude, Malaysia has not reached up to the level where offering sex tourism to international tourists but it is true that sex industry does exist. Sex industry in Malaysia is still under control whereby the religion; Islam are the root and still held strong by Muslims here and also supported respect by the non-Muslims citizen who lives here. In addition, Malaysian does not accept the culture that involved with free sex, which leads to curb this phenomenon from spreading widely. There are lots of aspects to be improved in the future to reduce this immoral activity in Malaysia. Hoping that Malaysia are free from the sex industry and eliminate country name from websites that promoting or listing Malaysia as a destination where they can find this immoral activity.

REFERENCES

Asianescorts.com (2013). Your #1 Escort Directory for Asia. Retrieved June 2, 2013 from http://www.asianescorts.com/#sort=r;name=;page=7.

Denzin, K.N. (2001). The Handbook of Qualitative Research (4th ed.). London: Sage.

Gore, R.M. (2009). Sex tourism trade throughout the world: Travellers taking trips for illicit purposes. Retrieved November 17, 2012 from http://sextourismtradethroughouttheworld.

Lee, P. (2013). Malaysians prefer home grown porn, says sex search engines. The Star Online. Retrieved May 10, 2013 from http://thestar.com.my/news/story.asp?file=/2013/6/4/nation/20130604094 11sec=nation.

Lim, L.L. (1998). The Sex Sector: The Economic and Social Bases of Prostitution in Southeast Asia. Geneva: International Labour Office.

Mount-Burton, S.H. (1995). International Tourism—Identity and Change. "Sex tourism and Traditional Australian Male Identity". London: Sage.

Over 600 000 seek for "sex buddies". (2013, May 12). Metro Ahad. p. 6.

Sagayam, A. (2006). Sex Trade a lucrative business in Malaysia. The Star Online. Retrieved March 17, 2013 from http://thestar.com.my/news/story.asp?file=/2006/7/27/nation/14963603&sec=nation.

Silverman, D. (2001). Interpreting qualitative data (2nd ed.). London: Sage.

Wall, G. & Mathieson, A. (2006). Tourism—Change, impacts and opportunities. Social consequences. Great Britain: Pearson.

World Health Organization (2001). Sex Work in Asia. Retrieved November 7, 2012 from http://www.who.int/whosis/whostat/EN_WHS2011_Full.pdf.

Hospitality and Tourism – Sumarjan et al. (eds)
 ISBN 978-1-138-00151-0

Common acceptable cuisine and national food identity

M.Z. Suhaimi, M.S. Mohd Zahari, Z. Othman & M.A. Hemdi
Universiti Teknologi MARA, Malaysia

ABSTRACT: Common acceptable cuisine usually discussed in the multicultural/ethnic nation as it represents the process of sharing cuisine among the ethnic groups. The common acceptable cuisine also considers as a precursor that takes place in the process of constructing the national food identity within the multicultural/ethnic countries. The cultural adaptation of cuisine through types of food is believed as the mechanism in enhancing the process of sharing the common acceptable cuisine with Malaysia without an exception. Through that, this study empirically investigates the general perception of the three major ethnic groups' chefs namely Malay, Chinese and Indian on the adaptation processes and their belief towards the formation of common acceptable cuisine in Malaysia. A remarkable finding emerged almost similar among three main ethnic chefs. The Malay, the Chinese and the Indian chefs are share a common view of the process of adaptation of ethnic cuisine through type of foods, and no doubt these chefs believed it giving advantages to the formation of common acceptable cuisine in the country.

Keywords: adaptation, ethnic cuisine, common acceptable cuisine, national food identity

1 INTRODUCTION

Besides religions, politics, and economy, food identity formation in the multi-cultural/ethnic countries has received significant attention among the academic scholars (Epp & Price, 2008; Fox & Ward, 2008). Scholars have argued that having a food identity, a nation can easily be recognized internationally (Spurrier, 2010) in term of food culture and image (Kwik, 2008) promoting gastronomic tourism (Kivela & Crotts, 2006) nation building and social integration (Thompson, 2011).

Clear examples can be looked at France, Italy, Spain, Thailand and Japan. The emerging of French food identity for instance, is not only creating a foundation in building the nation but, the cuisine itself is well established. Besides, the French food recognized and engaged in line with the development of global culinary and gastronomy (Pinkard, 2009).

Malaysia, as similar to other multi-cultural/ethnic nation, comprised of three major ethnic groups, namely Malay, Chinese, Indian and others who embodied their traditional ethnic identity is also struggling through various initiatives in establishing the country's national identity including nation food identity (Ishak et al., 2012; Manaf, 2008).

Due to the importance of having nation food identity, greater concern about this issue is receiving considerable attention from the Malaysian government through the former Minister of Tourism. Evidently, Dato Seri Ng Yen Yen stress on the contribution of the nation's culture and including the food identity to the destination branding for the country.

> *"The wide variety of culture and food in Malaysia can be an effective tool to demonstrate our nation's cultures and food identity to attract international tourists. The extensive ranges of our local dishes were very much a reflection of the cultural melting pot creating a unique Malaysian identity"*

In addition to the above notion, many argued that the precursor that take place in the process of constructing the nation food identity within ethnic groups in the multicultural/ethnic countries is reflecting all the way through sharing a common accepted cuisine (Appadurai, 1988; Spurrier, 2010). Looking at the context of ethnic relations in Malaysia and based on scholars notation (Phinney, 2003; Shalom & Horenzyk, 2004) it is strongly believed that adaptation of ethnic cuisine through types of foods will lead into the common acceptable cuisine for a country. In this sense, the adaptation of certain Malay, Chinese and Indian types of foods through preparation is believed will form into the common acceptable cuisine for a country. In line with that, this study empirically investigates the general perception of the three major ethnic groups' chefs namely Malay, Chinese and Indian on the adaptation processes and their belief towards the formation of common acceptable cuisine in Malaysia.

2 LITERATURE REVIEW

2.1 *Cuisines and the three major ethnic groups in Malaysia*

Apart from other cultural elements that have been introduced by governments such as the language, apparels, religious affiliation and others (Shamsul, 2011), cuisine is identified in giving significant roles in developing the social cohesion among the ethnic groups in Malaysia (Ishak et al., 2012; Jalis, 2008). This is based on the anecdote evidences whereby during the pre-independence of Malaysia each ethnic group centered their foods only for their own ethnic needs. There was also minimal cross cultural in understanding each other ethnic cuisine as well as other traditions and customs. Othman (2007) argue during that period Chinese restaurants for instance were found to be selling foods exclusively to the their community, such as pork, vegetables, mustard greens, mung beans, daikon radish, Chinese cabbage and many others. Likewise, Indian restaurants were also set up specifically to sell their own delicacies and savories, including curry, dhal, chapatti, idle and paratha to their fellow Indians while Malay restaurants and food were confined to the individual states or the same geographical location (Ramli & Ahmad, 2003).

However, with the growing numbers of modern ethnic restaurants in Malaysia after the post-independence era (Othman, 2007), restaurateurs started to incorporate various selection of menus ranging from diverse ethnic cuisines particularly the Malays, Chinese and Indians (Ramli & Ahmad, 2003). Obvious example can be seen in the concept of *Nasi Campur* (white rice served with various types of dishes). Several typical dishes from the three major ethnics such as ayam masak merah, asam pedas ikan pari, sambal udang petai, kangkung goreng belacan from the Malay ethnics, lemon chicken, sweet and sour fish, butter prawn, stir fried mixed vegetables, kai lan with soy sauce which represents the Chinese ethnics and mutton curry, tandoori chicken, dhal, rassam soup and many more symbolized to the Indian ethnics enhancing the concept of *Nasi Campur*. This is in line with what have been stated by Ramli and Ahmad (2003) that many ethnic restaurants offering such combinations of foods and this unique combination gradually created a significant assortment of Malaysian culinary potpourri.

2.2 *Cultural adaptation*

Cultural adaptation is a key element that enables humans to understand the other individuals' skills, knowledge, techniques and experiences. This process assists individuals to familiarize and adjust the new knowledge gain which leads or influence them to practice or develop the acquisition (Cordes, 2004). One of the notable findings is that the cultural adaptation process is a mechanism in reducing uncertainty and anxiety especially involving two different cultures (Baffoe, 2006). Auxiliary with this, different cultural groups yet have different experiences and challenges as they come into a particular culture. Li (2009) for instance indicates that different ethnic groups may face different situations in cultural adaptation even though the results may not always consistent, but previous researchers found ethnic/racial differences are still in control orientation.

Looking at Malaysia context, Hamid (2004) noted language is one of the cultural elements that best describes the process of cultural adaptation among the ethnic groups. In this sense, Bahasa Malaysia acts as the language instrument not only in the medium of education and the official language of administration but also in integrating the ethnic groups. He also argues that to force or encourage ethnic groups in Malaysia practicing one religion or one collective culture is impossible because this will be perceived as an attempt to eliminate the identification of certain ethnic groups. However, to encourage ethnic groups in Malaysia to accept and using Bahasa Malaysia as a communication tools is impossible, though it is likely to face many difficulties and obstacles. It is due to the fact that language is perceived as more universal in nature rather than religion and collective elements in culture (Hamid, 2004).

Besides language, cuisine as part of cultural elements representing each ethnic group in Malaysia start receiving the attention among academic scholars for its contribution of the cultural adaptation process (Henderson, 2009; Ishak et al., 2012). According to Fox and Ward (2008) the cultural adaptation of cuisine has typically been discussed in the multicultural/ethnic country perspectives since the differences in culture and customs are more pronounced.

2.3 *The concept of common acceptable cuisine*

As the complexity of each ethnic group that represents their collective identities including in the aspect of the cuisine, the nation requires to reach some common accepted food and cuisines to be portrayed in the global context. Cusack (2004) states that the multi-racial nation should have its own national cuisine, despite the globalization of culinary culture has added complexity to the relationship between food and nation. In this sense, national cuisine will have a complex and multiple origins and not static with one set of identities only (Cusack, 2004). The emergence of Mexican national cuisine can be seen as one of the examples. The complex culinary history in Mexico has donated to arrays of national identities was the most common culinary symbol for the Mexican nation was *mole poblano* (turkey in deep brown sauce). Towards this point of view, the concept of common acceptable cuisine among the multicultural/ethnic nation is considered important as it offers a collective sense of identity which benefits the country in various perspectives.

Mugalavai, Kiama and Omutimba (2012) posited the concept of common acceptable cuisine in the multicultural/ethnic country represents the process of sharing acceptable cuisine among the ethnic groups. Since each ethnic group embodied their own identity in term of food habits or preferences, some elements in the cuisine that are commonly accepted could facilitate the mutual understanding among the ethnic groups. In addition, Kifleysus (2004) noted that the concept of common acceptable cuisine can be achieved through the understanding of each ethnic belief, skills, ethical norms, and the social attitudes that practice within their cultural settings.

3 METHODOLOGY

3.1 *Sampling and instrument*

A descriptive research design using a quantitative approach through cross sectional study was applied. The sample populations were among ethnic chefs working in various foodservice establishments with the age range between 30 years old and above. Owing to the huge numbers of chefs in Malaysia those who work in Shah Alam, Petaling Jaya and Kuala Lumpur area were chosen as a sample.

Owing to geographical dispersion and large sample size, the Malay, Chinese and Indian chefs were identified through researchers' personal contact, culinary friends, Chefs Association of Malaysia (CAM) and Faculty of Hotel and Tourism Management UiTM alumni. On top of that, ethnic chefs were also identified through Facebook and Twitter.

As for the instrument, the questionnaire was divided into three main sections. Section A dealt with the respondents' demographic profiles. Section B elicits the overall perception of the Malay, Chinese and Indian chefs on the adaptation of ethnic cuisine through type of foods. Section C examines the beliefs of the Malay, Chinese and Indian chefs toward the formation of common acceptable cuisine resulted from the adaptation processes. Close-ended questions using a Likert scale ranging from 1 being 'strongly disagree' to 5 being 'strongly agree' was used in section B and C. The reliability and validity of the items were checked and confirmed through a pilot study which was carried out among the thirty selected chefs.

3.2 *Data collection*

The identified Malay, Chinese and Indian chefs were initially contacted via telephone and electronic mail to obtain permission and attain their willingness to participate in the research surveys. Owing to lack of time and working engagement some chefs acquired questionnaires to be sent via electronic mail.

Upon their agreement, the surveys were personally conducted by the researchers and research assistants with those chefs who agree to be surveyed. A survey questionnaire was also sent to those chefs who acquired questionnaire via electronic mail and reminder were given to submit within the stipulated period (2 weeks). In light of the positive feedback and the absence of any obvious problem with either the instrument or both processes, good responses with a total of 195 completed questionnaires were collected with 160 responses from self-administered and 35 responses via electronic mail.

4 RESULTS AND DISCUSSION

Out of 195 respondents, Malay constituted around 57.9%, as opposed to Chinese with 24.9% and Indian 15.4% of the total respondents. In regards to age group, similar frequency of 27.2% was received from respondents in the age range of 35–39 years and 40–44 years and followed by those in the age range between 45–49 years which accounted 17.4%. Respondents with the age range from 30–34 years old made up about 16.9%, and the lowest groups were among the 50 and above accounted 11.3% of total respondents. 30.8% of respondents passed the secondary school level in contrast to 69.2% who were graduated from colleges or universities.

4.1 *Overall perception on the adaptation of ethnic cuisine through type of foods*

The descriptive statistic looking at the mean score was used in examining overall perception on the adaptation of ethnic cuisine through type of foods (see Table 1).

The magnitude of the mean scores ranges from 3.29 to 3.75 indicates the majority of the chefs agreed with most of the items. As such, they agreed that easily in understanding each other ethnic types of food (M = 3.75). They also believed that many types of each other ethnic foods that can be adapted (M = 3.61). Slightly lower mean scores appeared on

Table 1. Type of foods.

Items	Indicator	Mean (M)	Std. deviation (SD)
Type of foods			
Easily in understanding each other ethnic type of foods	Tfo_1	3.75	1.155
I believe many types of each other ethnic foods that can be adapted	Tfo_2	3.61	1.001
Only type of foods that are accepted by each other ethnic can be adapted	Tfo_3	3.43	1.161
Only each other ethnic foods that are complying with religious requirements can be adapted	Tfo_4	3.48	1.100
Although many of each other ethnic meat dishes available, only those are suitable to a common taste can be adapted	Tfo_5	3.38	1.079
I believe most of each other ethnic vegetable dishes can be adapted	Tfo_6	3.43	1.116
I believe most of each other ethnic desserts can be adapted	Tfo_7	3.29	1.093

(Scale: 1 = Strongly Disagree, 2 = Disagree, 3 = Moderately Agree, 4 = Agree, 5 = Strongly Agree).
Note: (n = 195).

Table 2. Common acceptable cuisine.

Items	Indicator	Mean (M)	Std. deviation (SD)
Common acceptable cuisine			
I feel strongly attached with some of each ethnic cuisines in my culture	Cac_1	3.37	0.988
I believe the adaptation of the popular each other ethnic type of foods contributes to a common acceptable cuisine for Malaysian	Cac_2	3.56	0.974
I support the effort in promoting each other ethnic cuisines as common acceptable cuisines in Malaysia	Cac_3	3.79	0.886

(Scale: 1 = Strongly Disagree, 2 = Disagree, 3 = Moderately Agree, 4 = Agree, 5 = Strongly Agree).
Note: (n = 195).

two items. With religious restriction, it is not surprising that a majority of chefs believed the only type of foods that are accepted by each other ethnic (M = 3.43) and complying with religious requirements can be adapted (M = 3.48). In addition, the concern on certain foods especially for the meat based dishes shows slightly lower score rather than the vegetable dishes that can be adapted. It can be viewed at the level of agreement on the meat dishes that can be adapted (M = 3.38) compared to their level of agreement on the vegetable dishes that can be adapted (M = 3.43). Most of the respondents also moderately agreed on each other ethnic dessert that can be adapted (M = 3.29).

4.2 *Beliefs towards the formation of common acceptable cuisine*

Three items were used in measuring the three major ethnic groups' respondents on their beliefs towards the formation of common acceptable cuisine (see Table 2).

Results revealed that, each ethnic respondents moderately agreed that they strongly attached with some of each other ethnic cuisines (M = 3.37) and believed that the adaptation of the popular each other ethnic type of foods contribute to a common acceptable cuisine for Malaysian (M = 3.56). Having strong feelings on the preceding items it is not surprising that the majority of ethnic respondents are supporting the effort in promoting each other ethnic cuisines as common acceptable cuisines for Malaysia (M = 3.79).

This result indicates that most of the three major ethnic chefs, Malay, Chinese and Indian were having a common view of the process of adaptation of ethnic cuisine through type of foods and no doubt believed it giving advantages to the formation of common acceptable cuisine in the country.

4.3 *Comparison of common acceptable between ethnic groups*

Analysis using One-Way Analysis of Variance (ANOVA) with the Scheffé *post* hoc multiple comparison was undertaken to investigate the significant differences between ethnic groups on the beliefs of common acceptable cuisine. No, statistically significant differences were detected on any items. Ethnic respondents moderately agreed that they strongly attached with some of each other ethnic cuisines (3.39 for the Malay, 3.35 for Chinese and 3.33 for Indian and p = 0.945) and believed that the adaptation of the popular each other ethnic type of foods contribute to a common acceptable cuisine for Malaysian (3.51 for the Malay, 3.67 for Chinese and 3.53 for Indian and p = 0.614). It is not surprising that the majority of ethnic respondents is supporting the effort in promoting each other ethnic cuisines as common acceptable cuisines for Malaysia (4.02 for the Malay, 3.90 for Chinese and 3.75 for Indian and p = 0.236). What could be said from this section analysis and with no obvious ethnic dissimilarities appeared is that adaptation of the popular each other ethnic type of foods which involving time gradually contribute to a common acceptable cuisine in this country.

5 IMPLICATIONS AND CONCLUSIONS

Previous studies indicate food even in global phenomena influence every facet of human life including the routines, traditions, rituals, norms and the people's daily activities. The food influences the social-demographic, economic, education and life of the community for the betterment. Not only that, food is also acting in strengthening the integrative force, solidarity and social bonding and alliances among the communities. Most importantly and most remarkable finding is that adaptation on types of food creates a strong belief on the formation of common acceptable cuisine. This strengthening other researcher notion that the adaptations of food from various ethnic foods by ethnic group in multicultural/ethnic nation will lead to common acceptable cuisines and longitudinally forming what could be called national food identity (Fox & Ward, 2008).

Perhaps the ethnics' chefs and those individuals without realizing their adaptation of each other ethnic food not only are giving advantages for themselves but directly given the significant impact to the social and the nation. Adaptation increase the potpourri of acceptable food among the Malaysian, influence other public to practice and enhancing the Malaysia food image that will gradually be recognized locally and internationally. Therefore, if could be informed they should continuously practice what they are doing now as long maintaining the religious borders.

REFERENCES

Appadurai, A. (1988). How to make national cuisine: cookbooks in contemporary India. *Comparative Studies in Society and History,* 30(1), 3–24.

Baffoe, M. (2006). Negotiating two worlds: culture and cultural adaptation of immigrant and refugee youth. Quebec (Canadian) educational context, McGill University.

Cordes, C. (2004). The human adaptation for culture and its behavioral implications. *Journal of Bioeconomics,* 6(2), 143–163.

Cusack, I. (2004). Equatorial Guinea's national cuisine is simple and tasty: cuisine and the making of national culture. *Arizona Journal of Hispanic Cultural Studies,* 8, 131–148.

Epp, M., & Price, L. (2008). Family identity: a framework of identity interplay in consumption practices. *Journal of Consumer Research,* 35, 404–427.

Fox, N.J., & Ward, K.J. (2008). 'You are what you eat? Vegetarians, health and identity'. *Social Science and Medicine,* 66(12), 2585–95.

Hamid, Z. (2004). Bahasa Melayu sebagai bahasa penyatu dan bahasa pemisah warga Malaysia. *The 4th International Malaysian Studies Conference.*

Henderson, J.C. (2009). Tourism policy and cultural heritage in multi-ethnic societies: a view of Malaysia. *International Journal of Tourism Policy,* 2(1), 138–144.

Ishak, N., Zahari, M.S.M., Sharif, S.M., Muhammad, R., & Salleh, H.M. (2012). Acculturation, foodways and Malaysian food identity. In Artinah. et al. (Ed.), *Proceedings of the International Hospitality and Tourism Conference,* (pp. 359–363). Kuala Lumpur, Malaysia: Universiti Teknologi MARA.

Jalis, M.H. (2008). Acceptance level of Malaysian gastronomic tourism products among the western tourists. *Unpublished master thesis,* Shah Alam: Universiti Teknologi MARA.

Kifleyesus, A. (2004). *The construction of Ethiopian national cuisine,* University of Asmara.

Kivela, J. & Crotts, J.C. (2006). Tourism and gastronomy: Gastronomy's influence on how tourists experience a destination. *Journal of Hospitality & Tourism Research,* 30, 354.

Kwik, J.C. (2008). Traditional food knowledge: A case study of an immigrant Canadian "foodscape". *Environments,* 36(1), 59–74.

Li, Y. (2009). *One life, two worlds: a qualitative comparison of cultural adaptation, relationships and life experiences of Chinese and Colombian College students in the United States, Unpublished master thesis,* Omaha: University of Nebraska.

Manaf, Z.A. (2008). Establishing the national digital cultural heritage repository in Malaysia. *Library Review,* 57(7), 537–548.

Mugalavai, V.K., Kiama, F.W., & Omutimba, H.N. (2012). Using traditional cuisine contexts as a channel for inter-ethnic social integration in Kenya. *International Journal of Social Science Tomorrow,* 1(2).

Othman, Z. (2007). Service delivery system and customer patronization: a comparison of ethnic restaurants in Shah Alam. Unpublished master thesis, Shah Alam: Universiti Teknologi MARA.

Phinney, J. (2003). *Ethnic Identity and acculturation.* In Chun, K., Organista P. and Marin, G. (Eds.). Acculturation: advances in theory, measurement, and applied research. 63–82. Washington D.C: American Psychological Association.

Pinkard, S. (2009). A *revolution in taste: the rise of French cuisine, 1650–1800*. Cambridge: Cambridge University Press.

Ramli, A.S., & Ahmad, R. (2003). Factors influencing customers patronizing Mamak restaurants. *Proceeding of the 2003 Tourism Educators of Malaysia Conference.*

Shalom, U.B., & Horenczyk, G. (2004). Cultural identity and adaptation in an assimilative setting. Immigrant soldiers from the former Soviet Union in Israel. *International Journal of Intercultural Relations,* 28, 461–479.

Shamsul, A.B. (2011). *Modul hubungan etnik*. Shah Alam: Pusat Penerbitan Universiti (UPENA).

Thompson, J. (2011). *Gastronomic literature, modern cuisine and the development of French Bourgeois identity from 1800 to 1850*. History Honors Papers, paper 9. Retrieved from http://digitalcommons.conncoll.edu/histhp/9.

Hospitality and Tourism – Sumarjan et al. (eds)
© 2014 Taylor & Francis Group, London, ISBN 978-1-138-00151-0

Educational value of dark tourism: Tour operators perceived importance and action behavior

S.N.S. Syed Mahbob, M.S. Mohd Zahari, Z. Zainuddin & M.Z. Suhaimi
Universiti Teknologi MARA, Malaysia

ABSTRACT: Realizing the importance and its contribution, Malaysia without doubt perceived dark tourism able to generate revenues to the country's economy and efforts are undertaking in promoting dark tourism sites to the international tourists. This product, however, cannot aggressively be promoted without communal efforts from all bodies related to tourism including the tour operators. This paper empirically investigate the Malaysia local tour operators perceive the importance of educational values as part of dark tourism attributes and their actions behavior in promoting dark tourism sites. Majority of the local tour operators agreed that Malaysia has vast dark tourism sites, but some perceive slightly importance of it from educational values, therefore, do not yet aggressively transform their perceptions into action behavior in promoting the dark tourism sites. Visiting dark tourism sites are still getting less attention or promotion from some of the Malaysia local tour operators. These scenarios without doubt giving varying consequences and implications not only to the individual tour operators but t related tourism organizations and government authorities.

Keywords: action behavior, dark tourism, educational values, perceived importance, tour operators

1 INTRODUCTION

The rapid transformation of the tourism industry is significantly increasing the travelling pattern among the international tourists. Various reasons motivate travelers to travel, and this in turn, and in addition, to the conventional ones, create a range of new niche tourism products. New niche products are gradually accepted and support the existence of conventional products which tailored according to the tourists needs (Goel, Broder, Gabrilovich, & Pang, 2010; Novelli, 2005). Gastronomic tourism, culture based tourism, wild life tourism, green tourism and dark tourism without exception are examples of new niche tourism products. Apart from other tourism products dark tourism is starting to receive significant attention among countries and international tourists (Isaac, 2012).

Tarlow (2005) classified dark tourism is covered the visitation to places where tragedies or deaths have occurred while dealing with the legacy of humanity. Dark tourism products creating motivational tourists to travel to the places related to death and tragedies which have turned into new worldwide phenomenon (Stone & Sharpley, 2008). Keyes (2012) contended that dark tourism is playing a crucial part in constructing a national image, and emerge as vivacious part of the tourism industry which cannot be ignored by countries as an income generator.

Vietnam, Japan, Cambodia, and Ukraine just to name a few are getting the substantial income through dark tourism, in addition to other tourism products. The dark tourism is creating the unintentional demand and becomes part of tour companies' programs with the aims of making money by bringing the tourists to the dark sites around the world such as Auschwitz Concentration Camp in Poland, Tuol Sleng Genocide Museum in Cambodia, Ground Zero in New York, Hiroshima Peace Museum in Japan (Sharpley, 2009).

As with other countries, besides conventional and other new tourism products Malaysia without doubt perceived dark tourism as one the revenue to the country's economy. Various tourism sites have the mark as dark tourism sites such as Kelly Castle in Perak, Sabah memorial Park, Jerejak Island and Penang War Museum that has promoted as one of the tourist attractions (Braithwaite & Lee, 2008; Suhaini & Chai, 2010). Some efforts are undertaking by the government through Ministry of Tourism and other related agencies associated with the tourism industry to promote the dark tourism to the international tourists. Central and states government have come up with several planned in maintaining and upgrading the dark tourism sites (Ngo, 2012). Any new tourism products, however, cannot aggressively be promoted without communal efforts from all bodies related to tourism (Cooperative Research Centre, 2008). Strong collaboration and commitment from those parties including tour operators required in promoting the products.

In line with the above notion, questions arise especially related to Malaysian tour operators. How the local tour operators perceived importance of educational values as part of dark tourism attribute and how their behavioral actions toward this new emerging market? To be more specific and accurate, do the educational values as part of dark tourism attributes perceived importance to the local operators? In other words, does visiting the dark tourism sites be included in their tour itineraries? In understanding these questions, this paper empirically investigate the Malaysia local tour operators perceive the importance of educational values as part of dark tourism attributes and their actions toward promoting dark tourism sites as part of their tour packages.

This objective is further support with one hypothesis:

H1: There is a significant relationship between perceived importance of educational values as part of dark tourism attributes and tour operators' action behavior.

2 LITERATURE REVIEW

2.1 *Dark tourism and educational values*

The phenomenon of dark tourism has received significant attention among the researchers (Novelli, 2005), and it is worth mentioning that the long standing phenomenon of travelling to sites associated with deaths and disasters has existed since ancient times (Bittner, 2011; Stone, 2006). Stone (2005) claims people have long been drawn, purposefully or otherwise towards places, attractions or events that related with death, suffering, violence and disaster. Early examples of dark tourism could be found in Roman Coliseum, which considered as the first dark tourism attractions in the world. The patronage of Roman gladiatorial games involved the elements of death, torture and despair that consumed by the raucous spectators (Stone & Sharpley, 2008). These practices have been going on for centuries, but because of a transformation on how death and disaster handled; those activities used for more personal entertainment purposes (Chiles, 2010).

The visitation to sites related with death and disasters is recently growing and commercialized due to the increasing of tourism infrastructure and strong influenced from the media that provide up to date information related to the sites. This phenomenon is then well recognized and investigated under a range of different terms and definitions. Therefore, Stone and Sharpley (2008) mentioned the label of "dark tourism" has come into the academic discussion and media parlance that has generated momentous amount of research interest among the scholars.

Based on the literatures, cultural, historical and heritage and educational values are dark tourism attributes consider as the vital element in examining dark tourism phenomenon (Bujisic, 2011; Chiles, 2010; Moeller, 2006; Ozer, Ersoy, & Tuzunkan, 2012; Stone, 2006; Stone & Sharpley, 2008). For this paper, review is only focusing, focused on educational values. Stone (2012) revealed that regardless from the remembrance of history, tourists visiting the dark tourism sites can have the opportunity to congregate with the real situation and learn about life and death. In fact, visiting the war memorials can remind themselves of some larger human value that go beyond our day to day lives. Therefore, sites associated with the tragic history, war, death, murder can be encountered for educational purposes, and it should be recognized as educational tourism monuments so that it can be protected and conserved. This is the places for visitors to enhance their educational experience from connection of one's personal heritage (Biran et al., 2009).

2.2 *Action behavior in promoting dark tourism*

Marmor (2012) noted that an increase in visitation to the sites of genocide in recent years makes dark tourism as a big business among the tour operators. The popularity of such sites may be enhanced by the marketing and promotional activity of businesses or organizations that anxious to generate their profits through dark tourism industry (Seaton, 1996). Therefore, to attract the post-modern visitors, dark places must be packaged, promoted, priced and positioned, just like any other products or services. Many tourist agencies and small organizations offer tours around dark events and take their participants to dark locations, for example, the Ground Zero Tours and the Hurricane Katrina Tours. However, some state that bringing tourists around dark events seen as making money from despair, packaging pain for happiness, and business on dreadful human tragedies. Indeed, generating the profits from misery, packaging pain for pleasure, trading on terrible human tragedies regarded as profoundly inappropriate. There was an ethical and moral issue of profiting from loss among the victims. Therefore, it is necessary for the tour operator to promote peace and understanding the dark tourism sites to the visitors and tourists (Bujisic, 2011).

3 METHODOLOGY

3.1 *Sampling and instrument*

To discover the relationship between educational value and tour operator action behaviour, a quantitative approach applied for information gathering. The sample populations were among the selected and available tour operators that selling inbound and domestic tour packages within the Klang Valley and Kuala Lumpur area. A self-completed questionnaire designed, and the instruments adapted and modified from previous studies. The survey instrument comprised three sections. The first section designed to examine tours operators perception importance of dark tourism from educational value while the second section assessing the tour operators' action behaviour. Both sections used 7 point Likert scale which measuring tour operators level of agreement.

3.2 *Data collection process*

The telephone numbers and electronic mail addresses of 400 identified tour operators in selected areas of Klang Valley and Kuala Lumpur obtained from the Malaysian Association of Tour and Travel Agents (MATTA, 2012). The respective informants contacted via telephone and electronic mail in obtaining permission and willingness to participate in the research surveys. A brief explanation on the purpose of the study conducted and information obtained is strictly confidential, and anonymity preserved addressed.

The surveys personally conducted by the researchers and research assistant in 4 weeks during weekdays with tour operators that agree to be surveyed. Owing to of lack time and business engagement some tour operator acquired questionnaires to be sent via electronic mail. The remainder was given them to submit within the stipulated period (2 weeks). In light of the positive feedback and appropriate responses, a total of 120 completed questionnaires collected; 90 respondents from self-administered and another 30 respondents via electronic mail.

4 RESULTS AND ANALYSES

4.1 *Tour operator profiles*

Majority of tour operators participated were among the inbound compared to outbound and domestic tour operators. Most of them were in the tour businesses between 5 to 10 years and received the international tourists from Asian and the Western countries. The target markets were among international tourists and travel agencies.

4.2 *Perceived importance of educational values as part of dark tourism attribute*

The descriptive analysis undertaken looking at the perceived importance of educational values as part of dark tourism from the tour operators' perspective.

The result revealed most of the tour operators are slightly agreed that promoting the historical buildings and old monuments as part of dark tourism sites are important to provide a sense of learning values (M = 5.54; S.D = 1.491, item E1), enable the tourists learn about the collective artefacts and archaeology (M = 5.23, S.D = 1.301; item E2) and somewhat fulfil the tourists with knowledge about the country's history (M = 5.25; S.D = 1.451; item E3). The tour operators also slightly believed that promoting the historical buildings as part of dark tourism sites, provide the tourists with knowledge about the brutal history of the convict settlement (M = 5.17; S.D = 1.318; item E4) and slightly agreed dark tourism allow tourists to understand the tragedies that occurred during the past and empathize with the victims (M = 5.40; S.D = 1.387; item E5). With that impression, the tour operators also slightly believed the dark tourism sites conveying valuable messages for the benefits of future generation (M = 5.23, S.D = 1.369; item E6).

4.3 *Tour operators' action behaviour*

To know whether the tour operator's perception translated into their action behavior, this descriptive analysis is revealing the answer.

Table 1. Mean score of the educational values (n = 120).

No.	Item	Mean	S.D
1	EI	5.54	1.491
2	Promoting Dark tourism products are important to help tourists learn about collective artifacts and archaeology	5.23	1.301
3	It is important to fulfill tourist with knowledge about our country's history when visiting dark sites	5.25	1.318
4	Promoting dark tourism sites such old prisons enable visitors to learn about the brutal history of the convict settlement	5.17	1.318
5	The dark tourism sites such as a memorial park enables tourists to understand the tragedies that occurred during the past and empathize with the victims	5.40	1.387
6	Historical sites are important in conveying important messages and gaining knowledge of past events to the tourists	5.23	1.369

Strongly Disagree, 2 = Disagree, 3 = Somewhat Disagree, 4 = Neither Disagree or Agree, 5 = Somewhat Agree, 6 = Agree and 7 = Strongly Agree.

The majority of the tour operators just slightly or somewhat agreed that their company aggressively promoting dark tourism sites to the tourists (M = 5.06; S.D = 1.416; item F1) and slightly agree to visit the dark tourism sites are part of their company tour itinerary (M = 5.18, S.D = 1.424; item F2). With that feeling is it not surprising that the majority of the tour operators slightly agreed with the rest of the items as listed below:

- *Bringing more tourists to the dark sites are part of our company mission (M = 5.08; S.D = 1.12 item F4)*
- *Our company continuously promotes and encourages tourists to visit the historical buildings, old monuments and war museums (M = 5.03; S.D = 1.486; item F4)*
- *Owing to demand, our company keep promoting and offering more Malaysia dark tourism sites to tourists as it also as help to increase the company profits (M = 5.07; S.D = 1.175; item F5)*

Table 2. Mean score of the tour operators' action behavior (n = 120).

No.	Item	Mean	S.D
1	Our company aggressively promotes dark tourism sites to the tourists	5.06	1.416
2	Visiting dark tourism sites are part of our company tour itinerary	5.18	1.424
3	Bringing more tourists to the dark sites are part of our company mission	5.08	1.112
4	Our company continuously promotes and encourage tourists to visit the historical buildings, old monuments and war museums	5.03	1.486
5	Owing to demand, our company keeps promoting and offering more Malaysia dark tourism sites to tourists as it help to increase the company profits	5.07	1.175

Strongly Disagree, 2 = Disagree, 3 = Somewhat Disagree, 4 = Neither Disagree or Agree, 5 = Somewhat Agree, 6 = Agree and 7 = Strongly Agree.

Table 3. Results of multiple regressions of the education attribute on the tour operator's action behaviour.

Predictors	Model 1/std. β
Step 1: Model variables education	
R^2	0.51**
Adj. R^2	0.26
R^2 change	0.26
F-change	41.917

Note: *p < 0.05, **p < 0.01, ***p < 0.001.

Together, this result indicates that the perceived importance of dark tourism from educational value slightly contributed towards tour operator's action behavior in promoting the dark tourism sites to local or international tourists.

4.4 *The relationship between perceive the importance of educational values and tour operators' action behaviour*

In response to the hypothesis, standard multiple regression employed. The predictor variable comprised of educational value against the tour operator's action behaviour as a criterion variable. The educational value attribute was able to explain 26% ($R^2 = 0.26$; F-change = 41.917, $p < 0.01$**) of the variance on the tour operator's action behaviour. It evidently shows that educational attribute ($\beta = 0.51$, $p < 0.01$**) found to have a slight significant impact and positive influence on the tour operator's action behaviour.

5 THE IMPLICATIONS AND CONCLUSION

The result clearly show that the majority of the local tour operators agreed that Malaysia has vast dark tourism sites but perceive slightly importance of it from educational values, therefore, do not yet aggressively transform their perceptions into action behavior. In other word, visiting dark tourism sites, as opposed to cultural heritage, nature tourism, skyscrapers, city tour and shopping tourism are still getting less attention or promotion from some of the Malaysia local tour operators. Therefore and not to exaggerate that, visiting dark tourism sites probably are still not be included in the tour packages or as part of the tour company's itinerary for some of the local tour operators. This scenario probably due to the fact that promoting the dark tourism itself considered new in this country, and the market is still in the infant stage.

It becomes normal for any new tourism products slowly accepted, thus, received less effort from some of the tour operators. Less in business profit, less demand, fewer resources and many others may influence less initiative undertaken by them in organizing the dark tourism packages. However, results of this study also indicate that there are some tour operators who bringing tourists to visit such sites and making money out of it. Therefore, accusation of less interest among the tour operators could not solid reasons to base on but less of the information, supports, promotion and collaboration with the authorities may cause the causation, as well.

These without doubt are giving varying consequences and implications not only to the individual tour operators but related tourism organizations and government authorities. In this context, it is undeniable that the Government of Malaysia through Malaysia Tourism Promotion Board (MTPB) and Ministry of Culture, Arts and Heritage foresee dark tourism as promising products that significantly benefit the country. Support and promotion are among the pro-active activities undertaken by the government in upgrading and maintaining the infrastructures of the dark tourism sites, however, will be in vain or waste, if the information and clear mission of the efforts did not fully reach every local tour operator or other tourism organizations. This is not to say that there is no information reaches the operators, but it should continually be done and encourage agencies that responsibility with the dark tourism sites to disseminate information about the market which will provide monetary benefit to both parties.

As a conclusion, the collaboration and cooperation between tour operators and government related authorities, therefore, crucial and should be further strengthened if the "total tourism experience" among the international tourists in this country is to be achieved. Their commitment and cooperation are important in gear towards promoting the Malaysia dark tourism sites and become internationally recognized. Failure to develop such a commonality of approach may lead to this phrase.

"Despite the thousands of wonderful tourism products this country has it could be meaningless if there is no collective and continuous effort to utilize the available resources and promote them".

REFERENCES

Aydin, S., & Boz, M. (2009). The Importance of Alternative Types of Tourism in the Continuous Development of the Tourism Industry: Antalya Case. (Online) http://turizm-pazarlama.com.

Biran A., Poria Y., & Oren G. (2011). Sought experiences at (dark) heritage sites. *Annals of Tourism Research, 38*(3), 820–841.

Braithwaite, D., & Lee, Y.L. (2006). *Dark Tourism, Hate and Reconciliation: The Sandakan Experience.* International Institute for Peace through Tourism Occasional Paper, No. 8, March.

Brown, S., McDonagh, P., & Shultz, C. (2012). Dark marketing: ghost in the machine or skeleton in the cupboard. *European Business Review, 24*(3), 196–215.

Bujistic, M. (2011). Social implications and possibilities of post-war tourism development in Iraq when former enemy becomes a tourist. *The Journal of Tourism and Peace Research, 2*(2), 16–31.

Chiles, M.N. (2010). *The bonfire memorial experience: An exploration of the motivations, affects and experiences of site visitors.* Unpublished thesis, Texas A&M University.

Cooperative Research Centre (CRC). (2008). *Accessible Tourism Challenges and Opportunities. Research Informing Policy and Planning.* (Online) http://www.tourism.wa.gov.au.

Goel, S., Broder, A., Gabrilovich, E., & Pang, B. (2010). Anatomy of the long tail: Ordinary people with extraordinary tastes. In the proceeding of the third ACM International Conference on Web Search and Data Mining, New York, USA.

Isaac, R.K. (2012). Moving from a pilgrimage to dark tourism: leveraging tourism in Palestine. *Tourism, Culture & Communication*, 11, 149–164.

Keyes, M. (2012). War Tourism: Shaping memory and perception in Post-War Vietnam. *Department of Comparative Sociology.*

Malaysian Association of Tour & Travel Agents (MATTA). (2013). Members Directory.

Marmor, M. (2012, July). Dark Tourism: Caught Between Respect and Obscenity. Urban Times. (Online) http://urbantimes.co/magazine/.

Moeller, M. (2006.) *Battlefield tourism in South Africa with special reference to Isandlwana and Rorke's Drift Kwazulu-Natal.* Unpublished thesis, University of Pretoria.

Ngo, E. (2012, February). Ongoing Construction is to Improve Infrastructure, says Hamidah. *The Star.* (Online) http://www.thestar.com.my/story.

Novelli, M. (2005). *Niche tourism: Contemporary issues, trends and cases.* Oxford, UK. Butterworth-Heinemann Ltd.

Ozer, S.U., Ersoy, G.K., & Tuzunkan, D. (2012). Dark tourism in Gallipoli. *Procedia Social and Behavioral Sciences, 41*, 386–393.

Seaton, A.V. (1996). Guided by the dark: From Thanatopsis to Thanatourism. *International Journal of Heritage Studies, 2*(4), 234–244.

Sharpley, R. (2009). *Shedding light on dark tourism: An introduction.* In Sharpley. R, & Stone. P (Eds.), The darker side of travel: The theory and practice of dark tourism (pp. 3–22). Bristol: Channel View Publications.

Stone, P., & Sharpley, R. (2008). Consuming dark tourism: a thanatological perspective. *Annals of Tourism Research, 35*(2), 574–595.

Stone, P.R. (2012) Dark tourism and significant other death: towards a model of mortality mediation. *Annals of Tourism Research, 39*(3), 1565–1587.

Suhaini, I., & Chai, A. (2008). Challenges and opportunities for developing dark tourism in Malaysia: A pilot study on understanding visitor motivations at Jerejak Island, Penang. *Research Management.*

Tarlow, P. (2005) Dark tourism: The appealing 'dark' side of tourism and more. In, Novelli, M (Ed.), Niche tourism: Contemporary issues, trends and cases (pp. 47–57). Oxford: Elsevier.

Hospitality and Tourism – Sumarjan et al. (eds)
© 2014 Taylor & Francis Group, London, ISBN 978-1-138-00151-0

Acculturation of international students on Malay food acceptance

S.N. Amira & A. Zainal
Universiti Teknologi MARA, Malaysia

ABSTRACT: Looking into the tremendous growth of international students pursuing their higher education in Malaysia, it is a common for them to experienced culture shock symptom which is an excessive feeling on a situation that they are not familiar yet. Despite experiencing culture shock, international students are also vulnerable to the process of acculturation towards the host country. Therefore, this current study investigates the influence of acculturation among the international students towards their acceptance on Malay food in Malaysia universities. A total of 350 students from Asia, Middle East, America, Europe and Africa have participated in this survey and findings revealed that acculturation has a positive relationship with Malay food acceptance ($\beta = 0.262$ $p < 0.001$). As a result, study indicates that international students able to accept Malay food while staying in Malaysia although there is a difference in taste, texture and aroma.

Keywords: acculturation, international student, acceptance, Malay food

1 INTRODUCTION

Malaysia is on its way to establish its reputation as a regional hub for higher learning education, both in Asia and the world (The International College of Economics and Finance Monitor, 2013). The process of upgrading and establishing the local educational sector is in tandem with the current growth of student population around the world and therefore in order to propel the country's education sector, the Ministry of Education Malaysia has approved a grant enabling a local higher institution to offer places in their institution to 200,000 international students (The International College of Economics and Finance Monitor, 2013). With the growing numbers of international students in the world, the Director of Private Higher Education Institution revealed that, by the month of August 2011, numbers of foreign student pursuing their tertiary education have remarkably increased to 2% (University World News, 2011).

These significant increases revealed that Malaysia is now on the right path in positioning itself among the top ten countries in the world in providing an excellent education opportunity for the international community coupled with superior accommodation. Specifically, the diversifying group of international students may provide good opportunities for a cross cultural learning and communication but to certain extent the students himself may face many barriers that hampered during the learning experiences. By taking the international student diversity into account, this culturally heterogeneous group of international student has a tendency to experience various dilemmas in their educational ventures and social context (Barlett, 2002).

It is undeniable that moving from home country to Malaysia may be devastating and to a certain extent this feeling may influence the students to experience some degree of culture shock due to the uniqueness of the culture that they encountered. Experiencing a cultural shock can be described as an early adjustment to an unfamiliar environment (Pedersen, 1995), and furthermore culture shock which is an anxiety feeling which results of losing memorable sign and symbol of social intercourse which they do not carry at the level of conscious awareness may take place (Oberg, 1960).

Cultural differences among the international students may indirectly lead them to the isolation and distance from the local environment and community which may result in a lower level of confidence and feelings of disengagement. Berry (1997) explained that these changes (acculturation process) may involve biological (e.g. new eating habits) or economic alteration, whereby the changes commonly comprise of social and cultural changes such as changes in values, norms and habit.

Acculturation to a new society might occur on a different rates and level and research done by Seo, Phillips, and Kim (2012) revealed that some immigrant are able to accept a new culture relatively quickly whereas others requires more time and usually this process involves uncertainty avoidances which can be explained as the level of tolerances for uncertainty and ambiguity among the immigrants.

Thus, it is an opportunity for this current study to look upon the acculturation of the international students' on accepting the Malay food. A different contextual setting will be carried out in this study.

2 LITERATURE REVIEW AND HYPOTHESIS

2.1 *Acculturation*

Nowadays the number of students pursuing their tertiary education in different countries has somehow led to the acculturation process especially when the host country culture is different in terms of language, values and norms, religiosity, and food availability compared from their country of origin. It is mentioned that the changes experienced by the students indirectly alter their originality in order for them to suit with the current condition (Rudmin, 2003).

Phiney (1992) illustrated acculturation as a process involving two different categories of cultural and psychological changes which consists of alteration in group of customs, economic and political changes experienced by the person who came to contact with the different country throughout the process of migrating, studying or staying in a different community.

Acculturation process on psychological may influence the well being of a single person, their mental health and lead to their satisfaction when living in a new cultural context. Therefore, it is not surprising for the international students in Malaysia to experienced acculturation (Class, Castro, & Ramirez, 2011) since acculturation is a constant process happening throughout the life. According to Seo et al. (2012) this process will definitely influenced the students' ability to hold on to their originality due to the fact that the process begins with combination of two self regulating cultural in a host country (Class et al. 2011).

Research findings have shown that acculturation may happen in any situation but what really matters are their capabilities in acculturating and accepting new food (culture), belief and behaviour of the different group in the host country throughout the period of residing (Lee, Sobal, & Fronglilo, 2000).

2.2 *Malay food*

Malay food can be characterized by its use of varieties spices, herbs, coconut milk, shrimp paste and many other ingredients which can be found locally. The uses of ingredients which may cause distraction to many foreigners 'belacan' or shrimp paste, which is widely used in any 'sambaltumis' (cooked hot chilli dish) and popular condiment such as 'sambal belacan'. Rozin (2006) also posited that there are three components that differentiates Malay food from other ethnics like Chinese and Indians which is the uses of staple ingredients, the flavour principle (widely used of lemon grass, ginger, chillies, onion, herbs and spices) and lastly the cooking method which comprises of sautéing, boiling technique, steaming technique and much more.

The durian which has been known and consumed in southeastern Asia since prehistoric times, but has only been known to the western world for about 600 years can be linked to the Malay delicacy known as "tempoyak"; fermented durian flesh which is eaten as a condiment or cooked as a gravy for many types of fresh water fish. The prominent "tempoyak dish" variations found in the states of Perak, Pahang and Negri Sembilan are examples of authentic Malay food that need to presented as a food route (Zainal, Zali, & Kassim, 2010) so that foreigners will learn to appreciate these local food.

Based on the review of literature the following hypothesis was formulated:

H: Acculturation influences the acceptances of Malay food among the international students.

3 METHODOLOGY

3.1 *Participants and procedure*

The respondents of this study comprised of international students studying in Malaysia public and private universities located in Klang Valley. The study yielded 88% of response rate or 352 usable questionnaires.

A 30-items instrument used to measure the acculturation level was adapted from Bojanic and Xu (2006) and de Barcellos, Kügler, Grunert, Van Wezemael, Perez-Cueto, Ueland and Verbeke (2010). All items were measured using the 7-point Likert scale and the questionnaires were written in English and piloted prior to the actual study to ensure clarity and avoid ambiguity since most of the respondents are international students.

4 RESULTS AND FINDINGS

4.1 *Sample profile*

Data in the present study were gathered using multi-stage sampling with stratified and convenience sampling being done sequentially. 73.3% (N = 258) of the respondent were male while the

Table 1. Regression analysis for acculturation factor on Malay food acceptance.

Independent variable	Model summary 1
Step 1: Model variables acculturation	0.281
R^2	0.079
Adjusted R^2	0.76
F change	29.996
Sig. F change	0.000

remaining 26.7% (N = 94) were female with mean age of 20–29 years. 40.3% (N = 142) respondents has stayed in Malaysia for 1–2 years while 37.2% (N = 131) has stayed in Malaysia for 2–3 years. The remaining respondent stayed in Malaysia more than 3 years. 79% (N = 278) respondent were single and only 21% (N = 74) respondent married while studying in Malaysia.

4.2 *Results*

As illustrated in Table 1, total acculturation can explain as 7.9% of the variance in Malay food acceptance by the International students (R^2 = 7.9%; F Change = 29.996; $p < 0.001$). The beta (β) value for acculturation on Malay food acceptance ($\beta = 0.281$; $p < 0.01$) indicates that acculturation made a contribution on Malay food acceptance. Thus, the result shows that acculturation has positive significant influence on international students' Malay food acceptance. Thus the hypothesis is supported.

5 DISCUSSIONS

Based on the findings presented, it was observed that, acculturation has a significant relationship in influencing the acceptance of Malay food among the international students with a significant correlation $R^2 = 0.079$; beta value $\beta = 0.281$ ($p < 0.01$). This result is aligned with Wandel, Raberg, Kumar and Holmboe (2008) whereby in previous studies, Pakistanis and Sri-Lanka citizens who lived in Norway for the purpose of working and studying changed their meal pattern substantially according to the particular host country meal due to the changes in work patterns and climates consideration of the country that they lived in.

This result reveals that international students who are highly acculturate are capable of accepting the traditional Malay food as such "*sambal belacan*", "*masak lemak*" and "*cencaluk*" though the taste of the food is not familiar to them. In fact findings also revealed that moderately acculturate students has the potential to accept Malay food during the residing period but to a certain extent they are not willing to substitute their own homeland food with the local Malay food. The ability for them to accept Malay food is not too high due to the existence of the other international ethnic restaurants provided by the universities.

Findings also revealed that the length of living in the country has enabled the international students to acculturate with the local attitude and behaviour whereby they are willing to participate and join the Malay events held such as wedding ceremonies, 'Eidul Fitri' celebration and any other cultural events involving Malay food. This finding depicts that the longer the students live in the local environment the better the exposure to the local culture and therefore gradually influenced them to be involved and participate with the local community. This will help the students to engage and feel comfortable living with others and particularly the Malay community. Thus, it can be concluded that the direct involvement in a long term between international students and Malay community will enable the students to be highly acculturate with local community and help them to feel calm and content with the food and culture of the host country.

6 IMPLICATION

Findings from this study revealed that international students are capable of accepting Malay food although there is a distinctive characteristic of Malay food that are foreign to them; in terms of flavour, taste, aroma, appearance and texture which may not be favourable.

Furthermore, variation of food served during the three major meals (breakfast, lunch and dinner) also may influence them to at least try and accept the Malay food. Apart from that, constrain in terms of time, money and accessibility in searching for their country food has also significantly contributed to the process of acculturation towards accepting Malay food.

The finding also shows that local community in universities plays a major role in terms of convincing and influencing the international students to accept and eat Malay food. It is particularly a good decision for the universities to group the international students together with the Malay students especially during the residing period so that they will able to accept and experience the Malay food and in long terms they can be an international agent to promote our Malay food to their own country.

7 LIMITATIONS AND RECOMMENDATIONS

The findings gathered a high respondent rate but the distribution of respondents in terms of country of origin is disproportion whereby almost 143 of the respondent came from the Asia continent, 115 of them from the Middle East, 40 respondents from Europe, only 5 respondents from America, 2 from the Australasia and 47 of the respondents are from the other continent. In fact the students also confess that there.

Several directions for future research have also been identified based on the findings of the present study. Firstly, the theorized model's validity can be further tested using additional outcomes such as age and country of origin. Assimilation to the host country should also be taken into consideration. Thus, further study should look into this restriction and to achieve better findings and pattern of acculturation.

REFERENCES

Bartlett, K. (2002). *National report on international students in Canada 2000/2001*. Ottawa: The Canadian Bureau for International Education.

Berry, J.W. (1997). Acculturation and adaptation to new society. *International Migration, 30,* 303–332.

Bojanic, D., & Xu, Y. (2006). An investigation of acculturation and the dining out behaviour of Chinese living in United States. *International Journal of Hospitality Management, 25*(2), 211–226.

Class, L.M., Castro, F.G., & Ramirez, A.G. (2011). Conceptions of acculturation: A review and statement of critical issues. *Social Science & Medicine, 72*(9), 1555–1562.

De Barcellos, M.D., Kügler, J.O., Grunert, K.G., Van Wezemael, L., Perez-Cueto, F.J.A., Ueland, Ø. & Verbeke, W. (2010). European consumers' acceptance of beef processing technologies: a focus group study. *Innovative Food Science and Emerging Technologies, 11,* 721–732.

Lee, S.K., Sobal, J., & Frongillo, E.A. (2000). Acculturation and health in Korean American. *Social Sciene & Medicine, 51*(2), 159–173.

Lu, A. (2010, October 17). Tracking down fine Malay food The Star Online. Retrieved from sundaymetro@thestar.com.my.

Nummendal, M. & Hall, M. (2006). Local food and tourism: an investigation of the New Zealand South Island's bed and breakfast section's use and perception of local food. *Tourism Review International, 9,* 365–378.

Oberg, K. (1960). Culture shock & the problem of adjustment to new cultural environments. *Practical Anthropologist, 7,* 177–182.

Pedersen, P. (1995). The Five Stages of Culture Shock: Critical Incidents around the world. Westport CT: Greenwood Press.

Phiney, J.S. (1992). Ethnic identity in adolescents and adults: review of research. *Psychological Bulletin, 108*(3), 499–514.

Rozin, P. (2006). The integration of biological, social, cultural and psychological influences on food choice. The psychology of food choice. Oxfordshire: CABI.

Rudmin, F.W. (2003). Critical history of the acculturation psychology of assimilation, separation, integration, and marginalization. *Review of General Psychology, 7,* 3–37.

Seo, S., Phillip, W.M.J., Jang, J., & Kim, K. (2012). The effect of acculturation and uncertainty avoidance on foreign resident choice for Korean foods. *International Journal of Hospitality, 31*(3), 631–1036.

The International College of Economics and Finance (ICEF), (2013). Malaysia aim to be the sixth largest education by 2020. Retrieved from http://monitor.icef.com/2012/05/malaysia-aims-to-be-sixth-largest-education-exporter-by-2020/.

University World News, (2011, August 27). Malaysia: foreign students top 90,000. Retrieved from http://www.universityworldnews.com/article.php?story=20110819171628377.

Wandel, M., Raberg, M., Kumar, B., & Holmboe O.G. (2008). Changes in food habits after migration among south Asian settled in Oslo: the effect of demographic, socio economic and integration factor. *Appetite, 50,* 376–385.

Zainal, A., Zali, A.N., & Kassim, M.N. (2010). Malaysian gastronomy routes as a tourist destination. *Journal of Tourism, Hospitality and Culinary Arts,* 2(1), *15–24. University Teknologi MARA.*

Hospitality and Tourism – Sumarjan et al. (eds)
© 2014 Taylor & Francis Group, London, ISBN 978-1-138-00151-0

The impact of tourism towards the culture practice of indigenous people

N.A. Azhar Ismail, S.S.M. Mohamed Razak & M.A. Ismail
Universiti Teknologi MARA, Malaysia

ABSTRACT: Tourism in Malaysia viewed as a vehicle, which can produce an array of benefits for the local community in general, including the minority group of indigenous people. The main purpose of this study is to analyze the impact of tourism towards the culture practice of the indigenous people or Orang Asli in the state of Selangor, Malaysia. In order to achieve the purpose of study, a total of 400 questionnaires distributed to selected respondents in indigenous people's villages around Selangor, Malaysia. The findings revealed that tourism does have an impact towards indigenous people community in terms of economic, socio-cultural and environmental. However, despite the impacts of tourism, the indigenous people are still strongly practicing their culture inherited from their ancestors. Since cultural tourism is one of the main contributors to the tourism industry, it is imperative to investigate whether tourism development is coming hand in hand with culture preservation especially to this indigenous people as previous studies showed that tourists demand authentic experiences.

Keywords: tourism impact, Orang Asli, indigenous people's culture

1 INTRODUCTION

From January 2012 to September 2012, Tourism Malaysia (2012) has reported 18,153,643 tourists' arrival in Malaysia, generated more than RM50 billions' income for the country. Tourism is an industry with enormous economic impacts. It is also an industry that has many environmental and social consequences (Kim, 2002). Dyer, Aberdeen and Schuler (2003), in their study found that tourism development had brought various impacts to the local community who benefited in terms of job opportunities and increased cultural understanding, wealth and welfare. In Malaysia, tourism does not only impact the major community, but also the minor ethnic like the indigenous people. Hamzah (2004) mentioned that Malaysia is currently 'selling' its multiculturalism and cultural diversity representing all the major civilizations in Asia (Malays, Chinese and Indians) and other ethnic groups, projected by the promotion tag line, 'Malaysia Truly Asia'. Ministry of Tourism launched an Orang Asli tourism program in an effort to encourage more tourists to the country as well as to improve the livelihood of the community. The development of Orang Asli is not only due to government projects, but also a result of the increasing tourism activities in their area. Orang Asli community has gone through changes due to tourism activities; preference to be alone plenty of hard work, and modernization are some of the changes observed. Tourism has cause negative impact on the Orang Asli culture, losing their humble and hospitable nature (Ramachandran, 2009). Norlida, Redzuan, SitiHajar, Abdul Hamid and Doris (2012) mentioned in their study that the development of tourism in Cameron Highlands has had several impacts such as economic, social, cultural, and environment on the indigenous people within the surrounding areas. Preserving the authenticity of these indigenous people's culture believed to be important as authenticity of tourism products; which include cultural products as one of the many factors that attract tourist to a destination. Based on these mentioned issues, therefore, there is a need to discover whether tourism development within the Orang Asli or indigenous people's community in Malaysia is affecting their culture. Thus, this study carried out to identify the impact of tourism to the indigenous people and to assess the cultural activities practiced amongst the indigenous people. The findings later would provide whether or not tourism has negative impact on cultural practice among indigenous people. Should there be any, it can be a guide for the authority to improve their tourism development without eliminating the importance of culture preservation, which is paramount to the tourism industry.

From the academic perspective, a limited study carried out on tourism and cultural, in particular, on aspects of indigenous people. Therefore, the originality of this research could contribute to a new body of knowledge in Malaysia and extend the body of literature in heritage tourism study.

2 LITERATURE REVIEW

2.1 *Tourism impact*

Planning is essential to stimulate tourism development, and its sustainability, as without tourism planning, many unintended consequences may develop, causing tourist and resident dissatisfaction. These include damage to the natural environment, adverse impacts upon the cultural environment, and a decrease in potential economic benefits. Tourism is an industry with enormous economic impacts. It is also an industry that has many environmental and social consequences (Kim, 2002). Scholars have facilitated discussion on the issue of perceived economic, social, cultural and environmental impacts as a result of presence of tourism. As explained by Dyer et al. (2003), tourism development had brought various impacts to the local community who benefited in terms of job opportunities and increased cultural understanding, wealth and welfare. Income generated through tourism will help eradicate the shackles of poverty and social welfare and lead to more cultural pride and economic self-determination (Muqbil, 2009). However, not all of tourism impacts are always beneficial, for example, negative effects on demographic characteristics, social structures and relations, economic activity structures and sectoral dynamics, societal values and attitudes, built environment and land use, cultural heritage, and natural and environmental resources (Coccoossis & Mexa, 2004; Norlida et al., 2012).

2.2 *Economic impact*

Kuvan and Akan (2005) indicated that to mitigate economic difficulties many rural communities have adopted tourism as a new economic development strategy. The vast majority of studies in tourism economic impact have focused on employment opportunities, standard of living, revenue that a community derives from tourism activities and cost of living (Kim, 2002). Norlida et al. (2012) also found in their study that the indigenous people in Cameron Highland agreed that development of the tourism industry has led to increase in income, employment opportunities and family economic improvement, which is consistent with study by Brida, Osti and Faccioli (2011). However, there are also negative impacts in terms of economic due to the tourism industry. For example, Tosun (2002) and Weaver and Lawton (2001) in their study found that residents in several surveys indicated that there is an increase in the price of goods and services been provided, which is consistent with the study by Norlida et al. (2012). Norlida et al. (2012) finding revealed that development of tourism also caused increased in the prices or costs of houses.

2.3 *Socio-cultural impact*

Mercer (1994) found that indigenous people showed rejection to tourism by indicating that the occurrence of disruptions to the social life of indigenous communities as the number of tourism arrivals increased. Indigenous communities also felt that the influx of tourists could change and destroy their culture, especially if their settlements became a popular tourist attraction (Moowforth & Munt, 1998). Tourism also contributes to social ills such as begging, gambling, drug trafficking, prostitution, and family crisis and conflict between local residents and outsiders, as well as the uprooting of traditional society and causes deterioration of the traditional culture and customs of host countries (Ahmed & Krohn, 1992; Li, 2002), inconsistent with the finding by Norlida et al. (2012). Norlida et al. (2012) discovered that the increment of tourist arrivals to the local community settlements to improve their ability to talk about their culture to outsiders. Besides, the community also feels proud of their culture and has increased their interest in meeting outsiders, consistent with the study by Brida et al. (2011). Tourism may also have a broader impact on culture if it revives a culture because of renewed interest in indigenous tradition, especially on cross-cultural communication improvement and cultural revival (Bbus, 1999).

2.4 *Environmental impact*

Some people believe that tourism helps create a greater awareness and appreciation for the need to preserve the environment to capture its natural beauty for tourist purposes, and increase environmental infrastructure of the host country (Var & Kim, 1990). Tourism also thought to be a clean industry, without the pollution problems associated with other types of economic development (Kim, 2002). Brida et al. (2011) argued that tourism is an activity that contributes to the development of new environmental-oriented programs, although it contributes only marginally to the conservation of environmental assets. Norlida et al. (2012) in a study conducted in Malaysia stated that tourism development had an impact on the environment in terms of increased traffic congestion, solid waste disposal that pollutes the environment, as well as noise and water pollution consistent with other studies (Ahmed & Krohn, 1992; Andereck, 2005).

2.5 *Indigenous people*

The term "indigenous" has widely used recently in International Political discourse, Human Rights and Anthropology. United Nations (2009) in its factsheets states that Indigenous communities, peoples and nations are those, which, having historical continuity with pre-invasion and pre-colonial societies that developed on their territories, consider themselves distinct from other sectors of the societies now prevailing of those territories, or parts of them. They form at present non-dominant sectors of society and determine to preserve, develop and transmit to future generations their ancestral territories, and their ethnic identity, as the basis of their continued existence as peoples, in accordance with their own cultural patterns, social institutions and legal system.

2.6 *Indigenous people or Orang Asli in Malaysia*

Department of Social Welfare (1997) states that the Orang Asli is the indigenous minority people of Peninsular Malaysia. Orang Asli are not a homogenous race, which means that they have diverse cultures, traditions and ways of living, beliefs, and languages. There are 18 ethnic sub-groups of the Orang Asli, all of these subgroups depend on their ethnicity and the location (Nicholas, 2006). The Malaysian government has set up the Jabatan Kemajuan Orang Asli (JAKOA) with the responsibility to oversee the affairs of the Orang Asli. Some of the Orang Asli enjoyed modern facilities through the implementation of various development projects such as village resettlements programs, rural roads, provision of electricity and water supply, social amenities, as well as access to education.

2.7 *Culture practice*

Culture according Hofstede and McCrae (2007) define as the collective programming of mind that distinguishes one group or category of people from another. This stresses that culture is a collective, not individual attribute. Culture is also not directly visible but manifested in behaviours, as well as common to some but not to all people.

Guiso, Sapienza and Zingales (2006) defined culture as those customary beliefs and values that ethnic, religious and social groups transmit fairly unchanged from generation to generation whereas, Triandis (2000) argued that culture is to society what memory is to individuals. It refers to tools and ideas that are shared and transmitted to succeeding generations because they were once practical at some point of time.

3 METHODOLOGY

This study employed a descriptive research design. A questionnaire was developed for this study and comprises of three sections. The first section designed to measure the impact of tourism towards the indigenous people community, adapted from Norliza et al. (2012) looking on the economic impact, socio-cultural impact and environmental impact. The second section aimed to assess the culture practice of indigenous people, adapted from Hafazah and Amalina (2012). For both sections, respondents were to rate the items on a five-point Likert scale, ranging from 1 "strongly disagree" to 5 "strongly agree", and finally, Section 3 asked the for their respondents' demographic profiles. To ensure the reliability of the scales, a pilot test involving 30 personnel at the Jabatan Kemajuan Orang Asli (JAKOA) who involved in activities with indigenous people were conducted. Several modifications made based on the feedback from the pilot test.

The target population for this study was the indigenous people from two settlements in Selangor, which are actively involved in tourism activities. The two settlements are Kampung Bukit Kala, Batu 12, Gombak and Kampung Sungai Bumbun, Pulau Carey. A total of 346 respondents from Kampung Bukit Kala and 196 from Kampung Sungai Bumbunwere selected applying the convenience sampling.

4 RESULTS AND DISCUSSION

4.1 *Respondents' profile*

As presented in Table 1, there were 66.75% male and 33.25% female respondents. The majority of the respondents are in the age group of more than 36 years old (54.75%), 26–35 years of age (23.25%), 17–25 years (18%) and less than 16 years (4%).

Table 2 also illustrates the percentage and frequency of the demographic profile of respondents. In terms of education level, majority of the respondents 35% studied until From 3 (Lower Secondary), 34.5% studied until Primary 6, 27.8% had Form 5 (Middle Secondary) qualification and 2.8% studied until Form 6 (Higher Secondary).

4.2 *Reliability test*

As shown in Table 2, the Cronbach's Alpha coefficient value for all variables in the study revealed

Table 1. Demographic profile factors.

Items	Percentage (%)	Frequency
Gender		
Male	66.75	267
Female	33.25	133
Age		
More than 36 years	54.75	219
26–35 years	23.25	93
17–25 years	18	72
Less than 16 years	4	16
Education level		
Primary 6	34.5	138
Form 3	35	139
Form 5	27.8	111
Form 6	2.8	11

Table 2. Reliability coefficients for each section of the questionnaire.

Variables	No. of items	Cronbach alpha
Economic impact	5	0.700
Socio-cultural impact	10	0.682
Environmental impact	6	0.727
Culture practice	6	0.732

Table 3. Reported mean scores for tourism impact and culture practice (N = 400).

Variable	Mean (M)	Std. deviation (SD)
Economic impact	4.20	0.141
Socio-cultural impact	4.23	0.213
Environmental impact	2.22	0.129
Culture practice	4.74	0.206

a range of coefficient values from 0.682 to 0.732 accordingly.

4.3 *Analysis of tourism impact and culture practice*

Based on the illustration in Table 3, most of the respondents are sharing the same thoughts towards the items in the dimension. From the mean score table, respondents score (M = 4.20; SD = 0.141) for economic impact while (M = 4.23; SD = 0.213) for socio-cultural impact, and respondents score for environmental impact, respondents scored (M = 2.22; SD = 0.129). Whereas the analysis for culture practice of indigenous people showed a mean score of (M = 4.74; SD = 0.129).

Looking at the Table 4, the results of Pearson Correlation between tourism impacts (i.e., economic, socio-cultural and environmental) with culture practice of indigenous people were significant ranging from (r = 0.284 to 0.431, p = <0.01). The highest degree of relationship with culture practice is the economic impact (r = 0.431, p = <0.01), followed by relationship between culture practice with environmental impact (r = 0.378, p = <0.01), and finally, relationship between culture practice with socio-cultural impact (r = 0.284, p = <0.01).

Table 4. Pearson correlations matrix of study variables (N = 400).

		Culture practice
Economic impact	Pearson correlation	0.431**
	sig. (2-tailed)	0.000
Environmental impact	Pearson correlation	0.378**
	sig. (2-tailed)	0.000
Socio-cultural impact	Pearson correlation	0.284**
	sig. (2-tailed)	0.000

**Correlation is significant at the 0.01 level (2-tailed).

5 DISCUSSION

5.1 *Tourism impact*

In looking at the economic impact, respondents strongly agree that tourism increase their incomes and provide more jobs opportunities. This is in accordance with findings from the study done by Tosun (2002) and Norlida et al. (2012) which revealed that employment opportunities, increased income and family economic improvement as the positive impact of tourism.

Analysis on the socio-cultural impact of tourism, respondents strongly agree that tourism increases the number of outsiders coming to their area and tourism does not disturb their daily living. They also strongly supported that there is no disturbance in the practice of their own religion and culture, which opposed to the finding by Ahmed and Krohn (1992) that stated tourism causes deterioration of the traditional culture and customs. There is a strong disagreement from the respondents on tourism causes increment in vandalism and rest areas are becoming more congested, coherent with the result obtained by Norlida et al. (2012).

As for environmental impact, respondents came to an agreement on tourism helps to promote a cleaner environment and help to preserve the natural environment, consistent with the previous study (e.g., Kim, 2002; Norlida et al., 2012; Var & Kim, 1990). Respondents gave significantly low score, which portray strong disagreement on tourism causes noise, environment and water pollution, whereas, contrary results obtained in previous studies (Ahmed & Krohn, 1992; Andereck, 1995; Norlida et al., 2012).

5.2 *Culture practice of indigenous people*

The respondents strongly agree that they are still practicing the traditional marriage customs, funeral customs and birth and post-natal customs follow by a relatively strong agreement that they are still participating in community clean-up activities. Indigenous people in the studied area also agree that they still use the *dukun* or traditional medication, and they are still practicing customary dances as a custom compared to tourism purposes.

5.3 *Tourism impacts on the culture practice of indigenous people*

Significant positive relationships exist between all independent variables, namely economic, socio-cultural and environmental impact of tourism, with dependent variable, which is culture practice of indigenous people. Therefore, it can be concluded that tourism does give a positive impact towards culture practice of indigenous people, contradicts with the finding of (Moowforth & Munt, 1998; Ramachandran, 2009).

6 LIMITATIONS AND RECOMMENDATIONS FOR FUTURE RESEARCH

While the study offers a series of interesting and significant findings, there are limitations that need to be attended to. One of the possible limitations is related to the sample size as this study only focused on the indigenous people in the settlements that are actively involved in tourism in Selangor, which are Kampung Sungai Bumbun, Pulau Carey and Kampung Batu 12, Gombak. Since, the indigenous people are heterogeneous community, thus, the result of this study may not be applied conclusively to other community of indigenous people.

Hence, this study recommended several suggestions that could be uptake in future research, which may extend to selecting a different research setting in other indigenous people settlements in Malaysia other than in Selangor, such as in Pahang and Perak that actively involved in tourism activities. Besides, future researchers may have broader perspectives if the study conducted in Sabah and Sarawak, as in these states there many other ethnic with different culture practices. Plus, the Government of Malaysia is actively promoting these states as a tourism destination.

7 CONCLUSION

Result indicated that tourism does impact the indigenous people economically, socio-culturally and environmentally. There are positive and negative impacts of tourism towards the community in the studied area. This study also found that the indigenous people are still practicing traditional culture inherited from their ancestors. Finally, this study ruled out that tourism impact has a significant positive relationship with culture practice of indigenous people.

Therefore, it can be concluded that despite the impact of tourism which shown to bring development to this indigenous people community, such as improved income, increased job opportunities and better infrastructure, their culture practices not compromise. They are still strongly practicing their culture as customs and not due to tourism purposes only. This indicate a positive impact of tourism towards the indigenous people community, and the authority should take measures to ensure that there is a continuation of culture preservation among these indigenous people as it will benefit the tourism industry in Malaysia as a whole.

REFERENCES

Andereck, K.L., Valentine, K.M., Knopf, R.C. & Vogt, C.A. (2005), Residents' perceptions of community tourism impacts. *Annals of Tourism Research, 32*(4), 1056–76.

Brida, J.G., Osti, L., & Faccioli, M. (2011). Residents' perception and attitudes towards tourism impacts: A case study of the small rural community of Folgaria (Trentino—Italy). *Benchmarking an International Journal, 18*(3), 359–385.

Department of Orang Asli Affair. (2007). Annual Report of Department of Orang Asli Affair 2000. Kuala Lumpur: Jabatan Hal Ehwal Orang Asli.

Department of social welfare. (1997). A report of Orang Asli Children in School. Kuala Lumpur: Training Unit.

Dyer, P., Gursoy, D., Sharma, B., & Carter, J. (2007). Structural modeling of resident perceptions of tourism and associated development on the Sunshine Coast, Australia. *Tourism Management, 28*, 409–422.

Guiso, L., Sapienza, P., & Zingales, L. (2006). Does culture affect economic outcomes? Massachusetts: NBER Working Paper No. 11999, National Bureau of Economic Research.

Hafazah, A.K., & Amalina, H.H. (2012). The effect of a resettlement scheme on the social-cultural changes of the Temuan community. *Social and Behavioral Sciences, 42*, 362–373.

Hamzah, A. (2004). Policy and planning of the tourism industry in Malaysia. In the proceeding of the policy and planning of tourism product development in Asian countries.

Hofstede, G., & McCrae, R. (2007). Personality and culture revisited: linking traits and dimensions of culture. *Cross-Cultural Research, 38*(1), 52–88.

Kim, K. (2012). How does tourism in a community impact the quality of life of community residents? *Tourism Management, 1*(14), 1–14.

Kuvan, Y., & P. Akan. (2005). Residents' attitudes toward general and forest-related impacts of tourism: The case of Belek, Antalya. *Tourism Management, 26*(5), 691–706.

Li, Y. (2002). The impact of tourism in China on local communities. *Asian Studies Reviews, 26*(4), 471–486.

Moowforth, M., & Munt, I. (1998). Tourism and sustainability: New tourism in the third world. London: Routledge.

Muqbil, I. (2009). Indigenous tourism: More at stake than just "tourism". In the proceeding of the ITB Berlin 2009 Special Press Release: Indigenous Tourism.

Nicholas, C. (2006). The state of Orang Asli education and its problems. Consultancy report presented at Human Rights Commission of Malaysia (SUHAKAM). Kuala Lumpur.

Norlida, H.M.S., Redzuan, O., SitiHajar, M.I., Abdul Hamid, J., & Doris, P.S. (2012). *Journal of Tourism, Hospitality and Culinary Arts, 4*(1), 77–87.

Ramachandran, S. (2009). Responsible tourism: An inductive approach. *Integration and Dissemination, 4*, 41–50.

Tourism Malaysia (2012). Malaysia tourist arrivals by country of nationality September 2012. Tourism Malaysia Tourist Arrival Report. Retrieved April 12, 2013 from http://www.tourism.gov.my.

Triandis, H.C. (2000). Cultural syndromes and subjective well-being. Cambridge: The MIT Press.

United Nations. (2009). State of the world's indigenous peoples. New York: United Nations Publications. Retrieved November 10, 2012 from http://www.un.org/esa/socdev/unpfii/documents/SOWIP_web.pdf.

Hospitality and Tourism – Sumarjan et al. (eds)
© 2014 Taylor & Francis Group, London, ISBN 978-1-138-00151-0

The level of youth tourist satisfaction towards domestic tourism product

S.A. Jamal, N. Kamaruddin, S. Aliyas & N.A.A. Nik Hashim
Universiti Teknologi MARA, Malaysia

ABSTRACT: The focus of this study was to determine the level of satisfaction associated with the domestic tourism product among youth tourist in Malaysia. Quantitative study method was undertaken, with purposive sample of youth tourist age 15 to 30 that has experienced of using the tourism product being selected. Data that was collected via self-completed questionnaire was tested and analyzed by using the descriptive statistics. The result indicates that in association with the dimensions of tourism product, comprising; accommodation facility, transportation and attraction area, the youth tourist are satisfied with the domestic tourism product available for their usage with some exception to the price, tourism product and promotion activity that focus youth as target market. The findings provide valuable information and insights on the current trend of youth tourism in the country and thus strengthening the current literatures available for this area of study.

Keywords: domestic tourism, tourism product, tourist satisfaction, youth tourism

1 INTRODUCTION

Youth tourism is not a new phenomenon given that young people have always travelled around the world with their families, friends, in groups, or on their own for centuries (Taiyab, 2005). With the global rapid changes in the tourism industry, this particular market segment is becoming more visible among other leading tourism market segments such as Cultural Tourism and Adventure Tourism. Malaysian youth tourism market considered as a tourist between the age of 15 and 30 years old (Mohamed, Omar, & Lau, 2010). Youth tourism market includes all independent trips for periods of less than one year which are motivated, in part or full, by a desire to experience other cultures, build life experience and/or benefit from formal and informal learning opportunities outside one's usual environment (WTO, 2008). In order to ensure continuous growth of domestic youth tourism, the key issues relating to youngsters within this industry need to be given serious attention. Thus, issues related to the satisfaction of tourist are crucial as it can influence them to come back in the future.

Tourist can be considered as customers that have certain needs and demand that can only be supplied by the tourism players. Customer satisfaction is defined as an overall assessment of the performance of various attributes that constitute a product or a service (Bartikowski & Llosa, 2004). If the assessment is positive and regarded as well performed, the tourist will most likely feel satisfied with the attributes of the product that they have experienced. In tourism, tourist satisfaction usually influenced them to revisit besides passing information about the place they had visited to friends and family members informally through words of mouth.

The current conditions on the youth travel market generated a series of changes at the demands level, changes that will influence the elaboration and implementation of future strategies in the field (Moisa, 2010). However, the Malaysian tourism industry is still not organized to unlock the potential of youth tourism. Young people are an underestimated market for the tourism industry; it appears that it has specific behavioral characteristics and significant influence as far as demand for tourism is concerned. Nevertheless, young people, as a segment, are poorly understood because no specific discipline that studies youth behavior and no specific profession that deals with young people exist (Boukas, 2008). Furthermore, youth tourists are often described as inactive segments by tourism providers in developing countries. The seasonal and non-thematic nature of their travel undertakings has underprivileged them from being regarded as active participants and dominant segments in the tourism market (Habibah, Hamzah, Buang, Noor Alyanie, & Aimi Syairah, 2011). The growth in the number of youth tourists, and the increasing recognition of youth travel as a distinct market, has stimulated the development of services and products for this market (Taiyab, 2005). However, although youth tourism offers a great deal of opportunities for development, being at its infancy in the tourism sector of most developing countries, current knowledge on youth tourism provides limited understanding of the demand and travel pattern needed for future thematic planning of a country's tourism events and destinations (Habibah et al., 2011). Although various studies have explored the youth tourism market focusing on event tourism (Habibah et al., 2011), consumer behaviour (Labben & Johnson, 2004) perception towards tourism product (Boukas, 2008), and psychographic profiles (Reisinger & Mavondo, 2002), the current trend on the youth tourism in relation to level of satisfaction on domestic tourism product are yet to be explored among the Malaysian youth tourist. Given the increasing level of competitiveness, it is essential for tourism management authorities to understand the main factors causing satisfaction/dissatisfaction for the tourist (Massarat, 2012). This current study aims to fill this gap by exploring the youth satisfaction towards the domestic tourism product available for their consumption.

2 LITERATURE REVIEW

2.1 *Youth tourist*

Many industry experts believe that youth travel is the fastest growing travel market segment, with significant potential for long-term growth and economic impact to domestic tourism (Badaruddin et al., 2010). Today's young tourist has high level of curiosity and more explorative in nature as compared to the previous generation. In Malaysia, this youth tourism segment is has yet to position itself with other upstream market although there is vast potential in terms of making significant contribution to the domestic tourism market. According to Malaysia Current Population Estimates, the total population of Malaysia as of 2012 was 29.33 million, from that figure 11 million populations is represented by the youth (Malaysian Statistical Department, 2013). This shows that, with the growing number of youth population, it can become another market segment or niche market in Malaysian tourism industry, which need to be catered and seriously considered. Hence, the youth market presents an opportunity and a challenge to planners and policy makers to tap this lucrative source (UNWTO, 2012). In the long term, youth travel do not only limited to be focused on the domestic only. But, further cooperation with the neighboring country can further flourish this market within the Asian Region. In relation to the meaning of Youth Tourist, there is no exact definition given as many authors have used variety of terms to describe youth tourism (Mohamad et al., 2010). According to Richards (2008), at present, there is a wide range of definitions related to student and youth tourism in use, with little consistency in terms of age ranges. A concise definition of youth tourism is given by WYSE Travel, stating youth travel includes all independent trips for periods of less than one year by people aged 16–29 which are motivated, in part or in full, by a desire to experience other cultures, build life experience and/or benefit from formal and informal learning opportunities outside one's usual environment. In Malaysian context, the term youth is defined under the Youth Societies and Youth Development Act 2007 as a person not less than fifteen years and not more than forty years old. In addition, there exist no studies within Malaysian context that give proper definition of youth tourist.

In profiling the characteristics of Malaysian Youth Tourist, a study by Mohamed et al. (2010) was done to analyze the characteristics, patterns, behaviors and preference of Malaysian young tourist who travel domestically. This study revealed that youth is a potential market segment in domestic market that can be further serve by identifying their needs and demands. The Malaysian youth travel can be divided into three main segments which include students in high schools, students in higher education institution and working young tourist. A school-based trip is often conducted among schools where students are accompanied by the teachers in visiting the attraction area. This segment is limited in disposable income as they still depend on the budget from their parent to travel. Alternatively, higher education institution student are more independent and flexible. They possess their own budget during travel and enjoy leisure travel rather than for educational purposes. Therefore, the establishment of recreation clubs in schools and universities is a wise effort to encourage students to involve in tourism activities (Mohamed, 2005). Youth tourist in Malaysia is not any different from other type of tourism consumer. If their preferences and necessities are clearly identified, the players in the industry will be able to foretell and satisfy the customer needs and wants. By being satisfied, it will affect their post travelling behavior and intention.

2.2 *Satisfaction*

Satisfaction is the result of positive experience in using a particular product or services. Although visitor satisfaction is central concept in tourism (Danaher & Arweiler, 1996; Petrick, 2004), according to the specialty literature there are numerous and different definitions of the consumer's satisfaction, however some were particularly emphasized (Lache & Alexa, 2012). Hunt concept appreciates satisfaction as an evaluation based on the consumers' knowledge and experience, which, according to them, supposes a qualitative level for the respective product/service (Hunt & Morgan, 1994.). While the Howard Sheth concept–appreciates satisfaction/dissatisfaction as a psychological state, which reflects the adequacy/inadequacy of the effort made by the buyer, to acquire certain products and/or services.

Satisfaction results from the interaction of two factors in the individual, his aspiration or expectation level and his perception of the outcome. Presumably, if the performance of a product is good as expected, the buyer will be satisfied; if performance is less than expected, the buyer will be dissatisfied (Engledow, 1977). The satisfaction of the tourist services consumers is equally a psychological, emotional, affective and cognitive response, which is developed during the consumption, through the global evaluation of the multitudes of consumption services and experiences which make up the tourist product, and also by focusing the evaluative judgments towards the attributes of each and every service or experience (Lache & Alexa, 2012). Concerning its role in repurchase intention, tourist satisfaction level should be intensively studied for many practical reasons since costumers repurchase intention and loyalty are closely associated with their initial purchase.

2.3 *Tourism product*

Product or services is the backbone that drives the business of any industry. Product or services are created in response to the needs and demands of the potential customers. The development of tourism industry goes hand in hand with development of new products that can satisfy the consumer demands. In the relevant literature, there is no consensus definition of a tourism product, but there is a common understanding that such a product must appeal to travelers seeking either business or leisure activities (Jing, 2010). Gunn's model of the Tourist System denotes the tourism product as a complex consumptive experience that results from a process where tourists use multiple travel services during the course of their visit (information, transportation, accommodation, and attraction services) (Murphy, 2000). A study by O'Fallan (1994), categorize tourism product into three main categories; tourism experiences, place of product, and tourism products. The tourism experience is at the level of macro product including seeing, using, and experiences by tourists. As for the second category,

Figure 1. Tourism product as a total experience.

place of products is the destinations that tourists consume as a part of the tourism experience. And lastly, tourism product is defined as a generic product which includes accommodations, tourist attractions, restaurants, and souvenirs. Thus, for the purpose of this study tourism product can be defined as a generic product that comprises items such as accommodation facility, attraction area, transportation, restaurants and entertainment which can bring satisfaction to the tourist. In determining the dimension of tourism product, the researcher had applied the conceptual framework based from the concept by French et al. (1996). The concept is shown in Figure 1.

Based from the concept, tourism product is a sum of total experience by the tourist as a result of consumption of the product components from the beginning of their travel until the end of it. There are 10 components as shown in Figure 1. For the purpose of this current research, in replicating, this study in Malaysian context, only three components was selected to be dimensions of the study which are accommodation, transportation and attraction area. Thus, the dimension is referring to all the accommodation and transportation facility as well as the attraction area which is available to be used by youth within Malaysia.

3 METHODOLOGY

For this research, a descriptive research design using quantitative approach through cross sectional survey was selected as the time dimension involved one point in time only. This study had utilized the use of survey or self-administered questionnaire in collecting the data from respondents. From the total number of domestic tourist population in 2012, the tourist whose age between 15 to 30 years old was selected as respondents because they already have the experience using the domestic tourism product. Specifically, the purposive sampling was used in order to obtain information from the samples. In determining the appropriate number for the sample size, the researcher referred to the table of sample size for a given population size by Krejcie and Morgan (1970). Thus, the minimum sample size which was selected is 384. There were 500 questionnaires being distributed in order to minimize the unreliable data and increase the accuracy of data being gathered. After the screening process only 387 questionnaires was reliable for data analysis.

4 RESULT AND DISCUSSION

4.1 *Demographic profile*

The demographic analysis shows that, from total respondents of 387, 53.7% (n = 208) was presented by female respondents, while 46.3% by (n = 179) male respondents. From the data analysis the percentage of youth age 27 to 30 old has the higher percentage with 29.2% (n = 113). The percentage of respondents whose age between 23 to 26 and 19 to 22 is quite similar with 24.5% (n = 95) and 24.8% (n = 96) each. Lastly, the respondents age 15 to 18 represents 21.4% (n = 83) from the total 387 sample. For the race category, the Bumiputera represent the lowest percentage with only 3.9% (n = 15) as compared to other races. The Malays represent the highest percentage with 46.5% (n = 180), followed by Chinese with 27.9% (n = 108) and Indian with 21.7% (n = 84). The student represents the highest percentage from total sample of 387, with 53.2% (n = 206). However, if the percentage of working youth is being total up, it still cannot surpass the percentage by student category. Specifically the working youth represented by Administrative level with 18.6% (n = 72), Professional level with 15.5% (n = 60) and Technical level with 8.3% (n = 32). Only 3% (n = 1) of respondents represented by unemployed level. From 387 respondents, 36.7% (n = 142) were respondents with Degree qualification, 23.8% (n = 92) with Diploma qualification, 22% (n = 85) with secondary education, 10.9% (n = 42) with postgraduate qualification and 6.7% (n = 26) holding certificate. Based from this analysis, it can be inferred that most of the respondents have high level of literacy and ability to provide relevant responses while answering the survey. This is also proved by ability of respondents who able to express their opinion and level of awareness and concern on the youth tourism segment in Malaysia during the data collection.

4.2 *Descriptive analysis on level of satisfaction*

To identify the level of satisfaction of youth tourist towards the domestic tourism product the mean score was obtained using the descriptive statistics on three dimensions of the tourism product which are accommodation, transportation and attraction area.

Based from the mean score obtained for accommodation facility, it indicates that the score range from M = 2.2610 to M = 2.6822. However, the price affordability received a mean score that slightly near to neutral response. Looking at the post travel behavior the re-visit intention received mainly "strongly agree" as response M = 1.8889, while the satisfaction towards the accommodation facility received mainly "Agree" M = 2.4341 as response. This result indicates that youth tourists are satisfied with the accommodation facilities that are available for them during travelling.

In measuring the items for transportation facility, the items were analyzed in terms of the varieties of mode of transportation, price affordability, sufficiency of infrastructure, accessibility to rural areas, convenience, and adequacy of information facilities for transportation. The mean score range for these items is between M = 2.6098 and M = 2.2196. The mean score shows that the response received are mostly "agree".

For the post-travel behavior, the respondents agree that they will re-use the transportation facility that satisfied them during travelling with a mean score of M = 2.2196. As for the satisfaction level, the youth responded that they satisfied with the transportation facilities available for them for travelling, indicated by the mean score of M = 2.6357.

For Attraction Area dimension, the mean score received also revealed that the response received were mainly "agree" to the statement. However, the item for marketability of the attraction area to youth and availability of special tour package for youth received response which is nearing to "neutral" as the mean score received is M = 2.7106 and M = 2.695, respectively. The mean score also shows that the youth tourist mostly agree to re-visit the attraction area if they are satisfied, mean score (M = 2.1034). In terms of overall satisfaction the mean score was M = 2.2972, indicating most respondents given "agree" as answer.

5 IMPLICATION AND CONCLUSION

Although the findings revealed that the youth are being satisfied, there are certain areas where more attention is needed in order to further ensure the overall satisfaction of this market segment is achieved. The tourism product affordability and the promotional activities that focused more on the youth market need to be re-considered by the tourism players within the country, especially the accommodation and transportation providers and tour operators. Other than that, the policy maker also need to be more attentive at this segment and do not underestimate the potential that it can make towards the economy of the country. As a conclusion, as compared to other countries youth tourism in Malaysia is still at the premature level, thus to nurture this segment, further studies are needed to unlock the various aspects of this market within Malaysian context. The future research can look in the awareness on the importance of youth tourism market from tourism players' point of view. By looking at this, the researcher can identify the direction of this market and providing a hint to the policy maker on the current trend of youth tourism in Malaysia. All in all, this research had made contribution in unlocking the behavior of youth tourist in the country and the current trend that bring about new opportunities in the domestic tourism market.

REFERENCES

Bartikowski, Boris, & Sylvie Llosa (2004), Customer satisfaction measurement: Comparing four methods of attribute classifications. *Service Industries Journal, 24*(4), 67–82.

Boukas, N. (2008). *Cultural tourism, young people and destination perception: A case study of Delhpi, Greece.* (Unpublished master's thesis). University of Exeter, Devon, United Kingdom.

Danaher, P.J. & Arweiler, N. (1996). Customer satisfaction in the tourist industry: A case study of visitors to New Zealand. *Journal of Travel Research, 31*(1), 89–93.

Engledow, J.L. (1977). Was consumer satisfaction a pig in a poke? *Business Horizons, 20*(2), 87–94.

Gunn, C.A. (1988). Tourism Planning (2nd ed.). *New York: Taylor & Francis.*

Habibah, A., Hamzah J., Buang A., Noor Alyanie N.A., & Aimi Syairah, M.N. (2011). An inventory of youth tourist in event tourism. *World Applied Sciences Journal, 13*, 13–17.

Hunt, S.D. & Morgan, R.M., (1994). Theory of relationship marketing. Journal of Marketing, International Conference on Youth Tourism. India: WTO.

Jing, B.X. (2010). Perceptions of tourism products. *Tourism Management, 31*(5), 607–610.

Krejcie, R.V. & Morgan, D.W. (1970). Determining sample size for research activities. *Educational and Psychological Measurement, 30*, 607–610.

Labben, T.G. & Johnson, C. (2004). Developing the image of the young tourist: Details from Swiss festivals. *Tourism Review, 59*(1), 25–33.

Lache, C. & Alexa, E.L. (2012). Particular aspects of the tourist products consumers' loyalty. *Lucrări ştiinţifice seria Agronomie, 54*(1), 234–238.

Malaysian Tourism Satellite Account 2005–2011 (2013). Putrajaya: Department of Statistics, Malaysia.

Mohamed, B., Omar, S.I., & Lau, P.M. (2010). Malaysian youth market: A lucrative but overlooked tourist segment. *Malaysian Journal of Youth Studies, 3*, 1–20.

Mohamed, B., Y. Yusof & S.I. Omar. Eds. (2005). The psychographic clusters of domestic travellers. Penang: Universiti Sains Malaysia.

Moisa, C.O. (2010). Aspects of the youth travel demand. *Annales Universitatis Aoulensis Series Economica, 12*(2), 575–582.

Murphy P, Pritchard, M., & Smith, B. (2000). The destination product and its impact on traveller perceptions. *Tourism Management, 21*, 43–52.

O'Fallon, C.E. (1994). The role of central government in tourism: A public choice perspective. Lincoln University: Unpublished PhD thesis.

Petrick, J.F. (2004). The roles of quality, value, and satisfaction in predicting cruise passengers' behavioral intentions. *Journal of Travel Research, 42*, 397–407.

Reisinger, Y. & F. Mavondo (2002). Determinants of youth travel markets; perceptions of tourism destinations. *Tourism Analysis, 7*, 55–66.

Richards, G. (2008). *Youth travel matters: Understanding the globalphenomenon of youth travel.* Retrieved September 12, 2012, from http://www.sete.gr/files/Media/Ebook/2008/110314_Youth%20Travel%20Matters.pdf.

Hospitality and Tourism – Sumarjan et al. (eds)
© 2014 Taylor & Francis Group, London, ISBN 978-1-138-00151-0

Malaysia as a wedding destination: Perception and destination characteristics

M.H. Zamri, M.D. Darson, M.F. Wahab & L.K. Lim
Universiti Teknologi MARA, Malaysia

ABSTRACT: Wedding is the once-a-life event in people's life. Therefore, the purpose of this study is to identify what are the perceptions of international bridal tourist to choose Malaysia as their wedding destination and what are the most significant characteristics towards the arrival of international wedding bridal couples to Malaysia to organize their events. A qualitative approach was considered to be the most effective research design and semi structured interview was used in getting information from the informants. For this study, semi structured interviews with nine international bridal couples had been conducted through online and face to face interview. Findings clearly revealed that the uniqueness of Malaysian culture and the wedding venue is the main reason why these international bridal tourists choose Malaysia as a wedding destination. This study provides enormous benefits towards local event planners especially professional wedding planners the potential of leveraging the local resources as wedding tourism destination.

Keywords: bridal tourism, wedding destination, wedding events, wedding management

1 INTRODUCTION

Of all the significant events that would take place in one's life, nothing is more essential compared to one's a momentous day or wedding day. It is the moment that couples close a relationship with the promise to love each other for the rest of their lives. It may be a celebration by the beach at twilight, or in a grand ballroom filled with pink champagne roses and balloons. It may be a fairy-tale garden party, or romantic style on a private cruise witnessed only by the stars.

Wedding form part of personal or private events; it can be seen as the most complex to organize as it involves families and friends, catering, entertainment and accommodation (Shone & Parry, 2004). Thus, Shone and Parry (2004) suggested that special event can be categorized in leisure events (sport and recreation); cultural events (ceremonial, sacred, heritage, art and folklore); organizational events (commercial, political and charitable); and personal events (wedding, birthday and anniversaries). Strianese and Strianese (2003) stated that one of the biggest markets in the personalevent business comes from weddings. The objectives of this research paper are as following:

I. To seek perception of international tourist towards Malaysia as a wedding destination

II. To identify which of the characteristics are the most significant towards the arrival of international wedding bridal couples to Malaysia.

2 LITERATURE

The wedding management industry is still developing in Malaysia unlike other countries such as the United States and those in Europe, where public awareness of wedding planners is still at its developing stage. Nowadays, there are a growing number of modern couples who turn to wedding planners as they are too busy for the arrangement and they wanted a memorable wedding ceremony (Khoo, 2013). Wedding statistic 2011 in US shows that the average wedding cost per bridal couples is $25,631 per event. Market value is approximately $53.3 billion. Average number of guests is about 133–143 and the current trend of online market value is about $8.6 billion. The number of wedding businesses is in the range of 332,577–369,530 couples (The Wedding Report, 2013). In average wedding cost, items such as clothes & accessories, beauty and spa, entertainment, flower and decoration, gifts and favors, invitations, jewelry, photography and video, planner or consultant and lastly is venue, catering and rentals are part of the essential in wedding services.

Malaysia is blessed with many beautiful islands. Among them are Langkawi, Redang, Tioman,

Pangkor, Sipadan, Mabul and Perhentian islands, with beaches that inspire romance and clear blue waters that sweep soothingly to the shores. Weddings by the beach at sunset may be perfect, but couples can also look for a cruise wedding where they can sail away into the sunset as they exchange vows. This is another good innovation in the wedding industry, as well. Even though Malaysia is very much a tropical country holding a wedding in the cooler climes of the highlands, is an option, as well. For instance, Frasers Hill, Cameron Highlands and Genting Highlands are located on the main mountain range of Peninsular Malaysia with temperatures dropping between 16 and 21 degrees Celsius. Here, garden weddings or English wedding are the most chosen one, with the gently sloping green hills as a witness to the holy gathering.

With Malaysia's cultures and traditions, couples may want to exchange vows the way Malaysians themselves typically do it. Malay weddings will be treated the bride and groom being like a king and queen for the day, complete with the "royal" dais and traditional martial arts and cultural performances fit for the royal courts. Guests then take turns to "bless" the union with rosewater and fragrant potpourri. For a more oriental experience, couples can get dressed up in traditional Chinese wedding costumes and complete the Chinese tea ceremony with friends and family. Indian temple weddings are also a colorful affair as the bride and groom are celebrated with music and flowers by guests dressed in extravagant sarees of modern design. At the Sarawak Cultural Village, wedding ceremonies can be organized according to the traditional Iban, Bidayuh or Orang Ulu way. Similarly, the Monsopiad Cultural Village in Sabah caters to Kadazan styled weddings.

Wedding management is a high potential sector; it consists of nearly 70% of all the sectors in the market. Because people, especially Chinese, according to Chang, spend a lot for funerals, and weddings, the two most important occasions in their lives; thus, he believed that there is a huge demand for wedding planning (Wedding, 2010). Weddings and honeymoons contribute to local economic development in a number of ways. First, couples often use local logistics such services, relying on vendors such as florists, caterers, and transportation specialists whose businesses are in close distance to the wedding and reception. This expense stimulates the economy of the local community where the wedding is held. Secondly, weddings involve out-of-town guests who can be considered as tourists during their stay. Their expenditures on accommodation, food and beverage, gasoline, and entertainment pump new money into a region, thus boosting the local economy. Third, most couples will celebrate their newlywed status by taking a honeymoon. Now the newlyweds are the tourists, and their spending at a chosen destination will enhance the livelihood of the local businesses where they are vacationing. This combined expenditure of cash is significant; leading to healthy competition to attract engaged and newly married couples.

Weddings have several destination characteristics which are frequent in all types of services primarily in hospitality and tourism services. It is so much related towards this industry. Those characteristics are.

2.1 *Uniqueness*

When a human being involves in the production and consumption of services it will always move towards variability especially in tourism services. There is no such two service providers can come out with exactly the same quality of service because they are unique individuals. It is as same as the client or guests who might be having a different preference, expectations, needs and desires (George, 2004). The most important in wedding planning are the uniqueness of each event. This can lead to extraordinary theme and ideas just to ensure that the event is unforgettable once finished. These memories will create effect towards the bridal as well as their invitees. Therefore, every wedding is unique as the service providers; culture, guests, timing and locations are different (Shone & Parry, 2004).

2.2 *Wedding ambience*

Ambience is very important in the outcome of the wedding. For instance, superb food, great décor, smooth flow of the performances, smooth running of the technical team, great accessibility to the venue of the event and enough parking spaces for the guest to park their vehicles. However, these are not guaranteed the success of an event. Ambience is generally created by the crowd of guests or people who are attending the event. For example, the good company amongst friends and family may cause the great ambience that will ensure the success of the event without anything else needed. The reason why people attending an event are to experiences new situation and environment. However, that wedding planners must paying details attention and taking care of the needs of guests, in order to ensure that the wedding is a success (Shone & Parry, 2004).

3 METHODOLOGY

A qualitative approach through interview is online or face to face was considered to be the most effective research method. Semi structured interview was used to explore respondents experience, and views about Malaysia as potential wedding destination and what are the most significant characteristics of wedding event that attracts wedding tourist to Malaysia. The sample population for this study was derived from different part of the world. There are about nine respondents from four countries namely Philippines with six informants. While Nepal, Germany and Greece with one informant respectively. A set of interview questions was developed in order to obtain more insight on the perception of international tourist on Malaysia as a wedding destination.

4 FINDINGS

The information obtained from online and face to face interview session relating to the objective of the study is discussed in the subsequent series of analyses. In response to the objectives one which related to the perception of international tourist towards Malaysia as a wedding destination, there are six questions been asked as follows.

i. *How do you find the uniqueness of Malaysia as a wedding destination? Can Culture be described as one of the uniquenesses?*
ii. *In term of price, do you think it is worth it in term of money and services compared with other wedding destinations?*
iii. *How do you find Malaysia ambience for wedding event destination? Do you think the natural landscape in Malaysia for wedding purposes are appropriate?*
iv. *Does the ambience help to contribute in wedding photography?*
v. *How do you find the local wedding ritual? Do you think the Malaysia wedding ritual is an exotic experience?*
vi. *How the Malaysian does dishes suit with your wedding theme? Can the food mesmerize your guest?*

With regard to this question, all of the respondents agreed that Malaysia has a very strong potential to become leading international wedding destination. Majority of them mentioned about the uniqueness of the culture and beautiful sceneries. Some of the respondents were also mentioned about few places that they thought can be a potential wedding venue for their future wedding destination such as Langkawi, Penang and Cameron Highland. However, there are few informants who do not agree with the price range in the Malaysia wedding packages. From nine, informants, only five of them agreed that Malaysia wedding packages is attractive and affordable. One informant responded that Malaysia wedding packages were expensive, and another two informants have no opportunity yet to see packages.

With regard of the Malaysia ambiance for the wedding destination, most of the informant agreed that the tropical climate is the main reason for them to choose Malaysia as a wedding venue. This is in lined with the statement by MacCannell (1999), the more exotic and natural the landscapes presented, the more appealing it is for tourists who seek other experiences.

Ambiance plays a big role in wedding photography. It is another factor in wedding management. Based on the results, most of the informant agreed that the Malaysian ambiance is suitable for Asian theme wedding event. It is common for some wedding couples to experience something extraordinary. Some of the informants who lives in Kuala Lumpur had experience the local dishes. However, none of them experience the local wedding food. The wedding food, for example, the Malay wedding food is much richer in term of spices compared with the normal daily food. As time goes by, the new generations of Malaysia currently do not mind to incorporate the local wedding food with daily food. Many of the informants agreed that uniqueness of culture is the dominant factors that attract them to do wedding event in Malaysia. There are only one questions been asked as follows.

Informants agreed that uniqueness of the culture is the major contributor towards the arrival of bridal tourist to Malaysia. This is in lined with Shone and Parry (2004) who claimed that every wedding is unique as the service providers; culture, guests, timing and locations are different.

5 CONCLUSION

The summary of the preceding section clearly indicates that Malaysia has a bright potential to become an international wedding destination. In this sense, promoting Malaysian wedding packages with excellent services is highly demanded by the international bridal couples. In this sense, there are a few popular places such Penang, Langkawi, and Cameron highlands have the potential ability to shine as international wedding venue. It is due to the unique characteristics of the place such architecture, ambience and food. Despite this, findings of this study explicitly show that the mixed culture of Malaysian is also contributed as a reason why international tourists choose Malaysia as a wedding destination. Malaysia has its own standard when it comes to the multi-cultural concept. Be in Malaysia circle is similar as visiting all Asia countries. This is undoubtedly the most precious elements that Malaysia should preserve for the future betterment. The variety of cultural influences in Malaysia is the manifestation of centuries of immigration and trade with the Arab nations, China, and India. However, each culture remained untouched; it means no culture has truly been homogenized. Traditional temples are adjacent with churches while the mosque and Sikh temple exist side by side is typical scenery.

REFERENCES

Adler, H. & Chien, T.C. (2004). The wedding business: A method to boost food and beverage revenues in hotels. *Journal of Foodservice Business Research* 7(1), 117–125.

George, R. (2004). Marketing South African tourism. (2nd ed). Cape Town: Oxford. 415.

Khoo (2013). New Programme for a wedding planner wannabes. Retrieved May 28, 2013, from http://afterschool.my/new-programme-for-wedding-planner-wannabes/.

MacCannell, D. (1999). The Tourist: A new theory of the leisure class. Berkeley: UC Press.

Shone, A. & Parry, B. (2004). Successful event management. London: Thomson. 246.

Strianese, A.J. & Strianese, P.P. (2003). Dining room and banquet management (3rd. ed.) New York: Thompson Delmar Learning. 413.

Tassiopoulos, D. (2000). Event management: A professional and developmental approach. Cape Town: Juta. 494.

The Wedding Report, (2013). The wedding index, Retrieved June 11, 2013, from World Wide Web: http://www.theweddingreport.com/bz/index.php/the-wedding-/.

Wedding, (2010). The 7th Malaysia International Wedding and Banquet Exhibition. Retrieved 4th May, 2013 from World Wide Web: http://blog.wedding.com.my/archives/tag/wedding-exhibition.

Hospitality and Tourism – Sumarjan et al. (eds)
© 2014 Taylor & Francis Group, London, ISBN 978-1-138-00151-0

Indigenous peoples and their foods: A valuable contribution to tourism destination development—evidence from Lijiang, China

M. Gibson & C. Tang
Institute for Tourism Studies, Macao SAR, China

ABSTRACT: Indigenous peoples and their foods can be significant cultural attractions—a unique selling point, providing both ethnic enrichment and economic benefits to both local and national stakeholders. Tourists, on experiencing the 'authenticity' of this local food and culture, may have any pre-perceptions—where they exist, altered or strengthened as a result. Where such perceptions are absent these new experiences instead can help to create new insights. The success or failure of which, can in turn strongly influence future re-patronage intentions or through word of mouth, act as inspiration for others to visit. This paper seeks to identify the impact of the 'pull' of indigenous foods. Based on demographic profile and its strong indigenous culture, Lijiang in China is chosen as the case study. Questionnaire based quantitative research subjected to hierarchical cluster analysis is then used to identify response patterns to the Likert-scaled questions of experience and influence.

Keywords: foods, hospitality, indigenous, Lijiang, tourism

1 INTRODUCTION

There has always been a close relationship between both the hospitality and tourism industries. Indeed in the tourism destination, it is sometimes this hospitality—the provision of food and beverage services for instance, that can sometimes be a key determinant in the decision-making process. Yet despite this knowledge few studies directly examine this relationship, furthermore such research is especially lacking when it comes to indigenous peoples, their food and culture. The objective of this paper is to identify this impact in terms of both the destination image formation process and its role in the development of the tourism destination. In bridging this academic research gap this study also has practical merit with the goal of ultimately better aligning both industry's' needs whilst also contributing to the overall development of the local tourism industry. Considering the demographic profile of indigenous people and its current tourism development status, Lijiang in Yunan Province functions an appropriate case study for this research.

1.1 *Indigenous peoples*

Indigenous peoples, are groups also known as 'tribal peoples', 'ethnic' 'first' or 'native peoples', 'aboriginals', 'adivasi' and 'janajati' among others. They are also sometimes referred to using geographical or occupational terms such as hill people, hunter-gatherers, nomads, and peasants among others. In many if not most cases however, these groups are the ethnic minority descendants of those inhabitants of a country or geographical region who were displaced (removed from their traditional historical territories) or marginalized as a result of new arrivals at the time, of different cultures or ethnic origins, who then settled and eventually became the dominant population. These descendants then often chose to continue and cultivate their historical traditions and legacies for future generations.

Inherent in such lineage and irrespective of whether their status as indigenous peoples came about through natural or political means, indigenous groups have a rich, often ancient, social, economic and political set of traditions. These in turn are also often viewed as being both co- and independent of existing cultures whilst also in many cases simultaneously interdependent with both environmental and spiritual systems. This holistic view of life as well as their diversity and strong historical attachment ensure many indigenous peoples continuing importance in terms of world heritage (IFAD, 2013).

In terms of numbers, indigenous peoples are estimated to be more than 370 million in number, belong to 5000 or so groups, spread across 70–90 countries around the world from the Arctic to the South Pacific and constitute about 5% of the world's total population (United Nations, 2013). In this way it is not difficult to appreciate the value of these groups first as an entity unto themselves but also within the overall tourism product. Yet as groups they are not always easy to identify. This is not helped by the fact that there is no firm or widely accepted definition of "indigenous" peoples; not even in the reams of the United Nation's official literature. Instead a modern understanding of the term arises out of the consensus that such groups share certain common or at least similar characteristics among their numbers. In this sense they are more identified rather than defined –these include:

i. *Self-identification at the individual level and being accepted by the community as a whole;*
ii. *Historical continuity with pre-settler societies as well as strong links to territories and surrounding natural resources;*
iii. *Distinct social, economic or political systems often with unique language, cultural and belief systems;*
iv. *A strong resolve to maintain and ensure the continuance of their ancestral cultures as distinctive peoples and communities.*

From these criteria indigenous peoples are seen to retain their different and often unique traditional characteristics distinct from those of other segments of the dominant main-stream national populations in which they live. On a sour note though in some cases, the notion of being indigenous also sometimes carries with it, discriminative or negative connotations. In worst cases this can result in some indigenous peoples being amongst some of the world's most vulnerable, marginalised and disadvantaged groups of people (Cultural Survival, 2013). Another result of this persecution might serve to discourage some indigenous peoples from revealing their origins or even identifying with their roots at all.

1.2 *Destination image creation*

As a result of continuous and indeed sustained growth within the tourism sector consumers now face more choice than ever when it comes to tourism destinations. As a result, marketers of these destinations are now faced with big challenges in influencing or affecting consumer choices. In these circumstances and echoing the thoughts of many, Echtner and Ritchie (2003) observe that a more distinctive or differentiated and appealing image of the destination must be fostered if they are to appeal to consumers in this increasingly complex and competitive global market (Echtner & Ritchie, 2003).

Indeed the importance of 'destination image' or its perception was defined as long ago as the late 1970's by Crompton when he summarised the idea as comprising 'an attitudinal concept consisting of the sum of beliefs, ideas and impressions that a tourist holds of a destination' (Crompton, 1979, p.18). This still holds just as true as it was then, perhaps even more so since in today's competitive market. This was recognised by Gallarza, Saura, & Garcia (2002) when, echoing Cromptons words, suggested that the destination image continues to affect tourists' cognitive and subjective perceptions, their consequent behaviour and ultimately whether or not this becomes their destination of choice (Gallarza et al., 2002). In addition destination image remains just as relevant after the fact too as it influences satisfaction where tourists compare their preconceived perceptions with the actual reality encountered (Chon, 1990). Indeed Guthrie and Gale (1991) go as far as to argue that consumers' in fact are more motivated to act based on perceptions rather than actual reality. In this respect destination perception takes on just as importance as other more tangible resources.

So how are these perceptions formed? The formation of perceptions can be thought of as the product of complex cognitive mental development processes based on impressions extracted from a multitude of information streams (Reynolds, 1965). This might include such things as promotional literature (brochures and posters etc.), word of mouth and the opinions of others (family, friends and travel agent's suggestions); as well as from the general media (travel magazines, books, internet and TV etc.). The creation of this mental image does not end with the vacation either, perceptions gained before the trip are further modified with the experience of the actual trip (Echtner & Ritchie, 2003). This circular process is as important from the marketers view as it is from that of the consumer because positive tourism experiences are more likely to result in positive word of mouth. From these observations Gunn (1988) developed the seven phase travel experience model, stating that the image creation process comprises:

- *Phase 1: Accumulation of mental images about vacation experiences*
- *Phase 2: Modification of those images by further information*
- *Phase 3: Decision to take a vacation trip*
- *Phase 4: Travel to the destination*
- *Phase 5: Participation at the destination*
- *Phase 6: Return home*
- *Phase 7: Modification of images based on the vacation experience*

The three stages more closely associated with actual destination image formation can be identified among this group as occurring during phases 1, 2 and 7. In phases 1 and 2, destination perception is obtained through secondary sources while in phase 7 actual primary experiences are used to modify the destination image whether positively or negatively (Echtner & Ritchie, 2003).

Gunn (1988) suggests that the image formed in phase 1 is 'organic', and is more than likely not based solely on commercial and tourist sources of information alone but rather in combination with research, the opinions of friends and family, and the media among other sources. Also, while non-commercial information from the general media and education might actually be of limited value in the formulation the tourism destination image for individuals, historic, political, economic and social factors nevertheless can be very important influences in building the overall big picture of a particular destination. This is not surprising as a country's tourism image is often reflective of its national image (UNWTO, 1980).

After this initial image creation, commercial information including travel brochures and travel agency promotion then starts to influence these tentative early perceptions further in phase 2. This may serve to change, alter or even strengthen such perceptions (Echtner & Ritchie, 2003). Further building confidence in any decision making process suggests Beerli and Martin (2004), is the significant importance of word of mouth. In their view it is one of and the most believable and credible communication channels on offer. In this regard it is important that the reality of the destination is honestly portrayed and communicated in the market as such. Consequently and collectively, these image perception and modification processes as described in phases 1 and 2 are essential components of the consumers' decision tree. Further, the effectiveness of this process can be considered a measure of a particular destinations ability to influence the consumer's choice (Gunn, 1988).

In phase 7, images are further modified by actual experiences. This results in a more reasoned perception of the destination whereby positive experiences can translate into tangible benefits including recommendations and revisits (Gunn, 1988).

As a good framework of the destination image formation process, Gunn's model can be modified in this research to analyse the tourists' image building process to include experiences associated with indigenous peoples, in particular their food. Specifically, the extent to which positive food experiences are translated into word of mouth and/or the intention to re-patronage.

1.3 *Research concept*

The modified image formation processes as described in the research methodology provides a theoretical framework on which this research is based. This is applied to the comparative evaluation between perceived destination images before experiencing indigenous foods and the modified destination image afterwards. This specifically relates to phases 1, 2 and 6 of the modified image formation processes. This research will also examine which factor or factors relating to aspects of indigenous foods vis-à-vis preparation, presentation and the taste of the indigenous foods if any, that might significantly contribute to the destination image formation. This is relates to phases 4 and 7 of the modified image formation framework.

2 RESEARCH METHODOLOGY

2.1 *Justification and case study selection*

Lijiang is a prefecture-level city located in the northwest of Yunnan Province in the People's Republic of China with a total area of around 20,600 square kilometers. It was chosen for this study as the majority proportion of the 1.14 million population (about 58.1%) are indigenous peoples. More specifically, Lijiang contains a total of 22 indigenous races including the Naxi, Mosuo, Ni, Bai, Pumi, Lili, Miao, Zhuang peoples etc., with 12 of these groups having lived there for many generations (Lijiang City Council, 2013).

Lijiang has a long cultural history, since the Tang Dynasty the old town was considered the political, commercial and cultural center for the local indigenous people especially for ethnic Naxi's. Most notable among the Naxi people's history was the influence of the Mu family. the Mu family were the local ruling elite in Lijiang during the Ming and Qing Dynasty's. In fact so strong is the Naxi and the other Lijiang's indigenous peoples' legacy that local culture (including food), traditions and architecture have been significantly shaped by their influence through the ages (Lijiang City Council, 2013). What results is a rich cultural tapestry and in 2007 the United Nations recognised the old town of Lijiang as a UNESCO site of World Heritage. Coupled with this the local diverse natural landscape as well as the local romantic night life of the Lijiang's local environs and it is not difficult to understand why around 15.99 million tourists visited the area in 2012 generating around USD3.3 billion tourism revenue for the local economy. Not surprisingly tourism has now become one of the most important industries for Lijiang (Lijiang City Council, 2013).

Many indigenous peoples both contribute to and benefit from tourism development as a result they are increasingly eager to share their traditions, culture and hospitality to the masses. Indeed many work or operate businesses inside the food sector providing traditional local indigenous foods and snacks to locals and tourists alike. Among these foods famous dishes like the Naxi barbeque fish; Naxi barbeque pork; preserved ribs hotpot; Naxi pancake (aka Ba Ba); fried dragonfly; rice sausage; and chicken bean noodles are well known. Furthermore, many restaurants have gone that one step further by incorporating elements of the preparation of these foods as part of the overall dining experience. This might be considered a form of cultural promotion so far been well received by the tourists.

Yet despite the importance of indigenous peoples food and culture, to date there has been no direct research analyzing how such activities contribute to the tourism development of Lijiang. This is the justification of this research.

2.2 *Research objective and research questions*

The overall research objective of this paper is to identify those factors of indigenous foods that are worthy of note in the destination image formation process. Based on the demographic profile of the indigenous population and its current tourism development status, Lijiang in Yunan Province was chosen as an appropriate case study for this research. In achieving this research objective, the following research questions are developed:

- *To what extent indigenous foods in Lijiang contribute to the destination image formation?*
- *Which factor(s) related to the material, preparation, presentation and the taste of the indigenous food has the potential to significantly influence the tourists' perceived destination image?*
- *To what extent, does any subsequent image creation influence 'word of mouth' as well as intentions to revisit Lijiang prefecture?*

With the ability to identify basic patterns within target populations using the many varied statistical techniques, quantitative research by itself can elicit valuable information from a large number of participants (Neuman, 2006). To address the above objectives, a questionnaire-based quantitative approach is pursued.

The questionnaire design is based on Gunn's original framework of the image formation process. Following this model this research better evaluates the changes that take place between the perceived destination image creation and the modified destination image vis-à-vis experiencing the local indigenous food and culture. This following adaptation is modified from Gunn's original model:

- *Phase 1: Accumulation of mental image of the destination*
- *Phase 2: Modification of the image by further information*
- *Phase 3: Decision to take a vacation trip*
- *Phase 4: Travel to the destination*
- *Phase 5: Receiving additional stimulus from the material, preparation, presentation and the taste of the indigenous food*
- *Phase 6: Modification of images based on the experience of the indigenous food*
- *Phase 7: Return home (word of mouth effect or/ and revisit intention)*

Further strengthening these Likert-scaled questions consideration was given to theories of the AIDA (Attention, Interest, Desire and Action) and The Reasoned Action's (TRA) models.

To strengthen the results regarding representation of the target population, convenience sampling will be used to collect the data. With this in mind, as a great majority of visiting tourists in Lijiang stay overnight in hotels and hostels—these questionnaires will be distributed through these channels with the help of local employers.

The most recent version of the quantitative statistics software, SPSS, will be used to analyse the data. Hierarchical cluster analysis will be used to identify relatively uniform sub-samples differentiated by their response patterns. This hierarchical cluster analysis will be effective in highlighting the influence of the indigenous foods on consumer choices. Subsequent, comparison-of-means and chi-squared tests will then be used to evaluate how the clusters differ.

3 SIGNIFICANCE OF THE RESEARCH

As discussed, there has been very little literature directly analysing relationships between the food and tourism industries in the area of indigenous cultures and tourism destination development. This research will have substantial benefits both for Lijiang Province as well to other tourism destinations with significant indigenous populations. Additionally, it is hoped this research will reach the attention of other researchers with the potential to stimulating future research in this area.

REFERENCES

Beerli, A., & Martin, J.D. (2004). Factors influencing destination image. *Annals of Tourism Research, 31*(3), 657–681.

Chon, K.S. (1990). The role of destination image in tourism: A review and discussion. *The Tourist Review, 2*, 2–9.

Crompton, J.L. (1979). An assessment of the image of Mexico as a vacation destination and the influence of geographical location upon that image. *Journal of Travel Research Spring*, 18–23.

Cultural Survival (2013). *Cultural Survival—Who are Indigenous Peoples?* Retrieved on 07 April 2013, from http://www.culturalsurvival.org/node/10275.

Echtner, C.M., & Ritchie, J.R.B. (2003). The meaning and measurement of destination image. *The Journal of Tourism Studies, 14*(1), 37–48.

Gallarza, M.G., Saura, I.G., & Garcia, H.C. (2002). Destination image Towards a conceptual framework. *Annals of Tourism Research, 29*(1), 56–78.

Gunn, C. (1988). *Vacationscapes: Designing tourist regions*. New York: Van Nostrand Reinhold.

Guthrie, J., & Gale, P. (1991). *Positioning ski areas*. Paper presented at the New Horizons Conference Proceedings, Calgary.

IFAD. (2013). International Fund for Agricultural Development—Indigenous peoples: valuing, respecting and supporting diversity. Retrieved on 07 April 2013, from http://www.ifad.org/english/indigenous/.

Lijing City Council (2013). Introduction of Lijiang. Retrieved on 07 April 2013, from http://news.lijiang.gov.cn/others/article/2010-12/20/content_1304.htm.

Neuman, W.L. (2006). Social research methods: qualitative and quantitative approaches, Boston: Pearson.

Reynolds, T.J., & Guttman, J. (1984). Advertising is image management. *Journal of Advertising Research, 24*, 27–37.

United Nations (2013). Indigenous Peoples, Indigenous Voices. Retrieved on 07 April 2013, from http://www.un.org/esa/socdev/unpfii/documents/5session_factsheet1.pdf.

UNWTO (1980). *Tourist images*. Madrid: UNWTO.

Hospitality and Tourism – Sumarjan et al. (eds)
© 2014 Taylor & Francis Group, London, ISBN 978-1-138-00151-0

Molecular gastronomy movement and application in food

F. Faat & A. Zainal
Universiti Teknologi MARA, Malaysia

ABSTRACT: Technology is now being implemented in food which resulted in an emerging knowledge which implied science in the preparation of dishes. This category of knowledge is known as molecular gastronomy. As reported by Barham P., Cassi D., and The World's 50 Best Restaurant, restaurants which applied molecular knowledge such as El Celler de Can Roca, Noma, Fat duck, and Elbulli have been listed and won the best restaurants in the world. In developed countries, a phenomenon of molecular gastronomy is not something new in culinary realm, apparently in developing countries the understanding of this term and acceptance is still in the introductory stage. Perhaps, the application of molecular gastronomy in Malaysian cuisines could help increase the value of ethnic food and perception of how traditional food can be at par with molecular western dishes. This paper discusses the molecular gastronomy movement around the world and its application in food.

Keywords: culinary technology, molecular cuisine, molecular gastronomy

1 INTRODUCTION

Gastronomy comprised of multidisciplinary which consists the understanding of various social, cultures, historical components, literature, philosophy, economic religion, cooking, technology, and others with food as central axis (Santich, 2007). Living in this modern era and with the advancement of technology today, a new phenomenon of modern cuisine known as molecular gastronomy swamp almost entire of European country and it start to spread to all over the world. As mentioned by Gaya, Soler, Manes, Soriano, and Sebastia (2012), demand for molecular gastronomy restaurant has increased worldwide. In line with this, Cassi (2011) and This (2006) in European Molecular Biology Organization report (EMBO) and Barham et al. (2010) stated, a number of restaurant around the world have started to adopt a more scientific approach in their kitchen. As the result, these restaurants have acclaimed as being among the best in the world.

Notably, molecular movement was pioneered by Herve This and Kurti in 1988 and talented chefs, like Heston Blumenthal, Ferran Adria, Pierre Gagniare and Grant Achatz who were the first generation of chefs collaborated with scientist and applied the element of molecular gastronomy in their cooking. As mentioned by McGee (2004), Barham et al. (2010) and Cassi (2011), molecular gastronomy is collaboration between scientist and chef in creating and improvising food for a better understanding of food, especially on molecule reaction during cooking process.

Likewise, molecular gastronomy has been defined as the chemistry and physics behind the preparation of any dish (This, 2005a; This, 2006b; This & Kurti, 1994). Therefore, Barham et al. (2010) believed that molecular gastronomy is a scientific study of why some food taste varies which can make one food delicious and another one is not. This may be due to the ingredients and how they were grown, the manner in which the food was cooked and presented or the environment in which it was served. Throughout the molecular movement, varieties of food additives and equipment have been developed. This situation can be identified in most kitchens that applied molecular gastronomy in preparing food with surprise elements. For instance, the usage of algin and calcic in speherification, soy lecithin to stabilize the air in foam, sous vide 'under vacuum' cooking style, usage of espuma 'whipper cream' in creating splendid dishes.

It is important to embark on the element of acceptance prior to creation and introduction of new products and services. Accordingly a study by Yilmaz and Aydeniz (2012) inferred that the marketing success of food products dependent on many factors which includes the consumer sensory satisfaction. Clark (1998) stated when a person consumes the food and experienced the sensory characteristic, they are able to make a decision on whether or not the food is likeable or not.

Normally, the acceptance of food is measured in affective sensory test with potential consumer of the product or service (Kahkonen, 2000).

Specifically, acceptance refers to an individual experience, a feeling or an emotion with hedonic aspect (Cardello 1996). Overall, sensory consumer studies are used in product design to develop products, that are well accepted, with study in food and nutrition without exception. The difference is that, the focal point in food and nutrition studies are more towards the effect of various and distinctive factors related to food acceptance (Schutz, 1999).

2 THE APPLICATION OF SCIENCE TO FOOD

Do we normally question our food preparation process? Any enquiry lingering in our mind, for instance, why do we need to use this method, why this ingredients, and what to avoid in preparing food? To help in answering these questions, a better understanding in physic and chemical changes can solve the query.

Notably, food science is the discipline which consists of biology, physical science and engineering used to study the nature of foods, causes of their deterioration and principles underlying food processing (Yek & Struwe, 2008). Other application of food science is for the usage of food technology.

Since tools and technologies from food industry has been used in restaurant kitchen, varieties of techniques, equipment and dish has been developed. For instance, through the technique of spherification (a round pearls of gel with liquid center are construct by dropping a mixture of base with sodium alginate into calcium lactate bath) or reverse spherification (a round pearls of gel with liquid center are construct by dropping a mixture of base with calcium lactate into sodium alginate bath). With a contact with the calcium ions, the process of gelation that creates a very thin gel membrane is formed around a liquid.

Next, the usage of maltodextrin as ingredient in transforming oil to powder. "Powder oil" having a high affinity for water, pleasantly coats the mouth while releasing the potent flavor of the oil (Yek & Struwe, 2008). This powder is a staple ingredient in food manufacturing especially for dispersing dry ingredients; stabilize high fat ingredients and source of carbohydrate in energy drinks.

In line with the above, the innovation of thickener agent like xanthan gum (E415) creates a significant thickening and additionally it did not alter the flavor of the food. This assist culinarians from using traditional agent 'flour' to as it normally will dilute the flavor.

Another technique in molecular movement is the gelatinization process. Gelatine and pectin has been used widely in western culinary realms since decades ago. Gelatine produce pleasant gels like aspic, but the problem is that it starts to melt at 35 Celsius (Cassi, 2011). Thus, after Ferran Adria stumbled upon a common ingredient from Far East known as agar (melt at 85 Celcius) in western molecular. It has changed the value and perception of the westerners. Especially when Ferran Adria used agar in one of his famous creation called 'warm gelatine', the usage of agar in that particular recipe is to stabilize his warm beef gelatine dish.

A further study of gelatinization has created another gelling agent known as carrageenans, gellan gum (can be heated to 130 Celsius) and methylcellulose (thermoreversible). The way how methylcellulose works is when the mixture of food and the agent exposed with the heat (above 55 Celsius), it starts to transform into gel and as it cool, the food will be liquefied.

Similarly, the usage of liquid nitrogen is considered as a new technique. As mentioned by Ivanovic, Mikinac and Perman (2011) the temperature of the liquid is −196 Celsius and has been utilized for various industry purposes. This technique can reduce the crystallization of the product, due to the fastening process of freezing.

An executive chef of 'Salvador Dali of the kitchen', Ferran Adria was one the pioneer chef that popularized the movement by starting the greatest culinary revolution of the century. He uses the siphon (an equipment designed to make whipped cream by using carbon dioxide 'CO2' and nitro oxide 'NO2') to produce mousses and foam with unusual ingredients, such as vegetables, fruits, fish and meat (Adria, Soler, & Adria, 2010; Cassi, 2011) plus with his creation of fake caviar and spaghetti. Another responsible person is Chef Heston Blumenthal of Fat Duck Restaurant in United Kingdom (extensively used sous vide cooking in preparation of the food), Pierre Gagniare in France and Grant Achatz in United States.

3 MOLECULAR GASTRONOMY MOVEMENT

Previous literatures shows that molecular gastronomy is a culinary trend created by chefs and scientists in cooperation and it were defined as the scientific study of deliciousness (McGee, 2004; Mielby & Frost, 2010). This knowledge creates a new scientific approach in investigating culinary transformations, specifically the chemistry and physics behind preparation of food (This, 2006). As mentioned by Barham et al. (2010), molecular gastronomy encompasses a wide variety of topics, which includes understanding the effect of cooking method on the attributes of food, the investigation of new cooking method to improve and create new experiences in food, and the cerebral and sensorial interpretation of foods. It is shown that the practitioner of the molecular gastronomy movement seeks not only to delight the patron's palates but also evoke emotion and stimulate all other sense, making their food nothing short of performance.

It is worth noting that molecular cuisine is normally associated with high-end customer and luxury foods. Factors such as descriptive food names, the eating location and situation and social factor (up-scale restaurant) have been found to have an effect on the consumer perception (Carlsen, 2006; Gustafsson, 2004; Pierson, Reeve, & Creed, 1995) in Mielby and Frost (2010).

In addition, with the exceptional taste experience offered by the chefs in modernist cuisine, it could evoke a sensational moment. As mentioned by Svejenova, Mazza and Planellas (2007) and This (2002) in Mielby and Frost (2010) this new modern cuisine is highly scientific which often resulting in new combination of foods and application. Those dishes constructed often intend to challenge the eater by playing with the culinary mind set and being surprised. The early dish is scallop with liquorice milk, done by French two-star chef Pierre Gagniare.

4 PRODUCT DERIVATIVES

Basically, when a plate of food arrived on the table, our mind start to work by transmitting the data to the brain and interpret in systematic way. But, in surprising element the taste of the food is not exactly as they see. For instance, in Mielby and Frost (2010) experimental of creampuff, whereby the food they saw what seemed to be a classical creampuff and when eating the creampuff was meant to positively surprise with the small complex change in the taste and the deep pink color of the foam. Next, take a look at the dish of Alinea Restaurant 'Compressed watermelon mock sushi' and from Fat Duck Restaurant dish in 'orange and beetroot jelly', whereas the taste of those two jellies is contra from the color they represent. The beetroot color taste orange, while the orange jelly tastes beetroot (Blumenthal, 2009).

Another example of experimental cuisine is 'Pagens Classic', the results shows that customer when they were surprised when presented with the dish, since are unrecognized the food at all (Mielby & Frost, 2010). In addition, another one of the famous dish created by chef Heston called 'bacon and egg ice cream'. Uses of liquid nitrogen, which transforms the scramble egg into a scoop of ice cream (Blumenthal, 2009).

Next, another surprise dish is by replicating the ingredients of the food. For instance, signature dish by chef Thierry Marx called 'risotto de soja', the usage of bean sprout to imitate the size and appearance of rice and 'Golden egg', was made of crisp blown sugar and filled with savory egg filling and coated in a veneer of real gold leaf (Hill, 2009).

Then, the approach used by Thierry Marx Restaurant in combining the knowledge and transforms the classic cuisine 'The Chateau Cordeillan Bages' into a modern dish. The most influential person in searching true perfection of the food, recombination of classic into modern cuisine are Heston Blumenthal and Rene Redzepi. Supported by Petruzzeli and Savino, (2012) the rediscovery of old component is an innovation rule that has characterized the success of Redzepi and Noma.

With the high level of creativity and further exploration of molecular gastronomy by the chef at Moto Restaurant, has lead him to present unique ways in his menu by creating an edible menu and cutleries for the customer that could create astonishing experience.

On the other hand, to produce a new food, it is not necessary to use new tools and devices. It can be produced by using any equipment. However it rely on the level of knowledge and creativity of the chef. For example, the usage of molten glucose (glucose powder that is molten in the pan/pot) to replace the frying method without the use of fats and started the maillard reaction. Other example includes the usage of room temperature and alcohol in denature and coagulation of egg proteins, besides using mixture of salt, sugar and dry bean puree in denatures and coagulation of egg protein by an Italian chef Carlo Cracco.

5 CONCLUSION

Since 2002, restaurants like El Celler de Can Roca, Noma, Fat duck, The French Laundry, and Elbulli which applied molecular knowledge in cooking have been listed and won in The World's 50 Best Restaurant. Food science or molecular gastronomy does not only focus on texture and technology, it also comprises of important elements in providing good nutrition and health. Therefore, with the current molecular movement spreading in the western world, it is hoped that the idea will be well taken into the Malaysian restaurants that served authentic dishes. However, the underlying reason of the highlighting issue is still under investigation.

REFERENCES

Adria, F., Soler, J., Adria, A. (2010). A Day at elBulli. New York: Phaidon.

Barham, P., Skibsted, L.H., Bredie, W.L.P., Frost, M.B., Moller, P., Risbo, J., Snitkjaer, P., Mortensen, M.L. (2010). Molecular gastronomy: A new emerging scientific discipline. *Chemical Review.* 110(4), 2313–2365.

Blumenthal, H. (2009) The Fat Duck Cookbook. New York: Bloomsbury.

Carlsen, H.B. (2006). Er madikke bare noget der skal spises. In Bisgaard & Friberg (Eds.), Det aestetiskes aktualitet, 210–223. Copenhgen: Multivers.

Cardello, A.V. (1996). The role of the human senses in food acceptance. In (Eds) Meiselman, H.L. and Macfie, H.J.H., Food choice acceptance and consumption. 1–82. London: Blackie Academic & Professional.

Cassi, D. (2011). Science and cooking: the era of molecular gastronomy. *European Molecular Biology Organization Report.* 12(3).

Gaya, M.S., Soler, C., Manes, J., Soriano, J.M. and Sebastia, N. (2012). Nutritional assessment in tasting menu from molecular gastronomy restaurant. *International Food Research Journal.* 19(2), 787–789.

Gustafsson, I.B. (2004). Culinary arts and meal science—A new scientific research discipline. *Food Service Technology.* 4, 9–20.

Hill, B. (2009). Molecular Gastronomy: Research and Experience. Melbourne: International Specialized Skills Institute.

Institute Food Technology (2007). What is food science and technology? Retrieved on 18th April 2008 from http://www.ift.org.

Ivanovic, S., Mikinac, K., and Perman, L. (2011). Molecular Gastronomy in Function of Scientific Implementation in Practices. *Journal of Economics.* 2(2), 139–150.

Kahkonen, P. (2000). Consumer acceptance of reduced-fat-foods-the effects of product information. Unpublished Manuscript, University of Helsinki.

McGee, H. (2004). On Food and Cooking: The Science and Lore of the Kitchen, 2nd edition. New York: Scribner.

Mielby, L.H. and Frost, M.B. (2010). Expectation and surprise in molecular gastronomic meal. *Food Quality and Preference.* 21, 213–224.

Petruzzelli, A.M. and Savino, T. (2012). Search, Recombination, and Innovation: Lessons from Haute Cuisine. Long Range Planning.

Pierson, B.J., Reeve, W.G., and Creed, P.G. (1995).'The quality experience' in the food service industry. *Food Quality and Preference.* 6, 209–213.

Santich, B. (2007). The study of gastronomy: A catalyst cultural understanding. *The International Journal of the Humanities.* 5(6).

Schutz, H.G. (1999). Consumer data-sense and nonsense. *Food Quality Preference.* 10, 245–251.

Svejenova, S., Mazza, C., and Planellas, M. (2007). Cooking up change in haute cuisine. *Journal of Organizational Behavior.* 38, 539–561.

This, H., and Kurti, N. (1994). Physics and chemistry in the kitchen. *Scientific American.* 270, 44–50.

This, H., (2002) Molecular gastronomy. *Angewedte Chemie International Edition.* 41, 83–88.

This, H. (2005a). Modeling dishes and exploring culinary 'precision': the two issues of molecular gastronomy. *British Journal of Nutrition.* 93, 139–146.

This, H. (2005b) Molecular gastronomy. *Nature Materials.* 4, 5–7.

This, H. (2006). Food for Tomorrow? *European Molecular Biology Organization Report.* 7(11).

This, H. (2009). Molecular Gastronomy, a scentific look at cooking. *Accounts of Chemical Research.* 42(5), 575–583.

World's 50 best restaurant. (2013). Retrieved on 20 May 2013, from http://www.theworlds50best.com/2013list.html, 2013.

Yek, G.D. and Struwe, K. (2008). Deconstructing molecular gastronomy. *Food Technology.* 35–43.

Yilmaz, E. and Aydeniz, B. (2012). Sensory evalution and consumer perception of some commercial green table olives. *British Food Journal.* 114(8), 1084–1094.

Hospitality and Tourism – Sumarjan et al. (eds)
 ISBN 978-1-138-00151-0

The adaptations of Malay food knowledge among Malay generations in Kuala Lumpur, Malaysia

M.S. Mohd Sharif, M.S. Mohd Zahari & N. Mohd Nor
Universiti Teknologi MARA, Shah Alam, Malaysia

H. Mohd Salleh
Ministry of Sport and Youth, Malaysia

ABSTRACT: This study attempts to examine the adaptation of Malay traditional food knowledge within generations. Traditional food knowledge ideally refers to knowledge sharing of food tradition, which is being carried through the cooking skills, cooking techniques and recipes from generations. Traditional food knowledge refers as the unification of cultural identity, particularly for minority cultural groups. Food in all cultures is an expression of belief and linked to the culture pattern, where it precisely allows food to be a good indicator of identity, and it serves to define whom we are. Furthermore, foods are seen as a midpoint of cultural identity, which belongs to specific society or ethnic, and some elements of food preparation and consumption are the visible badges of identity. Therefore, it is crucial for every ethnic to preserve their food practices in order to maintain or retain the ethnic cultural identity. Qualitative method was adopted in the study by using pen-portrait and thematic analysis to comprehend the phenomena. Six sets of married women of two generations in a family were interviewed, and Malay village settlement area of Kuala Lumpur, Malaysia was chosen as the contextual setting. The respondents admitted that they adapted most of the knowledge of Malay traditional foods passed down by their mother in terms of preparation, method of cooking, equipment and cooking skills. The results also demonstrated that the emergence of modern equipments and utensils had directly altered the preparation of Malay traditional food. Working commitment and busy lifestyles are also identified contributing to limit the adaptation of Malay traditional food preparation among the young generation.

Keywords: adaptation, Malay food, Malay generation, food knowledge

1 INTRODUCTION

Traditional food knowledge ideally refers to knowledge sharing of food tradition which is being carried through the cooking skills, cooking techniques and recipes from generations. Traditional food knowledge refers as the unification of cultural identity, particularly for minority cultural groups. Food in all cultures is an expression of belief and linked to the culture pattern, where it precisely allows food to be a good indicator of identity, and it serves to define whom we are. Furthermore, foods are seen as a midpoint of cultural identity, which belongs to specific society or ethnic and some elements of food preparation and consumption are the visible badges of identity. Traditional food through it preparation and consumption are one of the best ways to understand a culture and ethnic identity (Verbeke & Lopez, 2005; Vu, 2009) and all traditional food-related activities, concepts and beliefs shared by a particular group of people gives a meaning to their identity, (Jaspal & Cinnirella, 2012; Niyogi, 2011) and this identity has been carried from one generation to the other generations through the transmission processes of food knowledge (Starr, 2010).

Food knowledge is a main tool for the individual to practice food production especially when it referred to traditional foods. Food knowledge can be obtained through several ways, but the most common is learned from the mother and family relatives when it involved family traditional foods. Some others may learn from community members when it involves a modern or the latest food trends. The individual of any age can learn about ethnic traditions in food practices especially during social or family gathering where they can exchange or gain food knowledge base on food served at the particular events (Kwik, 2008). Women are recognised as a transmission agent in delivering food knowledge to younger generations as they take charge of the family's food preparation (Chenhall, 2011). Conversely, due to women's involvement in the job-market last two decades have negatively influenced the transmission process of food knowledge to the young generation (Webb, 2011). According to UNESCO (2007) and the European Union (2007) there is a growing loss of food cultural heritage, owing to the lack of experienced senior generations to pass down the food skills and knowledge to the younger generation. For instance, a new generation of immigrants from Indonesia, India, China and Vietnam who settled in Canada faced an inability to produce the traditional food for their ethnic celebrations (Kemmer, 2000; Kwik, 2008; Nandy, 2004) and it also happened to Pakistani and Sri Lankan immigrant in Oslo, Norway (Wandela, Råberg, Kumar, & Holmboe-Ottesen, 2007).

This development had alerted scholars and non-government organization to uncover the main reason behind the scene and find the solution to retain or activate the transmission of food knowledge to the younger generation (Md. Sharif et al., 2012). Therefore, it is crucial for every ethnic to preserve their food knowledge practice in order to maintain the ethnic cultural identity. This study attempts to examine the adaptation of Malay traditional food knowledge within generations. Transmitting traditional food knowledge may contribute to personal skills that relate to the continuity of the ethnic and family food traditions that will further enhance the community capacity.

2 METHOD

This study specifically focuses on women with two generations in one family (mother age range from 50–70 years old and daughter age range from 30 to 40 years). Women are chosen as a focal gender in the study due to women is constantly alleged with food preparation in domestic settings (Chenhall, 2011). Six sets of respondents were involved in the study which located in "Kampong" areas near Kuala Lumpur owing of the high Malay population in these areas. For this study, the target population is specifically as mentioned above spotlight on the two generations between mothers and daughters (Malay ethnic). A set of open ended interview questions developed to guide the informants to share their knowledge and experience related to the adaptation process of Malay food knowledge within generations. The interview questions covered a wide range of information including demographic characteristics of respondents, Socio-economic status and general knowledge of the Malay traditional food, the transmission process including food preparation, methods of cooking and skills to generation practices. The questions were developed and adapted based on Kwik (2008) that conducted Traditional Food Knowledge (TFK) study towards Indonesian immigrants in Canada.

For data analysis, all interviews information was transcribed into a text using pen portrait analysis. This transcribing process formed a transcription of each interview. As a nature of a conversation, there are different intonation of voices which obviously indicate different types of expression and emotions. Therefore, the italic, question marks and exclamations are used to bring out interviewees voices, stressed or emphasized words. This is particularly important for further analysis as it represents the main ideas or keywords required by the study. In a simple way, pen portrait is like a storybook which tells the story from the very beginning to its climax (Merriam, 2009).

3 FINDINGS

3.1 *Respondents profiles*

Six sets of respondents (a set of mother and daughter) were interviewed in the study. Most of the interview process was done separately due to daughters were staying separated from their mother. Both respondents live within the Kuala Lumpur locality. Respondents under 'mother' category were aged between 54 to 70 years old and 'daughter' category was aged between 27 to 39 years old. It demonstrates that the average age gaps between mother and daughter is between 22 to 39 years. The respondent profiles are presented in Table 1.

Table 1. Respondents' profiles.

Respondent	Age	Education background	Occupation
Respondent 1			
Mother	59	Pri. school	Housewife
Daughter	37	College	Banker
Respondent 2			
Mother	62	Pri. school	Housewife
Daughter	39	Sec. school	Freelance
Respondent 3			
Mother	54	Sec. school	Retiree
Daughter	29	College	Ast. admin.
Respondent 4			
Mother	64	Pri. school	Housewife
Daughter	25	College	Supervisor
Respondent 5			
Mother	64	Sec. school	Retiree
Daughter	35	Bac. degree	Officer
Respondent 6			
Mother	70	Pri. school	Housewife
Daughter	37	Bac. degree	Teacher

Majority of the mothers' education backgrounds were up to secondary school while the 'daughter' was up to university level. This result essentially reflects the individual career undertaking where the profiles indicate that most 'mother' are a housewife whereas the 'daughter' is in the employment market.

3.2 *The adaptation process of Malay traditional food knowledge*

There are two different scenarios obtained from the interviewee. Most mothers confidently admitted they adapted most of the knowledge of Malay traditional foods passed down by their late mother in terms of preparation, method of cooking, equipment and cooking skills. In this sense, the originality of the style of cooking, ingredients is preserved without much alteration. A majority of them still preferred the original taste of rendang, masaklemak, curry and many others although some commercialization like pre mixed paste come to apparent in this present day. Evidently, the majority of them were not embarrassed with Malay traditional food commercial products.

Daughters, on the other hand, only adapted whatever convenient to them. Working routine is the reasons which limit them of adapting the traditional food knowledge passed down by their mothers which is similar addressed by Md. Nor et al., (2012). With this reason, the commercialization of Malay traditional products is the alternatives for them. In relation to this, *rendang* for instance is cooked using the commercial paste. In addition, some of the Malay traditional foods are sometimes ordered or buying from the restaurants or caterers. In sum, the level of adaptation on the Malay traditional food among the younger generation is slightly lessened compared to the older generation.

	Answer
Respondent 3	
Mother	*"Ohhhh ... till to date, I still follow my mother's cooking style. There is not many changes after all. Even though I am working."*
Daughter	*"Emmmmm .... yes, but not 100% because currently the ingredients are not easily to get especially typical herbs or spices. Aaaaa .... for instance, if there are no curry leaves, I don't see any problem for not adding in curry dish. So, I assume that it will not affect my dish after all and its ok!"*
Respondent 4	
Mother	*"I had observed my mother when she is busy preparing the dishes for wedding ceremony. Later, I followed her steps which I was called to prepare rendang .... Aaahhh, it was a bad experience, but it did not turn me down. Now, I have those experiences, and I'm confident in handling and prepare traditional main dishes for large quantities."*
Daughter	*"Yes, but not exactly perfect as my mother did. My family likes to eat Daging SalaiMasakLemakciliapi as you know .... It takes time for me to prepare them! Sometimes, I did not use charcoal, but I substitute to electric oven to smoke the beef."*
Respondent 5	
Mother	*"I try to prepare them as authentic as possible and always serve them to my children. Then, I notice that my daughter will try to replicate the dishes that I usually cooked, this will easily make them remember. When they are grown up and had children, they cook simple traditional dishes which required less preparation procedure."*
Daughter	*"She (mother) taught ... example ... masaklemak... she said "If you cook masaklemak.... You have to stir consistently under slow fire; otherwise the coconut milk will split (pecahlemak)" I will never forget her tips. Basically, every dish she would tell me why and for what."*

3.3 *Difficulties in adapting the Malay traditional food*

The majority of mothers involved in the interviewed ascertain that they did not quite facing any obstacles or difficulties in adapting the Malay traditional food either in preparation, equipment's, methods of cooking and skills although most of them were working. This is because the transfer of Malay traditional food knowledge normally occurred in their younger days. In other words, not much alteration or advancement took place on the styles of cooking, ingredients, equipment's with limited convenience ingredients available in the market. Most of them claimed that they preferred and used to the traditional ways although experiencing the process of modernization and advancement. In addition, mothers are attached with the methods or styles pass down by their mothers especially on the method of preparation. Despite that, they acknowledged the modern equipment helps them quicken the preparation. Unlike their mothers, who received some traditional Malay cooking knowledge from their mothers, most of the younger generations or daughters in the context of this study claimed of facing some hitches in adapting or following the traditional ways of Malay cooking.

	Answer
Respondent 1	
Mother	*"The most challenging part for the wood stove (dapurkayu) is to get some woods. While, charcoal stove … is easier to handle as compared to the wood stove. The elderly people used clay pots (belangga) for cooking… Especially sautéing (menumis)….. OOooooo …. the taste was really excellent!"*
Daughter	*"No. Because I love cooking, I am willing to learn masakasampedas. Then I started to like learning traditional dish like laukasampedas. To be more precise, I like to eat my grandmother's asampedas, masakpindang, masakkicapdaging.*
Respondent 3	
Mother	*"In traditional foods…. When people eat … they feel satisfied…. Like masakasam… but I prefer to use the old methods like using mortar to make chili past…. Fuhhh! The taste is so nice and if you use blender then the taste will slightly change or less …… Now blender is widely used and people have already forgotten the olden cooking methods. Perhaps our next generation will not be able to experience the authentic taste of our food culture."*
Daughter	*"Ha! Nenek's (grandmother) dishes were very detailed. She used fresh ingredients and various method of cooking. The challenge is… Whatever raw materials that she used, we must follow it exactly!. How to say …… ya…. If you cook it must use the same ingredients, and she insists that the dish should be similar taste as her dishes."*
Respondent 6	
Mother	*"Raw materials … nowadays we can get them easily ready made like peeled onion, shallots, garlic and others. Just buy from the market and blend it. Besides, chili paste and coconut milk (santan) are sold in a ready—pack. Hmm… (smile) if they are still lazy, I don't know what to say. Herbs and mix spices (rempahratus) are all ready and convenience… depend on them, either they want to cook or not."*
Daughter	*"Hahahaha… (laugh) Yes, Hahahaha….. OK, we had 'batugiling' (grinding stone) at home, I did it properly. My mother likes my chili paste and I did it nicely. Hehehe (giggle). So, since then I was always being called for grinding the chili and make it into a paste. The grind stone was pretty heavy. For your information, we have a blender but my mother still prefer 'failing'."*

Most of the daughters approved that modernization or advancement had not only altered some of the Malay traditional ingredients owing to the availability of commercial products, but also involved the food preparations and eating styles which therefore, altered the adaptation. Modernization through commercialization was claimed to have to change the ingredients and create pre-cooked product availability. Modernization also influences the equipment used in daily Malay traditional food preparation. The emergence of modern ovens, mixers, stoves, grinders, choppers and others for instance directly altered the preparation of Malay traditional food. Busy lifestyle also limits the adaptation of Malay traditional food preparation among the young generation.

4 DISCUSSION AND CONCLUSION

The result has revealed that the process of adapting the knowledge of Malay food did occur between mother and daughter. However the trend of knowledge transfer among Malay ethnic communities poses a declining pattern. Women's involvement especially among daughter's in the job market has limited their time spend on houseworks including cooking compared to women in the sixties. The majority of the daughter's highlighted that their main constraint in practicing Malay foods was the process is complicated and time consumes. Furthermore, western influenced and modern lifestyles are also identified to limit the adaptation of Malay food preparation among the young generation. As the result, fast foods, ready foods or convenience foods have become their priority to provide family meals. This scenario might slowly jeopardize the Malay food knowledge and practices, if there is no effort being taken to protect the food knowledge and practices of Malay foods.

In conclusion, an awareness and concern by ethnic community is necessary to ensure continuity of the ethnic culture and identity. Furthermore, the Malay cooking process has now been simplified with the emergence of convenience products, modern equipment and utensils and it should be seen as an advantage and motivations in adapting and continuing the family food tradition.

This study which focuses on the adaptation behaviour of the Malay ethnic food knowledge, indirectly have expanded and enhanced the gastronomic literature. Realizing that this particular area has yet thoroughly explored in Malaysian context. Therefore, research pertaining ethnic gastronomic adaptation should be further addressed.

ACKNOWLEDGMENT

This research was funded by the Ministry of Higher Education, Malaysia through Universiti Teknologi MARA under FRGS grant: 600-RMI/SSP/FRGS 5/3/Fsp (20/2011) and RAGS grant: 600-RMI/RAGS 5/3 (179/2012).

REFERENCES

Chenhall, C. (2011). Improving Cooking and Food Preparation Skills: A Synthesis of the Evidence to Inform Program and Policy Development (P.-C.P.H. Network, Trans.). Vancouver Canada: Public Health Agency of Canada.

European Union. (2007). *European Research on Traditional Foods*. Belgium: Publications.europa.eu.

Jaspal, R., & Cinnirella, M. (2012). The construction of ethnic identity: Insights from identity process theory. *Ethnicities*. doi: 10.1177/1468796811432689.

Kemmer, D. (2000). Tradition and Change in Domestic Roles and Food Preparation. *Sociology, 34*(2), 323–333. doi: 10.1177/s0038038500000201.

Kwik, J.C. (2008). Traditional food knowledge: A case study of an Immigrant Canadian "foodscape". *Environments, 36*(1), 59–74.

Md. Nor, N., Md. Sharif, M.S., Mohd. Zahari, M.S., Mohd. Salleh, H., Ishak, N., & Muhammad, R. (2012). The Transmission Modes of Malay Traditional Food Knowledge within Generations. *Procedia—Social and Behavioral Sciences, 50*(0), 79–88. doi: 10.1016/j.sbspro.2012.08.017.

Md. Sharif, M.S., Mohd. Zahari, M.S., Noor, N.M., Salleh, H.M., Ishak, N., & Muhammad, R. (2012). Traditional Food Knowledge (TFK) of Malay festive foods. In Z. et al (Ed.), *Current Issues in Hospitality and Tourism Research and Innovations* (pp. 365). London: Taylor & Francis Group.

Merriam, S.B. (2009). *Qualitative Research: A Guide to Design and Implementation*. San Francisco: Jossey-Bass.

Nandy, A. (2004). The Changing Popular Culture of Indian Food: Preliminary Notes. *South Asia Research, 24*(1), 9–19. doi: 10.1177/0262728004042760.

Niyogi, S. (2011). Bengali-American Fiction in Immigrant Identity Work. *Cultural Sociology, 5*(2), 243–262. doi: 10.1177/1749975510378208.

Starr, A. (2010). Local Food: A Social Movement? *Cultural Studies ↔ Critical Methodologies, 10*(6), 479–490. doi: 10.1177/1532708610372769.

UNESCO. (2007). Intangible Heritage/2003 Convention, 2011, from http://www.unesco.org/culture/ich/index.php?lg=en&pg=00022#art2.

Verbeke, W., & Lopez, G. (2005). A Longitudinal Application of the Theory of Reasoned Action to Women's Career Behavior. *British Food Journal,* 107(11), 832–840.

Vu, V. (2009). *The Changing Foodways of Vietnamese Americans in Orange County, California*. Ph.D, California State University, California.

Wandela, M., Råberg, M., Kumar, B., & Holmboe-Ottesen, G. (2007). Changes in food habits after migration among South Asians settled in Oslo: The effect of demographic, socio-economic and integration factors. [doi: DOI: 10.1016/j.appet.2007.09.003]. *Appetite, 50*(2–3), 376–385.

Webb, P. (2011). Family Values, Social Capital and Contradictions of American Modernity. *Theory, Culture & Society, 28*(4), 96–123. doi: 10.1177/0263276411404906.

Hospitality and Tourism – Sumarjan et al. (eds)

Food beliefs and practices among Malay women in postpartum period

M.S.Y. Kamaruddin, M.S.M. Zahari, R. Muhammad, A.F. Amir & A.A. Azdel
Universiti Teknologi MARA, Malaysia

ABSTRACT: Postpartum or confinement period generally perceived by most countries in the globe as an illness or disease thus requires special attention and treatment for women who experience it. However, due to technological and social advancement, the belief and practices are changing which therefore lead to different way of old generation and young generation mother perceive this confinement period. Using qualitative approach employing semi structured interview, young mother has more logical reasoning before accepting the food and practices in postpartum period as compared to old generation mother. The cultural belief and knowledge that is passed from generation to generation is now strengthened and supported by scientific proof as young generation mother is explained scientifically the relevancy of their ancestor in postpartum food practices and belief. Thus, the word "traditional practices" evolve with time and technology.

Keywords: food, foodways, belief, practices, Malay, postpartum, women

1 INTRODUCTION

In some part of the world, postpartum period is considered a disease for women because women are vulnerable to certain illness and also need to avoid almost any heavy tasks (Esther, 1993). Though it is considered illness, Stefanello, Nakano and Gomes (2008) noted postpartum phase is the time when women are given less attention about on their health especially in western countries.

In Malay community, the practices were strongly influenced by Hinduism when the Hindus brought into Malaysia in the early centuries until about 1400's (Ali & Chapman, 2007). They posited this traditional food and foodways knowledge practiced heretofore is the intersection between Hindu belief and Islamic religion. It is then altered due to the effect of modernization.

The advancement of medical facilities significantly promises good health care and treatment among mothers where many urban women nowadays prefer medical attention under the care of specialist in modern facilities (Ali & Chapman, 2007). Based on previous scholar where rural community hold the culture and custom strongly, urban area on the other hand seem to be more fragile in believing this food belief (Ali & Chapman, 2007). Additionally, they believed anything that has been passed from generation to generation about the food after childbirth nevertheless due to the evolution of social status whereby women seek for status equality (Counihan, 1999), this has driven women to seek a higher education ergo to question the traditional belief. Though there are several studies confinement period food of Malay, the paucity of research in determining the belief gap of this respective food and foodways still exist. Therefore, it is worth to comprehend and understand the old and young generation mother belief on food and foodways during the confinement period. By understanding both beliefs explain how cultural identity can change or redeveloped looking from the lens of postpartum food and foodways.

2 LITERATURE REVIEW

2.1 *Malay food and foodways during postpartum period*

The discourse pertaining to modernization has looked into many perspectives. There is the plethora of development of both constructive and disruptive argument on this area which many scholars have the different opinion. However, the corpus of modernization in relation to postpartum period specifically in sociological aspect is still lacking. In understanding the nature of this present study, this chapter will provide the body of literature combining argument of previous scholars from a different lens and area.

In Malay, postpartum period is called waktuberpantangselepasbersalin. Stefanello et al. (2008) assumed that the different forms of postpartum care are closely linked to cultural patterns that are meaningful in specific social settings. In addition, postpartum custom is rather a cultural setting with symbolic and practical functional meaning and is conducted by ritual of guidelines of quarantine and dietary (Chmielowska & Shih, 2007). Balwi (n.d) highlighted food is one of the crucial factors in determining the health including an ill mother during their post-partum care period. This is the period whereby women carefully need to select and eat food thus, commonly called "abstinence, diet or quarantine" (Stefanello et al., 2008). The careful choice of food is based on culture and belief as it is believed to affect the health of mother and also baby during weaning activity. Postpartum care has been constructed essentially within family (Stefanello et al., 2008) whereby the selection of food and food practices within this period in the olden days is based on the generation belief which being pass from mother to daughter.

Because in this confinement period when women are considered physically and symbolically ill (Stefanello et al., 2008), they adopt an array of food practices. The Malay also has their own set of rules of dietary during the postpartum period which was strongly influenced by Hinduism when the Hindus brought into Malaysia in the early centuries until about 1400 (Ali & Chapman, 2007). It is then accepted by the Malay as local culture until the Islam belief arrived (Ali & Chapman, 2007). Failing to follow the food practices would negatively influence the health of newborn (Barennes et al., 2007). Chmielowska & Shih (2007) listed it is considered taboo for women to do household duties in the kitchen.

This dietary is believed to be associated with medical methods that contained within Malay medical books known as kitab-kitabtib which the most popular called kitab Mujarrabat Melayu (Balwi, n.d, 2003). These medical books explained any physical and spiritual—or supernatural—disease that includes incantation and magic. In understanding how food affect the health of human, these books explained four elements of human which is soil, water, fire and wind. The balance of these four elements will have an effect on human health which if one of the elements dominates any of the three, human will go to an unhealthy state (Balwi, n.d). Many scholars agreed that women are vulnerable after delivering a baby and fall under the state of unhealthy (Ali & Chapman, 2007; Esther, 1993; Cheung, 1997; Matthey et al., 2002; Stefanello et al., 2008; Wong, 1994), and Balwi (2003) expounded food is the most significant factors to affect the balance of the human elements. Moreover, foodways (preparation up to consumption) believed to affect the human's health (Ali & Chapman, 2007).

In Malay food context, medicinal food derived from both flora and fauna. The derivation of food from flora obtained from roots, flowers and leaves either cooked or raw, called jamu. Food component derived from animal includes exotic ingredients such as hempedugajah, hempedusawa, lemaklandak, tulangsotongetc (Balwi, n.d.) to treat inner and outer disease. From both derivation of foods, it those can be categorized as cooling, windy, hot, sharp, poison or itchy in food confinement period (Shahar, Earland & Rahman, 2000). Traditionally, this conducted with the shamanistic methods. In addition, the effectiveness of food in treating the unhealthy condition (in this case, postpartum mother) must be followed with certain pantanglarang.

Rashiqah (2009) explained that Malay community still believe and follow pantanglarang even though it comes to the third millennium and ready to accept changes. Rahimi et al. (2003) reported that there is no significant difference in the belief of pantanglarang between women in the urban and rural area.

3 METHODOLOGY

The qualitative approach well suited in this context of study as suggested by Esterberg (2002) whereby belief and practice are subjective to an individual which therefore, this approach is appropriate to understand and scrutinize social process in the context. This also helps to increase the depth of understanding on how Malay women belief and practice their food and foodways during the forty crucial days of confinement period. As this study is looking on the belief and practice of food and foodways during the postpartum period, using a qualitative approach in this study justified. Young generation mother and old generation mother chosen as sample because to measure the actual impact of human behavior; precursor, antecedent or those who had experienced are able to give meaningful information than those who do not have the experience. Semi structured interviewing method employed for data collection process, and the interview process took approximately thirty to forty minutes.

4 FINDINGS

4.1 *Young generation food belief and practices*

In looking at the experience of practicing diet in the postpartum period, both Informant 1 and Informant 2 give almost an identical answer, whereby they have no idea what it is like during the first born baby. Informant 1 shared her experience:

> *"... I have two children now and the most challenging would definitely during my first baby was born. I have no idea after I delivered my first baby. I don't know any specific rules in doing things and even what to eat."*

Informant 2 recalled her memory showing how the act of ignorance impacts the psychology by not knowing the dietary practice during postpartum period:

"I am so lucky to have a supportive people around me which helped me undergone one month of postpartum period. I was totally blurred on what to eat and what cannot be eaten. I guess my ignorance towards postpartum period impact my psychology at that point of time. Serve me right [laughing]."

Based on the answer given by the informants below, it can be said that the place where a mother give a child a birth has a significant relationship on how they obtain their information regarding to food in-take but not food practices. Also, relatives who have experience especially mother and mother in law sharing their personal experience. Informant 2 did not explain directly about how hospital plays a role in constructing her first food intake during postpartum, but it is understood in a same notion as previous informant:

"During my time in hospital, I just ate whatever that is served by nurses and food that were brought by my mother and mother-in-law since I have no idea about food that is beneficial for my health."

Informant 1 demystified it:

"I have to admit that I obtained the information relating to postpartum guidelines firstly from nurses who took care of me. I do remember they gave me rice porridge, plain water and fish which I don't know what kind of fish was that in a one week I was at hospital."

"My mother and my mother-in-law also came to visit me and brought food that they said is nutritional which beneficial for my health and my baby."

While place and relatives can be said to have importance in constructing food intake among mothers, young generation mothers nowadays have a various range of reading material that enable them to strengthen their knowledge about food intake and practices. Informant 1 continued her explanation by elaborating it:

"But then, relying on them [mother and mother in law] would not be sufficed which I also search for books, articles and friends who had undergone the postpartum period. Yet, not all foods that were good for them (their mother) are not necessarily good for me."

Informant 2 shared her effort in obtaining information through her experience:

"During my time at home, I started to read any materials related to foods during postpartum period .... I started to read on health magazine, read on internet about food during postpartum period. It took me quite some time to absorb anything that is crucial for me to undergo the process."

Many previous research has outlined the food consumed during the postpartum period and looking on the nutritional value but not really into how the mother belief on the effect of the food. Informant 1 articulated what she 'needs' to eat as oppose to what she preferred:

"I mostly ate 'ikanharuan' (type of fish) and 'gamat' (sea cucumber) because it is believed to accelerate healing process. I don't simply eat anything because I was so scared if anything happened when I consume food that is not suitable for my baby [pause] and me .... I am a person who likes fried food and it is truly true for me when my mother said it is a tiresome month when I can only be allowed to eat food that are steamed, poached, boiled and soup. Everything was water based cooking method. No oil was allowed."

Based on the explanation given by Informant 2, she addressed that food during the confinement period does not have a pleasant taste which based on her personal taste bud and told some benefit of cooking with spices and herbs according to her mother's lesson. She exemplified it:

"Most of my foods during that time were rice and vegetables such as white radish, carrot and mustard green which I can tell you, the taste was not that good especially mustard green which it has a bitter aftertaste. I also eat ikanharuan, red meat and chicken. All foods that I have listed were mostly cooked with ginger and black pepper. My mother said that ginger is good to prevent you from bloating (masukangin or kembung) and black pepper is to regulate maintain your body temperature. At the end of every meal, I consumed orange and apples."

Informant 1 also explained how the food should be consumed based on temperature and hot and cold characteristic of the food:

"... my food consumption must be hot which immediately consumed after cooked. Do not even try to consume food that considered as 'cold' such as cucumber, spinach as it is believed to drop your body temperature."

4.2 *Old generation food belief and practices*

Informant 3 at one point of the interview session expressed her ambivalence feeling about the food intake as she compared to doctor's advices with relatives experience:

"Doctor also advices me to consume various type of food because each food offers specific nutritional value however doctor's advices mostly contradict with advices provided by my mom."

As an old generation mother who has vast experience about postpartum period because she had undergone normal childbirth and caesarean surgery, Informant 3 demystified it in a very systematic way about foods that are allowed and prohibited; have to and preferred:

"... I break the food into their type. For vegetables, I mostly ate pucukpegaga and pucukmaman (type of vegetables eaten raw), green mustard but only the leaves, not the stem, cauliflower, cabbage and mushroom. The vegetable that must be included in almost of my food is bitter gourd. Like it's name, it taste bitter and believe act as a cure. For meat, it needs to be cooked until dry with black pepper, ginger, garlic and shallot. Chicken is prohibited especially for caesarean operation because it can cause itchiness at the surgical part. For the fish, I can only eat fish that could not result in itchiness such as tenggiri, parang, bawal, ikanmerah and haruan which is also believed to accelerating healing process. As for fruit, I can eat tangerine and orange provided that it is ripe and sweet. I don't like sour food."

The reason of pineapple cannot be eaten during postpartum period clearly explained except for banana which cannot be explained by Informant 3 on the relevancy of restriction but has to follow the guideline.

"Never to eat pineapples because I was told that it has 'sharp' characteristic that will result in body aching. I don't know about that because I don't even dare to eat it."

"My mom really stressed out that never to eat pineapple and banana. ... I can't eat banana but pineapple will surely cause my womb to ache."

Further, Informant 3 added iced water would have repercussion for new mothers if they intended to consume it based on her mother's experience:

"... never drink iced water because it will decelerate the healing process. The other thing that my mom told me that the consequences that you will bear is not in the present but in the future in the age of forty and above whereby your body can easily become weak."

Informant 3 shared and added the ideology and belief on body condition as she grew older and having more children. Moreover, the reason of following strict diet is for her children safety:

"From the first child until the forth, I follow the rules accordingly. I follow the rules as strict as possible on my forth born child. Most of the veins around stomach will never recover; my mom said I was in a very vulnerable stage as compared to the first three children. Therefore, the tendency of body aching is considered high. Again, I strictly follow the rules for all my children."

5 CONCLUSIONS

The notable findings depicted that the belief of women in the postpartum period have slight changes. Those changes in Malay custom can be seen if it observed from the differences of conversation pattern between old and young generation mothers. They believed the number of the baby they have indicated the vulnerability of pain they will encounter. However, it is too early to hypothesize such statement but what can be postulated here is the postpartum diet that has set by predecessors is to reduce the postpartum pain endured by mothers, not to restore the condition before pregnancy. While literature stating most mothers still depend on their predecessors' rules and experience in determining food during the postpartum period, the finding of this study only partially agree with the statement. Old generation mothers are mostly following the traditional dietary guidelines of postpartum, but it is surprising there is also old generation mothers who have ambivalent feeling about the traditional practice which they compared with knowledge of doctors. It is irrefutable they rely on mother's experience, but they cannot deny the expertise of doctors advice to be credited in determining their health. What makes them follow the postpartum diet is for health but at the same time for beauty aspect.

ACKNOWLEDGEMENT

This paper is sponsored by Reseach Intensive Fund (RIF) Grant of Universiti Teknologi MARA Malaysia.

REFERENCES

Abdullah, R. & Beri, S. (1993). *Komuniti Orang Asli di Terengganu*. PercetakanYayasan Islam Terengganu.

Ali, A. & Chapman, P.H. (2007). Maternity services and the role of the traditional birth attendant, *bidan kampong*, in rural Malaysia. *Journal Public Health Management Practice, 13*(3), 278–286.

Balwi, M.K.M. (n.d.). Sains Perubatan Naturalistik-Melayu: Satu Kajian Awal. *Jurnal Kemanusiaan*. 34–41.
Balwi, M.K.M. (2003b). Ketamadunan Melayu Dan Sains: Satu Analisis AwalKe Atas Pencapaian Masyarakat Melayu Dalam Bidang Sains. *Jurnal Teknologi*, *39*(E), 47–61.
Barennes, H., Simmala, C., Odermatt, P., Thaybouavone, T., Vallee, J., Martinez-Ussel, B., Newton, P. & Strobel, M. (2007). Postpartum traditions and nutrition practices among urban Lao women and their infants in Vientiane, Lao PDR. *European Journal of Clinical Nutrition*, 1–9.
Belasco, W. (1999). Why food matters. *Culture & Agriculture*. *21*(1), 27–34.
Cheung, N.F. (1997). Chinese *zuoyuezi* (sitting in for the first month of the postnatal period) in Scotland. *Midwifery, 13*(2), 55–65.
Chmielowska, E. & Shih, F.S. (2007). Folk customs in modern society: 'tradition of *zuoyuezi*' in Taiwan: A physical anthropology perspective. Presented in EATS IV conference Stockholm 2007.
Counihan, C. (1999). *The anthropology of food and body: gender, meaning and power*. New York: Routledge.
Esterberg, K.G. (2002). Qualitative methods in social research. United States: McGraw Hill.
Esther, K. (1993). Recovering after childbirth in the Mixtec highlands (Mexico). *Medicaments et Aliments: L'approcheEthnopharmacologique*, 99–111.
Manderson, L. & Matthew, M. (1981). Vietnamese behavioral and dietary precaution during pregnancy. *Ecol Food Nutrition*, 11, 1–8.
Matthey, S., Panasetis, P. & Barnett, B. (2002). Adherence to cultural practices following childbirth in migrant Chinese women. *Health Care Women International*, *23*(6), 567–575.
O'Connor, A. (2008). *Maya foodways: A reflection of gender and ideology.* A thesis submitted in candicacy for special honors in the department of anthropology. Austin: University of Texas.
Rahimi, H., Fatimah, A.M., Rahimah, I., Sarah, Y. & Marlia, M.S. (2003). Adakahpantanglarangpemakanan di kalangan Orang Aslimempengaruhitahapkesihatanibudananak: pengalamandaerah Kuala Lipis. *Malaysian Journal of Public Health Medicine*, *3*(1), 73–77.
Rashiqah, I.A.R. (July 23, 2009). Pantanglarang Melayu: sesetengahnyaadabuktisaintifiktetapitidakdisedarimasyarakat. Berita Harian. 21.
Rearick, A.N. (2009). *"Food is something that we gather around": Foodways practices among Arab Americans in Columbus, Ohio*. The Ohio State University.
Shahar, S., Earland, J. & Rahman, S.A. (2000). Food intakes and habits of rural elderly Malays. *Asia Pacific Journal Clinical Nutrition*. *9*(2), 122–129.
Stefanello, J., Nakano, A.M.S., & Gomes, F.A. (2008). Beliefs and taboos related to the care after delivery: Their meaning for a women group. *Acta Paul Enferm*, 21(2), 275–281.
Wong, L.L. (1994). *Apart from the sesame oil chicken—the Chinese women 'doing the month'*. Taipei: Dou-Shin Ltd.

Hospitality and Tourism – Sumarjan et al. (eds)
© 2014 Taylor & Francis Group, London, ISBN 978-1-138-00151-0

Food heritage and nation food identity formation

A.M. Ramli, M.S. Mohd Zahari, N. Ishak & M.S. Mohd Sharif
Universiti Teknologi MARA, Malaysia

ABSTRACT: Identity formation is of one the crucial issues raised among multiracial and multicultural nations at all aspects of life covering distinctive societal, cultural as well as food origin. Food has been recognized as markers for identity regardless geographical, social and political divided populations. Arguably, transferring the traditional food knowledge, acculturation, assimilation, adaptations are the attributes that lead, create or form the foundation of food identity in the multicultural countries. Certification and gazetting the heritage and traditional food not only mechanisms to preserve but in the long run would strengthen and significantly manifest toward the national food cultural identity. This research note discusses the determinants that influence certification of food heritage in contributing towards the nation food formation which is still under investigation.

Keywords: determinant food heritage, food heritage, identity formation

1 INTRODUCTION

Identity formation is among the critical issues in multiracial and multicultural nations and many countries are still struggling in developing their identity. As identity formation would create a positive image of a nation, issues on how, where, when and why the identity be formed and measured including food in the pluralistic nations start to receive significant attention among the academic scholars (Lim, 2012; Vanhonacker, Verbeke & Guerrero, 2010). Arguably, Traditional Food Knowledge (TFK), acculturation, assimilation, adaptation and gazetting of traditional food heritage longitudinally are believed to create or form what could be called food identity in the multicultural countries (Laroche, Kim, Tomiuk, & Bélisle, 2005). In this sense, Traditional Food Knowledge (TFK) refers to a cultural tradition of sharing food, recipes and cooking skills and passing down that collective wisdom through generations (Md Nor, Sharif, Zahari, Salleh, Isha, & Muhammad, 2012) while acculturation is a process of learning, practice and customs of a culture, acquiring the capability to function within the dominant culture while retaining one's original culture (Ishak, Zahari, Sharif, & Muhammad, 2012).

The assimilation process creates new food identity as its hallmarks stamp out the originality of minor ethnic food by the dominant ones or by alteration in the consumption of the minor ethnic food in a new cultural environment without phase-out the original identity. Scholars deduced assimilation have effects at different level of generations and some researcher argued it process started with acculturation in the behaviour assimilation (Mukherji, 2005). Adaptation occurred when cultures can coexist within a given environment, retaining some (if not all) of their original/minority-cultural heritage while functioning within and acquiring the traits corresponding to the larger host/mainstream culture in which scholar debated that food adaptation would be one of the important process for formation of food identity (Cleveland, Laroche, Pons, & Kastoun, 2009).

DeSoucey (2010) claimed that certification and gazetting of gastronomic heritage significantly manifest the national cultural identity. Grasseni (2011) state that production, resources and cultural background factors give an image for Italian Cheese product within the country boundary and in the European countries. Therefore, gazetting of gastronomic heritage should not be ignored as one of the integral mechanism in the process of identity formation and recognition (Rodriguez, Schwartz, & Krauss Whitbourne, 2010). This research note conceptually discusses the relationship between certification and gazetting of food heritage and food identity formation. It has centrally focused on the determinants that influence certification of food heritage in contributing towards the formation of national identity.

2 LITERATURE REVIEW

2.1 *Food heritage*

When it comes to describing food heritage, McCoy (2012) argues that it is a rather ambiguous and amorphous term as it is dependent upon the place, time and individual. The researcher conveys the relationship of food heritage with agriculture, by rooting a community to its place and to its history and heritage, and of defining the cultural identity of a place, community, person, or region. Lyons (2012) reported heritage food consists of pure breed fruits, vegetables and livestock that naturally suited to the regional climate. It also fit perfectly in the intersection between culinary and heritage tourism which hold particular value in their ability to sustain and educate visitors on the local histories and creating a true sense of place. Chabrol and Muchnik (2011) looking food heritage from the traditional production of food like products production techniques and consumption skills and technology of local produce such as cheese, wine, rice, raw ham and chocolate. Food heritage which is associated with classical and traditional foods are continuously practicing by all generations without major altering of the original flavours (Wahid, Mohamed, & Sirat, 2009). Food heritage can be reflected from the environment history, belief, ideology and food technology of society in an era or period of time.

2.2 *Malaysia and food heritage*

Wahid et al. (2009) associate food heritage with classical and traditional foods which are continuously practicing by all generations without major altering of the original flavours. Food heritage can be reflected from the environment history, belief, ideology and food technology of society in an era or period of time. The Heritage Commissioner of National Heritage Department, Prof. Datuk Zuraina Majid states that food heritages are based on two categories. The first category refers synonymous or common foods which are part of our lives whereas the second consists of foods that are almost extinct in other words it were once part of our culture but are slowly dying out.

The first category refers to those food that represent the nation, places or ethnic groups (Mohammad & Chan, 2011) or food that are unique or signature dishes such *nasi dagang*, *masak lemak*, *assam laksa* and many others (Abby Lu, 2010). For the second category, food like *pulut kukus periuk kera* (glutinous rice cooked in monkey pot plants) and *ikan panggang tanah liat* (grilled fish wrapped in clay) are some examples that are slowly fading away owing to peculiarity in the method of cooking, recipes and technique of preparing those dishes (Wo, 2009) and some did not exist anymore especially the indigenous cuisines (Abdullatib, 2009).

This cultural heritage is difficult to preserve and measures than a physical object as it associated with values, beliefs, behaviours and rules of the society. The need the continuity and preserving food heritage is being considered as conditions of comparative advantage in maintaining a local food culture in the face of homogenizing pressures from the outside (Shariff, Mokhtar, & Zakaria, 2008) and continuation of preserving creating valued products especially the traditional cuisines. To date, there are more 100 types Malaysian favourite foods announced and registered as National food heritage. The food heritage has been divided into seven categories consists of rice, noodles, gravies and accompaniment dishes, appetizers, cake, porridge, dessert and drinks (Heritage, 2012). Each of the categories of the food selected based on preferential status with the purpose or preserving or not to be forgotten (Elis, 2009).

2.3 *Determinants of food heritage*

As described in the literatures food heritage determinants are associated with the historical elements, food characteristics, value of uniqueness, practice and integration element. In historical elements, Guerrero et al. (2009) stated dimension of traditional foods consist of elements of habit, natural, origin and locality. Habit and natural associated traditional food with something anchored in the past to the present, transmitted from one generation to another or food that has been consumed from the past or existed for a long time that has always been part of the consumers' life. Hjalager and Corigliano (2000) identified historic resources as focal points of food festivals and special events that attract tourists and local residents. They further asserted that the culture of the food and eating festivals can promote local culinary traditions, lifestyles, and gastronomic heritage (Guerrero et al., 2009). Lin, Pearson, and Cai (2011) revealed that the origin of food is the most important information in aiding international tourists to recognize the authenticity of a nation.

Staple, flavouring and preparation are the food characteristics closely related to the heritage. Belasco (2008); Rozin (2006) stated that staple foods or basic food that have unique value and very meaningful to communities ranging from meat and potatoes, stew and fufu (porridge) and many others could be classified as heritage. Flavouring, which has a distinctive way of seasoning dishes, unique flavour and combinations serve as important group "markers" closely associated with heritage. For instance, culinary identity in parts of China may expressed through the combination of soy sauce, garlic, and sesame oil, while a mix of garlic, tomato, and olive oil may signal "southern Italian. Specific cuisines favour or distinct manner of preparing food such as such as stir frying in

China; stewing for Mexico other food characteristics that linked with heritage. Guerrero et al. (2009) highlighted that taste as one of factors in recognizing the authenticity of a food product or cuisine. Karim et al. (2011) posited that staple food, cooking method and taste should be preserved and sustain as it representing identity of the community or ethnic and considered as country's food heritage and it the way representation of country food identity.

Guerrero et al. (2009) associate value of uniqueness as innovation with food that are new or unusual ingredients; new combinations of products; different processing systems or elaboration procedure, including packaging' coming from different origin or cultures' being presented and/or supplied in new ways; and always having temporary validity which has significant role in heritage.

In the practices and integration elements, cross-cultural process through the acculturation, assimilation, adaptation plays an important role which also closely associated with heritage (Kwik, 2008). The cross cultural process consists of food knowledge which referred to the cultural tradition of sharing food, recipes and cooking skills and techniques and passing down that collective wisdom through generations (Cleveland et al., 2009).

2.4 *Food identity formation*

Food holds the key to any culture and people make many assumptions about other people's diets based on who they claim to be and people also make assumptions about who people are, based on what they eat (Counihan & Van Esterik, 2012). Mintz (1996) noted that consumption is always conditioned by the meaning and the food we eat, and the manner in which we preserve, prepare, and serve it, all contain meaning at some level.

It is important for a pluralistic nation to have an identity and food besides other is another significant badge of identity (Schierup, 1988). Ruark (1999) claimed that food is one of the most common markers of identity. Ruark for instance state that food is more powerful than the sex drive, but people do not take notice of it. This is because we're fed three times a day.

Worth reiterating that scholars generally agreed that cross-culturing among the ethnic groups through acculturation, assimilation and adaptation processes will gradually create or form food identity in the multicultural countries (Fox & Ward, 2008). In addition, certification of the heritage and traditional food in the long run could strengthen the nation food identity formation (Manaf, 2008). DeSoucey (2010) on the hand claimed that gazetting of gastronomic heritage significantly manifest the national food cultural identity.

3 ISSUES THAT NEED FOR CERTIFICATION OF FOOD HERITAGE AND FOOD IDENTITY FORMATION

Parallel with other nations, Malaysia is also experiencing the urgency of having its own food identity when sharing food cultural background is becoming central issues between the neighboring countries. Countries like Singapore and Indonesia which are sharing a common historical roots and cultural heritage with Malaysia are disputing over some of the traditional food when each country pursues to validate those traditional foods as their identity. Owing to this issue and according to Chong (2012) each country is becoming more aggressive to defend and protect such heritage as theirs to safeguard the country as heritage is the core of a people's and country identity. Some clear examples can be seen on *Rendang, Nasi Lemak, Laksa* and many others (Jb, 2012). In promoting gastronomic tourism for the international tourists Singapore for instance insistently proclaimed some of the common Chinese, Indian, Malay and Peranakan food as their local heritage (Chaney & Ryan, 2012).

Not to be overstated that issues pertaining to food ownership or even worse on the Indonesia counterpart. A long debated history with the neighboring countries not only occurs in the territorial boundaries but also in cultural ownership like traditional dances, music instruments, arts and culinary treasures (Hussain, 2011). This issue to that extent sparks a nationalistic outrage protest, street demonstration and staged rallies which branding Malaysia as "a nation of thieves" by the irresponsible group. These resurgent disputes have given significant impact on the relationship between two countries which led to a discussion among the highest authorities from both countries (Chong, 2009).

Those incidences have open Malaysian eyes particularly the government on the importance to at least having our own cultures like oral tradition, languages, festive events, rites and beliefs, music and songs, the performing arts, traditional medicine, literature, traditional sports and games as well as traditional cuisine identity (Lim, 2012) although sharing the fundamental basis of it cannot be avoidable. Former Minister of Tourism, Dato Seri Ng Yen Yen for instance stressed the importance of having strong nation's culture and food identity that leads to the contribution of destination branding and image for our country.

"The extensive ranges of our local dishes were very much a reflection of the cultural melting pot creating a unique Malaysian identity. The wide variety of culture and food in Malaysia can be an effective

tool to demonstrate our nation's cultures and food identity to attract international tourists (Bernama, 2010b)".

Talking about food identity, scholars recognized the foundation that take place in the processes of constructing it within ethnic groups in the multicultural countries in particular is reflecting all the way through a sharing common accepted cuisine (Spurrier, 2010; Wilk, 2006). Cross-culturing of food among the ethnic groups through acculturation and assimilation processes gradually creates or form food identity in the multicultural countries (Helland, 2008). Worth reiterating that acculturation is denotes as a process of learning, practice and customs of a culture, acquiring the capability to function within the dominant culture while retaining one's original culture (Matsunaga, Hecht, Elek, & Ndiaye, 2010). Comparatively, although assimilation creates new food identity as it hallmarks to stamp out the originality of minor ethnic raises by the dominant ones therefore it arises several sensitive issues on preserving each ethnic cultural identity (Crocetti, Rubini, & Meeus, 2008). It is interesting to note that the adaptations of food from various ethnic groups by the dominant or majority ethnic group in multi-racial and cultural nations will also lead to common acceptable cuisines and longitudinally forming what could be called national food identity (Fox & Ward, 2008). Besides this, Manaf (2008) noted that gazetting the heritage and traditional food in the long run strengthening the nation food identity formation. Wo (2009) on the same note stressed that gazetting is not only mechanisms to preserve heritage food but importance element toward country identity formation therefore government should stimulate public awareness about it.

In line with what have been mentioned, Malaysian government presently is taking serious action in preserving it gastronomic treasures through various initiatives. Besides many initiation processes such the promoting the transferring of Traditional Food Knowledge (TFK), encouraging acculturation, assimilation and adaptation of culture within the ethnic group (Ishak et al., 2012; Md Nor et al., 2012), gazetting or certification heritage food by Department of National Heritage become one of the significant national agenda. In addition, almost all states also taking the same steps in preserving their heritage food (Bernama, 2010a). Despites these, some question arises. What are main purpose and the criteria used in selecting national heritage food? What are the main attributes used in determining heritage food? Is the criterion such history elements, food characteristics, value of uniqueness, practice and integration are considered? Which agencies responsible for the endorsement and most importantly what are the impacts of the endorsement or certification of those foods? These questions in fact need to be revealed and still under investigation.

REFERENCES

Abby Lu. (2010). Tracking down fine Malay food, thestar. Online: http://thestar.com.my/metro.

Abdullatib, S. (2009). 100 Makanan Tradisi, Kosmo Online. Online: http://kosmo.com.my/kosmo.

Belasco, W. (2008). Food: The key concepts. Oxford: Berg Publishers.

Bernama. (2010a). Memastikan Kesinambungan Warisan, Bernama. Online:http://www.kpkk.gov.my.

Bernama. (2010b). Yen Yen bent on tempting tourists with food. Online:http://updated.internalinsider.

Chabrol, D., & Muchnik, J. (2011). Consumer skills contribute to maintaining and diffusing heritage food products. Online: http://aof.revues.org.

Chaney, S., & Ryan, C. (2012). Analyzing the evolution of Singapore's World Gourmet Summit: An example of gastronomic tourism. International Journal of Hospitality Management, 31, 309–318.

Choi Tuck Wo. (2009, 23 April 2009). Dish branding, thestar. Online:http://thestar.com.my/lifestyle.

Chong, J.W. (2009). The Indonesia-Malaysia Dispute Over Shared Cultural Icons And Heritage. ILSP Law Journal, 177–186.

Chong, J.W. (2012). Mine, Yours or Ours? The Indonesia-Malaysia disputes over shared cultural heritage. SOJOURN: Journal of Social Issues in Southeast Asia, 27(1), 1–53.

Cleveland, M., Laroche, M., Pons, F., & Kastoun, R. (2009). Acculturation and consumption: Textures of cultural adaptation. International Journal of Intercultural Relations, 33(3), 196–212.

Counihan, C., & Van Esterik, P. (2012). Food and culture: A reader: Routledge.

Crocetti, E., Rubini, M., & Meeus, W. (2008). Capturing the dynamics of identity formation in various ethnic groups: Development and validation of a three-dimensional model. *Journal of Adolescence,* 31(2), 207–222.

DeSoucey, M. (2010). Gastronationalism. American Sociological Review, 75(3), 432–455.

Elis, S. (2009). Our rich 'food' heritage, Bernama. Online: http: //blis2.bernama.com.

Fox, N., & Ward, K.J. (2008). You are what you eat? Vegetarianism, health and identity. *Social Science & Medicine,* 66(12), 2585–2595.

Grasseni, C. (2011). Re-inventing food: Alpine cheese in the age of global heritage. Anthropology of food, 8. Online: http://aof.revues.org/6819.

Guerrero, L., Guardia, M.D., Xicola, J., Verbeke, W., Vanhonacker, F., Zakowska-Biemans, S., Hersleth, M. (2009). Consumer-driven defination of traditional food products and innovation in traditional foods. A qualitative cross-cultural study. *Appetite,* 52, 345–354.

Helland, S.H. (2008). Chinese Malaysian flavours: an anthropological study of food and identity formation in Penang. Master. University of Oslo.

Heritage. (2012). Senarai Warisan Kebangsaan-Warisan Makanan 2012 dan 2009 Jabatan Warisan Negara.

Hjalager, A.-M., & Corigliano, M.A. (2000). Food for tourists-determinants of an image. *International Journal of Tourism Research,* 2(4), 281–293.

Hussain, Z. (2011). Jakarta Cooks Up 'Rendang Diplomacy', The Jakarta Globe. Online: http://www.thejakartaglobe.com.

Ishak, N., Zahari, M.S.M., Sharif, M.S.M., & Muhammad, R. (2012). Acculturation, Foodways and Malaysian Food Identity. Paper presented at the Current Issues in Hospitality and Tourism and Innovations, Kuala Lumpur.

Jb, S.B. (2012). No need to patent 'rendang' dish: William Wongso, The Jakarta Post. Online: http://www.thejakartapost.com.

Kwik, J.C. (2008). Traditional Food Knowledge: Renewing Culture and Restoring Health. (Master), University of Waterloo.

Laroche, M., Kim, C., Tomiuk, M.A., & Bélisle, D. (2005). Similarities in Italian and Greek Multidimensional Ethnic Identity: Some Implications for Food Consumption. *Canadian Journal of Administrative Sciences,* 22(2), 143.

Lim, Y. (2012). KL central cultural makeover, The Star. Online:http://thestar.com.my.

Lin, Y.-C., Pearson, T.E., & Cai, L.A. (2011). Food as a form of destination identity: A tourism destination brand perspective. *Tourism and Hospitality Research,* 11(1), 30–48.

Lyons, L.T. (2012). The Food Heritage of Virginia An Untapped Asset of Community & Economic Development. University of Virginia. Online:http://www.virginia.edu.

Manaf, Z.A. (2008). Establishing the national digital cultural heritage repository in Malaysia. Library Review, 57(7), 537–548.

Matsunaga, M., Hecht, M.L., Elek, E., & Ndiaye, K. (2010). Ethnic Identity Development and Acculturation: A Longitudinal Analysis of Mexican-Heritage Youth in the Southwest United States. *Journal of Cross-Cultural Psychology*, 41(3), 410–427.

McCoy, L. (2012). Food Heritage Planning Proposals: Planning For Food Heritage Celebrations In Central Virginia. Online: http://www.virginia.edu.

Md Nor, N., Sharif, M.S.M., Zahari, M.S.M., Salleh, H.M., Isha, N., & Muhammad, R. (2012). The Transmission Modes of Malay Traditional Food Knowledge within Generations. Procedia—Social and Behavioral Sciences, 50(0), 79–88.

Mintz, S. (1996). Tasting Food. Tasting Freedom (Place, 1997), 13.

Mohammad, T., & Chan, J.K.L. (2011). Authenticity Representation of Malay Kelantan Ethnic Cuisine. Paper presented at the The 2nd International Research Symposium in Service Management, Yogyakarta, Indonesia.

Mukherji, J. (2005). Is Cultural Assimilation Related to Environmental Attitudes and Behaviors? *Advances in Consumer Research,* 32, 415.

Rodriguez, L., Schwartz, S.J., & Krauss Whitbourne, S. (2010). American Identity Revisited: The Relation Between National, Ethnic, and Personal Identity in a Multiethnic Sample of Emerging Adults. *Journal of Adolescent Research,* 25(2), 324–349.

Rozin, P. (2006). The integration of biological, social, cultural and psychological influences on food choice. In R. Shepherd & M. Raats (Eds.), The psychology of food choice (pp. 19–39).

Ruark, J.K. (1999). A Place at the Table. Chronicle of Higher Education, 9.

Schierup, C.-U. (1988). Integration?: Indvandrere, kultur og samfund: Billesø & Baltzer.

Shariff, N.M., Mokhtar, K., & Zakaria, Z. (2008). Issues in the Preservation of Traditional Cuisines: A Case Study in Northern Malaysia. International Journal of the Humanities, 6(6), 101–106.

Spurrier, C.T. (2010). Cassava, Coconut and Curry: Food and National Identity in Post-Colonial Fiji. (Master), University of South Carolina.

Vanhonacker, F., Verbeke, W., Guerrero, L., Claret, A., Contel, M., Scalvedi, L.,. Hersleth, M. (2010). How European consumers define the concept of traditional food: Evidence from a survey in six countries. Agribusiness, 26(4), 453–476.

Wahid, N.A., Mohamed, B., & Sirat, M. (2009). Heritage food tourism: bahulu attracts? Paper presented at the Proceedings of 2nd National Symposium on Tourism Research, Universiti Sains Malaysia, Penang, Malaysia 18 July 2009.

Wilk, R. (2006). Home cooking in the global village: Caribbean food from buccaneers to ecotourists. Oxford: Berg Publishers.

Wo, C.T. (2009, 23 April 2009). Dish branding, the star Online: http://thestar.com.my.

Hospitality and Tourism – Sumarjan et al. (eds)

The sport tourism perceived impacts: A sub-urban host community perspective

S.A. Syed Omor Junid, N. Aminudin & N.A. Mustapha
Universiti Teknologi MARA, Malaysia

ABSTRACT: The scale of sports tourism events are getting big and bigger in sub-urban area whereby the events are encroaching deeper into the lives of the host community. Previous researches on impacts of sports tourism events were carried out on urban communities, rather than sub-urban communities. Therefore, this conceptual paper seeks to investigate whether similar impacts are applicable to host community of sub-urban area. A quantitative method of self-administered questionnaire will be used in this study. The questionnaire will be divided into two sections; a demographic profile of the respondents and the sub-urban host community perceptions towards sporting event. A descriptive statistics and External Factor Analysis (EFA) will be used to analyze the data in this study. The context of this study is the sub-urban host community of Shah Alam in Selangor, with a plan of 400 questionnaires is distributed to the sub-urban host community who lives within 3 kilometres from Shah Alam Stadium. Findings from this study will provide sports event organizer an understanding on the local community perceptions towards sports event that is being held in their residential area and to support the development of sport events in Shah Alam.

Keywords: host community, impacts, sport tourism events

1 INTRODUCTION

Sport tourism is acknowledged as the world most favored leisure experiences (Ritchie & Adair, 2004). Sport tourism can be defined as a temporary movement from a person's settlement to pass their spare time by playing and watching games (Preuss et al., 2007). Therefore, individual who travelled with the purpose of watching or participating in sport activities and/or sport events is considered as performing sport tourism. According to Kurtzman and Zauhar, (2003) sport tourism events have the ability to attract tourists of which large percentages are spectators and it also attract non-residents whose primary purpose is to travel for participating or viewing sport events.

Internationally, events have been acknowledged as a strong component of sport tourism that attracts people from various locations (Gammon & Robinson, 2003; Getz, 1997; Penot, 2003; Sofield, 2003; Zauhar, 2004). A number of events that has a possibility to generate great number of visitors and expanses and these events have the potential to contribute to tourism development (Henderson et al., 2010). In Africa, events such as the 1995 Rugby World Cup, 2003 Cricket World Cup and 2010 Federation International de Football Association (FIFA) have contributed to the nation's economy development (Ntloko & Swart, 2008). Wide range of benefits produced to destination that played a host to sport tourism events, apart from economic development, it also create job vacancies for the local people, create a sense of community pride (Swart, 2001) and increase the number of sport facilities (Chao et al., 2011). Sport tourism events are greatly treasured and regarded owing to the fact that "tourism is a trillion dollar industry sport is a multi-billion dollar industry and has become a dominant force in the lives of millions of people globally" (Kurtzman, 2005). Nevertheless, there are studies that regard sport tourism events as a caused of negative impacts to the host community. In a study on sport tourism event impact in Taiwan, the study demonstrated negative impacts to the host community's environment (Chao et al., 2011).

In recent years, Malaysia has provided a variety of sports event and activities for tourist to pursue on their interest. There are a number of well-established international sports event that attract tourist to Malaysia are Formula One Grand Prix, Golf Maybank Malaysia Open, South East Asian Games (SEA Games), Thomas Cup Badminton Finals and Commonwealth Game (Yusof et al., 2009). Meanwhile local sport events such as the Malaysia Super League, Suzuki AFF Cup, the

Malaysia Hockey League, SUKMA Games (Sukan Malaysia) and Kuala Lumpur Marathon were also able to attract large crowd of tourists. When large number of participations in sport event exists, the impact will not only revolve around local economics but also environment and social of the region and host community (Ross, 2010; Ryan, 2001; Lin, 1995). The host community here refers to the individuals or residents who are staying at the event location or at close proximity to the event location and they are those who are likely to understand the event and impacts better (Burker, Page & Meyer, 2002; Delamere, 2001). Thus, consideration on the host community perceptions towards sport tourism events can be regard as initiatives to understand the ways which local residents react to the hosting events and its impact towards their society. Therefore, the purpose of this study is to investigate the sub-urban host community perceived impacts towards sport events held in Shah Alam Stadium and whether similar impacts are applicable to be applied to the sub-urban host community. The host community of this study will be residents who lives within 3 kilometres range from Shah Alam Stadium; in the area of section 9, section 13, Taman TunDr Ismail Jaya (TTDIJ) and BatuTiga.

2 LITERATURE REVIEW

2.1 *Sport tourism*

Standevan and Knop (1999) is defined as all forms of active and passive involvement in sporting activity, participated in casually or in an organized way for non-commercial or business/commercial reasons that necessitate travel away from home and work locality. Preuss et al. (2007) defined sport tourism as a temporary movement from a person's settlement to pass their free time playing and watching games and so forth. Nevertheless, Tassiopoulosa and Haydamb (2007) classified sports tourism into two; first is for advantages for tourism growth and the other one is sport based activity. The author added, since participation in sport activities appears in diverse motives (such as leisure, competition, and travel with the purpose to watch premier competitions or attractions), the genuine assets of sport and tourism amenities is an important function that attracts tourist to the host city, region or country.

2.2 *Resident's and host community attitude and perception*

The host community is referred to the citizens or society who lives at the location of the events or those who lives in a close range to the event location and they are the ones who could provide a better understanding of the event and impacts (Delamere, 2001; Burker, Page & Meyer, 2002). In tourism-related activity, it is essential to have an understanding on resident's attitudes and perceptions as they are the ones who have a strong moral case to be heard and given attention to (Weaver & Lawton, 2013). Residents have the ability to influence their relationships with tourists and shareholders, permitting tourism to thrive by either support or damaging it through resistance or sabotage (Deery, Jago, & Fredline, 2012). Even so, different host community may perceive tourism impacts differently (Hritz & Ross, 2010). Krippendorf (1987) suggested that the opinions, viewpoint, and expectations about tourism can be unlikely similar depending on the broad population or professional group that is being studied.

Various factors have been found to influence residents' attitudes towards tourism. Some of the factors involve the distance of venue or the tourist attraction with resident's home (Murphy & Andresson, 1988), the category of tourism (Dogan, 1989), respondents demographic background (Sharma & Dyer, 2009; Chen, 2000; Fredline & Faulkner, 2000; Lankford, 1991) and the length residents had stay in the community (Allen et al., 1988). A positive point of view towards tourism was found to have an association with gender (female), income levels, employment, education levels and the neighbourhood (the location where people reside) (Allen et al., 1993; Inbakaran & Jackson, 2006).

In a study on resident clusters by Inbakaran and Jackson (2006), the study identified a negative outlook in tourism expansion to be related to residents life cycle stage (age) and the distance where the major tourism attraction is positioned. Canosa et al. (2001, p. 52) also found a relationship between negative attitude of host community and levels of tourism, whereby tourism is perceived as a mean for creating socio-cultural changes to the residents. Krippendorf (1987) suggested that the opinions, viewpoint, and expectations about tourism can be unlikely similar depending on the broad population or professional group that is being studied.

Four activity categories were selected (Connaughton, 2012). The activity categories are:

1. *Professional and College Sports Teams. This category includes all the regularly scheduled major, minor, and college teams, including the Carolina Panthers, Charlotte Bobcats, Charlotte Knights, Charlotte Checkers, UNC Charlotte Men's Basketball, and JCSU Men's Football and Basketball.*
2. *Special Events. This category includes all the major professional and college sports events that occur on an individual basis in the Charlotte*

region during the year. These events include both NASCAR and non-NASCAR races at Charlotte Motor Speedway, PGA Tournament (estimate), the ACC Championship, the Belk Bowl, the Gold Cup, and the CIAA Basketball Tournament.

iii. *Amateur and Youth Events. This category includes amateur tournaments held in the Charlotte region that involve attendance by primarily out-of-town visitors. Most events have no ticketing fee and have little or no on-site spending. Most of the economic impact of this type of sports activity occurs off-site as a result of food and beverage spending and over-night accommodations.*

iv. *Sports Video and Media Productions. This category includes firms that engage in the production of sports related programming or distribution. This category includes ESPNU, SPEED, NASCAR Media Group, and Raycom Sports.*

2.3 *Social exchange theory*

Social Exchange Theory suggests that when an individual interact or reciprocate with others, they expect to "gain" some benefits from that person Blau (1964) and Goulner (1960) or it generates obligations between the people (Emerson, 1976). Therefore, the human relationships are created from cost-benefit analysis which creates "mutual obligations, reciprocity or repayment over time" (Cropanzano & Mitchel, 2005; Gouldner, 1960). The social exchange theory has been widely used "predominately" in post studies in assessing tourism impacts upon destination (Andereck & Vogt, 2000; Andriotis, 2005; Chen, 2001; Gursoy, et al., 2002; Harill, 2004; Vogt & Jun, 2004) and the theory allows an investigation of positive and negative resident's attitude in a destination to be carry out (Ap, 1992). Therefore in tourism context, it assumes that the host and visitors will exchange resources parties or more, likely, between residents and business community (organizer of the sport events). Both parties will be aware with the positive and negative outcome of tourism and they can either support or not the development of tourism depending on their perceptions and understanding on costs and benefits (Audriotis & Vaughan, 2003; Sutton, 1967). Harrill (2004) suggests that tourism development can take place only when the benefits outweigh the cost, for instance the economic benefits outweigh the cost of sharing social and environmental resources with tourist. Nevertheless, there was also an increase demand for accountability that requires even organizers to add value to the community's life where the event is held (Chalip et al., 2003).

2.4 *Malaysian Super League*

The Malaysian Super League (Liga Super Malaysia) is a Malaysian professional league for association football state teams and clubs. It is at the top flight of the Malaysian football league system and it is managed by Malaysia Super League Sdn Bhd, under supervision of the Football Association of Malaysia (FAM). The league is contested between 12 teams and operates on a system of promotion and relegation with the Malaysian Premier League. The 12 clubs participating in this top flight league need to pass a set of requirements and verification process, primarily related to professionalism and infrastructure feasibility.

It was previously known as the Semi-Pro League from 1989 to 1993 and the M-League (Liga-M) from 1993 to 2004. It is currently sponsored by Astro and therefore officially known as the Astro Malaysia Super League. The competition formed in 2004 following the decision by the Football Association of Malaysia (FAM) to privatize the league. As a result, MSL Sendirian Berhad (or MSL Proprietary Limited) was created to oversee the marketing aspects of the league.

2.5 *Economic benefits*

Sport tourism improves the individuals' life quality through tourist attraction and results in the economic welfare of the local communities (Swart & Bob, 2007). Tourists are a chief source of income and employment for the population of the hosting region (Batyk & Ski, 2009); that is why the world metropolitans consider sport tourism events as a unique opportunity for local marketing, because a potential economic advantage of being the host of a big sport event is to attract many spectators and tourists to the hosting city and consequently, it shall bring positive economic consequences (Preuss et al., 2007). Major event such as the Olympic Games focused on the long-term benefits that include cities reconstruction, new building, growth of tourism sector, public welfare improvement, job creation and local employment opportunities (Homafar et al., 2011).

3 ISSUES OF SPORT TOURISM EVENT

In Selangor, the Shah Alam Stadium is one of the sports complexes that located in a residential area. The stadium was opened in 1994 and it has been hosting various local sport events, concerts and exhibition (Wikipedia, 2013). Some of the sport events that was held are World Youth Championship matches in 1997, Chelsea FC Tour 2011 (Chelsea FC vs Harimau Malaya), and the

Malaysian Super League 2012. From there on Shah Alam continues to grow, from town to city. Event after event is being held at the stadium in order to utilize it. Event organizers, sponsorships, spectators, sports player and media come and go, but the community remains to witness the success or failure of events from their perspectives. In the case of Shah Alam, what residents perceived of event's, especially sports events impact on them remains to be investigated.

The recent football event, the Malaysia Super League which was held in Darul Makmur Stadium in Kuantan portrayed a negative view to the sport event in Malaysia. A chaotic scene was seen when the spectator's went out of control after second half of the game. Spectator's started to enter the football field and the crowd was difficult to control which caused the match to be stopped and postponed to another day. This raised a question on how does the host community perceived sport tourism event and its impact to the community.

For that reason, this study seeks to examine the perceptions of sports event being held in a city that was initially a well known quiet residential area until a national stadium was built, particularly local football event held in Shah Alam Stadium. The impacts of sport tourism events on host community can either be in positive or negative forms. Delamere (2001) claimed that understanding the host community perceptions towards sport events is essential because it may reduce the unwanted disruption of local community life and promote a balance between social and economic development in the area. Their perceptions may also determine the continuity of an event, since their disapproval of an event may lead to the cancellation or termination of an event.

The information obtained from this study will provide authorities such as state government of Selangor, Selangor Tourism and sport event organizers information's concerning the host community's perception towards sport event held in their community and whether they support the event or not. An understanding on the host community perspectives will help to enhance a good relationship between local community and authorities; by making they feel that their opinions and perspectives are important towards development in their residential area. A good image of sport events will increase the number of sport events and encourage the host community to participate in event such as through entrepreneurship. Apart from that, this research will contribute to the new body of knowledge in sport and tourism sector in Selangor as there has been few research conducted in the area of sub-urban host community perceived impact towards sport tourism event.

REFERENCES

Allen, L.R., Hafer, H.R., Long, P.T., & Perdue, R.R. (1993). "Rural residents attitudes toward recreation and tourism development", *Journal of Travel Research*, 31(4), 27–33.

Allen, L.R., Long, P.T., Perdue, R.R., & Kieselbach, S. (1988). "The impact of tourism development on residents' perceptions of community life", *Journal of Travel Research,* 27(1), 16–21.

Andereck, K.L., & Vogt, C.A. (2000). The relationship between residents' attitudes toward tourism and tourism development options. *Journal of Travel Research*, 39, 27–36.

Andriotis, K. (2005). Community groups' perceptions of and preferences for tourism development: Evidence from Crete. *Journal of Hospitality and Tourism Research*, 29, 67–90.

Andriotis, K., & Vaughan, R.D. (2003). Urban residents' attitudes toward tourism development: The case of Crete. *Journal of Travel Research*, 42, 172–185.

Ap, J. (1992). Residents' perceptions on tourism impacts. *Annals of Tourism Research*, 19, 665–690.

Batyk, I.M. & Ski, S.S. (2009). Tourism—common Cause. *Tourism Management*. 1–3.

Blau, P.M. (1964). *Exchange and power in social life*. New York: John Wiley & Sons.

Burker, M., Page, S.J. & Meyer, D. (2002). Evaluating the impact of the 2000 America's Cup on Auckland, New Zealand. *Event Management*, 7, 79–72.

Canosa, A., Brown, G. and Bassan, H. (2001), "Examining social relations between adolescent residents and tourists in an Italian coastal resort", *Journal of Tourism Studies*, 12(1), 50–59.

Chalip, L., Green, C., & Hill, B. (2003). Effects of sport event medial on destination image and intention to visit. *Journal of Sport Management*, 17, 214–234.

Chao, C.L; Chang, Y.C.; Chun, N.C.; Liu, Y.H., & Ni, K.L. (2011). A study of sport tourism's environmental impact: a case study of kenting traveling destination in Taiwan.

Chen, J.S. (2001). Assessing and visualizing tourism impacts from urban residents' perspectives. *Journal of Hospitality and Tourism Research*, 25, 235–250.

Connaughton, J.E. (2012). The economic impact of sports and sports events on the Charlotte MSA Economy Final Report.

Cropanzano, R., & Mitchell, M.S. (2005). Social exchange theory: An interdisciplinary review. *Journal of Management*, 31, 874–900.

Deery, M, Jago, L, & Fredlne, L (2012). Rethinking social impacts of tourism research: A new research agenda. *Tourism Management,* 33, pp. 64–73.

Delamere, T.A. (2001). Development of a scale to measure residents attitudes toward the social impacts of community festivals, part II: verification of the scale. *Event Management*, 7, 25–38.

Dogan, H.Z. (1989). Forms of adjustment: socio-cultural impacts of tourism. *Annals of Tourism Research,* 16(2), 216–236.

Emerson, R.M. (1976). Social exchange theory. *Annual Review of Sociology*, 2, 335–362.

Fredline, E., & Faulkner, B. (2000). Host community reactions: A cluster analysis. *Annals of Tourism Research*, 27, 763–784.

Gammon, S. & Robinson, T. (2003). Sport and tourism: A conceptual framework. *Journal of Sport Tourism*, 8(1), 21–26.

Getz, D. (1997). *Event Management and Event Tourism*. New York, NY: Cognizant Communication Corporation.

Gouldner, A.W. (1960). The norm of reciprocity: A preliminary statement. *American Sociological Review*, 25, 161–178.

Gursoy, D., Jurowski, C., & Uysal, M. (2002). Resident attitudes: A structural modeling approach. *Annals of Tourism Research*, 29, 79–105.

Harrill, R. (2004). Residents' attitudes toward tourism development: A literature review with implications for tourism planning. *Journal of Planning Literature*, 18(3), 251–266.

Henderson, J., Foo, K., Lim, H., & Yip, S. (2010). Sports events and tourism: The Singapore Formula One Grand Prix. *International Journal of Event and Festival Management*, 1, 60–73.

Homafar, F.; Honari, H.; Heidary, A.; Heidary, T., & Emami, A. (2011). The Role of Sport Tourism in Employment, Income and Economic Development. *Journal of Hospitality Management and Tourism*, 2(3), 34–37.

Hritz, N., & Ross, C. (2010). The Perceived Impacts of Sport Tourism: An Urban Host Community Perspective. *Journal of Sport Management*, 119–138.

Hua, K.P., Abdullah, N.D.T., & Radzuwan, R. (2010). Malaysian's women sport tourism pursue: Leisure constraints perceived and negotiation strategies adopted.

Inbakaran, R. and Jackson, M. (2006), "Resident attitudes inside Victoria's tourism product regions: A cluster analysis", *Journal of Hospitality and Tourism Management*, 13(1), 59–74.

Karim, H.A. (2008). The satisfaction of residents on community facilities in Shah Alam, Malaysia. *Journal of Asian Social Science*, 4(11).

Krippendorf, J. (1987). The holiday makers. Understanding the impact of leisure and travel. Oxford, UK: Butterworth Heinemann.

Kurtzman, J., 2005. Sport tourism categories. *J Sport Tourism*, 10(1), 15–20.

Kurtzman, J. & Zauhar, J. (2003) A wave in time: the sport tourism phenomenon. *Journal of Sport Tourism*, 8(1), 3–47.

Lankford, S.V. (1991), "An analysis of resident preferences, attitudes and opinions toward tourism and rural regional development in the Columbia River Gorge", PhD thesis, University of Oregon, Eugene, OR.

Malaysian Super League (2013). Wikipedia. Retrieved June 20, 2013, from http://en.wikipedia.org/wiki/Malaysia_SuperLeague.

Murphy, P.E. and Andressen, B. (1988), "Tourism development on Vancouver Island: an assessment of the core-periphery model", *The Professional Geographer*, 40, 32–42.

Penot, J. (2003). Sport tourism and tourism generated by sporting events. *Journal of Sport Tourism*, 8(2), 100–101.

Preuss H., Benoit S., Norm O. & Reilly K. (2007). "Profiling managing sport event visitors: The 2002 commonwealth games". 12, 5–23.

Ritchie, B., & Adair, D. (2004). Sport tourism: An introduction and overview. In B. Ritchie & D. Adair (Eds.), *Sport Tourism. Interrelationships, Impacts and Issues*, 1–29. Clevedon: Channel View.

Sharma, B., & Dyer, P. (2009). Residents' involvement in tourism and their perceptions of tourism impacts. Benchmarking. *An International Journal* 16(3), 351–371.

Sofield, T.B.H. (2003). Sport tourism: from binary division to quadripartite construct. *Journal of Sport Tourism*, 8(3), 144–146.

Standeven, J. & Knop, P.D. (1999). Sport tourism. Canada: Human Kinetics.

Sutton, W. (1967). Travel and understanding: Notes on the social structure of touring. *Journal of Comparative Sociology*, 8, 217–223.

Swart, K. 2001. A curriculum design for sport tourism studies in South Africa. Unpublished Doctoral Thesis, Illinois State University, United States of America.

Swart, K. & Ntloko, N.J. (2008). Sport Tourism Event Impacts on the Host Community: A Case Stud of Red Bull Big Wave Africa. *South African Journal for Research in Sport, Physical Education and Recreation*, 2(30), 79–93.

Swart, K. & U. Bob 2007. Listening to Community Voices.

Athlone and Green Point Residents' Views on the Location of the 2010 FIFA World Cup Stadium in Cape Town. 7th International Hamburg Symposium 'Sport and Economics' on the Economics of Top Performance in International Sports. Hamburg Chamber of Commerce, Hamburg.

Tassiopoulosa D. & Haydamb, N. (2007). "Golf tourists in South Africa: A demand-side study of a niche market in sports Tourism". *Tourism Management*. 29(5), 870–882.

Weaver, D.B. & Lawton, L.J. (2013). Resident perceptions of a contentious tourism event. *Tourism Management*, 165–175.

Vogt, C.A., & Jun, S.H. (2004). Residents' attitudes toward tourist market segments and tourism development in Valdez, Alaska: A comparison of residents' perceptions of tourist impact on the economy and quality of life. Proceedings from the 2004 Northwest Recreation Research Symposium, Newtown Square, Pennsylvania, 404–411.

Yusof, A., Fauzee, M.S.O., Shah, P.M., & Geok, S.K. (2009). Exploring Small-Scale Sport Event Tourism in Malaysia. *Research Journal of International Studies*, 47–58.

Zauhar, J. (2004). Historical perspective of sport tourism. *Journal of Sport Tourism*, 9(1), 5–101.

Hospitality and Tourism – Sumarjan et al. (eds)
© 2014 Taylor & Francis Group, London, ISBN 978-1-138-00151-0

Pork free restaurants in Malaysia: Issue of halal and muslim customer patronization

M.I. Haroun, M.S. Mohd Zahari, M.A. Hemdi & M.I. Zulkifly
Universiti Teknologi MARA, Malaysia

ABSTRACT: *Halal* is not only terms of religion but it governs cleanliness and purity and everything from the food that we consume to the business transactions that are performed in our daily lives. *Halal* is also closely associated with products and services including restaurant businesses and influences feeling, decision making, behavior and expectation towards patronizing the restaurant among the Muslim customer. As Malay Muslim customers are the highest markets in Malaysia the demand in ascertaining the products consumed are religiously fit or halal in the restaurants is obviously apparent. This research note discusses the issue of halal at the pork free restaurants in relation to Muslim customer patronzation which results are yet to be revealed.

Keywords: pork free restaurant, halal, muslim, customer, patronization

1 INTRODUCTION

In today's modern world, food is the factor that brings health and symbol of cultures of ethnics, communities and nations. It is not only an essential facet of human life but food is creating a social interaction and bonding between individual and kin groups. With that no doubt most of the food consumption behaviors are influenced by the social classes, gender, age, culture, race and religions. Of all these, religions are strongly believed to play a significant role in consumer attitudes and behavior in food purchasing decision and their eating habits especially among the Muslim (Bonne & Verbeke, 2008).

One of the massive issues related to food in the Muslim world today is dealing with Halal. This is due to Islam as the fastest growing religion and there are 2 billion Muslims scattered around 148 countries thus creating an increasing demand for *Halal* food products (Evans, 2005). The global food proprietors in fact are now shifting their attention to *Halal* food products in gaining a greater world market share.

In line with the above notion, *Halal* is a permissible and lawful act according to Shariah law which includes doing profitable business, consuming healthy food and beverages and other daily practices (Al-Qaradawi, 2001). The concept of *Halal* came from the verse Quran al-Baqarah: 168

> *"... oh mankind! Eat from the earth which is Halal and Toyyib and follow not the food steps of the devil. Lo! He is an open enemy for you".*

The *Halal* concept serves as a guideline for Muslims all over the world in all aspects of products and services (Burhan, 2000). It is a concept which encourages a Muslim to seek and use products, business and services that promote cleanliness in all as parts and parcels of a person's life. Thus, *Halal* food means that a product is safe for consumption, hygiene and wholesomeness. The bottom line is that *Halal* food is not just about what food is being prepared, served but how it is processed. Ahmad (2004) states that *Halal* is not only terms of religion but govern about cleanliness and purity and everything from the food we consume to the business transactions that are performed in our daily lives. In the restaurant businesses, products and services are closely associated therefore producing *Halal* food and environment play an important role that influence feeling, decision making, behavior and expectation towards patronizing the restaurant among the Muslim customers.

2 LITERATURE REVIEW

As a research note, the reviews of the literatures are centrally focused on the variables that relate to attributable that influencing the customers' decision making towards patronizing halal restaurants like knowledge, attitude and trust.

2.1 *Knowledge*

As with concepts of truth, belief and wisdom, there is no single definition on which scholars agree, but rather numerous theories about the nature of knowledge. Davenport and Prusak (1998) describe knowledge as

> ..... *a fluid mix of frame experience, important value, contextual information and expert insight that provides a framework for evaluation and incorporation of new experience and information.*

Generally, more than one type of knowledge exists (Abernathy, Dunlop, Hammond & Weil, 1999). DeLong and Fahey (2000) developed a useful framework to classify knowledge which distinguished human, social and structured knowledge. Human knowledge is what humans know or know how to do. For example, this can be based on the individual previous dining experience. Social knowledge is usually knowledge that arises out of relationships. An example of social knowledge is the way customers on different levels interact with other customers. Structured knowledge is rooted in the system, processes, rules and routines of an organization and is usually explicit knowledge. Of all these three types of knowledge, human knowledge is considered the most appropriate in this study as it relates to the experience of knowing something. In other words, knowledge is the confident understanding of a subject, potentially with the ability to use it for a specific purpose. Hence, knowledge in this research note refers to Malaysian Muslims knowledge towards *Halal* food and beverage products. As stated by Shaari and Mohd Arifin (2010) knowledge is one of the key ingredients that influence *Halal* purchase intention. Particularly, knowledge will be tested whether it is one of the predictor that influences the intention to patronize doubtful halal restaurants among the Malaysia Muslims customers.

2.2 *Attitude*

In simple term attitude is a settled way of thinking or feeling, typically reflected in a person's behavior. Attitudes are often described and measured because of its relationship to behaviors. Psychological literature of the 1970's purported a "theory of reasoned action" which, in simplified terms, states that one's beliefs shape one's attitudes which in turn can predict one's behavior toward which the belief and attitude are directed (Ajzen & Fisbein, 1980). In other word, when an individual or a person believes that performing a given behavior will lead to mostly positive outcomes, he she will hold a favorable attitude toward performing the behavior, while if she he believes that performing a given behavior will lead to mostly negative outcomes, he she will hold an unfavorable attitude. It is the person's favor or disfavor toward an action (Ajzen & Fishbein, 1980). Blackwell, Paul and James (2006) supported that attitude is the evaluation of performing a particular behavior involving the object, feeling whether tangible and intangible. Dabholkar (1994) the inclination or feeling that predisposes an individual to react favorably or unfavorably to a certain object, person, institution, or to any other aspect in his/her environment. For restaurant customers, this concept refers to a favorable or unfavorable assessment of behavior with regard patronizing the type of restaurants.

2.3 *Trusts*

From the popular beliefs, trust is related to believing for both logical and emotional act. Emotionally, it is where individual exposes to believing that they will or not take advantage or to experience and not experience against anything. Joey (2004) states trust is "that one believes in, and is willing to depend on, another party". The importance of trust has been widely analyzed from many literature streams (Heil, Bennis & Stephens, 2000; Luo, Zhang & Shim, 2010 Pavlou, 2002; Yousafzai, Pallister & Foxall, 2003). In the strategy and marketing literature, trust has been related to benefits such as richer information (Lo & Lie, 2008), competitive advantage (Barney and Hansen, 1994), firm performance (Garcia, Sanzo & Trespalacios, 2008) and the attainment of long-lasting and profitable relationships (Flavian, Guinaliu & Gurrea, 2006). From an economic perspective, trust reduces transaction costs (Bromiley & Cummings, 1995) and, in the organizational literature, trust is posited to operate as a governance mechanism (Bradach & Eccles, 1989). Overall, trust is the source of fundamental positive consequences. Trust also has been the focal point in relation to intention to patronize a restaurant. In this sense, trust is strongly based on the fact that the server or cook is a Muslim who is considered to be responsible for his acts toward God. In this sense, food and beverage sold by a Muslim is always to be trusted even when information about the *Halal* status is not directly available (Benkheira, 2002).

2.4 *Intention to patronize*

Intention is an indication of a person's willingness to perform the behavior or a person's probability dimension involving a relation between himself with some action and it is the immediate antecedent of behavior (Fishbein & Ajzen, 1975). Intention varies from time to time and as the time interval increases, the lower is the correlation between intention and action (Ajzen & Fishbein, 1980). Taylor and Todd (1995) found in their studies that intention is strong predictor of behavior. In addition, Armitage and

Conner (1999) stated that intention is recognize as the motivation for individuals to engage in a certain behavior. In this sense, customers who are positive towards the food and the service of the particular restaurants will be more likely to patronize that restaurant and vice versa.

3 PORK FREE RESTAURANT AND THE HALAL ISSUE

Malaysia is a multi-racial society, comprising of Malays, Chinese, Indians, and other ethnic group which Malay is considered a son of the soil who professes Islam. All Malays in Malaysia are Muslims and this is defined in Article 160 of the Constitution (Bhaskaran & Sukumaran, 2007) and others religions are free to be embraced by various ethnic groups. Islam as the national religion of the country sets the tone of life and consuming halal food or food products are one of the principles in Islamic teaching. As Malay Muslim is the highest market in this country they therefore demanding for more information to ascertain that the products are religiously fit for their consumption. They are interested to know whether the ingredients formulated in a given product, food, beverage and the business transaction conform to the Islamic law. Similar feeling goes to other value likes preparation and cleanliness in producing the food products (Dindyal, 2003) No exaggeration that religion is the important factor in shaping Muslim consumer attitudes and behaviors in general (Delener, 1994; Pettinger & Holdsworth, 2004) as well food purchasing and eating decisions (Blackwell & Miniard, 2001).

Despite the preceding notion, there were arguments that a number of basic products, foods, beverages and services available in the marketplace and restaurants did not conform to the system of belief of the Islamic teachings, which is the concept of "*Halal*". In the restaurants, although having a broad selection of restaurants and services either locally or internationally brand names, Muslim customers are facing on the issues of halal status. Some of the local restaurant brands appear to the public as 'Islamic' brands 'via their creative marketing strategies and labeling works (Shafie & Othman 2006). This is as an indirect signal to the primary target of Malaysian Muslim customers'. In line with this and owing to the multicultural nature of Malaysia, not all local restaurants are systematically *Halal* and with that a new 'Islamic' image known as Pork Free restaurants are rapidly emerging especially in Kuala Lumpur and Selangor (Shafie & Othman, 2006). Pork Free status is becoming a new marketing strategy in capturing the Muslim market.

Before proceeding into the problem statement, it is worth to slight touch on the categories of pork free restaurants in the context of this proposed study. The first type is that the restaurants serve pork free food with non alcoholic beverages and received *a Halal certificate* from JAKIM. The second type is that the restaurant serves pork free food, however as the restaurateur's wishes to cater to the needs of non Muslims therefore alcohol is served in the premise which will not receive halal certification from JAKIM. The third type is that the restaurant serves pork free or so called *Halal* food but operated by non-Muslims and without having *Halal* certification from JAKIM.

Having highlighted the types of pork free restaurants, many of non-Muslims restaurateurs' view that *Halal* food is the Pork Free therefore uses Pork Free as their *Halal* symbol to attract Malay Muslims to their restaurants although in actual fact and not to exaggerate that some of the main ingredients used are not *Halal.* With these, critical questions arise as to why some of Malaysian Muslims are happy patronizing the second and third type of pork free restaurant? Is these practices or action associate with their lack of halal knowledge which in turn leads to their attitude and trust to patronize those restaurants? These issues are therefore warrant and under investigation.

4 CONCLUSION

It is hoped the outcomes of this study will help the responsible authority and government to take necessary action or plan for the benefit of Muslim customers in particular with regard to halal food. In this context, if lack of understanding of halal food or the halal is being manipulated by restaurant operators using pork free as a tagline, the authority especially JAKIM can inject knowledge and help them in enriching their understanding and be more alert in this sensitive among the Muslim customers. However, if there are no changes after the initiative, the responsible authority can take a serious action against the restaurateurs' who transgress beyond normal acceptable practices. By doing this, all parties will benefit each other in developing and maintaining the continuous respects in the religious matters by each ethnic group.

REFERENCES

Abernathy, F.H., Dunlop, J.T., Hammond, J.H., and Weil, D. (1999). *A stitch in time*. Oxford. UK: Oxford University Press.

Ahmad, R.W. (2004). Halal perspective. *The Halal Journal,* 3, 23–25.

Ajzen, I. and Fishbein, M. (1980). *Understanding attitudes and predicting social behavior*. NJ: Prentice-Hall.

Al-Qaradawi, Y. n.d. (2001). *The lawful and the prohibited in Islam Al-Halal Wal-Haram Fil Islam*. Indianapolis, USA: American Trust Publications.

Armitage, C.J. and Conner, M. (1999). Distinguishing perceptions of control from self-efficacy: Predicting consumption of a low fat diet using the Theory of Planned Behavior. *Journal of Applied Social Psychology*, 29, 72–90.

Benkheira, M.H. (2002). *"Le rite a la lettre. Regime carne et normes religieuses."* In P. Bonte, A—M. Brisebarre, and A. Gokalp (Eds.). Paris, France: CNRS Editions.

Bhaskaran, S. and Sukumaran, N. (2007). National culture, business culture and management practices: Consequential relationships? *An International Journal, 14*(1), 54–67.

Blackwell, R.D. and Miniard, P.W. (2001). *Consumer Behavior*. Orlando, FL: Harcourt Inc.

Blackwell, R.D., Paul. W.M., and James, F.E. (2006). *Attributes of attitudes, consumer behavior*. New York, NY: Thomson Press.

Bradach, J.L. and Eccles, R.G. (1989). Price, authority, and trust: from ideal types to plural forms. *Annual Review of Sociology, 15*(1), 97–118.

Bromiley, P. and Cummings, L.L. (1995). *Transactions costs in organizations with trust*. In: Bies, R., Sheppard, B., Lewicki, R. (Eds.), Research on Negotiation in Organizations. JAI Press, Greenwich, CT.

Burhan, A. (2000). Consumer behavior and religiosity of Muslim community in Klang Valley: Consumer perception on halal and non halal in products and services. University Malaya.

Dabholkar, P.A. (1994). Incorporating choice into an attitudinal framework: analyzing models of mental comparison processes. *Journal of Consumer Research, 21*, 100–118.

Definitions. (2007). Definitions of "halal", "haram" and "musbooh". Forward, 11(38).

Delener, N. (1994). Religious contrasts in consumer decision behavior patterns: Their dimensions and marketing implications. *European Journal of Marketing, 28*(5), 36–53.

DeLong, D.W. and Fahey, L. (2000). Diagnosing culture barriers to knowledge management. *Academy of Management Executive, 14*(4), 113–128.

Dindyal, S. (2003). How personal factors, including culture and ethnicity, affect the choices and selection of food we make. *Internet Journal of Third World Medicine, 1*(2), 27–33.

Evans, A. (2005). Halal perspectives. *The Halal Journal, 1*(11), 25–28.

Fishbein, M. and Ajzen, I. (1975). Belief, attitude, intention, and behavior: An introduction to theory and research. Reading, MA: Addison-Wesley.

Flavian, C., Guinaliu, M. and Gurrea, R. (2006). The role played by perceived usability, satisfaction and consumer trust on website loyalty. *Information & Management, 43*, 1–14.

Garcia, N., Sanzo, M.J. and Trespalacios, J.A. (2008). New product internal performance and market performance: evidence from Spanish firms regarding the role of trust, inter functional integration, and innovation type. *Technovation, 28*(11), 713–725.

Heil, G., Bennis, W. and Stephens, D. (2000). Douglas McGregor. *Managing the Human Side of the Enterprise*. Wiley, New York.

Joey F.G. (2004). The theory of planned behavior and Internet purchasing. *Internet Research, 14*(3), 198–212.

Lo, S.K. and Lie, T. (2008). Selection of communication technologies—a perspective based on information richness theory and trust. *Technovation, 28*, 146–153.

Luo, X., Li, H., Zhang, J., and Shim, J.P. (2010). Examining multi-dimensional trust and multi-faceted risk in initial acceptance of emerging technologies: an empirical study of mobile banking services. *Decision Support Systems, 49*, 222–234.

Pavlou, P.A. (2003). Consumer acceptance of electronic commerce: integrating trust and risk with the technology acceptance model. *International Journal of Electronic Commerce, 7*(3), 101–134.

Pettinger, C. and Holdsworth, M. (2004). Psycho-social influences on food choice in Southern France and Central England. *Appetite, 42*(3), 307–316.

Shafie, S. and Othman, M.N. (2006). Halal certification: An international marketing issues and challenges. CTW Congress.

Taylor, S. and Todd, P.A. (1995). Understanding Information Technology Usage: A Test of Competing Models. *Information Systems Research, 6*(3), 144–176.

Yousafzai, S., Pallister, J.G., and Foxall, G.R. (2003). A proposed model of e-trust for electronic banking. *Technovation, 23*, 847–860.

Hospitality and Tourism – Sumarjan et al. (eds)
 ISBN 978-1-138-00151-0

Spa and wellness industry challenges: The perspectives of industry provider

S. Rasid & I. Azman
Universiti Teknologi MARA, Malaysia

ABSTRACT: The spa and wellness industry has become one of the main contributors of tourist arrival in the country. In order to sustain the business it is important to identify the challenges for future sustainability. This paper views the industry in Malaysia focusing on the challenges faced by the service providers especially in the Kuala Lumpur area. The paper aims to explore the current development of spa businesses in Malaysia, to identify the challenges faced by the spa and wellness providers, and to understand to the extent these challenges affect the growth of spa businesses in Malaysia.

Keywords: provider, spa challenges, spa and wellness

1 INTRODUCTION

The spa has been in existence for centuries, dating back to as early as 500 BC, from the Mesopotamians and Egyptians, to the Greeks and Romans, and later, the Japanese and the rest of the world (Dahlan, 2011). Various terms are used such as health tourism, spa tourism, wellness tourism, and medical tourism within the industry itself. However; spa and wellness tourism is a more familiar and well established concept within the health tourism phenomenon (Laing & Weiler 2008). It is known that spa therapies have been around since ancient times in many different forms that reflect the cultural, social and political milieu in which there are embedded (Cohen, 2008).

Asia is well-positioned to be a global leader in health and wellness tourism (Laing & Weiler, 2008). One of the reasons that enable Asia to become a global leader in health and wellness tourism is that many western tourists are seeking to return to nature and its elements when they are living in an increasingly artificial world. Furthermore, they may be drawn to Eastern practices and cultures because in the eyes of the consumers, 'Asia offers an intact world with authentic, original, genuine and deep encounters' (Fuchs, 2003). India, Thailand, Singapore, and Malaysia are key Asian countries which are successfully operating in this market (Laing & Weiler, 2008).

2 LITERATURE REVIEW

2.1 *Health, spa and wellness terms*

Laing and Weiler (2008), stated that health tourism is 'the sum of all the relationship and phenomena resulting from a change in location and residence by people in order to promote, stabilize and as appropriate, restore physical, mental and social well-being while using health services and for whom the place where they are staying is neither principal of residence or work'.

However, the history of health tourism in Europe has tended to be based around spa and seawater treatments, it is only in the recent years that cosmetic and beauty treatment has become more popular, as well as more spiritual and psychological activities (Koncul, 2012).

Wellness is defined by Cohen (2008) as 'a multidimensional state of being describing the existence of positive health in an individual as exemplified by quality of life and a sense of well-being'. Furthermore, Laing and Weiler (2008) suggest that wellness tourism is a subset of health tourism. The spa has been in existence for centuries, dated back to as early as 500 BC, from the Mesopotamians and Egyptians, to the Greeks and Romans, and later, the Japanese and rest of the world. The word 'spa' in Latin means "health through water" (Dahlan, 2012). However, (Laing & Weiler (2008), defines spa more broadly as 'entities devoted to enhancing overall well-being through a variety of professional services that encourages the renewal of body, mind and spirit'. Yussof (2010) reported that spa has different categories or known as 'typologies' namely:

- Day/Club/Salon Spas: Facilities that offer a variety of spa services such as massage, facials, and body treatment. Club spas are similar to day spas but focus more on fitness as the main purposes. While salon spas more focus on beauty services such as hair, make-up and nails treatment.

- Destination Spas and Health Resort: Offers myriads of offering such as fitness activities, educational classes, weight loss program and nutrition counseling. Involve all guests to participate in the activities.
- Health Resort Spas: Typically paid for a la carte basis and generally complement a hotel stay or other activities at the resorts.
- Medical Spas: Operates under the supervision of healthcare professional, this spa offers comprehensive wellness and medical care with traditional or alternative treatments and therapies.

2.2 *External environment*

Environment consists of factors that are largely if not totally, external and beyond the control of individual industrial enterprise and their managements. These are essentially the 'givers' within which firms and their management must operate in a specific country and they vary, often greatly, from country to country (Pal, 2000). It is supported by Froneman (2004), who stated that external components also known as market environment and refers to the macro environment of a business entity. LeRoux (2009), further define business external environment as macro environment that contains variables and factors which is outside of the business that brings negative or positive impacts on the sustainable and growth of business firms that encourage the achievement of objectives.

Dhar (2003) stated that the external environment can be divided into macro and micro environment. The public, suppliers, marketing intermediaries and customers are under the micro environment while economic, political, legal, demography and socio culture are the components of macro environment.

In spa businesses, one of the micro environment challenges derived from the consumers or the market. Barnes and Patricia (2008) mentioned that spa and wellness consumers can be divided into two profiles known as "Wellness-focused, moderate-to-active spa goers" and "Sickness reactors, not active spa-goers" and the consumer will either fall in one of the classifications.

On the other hand, GMDC (2009) has conducted a comprehensive market survey of consumer behaviors related to health and wellness. The results defined three segments of health and Wellness consumers, which also similar with the consumer profiles as specified by Barnes and Patricia. The three segments of Spa and Wellness consumers namely:

- Periphery: Falls in the profile of "Sickness reactors, not active spa-goers". Entry level consumers which aspire to be more involved in health and wellness but their action contradicts with their aspirations.
- Mid-level: Falls in the profile of "Wellness-focused, moderate-to-active spa goers". Have a moderate involvement in health and spa. Uses and purchase large amount of health products. Experienced and knowledgeable and somehow concern on pricing and convenience.
- Core: Falls in the profile of "Wellness-focused, moderate-to-active spa goers". Have the most involvement in health and wellness and served as trendsetter to the other users. Their major life focus is health and wellness.

Apart from the micro environment, spa also faces political and legal challenges of the macro environment. Spa businesses in Malaysia is facing the challenge be legal; the industry provider need to operate within the law such as having valid operating license, following statutory regulations during set-up and hire only therapist that is qualified with proper work visas and documentation (Suleiman, 2011). Additionally, a proper Standard Operation guide and training manuals is necessary and the providers also need to offer all treatments required by international standard to qualify their operations as spa.

2.3 *Internal environment*

Internal environment is also known as the micro environment, a controllable component of management and influencing the output of the business (Froneman, 2004). He further added that the variables in the micro environment are linked to the macro environment in a certain level depending on the decision made by the management. Dhar (2003) categorized few components that fall into the internal environment including organizational structure, mission and objectives, and also quality of human resources.

As mentioned in previous studies, quality of human resources is a component of internal environment. Tourism-Review (2007) mentioned that Malaysian spa industry is facing shortage of skilled labor back in 2007. In fact, the local spa industry is deemed ill-reputable by public discouraging professionals from the field to work in the country. This situation forces the spa businesses to recruit workforce from foreign countries or either forced to downsize and even close down the operation (Suleiman, 2011).

2.4 *Company's internal environment*

Tourism-Review (2007), mentioned that Malaysian spa industry is facing shortage of skilled labor back in 2007 and the situation forces the spas to recruit labor and workforce from foreign countries. Furthermore, Suleiman (2011), reported that spa providers in Malaysia chose to hire foreign therapist is simply due to lack of local therapist to work in the industry forcing them to downsized and even closed down the operation.

3 ISSUES

Malaysia, being a multi-cultural country has a huge range of traditional spa treatments handed down from multiple generations (El-Refae, 2012). The spa and wellness industry is seeing an emerging pattern with a big potential for spa to be developed as a new attraction in the country's tourism industry. Malaysia is one of the Asian countries actively offering spa and wellness facilities experiencing rapid growth in the region for the last few years; it is reported that as of December 2012, the country's Ministry of Tourism has rated 170 spas nationwide, of which 31 spas have obtained a 5-star rating, while 40 others obtained a 4-star rating (Economic Transformation Program, 2012).

Spa industry plays a vital role in the Malaysian economy and becoming the backbone of health development in the country as people are seeing the importance of a healthier lifestyle through spa treatment (Yusoff, 2010). The Ministry of Tourism mentioned that the spa industry was expected to generate RM666 million in revenue while employing some 5,644 therapists by 2020 (Carvalho, 2012).

The government sees the importance for the spa industry in Malaysia to have local expertise and thus introducing Centre Of Excellence (COEs) to train and recruits locals to become spa therapists (ETP Annual Report, 2012). To be able to achieve the targeted revenue and sustain the business in the long term, there is a need to identify the challenges. According to Dhar (2003), a success and failure of a business is depending on the extent of environmental scanning which is referring to identifying challenges and threats that leads to policies formulation and implementation. Therefore, the study attempts to identify the challenges within the spa businesses operation from the perspective of the spa owners. Thus it provides spa owners on the insight on coping with the challenges as being mentioned by Bittel (2013), the best way to deal with challenges is to have the information and be prepared.

4 CONCLUSION

The spa industry has become one of the main contributors of tourist arrival in the country. Therefore, in order to sustain the business it is important to identify the challenges for future sustainability. The information collected from the study will help the industry provider in foreseeing the challenges that they are facing. Knowing the challenges of spa and wellness business, it will help the spa providers to cope with the ever changing needs of the industry and sustaining the business. Thus, improvements and necessary actions can be implemented in the spa industry which can lead to a higher profitability rate and have a positive impact on the tourism industry in the country.

From the academic aspect, the study can contribute to a new body of knowledge in the area especially in extending the body of literature. There is little research done from the perspective of spa owners in the spa industry. Therefore, the information from the study can help in adding more literature to the existing one. Furthermore, the study will provide a clear view on spa and wellness industry to help the public in understanding more regarding the industry and thus, can correct the misconception among public.

REFERENCES

Bittel, L.R. (2013). Challenges Faces Today's Business Leaders. *ASPIRA/Morgan Stanley Entreprepeurial Leadership Initiative, 1–14*. Retrieved from: http://www.aspira.org/files/documents/entrepreneurship/session_04%20ex-6.pdf.

Carvalho, M. (2012). Tourism Ministry seeks to regulate spa industry to weed out vice activities—Nation | The Star Online. Retrieved April 8, 2013, from http://thestar.com.my/news/story.asp?file=/2012/6/5/nation/20120605133118&sec=nation.

Cohen, M. (2008). Spa, wellness and human evolution. In Understanding the global spa industry: Spa management (pp. 2–25). Amsterdam: Elsevier/Butterworth-Heinemann.

Dahlan, N. (2012). Service Innovation in the Business Models of the Spa Industry in Malaysia. Retrieved from http://www.wbiconpro.com/707-Full%20Paper.pdf.

Dhar, P.K. (2003). Indian economy its growing dimensions PB. Ludhiana u.a. India: Kalyani Publ.

Economic Transformation Programme Annual Report 2012 (n.d.). Retrieved April 7, 2013, from http://etp.pemandu.gov.my/Tourism-@-Tourism_-_EPP_9a-;_Developing_local_expertise_-%E2%97%98-_better_regulating_the_spa_industry.aspx.

El-Refae, B.A. (2012). The Relationships between Service Quality, Satisfaction, and Behavioral Intentions of Malaysian Spa Center Customers. *International Journal of Business and Social Science,* 3(1), 198–205.

Fuchs, H.J. (2003). Fareast goes West—New Opportunities for Asian Brands in Europe. *Asia Pacific Journal of Marketing and Logistics,* 15(3), 20–33.

GMDC & The Hartman Group (2009). Consumer Shopping Habits for Wellness and Environmentally Conscious Lifestyles Study: Insights for Health, Beauty and Wellness. Retrieved from http://www.pacific.edu/Documents/school-pharmacy/acrobat/Consumer%20Shopping%20Habits%20for%20Wellness%20-%20Presentation.pdf.

Koncul, N. (2012). Wellness: A New Mode of Tourism. *Economic Research,* 25(2), 525–534.

Laing, J., & Weiler, B. (2008). Mind, Body and Spirit: Health and Wellness Tourism in Asia. In Asian tourism: Growth and change (pp. 379–389). Amsterdam: Elsevier.

Suleiman, R (2011, November 15). The Law Seems to Bend Towards Illegal Spa. The Star Online. Retrieved June 24, 2013, from http://thestar.com.my/news/story.asp?file=/2011/11/15/central/9886739&sec=central.

Tourism-Review, Malaysian Spa Industry Lacking Quality Staff, (2007, January 1). Retrieved June 24, 2013, from http://www.tourism-review.com/malaysian-spa-industry-lacking-quality-staff.tr.

Yussoff, Y.M. (2010). Service Innovation: Case Study on Spa and Med Beauty at the Saujana.

Hospitality and Tourism – Sumarjan et al. (eds)
© 2014 Taylor & Francis Group, London, ISBN 978-1-138-00151-0

1Malaysia menu: Issues on food quality attributes toward customer satisfaction

I. Muhamad Shah Kamal, A.T. Salim, M.A. Sperico & F. Sulaiman
Universiti Teknologi MARA, Malaysia

ABSTRACT: 1Malaysia menu was introduced in 2011 by the Prime Minister of Malaysia Datuk Seri Najib Tun Razak. The purpose of this menu is to serve the privileged Malaysians to affordable food with various nutritional values and quality. Despite 1Malaysia Menu to assure the quality to be at par with other food menu however it still create an issue as customers' are percieve to associate the lower priced menu with lower quality. In line with that, this research note discusses on the 1Malaysia Menu by reviewing the past studies on food quality attributes toward customer satisfaction and relates the issue to the contextual setting which the empirical study is being undertaken.

Keywords: 1Malaysia menu, customer satisfaction, food quality attributes

1 INTRODUCTION

Nowadays people would always talk about what is 1Malaysia. It seems like a hot topic of discussion by Malaysians with various points of view. 1Malaysia is created by our sixth and current Prime Minister Datuk Seri NajibTun Razak and was launched in Perak on 18th of October 2009 (Jalil, 2009).

The 1Malaysia concept is a guideline on how to achieve "Bangsa Malaysia" just like what he used as slogan with the tagline "People First, Performance Now". 1Malaysia concept is the idea about making all Malaysians of different races and religions to fully understand and respect each other's cultures, and to live together, work hand in hand and sharing the responsibilities to achieve a better future Malaysian in terms of social economy and political stability.

Various efforts are undertaking to achieve 1Malaysia concept such as 1Malaysia shop, 1Malaysia clinic and "1Malaysia Menu". Bernama (2011) reported that 1Malaysia shop aimed to provide the masses products with low-priced to offset the high cost of living. The prices are a lot cheaper compared to the other supermarkets. The 1Malaysia clinics meanwhile were established to serve the health services to the urban poor communities which are a laudable effort. As the cost of food and price of meals has gone up tremendously the Malaysia government is finding ways to lessen the citizen's everyday burden. Thus, the 1Malaysia Menu was launched on 7 July 2011 by Datuk Ismail Sabri Yaakob, the Minister of Domestic Trade, Cooperatives and Consumerism (Bernama, 2011).

The program had received encouraging response with 1004 restaurants participating in the program nationwide including public higher educational institutions such as Universiti Teknologi Mara, Universiti Malaysia and Universiti Utara Malaysia. 1Malaysia menu offers breakfast and lunch for just RM2 and RM4 respectively where people specifically with low income and students can really enjoy food (Berita Harian, 2011).

2 LITERATURE

2.1 *The development of 1Malaysia menu*

The Prime Minister of Malaysia Dato' Sri Najib bin Tun Abdul Razak has invented a new concept called 1Malaysia with the theme of "People First, Performance Now". Consequently, he has come out with various programs with the main objective to serve the low income people by offering services and products at affordable prices.

Some of the program are *Klinik 1Malaysia, Bantuan Rakyat 1Malaysia, Kedai Rakyat 1Malaysia* and *1Malaysia Menu*. In line with that, 1Malaysia Menu offers breakfast and lunch for just RM2 and RM4 respectively where people, specifically those with low income and students can really enjoy the food offerings. (Berita Harian, 2011).

This program is being set up in collaboration with Kementerian Perdagangan Dalam Negeri, Koperasi dan Kepenggunaan (KPDNKK) which involved restaurateurs, cafeteria owners and some of the food providers. All food providers were voluntarily involved in this program by offering 1Malaysia Menu to their customers.

In 2012, the program successfully attracted the involvement approximately 3,228 participations from the food premises. The government is targeting another 1,000 participation in 2013. Until last 19 April 2013, there were about 742 premises participated in this 1Malaysia Menu program (Harian Metro, 2013).

2.1.1 *Food quality*

Food quality has been generally accepted as a fundamental element of the overall restaurant experience (Kivela, Inbakaran, & Reece, 1999; Raajpoot, 2002; Sulek & Hensley, 2004). According to Peri (2006), food quality is a necessary condition to satisfy the needs and expectations of customers. Sulek and Hensley (2004) investigated the importance of food quality, physical settings, and service in a full-service restaurant and reported that food quality appeared to be the most significant predictor of customer satisfaction although food quality explained only 17% of repeat-patronage intentions.

Despite the importance of food quality in the restaurant business, there is no consensus on the individual attributes that constitute food quality. Sulek and Hensley (2004) lumped all food attributes into only one variable, food quality, whereas Kivela et al. (1999), who designed a model of dining satisfaction and return patronage, saw that food quality had many attributes: presentation, tastiness, menu item variety, and temperature. Even so, little has been done with the crucial attributes of food quality in relation to customer satisfaction and behavioral intention. A thorough review of the literature revealed that the general description of food quality among researchers were (a) freshness, (b) taste, (c) healthy options, (d) variety of menu, (e) presentations, and (f) temperature.

2.1.2 *Freshness*

Freshness usually refers to the fresh state of food and appears to be related to crispness, juiciness, and aroma (Péneau, Hoehn, Roth, Escher, & Nuessli, 2006). In previous studies, freshness of food has been cited as a crucial intrinsic quality cue of food (Acebrón & Dopico, 2000; Johns & Tyas, 1996; Kivela et al., 1999).

2.1.3 *Taste*

Taste regarded as a key attribute in food in the dining experience (Kivela et al., 1999). Many customers have become food savvy, so the taste of food in restaurants has become increasingly important (Cortese, 2003). Therefore, it is not surprising that shabby restaurants with gourmet cooking are packed with customers and taste is usually believed to influence restaurant customer satisfaction and future behaviour intentions (Kivela et al., 1999).

2.1.4 *Healthy option*

Kivela et al. (1999) noted the importance of healthy foods in restaurants and suggested nutritious food as one of the core properties in dining satisfaction and re-patronization. The more notable thing is that restaurant customers are increasingly interested in healthy menu items (Sulek & Hensley, 2004).

2.1.5 *Variety of menus*

Variety involves the number or the assortment of different menu items. Restaurants constantly develop new menus to entice diners, and many proactive restaurateurs have created an assortment of food and beverage offerings. In previous studies, menu item variety was a crucial attribute of food quality in creating dining satisfaction (Kivela et al., 1999; Raajpoot, 2002).

2.1.6 *Presentation*

Presentation refers to how attractively food is presented and decorated as a tangible cue for customer perception of quality. Kivela et al. (1999) pointed out that the presentation of food is a key food attribute in modelling dining satisfaction and return patronage.

In addition, Raajpoot (2002) described food presentation as one of the products and/or service factors in TANGSERV.

2.1.7 *Tempreture*

Temperature is also a sensory element of food quality (Johns & Tyas, 1996; Kivela et al., 1999). According to Delwiche (2004), temperature influenced how the perceived flavour of food was evaluated, interacting with other sensory properties such as taste, smell, and sight.

2.2 *Customer satisfaction*

Customer satisfaction is an essential indicator of a company's past, current, and future performance and, therefore, has long been a critical focus among marketing practitioners and scholars (Oliver, 1999). In modeling satisfaction, two general conceptualizations of satisfaction exist in the literature: transaction-specific satisfaction and cumulative satisfaction (Anderson & Fornell, 1993; Boulding, Kalra, Staelin, & Zeithaml, 1993).

Recently, the focus of satisfaction research seems to have moved to emotional responses, instead of purely cognitive information, informing behavioral outcomes. As an example, Oliver (1993) argued that consumer attributions about satisfaction with specific product characteristics lead to positive and negative affective responses, which influence overall satisfaction.

3 DISCUSSION

3.1 *Issue on 1Malaysia menu and food quality attributes towards customer satisfaction*

In today's competitive foodservice industry and increase in cost of living, a critical challenge for foodservice provider is to provide quality food with reasonable price. Food can build the relationship between the customer and the restaurant itself (Salman et al., 2012). Qin and Prybutok (2009) noted food characteristic is a consideration that will influence customer to make decision in selecting a restaurant.

Furthermore, they noted that there is a significant relationship between food and customer satisfaction. Sulek and Husley (2004) mentioned that satisfying customers should be the crucial objective of business in order to build repeat purchase in the highly competitive restaurant industry.

From the aforementioned notion, food service providers in Malaysia in particular are competing with each other to introduce 1Malaysia Menu to the customer at their premise as one of the marketing strategy to attract customer. As previously mentioned, the type of menu offered in 1Malaysia menu is RM2.00 for breakfast and RM4.00 for lunch. However, based on preliminary information gathered from the blogs, social network and informal interviewed with customers on food quality of 1Malaysia menu, most commented that quality of food offered is not value for the money although Datuk Seri Ismail Sabri Yaakob, Domestic Trade, Co-operatives and Consumerism Minister noted people should not have negative perceptions about cheap food as long as it affordable. To see whether notion hold true the empirical investigation on the 1Malaysia menu need to be investigated and answers yet to be revealed.

REFERENCES

Anderson, E.W., & Fornell, C. (1993). A customer satisfaction researchprospectus. In R.L. Oliver & T.R. Rust (Eds.), *Service Quality: New Directions in Theory and Pratice* (pp. 239–266). Newbury Park, CA: Sage.

Anderson, E.W. & Fornell, C. (2000). Foundations of the American customer satisfaction index, *Total Quality Management, 1* No. 7, 869–883.

Berita Harian Online (2011, July 18). *Sarapan RM2 di 728 premis.* Retrieved From http://www.bharian.com.my/.

Bernama (2011, December 22). *Kedai Rakyat 1Malaysia and Menu Rakyat at All Universities Nationwide.* Retrieved from http://www.malaysiandigest.com.

Boulding, W., Kalra, A., Staelin, R., & Zeithaml, V.A. (1993). A dynamic process model of service quality: From. expectations to behavioral intentions. *Journal of Marketing Research, 30*(1), 7–27.

Brumback, N. (1998). Inn style. *Restaurant Business, 97*(3), 47–59.

Corporate Communication Unit (2011). UUM Launches the 1Malaysia Menu. Retrieved November 18, 2011, from www.uum.edu.my.

Cortese, A. (2003, October 27). Fine dining? Just across the lobby: Some of the best new restaurants are popping up in hotels. *Business Week,* 130–131.

Delwiche, J. (2004). The impact of perceptual interactions on perceived flavor. *Food Quality and Preference, 15*(2), 137–146.

Harian Metro (2013, May 6). Menu Rakyat 1Malaysia Berjaya Tarikh Minat Pelanggan.

Jalil. A.K. (2009). *Konsep1Malaysia.* Retrieved December 16, 2011, from http://www.scribd.com.

Johns, N., & Tyas, P. (1996). I nvestigating of the perceived components of the meal experience, using perceptual gap methodology. *Progress in Tourism and Hospitality Research, 2*(1), 15–26.

Kivela, J., Inbakaran, R., & Reece, J. (1999). Consumer research in the restaurant environment, Part 1: A conceptual model of dining satisfaction and return. patronage. *International Journal of Contemporary Hospitality Management, 11*(5), 205–222.

Kivela, J., Inbakaran, R., & Reece, J. (2000). Consumer research in the restaurant environment, Part 3: analysis, finding and conclusions. *International Journal of Contemporary Hospitality Management, 12*(1), 13–31.

Namkung, Y., & Jang, S., (2007). Does food quality really matter in restaurants? It's impact on customer satisfaction and behavioral intentions. *Journal of Hospitality & Tourism Research, 31*, 387.

Oliver, R.L. (1993). Cognitive, affective and attribute bases of the satisfaction response. *Journal of Consumer Research, 20*(3), 418–430.

Oliver, R.L. (1999). Whence consumer loyalty? *Journal of Marketing, 63*(4), 33–44.

Péneau, S., Hoehn, E., Roth, H.-R., Escher, F., & Nuessli, J. (2006). Importance and consumer perception of freshness of apples. *Food Quality and Preference, 17*(1/2), 9–19.

Peri, C. (2006). The universe of food quality, *Food Quality and Preference, 17*(1/2), 3–8.

Qin, H. & Prybutok, V.R. Service quality, customer satisfaction, and behavioral intentions in fast-food restaurant, *International Journal of Quality and Service sciences,* 178–179.

Raajpoot, N.A. (2002). TANGSERV: A multiple item scale for measuring tangible quality in foodservice industry. *Journal of Foodservice Business Research, 5*(2), 109–127.

Salman. W. W.A., Sebi. N.M., Othman. L.R. & M. Yahyaudin. (2012). Measuring Customer Satisfaction. *The case of Kelantan Delights Restaurant.* Information Current Issues in Hospitality and Tourism Research and Innovation.

Soriano, D. (2002). Customers' expectation factor in restaurants: the situation in Spain. *Internatioanal Journal of Quality & Reliability Management. 19*, (8/9), 1055–1067.
Sulek, J.M., & Hensley, R.L. (2004). The relative importance of food, atmosphere, and fairness of wait. *Cornell Hote and Restaurant Administration Quarterly, 45*(3), 235–247.
Walker, J.R. (2008). *The restaurant: From concept to operation* (5th ed). John Wiley and Son.

Hospitality and Tourism – Sumarjan et al. (eds)

Resettlement effect on Penan's food and foodways

M.S.Y. Kamaruddin, M.A. Sperico, F. Sulaiman & M.S. Kamal
Universiti Teknologi MARA, Malaysia

ABSTRACT: Penan people have been a nomadic people but now some of them moving or migrating away from the origin site permanently. Settlement of Penan increased since 1960's due to encouragement from Sarawak government which offers them inducement such as build permanent house because the construction of state's dam project. This resettlement of Penan has impact on their food and foodways. The purpose of this study is to explore the changes after resettlement towards food and food practices in the Penan community. This study use qualitative approach in collecting, analyzing, and interpreting data conducted through interviews. In line with this, the changes of food intake and foodways of Orang Penan will be revealed.

Keywords: food, food practices, Penan people, resettlement

1 INTRODUCTION

Resettlement is situation where a group of people move to a new area from the original settlement (Akpanudoedehe, 2010). According to Iuchi (2010), resettlement is regularly used to define the human translation movement or an action migrates away from the origin site permanently. Maruyama (2003) thought the resettlement process of the indigenous communities encountered difficulties in facing a new life and are not familiar with their practices. Contrary, Jamal (1996) found that by moving to a new area, migrants' behavioural patterns and acceptance at the new place will be influenced by their origin culture.

Zeleke (2012) portrayed resettlements generally involve groups of people who move from their own area to a new area or moved from one area that not fixed to the fixed areas. Reasons of resettlement varies and it may include the followings: dam development projects, industrial, roads, political disasters, war and hunger.

Sarawak is surrounded with large tracts of tropical rain forest and has more than 40 sub-ethnic groups, each with its own distinct language, culture and lifestyle (Zahari, Kamaruddin, Kutut & Langgat, 2011). Language and their cultures are different from the other groups. Groups often maintain relationships with subsistence economies derived from hunting, fishing, nomadic or seasonal agricultural transition forest, or subsistence farmers (Kuhnlein & Receveur, 1996).

According to Koizumi (2005), one of the traditional hunter-gatherer groups in Borneo is the Penan. Brosius (2006) stated that settlement of Penan people increased since 1960's due to encouragement from Sarawak government which offer them inducement such as permanent houses and knowledge on farming.

It is been proven that because of resettlement, there have been changes reported on the socio-cultural aspects among indigenous people (Zaman, 2010). Maruyama (2003) mentioned that resettlement will effect and influence the social life and aboriginal relations during the duration of moving from the area of origin and how they solve their problem in the new environment. As the resettlement process is ongoing, the tribe encounter a loss of cultural inventory in some level of socio-cultural compromise includes a change in or loss behavioural outline, cultural or economic practices, social institution and symbols (Platter, 2004), as well as in food and food ways (Hitchcock, 2010). Thus, Orang Penan socio-culture elements such as food and its practices are believed have been altered (Garay, 2006).

The availability study of Orang Penan specifically towards food practices as the progression resettlement is still in exploration. It involves the type of food, utensils used, ingredients, method of cooking and food intake. With this, the present proposed study is trying to address the issue.

2 LITERATURE REVIEW

2.1 *Penan community in Malaysia*

Zahari et al. (2011) cited that the indigenous groups that reside in Sarawak are the Iban,

Bidayuh, Melayu, Melanau, Chinese, Sebob and Orang Ulu. Approximately 5.5 percent (5.5%) of the total population of Sarawak is Orang Ulu that is a combination of indigenous tribes like Kenyah tribes, and the tribes of the nearest like Kejang, Kejaman, Punan, Ukit and Penan. Penan refers themselves as 'Penan', but outsiders often confuse this with another ethnic term of reference, 'Punan'. On the basis of fieldwork among the 'Penan' of Baram in the early 1950's, Needham (1953) says that there are actually three distinct groups of peoples respectively known as 'Penan', 'Punan' and 'Punan Bah'. The difference between Penan and Punan is language.

Traditionally, the Penan has been a nomadic people. As nomads, they live in groups ranging in size from fiften to seventy five people (15 to 75), with a mean of 32 (Needham, 1972). According to Needham (1972), and Urquhart (1951) they live in three types of camp hut: main camp, temporary camp and traveling camp. The main camp is occupied for a variable period of one month to a year (Urquhart, 1951) before another main camp is built. The temporary migration is used by the Penan as they move from one area to another in search for food and jungle produce. As for the traveling camp, it is used by hunters on a hunting trip.

As in 1988, there were about 9,327 Penan in Sarawak residing in not less than at 78 settlements. Today it is estimated that their population might have been increased to 12,000 people. The vast majority of them occupy a large area of tropical rainforest in the watersheds of the Baram and Rejang, the largest rivers in Sarawak.

2.2 *Traditional food practices*

Ethnic food refers to food in terms of expression of attitudes, behaviours, values and cultural beliefs as a tradition in culture and heritage, religion and country of origin (Mora, 1998). In olden times, indigenous people create their traditional food by using natural resources that acquired from the forest such as a wild animal meat, fish, shoots, leaves, roots, stems, bark and others (Zahari et al., 2011).

Based on Kuhnlein, Erasmus and Spigelski (2009) study, traditional foods are the foods that indigenous people can access locally, and within their traditional knowledge and the natural environment. At the same time, it is supported by Langub (2008), mentioned that they make a plant of hill rice as their crops for major food as well as various kinds of fruit trees, sugar cane and vegetables as a side dish.

Pearsall (1996) found that there are a number of traditional practices and cultural values of the Penan people that had created the problem for government in order to solve the issue of their resettlement. First, Penan people believe that animals that they feed should not be eaten even once. Second, they will leave their settlements when one of them died. Lastly, they refused to give up their nomadic way of life even though they live in settlements. Their way of life is more like hunting and collecting forest harvest because they cannot be good farmers who spend a lot of time toiling under the scorching sun and plant rice. Kuhnlein and Receveur (1996) stated that dietary changes of native people around the world have raised the threat of maintaining the traditional food and traditional knowledge that needed to traditional food systems will arise.

Foley (2005) alleged community living in urban areas will lose access to traditional food more quickly than those living in rural municipalities because they are easy to enjoy access to traditional foods. The changes in Sarawak natives' food not only engage the preparation, equipment, materials, processing methods, but also the manners of eating and other related matters. The modernization is not just revolutionize in the preparation and eating traditional food or daily food practices but also affect in traditional events, festivals and occasions such as weddings.

Foley (2005) worried that changes in the food environment in adaptating to the current environment that is geared towards fast food has become a norm in the life of society. Nowadays, people usually rely on foods that are easy to be prepared at home. Zahari et al. (2011) concluded if these changes affecting the overall system in terms of how preparation of food and style of consumption, it may have a huge impact on the young and future generation.

3 METHODOLOGY

The qualitative approach was used to investigate the resettlement influences on food intake in Penan people routines' life. It also gives more advantages as it evaluates and come out with one understanding of detailed result because it contributes in deep of analysis plus the explanation (Mason, 1996) using semi-structured interview. The questions were designed in an interactive manner to clearly bring out and develop a relaxing, friendly atmosphere to gain high quality information through the interview session (Taylor, 2004). Selected seniors of Orang Penan with the age range between 40 and 70 years old were chosen as the sample to track the changes of food intake and foodways. They are believed to have experienced the transmission of the nomadic life to resettlement process. Samples for this method were selected among the Orang Penan senior citizens of new settlement in Sungai

Asap, Belaga district. This place is chosen in this study owing to its latest settlement of Orang Penan as the development of Bakun Hydropower Dam.

4 FINDINGS

4.1 *Penan community before resettlement*

Penan community came from semi-nomadic life in deep jungle. Their life depending on natural resource gathered from the jungle. This style of living has been practiced by their ancestors where they lived in a nomad life and shifted to semi-nomad. They were living in small groups of 20 people and usually build simple house on the ground using bamboos and big leaves. Once the natural resources is depleted, they will leave the place and move to new settlement with new replenishment.

The Penan's sustenance is based from hunting and gathering activities; as sucha group of adults hunt wild animals, gather fruits and greens while women will do the households and some gathering plants and herbs in viscinity area. In fact, there are not much activities and routines in Penan community. Their life is based on survival deeds. They go out to find sustenance and eat at night and the same routine continued from day to day.

4.2 *Penan community in modern days*

Resettlement has changed the routine of Penan people, where most of the adults are working at the plantation and construction sites. Nowadays, they have money to buy food and others necessity such as clothes and medical aids compared to the old barter system for their survival. Besides, the new resettlement has changed the Penan children routine. They were sent to school build by the government near to their place. They were no longer playing and following their adult to the jungle seeking food for family.

The new resettlement of Orang Penan affects them both positively and negatively. From positive perspective, the new style of living enable them to earn money for living. Contrarily, finding revealed that Penan tend to spend unnecessarily. As such, the excess money after buying food was being used to buy alcoholic drink and this contributes to negative side effect such as health condition. For such reason, this community need to adapt and learn the new environment. It just takes time to get things to synchronize and benefits them in general aspect.

4.3 *Penan's foodways pattern before resettlement*

Before permanent resettlement, the relationship between Penan community and jungle is so strong where jungle is their place to find sustenance in form of foods or money if the need to buy something. The parameter of hunting and gathering is unlimited; including their staple food, upland *'kanen'* or rice, *'ubi'* or tapioca and *'saguk'* sago. As for protein, they will hunt for wild boar, deer, bats, fish such *'empurau'*. The nomadic or semi-nomadic Penan hunt with *'keleput'* or blowpipes made from ironwood and darts dipped in poison that can kill a wild boar.

The Penan community takes their diet three times a day back before they settled at Long Berawan; breakfast, lunch and lastly dinner. Food consumption of the Penan community was in big volume due to the usage of energy on the day routines.

The method used in preparing food was also using sources from the jungle. Food is usually prepared by *'nisak'* roasting directly and some may roast in *'lepek'* or bamboo and *'da'un balek'* or banana leaves. Thus, there is no variety of cooking methods. In addition, the Penan is using their bare hands in preparing food as well as to consume it. Despite of their large diet, they only hunt and prepare food for a day consumption because they do not have any proper storage facilities. Due to adequate food preparation, Penan generates less food disposal except for inedible substances such as hard bones and elastic skins.

4.4 *Penan's food and foodways after resettlement*

Consequently the development of landscape brings to resettlement of Penan community in Long Berawan to Sungai Asap village. Although the natural resources are low, there are replacements for the community to find food by buying different from normal practices where everything is free and hunted. The village of Sungai Asap is accommodated by few mini-markets and hawkers from other ethnics also ease the Penan community. From there, the community able to buy their necessity such *'kanen'* (rice), can foods, eggs and so on. Besides, chicken and fresh water products become substitutes for Penan's source of protein. Hence the food availability is still abandon for the Penan although they have been removed from their ancestors' land.

Penan's food intake is big because of their routines and activities. This also remains the same even getting bigger as they nowadays working in plantation estates and construction sites. The availability of food source supplied by the market nearby until night time also contributes the large food intake. The diet now is based on what the market offers and sells including can and instant foods. The new resettlement lifestyle has shifted the Penan's food ways to a new level. In addition, cooking methods and utensils used are also changed.

They agreed that the new resettlement gives variety of food substance and cooking method for instance cooking using gas fire, rice cooker as well pans and pots where they are changing their old cooking methods. The Penan perceive the cooking and preparation become more convenient.

Despite of cooking preparations that have been diversified, ways of consuming the food also changed. The usage of spoons and forks are also being practiced by the Penan as they are having their meals and no longer use their bare hand. In relation of socializing with other ethnics, exposure of using refrigerator also influences the Penan community. This gives them a place to keep their excessive foods. Besides, the Penan can keep their raw food for future usage to avoid wastages of food. With regards to the new resettlement, socio-environment also has been altered to adapt with people surround. It has shifted the modernization of Penan foodways especially in food storage. The usage of jars and containers for example is one of the new things in storing dry food ingredients such as sugar, salt and onions.

Nowadays the community move to pastoralism stage which farming livestock such as chickens and duck. The leftover like rice, other dishes and fruits being used to feed the animals. The Penan community still keeps dog as accompanies whether to secure the house or when they go to jungle. The Penan also gives their leftover food to their dogs. They are still continuing the tradition of feeding their animal as a mean of disposing food rather than dispose as rubbish.

5 CONCLUSIONS

The new resettlement has given big impact to the Penan lifestyle especially to as proven by the finding of this study. The nomadic and semi-settled styles of life have limited the exposure as living in deep and thick jungle far away from economic development. Although the shifting of culture specifically in food and food practices have changed more or less of the traditional practices, it also gives more benefits and advantages to the community. It brings diversification in many matters including sources, cooking methods, food consumption and food storing. The preservation of culture can be done via sincere efforts on helping the community in building their new life with knowledge sharing. It can facilitate them in adapting this new lifestyle. All individual within the community, related NGO's, private sectors, agencies and government should be together to help and maintaining the Penan tribe to assimilate with their new resettlement with the developing environment.

REFERENCES

Akpanudoedehe, J.J., Agba, Ogaboh, A.M. & Ushie, E.M. (2010). Socio-Economic and Cultural Impacts of Resettlement on Bakassi People of Cross River State, Nigeria. *Studies in Sociology of Science. 1*(2), 50–62. Department of Sociology, University of Calabar, Calabar, Nigeria.

Brosius, J.P. (2006). Between Politic and Poetic: Narratives of Dispossession in Sarawak, East Malaysia. Reimagining Polotical Ecology. Duke University Press.

Foley, W. (2005). Tradition and change in urban indigenous food practices. *Postcolonial Studies, 8*(1), 25–44.

Garay, G. (2006). *The Literature Review of Penan Religion.* University of Nevada, Reno.

Gillette, B.M. (1997). Contemporary Chinese Muslims (Hui) remember ethnic conflict: Stories of the late 19th century "Hui Uprising" from Xian. Paper Presented at the Association for Asian Studies meeting, Chicago, United State of America.

Hitchcock R.K. (2010). *Resource rights and resettlement among the san of Botswana.* Uprooted: Dispossession in Africa (Winter 1998) 22.4. Retrieved from http://www.culturalsurvival.org/publiccation/cultural-survivalquarterly/Botswana/.

Iuchi, K. (2010). Redefining a Place to Live: Decisions, Planning Processes, and Outcomes of Resettlement after Disasters. University of Illinois at Urbana.

Jamal, A. (1996). Acculturation: The Symbolism of Ethnic Eating Among Contemporary British Consumers. *British Food Journal, 98*(10), 12–26.

Koizumi, M. (2005). Ethnobotany of the Penan Benalui of East Kalimantan, Indonesia: Difference of Ethnobotanical Knowledge Among Villagers of Long Belaka. Graduate School of Asian andAfrican Area Studies, Kyoto University.

Kuhnlein, H.V., Erasmus, B., & Spigelski, D. (2009). Indigenous People Food System: The Many Dimension of Culture, Diversity and Environment for Nutrion and Health. Quebec, Canada: FAO Publishing.

Kuhnlein, H.V. & Receveur, O. (1996). Dietary Change and Traditional FoodSystems of indigenous Peoples. *Annual Reviews Nutrition. 16,* 417–442.

Kumar, R. (1996). *Research methodology.* London: Sage Publication.

Langub, J. (2008). Penan and the Pulong Tau National Park: Historical Links and Contemporary Life. *Borneo Research Bulletin, 39,* 128–165.

Maruyama, J. (2003). The Impacts of Resettlement on Livelihood and Socia Relationships among the Central Kalahari San. *African Study Monographs, 24*(4), 223–245.

Mora, J.K. (1998). Understanding Multiculturalism: Cultures and Values, San Diego State University, 15 August [Online] Available: http://coe.sdsu.edu/people/jmora/CulturalValues/. [Accessed 26 Oct 2006].

Needham, J. (1953). *Penan and Punan.* The Sarawak Gazette 79:27.

Needham. J. (1972). *Penan, Ethnic groups of insular Southeast Asia,* Vol. 1, pp. 17680. New Haven: Human Relations Area Files Press.

Pearsall, A. (1996). *The Penan.* Social Alternatives, *9*(4), 25–26.

Platter, Denise, T. (1994). Forced Migration and Resettlement as a Strategy for Development: An Analysis of Implication Based on a Study of The New Halfa Scheme in Eastern Sudan. University of Pennsylvania.

Taylor, E.A. (2004). Perceptions of the bureaucratic nightmare of HACCP: A case study. *British Food Journal, 65*.

Urquhart, Ian A.N. (1951). Some notes on jungle Punans in Kapit District. *Sarawak Museum Journal 5*, 495–533.

Zahari, M.S.M., Kamaruddin, M.S.Y., Kutut, M.Z., and Langgat, J. (2011). The Level of Alteration of Ethics Native Food: (A Case of Sarawak, Malaysia). *International Journal of Humanities and Social Science, 1*(6), 137–144.

Zaman. M.N. (2010). *Land Acquisition and Compensation in Involuntary Resettlement.* Land and Resources (Winter 1990) 14.4. Retrieved from http://www.culturalsurvival.org/ourpublications/csq/article/.

Zeleke, T. (2012). Resettlement and Sustainable Food Security: A Comparative Study of Inter-Zonal and Intra-Zonal Resettlement Schemes and Host Communities in Dawuro Zone, Southern Nations, Nationalities and People Region, Ethiopia. *Journal of Sustainable Development in Africa, 14*(2), 125–148.

Hospitality and Tourism – Sumarjan et al. (eds)
© 2014 Taylor & Francis Group, London, ISBN 978-1-138-00151-0

The differences of food consumption and practices during confinement between urban and rural Malay women

S.A. Yusof, S. Zulkifli, S.S.H. Shaidan & M.S.Y. Kamaruddin
Universiti Teknologi MARA, Malaysia

ABSTRACT: This paper is an exploratory study using a qualitative research interview to explore the differences of food consumption and practices during confinement between urban and rural Malay women. The informants consists of ten Malay women; 5 from urban area while the remaining from rural areas in the state of Selangor. The finding revealed that traditional confinement norms are still practiced in rural as well as urban areas. They also share similar belief about food categories during confinement period that include hot, cold, windy, itchy, poisonous and sharp. In terms of food practices, the only difference between both areas is the amout of practices being applied during confinement period.

Keywords: confinement, food consumption, food practices, rural, urban

1 INTRODUCTION

Confinement period refers to a phase for women after giving birth and an overview from the literature shows that the practices may vary among different ethnicity and cultural norms. Women will undergo this phase for a certain period of time. Regardless of its duration, the main purpose of confinement period is to heal the body, soul and psychology of the mother after a long and challenging moment of pregnancy (Mohamed & Bidin, 2012, Tee et al., n.d.; Warren, 2010). As for their food consumption during this stage, it differs between places (Hafez et al., 2010; Zahhura et al., 2012) and culture (Chmielowska et al., 2007; Freimer et al., 1983). Different people have different preferences for dietary intake during confinement and this is influenced by their culture.

This study focuses on Malay women in urban and rural areas. Previously, many researches have been done to examine food consumption during confinement but not much research looking from the perspectives of two societies between urban and rural areas. Based on comparison from different readings, it is apparent that the lifestyle beliefs, habits, as well as their food intake among both societies are different. Hence, there is an eventuality that their food consumption during confinement may be different as well. Early assumption shows that women in rural areas still continue practicing traditional methods when handling confinement period because they strongly hold on to the custom of their ancestors. Meanwhile, women in urban areas are prone to modern practices. Therefore, this study aims to explore the differences of food consumption and practices during confinement between rural and urban Malay women.

2 LITERATURE REVIEW

2.1 *Confinement period*

The confinement period varies within place and culture. For instance, women in Russia undergo this period for 40 days, Thai women for 30 days and Malay Women for 40 days (Mohamed & Bidin, 2012). According to Poh et al. (2005), he said that most of the Chinese women (82%), having 30 days of confinement period. Another author Simbar et al. (2004), said that there are three stages which are immediate, refer to the first 24 hours after giving birth, early postpartum refer to the first week and late postpartum which refer to the first six weeks after childbirth. In contrast, Lamxay et al. (2011) mentioned that in Southeast Asian cultures the women undergo their confinement period within 10 to 45 days. It is concluded that, based on different researchers, Malay culture practices their confinement period within 40 to 48 days (Zamani, 2001 & Rohaini, n.d.).

2.2 *Food during confinement*

In general, it is believed that every food has its own properties. The properties will somehow bring different effects towards women in postpartum period. As stated in the research by Shahar et al. (2000), according to rural elderly Malays food habits, there are eight properties of food which are cooling, windy, hot, sharp, poison, itchy, fatty and beverages. In contradiction, Mohamed et al. (2012) said that there are only six categories of food such as hot, cold, poisonous, sharp, windy and itchy. Despite having different views on the number of categories of food, these researchers do have the same opinion on the examples of food in the six categories.

In Chinese culture, they believed everything in the world is structured by the elements of *yin* and *yang*. They practice their diet by balancing the *yin* and *yang* elements which is 'hot' and 'cold' food (Thagard, 2006). Ginger, sesame seed oil and rice wine are considered as hot food. Meanwhile, most vegetables and fruits are considered cold foods such as winter melon, rambutan, papaya, cucumber, mushroom, and *kangkung* (Poh et al., 2005). With an imbalance of *yin* and *yang*, they are prone to diseases which can be recovered using herbs (Thagard, 2006). On the other hand, some researchers said that Chinese people believe that there are five categories of food such as sour, spicy, cold, sharp, and toxic (Tee et al., n.d.).

For Malays, drinking *air akar kayu* is a common practice (Zamani, 2001). He further added, *air akar kayu* is an energized drink made from remedial plants where it can be produced by boiling the plant's wood and should be drink while it is still warm. He also said that, besides *air akar kayu, kacip fatimah*, is also widely consumed herb.

2.3 *Food practices during confinement*

According to a review article based on Zamani (2001), traditional practices during postpartum is still widely popular in Malaysia. He also mentioned, 80% of the world's population depends on traditional remedy to maintain their health. He further added, these practices have been passed down from generation to generation and still being strongly followed. This is due to their practical experiences and observations which uses plants, and mineral substance. Women body are said to be cold after giving birth, so it is advisable for them to have warm drinks and eat 'hot' foods during confinement period such as *kacip fatimah*, tonic drinks, *jamu* and *makjun* (traditional medicine made from herbs and spices). Tonic drinks are meant to regenerate their energy, smoothens blood flow and prevent illness (Ramli, et al., 2010; Zamani, 2001). Although most seafood are discouraged to be eaten, but *ikan haruan* is commonly encouraged as it has a healing effect towards the dermal wounds especially for caesarean women. Cooking methods are also part of the considerations for food restrictions as they are not allowed to eat fried and oily foods (Zamani, 2001).

Despite having all the traditional practices, some researchers have their own findings on modern practices stating that their food choices are influenced by diversity of individual and environment factors. According to Mirsanjari et al. (2012), in terms of the individual factor, knowledge is useful in enhancing their dietary behavior by assuming that they are open to new information. Consequently, this may result in changes in attitude and practices (Mirsanjari et al., 2012). Similarly stated by Thi et al. (n.d.), mothers with advanced education background are less expected to follow the traditional practices and they also require scientific evidence in order to follow the certain restrictions. Chmielowska et al. (2007) also mentioned that it is essential for a mother to have a nutritious and balanced diet in order to regain their health. Based on Poh et al. 2005, majority of the Chinese women are practicing *zuo yuezi* (Chinese traditional method), but they also mentioned that the practices should be well incorporate with the modern method in order to achieve a balance diet. Tee et al. (n.d.) mentioned that there is a need to make them aware of the importance of the nutritional needs.

According to the Ministry of Health Malaysia's website, they are promoting a variety of care that combines traditional medicine and modern medicine for the mother's well being. These services may help doctors to make use of traditional beliefs to encourage health in the confinement period as well as to avoid potentially harmful beliefs or practices. Nevertheless, many still practicing the traditional way that has been brought down from generation to generation.

3 METHODOLOGY

The study is an exploratory study to show the differences in food consumption among Malays during confinement in urban and rural areas. It is in nature of an exploratory study that a qualitative approach is being used whereby, a qualitative research interview has been used to collect information. The in depth interview is done through a face-to-face and telephone interview. The purpose of using the interview method is to see the whole idea of the topic from the interviewees' perspective and understanding on how the idea was developed (Cassell et al., 2004).

The study uses a nonprobability sampling where Judgement Sampling is being used. Since this study is about confinement, only women that have gone through confinement period were selected as samples. The informants were taken from urban and rural areas in Selangor. Since the study is a comparative study, it is more relevant to compare the samples of the same state and culture.

The samples consist of 10 individuals where five informants are from rural area and another five informants are from urban area. The age group varies from 20's to early 60's. Besides the age factor, the numbers of children of the informants are also being considered.

4 FINDINGS

In general, we can see that the age groups of the informants are between 25 to 63 years old. Most of them have numbers of children, 1 until 7. The majority of their confinement period is between 42 to 90 days. Most of them follow food restrictions; however there are some who believes otherwise. However, all of them are believed to follow almost the same food practices. In overall, the respondents follow their food consumption and food practices until the end of their confinement period.

4.1 *Food consumption*

4.1.1 *Urban*

All the urban informants believe that food can be divided into six categories which are hot, cold, windy, poisonous, spicy, and also itchy, except for the informant 1U.

> *"I don't believe in all the food restriction because if a person is already allergic to the particular food, of course she will experience negative effects towards the body during confinement. I only follow certain restrictions just to avoid hurting my mother's and mother in law's feelings."*– Informant 1U

As informant 1U is a doctor herself, it shows that she holds her beliefs based on scientific evidence. For other informants (2U, 3U, 4U, 5U), they believed that out of all these categories, only hot food shall be consumed, as for the rest, it shall be avoided. This is because, during confinement, their body states are cold, so they need to consume hot food because it will give a good blood circulation.

> *"I will eat food while it is still hot in order to avoid muscle aches. I eat hot rice mixed together with ginger and black pepper most of the time to remove air from body. Besides that, I eat 'jamu' which is in powdered form together with 'telur ayam kampung' to give heat to the body, a lot of vegetables such as mustard leaves, cabbage, parsnip, and kailan to avoid constipation and increase milk production. I only consume chicken and certain types of fish which is not poisonous such as mackerel, seabass and many more. As for fruits, I take green apples and oranges but I avoid bananas because it is said to be cold. As for drinks, I am only allowed to drink coffee and milo."* – Informant 2U
>
> *"I eat all vegetables except for the cold ones such as cucumber and cabbage. I don't eat fruits at all, but I eat chicken, fish that does not bring itchiness, 'jamu' in capsule form and also I drink sufficient water. Besides that, I also take 'ikan haruan' essence which is sold in the market."* – Informant 4U
>
> *"I consume a lot of fish which are mostly expensive ones such as salmon, 'jenahak', and 'senangin'. The fish are all cooked with a lot of ginger, black*

Table 1.

Urban					
Respondent no.	Age	Education level	Working status	No. of children	Confinement period
1U	30	Degree	Working	1	90
2U	50	SPM	Working	4	42
3U	63	SPM	Working	7	44
4U	31	Degree	Working	2	44
5U	34	Degree	Working	1	44

Table 2.

Rural				
Respondent no.	Age	Education level	Working status	No. of children
1R	55	SPM	Not working	3
2R	28	Master	Working	1
3R	34	SPM	Working	1
4R	24	Degree	Not working	1
5R	52	SPM	Not working	4

pepper, and turmeric to heal internal wounds. As for vegetables, I eat the ones which is not windy such as parsnip, mustard leaves and 'pegaga'. Whenever I eat rice, it must be in a hot temperature and avoid spicy food that can cause air which affects both mother and baby. I do take 'jamu' which is for the womb to shrink back to its normal size." – Informant 5U

Taking from the wider perspective, it shows that the informants do believe in food consumptions which consist of certain restrictions and avoidances.

4.1.2 *Rural*

In rural areas, all the informants (1R, 2R, 3R, 4R, 5R) believe in all the food categories. Somehow all of them have different pattern in food consumption.

"I eat rice when it is still hot, then I eat fishes which are not itchy such as 'senangin', 'parang', 'bawal putih' and mackerel. Somehow, I don't eat chicken and beef but I eat vegetables such as spinach, parsnip, carrot, mustard leaves and potatoes. I drink plain water, coffee, milo, horlicks, and avoid tea because it is said to be windy ... I avoid taking spicy food which comes specifically from chillies but I do eat spices such as black pepper in a large amount than usual." – Informant 1R

"I eat all, chicken, beef, and fish such as red snapper and 'jenahak'. As for vegetables, I only eat mustard leaves and parsnip. For fruits, I only take oranges. For drinks, I take only orange juice, dates juice and also blended daun 'sirih', guava leaves and warm water. Seafood is strictly avoided because it brings itchiness and 'jamu' is also avoided because both give negative effects towards the breastfed baby."– Informant 2R

4.2 *Food practices*

4.2.1 *Urban*

Some of the informants do have similarities in terms of food practices. Some may practice *bengkung* (2U, 3U, 4U, 5U), *param* (2U, 3U, 4U), *pilis* (4U, 5U), herbs bathing (1U, 2U, 3U), and massaging (1U, 4U, 5U).

"During confinement, I use 'bengkung' as to slim down the abdomen area. It is normally used with lime, fresh turmeric and 'kapur' being spread on the stomach before wrapping with white cloth. Normally it is practiced until the end of the confinement period." – Informants 2U, 3U, 4U, 5U

"... I practice 'param' which uses uncooked rice which has been pounded with some herbs and being spread all over the body. This is to provide heat towards the body." – Informants 2U, 3U, 4U

"... I normally practice 'pili's where it uses mixed herbs and spices that will be spread on the forehead. The function is to remove air and avoid headache" – Informants 4U, 5U. *"Somehow, I don't really do it daily, I cannot stand the smell ..."* – Informant 5U

"I've got no choice, but to follow this practice every day as I've been taken care of my mother-in-law ..."– Informant 4U

"Normally, every morning and evening I will have herbs bath for the whole confinement period as this rejuvenates, gives freshness and provides good smell to the body" – Informants 1U, 2U, 3U

"... massaging has been part of my daily routine. It uses oil which has been extracted from plants. It gives me good blood circulation as well as relieve any muscle aches"– Informants 1U, 4U, 5U. *"... I normally have a massage lady at home every day that comes to massage me ..."*– Informant 5U

All the informants do have some sort of practices during confinement that they hold on to, but what determines the type of practices they do, depends on various factors such as knowledge, advices from family as well as from previous readings regarding confinement.

4.2.2 *Rural*

All of them (1R, 2R, 3R, 4R, 5R) do practice *bengkung* during their confinement period. However, not everyone share the same practices. As for *bertungku* and *berpilis* only informants 1R, 4R and 5R perform it. Then, informants 1R, 2R, 4R and 5R practice doing the same thing which is *berurut*. Besides that, informants 2R, 4R and 5R will normally have herb bath and only informants 2R and 4R take supplements by Shaklee which are vitamin B complex, vitamin C and multivitamin. Only informants 1R and 5R make it a habit eating *jamu akar kayu*. Meanwhile, informant 1R practices *berparam*.

"I practice 'bengkung' using 'kapur' and lime before wrapping it around my body throughout my confinement period" – Informants 1R, 2R, 3R, 4R, 5R. *"... I also put some ointment before wrapping myself ..."*– Informant 5R

"... I do 'bertungku' every day ..."– Informants 1R, 4R, 5R. *"... I normally do it before sunset ..."–Informant 4R. "... I'd prefer bertungku around after 2 pm ..."*– Informant 5R

"... for 'berpilis', I normally use herbs and spices which has been pounded into powder and just spread it onto my forehead to reduce headache" – Informants 1R, 5R. *"... I just practice 'berpilis' ... i don't even know what's being used in making the paste ..."*– Informant 4R

"... for 'berurut', normally oil extracted from plants are used to have a smooth blood circulation, remove air and also relieve muscle pain" – Informants 1R, 2R, 4R, 5R. *"... usually I practice berurut after bathing..."*– Informant 5R

"To remove unwanted smell, and give freshness to the body, I will have herb bath which uses various herbs such as 'serai wangi', 'daun senduduk putih',

'akar kayu', flowers and 'daun sirih' ..."– Informants 2R, 4R, 5R

"Besides that, I do take supplements from Shaklee ... vitamin C, vitamin B complex and multivitamin in order to aid in breastfeeding ... I take it because I don't take any 'jamu' ... got the idea from friends and internet sources."– Informants 2R, 4R

"... I drink only 'jamu akar kayu' ..."– Informant 1R. *"... I drink 'jamu akar kayu' because it comes with the package when I buy the 'bengkung' set ..."* – Informant 5R

"... and also 'berparam' ... made from rice flour and herbs which will be spread all over the body to prevent from excessive air and also to keep my body warm at times ..."– Informant 1R

5 CONCLUSIONS

Traditional confinement practices are still dominating among Malay women in both urban and rural areas in Selangor. Looking into the perspective of food consumption, both urban and rural give some sort of a difference due to their environment, knowledge, local beliefs and custom. Meanwhile, from the food practices perspective, not much difference can be seen. This could be due to their strong beliefs, as well as it is more practical and effective.

There are several limitations that can be found. Firstly, the research mainly focuses on Malay culture only which narrows down the perspective. Besides that, the geographical limitations also affect the findings where it is mainly located in one state which is Selangor. This will limit the variety of information that needs to be gathered. Furthermore, by having a small sample size, which consists of 5 urban and 5 rural, cannot represent a large number of populations. Thus, reduce the level of information precision. Lastly, the method used to gather data could also affect the findings. As for the research, a phone interview is less appropriate as compared to a face-to-face interview because body language, intonation and eye contact cannot be made. All these limitations that contribute towards the results cannot be generalized.

REFERENCES

Cassell, C., & Symone, G. (2004). *Essential Guide to Qualitative Method in Organizational Research.* Sage Publication: London.

Chew, K.S., Tan, T.W., & Ooi, Y.T. (2011). Influence of Chinese cultural health beliefs among Malaysian Chinese in a suburban population: A survey. *Singapore Medical Journal,* 52(4), 252.

Chmielowska, E., & Shih, F.S. (2007). Folk Customs in Modern Society: 'Tradition of Zuoyuezi' in Taiwan: An Physical Anthropology Perspective.

Freimer, N., & Echenberg, D. (1983). Cultural Variation-Nutritional and Clinical Implications. *The Western Journal of Medicine,* 6, 139.

Hafez, S.K., & Yakout, S.M. (2010). Early postpartum dietary practices among a group of Saudi women. *Journal of American Science,* 6(11).

Lamxay, V., Boer, H.J.D., & Bjork, L. (2011). Traditions and plant use during pregnancy, childbirth and postpartum recovery by the Kry ethnic group in Loa PDR. *Journal of Ethnobiology and Ethnomedicine,* 7–14.

Ministry of Health Malaysia, (n.d.). Tradiitonal and Complementary Medicine Division. Retrieved May 13, 2013, from http://tcm.moh.gov.my/v4/modules/mastop_publish/?tac=30.

Mirsanjari, M., Muda, W.A.M., Ahmad, A., & Shukri, M. (2012). Dietary Knowledge and Behaviors in a Sample of Malay Pregnant Women. *International Annual Symposium on Sustainability Science and Management,* 108.

Mohamed, A., & Bidin, A.A. (2012). Penjagaan kesihatan wanita semasa dalam pantang: Amalan dan kepercayaan. *Malaysia Journal of Society and Space,* 7, 20–31.

Poh, B.K., Wong, Y. Peng., & Karim, N.A. (2005). Postpartum dietary intakes and food taboos among Chinese Women attending maternal and child health clinics and maternity hospital, Kuala Lumpur. *Malaysia Journal Nutrition,* 11(1), 1–21.

Rohaini, I. (n.d.). Postnatal treatment: The Malay way. Hospital. *The International Medical Journal,* 9(1).

Sekaran, U., & Bougie, R. (2009). Research methods for business: A skill building approach. John Wiley & Sons: United Kingdom.

Shahar, S., Earland, J., & Rahman, S.A. (2000). Food intakes and habits of rural elderly Malays. *Asia Pacific Journal Clinic Nutrition,* 9(2), 122–129.

Tee, E.S., Kandiah, M., & Hanis, H. (n.d.). Food consumption and habits in rural Malaysian villages recent finding and some thoughts for improvement.

Thi, L.M., Pasandarntorn, W., & Rauyajin, O. (n.d.). Traditional postpartum practices among Vietnamese mothers in Anthi district, HungYen province.

Zamani, A. (2001). Traditional Practices In Postnatal Care: The Malay Community In Malaysia. TSJM, 2.

Zahhura, S.S.A., Nilan, P., & Germov, J. (2012). Food restrictions during pregnancy among indigenous temiar women in Peninsular Malaysia. *Malaysia Journal Nutrition,* 18(2), 243–253.

Hospitality and Tourism – Sumarjan et al. (eds)
© 2014 Taylor & Francis Group, London, ISBN 978-1-138-00151-0

Chinese cooking methods terminologies: Understanding and usage among young Chinese culinarians

K. Mohamad Abdullah, M.S. Mohd Zahari, R. Muhammad & M.A. Hemdi
Universiti Teknologi MARA, Malaysia

ABSTRACT: Chinese as one of the major ethnic in Malaysia is having vast and valuable cooking methods terminologies which have been created and used since the olden days. Owing to the lack of understanding and usage, much of the Chinese cooking terminology however are believed to be gradually fading out, especially among the younger generation. This study empirically investigates the level of understanding of Chinese cooking methods terminologies and its impact to the usage among the young Chinese culinarians in their daily cooking activities. The results revealed that, the majority of the young Chinese culinarians has a relatively poor understanding and usage of the terminologies. The efforts to increase the understanding and the usage of the terminologies, as well as preserving the Chinese heritage, language of foods, especially among the Chinese culinarians should be undertaken by all parties and the relevant authorities.

Keywords: Chinese, cooking methods, terminologies, understanding, usage

1 INTRODUCTION

Language plays a paramount role as a medium of communication that mirrors one's identity, elaborating and transmitting culture (Jiang, 2000). Lazear (1997) stated that language is like anthropomorphic organisms and the existence of it is dependent on the communicative activities and behaviours of their speakers. Within the Chinese community, language is more than simply a tool for communication but a living fossil carrying a culture and witness to the Chinese history and truly represents the regional culture (Li, Hong, Zheng, & Chua, 2010). Changed of dynasties, waves of migration, life of people and other aspects of history and culture are all recorded or reflected in the language. Since language is also an element of culture, protecting language means protecting the diversity of cultures (Ko, 2010). In other words, protecting the language is to preserve the richness and diversity of cultural heritage and thing behind them.

Within culinary perspectives, not to exaggerate that Chinese is one of the major ethnic groups in Malaysia having numerous, vast and valuable food terminologies which represent one of the language aspects that was created and used since the olden days (Lee, 1995). However, Li, et al. (2004) states that the biggest single threat to the Chinese food language is this modern culture and society is marginalizing tradition of the past. The richly diverse cultural and family traditions and the intergenerational wisdom that is imbedded in them are gradually faded. According to Jiang (2000), Chinese food terminologies are gradually reducing, the number of speakers has been decreasing and the circumstance of usage has been deteriorating in the past few years. This phenomenon occurs not only among the ordinary Chinese youngsters but also goes to young Chinese professional chefs and culinarians. To date, studies discovered low understanding and usage among the young Malay culinarian on Malay foods terminologies (Mohd Zahari, 2012). The question arises is the same phenomenon happening among the young Chinese culinarians? With the emphasized issues and the study gaps, this study aims to examine: a) general perceptions of Chinese food terminologies b) The level of understanding of Chinese cooking method terminologies; c) The level of usage of Chinese cooking methods terminologies; and d) the relationship between Chinese Food Terminologies Attributes and the Usage of it among the young Chinese culinarians.

2 LITERATURE REVIEW

2.1 *Food terminologies*

Food is a language with a structure and meaning that through its own words is related to individual feelings and beliefs about human relations as well as one's own identity. Food language is commonly expressed through its numerous terminologies (Mills, 2007). Food terminologies are used in describing its various operations and understanding of it is essential to the people of a particular country, especially those individuals who are involved in the culinary fields (Cook, 2010). Zhifang (2002) posited that food terminologies are the standardized way to communicate and defined cooking techniques or methods of preparation; therefore using the right terminologies is important in the culinary field. He added that ethnic cuisines often have their own vocabulary and terminologies which are unique to their culinary traditions. According to Collu (1991) food terminology could have one set of terms for everyday cooking and another for ceremonial cooking, therefore without food terminologies, secrets of cooking, techniques and skills may not be transferred and people may not acquire the knowledge of cooking through the generations, and cultural continuity will not exist (Lowinsky, 1992). On the other hand, terminologies of food may not be static but constantly growing and the changes which occur signify cultural swings in the way the terminologies are used and practiced by people (Matt, 2010).

2.2 *Chinese food terminologies*

Different cultures tend to have their own unique ways of cooking and each of them has different terminologies which often come from historical necessities (Kamaruddin, et al. 2012). As mentioned by Grant, (1999), Chinese, as the major ethnic group in Malaysia also possess their own food terminologies, from the preparations, methods of cooking to numerous unique of food names. He also claimed that Chinese food terminologies have been shaped by cultural transmission over generations and the average Chinese parents would usually pass the skills and processes of cooking on to their children through the terminologies that act as the medium of transmission, which occurs not only during daily cooking activities and traditional events, but also during wedding ceremonies (Kassim, 2008).

Some of the Chinese cooking ingredient terminologies includes *wǔ xiāng* (five spice seasoned-incorporating the five basic flavors of Chinese cooking), hóngfǔrǔ (red fermented tofu) and *dòuchǐ* (fermented black beans). Another important aspect of Chinese food terminologies is the traditional cooking equipments and utensils that have been used for thousands of years (Li, et al. 2010). Some of them may be disappearing in the modern world and some of them can only be seen in rural areas of Chinese families. Although some equipments may look rather unusual, these tools help tremendously in the preparation of some of the most delicious Chinese dishes (Lee, 1995). Despite the appearance of modern appliances, several types are still relevant, such as *qì guō* (bamboo steamer), *dǐng* (ancient cooking cauldron with two looped handles), *suǎn* (ancient utensil for washing rice), *tiě guō* (iron cooking pot), *qiēròu dāo* (meat cleaver) and many others.

2.3 *Chinese cooking methods terminologies*

There are so many specific techniques in Chinese cooking, apart from the well-known "stir-fry". According to Newman (1999), numerous methods of cooking terminologies are used in Chinese cooking which consists of dry, moist and combination methods. One of the earliest cooking methods is to broil called *zhi*, all that needed is to put meat over the fire. Jīan (Cooking in a pan with a light coating of oil or liquid and allowing the food to brown), *Xūn* (Cooking in direct heat with smoke and the source of the smoke is typically sugar or tea) and Kǎo (cooking by hot air through convection or broiling) are examples of terminologies for dry-heat cooking methods. On the other hand, moist-heat cooking methods include the terms such as *Shāo* (braising ingredients over medium heat in a small amount of sauce or broth and simmering for a short period of time until completion), *Zhēng* or *Xún* (Steaming food to completion over boiling water) and *Áo* (Cooking slowly to extract nutrients into the simmering liquid, used to describe the brewing process in Chinese herbology with the intention of using only the decocted brew) (Lum, 2005). For combination methods includes *Mēn*, (Stir-frying the ingredients until partially cooked and simmer with broth until the broth is fully reduced and ingredients are fully cooked) and *Hùi* (Quick precooking in hot water and finished by stir-frying). Clearly understanding the terminologies of the most used techniques can enhance culinary understandings (Yoshino, 2010).

3 METHODOLOGY

As this study examines the understanding and usage of Chinese cooking methods terminologies among young Chinese culinarians, a quantitative research approach is considered to be the most suitable method to be used. To base upon actual experience, the information is required for this study are obtained through self-administered questionnaire with young Chinese clearance who are working in the food industry or culinary field in Kelang Valley (Shah Alam, Petaling Jaya, Subang Jaya and Kelang), with the age range from 22 till 30 years old, who previously have undergone formal culinary education.

A survey questionnaire comprised of three sections. Section A elicited the general perception of

Chinese food terminologies among young Chinese culinarians. Respondents were required to translate their views on five types Likert scales ranging from 1 = Strongly Disagree, 2 = Disagree, 3 = Neither, 4 = Agree and 5 = Strongly agree. The understanding of Chinese cooking methods terminologies were directly investigated in Section B using the interval scale ranging from 1 = Very poor, 2 = Poor, 3 = Fair, 4 = Good and 5 = Very Good. Finally, the levels of usage of Chinese cooking methods terminologies were asked in Section C using five-point interval scales with 1 = Never, 2 = Rare, 3 = Sometimes, 4 = Often, and 5 = Very often.

The survey questionnaire was personally administered by the researchers. Two weeks were spent for the survey and in light with positive feedback 200 responses were obtained.

4 ANALYSIS AND RESULT

4.1 *General perceptions of Chinese food terminologies*

Frequencies showed that 64 percent (n = 128) of respondents were males compared to 36 percent of (n = 72) females. Descriptive analysis was then undertaken looking at general perceptions of Chinese culinarians toward Chinese food terminologies. Based on Table 1, the magnitude of mean scores ranging from 4.05 to 4.43 indicates that the majority of young Chinese culinarians agreed with all items. They agreed that Chinese are rich in cooking methods terminologies. They were certain that the Chinese recipes, techniques and secret cooking skills can be learnt through its food terminologies. They felt that the understanding and the usage of Chinese food terminologies is very important for Chinese culinarians as regular usage could preserve the Chinese culinary heritage. With that feeling, they felt that it is important to pass down the Chinese food terminologies to the young generation and to teach and use Chinese food terminologies, apart from English and French, in culinary institutions.

4.2 *The level of understanding of Chinese cooking method terminologies*

The results of the descriptive statistics in Table 2 revealed that, although Chinese is loaded with cooking methods terminologies, the young

Table 1. Mean score of the general perceptions of Chinese food terminologies.

Items	M	S.D
The Chinese are rich in valuable and unique food terminologies.	4.26	0.652
The Chinese are rich in cooking method terminologies.	4.05	0.693
Chinese recipes, techniques, secret cooking skills can be learnt through the food terminologies.	4.06	0.685
Understanding and usage of Chinese food terminologies are very important to Chinese Malaysian culinarians.	4.30	0.671
Regular usage of the Chinese food terminologies by Chinese culinarians will help to preserve the Chinese culinary heritage.	4.43	0.705
Important to pass down the Chinese food terminologies to the young generation.	4.43	0.733
Important to teach and use the Chinese food terminologies, apart from English and French terminologies, in culinary institutions.	4.24	0.877

Scale: 1 = Strongly Disagree, 2 = Disagree, 3 = Neither, 4 = Agree, 5 = Strongly Agree.

Table 2. Individuals' mean score for understanding of Chinese cooking methods terminologies.

Items	M	S.D
My knowledge of Chinese cooking method terminologies in general.	2.06	0.799
My understanding of Chinese dry-heat cooking method terminologies.	2.21	0.866
My understanding of Chinese moist-heat cooking method terminologies.	2.25	0.996
My understanding of Chinese combination cooking method terminologies.	2.18	1.205
My understanding of the method of *Wēi.*	2.10	1.096
My understanding of the method of *Xūn.*	2.26	0.983
My understanding of the method of *Mēn.*	2.28	1.162
My understanding of the method of *Lǔ.*	2.38	0.869

Scale: 1 = Very poor, 2 = Poor, 3 = Fair, 4 = Good, 5 = Very good.

Chinese culinarians still had a fair understanding of them in general. This can be seen through the moderate mean scores given to the understanding of Chinese dry-heat, moist-heat and combination cooking method terminologies. These results were further supported by the moderate mean scores given to some examples of Chinese cooking methods terminologies like *Wēi* (Slowly cooking in a ceramic vessel of broth and other ingredients by placing it in or close to hot embers), *Xūn* (Cooking indirect heat with smoke and the source of the smoke is typically sugar or tea). *Mēn* (Stir-frying the ingredients until partially cooked then cover and simmer with broth until broth is fully reduced and ingredients are fully cooked) and *Lǔ* (Cooking over prolonged and constant heat with the ingredients completely immersed in a strongly flavored soy sauce based broth).

4.3 *The level of usage of Chinese cooking methods terminologies*

Owing to the moderate level of understanding of the Chinese cooking methods terminologies, therefore it is not surprising that the levels of usage of Chinese cooking methods terminologies in daily cooking activities among the young Chinese culinarians were also relatively low.

Referring to Table 3, the majority of young Chinese culinarians admitted that they only used a few Chinese cooking methods terminologies in their daily cooking activities and this possibly reflected the relatively low usage of Chinese cooking methods terminologies among their working colleagues, during their culinary studies or culinary exposure as well as in recipes and menu writing.

Table 3. Individual mean scores of usage of chinese cooking terminologies.

Items	M	S.D
Level of usage of Chinese cooking methods terminologies in general.	2.24	0.920
Level of usage of Chinese cooking methods terminologies during my culinary studies or exposure.	2.33	0.987
Level of usage of Chinese cooking methods terminologies in recipes and menu writing.	2.39	0.925
The usage of Chinese cooking methods terminologies among my working colleagues.	2.22	0.938
The usage of Chinese cooking method terminologies in my daily cooking activities.	2.21	0.905

Scale: 1 = Never, 2 = Few, 3 = Sometime, 4 = Often, 5 = Very often.

Table 4. Multiple regression results.

	Usage of Chinese food terminologies
	Beta coefficients and significance levels
Cooking methods	0.39**
R^2	0.15

Note: *p < 0.05, **p < 0.001, ***p < 0.000.

4.4 *Relationship between Chinese food terminologies attributes and the usage*

The relationship between Chinese cooking methods terminologies and the usage of it among the young Chinese culinarians was confirmed with standard multiple regression. The result in Table 4 signifies that Chinese cooking methods terminologies was able to explained 15 percent ($R^2 = 0.15$, $p < 0.001$) of the variance in the usage of it. Understanding of cooking methods terminologies ($\beta = 0.39$, $p = 0.001$) had a relatively low impact on the usage. This result indicates that the understanding of Chinese cooking methods terminologies had a relatively small significant influence on the usage among the young Chinese culinarians. This results in fact similar to Zahari et al. (2012) which states that the less the culinarians understand the food terminologies, the fewer the terminologies used in their daily cooking activities.

5 IMPLICATION, RECOMMENDATION AND SUGGESTION

The study findings obviously indicate that preserving the Chinese food language is critically important as it help to enrich and preserved the Chinese culture, tradition and language at the same time. However, it is likely that the influence of other foreign food languages such as English and French among the young Chinese culinarians reduces the awareness of the importance of Chinese food terminologies in the culinary field. The advancement of processed foods, which enhances the varieties of new ingredients and products of Chinese food becomes a growing concern as many believe that this is leading to the unwillingness to learn and use traditional Chinese food terminologies.

In this sense, transferring the Chinese food terminologies and knowledge from generation to generation is the most effective way in achieving the preservation of it. It is suggested that every Chinese mother should encourage their children to help them in the kitchen, introduce the name of the foods, ingredients and the way of preparing it. The Chinese food terminologies should also be taught and introduced in culinary institutions, apart from English and French food terminologies, so that the culinary students will be aware of and understand their functions as well as the meaning of the terminologies. In addition, the relevant authorities, either in the public or private sector should help to increase the awareness and disseminate the information on the importance of preserving the Chinese heritage, language and foods through the media and food events. Such actions would at least encourage the young generations, especially the culinarians, to better understand and appreciate the Chinese foods language and heritage. Increasing the use of traditional Chinese food terminologies in recipe books as a source of references for future generations could also be done.

As a conclusion, knowledge and understanding of Chinese food terminologies are very important, especially among the young generation. This is because, the more Chinese food terminologies are understood, the better people will be able to communicate and use the terminologies in their daily life and at the same time will be able to sustain and constantly use the terminologies in the future.

REFERENCES

Collu, G. (1991). *The language of food in the fiction of Barbara PYM.* Unpublished master dissertation, McGill University Montreal.

Cook, G. (2010). *Sweet talking: Food, language, and democracy*. Open University, United Kingdom: Cambridge University Press.

Grant, D. (1999). Teaching Chinese culture through literature and food free press, New York, 131.

Jiang, W. (2000). The relationship between culture and language. *Journal of ELT, 54*(4), 328–332.

Kamaruddin, M.S.Y., Bakar, S.K.A., Mohd Zahari, M.S., Bakhtiar, M.F.S. (2012). Routine of cooking practices among Malaysian Working women. *Proceeding of the International Hospitality and Tourism Conference, Kuala Lumpur:* Universiti Teknologi Mara, Malaysia.

Kassim, A. (2008). Malay language as a foreign language and the Singapore's. *GEMA Online Journal of Language Studies, 89*(1), 47–56.

Ko, W.H. (2010). Evaluating effective culinary learning of food and beverage department students at technology universities. *Journal on World Transactions on Engineering and Technology Education, 8*(4).

Lazear, E.P. (1997). Culture and language. *Journal of Political Economy.* 1–50.

Lee, J. (1995). What is Real Chinese Food? *International Journal on Multicultural Societies* (IJMS), 2(3), 17–21.

Li, G., Hong, R., Zheng, Y.T., & Chua, T.S. (2010). Learning Cooking Techniques from YouTube. Springer-Verlag Berlin Heidelberg. 713–718.

Li, J.R. & Hsieh Y.H. (2004). Traditional Chinese food technology and cuisine. *Journal of Asia Pacific, 13*(2), 147–155.

Lowinsky, N.R. (1992). *Stories from the motherline: Reclaiming the mother-daughter bond, finding our feminine souls.* Los Angeles: Jeremy P. Tarcher, Inc.

Lum, F. (2005). Why is a terminology important? *Journal of American Academy of Ophthalmology,* (112), 2–12.

Matt, N. (2010). *Modernist cuisine, the art and science of cooking: The language of food.* (On-line) Retrieved from http://modernistcuisine.com/2010/10/the-language-of-food.

Mills, L.S. (2007). *Culinary terms defined.* Lehigh Carbon Community College: DM Feature.

Newman, J.M. (1999). Chinese cooking techniques. *International Food Research Journal,* 6(3), 25–27.

Yoshino, K. (2010). Malaysian cuisine: A case of neglected culinary globalization. *Proceeding In Globalization, Food and Social Identities in the Asia Pacific Region Conference,* Tokyo: Sophia University Institute of Comparative Culture.

Zahari, M.M.S., Abdullah, M.K., Azman, I., Kutut, M.Z., & Suhaimi, Z. (2012). Malay food terminologies: Understanding and usage. *Journal of Science Record, 14*(11), 118–130.

Zhifang, Z. (2002). Why are the native languages of the Chinese Malaysians in decline? *Journal of Educational Philosophy and Theory, 34(2), 161–170.*

Technology, advancement and innovation in hospitality and tourism

Hospitality and Tourism – Sumarjan et al. (eds)
© 2014 Taylor & Francis Group, London, ISBN 978-1-138-00151-0

Assessing consumer behavior from a main website of tourism product

M. Mohamed Razak, N. Aminudin, A.F. Amir & M.A. Ismail
Universiti Teknologi MARA, Malaysia

ABSTRACT: Website is one of the main promotional items in marketing for most of tourism products. In fact, governments all over the world use website as one of their marketing tools. To develop one that represents a national tourism product would not only be costly, nonetheless its effectiveness as compared to other non-official website remains to be investigated. Past researches have proven that the internet is an important influence on consumers' behavior. Therefore by using the Technology Acceptance Model (TAM) model, this study examines consumer's perceived ease of use and perceived usefulness of a website in determining customers' intention in visiting a tourism attraction. Questionnaires were distributed to 400 participants that visited homestays in Selangor. Descriptive statistics and correlation test were used to analyze the data. The information obtained provides relevant authorities with knowledge on the impact of website towards tourists' behavior and the development of tourism industry. Findings clearly revealed that TAM attributes positively influenced consumers' visit intention.

Keywords: customer behavior, homestay website, Technology Acceptance Model (TAM)

1 INTRODUCTION

The ICT driven re-engineering has gradually generated a new paradigm-shift, altering an industry structure and developing a whole range of opportunities and threats. In tourism industry, the development of ICT contributes to the escalation of tourism related activities to be a major evolving sector in Malaysia (Buhalis & O'Connor, 2005).

The Internet is changing the tourism industry structure by altering barriers to entry, minimizing switching costs, revolutionizing distribution channels, and facilitating price transparency, while enhancing efficiency. Thus, many of travel organization use the internet as the medium in promoting and marketing their products and services (Buhalis & Law, 2008). Yeoman and McMahon-Beattie (2006) predicted that by 2015, the majority of consumers will purchase holidays through the internet, and that the digital society will change their purchase behavior.

According to Ibrahim and Abdul Razzaq (2010), in Malaysia homestay program is supervised by the Ministry of Tourism (MOTOUR). They are responsible for marketing and promotion of homestay products that includes preparing a directory in Malaysia through www.go2homestay.com.mywebsite. It is also an initiative taken by the government to market the culture of Malaysia. This website is an official webpage for homestay program in Malaysia endorsed by the Ministry of Tourism Malaysia (go2homestay.com.my, 2012).

Since the internet has become an important medium for information searching and travel planning, more study is needed to understand the influence of website characteristics; this includes website's ease of use as well as its usefulness for vacation planning (Kaplanidou & Vogt, 2006). Hu and Hyun (2008) stated that due to the relative newness of this medium, there are still numerous problems preventing tourists from taking full advantage of it. Trip planning on the web is a frustrating experience because of several factors including technical difficulties such as website usability, lack of integration with brick and mortar tourism firms, credit card security and privacy issue.

Previous studies have focused on the relationship between the effectiveness of a website and purchase intention (Ranaweera, Bansal & McDougall, 2008; Thongpapani & Ashraf, 2011; Woodside, Vicente & Duque, 2011). Nevertheless, examining website usefulness and ease of use as a function of specific website characteristics and its impact on intentions to visit the destination is still at infancy stage (Kaplanidou & Vogt, 2006). Therefore by using the Technology Acceptance Model (TAM) model, this study investigates consumers' perceived ease of use and perceived usefulness of website in inducing consumers' visit intention to homestay.

2 LITERATURE REVIEW

2.1 *Technology acceptance model*

TAM proposed that the perception of belief about an innovation is influential in the development of attitude that ultimately effecting system utilization behavior (Davis, 1989). This model adapts the belief-attitude-intention-behavior relationship to an IT user acceptance model (Hu et al., 2012). As mentioned by Dishaw and Strong (1999), TAM identifies perceived ease of use, and perceived usefulness as key independent variables. This model hypothesizes that a system in use is directly determined by behavioral intention to use, which is in turn influenced by users' attitudes toward using the system and the perceived usefulness of the system (Yaghoubi & Bahmani, 2010).

2.2 *Perceived ease of use*

Davis (1989) defined perceived ease of use as the degree to which system would be free of effort. Usability studies on online stores have appear in website architecture, design, and various navigation processes to forecast how easy it is for users to attain what they want to do (Lohse & Spiller, 1998; Palmer, 2002). Chen, Gillenson and Sherrell (2004) indicated that poorly designed website processes have an adverse influence on the website's perceived ease of use. According to Vora (1998), other usability difficulty faced by web users are broken links and opposition in the web-site design. Rationale for these problems is due to lacking of design methodology and guidelines. To evade these problems, the designer should recognize the goal of the website, be constant in website design style, and understanding the user as well as the use environment (Vora, 1998).

2.3 *Perceived usefulness*

Davis (1989) defined perceived usefulness as the degree to which a person believes that using the new technology would enhance his or her job performance. Perceived usefulness can also be defined as the degree to which a customer believes that a particular website has effectively helped them in product information acquisition process to make informed purchase decision Yi and Jiang (2007). Koufaris (2002) asserted that the perception of perceived usefulness of using a website is vital, as it will develop tourist behavior towards the tourism website. Consumers perceive a tourism website as useful when they browse the website to gain particular information pertaining to the destination of interest.

2.4 *Consumer visit intention*

The Theory Reasoned Action Model (TRA) and its successor, the Theory of Planned Behavior (TPB) (Azjen, 1991), are well known, and have been extensively employed in studies relating to identifying behaviors (Azjen & Feishbein, 1980). In general, these theories (TRA, TAM) stated that a behavior is determined by an intention to execute the behavior. According to Dishaw and Strong (1999), TAM diverges from TRA in two key ways. First, it specifies usefulness and ease of use as the two external variables or values that concluded attitude toward an IT, intention to visit, and actual visitation. Thus, TAM does not require to be personalized to each behavior, as long as that behavior is use of IT. Second, TAM does not consist of the subjective norms assemble in TRA.

The TAM application to travel websites analyzes how helpful websites are in providing information trip planning to online travelers and how the ease of use and usefulness of the website influence intention to travel to a destination. Destination website can offer an immediate source of information to potential travelers to satisfy the goal of attaining functional and aesthetic trip information as a preliminary step to the trip intention phase (Kaplanidou & Vogt, 2006).

3 RESEARCH METHODOLOGY

3.1 *Sampling and instrumentation*

A non-probability sampling technique through convenience sampling was applied for data collection. Local and international tourists participating in homestay program were selected as samples; they were the individuals who may be using www.go2homestay.com website for information gathering and travel arrangement. 372 samples were deemed reasonable for this study; this was depicted from Krejcie and Morgan's (1970) table of sample size representing about 30,000 people. The questionnaire was developed in dual languages English and Malay and divided into four sections; section A uncover respondents profile; section B identifies perceived ease of use; section C measure the respondent's perceived usefulness of the homestay website; and section D elicits customer's behavior towards the homestay website. 5 point Likert scale were used in the instrument from 1 (strongly disagree) to 5 (strongly agree).

3.2 *Pilot study and data collection method*

A pilot study consisting 30 tourists were gathered at Selangor Tourism Centre prior to actual data collection. Following analysis of the pilot study, minor alterations were made and a total of 400 questionnaires were distributed in the actual data collection. In light of the positive responses, a total of 372 completed questionnaires were retrieved for analysis.

4 RESULT AND DISCUSSION

This section reports the overall mean score obtained with regard to respondents' perceived ease of use on the website. Viewing at the pattern of the mean score in Table 1, most of the respondents have similar feelings for all items within the variable. The magnitudes mean score of 3.83 indicating that the website design is perceived ease for use among respondents.

This section explains the mean score of perceived usefulness by respondents. Looking at Table 2, overall mean score was 3.64; as light inclination towards agreement indicating that perceived usefulness is important in inducing them to visit the homestay.

Table 3 views respondents' visit intention towards homestay. Overall mean score was 3.56; this means that the information seeking experience motivates them to visit homestay.

Looking at Table 4, the correlations between variables were statistically significant ranging from $r = 0.26$ ($p < 0.00$) to $r = 0.57$ ($p < 0.00$). The highest degree of relationship was between perceived usefulness and respondents' visit intention ($r = 0.57^{**}$, $p < 0.00$). In other cases, weak correlation appeared between perceived ease of use and consumers' visit intention ($r = 0.26^{**}$, $p < 0.00$) and also between perceived ease of use and perceived usefulness ($r = 0.39^{**}$, $p < 0.00$). Despite all these, the correlation between TAM attributes and respondents' visit intention were significant and positive.

Table 1. Overall mean score of perceived ease of use (N = 372).

Variable	Min	Max	Mean	Std. deviation
Perceived ease of use	2.25	4.75	3.838	0.58893

Table 2. Overall mean score of perceived usefulness (N = 372).

Variable	Min	Max	Mean	Std. deviation
Perceived usefulness	1.90	4.60	3.6484	0.6613

Table 3. Overall mean score of consumers' visit intention (N = 372).

Variable	Min	Max	Mean	Std. deviation
Intention to visit	1.33	5.00	3.5663	0.89125

Table 4. Pearson correlation matrix for study variable.

Variable	EOU (Perceived ease of use)	USE (Perceived usefulness)	INT (Consumers visit intention)
EOU	1	0.390**	0.261**
USE		1	0.571**
INT			1

5 LIMITATIONS AND FUTURE RESEARCH

Although this study highlighted a series of significant findings, some of the limitations have been identified and appropriate attention should be given in future research. The first limitation was relating to the sample size. This study only focuses on the consumers around Selangor and Kuala Lumpur, hence the result may only be generalized in a study contextual setting. In order to acquire a comprehensive result that can be generalized, it is important to reach more varied population. Thus, in future research it is recommended that the researcher focus on consumers from other states in Malaysia. In other words, the study should be done in diverse cities with dissimilar geographic characteristics, so that the result could be generalized to a broader geographical area.

The second limitation of this study was the research design. As for data collection, quantitative approach was used whereby the questionnaires have been distributed to the respondents. Even though this approach is less time consuming yet able to reach more respondents at a lower cost, it might not weight on the trustworthiness of the respondents in answering questionnaires. Despite the above notion, it is recommended for future researcher to use a mix mode approach; this method can provide further information beyond the scope of existing questionnaire. Additionally, further research can also explore the influence of variables such as information credibility and trust on perceptions of website ease of use and usefulness towards behavioral intentions. Therefore, this can provide assistance to the managerial side on how to research first-time and repeat consumers' perception of TAM attributes and visitation intentions.

6 IMPLICATIONS, RECOMMENDATION AND CONCLUSION

This study investigates consumer's perceived ease of use and usefulness of website in inducing their visit intention to homestay. Specifically, this study identified the attributes that strongly influence their visit intention. As for theoretical implications, this study provides an understanding about consumer's perceived ease of use and perceived usefulness towards behavioral intention. For practical implications, this study offers opportunity to relevant authorities and homestay operators to develop a better well designed website thus enhancing the image of SMEs' business. It is recommended that the incorporate application allowing consumers to do online. It is encouraged for website builders to use this technology by highlighting the effectiveness of the search tools, simplicity and friendliness of the online applications.

As a conclusion, this study contributes to the tourism literature by testing the TAM model for website, findings explained not only the determinants of individual's intentions to visit homestay but also the formation of these intentions based on the easiness and usability of the website. Results from this study will add upon values to website operators about the importance to create an overall favorable browsing environment to consumers.

REFERENCES

Ajzen, I. (1991). Theory of planned behavior. *Organizational behavior and human decision processes*. 50, 179–211.

Ajzen, I. & Fishbein, M. (1980). Understanding attitudes and predicting social behavior.

Buhalis, D. & Law, R. (2008). Progress in information technology and tourism management: 20 years on and 10 years after the Internet. *Tourism Management*, *29*(4), 609–623.

Buhalis, D. & O'Connor, P. (2005). Information communication technology. *Tourism Recreation Research*, *30*(3), 7–16.

Chen, L., M. Gillenson, & Sherrell, D. (2004). Consumer acceptance of virtual stores: A theoretical model and critical success factors for virtual stores. *Database for Advances in Information Systems*, *35*(2), 8–31.

Davis, F. (1989). Perceived usefulness, perceived ease of use and user acceptance of information technology. *MIS Quarterly*, *13*(3), 319–40.

Dishaw, M.T. & D.M. Strong. (1999). Extending the technology acceptance model with task-technology fit constructs. *Information & Management 36*(1), 9–21.

Go 2 homestay. (2012). Retrieved December 13, 2012, from http://www.go2homestay.com/about-us/.

Hu, P.J., Chau, Y.K., Sheng, O.R., & Tam, K.Y. (2012). Examining the technology acceptance model using physician acceptance of telemedicine technology. *Journal of Management Information Systems*, *16*(2), 91–112.

Ibrahim, Y. & Abdul Razzaq, A.R. (2010). Homestay programme and rural community development in Malaysia. *Journal of Ritsumeikan Social Sciences and Humanities*, 2, 7–24.

Kaplanidou, K. & Vogt, C. (2006). A structural analysis of destination travel intentions as a function of web site features. *Journal of Travel Research*, *45*(2), 204–216.

Koufaris, M. (2002). Applying the technology acceptance model and flow theory yo online consumer behavior. *Information Systems Research*, *12*(4), 205–223.

Krejcie, R. & Morgan, D (1970). Determining sample size for research activities. *Educational and Psychological Measurement*, 30, 607–601.

Lohse, G.L. & Spiller, P. (1998). Electronic shopping: The effect of customer interfaces on traffic and sales. *Communications of the ACM*, *41*(7), 81–87.

Palmer, J. (2002). Web site usability, design, and performance metrics. *Information Systems Research*, *13*(2), 151–167.

Ranaweera, C., H. Bansal, & G. McDougall. (2008). Web satisfaction and purchase intentions: Impact of personality characteristics during initial website visit. *Managing Service Quality*, *18*(4), 329–348.

Thongpapani, N. & Ashraf, A.R. (2011). Enhancing online performance through website content and personalization. *Journal of Computer Information Systems*, *52*(1), 3–13.

Vora, P. (1998). Human factors methodology for designing Web sites. In human factors in Web development, eds. C. Forsythe, E. Grose and J. Ranter. Mahwah, NJ: Lawrence Erlbaum Associates, Inc. Publishers.

Woodside, A.G., Vicente, R.M., & Duque, M. (2011). Tourism's destination dominance and marketing website usefulness. *International Journal of Contemporary Hospitality Management*, *23*(4), 552–564.

Yaghoubi, N.M., and Bahmani, E. (2010). Factors affecting the adoption of online banking: An integration of technology acceptance model and theory of planned behavior. *International Journal of Business and Management*, *5*(9), 159–165.

Yeoman, I. & McMahon-Beattie, U. (2006). Premium pricing and luxury. *Journal of Revenue & Pricing Management*, *4*(4), 319–328.

Yi, C. and Jiang, Z. (2007) The antecedents of online consumers' perceived usefulness of website: A protocol analysis approach. *12th International Conference on Human-Computer Interaction*, Beijing, China.

Hospitality and Tourism – Sumarjan et al. (eds)
 ISBN 978-1-138-00151-0

Hotel internet marketing channels and purchase decision

A.M. Amrahi, S.M. Radzi & S.K.A. Nordin
Universiti Teknologi MARA, Malaysia

ABSTRACT: This paper aims to investigate the relationship between hotel internet marketing channels and customer purchase decision of 5 star hotel in Kuala Lumpur. A quantitative approach used in this paper where instrumentation design based from previous research methodology. Questionnaire distributed to all 5 star hotels in Kuala Lumpur official Facebook fan page using an online survey provider. From the findings, there is a significant relationship between hotel internet marketing channels and purchase decision. This paper only focused on 5 star hotels in Kuala Lumpur, and if this study can be done towards 5 star hotels in Malaysia could be beneficial. This paper provides not only knowledge about the channels, but the practitioner also can know how the customer perceived when they get promotion from them.

Keywords: internet, direct marketing channel, indirect marketing channel, purchase decision, hotel industry

1 INTRODUCTION

Today, Internet had become a new trend of communication in the world. Many people and organizations use the internet as their communication channel because of the cost that is cheaper than traditional ways. Internet is an effective, efficient, and low-cost marketing channel. It shows that the Internet revolution has introduced a wide range of new marketing tools, which are accessible and affordable for smaller organizations (Sigala, 2003). In the hotel industry, hotels reported of competing each other to give the best products and services to the customers by using technology in marketing their products (Aziz, Radzi, Zahari, & Ahmad, 2011).

Moreover, how the hotel organization want to remain competitive in the market as the number of hotel keep on increasing every year (MAH, 2012). They need to compete with the competitor to ensure the customer choose their hotel rather than a competitor. By having a good internet marketing channels (direct and indirect marketing channels), is one ways for hotel organization to be competitive. Nevertheless, there is some argument from previous literature reveal that the increasing of false information in the internet also can affect online users. Hensens, Struwig and Dayan (2010) mentioned that, the success of indirect marketing channel (TripAdvisor) has raised questions with hoteliers, the travel trade, and rating bodies as to how the information provided by this new platform must be viewed in relation to conventional ratings. In addition, Schaffer and Mills (2004) stated that consumer fears to purchase a product online and lack of trust toward online marketing because of cybercrime incidences globally.

Based on statements and arguments from previous researchers, how the hotelier is able to face the challenges as the demand increasing every year and also the cybercrime also increasing. May be the problems will affect the customer purchase behavior. Some question may rise whether this marketing channels related with purchase decision. Therefore, this paper concentrates on investigating the relationship between hotel internet marketing channels (direct marketing channel and indirect marketing channel). From the relationship, the researcher also wants to examine which channel (direct and indirect marketing channel) that contributes more toward the relationship to customer purchase decision.

2 LITERATURE REVIEW

2.1 *Internet marketing*

Internet Marketing (IM) can be defined as digital marketing, web marketing, online marketing, or e-marketing. As the name states, it is the advertising of products or services over the Internet (Yazdanifard, Venpin, Yusoff & Islam, 2011). Moreover, the authors also stated that internet marketing also implies marketing through the wireless media and e-mail. Besides that, internet marketing also refers to a set of powerful tools used for promoting the product or services through the internet.

2.2 *Internet as marketing tools*

Currently, internet been used worldwide as a marketing tool. Many organizations will use the internet as a medium for them to promote their products or services. Internet not only used for promoting the products and services only, but it also used to build a relationship with the customers. As stated by Geiger and Martin (1999), in theoretically, the internet offers a unique chance for marketers to build up and remain relationships with their customers. It can be assumed that, information technology has changed the ways of decision making in the organizations. Moreover, Poon and Swatman (1997) also stated that the Internet allows small businesses to communicate as widely with individuals or trading partners as any large corporation can.

2.3 *Hotel internet marketing channel*

There are 2 dimensions in hotel internet marketing channel which are direct and indirect marketing channel in the hotel.

2.3.1 *Direct marketing channel*

Direct marketing can be defined as the direct interaction between buyers and seller. According to Shehzadi, Khan and Baig (2010), direct marketing defined as having a direct and immediate effect on the audiences. It means that the buyer will directly have contact with the seller or advertiser. Smith and Taylor (2004) defined direct marketing as the distribution of information, products or services through any advertising medium that invites the individual to response directly to the advertiser.

2.3.1.1 Hotel website

Website is a collection of one or more web pages grouped under the same domain name. Usually website will have a homepage where this page has a link between other pages. According to Bainbridge (2003), website in the hotel industry has a few categories which are a *travel agency website*: a website that offers more products than simple hotel booking functionality, *hotel booking agency website*: a website that takes hotel bookings/reservations as its primary business and *Hotel chain website*: a website specifically for a hotel chain. In previous research, there are a few authors stated that website is one of the direct marketing channel. Stated by O'connor and Murphy (2008), they conclude that direct Internet channels where they referred to the hotel's website are cheaper than indirect channels. Its show that hotel website become a marketing channel that represent the hotel directly. Moreover, Starkov and Price (2003) mentioned in *Online Travelers Prefer Booking Directly on the Hotel Website*, all Internet bookings in hospitality will be direct to the consumer which is via hotel-owned websites.

2.3.1.2 Email marketing

Email marketing is the use of email to deliver permission-based communications to build relationships, increase sales and improve customer retention. As stated by Clarke, Flaherty, and Zugelder (2005), The U.S. Federal Trade Commission (FTC) defines a commercial electronic mail message as any electronic mail message where the primary purpose is the commercial advertisement or promotion of a commercial product or service. Price and Starkov (2004) explained that email marketing is a powerful direct-to-consumer distribution and marketing tool. Moreover, email marketing allows hoteliers to engage clients in strong, personalized and mutually beneficial interactive relationships, increases conversions, and sells more efficiently. Email marketing become one of the most tools that been used widely by many organization especially hotel.

2.3.2 *Indirect marketing channel*

An indirect marketing channel consists of a chain of businesses (or intermediaries) involved in the distribution process, each passing the product down until it finally reaches the end-user. An indirect marketing channel might using a retailer to get to the end consumer, or it might using both a wholesaler and a retailer or even using an agent/broker/sales representative, a wholesaler and a retailer (Culler, 2012). Furthermore, according to Shehzadi et al. (2010), indirect marketing channel is a communication by means of stealth methods that do not directly make an individual realize that they being targeted.

2.3.2.1 Online travel agent

Travel agency is the most common intermediaries channel that be used by the hotel to offer products and services. Usually, travel agency will represent the hotel to cater the potential market and convert them to the buyer. This supported by O'Connor (2008), where most hotels used intermediaries such as travel agents, tour operators, or incentive houses. Increasingly, hotels are making use of the growing range of electronic-distribution channels. Travel Agencies (OTAs), also referred as third-party online intermediaries or TPI. According to Cheung and Lam (2009), travel agencies exist primarily as intermediaries between suppliers of travel services, such as airlines and hotels, and customers. Unlike distributors in many industries, travel agencies do not deal with physical products but with information.

2.3.2.2 Review site

The growth of internet user slightly affects the way of communication. By using the internet, they easily can share or review something that they had experienced. Like hotel guest, by using internet, they easily can share review based on their experience during a stay in the hotel. Review site can be defined as activities, practices and behaviors, among communities of people who gather online to share information, knowledge and opinions using conversational media (Hensens, Struwig, & Dayan, 2010). It can be assumed the review site is a place online where the user will able to write a review or share some information to other user. The authors also stated, review site is a web-based platform that focus on traveler reviews and provide independent platforms for guest feedback. The increasing adoption of online hotel reviews provides travelers greater and easier access to hotel-related information. The author also stressed that this growing form of electronic word-of-mouth platform offers travelers travel-related information from other travelers and makes for better more informed decisions (Kusumasondjaja, 2012). It shows that this review site provides information such as hotel information from past buyer and this information can be used by other traveler.

2.4 *Purchase decision*

Purchase decision is the action when customer reacts to certain products or services. This decision will affect the customer whether they want to buy or not to buy the products or services. Brey (2008) define the purchase decision as the study of the processes involved when individuals or groups select, purchase, use or dispose of products, services, ideas or experiences to satisfy needs and desires.

2.5 *Relationship between direct marketing channels and purchase decision*

In the previous study, some author discussed the relationship between the variable indirect ways. Anh and Assenov (2010) mentioned that the website of destinations should provide useful, complete, and relevant information to potential users. This is because providing useful and relevant information on the web is an essential feature that affects online users' intention to purchase. The statement shows that the website, which is referring as direct marketing channel will affect the customer intention to purchase.

2.6 *Relationship between indirect marketing channels and purchase decision*

Previous literature reveals that there is a relationship between indirect marketing channel and purchase decision. Zhu and Zhang (2009) stated that, with the Internet's growing popularity, online consumer reviews have become an influential resource for consumers seeking to discover product quality. It can be assumed that online consumer review which is referring to review side important for the customer to seeking information or products. Moreover, some firms even strategically manipulate online reviews in an effort to influence consumers' purchase decisions (Dellarocas, 2006). This can be assumed that an indirect marketing channel related to customer purchase decision.

3 RESEARCH METHODOLOGY

This study is a quantitative in nature. Descriptive study approach used to describe the variables. Moreover, Correlational approach used to identify the relationship between the variable which is hotel internet marketing and purchase decision. Online survey conducted to gathered data from respondents. All the respondents' data conveniently collected from hotel facebook page. There are 20 Five star hotels in Kuala Lumpur that have an official fan page. Initially, the link to the survey posted to all hotel fan page with permission. Next, the link can be access by all fans from the facebook page. Moreover, the respondents were approached personally through facebook private message then link to the survey given to them to answer after they agreed to participate to answer the survey. All the questions locked to avoid any missing answer from respondents and the link cannot be access after a certain period. The questionnaire consists of 4 sections, and 5point Likert scale used for 3 sections.

4 RESULTS AND FINDINGS

4.1 *Relationship between hotel internet marketing channels (direct and indirect) and purchase decision*

Based from the result, hotel internet marketing channels (direct and indirect marketing channels) able to explain 73% of the variation in the purchase decision while the remaining 27% of variation from other variable. The result showed that the hotel internet marketing channel significantly ($R = 0.73$, $P < 0.001$) contributed to purchase decision. Based on the importance and contribution, indirect marketing channel were more importance and contributed ($\beta = 0.704$) more to the model. As both direct and indirect marketing channel found to be significant to purchase decision, it shows that there was a relationship between hotel internet marketing channels and purchase decision. From the results also, indirect marketing channel contributed more ($\beta = 0.703$) compared to direct marketing channel ($\beta = 0.173$) towards the model.

Table 1. Results of multiple regressions: Hotel internet marketing channels (direct and indirect) and purchase decision.

	B	SE B	β
Direct marketing channel	0.184	0.051	0.173*
Indirect marketing channel	0.908	0.062	0.704*

Dependent variable: Purchase decision.
Note: $R = 0.73$, $*P < 0.001$.

A finding from the previous study shows that direct marketing channel (website) is important towards customer purchase decision. This channel (website) should have some criteria such as information provision, website design, online customer service and security. All these criteria are important to ensure customer buy the product offer (Anh & Assenov, 2010). Other than that a finding from Lim and Hall (2008), proven that direct marketing channels also influence customer purchase decision by providing the cheapest price information compared to other channels.

Moreover, 35% of travelers change their choice of hotel after browsing review site (Channel Advisor, 2010). This shows that before customer make their decision to buy any product; they tend to read review from other or experienced customer about the product. They believed that review from experienced customer will help them to make a better decision. Nevertheless, previous study revealed that majority customers prefer to use indirect marketing channels such as TripAdvisor or travel agent to book a room (Dickinger, 2010). It can be assumed that customer prefer to use indirect marketing channel to make any purchase.

5 STUDY LIMITATIONS

Some limitations highlighted in this study. First limitation is the sample used in the study. This study only focus respondent from hotel fan page only which not represent other online customer. Maybe it can be beneficial if to use other online medium to collect data and not only focus on facebook. Next, area of this study only focuses to 5 star hotels in Kuala Lumpur. The area not represents Malaysia as a whole population, and it just focus on the customer for 5 star hotels only. Other than that, this study only focuses on hotel rather than focuses on hotel and also resort.

6 SUGGESTIONS FOR FUTURE RESEARCH

It recommended for future research to conduct this study to whole Malaysia. Not only focuses on the respondent in facebook, future researcher can use other medium such as a review site (Tripadvisor), forum, website and other to gather data. This study can be expanding to entire Malaysia rather than focusing in Kuala Lumpur area only. Next, try to include 3 or 4 stars hotel because different rating of the hotel will give a different result. Lastly, rather than focus in the hotel only, try to conducting this study in resort. Different hotel will give a different result from respondents.

7 CONCLUSION

In conclusion, there is a relationship between hotel internet marketing channels and purchase decision. This result aligns with the previous study that also tries to investigate the relationship between the variable. The result also indicates that online marketing channel is vital to ensure customer buy the product or services of any hotel where indirect marketing channel is preferable compared to direct marketing channel. This indicates that hotel internet marketing channel is significant for the customer before they decide to purchase and practitioner can use this opportunity to use all the online channel to be more competitive in the market and gain more profit.

REFERENCES

Anh, L.T.P. & Assenov, L. (2010). Demand for online travel services in Vietnam. *Journal of Tourism, Hospitality & Culinary Arts*, 1(1), 69–77.

Aziz, A.A, Radzi, S.M., Zahari, M.S.M, & Ahmad, N.A. (2011). Hotel website dimension: Analyzing customers' preferences. *Journal of Tourism, Hospitality & Culinary Arts, 3*, 85–108.

Bainbridge, A. (2013, Mac 19). Hotel Booking Process. Retrieved from http://www.travelucd.com.

Brey, J.P. (2008). Consumer behaviour theory: Approaches And Models. Discussion paper. *Journal of marketing*, 69, 133–152.

Channel Advisor (2013, June 6th). Global Consumer Shopping Habit. Retrieve from http://go.channeladvisor.com/rs/channeladvisor/images/us-ebook-consumer-survey-2011.pdf.

Cheung, R. & Lam, P. (2009). How travel agency survive in e-business world. Communication of the IBIMA, 10, 85–91.

Clarke, I., Flaherty, T., & Zugelder, M. (2005). The CANSPAM act: New rules for sending commercial e-mail messages and implications for the sales force. *Industrial Marketing Management, 34*, 399–405.

Culler, S. (2012, November 7). Distribution. Retrieved from http://hortmgt.gomez.dyson.cornell.edu/PDF/Current%20Project/Marketing%20Modules/Distribution%20Module.pdf.

Dellarocas, C. (2003). The digitization of word of mouth: Promise and challenges of online feedback mechanisms. *Management Science, 49*(10), 1401–1424.

Dickinger, A. (2010). The trustworthiness of online channels for experience and goal-directed search tasks. *Journal of Travel Research, 50*(4), 378–391.

Geiger, S., & Martin, S (1999). The internet as a relationship marketing tool: Some evidence from Irish companies. *Irish Marketing Review, 12*(2), 24–36.

Hensens, W., Struwig, M., & Dayan, O. (2010). Review criteria on tripadvisor compared to conventional hotel-rating systems to assess hotel quality. *Eurochrie annual conference, 1*(1), 1–5.

Kusumasondjaja, S. (2012). Credibility of online reviews and initial online trust in hotel services: The roles of similarity and review quality. *Journal of Vacation Marketing, 18*(3), 185–195.

Lim, W.M., & Hall, M.J. (2008). Pricing consistency across direct and indirect distribution channels in South West UK hotels. *Journal of Vacation Marketing, 14*(4), 331–344.

Malaysia Association Hotel (2012, December 19), Registered Hotel December 2011. Retrieved from http://www.hotels.org.my/images/pdf/hotel_statistic/Registered_Hotels_Dec_2011.pdf.

O'connor, P. (2008). E-mail marketing by international hotel chains: An industry-practices update. *Cornell Hospitality Quarterly, 49*(1), 42–52.

O'connor, P. & Murphy, J. (2008). Hotel yield management practices across multiple electronic distribution channels. *Information Technology & Tourism, 10*(2), 161–172.

Poon, S. & Swatman, P.M.C. (1997). Small business use of the internet: Findings from Australian case studies. *International Marketing Review, 14*(5), 385–402.

Price, J. & Starkov, M. (2004). Developing an email Marketing Strategy in hospitality. Retrieved (2012, December 27) from Hebs Digital website: http://www.hebsdigital.com/articles/pdf/2004/Sep%2004%20HeBS%20Article%20%202004%20%20Developing%20an%20Email%20Marketing%20Strategy%20in%20Hospitality.pdf.

Schaffer, S. & Mills, J.E. (2004). Information and communication technologies in tourism 2004. The 11th ENTER International Conference in Cairo, 13(1), 413–423.

Shehzadi, I., Khan, S., & Baiq, E. (2010). Defining indirect marketing. *European Journal of Economics, Finance and Administrative Sciences, 21*(1), 1–5.

Sigala, M. (2003). Developing and benchmarking internet marketing strategies in the hotel sector in Greece. *Journal of Hospitality & Tourism Research,* 27(4), 375–401.

Smith, P.R. & Jonathan, T. (2004). *Marketing communications: An integrated approach. 4th edition.* London and Philadelphia: Kogan Page Limited.

Starkov, M. & Price, J. (2003). Online travelers prefer booking directly on the hotel website. Retrieve (2012, December 6) from Hebs Digital website: http://www.hebsdigital.com/articles/pdf/2003/Apr%2003%20HeBS%20Article%20-%202003%20-%20Booking%20Directly%20vs%20Intermediaries.pdf.

Yazdanifard, R., Venpin, M., Yusoff, W.F., & Islam, M.R. (2011). Internet marketing: The new era of innovation in e-commerce. *International Conference on Software and Computer Applications,* 9, 192–197.

Hospitality and Tourism – Sumarjan et al. (eds)
 ISBN 978-1-138-00151-0

The mediating role of brand image and awareness towards smart card usage in restaurants

S. Zurena @ Rena, M.S. Mohd Zahari & O. Ida Rosmini
Universiti Teknologi MARA, Malaysia

ABSTRACT: The availability range of payment products and services in the market today including the smart card is in response to diverse market demands of the changing behavior and lifestyle of the customers which impacting their actual usage behavior. Brand image and awareness are considering to play significant roles towards the customer usage of the smart card. Using the Technology Acceptance Model as the initial framework, this paper empirically examine the mediating role of brand image and awareness towards customer responsiveness and actual usage of Malaysia "Touch 'n Go" smart card in the restaurant sector. Through restaurant customers' surveyed in the Klang Valley, Malaysia, it was discovered that both brand image and awareness have significant customers causal effects on the actual usage of Malaysia "Touch 'n Go" smart card in the restaurants. Both attributable are having strong and positive effect on customers to accept and use the smart cards and therefore no exaggeration to say that smart card is becoming a necessity and one of the most important elements in this modern technology for customers and businesses in combating with the economic and financial needs of the country.

Keywords: actual usage, awareness, brand image, smart card

1 INTRODUCTION

Driven by advances and support of creative technologies, the range of payment products and services in the market expanded in response to diverse market demands which need to be accepted by both customers and merchants (Chakravorti & Rosan, 2006; Milne, 2005; Rochet & Tiroley, 2004). Based on the e-commerce and development report prepared by the United Nations Conference on Trade and Development (UNCTAD), the estimated total world e-commerce value in 2002 was about $2.3 trillion, and it was predicted to reach $12.8 trillion by 2006 (E-commerce and Development Report, 2012). Malaysia with no exception is embarking on the electronic payments and this is becoming a part of national larger national agenda to increase the efficiency of the nation's payment systems which would ultimately improve the competitiveness of the economy.

According to Bank Negara Malaysia Governor, Tan Sri Dr Zeti Akhtar Aziz, in the next 10 years, the Bank Negara aims to increase the number of e-payment transactions per capita from 44 transactions in 2010 to 200 transactions by the end of 2020, which is comparable to the e-payment transactions per capita in the developed countries (Bank Negara, 2011). In line with that statement, banks such as Hong Leong Bank has launched the Hello Kitty Debit Card, Malayan Banking through it Manchester United Debit Card to target the fans of famous characters and sports teams and Touch 'n Go Sendirian Berhad with no exception. Initially offering a contact less payment card for use on all the highways as the sole electronic toll collection operator, Touch 'n Go continually expanding its functionality. This card offered a wide range of applications including public transport, theme parks, parking lots, medical and health, movie, and restaurants as well. The earlier inventions in smart cards such as Touch n' Go, was still not prominent in restaurant usage, and did not gain support from the majority of customers although to the young and technologically-savvycustomers. Limited restaurants acceptance and low awareness of smart cards amongst customers were major constraints to its growth in Malaysia.

In order for the restaurants to closely monitor and conform to the needs and expectations of its customer, it is equally essential for them to build frequent interactions and exchanges of information between firms and their customers (i.e., ordering, service delivery, payment) and among various functional areas of restaurants (i.e., front to back-of-the-house communications, human resource management). To optimize such interactions, restaurants deployed technologies like point-of-sale and electronic payment systems

(Cobanoglu & DeMicco, 2007) as of lodging and retail with biometric systems that promise substantial benefits to firms and consumers (Lam, 2007; Pons, 2006; Tsai, 2007).

This evolution kept scholars and non government organization to expose and encourage service providers, restaurant operators and customers as the end users to jointly accept and use the system. It is important therefore for the restaurants to continually striving in the increasing competitive technology market environment. Brand image and awareness of the technology gazetted is considering to play significant roles towards the customer usage. In other words, brand image influence consumers to pick the right choice among the alternative brands or products. Using the Technology Acceptance Model as the initial framework, this study empirically examines the mediating role of brand image and awareness towards customer responsiveness (perceived ease of use; perceived usefulness; perceived trust) and actual usage of Malaysia "Touch 'n Go" smart card in the restaurant sector. In short, two hypotheses are to be tested:

H1: Brand Image mediates the relationship between customer responsiveness and actual usage

H2: Awareness mediates the relationship between customer responsiveness and actual usage.

2 LITERATURE REVIEW

2.1 *Technology acceptance model*

The TAM posits that perceived ease of use and perceived usefulness are highly relevant to the technology acceptance behavioral intentions. Perceived usefulness is the extent to which a person believes that using a particular system will enhance his or her performance, while perceived ease of use is the extent to which a person believes that using a particular system will be free of effort (Davis et al., 1989). As the original TAM was designed to predict technology adoption in work-related settings (Davis, 1989), the model is used in testing the mediating variable slike brand image and awareness to capture the context of actual customer usage of smart card in restaurants.

2.2 *Perceived trust*

Chen and Tan (2004) stated that trust has a positive impact on the users' attitude, purchasing intentions, and purchasing behaviors. Customers should trust the payment system adopted by the other user. The existing literature points out that high level of user confidence and trust is a contributing factor for the successful adoption of e-payment systems (Kurnia & Benjamin, 2007).

2.3 *Brand image*

Dobni and Zinkhan (1990), Martinez and Pina (2010) defined brand image as an important aspect of marketing activities; branding and market offering with varied definition and approaches to its conceptualization. Hsieh, Pan and Setiono (2004) argued, it helps consumer in recognizing their needs and satisfaction regarding the brand and it also distinguishes the brand from other rivals motivating customers to buy the brand. It contributes to the consumer's decision whether or not the brand is the one for him/her (Dolich, 1969) and it influences consumers' subsequent buying behavior (Fishbein, 1967; Johnson & Puto, 1987). Brand image also plays key roles for consumers to pick up the right choice among the alternative brands or product. Mangold and Miles (2007) stated that restaurant organizations can achieve a sustainable competitive advantage by properly and positively building the brand image. Well-developed branding benefits the restaurants through such things as higher levels of service quality, customer satisfaction, repeat purchase behavior, and lower employee turnover.

2.4 *Awareness*

Consumers go through "a series of process in knowledge, conviction, decision and confirmation" before they are ready to adopt a new product or service. The adoption or rejection of an innovation begins when "the consumer becomes aware of the innovation" (Rogers & Shoemaker, 1971; Sathye, 1999). Hence, for adoption of an innovation, it is necessary that the suppliers offering this service make the consumers aware about the availability of such a product and explain how it adds value relative to other products of its own or that of the competitors. Consumers must become aware of the new brand or technology in fact an important characteristic for any adoption of innovative service or product is creating awareness among the consumers about the service/product including restaurant (Sathye, 1999).

3 METHOD

This study employed a quantitative research design with a structured questionnaire as an instrument for primary data collection. The tool used to gather the information from the restaurant customers was self-reported questionnaire. The customers dining from five different restaurants in the Klang Valley were focused. Touch 'n Go participating restaurants namely Baskin Robins, Burger King, Dunkin Donuts, Starbucks, and The Chicken Rice Shop. Before the survey the corporate headquarters of these restaurant chains were contacted to obtain permission to survey their customers. Through self administering a total of 100 sets of questionnaires were distributed to the stipulated restaurants. Respondents or participants were asked to rate their valuable opinions and feedbacks on their response to Touch 'n Go smart card and how it influenced their actual usage. The data were analyzed using IBM SPSS Statistics version 21 for Windows. With no obvious problem the actual 100 questionnaires of the pilot were retained for analysis.

4 RESULTS AND DISCUSSION

Bootstrap analysis known as INDIRECT, implemented by Preacher and Hayes (2008) was used to assess the meditational model of *Brand Image* and *Awareness* as mediators of the relationship between *Customer Responsiveness* and *Actual Usage*. Table 1 indicates the summary of mediation results for both hypotheses (*H*1 and *H*2).

4.1 *Hypothesis testing*

H1: Brand Image mediates the relationship between customer responsiveness and actual usage

Table 1 indicates that the bootstrap estimation is based on 5,000 bootstrap samples. The mean indirect effect from the bootstrap analysis was positive and significant ($a \times b = 0.65$), with a 95% bias corrected and accelerated confidence interval excluding zero (0.33 to 1.10). In the indirect path, the customers who indicated high agreement level of customer responsiveness, were more likely to agree on the brand image ($a = 0.70$); while holding constant for customer responsiveness, the customers who had high agreement level on brand image, will also have a high level of actual usage ($b = 0.93$).

Table 1. Summary of mediation results (5,000 bootstrap samples, $N = 100$).

Independent variable (iv)	Customer responsiveness	Customer responsiveness
Mediating variable (M)	Brand image	Awareness
Dependent variable (DV)	Actual usage	Actual usage
Effect of IV on M (a)	0.70***	0.74***
Effect of M on DV (b)	0.93***	0.90***
Direct effect (c')	–0.42	–0.42
Indirect effect (a*b)	0.65	0.67
95% CI for a*b	0.33, 1.10	0.31, 1.07
Total effect (c)	0.90***	0.90***

Based on the significant mean indirect effect, H1 is accepted and it can be concluded that brand image mediates the relationship between customer responsiveness and actual usage. This is in line with Sondoh et al., (2007) which say that a positive brand image can influence customer loyalty so that they would repurchase more product or service and recommend others to buy.

H2: Awareness mediates the relationship between customer responsiveness and actual usage

The same scenario happened in the second hypothesis where the indirect effect was significant. The mean indirect effect from the bootstrap analysis in Table 1 was positive and significant ($a \times b = 0.67$), with a 95% bias corrected and accelerated confidence interval excluding zero (0.31 to 1.07). In this path, similar interpretation can be made where customers who indicated high agreement level of customer responsiveness, were more likely to have high level of awareness ($a = 0.74$); while holding constant for customer responsiveness, the customers who had high level of awareness, will also have a high level of actual usage ($b = 0.90$).

As a result, the hypothesis is accepted and awareness does mediate the relationship between customer responsiveness and actual usage.

5 CONCLUSION

The results of this study indicate that TAM with brand image and awareness as the mediator are useful in explaining restaurant customers' actual usage. Both brand image and awareness have significant causal effects on the actual usage of Malaysia "Touch 'n Go" smart card in the restaurants. This evidence is in line with Mangold and Miles (2007) that well-developed internal branding benefits the restaurants through such things the higher levels of service quality, customer satisfaction, repeat purchase behavior, and lower employee turnover could achieve. Therefore, it is no exaggeration to say that smart card is becoming a necessity and one of the most important elements in this modern technology for customers and businesses in combating with the economic and financial needs of the country. Smart cards may also represent a viable solution to the restaurant industry's technology needs and having a payment system could facilitate the efficient movement of funds and financial development and the growth of the country's economy.

REFERENCES

Bank Negara Malaysia, (2011). *Central Bank of Malaysia—Financial Sector Blueprint. 2011–2020.* http://www.bnm.gov.my ISBN: 978-983-9586-47-3.

Chakravorti, S. & Roson, R. (2006). 'Platform competition in two-sided markets: The case of payment networks', *Review of Network Economics, 5*(1), 118–143.
Chen, L.D. & Tan, J. (2004). Technology adaptation in e-commerce: Key determinants of virtual stores acceptance. *European Management Journal, 22*(1), 74–86.
Cobanoglu, C. & DeMicco, F. (2007). To be secure or not to be: This is the question! A critical look at hotel's network security. *International Journal of Hospitality and Tourism Administration, 8*(1), 43–59.
Davis, F.D. (1989). Perceived usefulness, perceived ease of use, and user acceptance of computer technology, *MIS Quaterly, 13*(3), 319–40.
Davis, F.D., Bagozzi, R.P., & Warshaw, P.R. (1989). User acceptance of computer technology: A comparison of two theoretical models. *Management Science, 35*(8), 982–1003.
Dolich, I.J. (1969). Congruence relationships between self images and product brands. *Journal of Marketing Research, 6*(1), 80–4.
Fishbein, M. (1967). Attitude and the prediction of behavior, in Fishbein, M. (Ed.). *Attitude Theory and Measurement,* (pp. 477–92). New York, NY: John Wiley & Sons.
Hsieh, M.H., Pan, S.L., & Setiono, R. (2004). Product, corporate-, and country-image dimensions and purchase behavior: a multicountry analysis. *Academy of Marketing Science, 32*(3), 251–70.
Johnson, M.D. & Puto, C.P. (1987). A review of consumer judgement and choice, in Houston, M. (Ed.), *Review of Marketing, American Marketing Association*, Chicago, IL, pp. 236–92.
Kurnia, S. & Benjamin, L. (2007). Exploring the reasons for a failure of electronic payment systems a case study of an Australian company. *Journal of Research and Practice in Information Technology, 39*(4), 34–67.
Lam, T. (2007). The applications of biometrics to payment verification. In W.S. Coats, A. Bagdasarian, T.J. Helou, & T. Lam (Eds.), *The practitioner's guide to biometrics* (pp. 147–162). Chicago, IL: American Bar Association.
Mangold, W.G. & Miles, S.J. (2007). The employee brand: Is yours an all-star? *Business Horizons, 50*(5), 423–433.
Martínez, E.Y. & Pina, J.M. (2010). Consumer Responses to Brand Extensions: A Comprehensive Model, *European Journal of Marketing, 44*(7/8), 1182–1205.
Milne A. (2005). What's in it for us? Network effects and bank payment innovation', *Bank of Finland Research Discussion Papers*, 16.
Pons, A. (2006). Biometric marketing: Targeting the online consumer. *Communications of the ACM, 49*(8), 61–65.
Preacher, K.J. & Hayes, A.F. (2008). Asymptotic and resampling strategies for assessing and comparing indirect effects in multiple mediator models. *Behavior Research Methods, 40,* 879–891.
Rochet, J.C. & Tiroley, J. (2004). Platform competition in two-sided markets. *Journal of the European Economic Association, 1*(4), 990–1029.
Rogers, E.M. & Shoemaker, F. (1971). *Communications in Innovation,* Free Press, New York, NY.
Sathye, M. (1999). Adoption of internet banking by Australian consumers: An empirical investigation, *International Journal of Bank Marketing, 17*(7), 324–334.
Sondoh, S.L., Omar, M.W., Wahid, N.A., Ismail, I. & Haru, A. (2007). The effect of brand image on overall satisfaction and loyalty intention in the context of color cosmetic. *Journal Asian Academy of Management, 12*(1), 83–107.
Tsai, R.C. (2007). Biometrics in the private sector: Trends and case studies. In W.S. Coats, A. Bagdasarian, T.J. Helou, & T. Lam (Eds.), *The practitioner's guide tobiometrics* (pp. 139–146). Chicago, IL: American Bar Association.
United Nations Conference on trade and development (2012). *E-commerce and Development Report, 2001 (UNCTAD/SDTE/ECB/2), New York: United Nations.*

Hospitality and Tourism – Sumarjan et al. (eds)
 ISBN 978-1-138-00151-0

A conceptual model for IT and website adoption in the tourism industry

S. Salavati
Azad University , Tonekabon Branch, Iran

N.H. Hashim, N. Khani & M. Bahrami
Universiti Teknologi, Malaysia

ABSTRACT: This paper proposes factors influencing website adoption in the tourism organization. Website adoption in this study includes both the acceptance and implementation of websites by tourism organizations to communicate and deliver their services. Review of website adoption literature identifies two gaps in adoption of website in the tourism industry. Based on these gaps, this study applies Rogers' Diffusion of Innovation and Technology-Organization-Environment models and two individual technology adoption theories, Theory of Planned Behavior and the Technology Acceptance Model to introduce its conceptual model. In addition, Institutional Theory is applied to identify the influence of external factors on innovation adoption.

Keywords: innovation, tourism, website

1 INTRODUCTION

Organizations may decide to adopt and implement IT from diverse motivations such as improving organizational process and flexibility (Sambamurthy, Bharadwaj, & Grover, 2003), providing the satisfaction of suppliers/customers (Chang & Wong, 2010) and discovering new business opportunities and access to market information (Tan, Chong, Lin, & Eze, 2009). In spite of many advantages of IT to businesses, studies reported several unsuccessful IT adoptions. In this regard, questions on how to apply the IT into business activities (Gregory, Wang, & DiPietro, 2010); which variables are related to better adoption of this technology in a certain industry (Law & Cheung, 2006); and what theories help to predict and explain the successful adoption in this industry remain crucial (Murphy, Schegg, & Olaru, 2006).

Organizations is a complex process that is influenced by both external and internal business environments. In this study, website adoption is particularly defined as a process including both accepting and implementing of a website by tourism organizations to communicate, and deliver their services such as booking, reservation, and payment. Despite the significance of IT adoption and the vast literature, knowledge about the organizational IT adoption is still limited (Hameed, Counsell, & Swift, 2012), particularly those investigating the implementation of the technology following acceptance. Review of the literature on IT and website adoption reveals that knowledge in this area is either case specific or fragmented, limiting our understanding of the adoption behavior of organizations (El-Gohary, 2012).

In addition, there is a call to develop a model that includes comprehensive factors related to the adoption of IT by the hospitality industry (El-Gohary, 2012). Previous models and theories on IT adoption have two shortcomings. First, the lack of an integrated perspective to explain the influence of both organizational and individual components of IT innovation adoption collectively (Hameed et al., 2012), and second, limited consideration of external factors influencing IT adoption process.

Earlier studies on IT adoption examined adoption at the individual level before expanding it to the organizational level. In tourism, some studies on IT in organizations considered website performance and the rate of acceptance in order to describe website adoption (Salavati & Hashim, 2011), while others investigated factors influencing website adoption (e.g., Hashim, Murphy, Purchase, & O'Connor, 2010). These studies applied different models and theories or combined related models and theories to explain organizational adoption (Hameed et al., 2012).

To explain the organizational adoption process researchers sometimes combine existing models. Technology-Organization-Environment framework (TOE) (Tornatzky & Fleischer, 1990) is a model, which is broadly used to explain organizational IT innovation adoption. Based on TOE, the process by which an organization adopts and implements an innovation may be affected by technological, organizational, and environmental context (Tornatzky & Fleischer, 1990). Although previous studies employed TOE to introduce the determinants influencing IT, it is not able to explain the individuals' behavior in accepting new technology.

In addition to TOE, Diffusion of Innovations (DOI) (Rogers, 1962) is a widely used theory, which explains not only organizational but also individual innovation adoption. Nonetheless, DOI has some limitations when it comes to explaining organizational innovation adoption such as inadequate constructs to deal with collective adoption behaviors (Lyytinen & Damsgaard, 2001). Likewise, there is also a lack of discussion on post-adoption behavior such as the implementation of the technology (Zaltman, Duncan, & Holbeck, 1984) and how attitude evolves into accept or reject decisions, and how innovation attributes fit into this process at the organizational level (Karahanna, Straub, & Chervany, 1999).

Individual level measures user acceptance and the actual use of innovation (Hameed et al., 2012). Most organizational IT adoption studies have considered only a single individual, usually the CEO or CIO. The innovation adoption process in an organization is considered to be successful if the innovation is accepted and integrated into the organization by its employees who use apply the innovation over time (Hameed et al., 2012). Rogers (2003) stated that an innovation can be initially accepted and then rejected (p. 31). An organization's top managers may accept the innovation, only to have the employees reject it during the implementation process. Hence, in investigating the process by which an organization adopts innovation, combining information from top management and employees would be important as it would allow a more in depth understanding of the factors which hinders or facilitate innovation adoption in organizations (Sellin, 1990).

Based on the above discussion, this research proposes a model to explain IT adoption within the hospitality industry. This model combines TOE and DOI and two most frequently used individual IT adoption models to explain individual and organizational factors influencing website adoption. These two individual models include Theory of Planned Behavior (TPB) (Ajzen, 1991), and the Technology Acceptance Model (TAM) (Davis, 1989), which are employed to explain the behavior employees in accepting website.

The literature has also acknowledged that most existing models and theories are limited in their ability to explain organizational external pressures on the adoption of IT innovation (Zhang & Dhaliwal, 2009). In an attempt to investigate the competitiveness and rapid rates of change in an organization's environment, this study uses Institutional Theory (INT) (DiMaggio & Powell, 1983) to identify the influence of environmental pressures on the organization's IT innovation adoption. INT provides a useful research view into the study of IT adoption within organizations (Zhang & Dhaliwal, 2009).

2 THE RESEARCH MODEL

This section deals with the dissertation conceptual model. The approach of this paper is to use an interactive perspective which takes into account both organizational and individual behavior characteristics. The literature on IT adoption indicates that most scholars conduct their research by integrating theories utilized in information systems with a framework that covers the contextual antecedents (Hameed et al., 2012). For the purpose of this study, organizational adoption process previous models need further adjustment to characterize (Hameed et al., 2012).

The Rogers Innovation-Decision Process model is chosen as the foundation model in this study, and the other four models, TAM, TPB, ITN and TOE, are also used to adjust and to improve the research model. Figure 1 illustrates this study conceptual model.

As shown in Figure 1, the innovation adoption process by an organization is divided into two sub-processes: acceptance process and implementation process. The acceptance process refers to a sequence of events initiating with adopter awareness of an innovation leading to adopting or rejecting of an innovation. The adopter here is mainly refer to the organization CEO. By adopting and entering the innovation into the organization the acceptance process is over and the implementation process will initiate. Thus, the implementation process refers to the process in which an innovation uses regularly for targeted purposes.

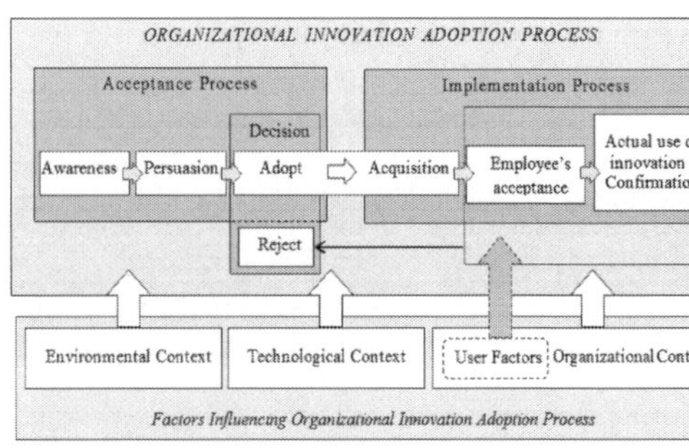

Figure 1. The research conceptual model.

In some cases, adopters may adopt the innovation, but for some reasons do not continue to use, and in practice, they would not fall into the implementation process. For instance, hotels may establish a website, but never utilized it for the purpose of their business.

In fact, this study supposes that after entering an innovation into an organization, the acceptance of its staffs plays a significant role to implement this new technology. In this regard, the users' characteristics is shown as an important independent variable related to adoption of website by tourism organization separately in the model. The following introduces the factors influencing IT and website adoption followed by an operational model.

2.1 *Factors influencing it adoption*

Studies on organizational innovation adoption have identified a large number of variables as predictors of innovation adoption. For instance, Jeyaraj, Rottman, and Lacity (2006) in a review of empirical studies conducted on innovation adoption between 1992–2003 found out hundred related variables as predictors of organizational adoption. Similarly, Hameed et al. (2012) in their study found more than one hundred and twenty factors related to the organizational IT adoption, which have been used in the previous studies between the years 1990–2010. Table 1 lists the most frequently factors that have been cited by scholars and that have been mainly used to predict IT innovation adoption. Regarding to the TOE framework, the factors are classified into three major categories including organizational context, environmental context and technological context. This classification is almost similar to it has been used by a number of studies especially those studying the innovation adoption by small organizations (Li, Lai, & Wang, 2010). However, there is a major difference in this classification in the organization category. As shown in Table 1, this category itself is divided into individual factors and organizational factors. The individual factors here are considered as the factors related to the website user and Chief Executive Officers (CEOs).

This study identified 21 factors related to the hospitality website adoption context based on the five selected models. Table 1 summarizes these factors.

2.1.1 *Organizational context factors*

The most frequently examined factors in organizational IT adoption are the attributes related to the organizational context (Hameed et al., 2012). As mentioned earlier, for the purpose of this paper, organizational context determinants include both individuals and organizational factors, which explain with details in the following.

2.1.1.1 Individual factors

Studies on organizational innovation adoption stated that individual attributes have a significant effect on the innovativeness of an organization

Table 1. Factors influencing on IT adoption and their supportive models.

Context	Sub-categories	Factors	Model
Organizational context	*Individual factors* Users attributes	Educational level	DOI
		IT knowledge	DOI
		Perceived usefulness	TAM
		Perceived ease of use	DOI; TAM
		Subjective norm	TPB
	CEO attributes	Educational level	DOI
		IT knowledge	DOI
		Attitude toward change	DOI
		Perceived risk	DOI
	Organizational factors	Organization size	TOE
		Organization age	TOE
		Organizational type	DOI
Environmental context		Coercive pressure	TOE; INT
		Imitative pressure	INT
		Normative pressure	INT
Technological context		Relative advantage	DOI
		Compatibility	DOI
		Trialability	DOI
		Cost	DOI
		Observability	DOI
		Complexity	DOI

(Hausman, 2005). In this paper, these factors are divided into two groups; the user's attributes and CEOs attributes. Users refer to the hospitality organization staffs who involve with the website or, in general, to the IT activities. CEOs refer to the manager(s) who make strategic decisions for an organization.

The acceptance of IT by users has been the subject of some previous studies (Hameed et al., 2012). While an innovation enter into an organization, the acceptance of users plays a significant role to implement it. The constructs of DOI, TAM and TPB contribute most towards user acceptance characteristics. Perceived Usefulness, Perceived Ease of use, Subjective Norm, Users' IT Knowledge, Educational level, and Age are the six related user attributes.

Studies have emphasized that the inherent attributes and the ability of managers/owner, as being the key determinants of the nature and level of business conducted. Heart and Pliskin (2002) stated that the lack of these factors in small and medium-sized hospitality stakeholders is a significant barrier of adoption of IT in this industry.

2.1.1.2 Organizational factors

Organizations characteristics are the most frequently attributes examined in adoption of IT (Hameed, et al., 2012). Organization size, age and type are considered as the most examined organizational factors influence the organization innovation adoption process. Many studies on tourism organization considered to this factors. For instance, the number of room is considered as the size of hotels or independence or affiliated branches of hotels is considered as the type of tourism organizations. Previous studies mentioned that the organizational factors have a direct influence on the IT and website innovation adoption (El-Gohary, 2012; Salavati & Hashim, 2011).

2.1.2 *Environmental context factors*

Consistent with the TOE framework, factors that relate to environmental context influence innovation adoption of an organization. Previous studies indicated that the environment plays a significant role in IT adoption (Tornatzky and Fleischer, 1990). Recently, a few previous IT studies employed institutional theory to better understanding the effects of external environmental pressures on organizational adoption of IT. Review of these studies suggests that organization's external environmental pressures (e.g., competitors and competition intensity, and customers) are found to be significant influencing an organization's Internet and IT adoption. These studies indicate that three types of pressures in an organizational external environment can affect organizational IT

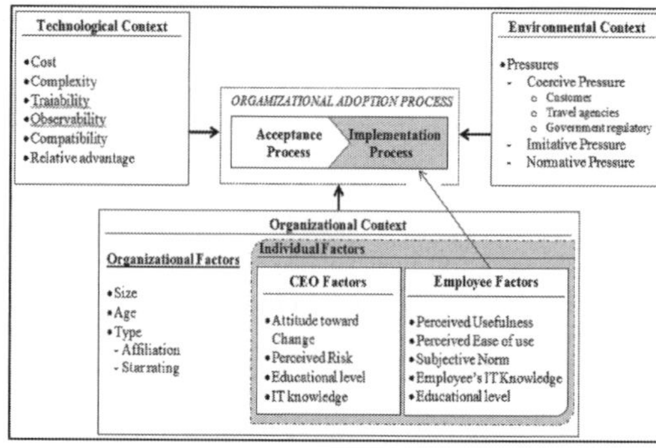

Figure 2. The conceptual model with details of its factors.

adoption: normative pressures, coercive pressures and imitative pressures (DiMaggio and Powell, 1983). Hence, according to institutional theory, these three pressures are employed in this study as environmental factors that may influence organizational website adoption.

2.1.3 *Technological context factors*

Previous studies on IT adoption suggest that the technical context of an organization typically influences organizational adoption of IT innovation (Chong and Pervan, 2007). Considering DOI and TOE, this study proposes three innovation characteristics on the adoption of tourism organization website: relative advantage, compatibility, trialability, observability, complexity, and cost.

3 THE PROPOSED MODEL WITH DETAILS

According to the above, Figure 2 detail out this study proposed model in terms of factors influencing the two-stage of the adoption processes: acceptance and implementation. Review of organizational IT adoption studies identified factors influencing organizational IT adoption decision in three contexts: technological, organizational and environmental. However, following the acceptance of the innovation in the organization, employee acceptance plays an importance role in utilizing and implementing of an innovation.

4 CONCLUSION AND FUTURE RESEARCH

In an attempt, this study developed a conceptual model to explain IT innovation and website adoption in the tourism industry. We combined the most frequently used theoretical model and framework to build this research model. The proposed model tests the Innovation adoption process into two processes; acceptance and implementation process. It also explains two levels of analysis (organizational and individual) on a tourism organization behavior.

In general, five previous and famous models and framework (DOI, TAM, TPB, TOE and INT) employed to construct the research proposed model. TOE provides a suitable framework combining the variables influencing organization adoption process. The combination of TAM and TPB explain user acceptance behavior. However, DOI mainly describes the technological factors influencing adoption process. To have a better explanation of environmental context, this study employed INT to describe a tourism organization pressure in adopting website.

Future studies may test the proposed model in terms of examining the significant of each suggested factor. They also can develop new attribute(s) to the model by organizing interview from both decision makers and innovation users inside the organization.

REFERENCES

Ajzen, I. (1991). The theory of planned behavior. *Organizational Behavior and Human Decision Processes*, *50*(1), 179–211.

Chang, H.H. & Wong, K.H. (2010). Adoption of e-procurement and participation of e-market place on firm performance: *Trust as a moderator. Information and Management*, *47*(5–6), 262–270.

Chong, S. & Pervan, G. (2007). Factors influencing the extent of deployment of electronic commerce for small- and medium-sized enterprises. *Journal of Electronic Commerce in Organizations*, *5*(1), 1–29.

Davis, F.D. (1989). Perceived usefulness, perceived ease of use, and user acceptance of information technology. *MIS Quarterly, 13*(3), 319–340.

DiMaggio, P.J. & Powell, W.W. (1983). The iron cage revisited: Institutional isomorphism and collective rationality in organizational fields. *Am Social Rev*, *48*(2), 147–160.

El-Gohary, H. (2012). Factors affecting e-marketing adoption and implementation in tourism firms: An empirical investigation of Egyptian small tourism organisations. *Tourism Management*, *33*(5), 1256–1269.

Gregory, A., Wang, Y.R., & DiPietro, R.B. (2010). Towards a functional model of website evaluation: A case study of casual dining restaurants. *Worldwide Hospitality and Tourism Themes*, *2*(1), 68–85.

Hameed, M.A., Counsell, S., & Swift, S. (2012). A conceptual model for the process of IT innovation adoption in organizations. *Journal of Engineering and Technology Management*, *29*(3), 358–390.

Hashim, N.H., Murphy, J., Purchase, S., & O'Connor, P. (2010). Website and email adoption by Malaysian hotels. *International Journal of Hospitality Management*, *29*(1), 194–196.

Hausman, A. (2005). Innovativeness among small businesses: theory and propositions for future research. *Industrial Marketing Management*, *34*(8), 773–782.

Heart, T., & Pliskin, N. (2002). Business-to-business e-commerce of information systems: two cases of ASP-To-SME e-rental. *INFOR*, *40*(1), 23–34.

Jeyaraj, A., Rottman, J., & Lacity, M. (2006). A review of the predictors, linkages, and biases in IT innovation adoption research. *Journal of Information Technology*, *21*(1), 1–23.

Karahanna, E., Straub, D.W., & Chervany, N.L. (1999). Information technology adoption across time: Across sectional comparison of pre-adoption and post-adoption beliefs. *MIS Quarterly*, *23*(2), 183–214.

Law, R. & Cheung, C. (2006). A study of the perceived importance of the overall website quality of different classes of hotels. *International Journal of Hospitality Management*, *25*(3), 525–531.

Li, D., Lai, F. & Wang, J. (2010). E-business assimilation in China's international trade firms: the technology organization environment framework. *Journal of Global Information Technology*, *18*(1), 39–65.

Lyytinen, K. & Damsgaard, J. (2001). What's wrong with the diffusion of innovation theory? The case of a complex and networked technology. Paper presented at the Conference on Diffusing Software Products and Process Innovations, Banff, Canada.

Murphy, J., Schegg, R., & Olaru, D. (2006). Investigating the evolution of hotel internet adoption. *Information Technology and Tourism*, *8*(1), 161–177.

Rogers, E.M. (1962). Diffusion of Innovations. New York: The Free Press.

Rogers, E.M. (2003). Diffusion of innovations (5th ed.). New York: The Free Pressd.

Salavati, S. & Hashim, N.H. (2011). An investigation on website adoption and performance on Iranian hotels. Paper presented at the International Conference on Business and Economic Research (2nd ICBER 2011), Langkawi, Malaysia.

Sambamurthy, V., Bharadwaj, A., & Grover, V. (2003). Shaping agility through digital options: Reconceptualizing the role of information technology in contemporary firms. *MIS Quarterly*, *27*(2), 237–263.

Sellin, N. (1990). On an aggregation bias. *Journal of Educational Research*, *14*(3), 257–268.

Tan, K.S., Chong, S.C., Lin, B., & Eze, U.C. (2009). Internet-based ICT adoption: Evidence from Malaysian SMEs. *Industrial Management and Data Systems*, *109*(2), 224–244.

Tornatzky, L. & Fleischer, M. (1990). The process of technology innovation. Lexington: Lexington Books.

Wang, Y. & Qualls, W. (2007). Towards a theoretical model of technology adoption in hospitality organizations. *Hospitality Management*, *26*(1), 560–573.

Zaltman, G., Duncan, R., & Holbeck, J. (1984). Innovations and Organizations. Florida: Krieger: Malamar.

Zhang, C. & Dhaliwal, J. (2009). An investigation of resource-based and institutional theoretic factors in technology adoption for operations and supply chain management. *International Journal of Production Economics*, *120*(1), 252–269.

Hospitality and Tourism – Sumarjan et al. (eds)
 ISBN 978-1-138-00151-0

Express marketing of tourism destinations using Instagram in social media networking

H. Hanan
Kuala Lumpur Metropolitan University College (KLMUC), Malaysia

N. Putit
Faculty of Hotel and Tourism Management, Sarawak, Malaysia

ABSTRACT: *"A picture is worth a thousand words"*. Considering the massive and growing popularity of digital media, the varied use of picture-based social media tools or applications such as *Instagram, Pinterest, Snapbucket, Twitter, Tumblr* and others picture-based social media in promoting and marketing tourism destinations has been seen as the current trends in the tourism industry. These social media applications help to raise awareness of any potential place of interests and allow public sharing of photos and comments that could potentially be the "make it" or "break it" factor when it comes to promoting and express marketing of tourist destinations. As one of the top most used picture-based social media applications in the world, Instagram is seen as a trendy tool in digital tourism where snapshots and comments can be shared all across one's social media channels within seconds and anybody could be 'the tourism expert' in expressing their likes or dislikes towards a place of interests. Following this current social media phenomenon, most tourism promotional activities use Instagram pro actively in engaging and attracting their potential clientele as this form of online express marketing also helps them save a great deal of media promotional budget. This research provides valuable information regarding express marketing of tourism destinations via the integration of social media network and addresses the strategic role and impact of Instagram in maximizing the destination popularity whilst strengthening the tourism operators' brands and successfully reducing the marketing campaign cost.

Keywords: destination image, express marketing, marketing, social media network, tourism destination

1 INTRODUCTION

Rapid growth of the use of the social media network creates new phenomenon in promoting and creating awareness on the existence of the tourism destination. The process of posting the photo of the destination on the social media networking creates interest among the user and the perception of *"A picture is worth a thousand words"*.

According to Duggan and Brenner (2013), the Internet users age 50 and below are the one mostly like to use the social media networking and 83% respondents are from age 18–29 years old. The influence of social media network creates a new phenomenon and new wave to do transformation on marketing tools which focus on the use of new technology. The changes of technology that lead to change of demands by the customer towards the tourist destination need to be balance in order to create the good destination image for that particular place.

This research provides valuable information regarding express marketing of tourism destinations via the integration of social media network and addresses the strategic role and impact of Instagram in maximizing the destination popularity whilst strengthening the tourism operators' brands and successfully reducing the marketing campaign cost.

2 LITERATURE REVIEW

2.1 *Express marketing*

Marketing is a method that been used to promote the products and services that been produce by certain organization. In order to look clear definition on the process and function of marketing, American Marketing Association had come out with clear definition on it. Marketing is the set of institutions, activities and processes for creating, delivering, communicating and exchanging offerings that represent the value for the clients, customers, partners and society at large (American Marketing Association, 2007). According to the definition done by

American Marketing Association (AMA), marketing is a must in order to introduce the products and services been offered by using several methods and tools of delivering the message. Express marketing is the marketing tool that uses the current internet networking technology. It works on the social media network such as the Facebook and twitter which can promote the products or services within second. Express marketing is a publicity method that focuses more on the feeling and experience of the customers towards the product or services. Kotler and Zaltman (1971) stated that the publicity is any unpaid form of non-personal promotion and presentation of ideas, products or services where the sponsor is unidentified.

From the definition of publicity by Kotler and Zaltman (1971), express marketing can be turn into one element of the publicity in marketing which any person can be the sponsors of the ideas, products or services that can present it rapidly in the social media network. The use of Instagram will create the impression among the public on the picture that been post by unidentified sponsors. According to iContact.com (2013), Instagram is a free social network based photo-sharing program that enables users with a valid account to immediately take, apply a digital has tag or filter, and instantly share a photo on their profile page in Facebook or twitter.

Traditionally, postcards create impression towards the holiday destination but the impressions have been edited by the professional photo editor. The glossy image of man-made and natural attractions, as well as abstract and iconic description of the destination are produced and designed so tourists will have evidence for their travel and in the same time the travelers send the postcards to relatives and friends to validate their trip, either by sharing the experience at the destination or to let others know that they are thinking about them (Milman, 2012).

This free social network program can be use to promote the tourist destination by creating the mood and impression towards the destination too. The function of postcards as the traditional tools of marketing continues by the photo sharing in Instagram which the traveler can upload easier and share the experience on the place of interest that lead to the development of express marketing for the tourist destination. The digital technology is more efficient and cost lesser when promoting the tourist attraction. However, there are risks involve which based on personal view and experience which may different from one person to another person. Positive image may lead to the success in publicity but the negative impression lead to bad image of the tourist destination.

2.2 *Development of social media network*

Currently, social media networking are rapidly growing by the increasing number of users worldwide. Social media can be generally defined as internet based applications that focus on the consumer-generated content which encompasses "the media impressions created by the consumers, usually informed by relevant experience and shared or archived online for easy access by other impressionable consumers" (Blackshaw, 2006; Zheng & Gretzel, 2010). The combination of new technologies and social networking in daily life of users especially the young people is a fact and the introducing then new media create the new possibilities, opportunities and challenges for those who are involved in the tourism sector (Bizirgianni & Dionysopoulou, 2013).

Bosari (2012) stated that there are several facts on the role of social media currently playing in the business world based on 2012 Social Media Marketing Industry Report. 94% of all businesses with the marketing department used social media as their main marketing platform, almost 60% of marketers are focusing on the social media marketing development and maintenance, 43% of people from age 20 to 29 are spending more than 10 hours a week on social media sites, 85% of all businesses that have dedicated social media platform as marketing strategy reported an increase in market exposure, and 58% businesses used social media marketing for over 3 years reported an increase in sales over that period. Based on this fact, pattern of the marketing strategy had change from the traditional marketing method that need to pay the money for promoting the products and services into the modern marketing method which the use of social media network to promote the product and services with less cost.

Deciding the best social media network for promoting the products and services for an organization are crucial in order to plan for the strategic marketing planning and Eckerle (2013) had come out with four (4) tips for developing a winning social media strategy:

1. Decide if social media the been chosen is right for your organization
2. Match the medium with the message that your company want to deliver to your target customer
3. Explore the aspects of every medium
4. Track growth and measure the level of success for choosing the medium

The four (4) tips done by Eckerle (2013) showing that the chosen of the right social media as a medium of delivering the message is really important in order to make sure the success of the marketing strategy that been set up earlier. Posting the photo of the destination that been visited by tourists and travelers indirectly create the new tools and method in promoting the destination. The development of express marketing through posting photo on the social media network influence the new interest from the social media user to travel and this new method only use the publicity method. Even though it is hard to measure the level of success for this method but the effect of this medium can be seen through the comment and sharing the photo on the social media network.

The social media network keeps on updating base on the demand for the digital technology and program. Creative and innovative marketing strategy that follow the current trend and demand influence the decision making among potential visitors and tourists towards buying the tourism products. The strategies and tools for communicating with the customer have changed significantly with the emergence of the phenomenon known as social media, also referred to as consumer-generated media and this form of media describe the variety of new online information sources created, initiated, circulated and used by consumers intent on educating each other about services, products, brands issues and personalities (Blackshaw & Nazzaro, 2004; Mangold & Fauld, 2009).

The 21st century is the witnessing an explosion of the internet-based message transmitted through the social media network and it becoming a major factor in influencing various aspects of customer behavior including information acquisition, awareness, attitudes, opinions, purchase behavior and post-purchase communication and evaluation (Mangold & Fauld, 2009). The changes of the new technology lead to the influence of new demand for tourism product and it leads to the high perception towards the services done by the Destination Management Organization (DMO).

2.3 *Express marketing on tourism destination image*

The aim of improving a destination's marketing effectiveness and efficiency involve adopting the appropriate approaches and using the adequate tools and techniques by Destination Marketing Organizations (DMOs) and it need adopt and implement a strong focus on strategic marketing due to the current demanding tourism market (Soteriades, 2012). Express marketing by using the Instagram create the good and bad impression based on the personal experience towards the tourist destination. Rapid posting on the social media network such as Facebook and twitter make the viewer know about the destination. The photo that had been taken creates its own impression to the other person and it more on sharing the experience on that particular destination. Before the existence of Instagram, traveler and tourist tend to buy post card that shows the destination that they visit. With the introduction of the new technology, public tend to use it for expressing their feeling towards the destination.

The stronger positive image of the tourist destination will have a higher probability of being included and chosen in the process of decision making (Alhemoud & Armstrong, 1996; Bigne, Sanchez, & Sanchez, 2001; Echtner & Ritchie, 1991; Johnson & Thomas, 1992; Telisman-Kosuta, 1994). Express marketing through the change of technology will introduce the destination. The photo that had been transferred directly to the social media network will influence the image of the destination by creating impression. Express posting within a second on the social media network will create fast impression toward the destination and it influence first impression for those who do not know much about the destination.

The uniqueness and the art of snapping the picture can create the certain kind of emotion and feeling toward the destination. The environment and ambiance of the destination will make the destination more attractive in the picture. In addition of the Instagram, the new program which is Instaphoto had been launch whereby it provide the detail on the correct longitude and latitude of the destination that make this program more effective. The details on the longitude and latitude of the destination will guide the potential traveler to find the destination using the GPS system. The system increase the motivation to travel whereby the system lead to the easiest way of finding the destination besides showing the current traffic situation on that destination.

3 CONCLUSION

Demand on the products and services keep changing based on the influence of the current technology that been introduce. The use of social media networking is one of the best ideas for the marketers in order to tackle the worldwide market for promoting the product and services that been product. The new product and services that been produce can create the new demand on it based on the influence of the photo snapping by the unidentified sponsors on the social media network. Increase number of social network user worldwide influence the new demand on tourism products and services that been provide by the destination management organization. Technology use had reduced the human touch on the services but the expectation from the potential customers is high towards it.

The image on the Instaplace makes the customer wondering on the services on the destination. Balance technology and human touch will lead to the success of the destination marketing and create good destination image for that particular tourism destination. Increase on the destination image lead to the loyalty of tourists visited the destination and it became a 'must visit destination' for the tourists to perceive.

REFERENCES

American Marketing Association. (2007). Definition of marketing. Retrieved May 25th, 2013 from http://www.marketingpower.com/aboutama/pages/definitionofmarketing.aspx.

Bigne, J.E., Sánchez, M.I., & Sanchez, J. (2001). Tourism image, evolution variable and after purchase behavior: inter-related. *Tourism management, 22*, 607–616.

Bizirgianni, I., & Dionysopoulou, P. (2013). The influence of tourist trends of youth tourism through social media (SM) and information and communication technologies (ICTs). *Procedia-social and behavior sciences, 73*, 652–660.

Bosari, J. (2012). The developing role of social media in the modern business world. Retrieved May 25th, 2013 from http://www.forbes.com/sites/moneywisewomen/2012/08/08/the-developing-role-of-social-media-in-the-modern-business-world/.

Duggan, M. & Brenner, J. (2013). The demographics of social media user—2012. Retrieved May 21st, 2013 from http://pewinternet.org/Reports/2013/Social-media-sers.aspx.

Eckerle, C. (2013). Social media marketing: 4 tips for developing a winning social media strategy. Retrieved May 26th, 2013 from http://www.marketingsherpa.com/article/wrap-up/social-media-strategy-tips-webinar#.

iContact. (2013). Instagram. Retrieved May 26th, 2013 from http://www.icontact.com/define/instagram/.

Kotler, P. & Zaltman, G. (1971). Social marketing: An approach to planned social change. *Journal of marketing, 35*, 8–12.

Mangold, W.G. & Fauld, D.V. (2009). Social media: The new hybrid element of promotion mix. *Business horizons, 52*, 357–365.

Milman, A. (2012). Postcards as representation of a destination image: the case of Berlin. *Journal of vacation marketing, 18*(2), 157–170.

Soteriades, M. (2012). Tourism destination marketing: approaches improving effectiveness and efficiency. *Journal of hospitality and tourism technology, 3*(2), 107–120.

Zheng, X. & Gretzel, U. (2010). Role of social media in online travel information search. *Tourism management, 31*(2), 179–188.

Hospitality and Tourism – Sumarjan et al. (eds)
 ISBN 978-1-138-00151-0

A pilot study on effects of e-commerce and internet technology acceptance on tourist satisfaction in Malaysia

T. Sajjad, A. Zainal & S.M. Radzi
Universiti Teknologi MARA, Malaysia

ABSTRACT: Every year the number of international tourist arrivals to Malaysia is showing an upward trend. The significance of the internet as a commercial channel in the context of tourist information search and provision has been well recognized. It is noted that providing a means of integrating travel businesses with service providers and distributors is the crucial first step towards developing a successful e-commerce business model and making tourists satisfied. Thus, the purpose of this study is to assess the effects of e-commerce and internet technology on tourist destination satisfaction in Malaysia. The data were collected through self-administered survey questionnaire at Kuala Lumpur International Airport (KLIA). 50 tourists from 17 countries participated in this study. Accordingly, the results of this initial study indicated internet technology has significant effect on tourist destination satisfaction as compared to e-commerce.

Keywords: e-commerce competitive advantage attributes, internet technology acceptance, tourist destination satisfaction

1 INTRODUCTION

Information search tends to be more intense before arrival at the destination than during the travelling period (Mansfeld, 1992; Peres, Correia, & Moital, 2011; Sheldon, 1997). At this step, tourists decide upon whether to travel to the destination and make preliminary decisions regarding what to do once there. However, there is a growing recognition of the role of information search during travel. However, web-based virtual environment brings more chances to access the experiential information through virtual communities.

As tourism plays an important function in the Malaysian economy, this current study aims to explore the effect of e-commerce and the internet-technology in particular for the future competitiveness of the destination satisfaction in tourist industry. Furthermore, this paper explores the current situation of e-tourism developments in the Malaysian tourism industry.

1.1 *Information technology*

The significance of the internet as a commercial channel in the context of tourist information search and provision has been well recognized in previous studies (Bonn, Furr, & Susskind, 1999; Marcussen, 1997). It is noted that providing computer systems that integrate travel businesses with suppliers, distributors, and retailers is the crucial first step toward developing a successful e-commerce business model and make tourists satisfied. Internet technology can also contribute to minimizing transaction costs, lowering uncertainty, sharing market information, and aiding in distribution channel efficiency. Studies have noted that the main reason for many travellers to use the Internet for information search and subsequently making purchase offline is online privacy (Buhalis & Law, 2008). According to Kim and Kim (2004), the price, safety, and convenience are the three significant reasons for both online and offline customers. The internet serves as a new communication media and distribution channel for e-travellers and suppliers of travels services and products (Law, Leung, & Wong, 2004).

Hence, the purpose of this study is to relate e-commerce and internet technology acceptance in one model inclusively given the current and potential importance of these two dimensions as a source of information in the tourism sector, it appears necessary for this current study to analyze further the consequences of tourist destination satisfaction in Malaysia.

2 LITERATURE

2.1 *Background*

One of the critically important factors to any destination is the sufficient provision of information to virtual visitors before their visit, and to virtual and face-to-face visitors during their visit. According to

Buhalis (1998, 2003) information providers and consumers are still using of many traditional information resources, also Information and Communication Technologies (ICTs) are having a profound result on the diffusion of tourism information (Mistilis & D'Ambra, 2010). Referring to Mistilis and D'Ambra (2008, 2010) for instance, in view of travel within Australia, an upward number of visitors, domestic and international, are using the Internet to look for information and/or to make reservations. Therefore, the Internet played for these international visitors a greater role in travel planning once the decision to visit Australia had been made rather than in influencing the choice of destination.

2.2 *E-commerce competitive advantage attributes*

According to Yang et al. (2012) E-commerce (Electronic Commerce/EC) means commercial activities that are finished completely or partly online, and the products sold on the Internet are physical goods, services or information. Generally, e-commerce is rolling for the travel industry and more and more competitors have come into the field (Yang et al., 2012). Due to the increase in applied IT in the travel industry more traditional travel intermediaries are adopting IT to provide added value to travel consumers. IT has thus become a vital competitive advantage for travel agencies, as they use e-commerce development strategies to retain performance advantages. Therefore, e-commerce creates opportunities for there-intermediation of the on-line travel agency. The four competitive attributes of e-commerce in this study are product, brand, service and price (Tsai et al., 2005).

According to Yang, Chou, and Liu, (2012) when a travel company wants to make a plan connected to e-commerce, they must consider their product planning in relation to positive content and value of the product, it will help the customers to understand the content and value of the product as customers expect. The companies must recognize how to shift virtual commodities into first-rate service. When companies design their websites, they must certify the websites are easy to use, offer a smooth browsing experience and have rich content.

For instance, referring to the statistics from the MIC (Market Intelligence Center) in 2009, the e-commerce market in Taiwan reached 311.6 million NT, with travel accounting for 52%, most of which was air tickets sold (Yang et al., 2012).

2.3 *Internet technology acceptance*

According to Matijaševic et al. (2008) travel products connect a higher level of differentiation and higher level of involvement, intangibility than other tangible consumer goods. Because of the Internet the environment has transformed into two dimensions: first, there has been a raise in the awareness of the travel options available, and second, the provision of user friendly software that provides direct contact to relevant information. These two factors are powerful in attracting consumers to adopt online purchasing. The role of Internet as an information source to enhance tourism volume is vital and it led to attracting new visitors.

This study shows that the user finds the system useful for achieving a specific result. It is distinct as "the prospective user's subjective probability that using a specific application system will increase his or her job performance within an organizational context" (Davis et al., 1989; Castaneda et al., 2009). Perceived usefulness is a multi-dimensional concept related to the issues of working speed, work efficiency and effectiveness, making work easier, and other practical considerations. Perceived ease of use is defined as "the degree to which the prospective user expects the target system to be free of effort" (Davis et al., 1989; Castaneda et al., 2009).

2.4 *Tourist destination satisfaction*

According to Yoon and Uysal (2005), satisfaction is playing a significant role in planning marketable tourism products and services. Improvements in tourist satisfaction have come about through better management for tourists, regarding better reservations systems, better signage, new customer care courses, and the installation of other information provision (Augustyn & Knowles, 2000). Tse and Wilton's (1988) perceived overall performance model, consumer dissatisfaction is a function only of actual performance and is unrelated to consumer expectations. In other words, actual performance and initial expectations are considered independently.

2.5 *Theoretical framework and hypothesis*

Based on the literature, the following hypotheses were formulated.

H1: There is a significant effect between E-commerce competitive advantage and tourist satisfaction.
H2: Higher internet technology leads to higher tourist satisfaction with the selected holiday destination.

3 METHODOLOGY

3.1 *Participants and procedures*

The respondents of this pilot study comprised of 50 tourists who had visited Malaysia. All respondents have used internet to seek information before coming to Malaysia. The data were collected through self-administered survey questionnaire at Kuala Lumpur International Airport (KLIA).

3.2 *Measures*

A close-ended questionnaire using a numerical 7-point Likert Scale was utilized. The dimensions of e-commerce competitive advantage were product, brand, service and price. While the dimensions of Internet Technology acceptance were usefulness and easiness.

3.3 *Data analyses*

The regression analyses were conducted to assess the relationship between the e-commerce and internet technology towards the criterion variable of tourist satisfaction.

4 RESULTS

Tables 1 and 2 illustrate the reliability test and the results of e-commerce and internet technology to tourist satisfaction in Malaysia. The results of this initial study indicated internet technology has significant effect on tourist destination satisfaction as compared to e-commerce.

Table 1. Reliability coefficient of the variables.

Variables	Number of items	Cronbach's alpha coefficient
E-Commerce		
Product	3	0.817
Brand	3	0.660
Service	3	0.604
Price	3	0.841
Internet technology		
Usefulness	4	0.939
Easiness	4	0.775
Tourist satisfaction		
Perceived overall performance	3	0.936
Total	23	

Table 2. Pearson correlation matrix for sub variables of e-commerce and internet technology to tourist satisfaction.

Step	Measures	β
1	E-commerce	0.438*
1a	Product	0.482**
1b	Brand	0.235
1c	Service	0.245
1d	Price	0.411*
2	Internet technology	0.664**
2a	Usefulness	0.677**
2b	Easiness	0.522**

*Correlation is significant at the 0.05 level (2-tailed).
**Correlation is significant at the 0.01 level (2-tailed).

5 DISCUSSION

As shown in Table 1, the Cronbach's alpha coefficient for e-commerce includes 4 dimensions, such as: product (3 items) had a value 0.817 and price (3 items) had a value 0.841, values for the internal consistency were considered very good and accepted; while brand (3 items) had a value 0.660, and service (3 items) had a value 0.604, for these two dimensions the Cronbach's alpha values were considered moderate but accepted. Internet technology includes 2 dimensions: the usefulness had a value 0.939; therefore, the Cronbach's alpha value was considered excellent and accepted. The dimension of easiness had a value 0.775; therefore the Cronbach's alpha value was considered good and accepted (Blaikie, 2003; Hinton, 2004). The dependent variable of tourist satisfaction includes one dimension that perceived overall performance with a value of 0.936. Therefore, according to the strength of the rules of thumb this value is excellent internal consistent reliability.

The Pearson Correlation showed that Internet technology had a significantly positive effect on tourist satisfaction as compared to e-commerce. The two dimensions of Internet technology such as usefulness ($\beta = 0.677$), easiness ($\beta = 0.522$) had the largest influence and significant unique contribution on tourist destination. While e-commerce with four dimensions such as product ($\beta = 0.482$), brand ($\beta = 0.235$), service ($\beta = 0.245$), and price ($\beta = 0.411$) had moderate influence on tourist satisfaction. In determining the strength of the relationship the correlation between Internet technology and tourist satisfaction ($\beta = 0.664$) is significant and positive and E-commerce is ($\beta = 0.438$).

6 IMPLICATIONS

This preliminary study is able to assist travel agencies, tour operators and tourism industry in Malaysia to become aware of the critical factors such as e-commerce competitive advantage attributes and Internet technology that can have a significantly unique contribution and positive influence to tourist destination satisfaction in Malaysia.

The effect of this study contributes several practical and theoretical implications. In particular, this study confirmed the relationship between e-commerce and internet technology to tourist destination satisfaction. Although this is a preliminary study it provides definite evidence that Internet technology is playing an important role to tourist destination satisfaction at present and for the near future and e-commerce competitive advantage at the same time can look as an influential aspect to tourist satisfaction in destination.

7 CONCLUSION

Tourism is changing rapidly as nature, heritage, and recreational destinations become more important, and as conventional tourism is forced to meet tougher environmental requirements. This presents a challenge to government and private enterprise to develop new approaches to the tourism market. Successful Malaysian tourism must benefit local population economically and culturally to give them incentives to protect the national resources which create the attraction. According to statistics the number of international tourist arrivals to Malaysia shows an upward line chart. Therefore, understanding the expectations of tourist arrivals to Malaysia is important to tourist image formation.

In line with this, the internet technology is playing a very important role for promoting a tourist destination's exciting activities and cultural images. Additionally, tourism businesses need to become more flexible, more efficient and quicker in responding to travelers' requests. Therefore, the importance of the e-commerce and internet as a source of information in the tourism industry should be emphasized. In particular recent studies have highlighted the role of internet in tourist image formation. Within the Malaysian hospitality and tourism industry, this study would be able to develop the understanding of the e-commerce together with internet technology actual acceptance among users in order to choose Malaysia as their holiday destination.

REFERENCES

Augustyn, M. & Knowles, T. (2000). Performance of tourism partnerships: A focus on York. *Tourism Management*, 21, 341–351.

Blaikie, N. (2003). Analyzing Quantitative Data: From Description to Explanation. London: Sage.

Bonn, M.A., Furr, H.L. & Susskind, A.M. (1999). Predicting a behavioural profile for pleasure travellers on the basis of internet use segmentation. *Journal of Travel Research*, *37*(4), 333–340.

Buhalis, D. & Law, R. (2008). Progress in information technology and tourism management: 20 years on and 10 years after the internet the state of e-Tourism research. Tourism Management, *29*(4), 609–623.

Buhalis, D. (1998). Strategic use of information technologies in the tourism industry. *Tourism Management*, *19*(5), 409–421.

Buhalis, D. (2003). *eTourism*. London: Prentice Hall.

Castañeda, J.A., Frías, D.M. & Rodríguez, M.A. (2009). Antecedents of internet acceptance and use as an information source by tourists. *Emerald Group Publishing Limited, 33*(3), 548–567.

Davis, F.D., Bagozzi, R.P., & Warshaw, P.R. (1989). User acceptance of computer technology: A comparison of two theoretical models. *Management Science*, *35*(8), 982–1003.

Hinton, P.R. (2004). *Statistic Explained*, (2nd ed.). London: Routledge.

Kim, W.G. & Kim, D.J. (2004). Factors affecting online hotel reservation intention between online and offline customers. *International Journal of Hospitality Management*, *23*(4):381–395.

Law, R., Leung, K. & Wong, J. (2004). The impact of the Internet on travel agencies. *International Journal of Contemporary Hospitality Management*, *16*(2), 100–7.

Mansfeld, Y. (1992). From motivation to actual travel, *Annals of Tourism Research*, *19*, 399–419.

Marcussen, C.H. (1997). Marketing Europe tourism products via internet/WWW. *Journal of Travel and Tourism Marketing*, *6*(3/4), 23–24.

Matijaševic, M.S., Krce, M.B. & Korona, Z.L. (2008). Symbols in visual presentation of Croatian marinas on the internet. Proceedings 4th International Conference. An Enterprise Odyssey: Tourism-Governance and Entrepreneurship, Croatia, 941–954.

Mistilis, N. & D'Ambra, J. (2008). The visitor experience and perception of information quality at the Sydney visitor information centre. *The Journal of Travel & Tourism Marketing*, *24*(1), 35–46.

Mistilis, N. & D'Ambra, J. (2010). Assessing the e-capability of visitor information centers. *Journal of Travel Research*, *49*(2), 206–215.

Peres, R., Correia, A., & Moital, M. (2011). The indicators of intention to adopt mobile electronic tourist guides. *Journal of Hospitality and Tourism Technology*, *2*(2), 120–138.

Sheldon, P. 1997. Tourism Information Technology. CABI, Wallingford.

Tsai, H.T., Huang, L., & Lin, C.G. (2005). Emerging e-commerce development model for Taiwanese travel agencies. *Tourism Management*, *26*, 787–796.

Tse, D.K. & Wilton, P.C. (1988). Models of consumer satisfaction formation: An extension. *Journal of Marketing Research, 25*, 204–212.

Yang, D.J., Chou, D.H. & Liu, J. (2012). A study of key success factors when applying e-commerce to the travel industry. *International Journal of Business and Social Science*, *3*(8), 114–119.

Yoon, Y. & Uysal, M. (2005). An examination of the effects of motivation and satisfaction on destination loyalty: A structural model. *Tourism Management*, *26*, 45–56.

Hospitality and Tourism – Sumarjan et al. (eds)
© 2014 Taylor & Francis Group, London, ISBN 978-1-138-00151-0

Electronic word-of-mouth, trust and purchase decisions in five-star hotels

R.N. Ningkan, H. Hashim & H. Ismail
Universiti Teknologi MARA, Malaysia

ABSTRACT: The increase in the availability and accessibility to websites related to customer feedback makes electronic word-of-mouth an important marketing tool for many hospitality organisations. The electronic word-of-mouth process is essentially a communication between the users, ranging from a casual topic to the advocacy of a particular customer brand. This research note discusses the development and use of electronic word-of-mouth systems and levels of trust as they relate to the final purchase decisions of customers. It outlines issues that form part of the author's current empirical study.

Keywords: consumer, customer, electronic word-of-mouth, purchase decision, trust

1 INTRODUCTION

The emergence and growth of social media and travel feedback websites in the last two decades make it much easier now for the consumer to access a wide variety of information and to exchange information with other consumers. According to Hennig-Thurau et al. (2004), electronic word-of-mouth refers to the feedback made by a potential or existing consumer regarding a specific service or product. Liu (2006) further stresses the informal aspect of such communication. Ratnasingam (2012) added that first-time customers who have not yet seen or visited a hotel are much more likely to seek out information on the hotel in question before deciding to proceed with their purchase. Consumers' choices, experiences and evaluations of hotels are complex areas with many factors coming into play. Further, Hanlan et al. (2005) noted that the consumer will evaluate product performance based on their expectations and post-experience satisfaction levels.

Trust in the integrity of websites remains a significant issue. Trust can be defined as a psychological state in which consumers believe in the competence, integrity and benevolence of the e-vendor and in the predictability that the e-vendor will meet the obligation of the transaction (Grefen et al., 2003). However, Zhu et al. (2011) noted that online consumers may expose themselves to the loss. Building trust can reduce the perception of risk by consumers (Somali et al., 2009), and impact positively on those who may be reluctant to pursue an online purchase (Gefen, 2000).

Online purchase intention is an important factor of real-life behavior. It refers to the assessment criteria used by consumers, based on the quality of a website, the information given, and the product review evaluations (Poddar et al., 2009). The intention to purchase a product can be defined in terms of the probability of a potential consumer to complete the purchase of the product (Sam & Tahir, 2009). Hellier et al. (2003) note that the intention to repurchase is an individual judgment regarding the buying again of a product from the same company by taking into account his or her current situation and other information.

2 LITERATURE REVIEW

2.1 *Electronic word-of-mouth*

The growth of the internet in the Web 2.0 era has resulted in more individuals being able to manipulate data and create content according to their own needs, with such techniques leading naturally to the evolution of electronic word-of-mouth systems (Alexandru, 2010). Kelly (2007) noted that electronic word-of-mouth can be seen as both a consultation and an advertisement, by potential customers reading the ratings and recommendations before making a final online purchase decision. Indeed, it has created an environment whereby online information providers such as tourist boards, hotel and resort websites, travel agents, bloggers and magazines actively compete with each other for the attention of searchers and ultimately purchasers of products (Dickinger, 2010).

Cheung and Thadani (2012) further argued that the ongoing improvement of internet technology systems offers a fertile base for the continued development of electronic word-of-mouth communication. The relevance of this development can be construed both in terms of consumers' needs and equates its impact on the hospitality industry in relation to the retention of existing customers and the gaining of new customers. Doh and Hwang (2009) noted that positive feedback from the other internet users brought a positive image to the five-star hotel from the point of view of customers intending to purchase a hotel room. In contrast, Park et al. (2007) reported that negative electronic word-of-mouth statements have a greater impact compared to the positive word-of-mouth statements. In other words, the negative statements from other consumers can bring a negative effect towards the product or service.

Alexandru (2010) noted that electronic word-of-mouth can give an advantage to those companies that try to use it, but that not all companies will receive positive word-of-mouth statements from consumers. According to Lee et al. (2012), independently-minded and self-controlled consumers tend to like to help other internet users by posting and making available information on particular topics and subjects. They also noted that self-control must be understood by the other users in order to ensure the consumers take part in electronic word-of-mouth processes and systems. The presence of product reviews from other consumers enhances consumers' searches and purchase behaviours (Huang et al., 2009).

2.2 *Role of trust*

As can be seen from the above notations, trust remains one of the most significant factors that impacts on a consumer making a final purchasing decision. Trust is based on a mental model of cognitive and emotional elements (Rusman et al., 2007). In the case of online hotel reservations, customers make themselves vulnerable to the actions of the hoteliers, based upon the expectation that the hotelier will provide the service as promised on the website (Ratnasingam, 2012). Trust is the consumer confidence, belief or expectation in the seller providing the service as described.

There appears to be no agreed consensus on how trustworthy consumers feel social media and feedback websites are. The Edelman Trust (2010) found that consumers were increasingly losing trust in one another when it came to providing credible information about products or companies. Between 2008 and 2010, confidence by consumers in the online peer reviews of products or services dropped by almost half, with only 25% of respondents trusting their peers online for credible information. Interestingly, and in contrast, the Edelman Trust (2012) reported an improvement in the level of trust that consumers reported in relation to online peer reviews and social media sites. While trust overall in governments, business and NGOs declined in the twelve months to 2012, consumer trust in the media, and in particular in social media, increased to just under 50% (Edelman Trust, 2012).

2.3 *Future purchasing decisions*

According to Gretzel and Yoo (2008), three-quarters of travellers have considered online consumer reviews as an information source when planning their trips. The impact of electronic word-of-mouth feedback can act both directly and indirectly on potential consumers and their intention to purchase a hotel room (Kim et al., 2004). According to Chang et al. (2010), the effect of recommendations by other experienced users will lead to a greater motivation to purchase a service or a product online. Furthermore, Xiaofen and Yiling's (2009) study shows that a positive recommendation by electronic word-of-mouth will increase the willingness to purchase online compared to advertising and other recommendation systems. The electronic word-of-mouth effect appears to be greater than other types of recommendations, including even advertising (Xiaofen & Yiling, 2009).

However, there are different levels of expertise with some internet users being seen as experts in analysis of the feedback left on websites compared to others who are new to the environment. Park and Kim (2008) found that an expert user has strong opinions in terms of reviewing purchase intentions compared to inexperienced users; the less experienced user tends to focus on positive feedback. Park and Kim (2008) argued that in these cases there is a positive effect of the quality of reviews on purchasing intentions: the purchasing intention towards the product or service increases in line with the higher number of positive feedback statements (Park et al., 2007).

3 ISSUES RELATING TO ELECTRONIC WORD-OF MOUTH

According to Text 100's Digital Index (2012), traveller demands for information from advice-based websites have become a vital factor in the hospitality industry. The study demonstrated the significance of social media websites prior to making a purchasing decision. Over half of the respondents have shared their photographs, videos, and blogs on their vacations. In addition, the study commented on the need for the travel industry to remain up-to-date with new technology, given the rapid development in mobile computing and telephony systems that enables consumers to access online advice-websites at any time they have access to the internet. These social media and travel feedback websites have had a huge impact on both consumers and the hospitality industry (Racherla et al., 2013).

In addition, recent research reported that each year, millions of potential hotel visitors consult travel review websites (Tuominen, 2011). It has been estimated that 88% of potential consumers have their choices affected by what they read on such websites (European Travel Commission, 2009, cited in Tuominen, 2011). Existing research has not focused on how significant such travel review websites are in consumers' decisions to purchase a five-star hotel rooms in Malaysia, and the level of trust that these consumers feel in making full use of social media and travel feedback websites. In line with that, addressing these issues will be the aim of the author's future research.

REFERENCES

Alexandru, B. (2010). The art and science of word-of-mouth and electronic word of-mouth. *Fascicle of Management and Technologicl Engineering*, *9*(19), 4.7–4.16.

Chang, C.C. & Chin Y. (2010). The impact of recommendation sources on online purchase intentions: The moderating effects of gender and perceived risk. *World Academy of Science, Engineering and Technology*, *42*, 111–114.

Cheung, C.M.K. & Thadani, D.R. (2012). The impact of electronic word-of-mouth communication: A literature analysis and integrative model. *Decision Support System, 54*, 461–470.

Dickinger, A. (2010). The trustworthiness of online channels for experience and goal—directed search tasks. *Journal of Travel Research*, *50*(4), 378–391.

Doh, S.J. & Hwang, J.S. (2009). How consumers evaluate EWOM (electronic word-of-mouth) messages. *Cyber psychology & behaviour, 12*(2), 193–197.

Edelman Trust (2010). Edelman trust. Retrieved on 4 September 2012 from http://www.edelman.com/trust/2010/docs/2010_barometer_executie_summary.pdf.

Edelman Trust (2012). Edelman trust. Retrieved on 22 May 2012 from http://www.edelman.com/wp-content/uploads/2012/01/Final-Brochure-1.16.pdf.

European Travel Commission (as cited in Tuominen, P., 2011) The influence of tripadvisor consumer-generated travel reviews on hotel performance. Retrieved on 23 July 2012 fromhttps://uhra.herts.ac.uk/dspace/bitstream/2299/7612/1/S128.pdf.

Gefen, D. (2000). E-commerce: The role f familiarity and trust. *The International Journal of Management Science, 28*, 725–737.

Grefen, D., Karahanna, E., & Straub, D.W. (2003). Inexperience and experience with online stores: The importance of TAM and Trust. *IEEE Transactions on Engineering Management*, *50*(3), 307–321.

Gretzel, U. & Yoo, K.H. (2008). Use and impact of online travel reviews. *Information and Communication Technologies in Tourism*, 35–46.

Hanlan, J, Fuller, D., & Wilde, S.J.C. (2005). Destination decision making and consumer demands: Identifying critical factors. center for enterprise development and research occasional, Paper, No. 2, Centre for Regional Tourism Research, Southern Cross University, Coffs Harbour, NSW.

Hellier, P.K., Geursen, G.M., Carr, R.A. & Rickard, J.A. (2003). Customer repurchase intention: A General structural equation model. *European Journal of Marketing*, *37*, 1762–1800.

Hennig-Thurau, T., Gwinner, K.P., Walsh, G., & Gremler, D.D. (2004). Electronic word-of-mouth via consumer-opinion platforms: What motivates consumer to articulate themselves on the internet*? Journal of Interactive Marketing, 18*, 38–52.

Huang, P., Lurie, N., & Mitra, S. (2009). Searching for experience on the web: An empirical examination of consumer behavior for search and experience goods. Journal of Marketing, *73*(2), 55.

Kelly, L. (2007). Beyond buzz: The next generation of word-of-mouth marketing. New York: AMACOM.

Kim, J.I., Lee, H.C., & Kim, H.J. (2004). Factors affecting online search intention and online purchase intention. *Seoul Journal Business*, *10*(2), 28–48.

Lee, D., Kim, H.S., & Kim J.K. (2012). The role of self-contrual in consumers: Electronic of mouth (Ewom) in social networking sites: A social cognitive approach. *Computers in Human Behavior*, *28*, 1054–1062.

Liu, Y. (2006). Word-of-mouth for movies: Its dynamics and impact on box office revenue. *Journal of Marketing*, *70*(3), 74–89.

Park, D.H. Lee, J.,& Han, I (2007). The effect of on-line consumer reviews on consumer purchasing intention: The moderating role of involvement. *International Journal of Electronic Commerce*, *11*(4), 436.

Park, D., & Kim S. (2008). The effects of consumer knowledge on message processing of electronic word-of-mouth via online consumer reviews. *Electronic Research and Application*, *7*, 399–410.

Poddar, Amit, Naveen D.,& Yujie Wei (2009). Website orientations, website quality and purchase intention: The role of website personality. *Journal of Business*, *62*(4), 441–450.

Racherla, P., Connolly, D.J., & Christodoulidou, N. (2013). What determines consumers' ratings of service providers? An exploratory study of online traveler reviews. *Journal of Hospitality Marketing & Management*, *22*(2), 135–161.

Ratnasingam, P. (2012). Consumer's trust indicators in the online hotel booking decision. *Humanities and technology*, *2*(2), 192–198.

Rusman, E., Van B.J., & Koper, R. (2007). Theoretical framework for the design and development of a personal identity profile fostering interpersonal trust in virtual project team. Retrieved on 6 November 2012 from http://lnx-hrl-075v.web.pwo.ou.nl/bitstream/1820/943/7/%20sid%20workshopvirtual%20team%20perceived%20trustworthiness_adapted.pdf.

Sam, M.F.M. & Tahir, M.N.H. (2009). Website quality and consumer online purchase intention of air ticket. *International Journal of Basic & Applied Sciences*, *9*(10), 20–25.

Somali, S.A., Gholami, R., & Clegg, B. (2009). An investigation into the acceptance of online banking in Saudi Arabia. *Technovation*, *29*, 130–141.

Text100's Digital Index (2012). Text100's digital index: Travel and tourism study. Retrieved on 10 October 2012 from http://info.text100.com/Portals/162056/docs/Text100DigitalIndextTravelTourismStud-TOP10Final.pdf.

Tuominen P. (2011). The influence of tripadvisor consumer-generated travel reviews on hotel performance. Paper Presented at the 19th Annual Frontiers in Service Conference. Retrieved on 22 May 2012 from https://uhra.herts.ac.uk/dspace/bitstream/2299/7612/1/S128.pdf.

Xiaofen, J. & Yiling, Z. (2009). The impacts of online word-of-mouth on consumer's buying intention on apparel: An empirical study. *International Symposium on Web Information Systems and Applications*, 24–8.

Zhu, S.D., Lee, Z.C., & O'Neal, S.G (2011). Mr. Risk! Please trust me: Trust antecedents that increase online consumer purchase intention. *Journal of Internet Banking and Commerce*, *16*(3), 1–23.

Hospitality and Tourism – Sumarjan et al. (eds)
© 2014 Taylor & Francis Group, London, ISBN 978-1-138-00151-0

Five-star hotel in-room entertainments: The relationship of technology readiness index, perceived ease of use and perceived usefulness

F.H. Ramly & A.A. Azdel
Universiti Teknologi MARA, Malaysia

ABSTRACT: Technology in services has been a part of a person's life thus technology-based product becoming unavoidable in a person's life. Hotel have spent a tremendous amount of money to install and upgrade new technology in their property in order for them to satisfy the customer. Hence, in-room entertainment technologies services are one of the revenue producing opportunity for hotel which allows for guest-customized experiences. However, there are certain people who able to adapt to technology changes and some do not. Thus it is beneficial to understand the reason why some people able to adapt and some were restrict the usage of technology in their daily life. These differences can be further investigated using TRI which the ability of consumer to adapt and use the new technology for them to reach their goals. This study will use the models of Technology Readiness Index (TRI), perceived ease of use and perceived usefulness to understand further the relationship between readiness and the acceptance of the user.

Keywords: in-room technology, perceived ease of use, perceived usefulness, technology readiness index

1 INTRODUCTION

Technology has been an integral part of people life and technology-based products as well as services are becoming increasingly unavoidable (Mick & Fournier, 1998; Meuter et al., 2005). Even though people know the benefits of using the technology; some people choose to ignore, refuse or delay their adoption of technology (Mick & Fournier, 1998). Thus, it is beneficial to understand the reason why some people able to adapt and some restrict the usage of technology in their daily life.

Additionally, the infusions of advanced information technology in hotel management system help the hotel to improve their service quality and service efficiency which in the end it helps in lowering the management cost and enhances customer satisfaction (Ying, 2013). However, there are a certain group of people who doesn't use technology at all. Some of the customers do take real advantages and interested in the improvement of hotel technology however there are still some guests who don't want to take part and are not willing to pay for technological innovations. Rogers (1983) stated that innovations that are perceived by receivers as having a greater relative advantage, trialability, compatibility, observability and less complexity will be adopted more rapidly compared to other innovations without these qualities.

Hence, this study will use the models of Technology Readiness Index (TRI), perceived ease of use and perceived usefulness to understand further the relationship between readiness and the acceptance of the user. The effective use of information technology enables hotel industry to make significant improvement in enlarging its operational efficiency (Kapiki, 2012). Walczuch, Lemmink & Streukens (2007) said the difference in personality show the difference in the adoption process of IT and this difference explained by using the influence of personality of the user.

2 LITERATURE REVIEW

2.1 *Technology readiness index*

The term technology readiness discusses about the person's tendency to embrace and use new technologies for accomplishing goals both in home life and at work (Parasuraman, 2000). Technology readiness comprises four dimensions which are optimism, innovativeness, discomfort and insecurity. Optimism relates to a positive view about a technology and a belief that technology offers people increased over control, flexibility and efficiency in life (Parasuraman, 2000). Innovativeness often refers to a person's tendency to try out new things (Parasuraman, 2000). Customers that have a high tendency of innovativeness tend to feel more comfortable to use the new technology product and require little proof on the outcomes of the technology (Agarwal & Prasad, 1998). Insecurity involves the distrust of technology for security and privacy reason (Parasuraman, 2000). Insecurity customer tends to have negative input towards new technology thus it reduces the ability for them to try and avoid new technology (Meuter et al., 2005). Thus people who have discomfort may use the product or services less frequently. However, basic functional products are less complex and require less knowledge to use them.

TRI can help managers to understand the importance of offering technology and tailor he need and preferences toward different segment based on their comfort with the usage of technology (Victorino, Karniouchina, & Verma, 2009). However, there are a certain group of travelers that will be discouraged from staying at the hotel with the reason of hotel implementing new technology (Withiam, 2013). Therefore, it is important for the hotel to ensure that the customer has positive experience and feedback while using the technology (Badri, Al Rashedi, & Mohaidat, 2013). Moreover, it is advisable for the hotel to pay attention toward managing customer feedback, it is recommended that the hotel need to continuously providing face-to-face interaction rather than depend on various technology-based interfaces (Victorino, Karniouchina, & Verma, 2009). In line with that notion, technology adoption in the hospitality industry can be said as one of the biggest opportunity yet it still pose as one of the biggest challenges for the management therefore adoption of the new technology requires insight and inputs from a strong research initiatives and organizational commitment (DiPietro & Wang, 2010).

2.2 *Perceived ease of use and perceived usefulness*

The perceived ease of use refers to "the degree to which a person believes that using a particular system would be free of effort" (Davis, 1989). This can be said as if the user were given an application that is useful but at the same time they believe that the systems is too hard to use and the benefit of using the application is out-weighed by the effort of using the application. Meanwhile the perceived usefulness is defined as "the degree to which a person believes that using a particular system would enhance his or her job performance" (Davis, 1989). The act of people that tend to use or not to use an application to the degree they believe it will help them perform their job better. In addition, it can be said that the usefulness of the application is influenced by the perceived ease of use (Davis, 1989).

Although demand for using the Information Technology (IT) is booming in the hospitality sector, how does this application of technology able to utilize its operation and function to produce maximum customer values as opposed to differences among customers personal use of computer and their familiarity (Lai, Huang, Lu, & Chang, 2013). Therefore, it is important to ensure that the customer to have a more positive attitude toward the use of technology as to exploit the benefit of IT tool in the hotel industry (Moses, Wong, Bakar, & Mahmud, 2013). Moreover, Wang (2013) stated that the actual use of technology can be affected by the attitude of the customer toward adopting the technology. The researcher also added that if the business enterprise has a strong ability to influence the customer attitude then it will have better performance. Hence, it is agreed that if the customers think that the IT is useful therefore the customers are ready to adopt it.

2.3 *In-room technology*

Hotel often benefit from utilizing the technology as a value-added amenities to help in promoting the differences in hotel, enhancing guest satisfaction and building loyalty among guest (Cobanoglu, Ryan, & Beck, 1999). Some studies agree that it help the hotel to improve their guest experience and enhanced guest satisfaction provide major advantages of hotel technology applications (Brewer, Kim, Schrier, & Farrish, 2008).

In the hospitality industry, the main contributor to hotel revenues comes from room sales. Some hospitality industry experts believe if the manager emphasize about the importance of in-room technologies as the traveling public continues to become more technologically savvy (Higley, 2007; Munyan, 2008; Squires, 2008). In-room technology were designed to provide a more comfortable and safe environment which may include minibars, electronic locks and safes, alarm clocks, desktop computers, entertainment systems, climate control systems, security systems and others (Collin & Cobanoglu, 2008).

Too (2010) added with the increasing number in wired and wireless technology it is hard for the hotel industry to keep up and provide such sophisticated system in the hotel room. The author also added, in such a complex and dynamic environment the capabilities for the hotel to decide what system and infrastructures to invest in are real challenges.

3 DISCUSSION

3.1 *Issue*

The improvement in multimedia entertainment over the last few years has resulted in an increased acceptance by consumers and this behavior has shown the importance of providing in-room entertainment in the hotel (Boukis, 2007). In-room entertainment technologies services are one of the revenue producing opportunity for hotel which allows for guest-customized experiences. Although installing this technology helps to increase customer satisfaction however it is important for the hotel industry to understand that not all technology impact guest satisfaction equally (Cobanoglu, Berezina, Kasavana, & Erdem, 2011).

Godoeand Johansen (2012) added that when introducing new technology system, the hotel needs to test the system before it is being adopted by the hotel as it will help the hotel to take necessary step before commencing the implementation of the technology. The author also added that in order to understand the acceptance toward the adoption of technology, it is agreeable to say that personality dimension can influence the way people interact with experience and use new technology.

Previous research examined on how consumer traits influence technology acceptance (Dabholkar & Bagozzi, 2002; Ellen et al., 1991; Hill et al., 1987). However the role of TRI within the TAM has been under-explored. Jiun-Sheng and Hsing-Chi (2011) verified that it is important to incorporate TRI into TAM because it plays an important role in consumer technology adoption.

4 CONCLUSIONS

The findings from this study are expected to contribute to the extension of knowledge on technology readiness and technology acceptance model which focus on the hotel industry. Specifically, this study will extend the knowledge in the hotel industry and internet marketing. From the academic perspective, this study may contribute to the body of knowledge in the area of hotel technology.

Although all instruments were adapted from the previous researchers, this is an opportunity to examine the instrument applicability in Malaysia environment. The finding of this study is relevant to hotel managers in forming strategies to retain guest loyalty. The benefit of this study is beneficial for the hotel which implemented the in-room technology as this study would provide an insight on the behavior of the guest who prefers in-room technology. Such knowledge can benefit the managers to fine-tune their product positioning and communication strategies to match with the TRI profiles of potential customer in different stages of the product life cycle.

REFERENCES

Agarwal, R., Prasad, J. (1998). A conceptual and operational definition of personal innovativeness in the domain of information technology. *Information System Research*, 9(2), 204–15.

Badri, M., Al Rashedi, A., & Mohaidat, J. (2013). School teachers' technology readiness: An empirical study applying readiness factors and teacher type categorization. International Conference on Information, Business and Education Technology (ICIBIT 2013).

Brewer, P., Kim, J., Schrier, T., & Farrish, J. (2008). Current and future technology use in the hospitality industry. American Hotel and Lodging Association. Retrieved from http://www.ahla.com/membersonly/content.aspx?id=5964.

Cobanoglu, C., Berezina, K., Kasavana, M.L., & Erdem, M. (2011). The impact of technology amenities on hotel guest overall satisfaction. *Journal of Quality Assurance in Hospitality & Tourism,* 12(4), 272–288.

Cobanoglu, C., Ryan, B., & Beck, J. (1999). The impact of technology in lodging properties. In International Council on Hotel, Restaurant, and Institutional Education Annual Convention Proceedings (pp. 34–39). American Hotel and Lodging Association (AH & LA).

Collins, G.R. & Cobanoglu, C. (2008). Hospitality information technology: Learning how touse it (6th ed.). Dubuque, IA: Kendall/Hunt Publishing Company.

Davis, F.D. (1989). Perceived usefulness, perceived ease of use, and user acceptance of information technology. *MIS Quarterly*, 319–340.

Higley, J. (2007). Keep technology working, make guests happy Questex Media Group.

Jiun-Sheng Chris Lin, & Pei-ling Hsieh, (2006). The role of technology readiness in customers' perception and adoption of self-service technologies, *International Journal of Service Industry Management*, 17(5), 497–517.

Kapiki, S. (2012). Current and future trends in tourism and hospitality: The case of Greece. *International Journal of Economic Practices and Theories*, 2(1).

Lai, Y.H., Huang, H.C., Lu, R.S., & Chang, C.M. (2013). The effects of website trust, perceived ease of use, and perceived usefulness on consumers' online booking intention: Evidence from Taiwan B & B sector. *Life Science Journal*, 10(2).

Mick, D. & Fournier S. (1998). Paradoxes of technology: Consumer cognizance, emotions, and coping strategies. *Journal of Consumer Research*, 25(2), 123–43.

Meuter M.L., Ostrom A.L., Bitner M.J., & Roundtree R. (2005). The influence of technology anxiety on consumer use and experiences with self-service technologies. *Journal of Business Research*, 56(11), 899–906.

Moses, P., Wong, S.L., Bakar, K.A., & Mahmud, R. (2013). Perceived usefulness and perceived ease of use: Antecedents of attitude towards laptop use among science and mathematics teachers in Malaysia. *The Asia-Pacific Education Researcher*, 1–7.

Munyan, R. (2008). Technology in the next generation of hotels. *Lodging Hospitality*, 64(16), 78–88.

Parasuraman, A. (2000). Technology readiness index (TRI): A multiple-item scale to measure readiness to embrace new technologies. *Journal of Service Research*, 2, 307–320.

Rogers, E.M. (1983). *Diffusion of Innovations,* 3rd Ed, The Free Press, London.

Squires, M. (2008). Technology changes lodging workforce. *Lodging Hospitality*, 64(16), 89–94.

Too, C.W.P. (2010). Technology enhancement in hotel guestroom.

Walczuch, R., Lemmink, J., & bStreukens, S. (2007). The effect of service employees' technology readiness on technology acceptance. *Information & Management*, 44(2), 206–215.

Victorino, L., Karniouchina, E., & Verma, R. (2009). Exploring the use of the abbreviated technology readiness index for hotel customer segmentation. *Cornell Hospitality Quarterly*, 50(3), 342–359.

Wang, Y. (2013). The Influence factors of users' attitude to adopt mobile business. International Conference on Information, Business and Education Technology (ICIBIT 2013).

Withiam, G. (2013). Tech-savvy guests spend more. Cornell Center for Hospitality Research Retrieved from http://hospitalitytechnology.edgl.com/top-stories/Tech-savvy-guests-spend-more54918.

Hospitality and Tourism – Sumarjan et al. (eds)
 ISBN 978-1-138-00151-0

Motivational function and online community participation towards brand commitment: A case of Grand Blue Wave hotel Shah Alam

N.S. Shariffuddin, N.A. Ahmad, N.A. Ishak & P.S. Jalal
Universiti Teknologi MARA, Malaysia

ABSTRACT: Due to advance technology, social media has become the right way to gather information, communication, knowledge and entertainment. Thus, Facebook becomes a world trend or "King" of Social Media and becoming a popular tool for marketing. In the hospitality industry, social media helps them to market their business. Thus, building good relationship with online communities, hotel can generate strong commitment to their brands. The marketers have to consider which motivational functions that influence consumer to participate in online communities. This study provides a deep comprehension of consumer behavior in social media and the reasons of why they get involved and participate in online communities.

Keywords: brand commitment, Facebook, motivational functions, online community participation, social media

1 INTRODUCTION

Nowadays, owing to the advancement of technology, Naveed (2012) noted that social media can be considered as an exact mode for the consumer to get the right information at the right time. In fact, Paris, Lee and Seery (2010) explained that social media can be defined as a second generation of Web development and design, with the aims to facilitate communication, secure information sharing, interoperability, and collaboration on the World Wide Web. Hence, social media can be considered as a 21st century version of networking that uses the internet as a medium to reach out (Seth, 2012).

Since social media reach huge numbers of people far and wide, Hartshorn (2010) testified that it has appeared as a very effective business instrument to involve with consumers and in this manner can build a brand name by continuous and prompt correspondence (Seth, 2012). According to Nelsonwire (2010) stated that consumers participate on social media website more than 5.5 hours per day. Lanz, Fischhof and Lee (2010) added that there are several social media famous sites such as Facebook, Twitter, MySpace, LinkedIn, YouTube, and Tagged.

Currently, Facebook is the world leader Social Network (Krahl, 2013) and statistic shows that in April 14, 2013 Facebook had a total number of 978 million active users all over the world (social times. me, 2013) and 680 million monthly active users who used Facebook mobile products as of December 31, 2012 (facebook.com, 2013). With such a huge user base, Facebook is becoming a popular instrument for public relations and advertising professionals to reach mass audiences and Facebook fan pages allow brands to create an online community of brand users on the social networking site (Bushlow, 2012).

In line with the hospitality industry, Lanz, Fischhof and Lee (2010) noted that most successful hotels are finding innovative ways to integrate social media with traditional marketing techniques stay forward in the race. Hotel can generate strong commitment to their brands by constructing a good relationship with online communities. It is supported by Seth (2012) consumers will spend more, get more referrals, and have higher repeat clientele when hotels engage their customers via social media. In addition, hotels can track the number of conversations taking place over a new product or service offered by them and can immediately respond to reviews, comments and feedback (Seth, 2012). Thus enable them to overcome the problems faster and put an effort to improve future brand performance.

2 LITERATURE REVIEW

2.1 *Overview of Facebook*

Facebook is perhaps the "king" of social networking sites, with its over almost 1 billion users globally (Facebook, 2011). Facebook is an SNS developed in 2004 by former Harvard undergraduate student Mark Zuckerberg, which allows users to add friends, send messages, and update personal profiles in order to notify friends and peers about themselves (Quan-Haase & Young, 2010). Furthermore, Facebook users also can join groups to discuss anything that they're interested, play online games and also create events and invite their friends to attend. Facebook has become more and more useful for their users as they can do many things in the same place (Krahl, 2013). According to internetworldstats.com (2013) in Asia Internet use, Population Data, and Facebook Statistic shows that in Malaysia there are 17,723,000 internet users in June 30, 2012 and 13,589,520 Facebook users in December 31, 2012.

Bushelow (2012) argued that the like button is a powerful tool because after it is pressed, Facebook shows the individual's entire network what he or she has just liked, spreading information and affiliations in a viral manner. Although this may be true, SocialMediaExaminer (2011) noted that an increasing number of companies have adopted Facebook pages as a marketing tool, yet many Facebook marketers still struggle to understand how they can best create business value through the medium. Thus, practitioners often discuss the Return-On-Investment (ROI) of social media, indicating the difficulty to substantiate the impact of social media marketing on sales.

2.2 *Social media and online community marketing*

With consumers becoming more immune to conventional advertising and communication channels, the hospitality industry should turn to social media (such as LinkedIn, Facebook, Twitter) as a means of reaching out to their audience (Lim, 2010). Social media are innovative Web-based applications in online marketing (Yang, Kim, & Dhalwani, 2008). Social media promote people to provide personal information. As a new marketing channel, online communities allow marketers to collect all information about potential or current consumers from their profiles, infer consumers' needs and preferences based on their history of community usage, and obtain direct replies from consumers (Sigala, 2003).

According to Spangler, Hartzel and Gal-Or (2006), social media provide companies with various marketing opportunities by enabling them to build online communities, negative outcomes may arise with regard to privacy concerns. Social media assists marketers to advertise their new offerings to targeted consumers (Chung & Buhalis, 2008).

Kang (2011), noted that by monitoring content posted by community members, marketers can achieve a high level of customization and can obtain an in-depth understanding of each consumer's needs, using this information to develop new products or services. In the year 2010, The Clockwork white paper has listed eight in which hotel brands are using Facebook: to promote special offers, to promote special rates for Facebook fans, as a competitive tool, to promote portfolio of the hotel properties, for surveys, to promote events, to enable booking, and to promote the latest news.

2.3 *Online community participation*

Participation in online communities can be characterized as passive or active (Kang, 2011). In relation, active community members are those who interact with other members as opposed to those who merely observe information (Madupu, 2006). Passive referred as lurkers or free riders members browse online communities but rarely become involved in community activities (Preece, Nonnecke, & Andrews, 2004). Moreover, lurkers tend to pursue their own goals and merely take advantage of the benefits of the communities (Ridings, Gefen, & Arineze, 2006). In contrast, Wang & Yu (2012) argued that the term "lurker" may not be suitable in many contexts; given that those individuals do engage in certain activities in online communities, and these activities can also be important to online communities.

When the community member actively sharing their information and experience by posting photos, videos, and comments at hotel brand's Facebook fan page it will attract new member and maintain the relationship with the existing member. In addition Muniz and O'Guinn (2001) stated that, members' active participation enhances their knowledge regarding brands and products and thus Flavián and Guinalíu (2006) noted that active member enables them to offer suggestions to solve problems with product usage and help each other make purchasing decisions.

Brown and Duguid (1991) testified online communities allow individuals to interact with each other. In addition, Wagner (2004) said they can create and share knowledge through dialog with questions and answer. However, Phang, Kankanhalli and Sabherwal (2009) argued that the value of online communities depends on the ongoing participation in terms of two key activities: i.e., knowledge seeking and contribution. Moreover, according to Kang (2011) online community marketers must understand the consumers' motivations to participate relative to what they desire to receive from online communities. To attract new members and encourage existing members' active participation the marketers must study and well understand about consumer behavior and what the benefits that participant seeks.

2.4 *Motivational functions*

Functional benefits: According to Peter, Olson and Grunert (1999), functional benefit is one that increases the ease and/or efficiency of completing transactions (i.e., purchasing products and services) and exchanging information (i.e., information gathering and sharing). Interact with other community members can facilitate purchasing decisions is one of the functional benefits of an online community (Armstrong & Hagel, 1996).

There are several arguments from the past studies about the relationship between functional benefit as the motivational function in online community participation. The studies reveal that Wang and Fesenmaier (2004) found functional benefits have a negative impact on members' participation, and Hwang and Cho (2005) said there is no significant relationship between functional benefits and members' community activities. However, Chung and Buhalis (2008) found a positive relationship between members' information acquisition and their participation. The purpose of consumer participate in online community is to gather and share the information even though they don't have specific plan to travel and stay which hotels.

Social benefits are derived networking, collaborating with and forming relationship with other community members for various product- related, personal or professional reasons (Dholakia, Blazevic, Wiertz, & Algsheimer, 2009). Nambisan and Baron (2007) proposed the social integrative benefits; deriving from the social and relational ties that develop over time among the participants. Community members help and support each other by exchanging ideas and opinions of interest, answering other members' questions, and introducing new topics for discussion (Dholakia et al., 2009) and (Madupu, 2006).

Frequent participation in community activities engenders trust (Wang & Fesenmaier, 2004), which in turn leads to flourishing relationships among community members and with organizers of the community (Preece, 2000). In addition Dholakia et al. (2009) stated that maintaining interpersonal connectivity denotes social benefits derived from establishing and maintain contact with other people (e.g. social support, friendship an intimacy). When online community recognizes that they share mutual interests with other members, they tend to increase the number and length of visits to hotel brand's Facebook fan page, and to actively participate in online community activities (Hwang & Cho, 2005).

The psychological benefits are derived from feeling connected to community members, and include identity expression through the community, a sense of belonging to the community, and a sense of affiliation with other members (Bressler & Grantham, 2000). As members gain such knowledge about their online communities, they come to understand the community and feel a strong sense of belongings and affiliation, which in turn develops a permanent sense of identification (Wang & Fesenmaier, 2004). Hence, this psychological dependence makes members feel confident and positive about their interactions, a psychological benefit that encourages members to increase their participation (Kang, 2011).

Hedonic benefits are something bringing the participant positive reactions and enjoyment (Nambisan & Baron, 2007). In this situation, community members do not seek for information whereas they seek for entertainment e.g., playing games, and contest. Kang (2011) conclude that if participating in an online community is perceived as fun or entertaining, members are more likely to visit the community and to spend more time visiting it.

In hospitality research, monetary benefits have been considered as part of individualized services that fulfill the consumers' specific needs (i.e., special treatment benefits) (Lee, Ahn, & Kim, 2008). To attract new members and benefit online community members, these businesses tend to offer special promotions and coupons (Treadaway & Smith, 2010). By promoting special rates of rooms in hotel brands Facebook fan page, the marketers able to attract community member to participate thus increase the room sales.

2.5 *Brand commitment of online community members*

Brand commitment is defined as strong and positive feelings of attachment to a brand (Beatty & Kahle, 1988). Online communities are also found to affect customer loyalty and purchase intention (Kim et al., 2004). Donio, Massari and Passiante (2006) suggested that satisfaction and commitment are indicators of brand loyalty. Consumers who are highly committed to a specific brand evaluate competing brands less positively or avoid considering competitors' brands when making purchasing decisions (Ahluwalia, Burnkrant, & Unnava, 2000). They tend to defend their favorable attitudes toward brands when perceiving a threat such as unfavorable information about their preferred brands or favorable information about competing brands (Chaiken, Liberman, & Eagly, 1989). Consumers who perceive such threats tend to secure their positive attitudinal position toward their preferred brands by searching for favorable information about their brand (Jain & Maheswaran, 2000).

3 ISSUE

To attract the online community to participate actively, Wasko and Faraj (2000) said that the business must understand consumers' motivation first with regard to the benefits that they expect in the return of participation. Wang, Yu and Fesenmaier (2002) identified four categories of benefits: functional, social, hedonist, and psychological, and found that these benefits bear an impact on whether members participate actively or passively. Conversely, Kang (2011) listed previous studies had found different outcomes from the relationships between participation and brand commitment in the literature review. In fact, Antikainen (2007) and Dholakia, Bagozzi, and Pearo (2004) mention that, if companies can offer the desired benefits such as information and a sense of belonging, the companies will be able to attract new customers, build relationships with their customer, and inspire them to visit again.

However, this issue is raising a critical question as to why less participation among the online community in Grand Blue Wave Hotel Facebook Fanpage and what are the underlying reason of this question. In June 6, 2013, only 760 persons like Grand Blue Wave Hotel Fan page and as compared to the hotel competitor which is Concorde Shah Alam received 3,693 like for their fan page. Based on personal observation throughout Grand Blue Wave Facebook Fanpage, it shows that their numbers of fans are slightly behind from their competitor because of lack of online community participation from Facebook user.

Therefore, it is worth to investigate and revealed the causation. Besides that, the previous study by Kang (2011) about social media marketing in hospitality industry conducted in United State. Thus, he added that in the hospitality industry, yet studies concerning about online communities has been infrequently conducted. For this reason, this study being carried out in order to strengthen the theoretical and empirical the role of motivational functions influencing online community participation for hospitality and the impact towards commitment hotels' brands in Malaysia. In addition, Lim (2010) stated that if social media can become the new "tool" for effective branding and marketing purposes, there is a need to explore and discover the social media.

4 CONCLUSION

In hospitality industries, social media is another initiative of marketing tools. They have become a major factor in influencing various aspects of consumer behavior including recognition needs, information acquisition, opinions and attitudes, purchase behavior, and post-purchase communication and most importantly evaluation (Mangold & Faulds, 2009). Via Facebook, the hospitality companies can easily access their customer's feedback. For that reason, the manager can know what customers' preferences are. It helps a lot to improve hotel product or services even better. Besides that, it is the easiest way to spread the information regarding the new products or services.

From a theoretical perspective, the study provides a deep comprehension of consumer behavior in social media and the reasons of why they get involved and participate in online communities. From a practical perspective, to keep online communities active, online marketers can establish sustainable marketing strategies by identifying what benefits that community member's aspect for in participating, and thus can retain members who are willing to be involved in a long-term relationship with the community (Kang, 2011).

REFERENCES

Ahluwalia, R., Burnkrant, R.E., & Unnava, H.R. (2000). Consumer response to negative publicity: The moderating role of commitment. *Journal of Marketing Research, 37*, 203–214.

Antikainen, M. (2007). *The attraction of company online communities: A multiple case study*. Public Discussion in University of Tampere, Department of Management Studies, Finland.

Armstrong, A.G., & Hagel, J. (1996). The real value of online communities. *Harvard Business Review, 74*(3), 134–141.

Beatty, S.E. & Kahle, L.R. (1988). Alternative hierarchies of the attitude-behavior relationship: The impact of brand commitment and habit. *Journal of the Academy of Marketing Science, 16*(2), 1–10.

Bressler, S.E. & Grantham, C.E. (2000). *Communities of commerce: Building internet business communities to accelerate growth, minimize risk, and increase customer loyalty*. New York: McGraw Hill.

Brown, J.S. & Duguid, P. (1991). Organizational learning and communities-of-practices: Toward a unified view of working, learning, and innovation. *Organizationn Science, 2*(1).

Bushlow, E.E. (2012). Facebook pages and benefits to brands. *Strategic Communications Elon University.*

Chaiken, S., Liberman, A., & Eagly, A.H. (1989). Heuristic and systematic information processing within and beyond the persuasion context. In J.S. Uleman & J.A. Bargh (Eds.), *Unintended thought* (pp. 212–252). New York: Guilford.

Chung, J.Y. & Buhalis, D. (2008). Information needs in online social networks. *Information Technology & Tourism, 10*, 267–281.

Clockwork White Paper: Travel and Tourism. (2010). Facebook andhotels: What should you be doing? *Clockwork marketing.*

Dholakia, U.M., Bagozzi, R.P., & Pearo, L.K. (2004). A social influence model of consumer participation in network and small group based virtual communities. *International Journal of Research in Marketing, 21,* 241–163.

Dholakia, U.M., Blazevic, V., Wiertz, C., & Algsheimer, R. (2009). Communal service delivery: How customers benefit from participation in firm-hosted virtual P3 communities. *Journal of Service Research, 12*(2), 208–226.

Donio, J., Massari, P., & Passiante G. (2006). Customer satisfaction and loyalty in a digital environment: An empirical test. *Journal of Consumer Marketing, 23*(7).

Facebook. (2011). Newsroom. Retrieved from: http://newsroom.fb.com/content/default.aspx?NewsAreaId=22.

Facebook. (2013). Facebook for business. Read 19.2.2013. http://www.facebook.com/business.

Flavián, C. & Guinalíu, M. (2006). Consumer trust, perceived security and privacy policy. Three basic elements of loyalty to a web site. *Industrial Management & Data Systems, 106*(5), 601–620.

Hartshorn, S. (2010, May 4). *5 Differences Between Social Media and Social Networking.* Retrieved from: http://www.socialmediatoday.com/SMC/194754.

Hwang, Y.H. & Cho, Y.H. (2005). The influence of online community's functions on members' attitude toward the community and off-line meetings. *Korea Tourism and Leisure Research, 17*(4), 141–159.

Internetworldstats.com. (2013). Asia internet use, population data, and Facebook Statistic Retrieved from: http://www.internetworldstats.com/stats3.htm#asia.

Jain, S.P. & Maheswaran, D (2000). Motivated reasoning: A depth-of-processing perspective. *Journal of Consumer Research, 27,* 358–371.

Kang, J. (2011). Social media marketing in the hospitality industry: The role of benefits in increasing brand community participation and the impact of participation on consumer trust and commitment toward hotel and restaurant brands. *Research for Hospitality Management. Iowa State University.* UMI Number: 3494053.

Kim, W.G., Lee, C., & Hiemstra, S.J. (2004). Effects of an online virtual community on customer loyalty and travel product purchases. *Tourism Management, 25,* 343–355.

Kerpen, D. (2011). *Likeable social media.* New York: McGraw-Hill.

Krahl, A. (2013). Social Media and Internet Tools In Hospitality Marketing. *Bachelor's thesis March 2013 Degree Programme in Tourism.*

Lanz, L., Fischhof, B., & Lee, R. (2010). *How are hotels embracing social media in 2010? Examples of how to start engaging.* New York: HVS sales and marketing services.

Lee, Y., Ahn, W., & Kim, K. (2008). A study on the moderating role of alternative attractiveness in the relationship between relational benefits and customer loyalty. *International Journal of Hospitality & Tourism Administration, 9*(1), 52–70.

Lim, W. (2010). The Effect of Social Media Networks in the Hospitality Industry. *UNLV Theses/Dissertations/Professional Papers/Capstones.* Paper 693.

Madupu, V. (2006). Online brand community participation: Antecedents and consequences. *(Doctoral dissertation). Available from ProQuest Dissertations and Theses database.* (UMI No. 3230964).

Malaysian Digital Association, (2012). Malaysian Website Rankings for November 2011, *Malaysia Malaysian Digital Association Reports.* January 30, 2012—Kuala Lumpur.

Malmivaara, T. (2011). *Motivations behind liking: Implications of Facebook brand community behavior on purchase intentions.* Retrieved from Social Sciences Citation Index.

Mangold, W.G. & Faulds, D.J. (2011). Social media: The new hybrid element of the promotion mix. *Business Horizons, 52,* 357–365.

Muniz, A.M. & O'Guinn, T.C. (2001). Brand community. *Journal of Consumer Research, 27*(4), 412–432.

Nambisan, S. & Baron, R. (2007), Interactions in virtual customer environments: Implication for product support and customer relationship management. *Journal of Interactive Marketing, 21,* 2.

Nelsonwire. (2010). Led by Facebook, Twitter, global time spent on social media sites up 82% year over year. Website:http://blog.nielsen.com/nielsenwire/global/ledby-facebook-twitter-global-time-spent-on-social-media-sites-up-82-year-over-year/.

Paris, C.M., Lee W., & Seery, P. (2010). The role of social media in promoting special events: acceptance of Facebook Events. *Information and Communication Technologies in Tourism, 14,* 531–541.

Peter, J.P., Olson, J.C., & Grunert, K.G. (1999). *Consumer behavior and marketing strategy, European edition.* Berkshire: McGraw-Hill.

Phang, C.W., Kankanhalli, A., & Sabherwal, R. (2009). Usability and sociability in online communities: A comparative study of knowledge seeking and contribution. *Journal of the Association for Information Systems, 10*(10), 721–741.

Preece, J. (2000). *Online communities: Desingning usability, supporting sociability.* Chichester: Wiley.

Preece, J., Nonnecke, B., & Andrews, D. (2004). The top five reasons for lurking: Improving community experiences for everyone. *Computers in Human Behavior, 20*(2), 201–223.

Quan-Haase, A. & Young, A.L. (2010). Uses and gratifications of social media: A comparison of Facebook and instant messaging. *Bulletin of Science, Technology and Society, 30*(5), 350–361.

Ridings, C., Gefen, D., & Arinze, B. (2006). Psychological barriers: Lurker and poster motivation and behavior in online communities. *Communications of the Association for Information Systems, 18,* 329–354.

Seth, G. (2012). The effects of social media networks in the hospitality industry. *University of Nevada, Las Vegas.* Paper 693.

Sigala, M. (2003). Developing and benchmarking Internet marketing strategies in the hotel sector in Greece. *Journal of Hospitality & Tourism Research, 27*(4), 375–401.

Socialmediaexaminer.com. (2011). Social media marketing Report 2011, retrieved November 8th 2011. Retrievedfrom: http://www.socialmediaexaminer.com/SocialMediaMarketingReport2011.pdf.

Socialtimes.me. (2013). Social media global info center. *Facebook Statistics*. Retrived from: http://socialtimes.me/.

Spangler, W.E., Hartzel, K.S., & Gal-Or, M. (2006). Exploring the privacy implications of addressable advertising and viewer profiling. *Communications of the ACM*, *49*, 119–123.

Treadaway, C. & Smith, M. (2010). *Facebook marketing: An hour a day.* Indianapolis, Indiana: Wiley Publishing, Inc.

Wagner, C. (2004). Wiki: A technology for conversational knowledge management and group collaboration. *Communications of the Association for Information Systems*, *13*, 265–289.

Wang, Y.C., Yu, Q., & Fesenmaier, D. (2002). Defining the virtual tourist community: Implications for tourism marketing. *Tourism Management*, *23*(4), 407–417.

Hospitality and Tourism – Sumarjan et al. (eds)

Food review (blog) and customer purchase intention behaviour: A case study on Friedchillies website

A. Zaileen Elina, W.I. Wan Ruzanna & A. Roslina
Universiti Teknologi MARA, Malaysia

ABSTRACT: Malaysia is known as a 'food heaven' whereby everyone can find different type of food not only from 3 different major races which is Malay, Chinese and Indian but also the other minorities and food available almost 24 hours a day. This wide range of choices of food has led to the problem of choosing what food should we eat and where the best place to have it. The technology advancement help to the process of sharing information not only through the picture post but also food blog and food review online. This research note discusses on the food review (food blog) by reviewing the past studies and its attribute which relates to the customer purchase intention behaviour where the empirical study is still under investigation.

Keywords: attributable, customer purchase intention behavior, food review

1 INTRODUCTION

"Jalan-JalanCariMakan" or JJCM on TV3 was formerly hosted by Tengku Maria Tengku Sabri is one of the first television show that gives out review on food in restaurants in Malaysia. A high-rating and outstanding feedback about the show from viewers has sustained this show for more than six seasons with several different hosts, hosting the show. This show opens an opportunity to other television stations to come up with the similar programmes providing viewers with restaurant food and an idea of new places to dine and try such as "Sarapan di mana?" and "Warung kita". Besides that, newspaper articles, magazines and the current blogs are another form of food review which has been widely used.

According to Cox and Blake (2011), website or blog has become one of the important sources of information seekers these days. Though the main idea of having a blog is for 'blogger to express their feelings, voice, opinions and share experiences and idea (Herring et al., 2005) it's also has been used as a form of marketing and promotion by the companies. Some bloggers gain income through their writings. Currently, countless blogs are available either on fashions, films, music's, tourisms and not to forget, food blog. Various blogs on food found on the internet which shared the recipes, selling cupcakes or cookies as well as blog reviews of food and restaurants. Some of the food blogs which focus on food is a personal blog that shared its experiences. There are also official websites that owned by a company which tries to promote food available in Malaysia from the local street stall such as 'goreng pisang' and 'ubi ketuk' to upscale restaurant menus such as grilled Salmon and Sushi. The food reviews posted at www.friedchillies.com for instance were mostly arisen through the experiences of the authors and there were more than five different authors who worked full time as well as part time sharing their experiences with food. The actual idea of having this food review is to share of good places to dine in Malaysia not only for locals but foreigners. This shall give the readers varieties of food and maybe new places to try on.

2 LITERATURE

This review is divided into three sections. The first section is reporting and reviewing on the relation between the trustworthiness in the blog. The importance of the trustworthiness of the information shared is the central focus of this section. The online purchase intention is thoroughly discussed in the second section. Each of elements which relate and give impact towards customer purchase intention is also reported. The third section discusses on the products and the customer purchase intention behaviour.

2.1 *Blog trustworthiness*

Many researchers discussed on the trustworthiness of the information shared through online

(Chang et al., 2011; Lee et al., 2001; Ryu, 2003). According to Tripp et al. (1994), "trustworthiness is the degree to which the audience perceives that the communicator intend to convey a valid statement". Hardin (2002) stated that trust involves the reasons for which the relevant party is considered to be trustworthy. Meanwhile, Steward (2003) states that trust can be transferred through internet which at the same time gain the reader loyalty. Trust also relates "that the blogger credibility plays a positive role in building relational trust" (Yang & Lim 2009). Chang et al. (2011) confirmed that the trustworthiness of the content in the internet has the positive impact towards the e-WOM.

According to Ahn et al. (2012) the online repute can be considered as a main factor for the organization to gain trust. Ahuja and Mercury (2010) posited that it will be easier for the well known company or high reputation company to increase trust. On the other hand, Droge et al. (2010) noted that the blog has been used as an important information distribution channel and as a knowledge-sharing tool whereby this shall be used in order to gain trust. Nevertheless, the absence of trust will force the information seekers to judge the legitimacy of the source information (Tormala & Clarkson, 2008).

The trustworthiness also shall be gained through the writing style used in the blog. According to Ahn et al. (2012), one of the important criteria that might lead to the customer trustworthiness is the quality in writing. The blogger writing style may gain the readers trust and influence the customer purchase intention. In line with this statement, both the reputation and writing style which leads to trustworthiness shall lead to the behaviour intention or purchase intention.

2.2 *Blog involvement*

Several studies related to the blog have been done by many researchers (Bull et al., 2010; Du & Wagner, 2006; Duke, 2009; Wyld, 2008; Zhang & Han, 2007). According to Wu and Lee (2012) blogs are a new sharing method which allows the bloggers and readers to communicate and interact with one another based on their topic of interest. Hsu and Lin (2008) added that the blog-hosting provider should develop tools that require minimum effort to influence positive visitors believe. Kent (2008) argues the strength of the blog is included persuasion ability.

Bouhlel et al. (2010) looked at the impact of blog on consumers' behaviour. They noted that the perceived credibility has a positive effect towards the blog. They also noted that positive comment from the online shoppers past experience has increased the blog credibility and purchase desire. On top of that Chen (2008) agreed that a customer comment shall influence the decision making. Chang et al. (2011) state that the discussion, interaction and participation in the blog has a positive impact towards online purchase intention compared to the traditional websites. Meanwhile, the attitude towards the blog had a positive impact on the purchase intention which include the content display and information that came from the reliable resources (Bouhlel et al., 2010).

According to Wu and Lee (2012), there is a significant relationship between the interaction with the blog viewers and the bloggers towards the customer purchase intention. Communication through a suggestion from users shall help to encourage customer trust as well as willingness to purchase and there is a strong relationship between power and posting of similar opinion expressed on blogs and the behaviour (Tan et al., 2011).

2.3 *Blog attitudes*

Product blog and attitudes become a general subject in describing about the blog. Most of blog appeared with the very best attitudes with the same descriptive meaning of attracting the reader. Customer purchase behaviour has been influenced by two main components that have been known as the trust of power blog which includes quality of information and the blog reputation and size (Ahnet et al., 2012). People's viewpoints and external environment have a potential influence towards purchase behaviour where it becomes one of element in product attitudes. With regards to people, they do commit with different attitudes, character and opinion whereby all these major factors slightly influenced towards their intention. The wisdom power of words have been directly influenced the potential of a blog reader in purchase behaviour. It shows that marketing that used people as medium have a big impact and upon giving best results towards the purchase intention and behaviour. Thus, when it concerns to the blog attitudes, it always portrays an image about the characteristic of the blog and website itself. The blog layout, colour, or even the word used in the persistence of describing the main component of blog product has been clarified as an attitude which leads to the reader's attention. Bouhlelet et al. (2012) claimed that the quality of the product information provided by online intermediaries has a positive impact on product attitude.

It has been proven that when the consumer is highly involved in with blog usage they will more frequently gather relevant product knowledge through blog interaction (Wei, 2012). Behaviour toward product may stimulate the higher level of purchase intention. Describing the blog attitudes through the writing styles could catch the reader's intention. Although reader attention has highly influenced by the blog attributes such as attractive layout and design, advertising promoted and pictures show on the blog the exquisiteness of the writing about the food and product promoted should not be overlooked.

2.4 *Issue on blog review*

Information technology and foodservice might seem to be unrelated back then, but in today's world it has been almost like a backbone to one another as they support each other to succeed in the industry. Through information technology, the marketing strategy can be done widely without any barriers. Since we are in the 21st century era, where technology has revolved from wants to needs, and information is "right at your fingertips", all the information needed can be added within a minute either through websites, newspaper online, or blogs. Food review through blogs for examples has opened an opportunity for the capitalist out there to gain some profit through their posting. This on the other hand food blog directly helps the customer to make up their mind and maybe make a decision on which restaurant to choose. Therefore, the reliability of the information posted through this channel needs to be well sort out and digest by the reader in order to make sure they do not receive false fact. Bearing this point in mind, it raises critical questions to whether all the information shared through the blog can be trusted particularly relating to food. On the other hand, the consistency of blog involvement and product attitude need to be confirmed either it has direct impact towards customer purchase intention. The underlying reasons of the highlighting issue are still under investigation.

REFERENCES

Ahn S.H., Jang K.H., & Lee C.H. (2012). A study on the influence of the characteristics of power blogs and the user's trust on the purchasing intention. *2012 3rd International Conference on e-Education, e-Business, e-Management and e-Learning,* 27.

Ahuja, V. & Mercury, Y. (2010). Corporate blogs as e-crm tools: Building consumer engagement through content management. *Journal of Database Marketing and Customer Strategy Management*, 17(2), 91–105.

Ba, S. & Pavlou, P. (2002). Evidence of the effect of trust building technology in electronic markets: Price premiums and buyer behavior. *MIS Quarterly*, 26(3), 243.

Bouhlel O., Mzoughi N., Ghachem M.S., & Negra A. (2010). Online purchase intention: understanding the blogosphere effect. *International Journal of e-Business Management*, 4(2), 37–51.

Chang, L.Y., Lee Y.J., & Huang, C.L. (2011). The influence of e-word-of-mouth on the customer's purchase decision: A case of body care products.

Chen, Y.F. (2008). Herd behavior in purchasing books online, *Computers in Human Behavior*, 24(5), 1977–1992.

Cox, A.M. & Blake, M.K. (2011). Information and food blogging as serious leisure, *Aslib Proceeding*, 63(2/3), 204–220.

Demopoulos, T. (2007). *Blogging and podcasting*. Chicago: Kaplan Publishing.

Drodge, C., Stanko, M., & Pollitte, W. (2010). Lead users and early adopters on the web: The role of new technology product blogs, *Journal Product Innovation Management*, 27(1), 66–82.

Du, S.H., & Wagner, C. (2006). Weblog success: Exploring the role of technology. *International Journal of Human Computer Studies*, 64(9), 789–798.

Duke, S. (2009). Educating public relations students to enter the Blogosphere: Results of a Delphy study. *Journalism & Mass Communication Educator*, 63(4), 37.

Edelman, Intelliseek. (2005). Trust media: How real people are finally being heard (website).

Ghazisaeedi. M., Steyn. P.G., & Heerden. G.V. (2012). Trustworthiness of product review blogs: A source trustworthiness scale validation, *African Journal of Business Management*, 6(25), 7498–7508.

Hardin, R. (2002). Trust and trustworthiness. New York: Russell. Sage Foundation.

Herring. S.C., Scheidt, L.A., Wright, E. & Bonus, S. (2005). Weblogs as a Bridging Genre, *Information Technology & People*, 18(2), 142–171.

Hsu, C.L. & Lin, J.C.C. (2008). Acceptance of blog usage: The role of technology acceptance, social influence and knowledge sharing motivation. *Information & Management*, 45, 65–74.

Kent, M.L. (2008). Critical analysis of blogging in public relations, *Public Relation Review*, 34(1), 32–40.

Lee. C.Y., Jung. S.R., & Shin. G.H. (2001). A factor analysis for trustworthiness among web users. *The Korea Society of Management Information System Conference*, 663–673.

Ryu, I.C. (2003). Factors influencing the consumer trust and mediating roles of trust on purchasing intention in B2C electronic. *The Journal of MIS Research*, 13(4), 49–72.

Sekaran, U. & Bougie, R. (2010). *Research Methods of Business: A Skill Building Approach (5th Ed)*, John Wiley & Sons Ltd. 7.

Sneiderman, B. (2000). Designing trust in to online experiences. *Communications of the ACM*, 43(12), 57–59.

Steward, K.J. (2003). *Trust Transfer on the World Wide Web. Organization Science*, 14(1), 5–17.

Tan, L.K.W., Na, J.C., & Theng, Y.L. (2011). Influence detection between blog posts through blog features, content analysis, and community identity, *Online Information Review*, 35(3), 425–442.

Tormala, Z.L. & Clarkson, J.J. (2008). Source of trustworthiness and information processing in multiple message situations: A contextual analysis social cognition, 26(3), 357–367.

Tripp, C., Jensen T., & Carlson L. (1994). The effects of multiple product endorsement by celebrities on consumers' attitudes and intentions. *Journal of Consumer Respond*, 20(4), 535–547.

Wu, W.L. & Lee, Y.C. (2012). The effect of blog trustworthiness, product attitude, and blog involvement on purchase intention. *International Journal of Management & Information System*, 16(3), 265–276.

Wyld, D.C. (2008). Management 2.0: A primer on blogging for executives. *Management Research News*, 31(6), 448–483.

Yang, S.U. & Lim, J.S. (2009). The Effect of blog-mediated public relations (BMPR) on relation trust. *Journal of Public Relation Resolution*, 21(3), 341–359.

Zhang, Y. & Han, Y. (2007). Discussion on the influence of web blog on organization and individual. *Asian Social Science*, 3(2).

Hospitality and Tourism – Sumarjan et al. (eds)
 ISBN 978-1-138-00151-0

Customer satisfaction and repurchase intention towards Darul Rahmat chicken

R. Muhammad & S.A. Othman
Universiti Teknologi MARA, Malaysia

ABSTRACT: The fast food industry in Malaysia is facing increasing competitive challenges as with other industries around the world. There are over 2,500 fast food outlets in the country representing more than 240 products (Mohamed et al., 2012). Darul Rahmat Chicken is one of the local fast food that has been rebranding their name from the Radix Fried Chicken since May 2012. The main objective of the creation of the Darul Rahmat Chicken or D'RC to give consumers the opportunity to choose foods that are prepared to meet halal and organic methods ranging from chicken farming system that will be used in the D'RC. The purpose of this study: 1) to determine the attributes that influences the level of customer satisfaction towards D'RC, 2) to examine the level of customer's satisfaction towards D'RC, and 3) to identify whether the customer intends to repurchase D'RC products. This study will be conducted in D'RC Shah Alam and are aimed to identify the significant relationships between customer satisfaction and customers repurchase intention. Understanding the impacts from the previous study will lead to the empirical investigation in D'RC issues.

Keywords: customer satisfaction, Darul Rahmat Chicken, fast food, repurchase intention

1 INTRODUCTION

In Malaysia, fast food outlets such as McDonald's, KFC, and Pizza Hut are undergoing dramatic transformations and are experiencing heightened competition (Yap & Kew, 2007). A fast-food restaurant is characterized as quick services of a franchised restaurant chain, which supplied the food quickly after ordering and minimal service offering for dining and takeout facilities (Elliot & Reed, 1999). Nowadays, almost all the Fast Food Restaurants (FFRs) focus on several ways to increase their service quality in order to increase the level of satisfaction among their customers and thus increase their purchase intentions as well as loyalty (Qin & Prybutok, 2008; Gillbert, Veloutsou, Goode, & Moutinho, 2004; Kara, Kaynak, & Kucukemiroglu, 1995). When FFRs are able to achieve or exceed the expectations of customers, the customer will be satisfied with the service. The fast food industry is especially attractive for such research on customer satisfaction, because it is a worldwide phenomenon and growing rapidly (Ball, 1992).

There are many fast food restaurants in Malaysia. To name a few, McDonalds, Pizza Hut, KFC, Subway, A&W are some of the examples. Radix Fried Chicken is also considered a new fast food restaurant that was founded by Mr. Haji Ismail Bin Haji Ahmad, the Managing Director of the Herba Penawar Al-Wahida or HPA. Radix Fried Chicken began its operations on 1 June 2008. Now Radix Fried Chicken or RFC already has branches in other country such as Saudi Arabia. In addition, Radix Fried Chicken is also famous for its coffee known as Radix coffee. RFC is a local fast food as an alternative of this kind of restaurant that was introduced by overseas companies. The main objective of the creation of the RFC to give consumers the opportunity to choose foods that are prepared to meet halal and organic methods ranging from chicken farming system that will be used in RFC restaurant chain.

The beginning of an idea occurs after exposure in the Malaysian media on the level of doubt broiler production. On March 2008 Managing Director conventional, HPA met in the East Regional Telecom Training Center, Marang, Terengganu. Several experienced partners in poultry meat, including a vet (chicken farming) Dr. Saari b. Husin (DAHP, DVM—UPM) are present. Results taken from discussions with Dr. Saari to organize and develop large-scale farming system began formulating organic ration formulation (independent basic material of dubious food formulations in the context of Islamic law, particularly in dietary protein sources, free of chemicals such as antibiotics and "growth promotant").

On May 2013, Darul Rahmat Chicken was re-brand from the RFC due to the management issues. Darul Rahmat Chicken was managed by the Darul Rahmat Sdn. Bhd. and it has 4 franchises in Klang Valley area which situated at Shah Alam, Sungai Besi, Bandar Baru Bangi and Setiawangsa. Darul Rahmat Chicken or D'RC still maintained their fast food service with the organic chicken on the menu. Nevertheless, in a mature and slow-changing sector in Malaysia, customers are being served with ordinary food with most of the sellers are lacking in terms of creativity and new ideas.

This is the reason why the customers see there is a little impact in the innovation towards their decision to purchase fast food restaurants such as a DR'C restaurant to the customer. The innovation to the customers is at no difference between the competing fast food providers serving them in an almost similar kind of environment. For example, Subway Restaurant and O'Briens Irish Sandwich Bars for a product of sandwich while Mc Donald's and Burger King for a product of a kind of burger. On the other hand, Malaysian local restaurant, Radix Fried Chicken is offering almost similar business settings with Kentucky Fried Chicken selling a kind of similar product of fried chicken.

In a Shah Alam area itself, the competition is high between McDonald's, KFC, Subway and A&W therefore D'RC should maintain high service quality to compete with this establish brand. Yap and Kew (2011) noted that quality and customer satisfaction have long been recognized as playing a crucial role for success and survival in today's competitive market. In addition, consumers are seeking quality, value and desirable environment away from the daily life pressures (Soriano, 2002). With the growing affluent and educated society, consumers nowadays are demanding safe and better food quality to consume. Factors such as safety, freshness, storage, preparation, consumption, price, packaging, and place of purchase are the factors that may create the consumers' purchase intentions towards their food purchasing behaviour.

Quality and safety are two important elements in consumer food perceptions and decision making associated with food choice (Grunert, 2005, Röhr et al., 2005). Price is not the prime factor and those sellers who are able to serve good quality food will always win the customer. Nevertheless, consumers always are seeking a better value for their money (Klara, 2001). Food quality and fresh ingredients are the most important reasons why customers return to the restaurant (Brumback, 1998).

In contrast, failure to deliver quality food can only threaten marketers' long term profitable relationship with the customer. The future of the food product will be short-lived and the consumer acceptance will fall drastically. As the consumer perception is getting lower, it would seem reasonable to suppose that that the overall chances to induce the consumer intention to purchase the food product will definitely be an uphill task to the marketer. Significance to this, the researcher would like to find out to what extent D'RC can satisfy their customers and this issue is still questionable due the company is still new and did not well establish. Therefore this study will explore this area of interest.

2 LITERATURE REVIEW

2.1 *Customer satisfaction*

Satisfied customers are central to the long term success of the business, and their linkage between customer satisfaction and customer behaviour has been well documented in marketing literature (Gilbert et al., 2004). Customer satisfaction is an essential indicator of a company's past, current, and future performance and, therefore, has long been a critical focus among marketing practitioners and scholars (Oliver, 1999). Customer satisfaction is defined as a post-purchase evaluative judgment concerning a specific buying decision (Homburg & Giering, 2001).

In modeling satisfaction, two general conceptualizations of satisfaction exist in the literature: transaction-specific satisfaction and cumulative satisfaction (Anderson & Fornell, 1993; Boulding, Kalra, Stalin, & Zeithaml, 1993). In the transaction-specific viewpoint, satisfaction is transient: How happy is a customer with a product or service at a certain point in time (Cronin & Taylor, 1992). Alternatively, cumulative satisfaction is a customer's evaluation of the total consumption experience with a product or service to date, which directly affects post purchase phenomena such as attitude change, repeat purchase, and brand loyalty (Johnson & Fornell, 1991).

For decades, customer satisfaction has been one of the key global constructs predicting customers' behavioural intentions and actual behaviour (Oliver, 1999; Westbrook & Oliver, 1991; Yi, 1991). Customer satisfaction has a positive effect on attitude change in the post purchase period (Oliver, 1999; Westbrook & Oliver, 1991; Yi, 1991), and these positive attitudes do enhance repurchase intentions (Yi, 1991).

Furthermore, previous researchers suggest that it is important to include affective variables in the modelling satisfaction evaluation along with cognitive variables, especially when consumers are involved in a service setting (Pfaff, 1997). More recently, the focus of satisfaction research seems to have moved to emotional responses, instead of purely cognitive informing behavioural outcomes (Namkung & Jang, 2007). As an example, Oliver (1993) argued that consumer attributions about satisfaction with specific product characteristics lead to positive and negative responses, which influence overall satisfaction.

2.2 *Food quality*

Food quality has been generally accepted as a fundamental element of overall restaurant experience (Kivela et al., 1999, Namkung & Jang, 2007; Rajpoot, 2002; Sulek & Hensley, 2004). According to Peri (2006), food quality is a necessary condition to satisfy the needs and expectations of customers. Sleek and Hensley (2004) investigated the relative importance of food quality, physical settings, and service in a full-service restaurant and reported that food quality appeared to be the most significant predictor of customer satisfaction although the food quality explained only 17% of repeat-patronage intentions. Susskind and Chan (2000) also found the food quality was one significant determinant of customer assessments of restaurants in the Toronto, Canada, and restaurant market.

Despite the importance of food quality in the restaurant business, there is no consensus on the individual attributes that constitute food quality. Sulek and Hensley (2004) lumped all food attributes into only one variable, food quality, whereas Kivela et al. (1999), who designed a model of dining satisfaction and return patronage, saw that food quality had many attributes: presentation, tastiness, menu item variety, and temperature. Even so, little has been done with the crucial attributes of food quality in relation to customer satisfaction and repurchase intention.

2.3 *Service quality*

Ghobadian et al. (1994) posit that most of the service quality definitions fall within the "customer led" category. Juran (1999) elaborates the definition of customer led quality as "features of products which meet customers' needs and thereby provide customer satisfaction." As service quality relates to meeting customers' needs, we will be looking at "perceived service quality" in order to understand consumers (Arnauld et al., 2002). Grönroos (1984) and Parasuraman et al. (1985) looks at perceived quality of service as the difference between customers' expectation and their perceptions of the actual service received. Other researchers look at the perceived service quality as an attitude. Arnauld et al. (2002) defined perceived quality "whether in reference to a product or service" as "the consumers' evaluative judgment about an entity's overall excellence or superiority in providing desired benefits" (p. 327). Hoffman and Bateson (2001) define service quality as an attitude "formed by a long-term, overall evaluation of a performance". Attitude is defined as "a consumer's overall, enduring evaluation of a concept or object, such as a person, a brand, or a service" (Arnauld et al., 2002). Service quality as "an attitude" is consistent with the views of Parasuraman et al. (1988), Cronin and Taylor (1992) and Sureshchandar et al. (2002). The basis of the view is elaborated by the latter:

> *"As perceived service quality portrays a general, overall appraisal of service i.e. a global value judgment on the superiority of the overall service, it is viewed as similar to attitude." (p. 364) Khalidah Abu, Proceeding of IBBC 2004-635. Feinburg and de Ruyter (1995) pointed the importance of adapting the definition of service quality in different cultures. Ueltschy and Krampf (2001) contended that differences in culture affect measures of quality in a service sector. They encapsulated service quality measures as "culturally sensitive" and "may not perform properly or comparatively in a culturally diverse group domestically or abroad".*

Moreover, since perceived levels of service in restaurants are based on the relationship between customers and service providers (Nikolich & Sparks, 1995), customer's perceptions and evaluation of service quality may be highly dependent upon the service provider's performance during service delivery. Reliable, prompt, and assured service can be considered intangible cues that influence satisfaction and post-consumption behaviours (Brady & Robertson, 2001). Thus, in this study service quality refers to the level of service provided by restaurant employees.

2.4 *Atmosphere and novelty*

The atmosphere of a service setting has been deemed essential in determining customer satisfaction (Bitner, 1990; 1992). It has also been established that atmosphere quality can be influential in determining repeat patronage (Wakefied & Blodget, 1994). Novelty seeking behaviour relates to a person's desire to seek out novel experiences, or new sources of stimulation. It is conceptualized as a willingness to try new things (Hirschman, 1984). It is thought that consumers' novelty seeking drive can have implications in buyer choice and brand loyalty within the service industries and this phenomenon influences purchase behaviour. It has been found that novelty seeking in a motivator in desiring to travel (Lee & Crompton, 1992). In addition, it has been found that people who visit novel travel locations are less likely to return to the same location again (Bello & Etzel, 1985).

2.5 *Brand preference*

Customer satisfaction can influence attitudinal change (e.g. service and supplier preference) which in turn affects repurchase intention (Innis, 1991; Oliver, 1980; Oliver & Bearden, 1985; Phillip et al., 2002; Stauss & Neuhaus, 1997). A high level of satisfaction is likely to increase the probability that the brand in question will be retained in the customer's consideration set and will increase the customer's preference for the brand (Westbrook and Oliver, 1981). The effect of brand preference on willingness to buy has rarely been examined (Dodds et al., 1991). Encouraging approaches to be more precise specification of customer choice behavior are provided by developments in consideration set theory by Kardes et al. (1993), Roberts and Lattin (1991, 1997) and Shocker et al. (1991).

2.6 *Repurchase intention*

Satisfaction may be merely a judgment with cognitive and affective response to the target whereas the repurchase intention also involves a behavioural component (Kim & Ok, 2009; Mittal, Ross, & Baldasare, 1998). Thus, to best predict repurchase intentions, other variables than satisfaction should be considered such as brand preference and attitude. Improving current customers' repurchase intentions is economically more advantageous than constantly seeking new customers (Kim & Ok, 2009; Reichheld & Sassers, 1990; Sheth & Partvitiyar, 1995).

REFERENCES

Anderson, E.W., & Fornell, C. (1993). A customer satisfaction research prospectus. In R.L. Oliver & T.R. Rust (Eds.), *Service Quality: New directions in theory and practice* (pp. 239–266). Newbury Park, CA: Sage.

Arnauld, E.J., Price, L.L., & Zinkhan, G.M. (2002). *Consumers, New York, NY: Mc-Graw-Hill Higher Education.*

Ball, S. (Ed.) (1992). Fast food operation and their management, *Stanley Theories Ltd, London.*

Bello, D.C. & Etzel, M.J. (1985). The role of novelty in the pleasure travel experience. *Journal of Travel Research,* Summer, 20–26.

Bitner, M. (1990). Evaluating service encounters: The effects of physical surroundings and employee responses. *The Journal of Marketing*, *54*(2), 69–82.

Bitner, M.J. (1992), Servicescapes: The impact of physical surroundings on customers and employees. *Journal of Marketing*, *56*, 57–71.

Boulding, W., Kalra, A., Staelin, R., & Zeithaml, V.A. (1993). A dynamic process model of service quality: From expectations to behavioral intentions. *Journal of Marketing Research, 30*(1), 7–27.

Brady, M.K. & Robertson, C.J. (2001). Searching for concensus on the antecedent role of service quality and satisfaction: An exploratory cross-national study. *Journal of Marketing Research, 30*(1), 7–27.

Cronin, J. & Taylor, S. (1992). Measuring service quality: A reexamination and extension. *Journal of Marketing, 56,* (July), 55–68.

Dodds, W.B., Monroe, K.B., & Grewal, D. (1991). Effect of price, brand and store information on buyers' product evaluations. *Journal of Marketing Research, 28*(3), 307–319.

Elliot, P. & Reed, R. (1999). The valuation of fast-food outlets: Analysis, methodology, and reliability. *The Appraisal Journal, 67*, 359–369.

Gillbert, G.R., Veloutsou, C., Goode, M.M.H., & Moutinho, L. (2004). Measuring customer satisfaction in the fast food industry: A cross-national approach. *Journal of Services Marketing, 18*(5), 371–383.

Ghobadian, A., Speller, S., & Jones, M. (1994). Service quality: Concepts and models, *International Journal of Quality & Reliability Management*, *11*(9), 43–66.

Grönroos, C. (1984). A service quality model and its marketing implications. *European Journal of Marketing, 18*(4), 36–44.

Hirschman, A. (1984): A dissenter's confession: The strategy of economic development revisited, in G. Meier and D. Seers (eds.). *Pioneers in Development*, New York, World Bank.

Hoffman, K.D. & Bateson, J.E. (2001). Essentials of service marketing: concepts, strategies, and cases (p. 324). *(2nd Ed.). Australia: South Western Thompson Learning.*

Homburg, C. & Giering, A. (2001). Personal characteristics as moderators of the relationship between customer satisfaction and loyalty: An empirical analysis. *Psychology & Marketing*, *18*(1), 43–66.

Innis, H.A. (1991). *The bias of communication. Toronto: University of Toronto Press.*

Johnson, M.D. & Fornell, C. (1991). A framework for comparing customer satisfaction across individuals and product categories. *Journal of Economic Psychology, 12*(2), 267–286.

Kara, A., Kaynak, E., & Kucukemiroglu, O. (1995). Marketing strategies for fast-food restaurants: A customer view. *International Journal of Contemporary Hospitality Management*, *9*(3), 248–256.

Kardes, F.R., Kalyanaram, G., Chandrashekaran, M., & Dornoff, R.J. (1993). Brand retrieval, consideration set composition, consumer choice, and the pioneering advantage. *Journal of Consumer Research, 20*(1), 62–75.

Kim, W. & Ok, C. (2009). The effects of relational benefits on customer's perception of favorable inequity, affective commitment, and repurchase intention in full-service restaurants. *Journal of Hospitality and Tourism Research, 33*, 227.

Kivela, J., Inbakaran, R., & Reece, J. (1999). Consumer research in the restaurant environment, Part 1: A conceptual model of dining satisfaction and return patronage. *International Journal of Contemporary Hospitality Management*, *11*(5), 205–222.

Lee, T.H., & John L., Crompton. (1992), Measuring Novelty Seeking in Tourism. *Annals of Tourism Research, 19*, 732–751.

Mittal, V., Ross, W.T, Jr., & Baldsare, P.M. (1998). The asymmetric impact of negative and positive attribute-level performance on overall satisfaction and repurchase intentions. *Journal of Marketing*, *62*(1), 33–47.

Namkung, Y. & Jang, S.S. (2007). Does food quality really matter in restaurants? Its impact on customer satisfaction and behavioral intentions. *Journal of Hospitality and Tourism Research, 31*(3), 387–410.

Nikolich, M. & Sparks, B. (1995). The hospitality service encounter: the role of communication. *Hospitality Research Journal, 19*(2), 43–56.

Oliver, R.L. (1980). A cognitive model of the antecedents and consequences of satisfaction decisions. *Journal of Marketing Research*, 460–469.

Oliver, R.L. (1993). Cognitive, affective, and attribute bases of the satisfaction response. *Journal of Consumer Research, 20*, 418–430.

Oliver, R.L. (1999). The red queen rules. *Journal of Business Strategy, 20*(3), 8.

Oliver, R.L. & Bearden, W. (1985). Disconfirmation processes and consumer evaluations in product usage. *Journal of Business Research, 31*, 235–246 .

Parasuraman, A., Zeithaml, V.A., & Berry, L.L. (1988). SERVQUAL: A multiple-item scale for measuring consumer perceptions of service quality. *Journal of Retailing, 64*(1), Spring, 12–40.

Parasuraman, A., Zeithaml, V.A., & Berry, L.L. (1985). A conceptual model of service quality and its implications for future research. *Journal of Marketing, 49*(Fall), 41–50.

Pfaff, M. (1997). The index of consumer satisfaction: Measurement problems and opportunities. In H.K Hunt (Ed.), *Conceptualization and measurement of consumer satisfaction and dissatisfaction* (36–71). Cambridge, MA: Marketing Science Institute.

Phillip K. Hellier, Gus M. Geursen, Rodney A. Carr., & John A. Rickard. (2002). Customer repurchase intention: A general structural equation model. *European Journal of Marketing, 37*(11/12), 1762–1800.

Qin, H. & Prybutok, V.R. (2008). Determinants of customer-perceived service quality in fast food restaurants and their relationship to customer satisfaction and behavioral intentions. *Quality Management Journal, 15*(2), 35–50.

Rajpoot, N.A. (2002). TANGSERV: A multiple item scale for measuring tangible quality in foodservice industry. *Journal of Foodservice Business Research*, *5*(2), 109–127.

Reichheld, F.F. & Sasser, W.E., Jr. (1990). Zero defections: Quality comes to service. *Harvard Business Review*, *68*(5), 105–111.

Roberts, J.H. & Lattin, J.M. (1991). Development and testing of a model of consideration set composition. *Journal of Marketing Research, 28*(4), 429–440.

Roberts, J.H. & Lattin, J.M. (1997). Consideration: Review of research and prospects for future insights, *Journal of Marketing Research, 34*(3), 406–410.

Shocker, A.D., Ben-Akiva, M., Boccara, B., & Nedungadi, P. (1991). Consideration set influences on consumer decision-making and choice: Issues, models, and suggestions. *Merketing Letters*, *2*(3), 181–197.

Stauss, B. & Neuhaus, P. (1997). The qualitative satisfaction model. *International Journal of Service Industry Management*, *8*(3), 236–249.

Sulek, J.M. & Hensley, R.L. (2004). The relative importance of food, atmosphere, and fairnness of wait. *Cornell Hotel and Restaurant Administration Quarterly*, *45*(3), 235–247.

Sureshchandar, G.S., Rajendran, C., & Anantharaman, R.N. (2002). The relationship between service quality and customer satisfaction: A factor specific approach. *Journal of Services Marketing, 16*(1), 9–34.

Susskind, A.M. & Chan, E.K. (2000). How restaurant features affect check averages. *Cornell Hotel and Restaurant Administration Quarterly*, *41*(6), 56–63.

Wakefield, K.L. & Blodgett, J.G. (1996). The effect of the servicescape on customers' behavioral intentions in leisure service settings. *Journal of Services Marketing*, *10*(6), 45–61.

Westbrook, R.A. & Oliver, R.L., (1981), Developing better measures of consumer satisfaction: Some preliminary results. *Advances in Consumer Research, 8*(1), 94–99.

Yap, S.F. & Kew, M.L. (2007). Service quality and customer satisfaction: Antecedents of customer repatronage intentions. *Sunway Academic Journal, 4*, 60–72.

Yi, Y. (1991). A critical review of consumer satisfaction. In V.A Zeithaml (Ed.). Review of marketing (pp. 68–123). *Chicago: American Marketing Association.*

Green hospitality and tourism

Hospitality and Tourism – Sumarjan et al. (eds)
© 2014 Taylor & Francis Group, London, ISBN 978-1-138-00151-0

Consumers' awareness towards restaurant green practices and its influence on behavioural intentions

A.L. Mohd Rasdi, F.S.C. Abdullah, A.A. Sukur & M.M. Razak
Universiti Teknologi MARA, Malaysia

ABSTRACT: Consumers are becoming more conscious in purchasing products and services which are environmentally friendly and this leads to the increasing demand for 'green' products and services. Companies across all sectors, hospitality sector included, try to make great strides toward instituting green practices and earning green certifications. Restaurants in Malaysia are following this trend and trying to implement green practices. This study investigated consumers' awareness towards restaurant green practices and its influence on behaviour intentions. Questionnaires were distributed to a convenience sample of 400 shoppers at some well-known hypermarkets and supermarkets in Shah Alam, such as Tesco, Giant, Mydin and SACC Mall. The results show that, consumers almost agreed about being aware of restaurant green practices. The findings of this study can help restaurant operators better understand the consumers' demands and possible behaviour towards green practices.

Keywords: behavioural intention, consumers' awareness, restaurant green practices

1 INTRODUCTION

Wang (2012) mentioned that, consumers are becoming more conscious in purchasing products and services that are environmentally friendly and this indirectly leads to the increasing demand for 'green products and services'. Consumers are also getting behind the idea of being green and are more concerned about climate change. The industries had invested enormous efforts to produce and encourage eco-friendly goods. Thus, many companies started developing environmentally friendly packaging and putting in numerous efforts to keep-in step with the environmental movement in order to be more socially responsive in addressing pollution and waste disposal (Chen & Chai, 2010).

The hospitality sector is also making great strides toward instituting green practices and earning green certifications. Tourism industry which includes hospitality, transportation, and food and beverage sectors are the biggest industries and often rely on the integrity of the environment. Chou, Chen and Wang (2012) stated that, green practices in the tourism and hospitality industry can be mainly traced back to the idea of sustainable tourism and have been broadly addressed.

According to Jeong and Jang (2010), the focus of a friendly environment has also encouraged the restaurant sector to adjust their services as well as to meet the changing expectations and demand from the customers. Thus, restaurant owners have to implement green practices which include energy efficiency, water conservation, pollution prevention, environmental health, recycle and reuse programs, green purchasing, green materials, sustainable foods and green designs of building and space (Chou et al., 2012).

However, little information exists about the need for green practices in the restaurant sector since many researchers had more interest and focus on studying green practices in the lodging sectors (Schubert, 2008). Besides, there is very limited information available on how green practices are perceived by restaurant customers especially in Malaysia. Schubert (2010) explained that the low awareness of effective procedures for food service practices causes restaurants to face difficulties when implementing green practices.

Restaurant operators also hesitate to implement 'green practices' due to the lack of knowledge available on consumers' behavioural intentions regarding 'green' restaurant, including a plan to patronize the restaurant and willingness to pay more. Hu et al. (2010) suggested that researchers should investigate consumers' patronage intentions towards green restaurants from other countries which give differences in values and cultures.

In Malaysia, the green trend is still new-fangled. Chen and Chai (2010) recommended that further study should include a cross section of the Malaysian population which focuses on knowledge of consumer on the specific type of green products in the market. This study investigated consumer awareness levels of restaurant green practices. The researcher also explored on consumers' intention to pay premium prices for these green products or services in Malaysia context.

2 LITERATURE REVIEW

2.1 *Restaurant green practices*

Jang et al. (2011) defined green restaurant as one that offers a variety of green food menu items that use locally grown or organic certified food as well as one that implements green practices such as recycling program, efficient use of energy and water, and reduction of solid waste. Based on Tan and Yeap (2012), when restaurants engage in green practices, such practices must be aimed to effectively reduce the environmental and social problems that arise directly or indirectly from their operations. Namkung and Jang (2012) further explained that, in restaurants, various green practices can be applied; this includes energy efficiency, water efficiency, recycling, and sustainable food. For example, upscale casual restaurants serving organic food and drinks, while in fast food restaurants, recycled products are used instead of Styrofoam to-go containers.

2.2 *Consumers' awareness*

Yaacob and Zakaria (2011) argued that generally, awareness comprises a human's perception and cognitive action to a condition or event. Awareness is an ability to be conscious of the feel or perceive and does not necessarily imply understanding. Companies try to expand consumers' knowledge of the product and its environmental attributes by raising awareness about green products in order to encourage their purchasing behaviour. From their study regarding customers' awareness, perception and future prospects of green products in Pahang, Malaysia, the respondents were found to usually get the information regarding green products from newspapers and television. Rashid (2009) investigated awareness of the Eco-label in Malaysia's green marketing initiative and found that, within the Malaysia context, awareness of green products is still low. However, this study did not investigate Malaysian consumers' reaction to the use of eco-label in their purchase intention for products with environmentally friendly features.

2.3 *Customers' behavioural intention towards restaurant green practices*

Rezai et al. (2012) investigated on the consumers' awareness and consumption intention towards green food consumption in Malaysia. The result showed that consumers are more concerned about their health, food safety, protection of the environment and animal welfare. Most of the consumers are aware of the fact that green foods are healthier, safe to consume and help to protect the environment under the principle of sustainable agricultural development. The findings indicate that the majority of the respondents had aware and had a positive perception towards green foods in Malaysia due to the rapid change in food consumption patterns and consumers' lifestyles. The majority of the respondents will have intention to purchase green foods in the near future.

From the restaurant industry context, Hu et al. (2010) explored the dynamic of green restaurant patronage. It was found that consumer knowledge of sustainable restaurant practices and environmental concerns were significant determinants of the intention to patronize green restaurants. The results from this study also suggested there is a significant relationship between environmental concerns and consumers' intentions to patronize a green restaurant. Most respondents of this study indicated their willingness to pay more for a green restaurant. Lien et al. (2012) argued that green consumption attitudes, green subjective norms, and green perceived behavioural control of consumers had a significant and positive influence on their green consumption behavioural intention with regard to patronizing of green restaurant.

3 METHODOLOGY

A descriptive research design with a quantitative approach was used for data collection and self-completion survey was developed in order to obtain required information from the consumers. The data collection was conducted at well-known hypermarkets and supermarkets in Shah Alam which are Tesco, Giant, Mydin and also Shah Alam City Center Mall. The respondents were conveniently selected at these hypermarkets and supermarkets. Tough only 384 samples were needed, 500 questionnaires were distributed in case of incomplete questionnaire.

The respondents were approached at the entry and exit doors as well as the parking areas of the hypermarkets and supermarkets to answer the questionnaires. The researcher also approached consumers who were seated on benches, in hypermarkets and supermarkets, waiting for their family members to finish shopping. The questionnaires prepared in two languages, English and Malay language, were distributed to the respondents and collected immediately upon completion.

4 ANALYSIS AND FINDINGS

Consumers were slightly aware on the restaurant green practices. Table 1 listed overall mean score of consumers' awareness level of restaurant green practices. From the results, majority of the consumers were aware on the restaurant green practices with social focused, followed by environmental focused and lastly food focused. It can be assumed that even though they were familiar with green practices, but they were still lacking on the knowledge towards restaurant green practices.

Consumers do have intention to patronize and willingness to pay more to restaurant which practices green. However, the different value of mean scores means that their intention to patronize were higher than their willingness to pay more. Besides that, from this result, it shows that, small numbers of respondents were willing to pay more for restaurant green practices.

The results show that the there was a moderate correlation between the consumer awareness level of restaurant green practices and intention to patronize. However, there was a fragile relationship between consumers' awareness level and willingness to pay more for restaurant with green practices.

Table 1. Overall mean score of consumers' awareness level of restaurant green practices (N = 384).

Variables	Mean (M)	Std. deviation
Food focused	3.85	0.6263
Environmental focused	3.88	0.6097
Social focused	3.94	0.6841

(Descriptive statistics).

Table 2. Overall mean scores of consumers' behavioural intention (N = 384).

Variables	Mean (M)	Std. deviation
Intention to patronize	3.925	0.75056
Willing to pay more	3.462	0.93496

(Descriptive statistics).

Table 3. Summarized of correlations between all variables.

	Intention to patronize at restaurant green practices	Willingness to pay more for restaurant which practices green
Consumers' awareness of restaurant green practices	Moderate relationship (r = 0.601)	Weak relationship (r = 0.388)

5 LIMITATION AND RECOMMENDATION

As this study was only focused on the consumers in Shah Alam, Selangor, which is considered an urban area. Therefore, in future research, it is recommended that the researcher focus on consumers from other states in Malaysia. The study should be done in various cities with different geographic characteristics. In fact, various areas of the country might have varied in consumers' awareness and behavioural intention towards restaurant green practices.

As for theoretical implication, this study provides an in-depth knowledge and a better understanding toward Shah Alam's consumers' awareness of restaurant green practices and its influence on behavioural intention. Since there is a lack of study that has been done in relation to restaurant green practices, therefore, from this result, one can achieve more collections of references for future study. From the results which show that consumers are slightly agreed have intention to patronize restaurant green practices, therefore, restaurant operators have opportunities to create a new market niche which can enhance their business image and generate competitive advantage.

As the results show that the majority of the respondents slightly agree that they were aware about restaurant green practices and almost agree had the intention to patronize at restaurant green practices, it is recommended for restaurants that have implemented one or more green practices should proudly advertise the kind of practices they have adopted.

6 CONCLUSION

Restaurant operators should advertise for their involvement in green practices in order to increase levels of consumers' awareness. All those attributes impacted the consumers' behaviour such as intention to patronize and willingness to pay more for restaurant green practice. Specifically, public advertising and brochures may increase consumers' awareness and consequently affect their behavioural intention.

REFERENCES

Chen, T.B. & Chai, T.L. (2010). Attitude towards the environment and green products. *Management Science and Engineering*, *4*(2), 27–39.

Chou, C.J., Chen, K.S., & Wang, Y.Y. (2012). Green practices in the restaurant industry from an innovation adoption perspective: evidence from Taiwan. *International Journal of Hospitality Management, 31*, 703–711.

Hu, H.H., Parsa, H.G., & Self, J. (2010). The Dynamics of Green Restaurant Patronage. *Cornell Hospitality Quarterly,* 51(3), 344–362.

Jang, Y.J., Kim, W.G., & Bonn, M.A. (2011). Generation Y consumers' selection attributes and behavioral intentions concerning green restaurants. *International Journal of Hospitality Management, 30*, 803–811.

Jeong, E.H. & Jang, S.C. (2010). Effects of restaurant green practices: which practices are important and effective? *Caesar Hospitality Research Summit.*

Lien, C.Y., Huang, C.W., & Chang, H.J. (2012). The influence of green consumption cognition of consumers on behavioural intention: A case study of the restaurant service industry. *African Journal of Business Management, 6*(26), 7888–7895.

Namkung, Y. & Jang, S.C. (2012). Effects of restaurant green practices on brand equity formation: Do green practices really matter. *International Journal of Hospitality Management.*

Rashid, N. (2009). Awareness of Eco-label in Malaysia's Green Marketing Initiative. *International Journal of Business and Management, 4*(8), 132–141.

Rezai, G., Teng, P.K., Mohamed, Z., & Shamsudin, M.N. (2012). Consumers' awareness and consumption intention towards green foods. *African Journal of Business Management, 6*(12), 4496–4503.

Schubert, F. (2008). *Exploring and predicting consumers' attitudes and behaviors towards green restaurants.* Unpublished master's thesis, The Ohio State University.

Schubert, F., Kandampully, J., Solnet, D., & Kralj, A. (2010). Exploring consumer perceptions of green restaurants in the US. *Tourism and Hospitality Research, 10*(4), 286–300.

Tan, B.C. & Yeap, P.F. (2012). What drives green restaurant patronage intention? *International Journal of Business and Management, 7*(2).

Wang, R. (2012). Investigations of important and effective effects of green practices in restaurants. *Social and Behavioral Sciences, 40*, 94–98.

Yaacob, M.R. & Zakaria, A. (2011). Customers' awareness, perception and future prospects of green products in Pahang, Malaysia. *The Journal of Commerce, 3*(2).

Hospitality and Tourism – Sumarjan et al. (eds)
 ISBN 978-1-138-00151-0

Scuba divers' underwater responsible behavior: A comparison of divers' specialization

S.Z.Z. Abidin, N. Salim & M. Badaruddin
Universiti Sains, Malaysia

ABSTRACT: The uniqueness and breathtaking scenery of coral reefs are attracting an increase of divers worldwide, contributing to growing concerns about coral damages. Sustainability of recreational diving is highly dependent upon the health of reef ecosystems. The diver can reduce their impacts towards coral reefs through responsible diving behaviours. The purpose of this study is to examine SCUBA divers' specialization in relationship with underwater responsible behaviour in Perhentian Island. Data were collected using convenience sampling among three specialization groups of SCUBA divers. The results reveal that contact between divers and coral reefs mostly happened among beginners and during underwater photography. This survey confirmed that divers' norms did not differ by the specialization level except for losing buoyancy control and touching sea turtle. This study not only essential to promote responsible diving behaviour while engaging in marine parks, but also to preserve the longevity of coral reef health. Although further research is necessary to validate these findings, this study serves as a preliminary study for future research on social aspects of dive site planning and management.

Keywords: coral reefs, divers' specialization, Scuba diving, underwater responsible behaviour

1 INTRODUCTION

The uniqueness and breathtaking scenery of coral reefs is attracting an increase of divers worldwide. Coral reefs are one of the most diverse, productive and ecologically valuable environments which provide a number of notable benefits to humankind. Recent regional surveys have confirmed the extent of reef degradation. The causes of this destruction include increased ocean temperatures, overfishing, the use of destructive fishing methods, direct channelled sea pollution and uncontrolled tourism activities (diving and snorkelling) (Gardner et al., 2003).

Dive tourism has marked up as a vast growing recreational activity worldwide including Malaysia. Malaysian government identifies dive tourism as one of the high yield components of marine tourism and is aggressively promoting tourism icons such as Sipadan Island and Tunku Abdul Rahman Marine Park as a world-class scuba diving destination. According to Professional Association of Diving Instructors (PADI) Diver Statistics (2013), there are approximately 135,710 certified divers registered with PADI. Popularity of recreational diving activity among tourists is due to the development of safe and affordable diving equipment, along with technological advances that enable access to remote SCUBA diving sites and the growing interest in learning about and experiencing natural environments (Dimmock 2009).

Due to its popularity, diving activity has become uncontrollable in some parts of the countries. Corals in Great Barrier Reef (GBR) encountered tremendous damages caused by excessive number of divers. Red Sea experienced an increase of dive tourism produced major impacts from reef infilling for resort constructions and direct contacts by divers on the reefs. In Malaysia, the scenario of the coral damage was quite similar to other foreign countries as islands like Redang, Pulau Payar as well as Perhentian experienced coral damage due to resort development and uncontrolled diving behaviour.

Divers play a crucial role in marine environmental conservation as their presence in these vulnerable ecosystems may harm coral communities. Tratalos and Austin (2001), through their research suggested that diving is having a significant impact in areas subject to high levels of usage. There was also more dead corals and coral rubble at the site experiencing greater increase in diving. Therefore, it is important to understand the nature and causes of impacts if management interventions are to be successful in their mitigation.

2 LITERATURE REVIEW

2.1 *Recreation specialization*

Recreational specialization as suggested by Bryan (1977) is defined as 'a continuum of behaviour from the general to the particular, reflected by equipment and skills used in the sport, and activity setting preferences'. Recreation specialization can be ordered based on a few characteristics: level of experience, individual's skill, relationships, equipment use, orientation and commitment towards the given activity (Thapa et al., 2006). Within the range, individuals are categorized for beginner to expert.

Previous research on level of specialization suggests that an increase in activity involvement leads to more positive attitudes toward environmental concerns. Level of experience and skills is related to environmental awareness and more prominence among divers on the natural preservation (Musa et al., 2011). Expert divers are more aware of specific environmental issues rather than general ones. They are conscious of negative impacts of their activities and favour low-impact practices. Therefore, they tend to show interest to be involved in reef monitoring projects during their dive trips (Thapa et al., 2005). However, a collaborative study by Centre for Island and Coastal Tourism (CENTICA), University of Kent and Tourism Planning Research Group (TPRG), Universiti Teknologi Malaysia, on the impacts of dive tourism to local community revealed that skilled divers bring more coral damage while using underwater camera due to their determination to get the best underwater shots (Kaur 2010).

Dearden (2006) identified that more specialized divers will want to see their activity continue in a sustainable way in the future. Therefore, most of the expert divers are more likely to support the rules, norms and procedures associated with the given activity to ensure future participation. Anderson & Loomis (2011), reporting the result from a qualitative analysis, showed that the number of physical contacts with corals declined with the increasing diver experience. Likewise, advanced diving training may contribute to fewer contacts with coral reefs.

Several on the impacts of diving activities have been conducted (Dearden et al., 2006; Gardner 2003; Harriott et al., 1997; Hawkins et al., 1999; Rouphael & Inglis 2001; Tratalos & Austin 2001; Thapa et al., 2006). Among those who studied divers' underwater behaviour are Barker and Roberts (2004), Anderson and Loomis (2011), Musa et al. (2011), and Ong and Musa (2012) but there has been little research on diver specialization. Specialization can provide important insights on a relationship between an activity and the environment it takes place in and, ultimately, the sustainability of the activity.

2.2 *Underwater behavior*

Normative research provides one way of understanding the behaviour of recreation settings. Anderson and Loomis (2011) stressed that social norm may influence human behaviour through the cognitive and emotional components of the obligations and sanction. As recreation activity takes place in a sensitive environment such as reef ecosystem, understanding the divers' shared norms is crucial since the collective behaviours have potentials to affect reef health and longevity of diving activity.

According to the standard requirements for recreational SCUBA diving proposed in ICRAN MAR 1 (2006), divers shall accordance with all applicable local and national laws and regulations on diving activity. The guidelines include avoiding interactions with marine life, maintaining an awareness of fins, adjusting buoyancy and securing equipment and cameras from contacting corals or stirring up sediments while diving. Divers are not allowed to touch the corals or other reef-dwelling organisms. Harassing or swim close to any marine life such as sea turtles is forbidden. Feeding fish and collecting shells and corals are strictly prohibited.

Divers could cause harm to corals via direct physical contact by touching or walking on them, some of the divers may even accidentally kick corals with their fin when they lose their buoyancy, or by stirring up benthic sediment, thereby subjecting coral polyps to increased sedimentation loads. Studies agreed that fins cause most damage to the corals followed by hands, knees and equipment (Rouphael & Inglis 2001; Barker & Robert 2004). Expert divers can mitigate impacts to coral reefs by moving through the water with proper buoyancy and fin techniques. Likewise, all divers can protect corals by avoiding physical contact with them (Thapa et al., 2005).

Rouphael and Inglis (2001) observed that around 70% of divers had one or more physical interactions with reef benthos, which is most contact, were by fin kick (58%) or deliberately holding onto corals (32%). Although the damage done by individuals is relatively small, there is some evidence can cause significant localized destruction of sensitive marine organisms (Harriott et al., 1997; Hawkins et al., 1999).

Several researchers have stressed that the damage on coral reefs can be minimize by changing the divers' behaviour (Hawkins et al., 1999). In an effort to alleviate negative impacts of divers on the reefs, Rouphael and Inglis (2002) suggested that it would be more realistic and efficient to identify and mitigate the causes of damaging behaviour of divers rather than using other methods, such as limiting the numbers of visitors to dive sites, promoting site substitution policies, practical use of artificial reef, educating low-impact recreational practices and rules and regulations enforcement. In response to this, the present study examines the diver specialization–behaviour relationship among divers for further understanding of the antecedents of underwater responsible behaviour.

3 METHODOLOGY

3.1 *Description of the study site*

Perhentian Islands consist of two main islands namely Perhentian Besar and Perhentian Kecil. These islands are belonged to Redang Island Marine Park. Most of the accommodations and dive shops can be found in Perhentian Besar while Perhentian Kecil is a fishing village inhabited by a small local population. Perhentian Island was chosen as the study site because of the high quality and range of dive opportunities in the area.

3.2 *Research instrument*

The survey questionnaire for this study consisted of four sections which measured variable of diving attitude, underwater behaviour, recreation specialization and demographic profile. The questionnaires were developed based on the previous literature and expert opinions. They were modified according to the study site and the target of population.

Behavioural norm questions were designed to measure the sanction components of norms and are based on previous approaches to measuring these concepts (Heywood 2002). Respondents were asked to rate how embarrassed they would feel if they engaged in each behaviour at a coral reef and others saw them. Applicable diving behaviours were selected based on work conducted by other researchers on diving (Anderson & Loomis, 2011; Thapa et al., 2005). Sanction responses were measured using a five-point Likert scale ranging from 1 (not at all embarrassed) to 5 (extremely embarrassed). Factors that lead to contact between divers and corals also measured using a five-point Likert scale ranging from 1 (not priority) to 5 (essential).

The diver specialization level was measured using a diver specialization index which designed based on the responses to 12 questions on items that have been linked to the degree of specialization in other activities (see Dearden et al., 2006), as well as personal knowledge of diving. The questions fell into three categories: (1) Past diving experience: the amount, variety and type of diving experience. (2) Investment in diving: both financially and time consumed. (3) Centrality of diving to life.

Table 1. Distribution of diver specialization group.

Diver specialization rating	Percentage	Divers
Beginner (0–2)	33.3	15
Intermediate (3–5)	31.1	14
Expert (6–8)	35.6	16

3.3 *The sampling method*

Questionnaires were randomly distributed to the divers in the dive shops and who just got back from diving. Divers were approached during in-person intercepts using the convenience sampling method. The respondents were first asked whether they volunteer to participate in the survey. Respondents took about 30 minutes to answer all the questions.

3.4 *Data analysis*

In this study, descriptive statistics including mean ratings were computed on the respondents' demographics, diver behaviour sanctions and factor that leading to contact between divers and corals.

Divers were grouped into subgroups based on their responses to the indicator questions using a binary scale. If they met the requirement for a particular indicator feature, they were given a score of one. If they did not, they were assigned a zero for that feature. The individual scores were tallied, and divers assigned to either the low, medium or highly specialized group (see Table 1).

Analysis of Variance (ANOVA) tests was used to assess whether significant differences occurred among the diver specialization groups in terms of their responsible underwater behaviour and factor causes leading to contact between divers and corals.

4 RESULT AND DISCUSSION

4.1 *Profile of respondents*

A total of 45 out of 50 of respondents completed the questionnaire in the pilot survey, representing a response rate of 90%. There were more male respondents (71.1%) than female respondents (28.9%). The majority of divers (57.8%) were young aged between 20 and 27 years. With regard to nationality, there was a higher percentage (66.7%) of foreign divers than domestic divers (33.4%).

A breakdown of non-Malaysian showed that they mainly originated from Europe, Oceania, North America and Middle East with 53.3%, 6.7% and 4.4% respectively. Educational background was not included, since all respondents (100%) achieved academic qualifications of a diploma or higher.

4.2 *Diver specialization*

Divers in the sample clustered towards the higher end of the specialization continuum, with 33.3% of divers falling into the 'high', 31.1% of divers falling into the 'medium' specialization levels and 35.6% of divers falling into the 'low' specialization levels (see Table 1).

4.3 *Diver underwater behaviour*

The mean score for all behaviour sanctions was relatively high. This clearly indicates that divers would feel moderately or very embarrassed if seen engaging in any of the behavioural sanctions as in Table 2. Divers reported that they would feel very to extremely embarrassed if seen engaging in two of nine behaviours, while other behaviours would lead to moderately high embarrassment levels.

There was a statistically significant different between specialization and underwater behaviour as determined by one-way ANOVA ($F = 5.745$, $p = 0.006$). A turkey post hoc revealed that lose buoyancy control was statistically significantly lower when compare between expert divers to the intermediate ($p = 0.039$) and beginners ($p = 0.013$). There were no statistically significant differences between beginner and intermediate ($p = 0.889$).

The results shows that there are significant differences between specialization and underwater behaviour in terms of touching the sea turtle ($F = 5.096$, $p = 0.011$). A turkey-post has revealed that touching a sea turtle was statistically significantly lower when compare between expert divers to the beginner ($p = 0.006$) but there were no statistically significant differences between expert and intermediate ($p = 0.380$), and between beginner and intermediate ($p = 0.231$).

Table 2. Diver behaviour sanctions in order of strength.

Items	Mean	SD	*n*
Break off pieces of live corals	4.62	0.75	45
Leave garbage on the sea floor	4.56	0.92	45
Remove shell from a reef	4.07	1.16	45
Take pieces of dead corals	4.04	1.15	45
Knowingly anchor a boat	3.93	1.28	43
Touch sea turtle	3.89	1.13	44
Touch corals with hands	3.76	1.19	45
Feed marine life	3.53	1.42	45
Lose buoyancy control	3.20	1.31	45

Table 3. The causes leading to contact between divers and corals (n = 44).

Variables	Mean	SD
Poor buoyancy control & fin technique	4.41	0.66
Poor swimming techniques	4.07	0.82
Ignorance about coral conservation	4.00	1.22
Walking on the corals	3.82	1.37
Incorrect weighting	3.68	0.93
When divers panic	3.48	1.11
Underwater photography	2.93	1.15

Table 3 shows diver perceptions towards contact between divers and corals based on priority. Divers reported that contact between divers and corals usually occur when an individual has poor buoyancy control and fin techniques. This situation usually occurs among the first-timers or beginners since they have less self-confidence and can easily panic when they do their first dive. Likewise, poor swimming techniques can also lead to contact between divers and coral reefs.

There was a statistically significant difference between specialization and factor that causes leading to contact between divers and corals, whereby when divers panic ($F = 7.895$, $p = 0.001$) and during underwater photography ($F = 0.562$, $p = 0.041$).

Divers indicated that they would feel a high level of embarrassment for breaking off pieces of live coral, leaving garbage on the sea floor, remove shell from a reef and taking pieces of dead corals. Behavioural sanctions were moderate for anchoring a boat on coral, swimming close to and touching the sea turtle and corals with hands, slight for feeding marine life and lose buoyancy control. Anchoring cause considerable and long lasting damage to coral communities. Likewise, between the two sanctions that differed by the specialization level, one directly relates to reef health, which is losing buoyancy control. However, the other sanctions, touching sea turtle, may have a less direct connection to reef health, but this action would scare this endangered species. Experience may also partly explain the greater embarrassment ratings for losing buoyancy control.

Findings from this study demonstrate that divers in the Perhentian Island represent a diversity of recreation specialization subgroups. While other reef ecosystems have seen the displacement of high expect divers by beginner (Dearden et al., 2006), it appears Perhentian Island coral reefs are visited by divers of many specialization levels, including expert specialization divers.

5 CONCLUSION AND RECOMMENDATIONS

Responsible diving behaviours are important to protect coral longevity. As this study used a pilot survey method, for the next time around, data collection should be more precise and cover other behaviour obligations that measure diver underwater behaviour. Future studies may want to explore irresponsible underwater behaviours among divers. In general, divers on Perhentian Island felt obligated to engage in behaviours that would protect coral reef resources.

The research reported here suggests that monitoring the specialization levels amongst divers can serve as an early warning system for management interventions and help address the lack of equipment and detailed knowledge required for biological monitoring. Understanding the social science dimensions of the dive experience will ultimately contribute to a more successful, sustainable dive industry, and a more effective exponent of incentive driven conservation at work.

Other approaches to minimize diving impact is to include provision of safe entry and exit points that allow divers to correct their buoyancy without causing damage before they proceed to more sensitive areas of the site, and by visiting particularly sensitive dive sites later in the trip, once divers have become familiar with their equipment and surroundings.

Lastly, although this study focused on divers in the Perhentian Island, the findings are likely applicable to other areas where recreation occurs in sensitive ecological environments. Responsible diving behaviours are an important component of coral reef protection throughout the world. The similarity in diver norms found across specialization levels in this study may relate to the training and certification requirements of Perhentian Island divers. Managers in other regions may benefit from considering how environmental protection messages can be incorporated into these programs.

ACKNOWLEDGEMENT

The authors would like to extend their appreciation to the following institutions that made this study and paper possible:

i. *Ministry of Higher Education, Malaysia for granting the Long Term Research Grant Scheme 2011 [LRGS Grant No. JPT.S (BPKI) 2000/09/01/015Jld.4 (67)]*

ii. *Universiti Sains Malaysia for granting the Research University Grant called Tourism Capacity and Impact Studies [Grant No. 1001/PTS/8660011].*

REFERENCES

Anderson, L.E. & Loomis, D.K. (2011). SCUBA diver specialization and behavior norms at coral reefs. *Coastal Management, 39*, 478–491.

Barker, N.H.L. & Roberts, C.M. (2004). Scuba diver behaviour and the management of diving impacts on coral reefs. *Biological Conservation, 120*, 481–489.

Bryan, H. (1977). Leisure value systems and recreational specialization: The case of trout fishermen. *Journal of Leisure Research, 22*, 233–257.

Dearden, P., M. Bennett, & Rollins R. (2006). Implications for coral reef conservation of diver specialization. *Environmental Conservation, 33*, 353–363.

Dimmock, K., & Wilson, E. (2009). Risking comfort? The impact of in-water constraints on recreational SCUBA diving. *Annals of Leisure Research,* 12(2), 173–194.

Gardner, T.I., Cote, J., Gill, A., Grant, & Watkinson A. (2003). Long-term region-wide declines in Caribbean corals. *Science, 301*, 958–960.

Harriott, V.J., Davis, D., & Banks, S.A. (1997). Recreational diving and its impact in marine protected areas in eastern Australia. *Journal of the Human Environment, 26*(3), 173–179.

Hawkins, J.P., Roberts, C.M., Van't Hof, T., De Meyer, K., Trantolos, J., & Aldam, C. (1999). Effects of recreational SCUBA diving on Caribbean coral and fish communities. *Conservation Biology, 13*, 888–897.

Heywood, J. (2002). The cognitive and emotional components of behavior norms in outdoor recreation. *Leisure Sciences, 24*, 271–281.

ICRAN MAR 1. (2006). Standard requirements for recreational scuba diving services in the Mesoamerican reef system. Retrieved June 5, 2013 from standards@coral.org.

Kaur, C.R. (2010). Dive tourism in Malaysia: Addressing the developmental impacts. Retrieved July 5, 2013 from http://www.bairdmaritime.com/index.php?option=com_content&view=article&id=6534:dive-tourism-in-malaysia-addressing-the-developmental-impacts-&catid=114:workboats&Itemid=209.

Musa, G., Seng, W.T., Thirumoorthi, T. & Abessi, M. (2011). The influence of SCUBA divers' personality, experience, and demographic profile on their underwater behaviour. *Tourism in Marine Environments, 7*, 1–14.

Ong, T.F., & Musa, G. (2012). Examining the influences of experience, personality and attitude on SCUBA divers' underwater behaviour: A structural equation model. *Tourism Management, 33*, 1521–1534.

PADI Diver Statistics. (2013). PADI worldwide certification history. Retrieved from June 2013 (updated February 2013). http://www.padi.com/SCUBA/about-padi/PADI-statistics/default.aspx.

Rouphael, A.B., & Inglis, G.J. (2002). Increase spatial and temporal variability on coral damage caused by receational scuba diving. *Ecology Application, 12*(2), 427–440.

Rouphael, T. & Inglis, G.J. (2001). Take only photographs and leave only footprints?: an experimental study of the impacts of underwater photographers on coral reef dive sites. *Biological Conservation, 100*, 281–287.

Thapa, B., Graefe, A.R., & Meyer, L.A. (2006). Specialization and marine based environmental behaviours among SCUBA divers. *Journal of Leisure Research, 38*(4), 601–615.

Thapa, B., Graefe, A.R., & Meyer, L. (2005). Moderator and mediator effects of SCUBA diving specialization on marine-based environmental knowledge-behaviour contingency. *Journal of Environmental Education, 37*(1), 53–67.

Tratalos, J.A. & Austin, T.J. (2001). Impacts of recreational SCUBA diving on coral communities of the Caribbean island of Grand Cayman. *Biological Conservation, 102*, 67–75.

Hospitality and Tourism – Sumarjan et al. (eds)
 ISBN 978-1-138-00151-0

A snapshot of environmental strategies in hotels: Locals' perspective in Malaysia

M.F.S. Bakhtiar, A.A. Azdel, M.S.Y. Kamaruddin & S.A. Talib
Universiti Teknologi MARA, Malaysia

ABSTRACT: Researches over the past decade have shown society's overwhelming concern for global environment; the same is evident in the hotel industry. Despite ongoing pressure to sustain with such demand, hoteliers need to strike a balance; they need to simultaneously ensure that environmental strategies applied are unobtrusive while keeping it visible to satisfy environmentally-conscious clientele. Hence, it is noteworthy to understand guests' actual perception about environmental strategies applied in hotels. Through self reported questionnaire, 288 valid respondents were gathered using convenience sampling method to unravel the issue. Various opinions were discovered about environmental strategies in hotels. Any environmental strategy implemented should be planned with care especially if it creates a limitation in individual's activity or preference. Findings showed respondents were inclined to refuse strategies compromising personal privacy, preferences and hygiene as well as other fundamental individual needs. They were unlikely to support strategies causing much inconvenience. Cutback and discrimination in service delivery are encouraged to be avoided. Strategies designed must not be perceived by guests as a cost saving approach benefiting the organization solely. Interestingly, respondents were unaffected by eco-friendly guestroom design. They were willing to tolerate the reuse, recycle and other environmental campaigns together with the green gadget implementation.

Keywords: environmental, strategies hotel, perception

1 INTRODUCTION

The ongoing global concern for environmental protection has led to changes in the way people live. The demand for environmental products and services are becoming more apparent; even in the hotel industry. There were studies attempting to capture hotel guests' changing attitude towards protection of the environment. According to Swanger, Benson & Paxson (2009), 90% of hotel guests prefer to stay in a hotel practicing environmental strategies. Hence, it has become important for hoteliers to understand the role environmental issues play to encourage guests' purchasing decision.

Creating a balanced environmental atmosphere in hotels is crucial; hoteliers must be able to simultaneously keep environmental strategies unobtrusive so that guests don't feel inconvenienced, but make them visible enough to satisfy environmentally-conscious clientele (White, 2010). Several studies have proven that unwarranted environmental strategies have caused guests to respond negatively when such policies impacted their experiences (El Dief & Font, 2012). Thus, environmental strategies shouldn't at any time compromise personal privacy, preferences and hygiene as well as other fundamental individual needs. According to Parasuraman, Zeithaml and Berry (1985) in the conceptual model of service quality, customers experiencing poor service will become dissatisfied and decided not to return in the future. Interestingly, the same is evident through studies in many of hospitality and service related industries including information technology, retailing, finance and even healthcare (Bakhtiar, Mohd Zahari, Azhar, & Kamaruddin, 2012; Abdul Aziz, Bakhtiar, Che Ahmat, Kamaruddin, & Ahmad, 2012; Choi, Cho, Lee, Lee & Kim, 2004).

Based on the above contemplation, it is crucial for hoteliers to understand how each environmental strategy applied within their establishments was perceived by current or potential guests. Albeit the fact many researchers have addressed this issue in the U.S. Europe and several Asian countries, minimal is established in Malaysia and countries with similar geographical, climate and cultural characteristics. Therefore, this study particularly examined the perception of Malaysian towards prevalent environmental practices in hotels.

2 LITERATURE REVIEWS

Many of environmental management practices in hotels are applied in operational areas including guest rooms, food and beverage facilities, conference and meeting areas, and other public spaces within the hotel. Environmental management in the hotel is typically categorized into the following segments; energy and water conservation as well as waste reduction and recycling management (Molina, Claver, Pareira & Tari, 2009; Bohdanowicz, 2006). The span of hotel environmental management is also built around back-of-house operations, organizational system and culture as well as external business relationship (Park, 2009).

2.1 *Energy management*

The hotel industry in general consumes considerable amount of electricity and fossil fuel energy in various operational areas (Park, 2009). The U.S Environmental Protection Agency (EPA) reported that the hotel industry can save up to $285 million just by reducing the amount of energy use by 10 percent (McLeish, 2007). The usage of energy efficient bulbs can save about 25% of energy costs, depending on the scale of the hotel (Bohdanowicz, 2006). Other strategies would include the use of energy efficient equipments certified by the EPA, implementing environmental building design and facility selection according to the specification by U.S Green Building Council (USGBC).

In addition to the strategies mentioned above, other countries also implement numerous energy management initiatives including application of renewable energy program such as the use of wind, solar, and run-of river power; adopting automated energy control system through key-card; replacing incandescent bulbs with fluorescent lighting; installing energy-efficient equipment in major operational areas such as laundry; using digital thermostats to control guestroom energy consumption; installing occupancy sensor to automatically turn off lights when guests leave their room (Kasimu et al., 2012; Millar & Baloglu, 2011; Park, 2009). Other energy management initiatives recalled from Bohdanowicz, Zientara and Novotna (2011), and Kasavana (2008) were reducing air-circulating equipment through implementing smoke-free policy; using waste heat from the power generator; using Energy Star-qualified products; installing triple-glazed windows or reflective glass to save energy for heating and cooling; replacing outdoor and exit signs with Light Emitting Diode (LED) signs; as well as designing parking areas under the hotel's building to reduce site footprint and lower cooling costs by eliminating blacktop paving heat effect.

2.2 *Water management*

According to Brodsky (2005), an average hotel consumes about 209 gallons of water per occupied room daily; only 34 gallons below a typical U.S household usage. The high cost consideration has driven many hotels to implement water conservation programs. For example, Marriott International introduced 'linen reuse program' that has been proven to save about 17% hot water consumption, by encouraging guests to reuse linens and towels during their stay in the hotel (Marriott International, 2007). A similar approach has been implemented by other hotel brands worldwide and it is evident in various studies including in Kasimu, Zaiton and Hassan (2012) and Bohdanowicz (2006). The benefits of this program extend beyond utility conservation; it also helps to minimize waste. Other commonly applied water conservation initiatives include installation of low-flow faucets and shower heads and low-water-volume toilets; infrared-activated faucets, sink aerators; regularly fixing toilet leaks; using water-efficient laundry equipment and dishwashers; placing water meters in guest rooms to track usage; and adopting water saving campaigns in kitchens as well as adopting water-efficient or xeric gardening techniques; and using treated wastewater in garden irrigation (Bohdanowicz 2011; Park, 2009).

2.3 *Waste reduction and recycling program*

Another area of environmental management in the hotel is waste reduction and recycling. The main purpose of waste management is to reduce quantity and toxicity of waste produced (Kirk, 2012). The biggest amount of waste produced by hotels derived from papers and food waste (Wastecare Corporation, 2013). The type of waste generated in hotels is different according to the operational functions it serves. For example, the food and beverage function produces a mixture of solid and organic wastes from the packaging and food waste, aluminum cans, glass bottles, corks and cooking oils (Park, 2009). Housekeeping generates general cleaning wastes including laundry chemicals, plastic and cardboard packaging. Other departmental functions such as sales and marketing, finance, administrative, and human resources produces similar solid waste such as toner cartridges, paper, cardboard waste, whereas the maintenance facility releases remaining types of solid wastes such as general fixtures and equipments (Radwan, Jones, & Minoli, 2012). The following are other commonly applied waste and recycling management strategies; placing recycling bins in all front and back-of-house areas; purchasing used or recycled-content products; adopting a donation program for leftover guest amenities, old furniture and appliances; and composting organic kitchen waste (Park, 2009). In addition, using refillable amenity dispensers; providing reusable items such as cloth napkins, glass cups, ceramic dished with food and beverage service; grinding guest soaps to use as laundry detergent for hotel uniforms; purchasing food items and cleaning chemicals in bulk containers; and recovering used cooking oil and food waste are also practiced in hotels.

2.4 *Other environmental programs in hotel*

Besides the three major areas of environmental management discussed earlier; there are several other practices designed to encourage eco-friendly hotels. The following suggestions were derived from Kasimu et al. (2012), and Park (2009) which includes utilizing eco-friendly cleaners and detergent; purchasing locally produced supplies; buying from environmentally responsible suppliers; establishing a formal channel to cooperate NGOs; establishing guests education programs; supporting local communities to enhance local environment; and incorporating environmental reporting in corporate control systems. Additional strategies indicated by Millar and Baloglu (2011), and Park (2009) were providing employees with environmental training; having a written environmental policy as an organizational culture; having a manager or a team in charge of environmental management; as well as pursuing green certifications from respective environmental agencies.

3 METHODOLOGY

In accessing respondents' perception about environmental practices in hotels, a quantitative research approach was employed using a self administered questionnaire. Following simple screening questions, local travelers at public airport terminals in the central region of Malaysia were chosen for data collection using the convenience sampling technique. A total of 288 valid respondents was gathered and analysed using descriptive analysis approach.

4 FINDINGS AND DISCUSSION

4.1 *Demographic profile of the respondents*

Before commencing into statistical analysis, all data were checked for irregularities, missing item, or unrealistic responses. Frequency test was used against all demographic variables (gender, age, education background and occupation together with their region of origin). Of the 288 respondents, 51.1 percent were male as opposed to 48.9% female. About half of respondents (47.6%) were between 21 to 30 years old, followed by 34.4% around 31 to 40, another 13.5% from 41 to 50, another 2.8% were under 20 while the remaining of 1.7% between 51 to 60 years old. Moving to education profile, almost all respondents with over 90% own at least undergraduate academic background. Since majority respondents are with higher level education, it is supposed that they may have some basic knowledge about environmental strategies hence able to respond well to the subject matter. Meanwhile, a review of respondent's occupation revealed that the majority (45.1%) were corporate professionals, followed by 24.7% working in the government sector, while the remaining were either self-employed or students. Since majority of respondents was classified as professionals, it is assumed that those individuals could be business travellers and have various hotels stay experiences. The last demographic component analyzed was respondents' region of origin. Based on the analysis, 89.9% respondents were from the central region (Kuala Lumpur, Selangor and Putrajaya) another 10.1% from the southern region (Johor, Melaka and Negeri Sembilan combined) while the rest were from north and east coast states in Malaysia.

4.2 *Perception on energy management program in hotels*

In response to the objective of the study, mean score was used to indicate respondents' perception on 3 categories of environmental practices; (1) energy and (2) water management as well as (3) waste reduction and environmental protection campaign. Respondents' perception about the energy management program is presented in Table 1. From the table, the majority of respondents was inclined towards agreeing with an energy management program in hotels with a mean score between 3.92 and 4.27. The use of key cards to enable electricity in guestroom received the highest score ($M = 4.27$, $SD = 0.740$).

Table 1. Mean score for perception of energy management strategies in hotels (n = 288).

Variables	*M*	*SD*
Air-conditioning activated by motion sensor along hallways	3.92	0.938
Use key-card to enable electricity usage in guestrooms	4.27	0.740
Adopt "smoke-free policy" within hotel to save energy	4.13	0.905
Installation of reflective coating at guestroom's window	4.24	0.770
Tubular skylight room design	4.15	0.768

Scale: 1 = Strongly Disagree, 2 = Disagree, 3 = Moderate, 4 = Slightly Agree, 5 = Totally Agree.

Contrarily, installation of automated (motion sensor) air-conditioning system along hallway got the lowest score. Being in a country with a warm tropical climate all year round; the idea of pursuing automated air-conditioning system may be perceived as operationally ineffective.

Although other energy management strategies were viewed positively, the standard deviation value of all items was high; between 0.740 and 0.938. This shows that not all of them were willing to accept suggested strategies evenly. As such, smokers might be less enthusiastic with the idea of "smoke free zone" if it runs the entire guest rooms and all public areas. Interestingly, respondents were not affected by eco-friendly guestroom design; installation of reflective coating and tubular skylight were perceived positively. Overall, it is important to design an energy management program with care especially if it involves limitation in individual's activity or preference. The decision to implement a strategy is encouraged to be made upon suitability to respective geographical setting. Additionally, alternative must always be made available to those unready to forgo their lifestyle for green practices.

4.3 *Perception on water management program in hotels*

Table 2 depicts respondent's mean score for water management strategies in hotels. As compared with energy management, the majority gave a moderate score for water management program in hotels. Although the range of mean score was only between 3.38 and 3.98; an in-depth analysis reveals that they were still inclined towards favouring 3 from the 4 strategies listed. Again, the standard deviation for all items was high between 0.877 and 1.189 indicating varies feedback among respondents. Installation of water efficient devices in toilet was perceived most acceptable (M = 3.98) as compared with others.

Table 2. Mean score for perception of water management strategies in hotels (n = 288).

Energy management		
Variables	*M*	*SD*
Not changing bed sheet	3.53	1.162
Not changing bath linen	3.38	1.189
Water efficient devices in all toilets/bathrooms	3.98	0.877
Reuse waste-water in hotel's toilet flushing system	3.91	1.126

Scale: 1 = Strongly Disagree, 2 = Disagree, 3 = Moderate, 4 = Slightly Agree, 5 = Totally Agree.

This may be probable since the strategy has no significant changes to the way such devices being used, thus there is none credible reason to completely refuse the strategy. Similar argument may be applied to application of waste water in the hotel's flushing system.

Contrarily, fewer respondents agreeing to not changing bed sheet during a guest's stay (M = 3.53). Even less agreed to not changing bath linens during their stay in the hotel. This may indicate that respondents were unlikely to compromise cleanliness and personal hygiene for meeting environmental goals. Based on analysis and pattern discussed about water management, it is worthwhile to mention that respondents were most likely to refuse strategies impacting fundamental individual needs (not changing bed sheet and bath linens) as opposed to installation of environmental conservation devices having no clear effect to them.

4.4 *Perception on waste reduction management and environmental protection campaign*

The mean score for perception on waste reduction management and environmental protection campaign in hotels is depicted Table 3. A glance through the figures revealed that majority respondents were inclined towards agreeing with all strategies listed. They favoured the idea of recycling bins in guestrooms and public areas (M = 4.09); heavy promotion about environmental awareness

Table 3. Mean score for perception of waste reduction and environmental campaign (n = 288).

Waste reduction and environmental campaign		
Variables	*M*	*SD*
Recycle bins in guestroom and public areas	4.09	0.803
Using refillable amenities (i.e. soap & shampoo) dispenser	3.93	0.819
Heavy environmental awareness promotion program	4.09	0.815
Reusable items in food & beverage service	4.01	0.855
Complimentary newspapers in public areas only	3.97	0.966
Special car park entitlement to hybrid vehicles driver	3.77	0.992
Replace hand-towel tissue paper with automated hand dryer	4.03	0.867
Brochure via email; printed copy upon request	4.05	0.797

Scale: 1 = Strongly Disagree, 2 = Disagree, 3 = Moderate, 4 = Slightly Agree, 5 = Totally Agree.

does not seem to bother them (M = 4.09); they were willing to accept the hotel's brochure and promotion via email while printing is made upon request (M = 4.05); replacing hand-towel tissue paper with automated hand dryer was perceived positively (M = 4.03), even using reusable items in food & beverage service was deemed acceptable (M = 4.01).

From another standpoint, the strategy to offer complimentary newspapers only in public areas as opposed to delivering it to each guestroom were minimally favourable (M = 3.97); maybe respondents perceived the ideas as a cost saving approach benefiting the organization rather than the guests or they were make inconvenienced as a result of supporting such strategy. In a similar gesture, using the refillable amenities dispenser was less favourable strategy (M = 3.93); as well as offering special parking entitlement to guest arriving with hybrid vehicles (M = 3.77). Based on the three final items described, maybe respondents were unlikely to accept a cutback and discrimination of service delivery to uphold hotel's environmental program.

5 CONCLUSION

Overall, this study revealed that not all respondents were willing to accept suggested strategies evenly. Therefore, any environmental management strategy implemented in a hotel should be planned with care especially if it involves limitation in individual's activity or preference. Based on the findings, it was clear that respondents were inclined to refuse strategies that compromise personal privacy, preferences and hygiene as well as other fundamental individual needs. They were unlikely to support if the strategy is perceived to cause much inconvenience to them. Furthermore, any cutback and discrimination in service delivery made in order to uphold the hotel's environmental program should be avoided; hence it is crucial for hoteliers to understand that the strategy implemented must not be viewed as a cost saving approach benefiting the organization. Interestingly, respondents were unaffected by eco-friendly guestroom design. They were willing to tolerate reuse, recycle and other environmental campaigns in addition to green gadget implementation positively.

ACKNOWLEDGEMENTS

This research is funded by Universiti Teknologi MARA through Research Intensive Fund (RIF) grant.

REFERENCES

Abdul Aziz, A., Bakhtiar, M.F.S., Che Ahmat, N.H., Kamaruddin, M.S.Y., & Ahmad, N.A. (2012). The usage of ICT applications in 5 star hotels in Kuala Lumpur, Malaysia. Current Issues in Hospitality and Tourism Research and Innovations, Taylor & Francis Group, London.

Bakhtiar, M.F.S., Mohd Zahari, M.S., Azhar, A.R. & Kamaruddin, M.S.Y. (2012). Hypermarket fresh foods' attributes towards customer satisfaction. Current Issues in Hospitality and Tourism Research and Innovations, Taylor & Francis Group, London.

Bohdanowicz, P. (2006). Environmental awareness and initiatives in the Swedish and Polish hotel industries survey results. *Hospitality Management, 25*, 662–682.

Bohdanowicz, P., Zientara, P., & Novotna, E. (2011). International hotel chains and environmental protection: An analysis of Hilton's we care programme (Europe, 2006–2008). *Journal of Sustainable Tourism, 19*(7), 797–816.

Brodsky, S. (2005). Water conservation crucial to energy savings. *Hotel & Motel Management, 220*(13), 12.

Choi, K.S., Cho, W.H., Lee, S., Lee, H., & Kim, C. (2004). The relationships among quality, value, satisfaction and behavioral intention in health care provider choice: a South Korean study. *Journal of Business Research, 57*(8), 913–921.

El Dief, M., & Font, X. (2012). Determinants of environmental management in the Red Sea hotels personal and organizational values and contextual variables. *Journal of Hospitality & Tourism Research, 36*(1), 115–137.

Kasavana, M. (2008). *Green hospitality*. Hospitality upgrade.

Kasimu, A., Zaiton, S., & Hassan, H. (2012). Hotels involvement in sustainable tourism practices in Klang Valley, Malaysia.

Kirk, D. (2012). *Environmental management for hotels*. Routledge.

Marriott International (2007). Marriott helps "Clean up the world". Retrieved, from the World Wide Web http://www.marriott.com/news/detail.mi?arrArticle=160342.

McLeish, B. (2007, November). Pairing green design with energy purchasing strategies. *Hospitality Construction. 2*(6), 52–54.

Millar, M.& Baloglu, S. (2011). Hotel guests' preferences for green guest room attributes. *Cornell Hospitality Quarterly, 52*(3), 302–311.

Molina-Azorín, J.F., Claver-Cortés, E., Pereira-Moliner, J., & Tarí, J.J. (2009). Environmental practices and firm performance: An empirical analysis in the Spanish hotel industry. *Journal of Cleaner Production, 17*(5), 516–524.

Parasuraman, A. Zeithaml, V., & Berry, L. (1985). A conceptual model of service quality and its implications for future research. *Journal of Marketing, 49*(4): 41–50.

Park, J. (2009). The relationship between top managers' environmental attitudes and environmental management in hotel companies (Doctoral dissertation, Virginia Polytechnic Institute and State University).

Radwan, H.R., Jones, E., & Minoli, D. (2012). Solid waste management in small hotels: A comparison of green and non-green small hotels in Wales. *Journal of Sustainable Tourism, 20*(4), 533–550.
Swanger, N., Benson, L.S., & Paxson, C. (2009). Ecotourism projects: Impact on environmental attitudes in introductory hospitality courses. *Journal of hospitality and tourism education, 21*(2), 24–29.
Wastecare Corporation (2013). Waste reduction and recycling tips for hotels, resorts and motels. Retrieved from http://www.wastecare.com/Articles/Waste_Reduction_Recycling_Tips_Hotels.htm.
White, M.C. (2010). For hotels, eco-friendly ideas await a friendlier economy. The New York Times. Retrieved from http://www.nytimes.com/2010/08017/business/7green.html.

Food service and food safety

Hospitality and Tourism – Sumarjan et al. (eds)
© 2014 Taylor & Francis Group, London, ISBN 978-1-138-00151-0

Knowledge, attitude and food safety practices among the food handlers: A case of Kota Kinabalu, Sabah, Malaysia

M. Sylvester, P.S. Toh & M.A.A. Bashir
Universiti Teknologi MARA, Malaysia

ABSTRACT: The rapid expansion of restaurants in Kota Kinabalu Sabah is currently due to higher disposable incomes of the consumers. This study aims to determine the level of knowledge on food safety among food handlers in Kota Kinabalu Sabah. Furthermore, it is also trying to determine the attitudes of food handlers towards food hygiene practices and perform an assessment on the aspect of whether the food hygiene practices are applied by food handlers. The sample populations selected for this study were restaurants' food handlers who are involved in the daily operation for the restaurant. The results were significant for food Safety knowledge toward food safety practices ($\beta = 0.35$, $p < 0.001$) and between food safety attitudes toward food safety practices ($\beta = 0.50$, $p < 0.05$). The findings of this study can be used by policy makers particularly the Kota Kinabalu City Hall in reviewing the current conditions of food premises. Through the findings, they can plan on what steps that need to be taken to address the recent situations of the foodservice premises. In addition, the outcomes of this study may perhaps increase the awareness of food handlers regarding the importance of hygiene in food premises.

Keywords: attitude, food handlers, food safety, knowledge, practice

1 INTRODUCTION

There is no doubt on the fact that food borne illnesses are very common in Malaysia. This is due to the fact that there are so many factors involved in the causes of food borne illnesses such as improper food handling, personal hygiene and cleanliness. This applies to numerous foodservice operations including in the restaurant sector (Abdul-Mutalib, Abdul-Rashid, Mustafa, Amin-Nordin, Hamat & Osman, 2012). Food hygiene is very important in every food service establishment because it relates to food safety. In conjunction with the notion, the hands of food service employees can be the source in the spread of foodborne diseases because of poor personal hygiene and cross-contaminations (Eihiri & Moris, 1996; Siow & Norrakiah, 2011).

Most of food handlers in local restaurant in Kota Kinabalu did not apply the food safety and food hygiene practices in their premises. These concerns could potentially develop into a health issue for the Malaysian citizens and tourists, in regards to food-borne diseases, thus leading to food borne outbreaks (Pang & Toh, 2008). Manask (2002) mentioned that in countries where the Foodservice industry is considered important and developed, such as in the United State, food borne outbreaks could be high. Yet, the food safety issues in the foodservice industries usually receive the minimum amount of attention and care (Pang & Toh, 2008).

2 LITERATURE REVIEW

Gibson et al. (2002) pointed out that that hygienic and safe food preparation together with the education of those who are involved in the food preparation, food processing and service of meals are important in reducing the risk of food borne illnesses. Campos et al. (2009) confirmed that infections caused by microorganisms can be reduced by maintaining a proper hand hygiene practices. Several studies have verified that the presence of microbes on food-handlers' hands and the likelihood for infected and unhygienic hands to become a contamination source was due to poor personal hygiene and improper hand washing practice (Aycicek et al., 2004; Lues & Tonder, 2007).

Martins, Hogg and Otero (2012) mentioned that the application of good hygiene practices in food businesses especially for food handlers is essential to protect consumers from food borne illnesses. Food handlers should have the necessary knowledge and skills to enable them to apply this practice. Education and work experience are important factors in healthy food handling. Incorrect practice such as improper and inefficient cleaning due to the lack of knowledge on hygiene may cause food contaminations (Martins et al., 2012).

Besides knowledge, attitude is also an important factor that may decrease the trend of food borne diseases. Howes et al. (1996) pointed out that there is a relationship between positive behaviour, attitudes and the continuation of education of food-handlers in terms of food safety and hygiene towards the maintenance of safe food handling practices. Safe and hygienic food handling are an important part of food handlers' responsibilities, and learning more about food hygiene is important for them. Food handlers are supposed to wear caps, masks and gloves reduce the risks of cross contaminations and food contaminations. In the study done by Tokuc et al. (2009), they confirmed that 94.5% of food handlers agreed that washing hands before handling food reduces the risk of contaminations.

In addition, Buccheri et al. (2007) also stated the importance of positive attitude toward safe storage practices involving temperature control and correct handling of food using adequate clothing and gloves. Despite the generally positive attitudes towards safe storage practices, correct handling of foods, and using adequate clothing and gloves revealed in the study, the respondents fared badly in answering questions on refrezing. Similar results have been obtained from previous studies (Askarian et al., 2004; Buccheri et al., 2007). Perlik (2002) confirmed that food safety practices are vital in any food service establishment. This is due to the fact that consumer food safety concerns in a food service establishment are entailed from the observable cues such as food hygiene and safety practices.

Dettenkofer and Spencer (2007) discovered in their study that self-reported behaviours revealed improper practices, such as sharing of utensils for raw and cooked foods and thawing of frozen food at room temperature. In addition, washing hands before and after touching unwrapped and raw food was not as generalized as expected in personnel who should have been constantly trained about hand hygiene. Hand hygiene, however, is considered most critical in the pathogen control than cleaning and disinfection of environmental surfaces.

According to Codex Alimentarius (2003), food handlers should tie back long hair and wear appropriate hair covering (hairnets or toques), cut their nails short and avoid wearing nail polish. During handling they should remove all jewellery and makeup. Moreover, Campos et al. (2009) found that there is a higher value in the present work, in which 81.5% of the handlers had long nails and/or wore nail polish and jewellery and 51.9% reported performing routine activities with respiratory or skin infections, in violation of current sanitary legislation norms.

3 METHOD

A sample of 300 respondents through convenience sampling was selected in three main areas in Kota Kinabalu Sabah, Malaysia which are Segam, Sinsuran and Kampung Air area. A self-completed questionnaire has been developed for gathering the required information for this study and conducted in March 2013. An identical five point Likert scale for every question in each section in the questionnaire as it allow the respondents to better express themselves, and at the same time increase the response rate and the quality of the data obtained (Babakus & Mangold, 1992).

The questions used in the study were based on and modified from previous research (Abdul-Mutalib et al., 2012; Buccheri et al., 2010; Jianu & Christ, 2012). A pilot study was undertaken among (30) respondents which are food handlers who are working in a local restaurant (independent own restaurant). Cronbach's alpha for each set of the questions range within acceptable limit (>0.7). Descriptive statistic was used to measure the internal consistency define by the mean and standard deviation for each item and inferential statistics is undertaken to determine the relationship between variables.

4 RESULT AND DISCUSSION

4.1 *Attitude*

Looking at the respondents' attitudes toward food safety, analyses of Table 1 clearly reveals the overall information regarding the respondents' attitude. Most of the respondents agreed with the notion that work area should be clean before they start working ($M = 4.17$, $SD = 0.5676$) and they should not handle or touch food with a wounded hand ($M = 4.06$, $SD = 0.7864$). Additionally, most of the respondents also disagreed that they should wash their hands before they start working ($M = 2.50$, $SD = 0.727$).

However, the respondents also agreed that a same towel can be used in many places in kitchen area and the dining room ($M = 4.16$, $SD = 0.7137$) and jewellery such as ring, watch can be worn while handling food ($M = 4.16$, $SD = 0.6448$). In a correct manner, any jewellery should not be worn by the food handler such as ring and watch during handling food as it is contributing to physical hazard in food. Furthermore, most of the respondents disagreed that separate utensils must be used when preparing raw and cooked food and they should not touch their hair, face, nose, etc while preparing food.

Table 1. Mean scores on respondent's attitude regarding food safety.

Respondent attitudes	*M*	*SD*
The work area should be clean before start working	4.17	0.56760
Hands should be washed before start working	2.50	0.727
Working with unclean hand should be avoided	3.16	0.9785
Smoking during work should be avoided	3.18	0.9318
Should not rub or touch our hair, face, and nose while working	2.55	0.9505
To clean many places in the kitchen and Dining room, same towel can be used	4.16	0.7137
It is not necessarily for cooked food and raw food to be kept separately	4.23	0.7508
To prepare raw or cooked food, separate utensils must be used	2.51	0.8157
Should not re-frozen food that already being defrosted	2.53	0.8590
Should not handled or touch food with a wounded hand	4.06	0.7864
Jewellery such as ring, watch can be worn while handling food	4.16	0.6448

Table 2. Mean score on respondents' knowledge regarding food safety.

Knowledge regarding food safety	*M*	*SD*
Food that is prepared in advance is most probably will contribute to food poisoning.	2.54	1.1999
Reheating food is most probably will contribute to food contamination.	2.40	1.0380
A false application of cleaning and sanitization process for equipment (refrigerator, slicing machine, mincer) increase the risk of foodborne disease to consumers.	4.44	3.0908
Before handling food, hands should be washed to reduce the risk of contamination.	3.70	1.1435
Wearing proper gloves while handling food reduce the risk of cross contamination.	2.74	1.1991
Proper use of cap/hair net, masks, and adequate clothing can minimize the risk of food contamination.	4.26	1.902
The importance to know the correct temperature of the refrigerator/freezer is to reduce the risk of food spoilage.	4.73	3.1385
Inappropriate storage of foods (raw, cooked and ready-to-eat) may cause a health hazard to consumers.	2.94	1.2538

In the study done by Tokuç et al. (2009), they confirmed that 94.5% of food handlers agreed that washing hands before handling food reduces the risk of contamination. Martins, Hogg, and Otero (2012) mentioned that the application of good hygiene attitude in food businesses especially for food handlers is essential to protect consumers from food borne illness which is to ensure that all utensils, cloth and equipment should be separated when handling raw and cooked food.

Food handlers should have a clean, tidy and proper appearance, without any skin infections, good dental hygiene, short finger nails and are not in the habit of biting nails, not wearing jewellery except wedding ring, wearing no make-up, work in clean shoes and uniform, and stick to good hygiene practices (Bas et al., 2004). A proper attitude during food preparation and serving is important not only to maintain the quality of food but also the safety of food in term of avoiding any possibilities of cross-contaminations from occurring. Cross-contaminations may lead to food borne illnesses if there is a lack of monitoring and controlling regarding food handlers' attitude in their working routine.

4.2 *Knowledge*

Analysis on respondents' knowledge regarding food is shown in Table 2. The frequency analysis reveals all mean scores of each item under knowledge dimension. From the analysis, most of the respondents disagreed with the statement that food prepared in advance most probably will contribute to food poisoning ($M = 2.54$, $SD = 1.1999$) and reheating food most probably will contribute to food contaminations ($M = 2.40$, $SD = 1.0380$). This is in line with Abdul-Mutalib et al. (2012) stated that reheating food may not lead to food contamination. However, most of the respondents disagreed that by wearing proper gloves while handling food reduces the risk of cross-contaminations ($M = 2.74$, $SD = 1.1991$). Based on this notion, food handlers should wear gloves in order to avoid bacterial contaminations directly to food. Angelillo et al. (2001) and Askarian et al. (2004) stated that the usage of gloves is important in reducing bacterial contaminations directly to food.

Apart from that, most of the respondents agreed that a proper use of cap or hair net, masks, and adequate clothing can minimize the risk of food contaminations ($M = 4.26$, $SD = 1.902$) and it is important to know the correct temperature of the refrigerator or freezer to reduce the risk of food spoilage ($M = 4.73$, $SD = 3.1385$). However, the respondents disagreed with the fact that inappropriate storage of foods (raw, cooked and ready-to-eat) may cause a health hazard to consumers (M = 2.94, SD = 1.2538). Improper storage of food item especially raw and ready-to-eat food may cause bacterial contaminations to ready-to-eat food. This condition may cause food borne illnesses. Lucca and Torres (2006) pointed out that all the problems related to handling, inadequate or insufficient storage and poor hygienic conditions, will lead to higher risk of contracting food-borne disease.

4.3 *Practice*

Analysis of respondents' practices toward food safety in Table 3, the result showed that most of the respondents disagreed that they should wash their hands before they start working and after using the washroom (M = 2.60, SD = 0.8537). Furthermore, they also do not wash their hands after coughing or sneezing. Siow and Norrakiah (2011) confirmed that besides poor hand sanitation, the lack of personal hygiene amongst food handlers is also one of the most commonly reported practices that contributes to the rise of food borne illnesses. Therefore, hand washing is important to reduce the chances of food contaminations from occurring. Moreover, the majority of them uses an apron to wipe their hands frequently (M = 4.48, SD = 0.6145) and use the same towel to clean kitchen area and also the dining room (M = 4.62, SD = 0.5242). Additionally, most of the respondents agreed that jewellery can be worn while working (M = 4.73, SD = 0.4997) and they do not separate cooking and raw food during food preparation (M = 4.60, SD = 0.7925).

Table 3. Mean score on respondents' practices.

Practices regarding food safety	(M)	(SD)
I clean and organize the workstation before I start working	4.21	0.6837
I wash my hand first before start working and after using the washroom	2.60	0.8537
I use a tissue when I cough and sneezing and wash hands after that	2.68	0.8908
I chew gum and eating while I'm working	3.06	0.8050
I smoke while I work	3.38	0.9377
I always use my apron as a cloth to wipe my hand frequently	4.48	0.6145
I use the same towel to clean all places in the kitchen and dining room	4.62	0.5242
I wear jewellery and watch while working	4.73	0.4997
I do not separate cooked food and raw food during preparation	4.60	0.7925
I always use separate utensils to prepare cooked food and raw food	2.79	1.383

Table 4. Result of multiple regressions of the attitude and knowledge on practices.

Variables	B	SE B	β	Sig
Constant	2.745	0.163		
Attitude	0.268	0.051	0.504	0.015
Knowledge	0.141	0.017	0.348	0.000

Note: $R^2 = 0.721$; $*p < 0.05$, $**p < 0.01$, $***p < 0.001$.

On the other hand, most of the respondents do not use separate utensils to prepare cooked and raw food (M = 2.79, SD = 1.383). These results indicate that the food handlers do not apply the basic condition of food safety and hygiene in food handling. Dettenkofer and Spencer (2007) discovered in their study that self-reported behaviours revealed improper practices, such as sharing of utensils for raw and cooked foods. In addition, washing hands before and after touching unwrapped raw food and also after using the washroom can be the source of cross-contaminations. Hand hygiene, however, is considered most critical in the pathogen control than cleaning and disinfection of environmental surfaces. Improvement of food workers' hand washing practice is, therefore, vital to reduce the incidence of food borne illnesses (Lynch, Elledge, Griffith, & Boatright, 2003).

4.4 *Multiple regressions for knowledge, attitude and practices*

As can be seen from Table 4, the model was significant where two (2) variables were able to explain 72.1% (R^2 = 0.721, F = 14.772, p < 0.001) of the variance in the practices. The outcomes revealed that attitude and knowledge significantly contribute to the prediction of food handlers' practices. It is evidently shown that attitude ($\beta = 0.50$, $p < 0.05$) had the most impact on the food handlers' practices, followed by knowledge ($\beta = 0.35$, $p < 0.001$). Each of the dimensions was found to significantly and positively influence on the food handlers' practices. It can be said that the assumption on the food handlers' practices are strongly supported.

5 CONCLUSION

In conclusion, although the results from the analysis are significant, several aspects regarding food safety and hygiene still need to be emphasized. Food safety is a critical necessity, and that it is an utmost importance in any food premises. Therefore, food handlers should be trained and be informed concerning on the aspect of food safety and hygiene. Restaurants' owners or managers should also contribute to this matter in terms of controlling and monitoring their workers. The involvement of government especially the local authorities are important to ensure all food premises run their operation in accordance to the rules and regulations of food safety and hygiene.

REFERENCES

Abdul-Mutalib, N.A., Abdul-Rashid, M.F., Mustafa, S., Amin-Nordin, S., Hamat, R.A., & Osman, M. (2012). Knowledge, attitude and practices regarding food hygiene and sanitation of food handlers in Kuala Pilah, Malaysia. *Food Control, 27*, 289–293.

Angelillo, I.F., Viggiani, M.A., Greco, R.M., & Rito, D., (2001). HACCP and food hygiene in hospital: Knowledge, attitudes and practices of food services staff in Calabria, Italy. *Infection Control Hospital Epidemiology, 22*, 1–7.

Askarian, M., Kabir, G., Aminbaig, M., Memish, Z., & Jafari, P. (2004). Knowledge, attitudes, and practices of food service staff regarding food hygiene in Shiraz, Iran. *Infection Control and Hospital Epidemiology, 25*, 16–20.

Ayçiçek, H., Aydogan, H., Küçükkaraaslan, A., Baysallar, M., & Baoustaoflu, A.C. (2004). Assessment of the bacterial contamination on hands of hospital food handlers. Food Control, *15*, 253–259.

Babakus, E. & Mangold, W.G. (1992). Adapting the SERVQUAL scale to hospital services: An empirical investigation. Health Services Research, *26*(6), 767–786.

Bas, M., Ersun, A.S., & Kivan, C.G. (2004). The evaluation of food hygiene knowledge, attitudes and practices of food handlers in food businesses in Turkey. *Food Control, 17*, 317–322.

Buccheri, C., Casuccio, A., Giammanco, S., Giammanco, M., La Guardia, M., & Mammina, C. (2007). Food safety in hospital: knowledge, attitudes and practices of nursing staff of two hospitals in Sicily, Italy. *BMC Health Services Research, 7*, 45–56.

Campos, A.K.C., Cardonha, Â.M.S., Pinheiro, L.B.G., Ferreira, N.R., Azevedo, P.R.M., & Stamford, T.L.M. (2009). Assessment of personal hygiene and practices of food handlers in municipal public schools of Natal, Brazil. *Food Control, 20*, 807–810.

Codex Alimentarius. (2003). Recommended international code of practice general principles of food hygiene. CAC/RCP 1–1969, Rev.

Dettenkofer, M., & Spencer, R.C. (2007). Importance of environmental decontamination: A critical view. *Journal of Hospital Infection, 65*, 55–57.

Ehiri, J.E. & Morris, G.P. (1996). Hygiene training and education of food handlers: Does it work? *Ecology of Food and Nutrition, 35*, 243–251.

Health and Urban Service Department. (2013). Public Health Division, Health and Urban Service Department. Retrive March 3rd, 2013, from http//:www.dbkk.sabah.gov.my.

Jianu, C. & Chis, C. (2012). Study on the hygiene knowledge of food handlers 409 working in small 410 and medium-sized companies in western Romania. *Food Control, 26*, 151–156.

Lucca, A. & Torres, E.A.F.S. (2006). Street-food: the Hygiene conditions of hot-dogs sold in São Paulo, Brazil. *Food* Control, *17*, 312–316.

Lues, J.F.R. & Tonder, I.V. (2007). The occurrence of indicator bacteria on hands and aprons of food handlers in the delicatessen sections of a retail group. *Food Control, 18*, 326–332.

Lynch, R.A., Elledge, B.L., Charles, P.H., Griffith, C.C., & Boatright, D.T. (2003). A comparison of food safety knowledge among restaurant managers, by source of training and experience, in Oklahoma County, Oklahoma. *Journal of Environmental Health, 66*, 9–14.

Manask, A.M. (2002). The Complete Guide to Food Service in Cultural Institutions. John Wiley and Sons, New York.

Martins, R.B., Hogg, T., & Otero, J.G. (2012). Food handlers' knowledge on food hygiene: the case of a catering company in Portugal. *Food Control, 23*, 184–190.

Pang, F. & Toh, P.S. (2008). Hawker food Industry: Food safety/public health strategies in Malaysia. *Nutrition & Food Science, 38*(1), 41–51.

Perlik, A. (2002). Feeding the faith. *Restaurants and Institutions, 12*(13), 40–48.

Siow, O.N. & Norrakiah, A.S. (2011). Assessment of knowledge, attitudes and practices (KAP) among food handlers at residential colleges and canteen regarding food safety. *Sains Malaysiana, 40*, 403–410.

Tokuç, B., Ekuklu, G., Berberoglu, U., Bilge, E., & Dedeler, H. (2009). Knowledge, attitudes and self-reported practices of food service staff regarding food hygiene in Edirne, Turkey. *Food Control, 20*(1), 565–568.

Hospitality and Tourism – Sumarjan et al. (eds)
© 2014 Taylor & Francis Group, London, ISBN 978-1-138-00151-0

Food related activities and Malaysian tour operators

N. Mohd Yusoff, M.S. Mohd Zahari, M.Z. Kutut & H.A. Hadi
Universiti Teknologi MARA, Malaysia

ABSTRACT: Malaysian gastronomic products namely food, beverages and food related activities are proven important to the tourism industry. Various efforts are implemented by government in promoting these products. Scholars argue that any tourism products could not successfully be promoted without communal efforts from all tourism players including tour operators. This study qualitatively explores the importance of food related activities as part of the gastronomic product to the Malaysian tour operators and their action behaviour in organizing the food tour activities. Tour operators clearly perceived the importance various food culture related activities like types of food festivals, night market, hawkers street stall, food exhibition, food bazaar including Ramadan Bazaar, Homestay and many others as part of the Malaysian gastronomic product. Some of the tour operators admitted food culture related activities are important to their business.

Keywords: food related activities, Malaysian, operators, tour

1 INTRODUCTION

In this modern era the importance of tourism products besides the conventional ones gradually shifted whereby tourist no longer expected to be passive, relaxed and looking at the sight but looking more on tangible activities (Boniface, 2003). The niche new products like sport, shopping, golf, MICE, gambling and entertainment, cruise, cultural, yacht, wildlife tourism and many others are the examples of tangible activities that lead tourism becoming a more interesting industry (Aydin & Boz, 2009). Besides those mentioned products, foods and beverages and food related activities which are part of gastronomic realm is realizing and significantly important for creative tourism that contributes substantial international revenues (Hjalager & Richards, 2002).

Goeldner and Ritchie (2006) deduce that gastronomic products and tourism cannot be separated from tourists as they will bring their taste bud together while travelling and in fact food and drink is one of the criterion or reason for travel (Boniface, 2003). Santich (2004) regard gastronomic tourism as a form of travelling not only focused on food, drink but food related activities like food culture and events while Chaney and Ryan, (2011) denote that gastronomic tourism is a subset of cultural tourism which reflect the culture and heritage of each country and destination. Santich (2004) also explains gastronomic tourism is about participation or engagement in one culture, including place and people of particular destination and even 8000–10000 years ago food, drink, tourism and culture is already seen as a platform featuring the destination image and interrelated with each other although not been given serious attention (Boniface, 2003).

Scholars contend gastronomy through food not only means as nourishment but increase tourist's experience towards it in a particular destination (Hjalager & Richards, 2002; Kivela & Crots, 2006). Gastronomy products are now used by the tour operators as a marketing tool, a pulling factor in attracting visitors and play a leading role in adding value to a destination (Jalis et al., 2009; Shenoy, 2005). The Intrepid Travel Agency (2004) reported that much of the tour operators in Australia, United States, Europe and some of Asian countries especially Thailand, China and Japan offers gastronomy packages on top the shopping and sightseeing. Holiday with cooking, food related activities and wine appreciation feature regularly in destinations like Tuscany as well as Provence in Europe, Melbourne and the Sydney Napa Valley in Australia and the Sonoma Valley in California. In fact, tour operators in New Zealand, Australia, Germany and United Kingdom are aggressively offering weekend gastronomy packages due to demand for a short and/or weekend holidays among food and wine lovers (Scarpato, 2002).

It is proven that many western countries' tour operators in particular perceived the importance of gastronomic tourism not only in creating the memorable experience for tourists (Goeldner & Ritchie, 2006) but would be the best predictor of future behaviour and choices among potential tourists (Conner & Armitage, 2002). Whether similar scenario is occurring among the Malaysian tour is not known. In other words, the extent to which local tour operators perceived the importance of gastronomic tourism products and how it implicate their behaviour action has not yet been widely researched. This study qualitatively explores the importance of food related activities as part of the gastronomic product to the Malaysian tour operators and their action behaviour in organizing the food tour activities among the international tourists.

2 LITERATURE REVIEW

2.1 *Gastronomy and tourism*

According to Pine and Gilmore (1999), the basis of the economy now is slightly shifting from delivering services to staging experiences. Food and beverage besides accommodation and other tourism activities are considered one of the most important elements in tourism and hospitality industry. The various types of food and beverages are viewed as an integral component of a travel experience. The quality of the basic elements of the tourism products will increasingly be experienced by the tourists including food and beverages, food culture practices and food heritage.

Goeldner and Ritchie (2006) stated that food and beverages cannot be ignored by tourists and some consider them as one of the most important elements when they are in the travelling mood. Richards (2002) contended that besides experiencing various tourism activities, tourists without doubt will encounter various types of food and beverages in the visited country. During the period of vacation tourists will also experience and consume varieties of food and beverages, as well as learn the food culture and heritage of that particular country (Hall & Mitchell, 2005; Santich, 1996) and all these food and beverages and food-related activities as part of gastronomy product are closely related to tourism.

2.2 *Malaysian gastronomic products*

Malaysia is one of the well known countries in South East Asia of having vast gastronomy products that can be offered to international tourists (Jalis et al., 2007). A multi-ethnic, multi-cultural and multilingual society comprises of Malay, Chinese, Indian and other ethnic groups whom live side by side has formed the unique gastronomy products with a combination of ingredients and taste ranging from mild to spicy and from plain to sweet beverages (Hutton, 2000; Jalis et al., 2007).

The styles of cooking, the equipments and the way of the food and beverages served are also fascinating. Beside food and beverages, the uniqueness of Malaysian food cultures (food-related activities and events) representing each ethnic group have become part and parcel of Malaysian gastronomy products (Jalis et al., 2007). Most of these are derived either from religious or cultural events such as Hari Raya (Malay), Chinese New Year (Chinese) and Deepavali (Indian). Another famous food-related activity that portray the excitement of the Malaysian food culture are the fare offered in open air night markets popularly known locally as pasar malam and Hawkers Street stalls (Jalis et al., 2007; Syed Amir, 1991).

In addition, by visiting these places tourists for sure will get an opportunity to watch closer the uniqueness of traditional skills practised by the hawkers in preparing foods. Malaysia is also rich with traditional eating manners which continuously inherit and practices from one generation to another. The most popular method of eating is to pile on the rice onto a plate with other side dishes heaped on top and tuck in heartily using the fingers of the right hand among the Malays and Indian whilst the Chinese eat their food by using chopsticks (Hutton, 2000; Jalis et al., 2007; Syed Amir, 1991).

3 METHODOLOGY

The qualitative approach through face to face interview was employed in this study as it provides in depth understanding on the issues investigated. In addition, qualitative are more subjective in nature and involve examining and reflecting on perception to gain an understanding of the social and human activities. The informants were among Malaysian tour operators that cater or organize the inbound and outbound packages. In tapping tour operators' view and action behaviour, a set of open ended interview questions was developed. With that, 10 selected tour operators in the Klang Valley (Kuala Lumpur Petaling Jaya and Shah Alam) were successfully interviewed. All the interviewed information was analysed using thematic analysis.

4 RESULT AND DISCUSSION

4.1 *Informant profiles*

Information's obtained, showed that all 10 CEO's of the tour companies had more than fifteenth year's experiences in travel and tour businesses especially dealing with international tourists and offering the inbound as well as outbound packages.

The tour companies involved have been classified according to time duration of operation, main products and activities, types of packages and type of customers. Cater the international tourists from various regions namely Asia, Middle East, Europe, Australia and Northern America make all the tour companies involved in this study are well-known among the local travel industry.

4.2 *Importance of food related activities as part of gastronomic products for tourism industry*

Tour operators agreed that food related activities or food culture is important gastronomic products. Various foods related activities like types of food festivals, night market, Hawker street stall, Ramadan Bazaar, Homestay and many others as food related activities that could be used to attract more international tourists exploring Malaysia as an interesting tourism destination. Tour operators believed that some tourists are eager to learn and explore food and other relater culture that can enhance their experience towards a destination. Some of the interesting verbatim answered as follows:

> *"Well, various foods related activities like food festivals, night market, hawkers street stall, Ramadan bazaar, Homestay are gastronomic products. These products are important and no doubt closely related to tourism industry"*
>
> *(Informant 1)*

> *"Not to exaggerate that some adventure tourists besides others travel to experience and learnt about the destination culture including food culture. So, our food related activities in addition to existing ones perhaps could be used to attract more international tourist... Ramadan Bazaar has proven to be one the food related events enjoyed by the international tourists"*
>
> *(Informant 5)*

> *"Malaysia having variety food related activities that could be a potential in attracting more tourists. Night markets, street hawkers' stalls, food festivals are some of the interesting food related activities."*
>
> *(Informant 7)*

Seeing to its importance and great potential of attracting international tourists the tour operators then believed that food related activities can expand the tourism industry and increase the local earnings. Food related activities in fact, can be a significant tool in allowing the international tourists explore the local culture and heritage. Not only Malaysian food and beverages, tour operators also believed that food related activities have its own significance in creating Malaysia tourism industry to become more attractive and interesting. Some of the tour operators verbatim answered:

> *"As mentioned earlier, food related activities are an important part of Malaysian gastronomic products thus owing to its potential these products could colour our tourism industry. We have such a diverse culture and ethnics which I believed tourists eager to learn and experiencing it..."*
>
> *(Informant 8)*

> *"Besides our local food and beverages, I strongly believed that our food related activities are having a great potential and making the tourism industry even more interesting"*
>
> *(Informant 10)*

4.3 *Tour operators' action behaviour*

Having touched on the perceived importance of food related activities as part of Malaysia gastronomic products, this section analysis revealed the tour operators' action behaviour in organising trips or tours toward food culture related activities. Majority of tour operators perceived food culture related activities also importance to their business and local tourism industry although some of it yet not aggressively be translated into positive action behaviour or part of their tour package. In other words, other than restaurant dining food culture related activities like visiting night market, food events and festival still not what could be called a must visit itinerary. Despite this, all those activities are gradually gaining demand and start to receive it popularity especially experiencing the Homestay program. Some of the tour operators admitted of bringing the international tourists to such program especially in sub-urban and rural areas. Some of the verbatim answers are as follows:

> *"Not to be pessimistic. Food related activities like visiting the night markets or pasar malam, food events still not our main activities or part of our main tour itinerary.. We have asked the tourists... besides dining at the Malaysian eating spot they prefer to visit the Homestay program compared to the night markets and food events"*
>
> *(Informant 3)*

> *"Even though food related activities are part of gastronomic products and seem important to the tourism industry, unfortunately not all them are include in our packages ... however some positive indication is there as some of the international tourists eager to learn such culture. Some of the groups asked for us to drop them at the China town"*
>
> *(Informant 8)*

"We still consider food related activities are important to our business even though those products are not our tour itinerary. But somehow, tourists visit street hawkers, pasar malam at Kampong Baru with their own accord and they enjoyed it. I guess these kinds of activities can be included in our tour itinerary and in fact we start to do so...... see how the feedback.... If they happy and no complaint about it..... Then why not ... most importantly money to our company"

(Informant 2)

5 IMPLICATION AND CONCLUSION

It is proven from this study that, tour operators clearly perceived the importance various food culture related activities like types of food festivals, night market, hawkers street stall, food exhibition, food bazaar including Ramadan Bazaar, Homestay and many others as part of the Malaysian gastronomic product. Some of the tour operators admitted that food culture related activity's importance to their business. In other words, although not as popular as restaurant dining activities, engaging on food related activities along with the uniqueness of vast Malaysian food and beverages is gradually receiving attention among the tour operators.

Bearing this point in mind, tour operators in this country especially those who directly deal with international tourists should increase their marketing strategy in promoting the local gastronomy products and incorporating food related activities in addition to the existing ones. To the researcher's knowledge, a few internationally known local tour agencies have started incorporating some food culture activities as part of their tourism activities or package itineraries. As a conclusion, this good indication should not be ignored but aggressively be utilized by tourism agencies, tour operators, government bodies that these products in addition to other tourism products able to generate and boost the economy not only to the operators but to all parties involved.

REFERENCES

Abdul Karim, N.I. (2012). Ramadan bazaar, international tourist's attraction and disseminating information behavior. Unpublished doctoral thesis, Universiti Teknologi Mara.

Aydin, S., & Boz, M. (2009). The importance of alternative types of tourism in the continuous development of the tourism industry: Antalya case. Retrieved February 1st, 2012 from http://turizm-pazarlama.com/?p=30.

Boniface, P. (2003). *Tasting tourism: Traveling for food and drink new directions in tourism analysis*. London: Ashgate Publishing.

Chaney, S., & Ryan, C. (2011). Analyzing the evaluation of Singapore's World Gourmet Summit: An example of gastronomic tourism. *International Journal of Hospitality Management*, *31*(2), 309–318.

Conner, M. & Armitage. C.J. (2002). The social psychology of food choice. Wallingford: CABI.

Frochot, I. (2003). An analysis of regional positioning and its associated food images in French tourism regional brochures. *Journal of Travel and Tourism Marketing*, *14*(3), 77–96.

Goeldner, C.R. & Ritchies, B.J. (2006). *Tourism: Principles, practices and philosophies* (10th Ed.). New Jersey: John Wiley and Sons, Inc.

Hall, M., & Sharples, L. (2003). *The consumption of experiences or the experience of consumption? An introduction to the tourism of taste*. Oxford: Butterworth-Heinemann.

Henderson, J.C. (2009). Food tourism reviewed. *British Food Journal*, *111*(4), 317–326.

Hjalager, A. & Richards, G. (2002). *Tourism and gastronomy*. London: Routledge.

Intrepid Travel. (2004). *Intrepid gourmet traveller.* Melbourne, Australia.

Jalis, M.H., Mohd Zahari, M.S., Zulkifly, M.I., & Othman, Z. (2009). Western tourists perception of Malaysian gastronomic products. *Asian Social Science Journal*, *5* (1), 25–36.

Jalis, M.H., Mohd Zahari, M.S., Zulkifly, M.I., & Othman, Z. (2007). Gastronomic products as tourist motivational factors: An opportunity to Malaysia. In proceeding of the 2007 Tourism Educators of Malaysia Conference.

Kivela, J. & Crotts, R. (2006). Tourism and gastronomy: Gastronomy's influence on how tourists experience a destination. *Journal of Hospitality and Tourism*, *3*(30), 354–377.

Mohamed, B. (2006). Image of Malaysia as an international destination. Retrieved July 17, 2007, from http://www.hpb.usm.my/tourism.

O'Halloran, R.M. & Deale, C.S. (2004, June). Food tourism: creating and positioning tourism supply chain. Paper and proceedings for Administrative Sciences Association of Canada, Laval City, Quebec Canada.

Santich, B. (2004). The study of gastronomy and its relevance to hospitality education and training. *International Journal of Hospitality Management*, *23*(1), 15–24.

Scarpato, R. (2002). *Gastronomy as a tourist product: The perspective of gastronomy studies*. London: Routledge.

Shenoy, S.S. (2005). Food tourism and the culinary tourist. Published doctoral thesis, Clemson University.

Tourism Malaysia. (2011). Events and festivals. Retrieved February 3rd, 2012, from http://www.tourism.gov.my/events/.

Tourism Malaysia. (2012). Events and festivals. Retrieved February 3rd, 2012, from http://www.tourism.gov.my/events/detail.php?id=1225.

Hospitality and Tourism – Sumarjan et al. (eds)
 ISBN 978-1-138-00151-0

Perceived leadership effectiveness on food handlers' spiritual intelligence in food service business

M. Saad & P.S. Toh
Universiti Teknologi MARA, Malaysia

M.A. Mohamed Adil
Academy of Contemporary Islamic Studies, Universiti Teknologi MARA, Malaysia
Accounting Research Institute, Universiti Teknologi MARA, Malaysia
International Institute of Advanced Islamic Studies, Malaysia

S.K. Ab Manan
Academy of Contemporary Islamic Studies, Universiti Teknologi MARA, Malaysia
Accounting Research Institute, Universiti Teknologi MARA, Malaysia

ABSTRACT: For centuries, scholars continue to discuss on how to success and sustain in good business. Thus, this study aims to examine the Perceived Leadership Effectiveness (PLE) connection on the food handlers' Spiritual Intelligence (SQ) in food service business. Data collected from 1,026 food handlers in 82 government-controlled food service businesses throughout Malaysia. Based on judgmental sampling, this cross-sectional study applied standard data analysis approaches of quantitative data. The quantitative findings revealed that the dominant participations were Muslims with moderate education background, engaged in the industry as non-permanent labourers with less than five years working experiences. Subsequently, simple regression analysis suggested that PLE had a significant relationship with the SQ dimension. While this study would discuss on the implication for theory and practice, it provides empirical evidence on individual differences that is likely to underpin sets of behaviours in food service industry and adds knowledge to the body of literature. This study limited to four criteria of integrated SQ scales (ISIS); holistic view, challenges, self-consciousness, and higher-consciousness, restricted to food handlers in the government-controlled of food service businesses and self-rated report. Replication of the criteria in ISIS or utilization of the other criteria may be beneficial.

Keywords: food service business, perceived leadership effectiveness, spiritual intelligence

1 INTRODUCTION

Being at a competitive edge, success in business should not solely be measured financially. Non-financial measurement is also significant as it could provide inspiring and meaningful working environment because employees are seeking beyond the economic reward from their job (Kinjerski & Skrypnek, 2004; Robert & David, 2005). One good example of non-financial measurement for success is having leadership effectiveness. Previous studies indicate that leadership effectiveness has an influence on the Spiritual Intelligence (SQ) of workers (Abdul Ghani, Naser, & Instsar, 2009).

In relating to this, SQ is a recent phenomenon concept that associates with how to success and sustain in good business. Especially in the stage of average organizational condition, the spiritual worker could experience and adapt great uncertainty by believing that things will turn out right for everyone towards responsible behaviour (Marques, 2006). The effective application of individual SQ at work is a stronger predictor of organizational effectiveness (Garcia-Zamor, 2003). Spirituality according to Emmons (2000) is intelligence due to five core abilities, namely: (a) Capacity for transcendence; (b) the ability to enter into heightened spiritual states of consciousness; (c) the ability to invest everyday activities, events, and relationships with a sense of the sacred; (d) the ability to utilize spiritual resources to solve problems in living; and (e) the capacity to engage in virtuous behaviors or to be virtuous (to show forgiveness; to express gratitude; to be humble; to display compassion) (p. 10).

Whilst research has demonstrated the importance of SQ, the food service operators are struggling in current business environment. Significant growth of food poisoning incident rates in 12 consecutive years has been increasing up to 25 times (Mazni, Toh, & Mohamed Azam, 2013). Apart from the aforementioned constraint, leaders need to address workers' potential that is their SQ, to achieve what they had aimed for the bottom line (Garcia-Zamor, 2003). Due to the interest to improve the food service business, this study proposed to determine the relationship between PLE and SQ.

2 LITERATURE REVIEW

2.1 *Leadership effectiveness*

Leadership effectiveness has become the prerequisite of management source to enhance the bottom line of the competitive business, in which in this study it would foresee the impact on the workers' spirituality level. At different levels of effectiveness, all leadership styles have a strong influence on performance, people, and managers at the organizational level (Mannion, 2009; McDermott, Kidney, & Flood, 2011; Syed Kamran et al., 2011). Full-range theory of leadership includes transformational, transactional, and laissez-faire leadership styles. Whilst leadership styles have been studied to administer the organizations effectively, transformational leadership and transactional are amongst the well-liked styles which integrate trait, behavioural, and contingency approach to attain organizational goals and predicted higher leadership effectiveness ratings (Abu Daud, 2009; Batool, 2013).

2.2 *Spiritual intelligence*

Spirituality can be classified as a form of intelligence to which the spiritual elements consist of: the research for and experience of sacred, meaning, higher-consciousness, and transcendence (Amram & Dryer, 2007; Tischler, Bibierman, & McKeage, 2002). SQ is an extension to the Emotional Intelligence (EQ) concept. According to George (2006), EQ is important to understand and control one's emotions and feeling, and at the same time it is subtle to the others' feelings. Conversely, SQ is the intangible inner attributes or deep awareness of oneself in which can be a ground for a decision-making based on logic and analysis. Therefore, SQ is believed to have higher intelligence compared to EQ (Tischler et al., 2002). Indeed, SQ is needed by modern organizations to renew themselves for a better management (Marques, 2006). Few empirical research such as Ayranci and Semercioz (2011) has linked the SQ with organizational context.

Ayranci and Semercioz (2011) studied Turkish top management's spiritual leadership, spirituality, and religiosity. The authors found that there were no common factors formed the three dimensions, and no statistical significant relationship was found between the spiritual leadership and the issues of spirituality and religiosity. However, their study revealed that there were positive relationships among the three dimensions.

A strong and positive relationship between spirituality in the workplace and job satisfaction measured from 100 employees in the insurance companies (Gupta, Kumar, & Singh, 2013). Gupta et al. (2013) suggested that the employer should look at more aspects to include the spiritual element than merely materialistic rewards for their workers.

In the study of finding "spirit at work," Kinjerski and Skrypnek (2004) collected data from 14 professionals through semi-structured interactive interviews and written open-ended surveys. The authors concluded the work environment provides a place to articulate creativity and total self when the "spirit at work" emerges at the individual and organizational level.

Western literature often debates that spirituality and religion are completely same phenomenon, overlapping, or totally in separately viewed (Fry, 2003; Gupta et al., 2013; Marques, 2006; Reave, 2005). However, Marque (2006) argued that spirituality as compared to religion drives human beings to higher awareness that interconnect to all living beings.

2.3 *Hypothesis testing*

In the quest for studying the success and sustaining in good business, it is crucial to examine leadership effectiveness influence on the workers' SQ. Leadership effectiveness and SQ are two distinct concepts, but immateriality derived from the spirituality may be incorporated into business (Ayranci & Semercioz, 2011). There were more than 150 studies that have indicated a "clear consistency between spiritual values and practices and effective leadership" (Reave, 2005, p. 655). However, the closer examination revealed that most the studies conducted were non-empirical in nature. The spiritual terms, in general, were related to leadership responsibility and type of leaderships, discussing the spirituality in leadership, or measuring leadership success in motivation, satisfaction, retention, follower ethics, and Organizational Citizenship Behaviour (OCB) (Brown, Treviño, & Harrison, 2005; Thompson, 2004; Wolf, 2004).

Empirically, the leadership effectiveness and workplace spirituality was attempted in the study at secondary schools in Northern Peninsular Malaysia (Abdul Ghani et al., 2009). Abdul Ghani et al. (2009) found that workplace spirituality aided leadership effectiveness and bring significant variance in teacher's satisfaction, commitment, and departmental performance.

In addition, most of the studies done viewed at the spiritual's elements from the Christianity perspectives. Adapting this study to the local setting with Islam being the largest practiced religion, the following hypothesis was posited:

H1: Leadership effectiveness is positively and significantly related to SQ.

3 METHODOLOGY

3.1 *Participants, measurement, and data analysis*

The unit of analysis for this study was individual operational food handlers working in 82 government-controlled food service businesses throughout Malaysia. Based on judgmental sampling, 1,640 questionnaires were distributed among food services. Six-point *Likert*-type scale measured each statement for the PLE and SQ items, where 1 represented "never or almost never" and 6 signified "always or almost always." The respondents were asked to indicate the effectiveness level of their immediate supervisor (PLE) and to self-report their level of SQ.

The approach of perceived effectiveness was used in past studies such as Lu and Yang (2010) and Filipe (2011) as leadership effectiveness has been measured by the followers' subjective evaluations (Reave, 2005). This study established questions items for the food handlers to perceive their food operator's leadership effectiveness. Five questions items were modified Bass and Avolio (1990) that found on Yukl (2008) leadership effectiveness survey. Meanwhile, 12 statements measured food handlers' SQ that have been adapted from Ayranci (2011). Out of 16 criteria, this study utilized only four criteria in the integrated SQ scales (ISIS) that achieved high Cronbach's Alpha. The food handlers' SQ was captured by the criteria of holistic view, challenges, self-consciousness, and higher-consciousness that have been adapted from Ayranci (2011). IBM SPSS software version 21 for Windows was used to analyze the data collected based on descriptive statistics and item total correlation analysis. Testing of the hypothesis was done using simple standard regression.

4 RESULTS AND DISCUSSION

4.1 *Profile of the respondents*

Of 1640 questionnaires, the response rate was 63%, which 1,026 questionnaires were returned and usable for the study. Table 1 shows the respondents' profile. The analysis of respondent's characteristics reported that female respondents slightly dominated the male by 10%. About 57% were below 30-years-old, and 75% had completed their secondary school as their highest education background. The majority respondents were Muslim (91%), in which encompassed 76% non-permanent food handlers and 69% of them worked at normal hours. Interestingly, more than 80% are experienced in the field; the majority of them had working experience lower than five years (81%). For minimum food safety requirements, more than 90% declared they had typhoid immunization and 98% agreed that the food handling course was useful for their work.

Table 1. Profile of respondents.

Demographic Variables	Categories	Frequencies	Percentage
Gender	Male	457	44.8
	Female	563	55.2
Age	<18 years	38	3.8
	18 to 30	539	53.6
	31 to 40	181	18.0
	41 to 50	159	15.8
	>50	89	8.8
Education	College/ University	75	7.4
	Secondary school	768	75.4
	Primary school	159	15.6
	Others	16	1.6
Religion	Muslim	931	90.7
	Others	86	8.4
Working Status	Permanent	244	24.3
	Non-permanent	761	75.8
Working Hours	<8 hours	101	10.0
	8 to 10 hours	698	69.2
	>10 hours	210	20.8
Working Experience	<1 year	325	35.2
	1 to 5 years	423	45.8
	6 to 10 years	128	13.9
	>10 years	48	5.2
Typhoid Immunization	Yes	929	90.5
	No	47	4.8
Food handling Course	Agree	976	97.7
	Disagree	23	2.3

Note: N = 1026; items are based on 6-point Likert-type scale (1 = "never" to 6 = "always").

4.2 *The effect of PLE on SQ*

Correlational analysis was used as initial examination on the relationships between the PLE and SQ. Table 2 gives details of the means, standard deviations, and intercorrelations among the two dimensions.

The summated means of all factors were above 5.0, indicating high practised of PLE and SQ. Specifically, food handlers seem to agree that leaders have high leadership effectiveness and consider themselves as having high SQ. The positive correlation ($r = 0.344$) obtained provides initial indication that there is a positive relationship between the two variables.

4.3 *Hypotheses testing*

Hypothesis 1 (H1) predicts that there is a linear relationship between the PLE and SQ. The regression estimates as shown in Table 3. Finding of the study indicates that there is enough statistical evidence to reject null hypothesis and accept the alternative hypothesis ($t = 11.67$; $p < 0.05$). Therefore, it can be concluded that there is a positive relationship between PLE and SQ.

The significant result replicates previous findings (e.g., Abdul Ghani et al., 2009; Tischler et al. 2002; that the leadership effectiveness has connected significantly with the SQ. Exhibited resilience through spirituality is a key aspect that has been described by the effectiveness of their leader in the workplace. The result seems implied that food handlers' human potential can be cultivated by emphasizing the role of the leadership effectiveness. On the other hand, current study revealed opposite outcome with Ayranci and Semercioz (2011), which they had found that the leadership effectiveness was negatively connected with the spirituality elements. Discrepancy may be due to the adapted SQ dimension with the local atmosphere in current study was perfectly matched with Muslim respondents.

Other important finding was that PLE only contributed about 12% whilst the means score of PLE and SQ demonstrated of high practice. The emergent results indicate that other factors might enhance the predictions of food handlers' SQ.

Table 2. Intercorrelations among study dimensions.

	Means	SD	1	2
Leadership effectiveness	5.2272	0.71757	–	0.344**
Spiritual intelligence	5.2702	0.67740	0.344**	–

Note: **Correlation is significant at the 0.01 level (2-tailed).

Table 3. Regression of PLE on SQ.

Coefficients[a]					
	Unstandardise coefficients				
Model	B	Std. error	Beta	t	Sig.
1 (constant)	3.572	0.147		24.311	0.000
	0.325	0.028	0.344	11.666	0.000

Note: [a]Dependent variable: SQ.
Adjusted $R^2 = 0.118$; $F_{1,1013}$ value = 136.085, ($p = 0.000$).

5 IMPLICATIONS OF THE STUDY FINDINGS

Paralleling leadership effectiveness to food handlers' SQ concludes that the effectiveness of the leadership plays an important role to determine the fate of their organization. Food safety remains a major public health responsibility, thus associating SQ to the current food service business environment could address the food handlers' motivational aspect in proactive behaviour towards good business practices. SQ dimension in the current study incorporates flexible statements, in which it contains universal values, including the religion of Islam. These flexible statements can add value to the SQ literature.

6 LIMITATIONS AND FUTURE RESEARCH

Self-report methods can gather behavioural factors and patterns of the practices including intelligence elements (Batool, 2013; Hawkshead, 2007), but this type of data collection may have been bias due to potential over-estimating adherence or self-awareness (Batool, 2013; J. George, Kong-David, & Steward, 2007), particularly when the unit of analysis is individuals. Due to this, dyad as the unit of analysis (manager-worker) is proposed to reduce the bias issue. Besides, the highly practised of the SQ may consider other 12 criteria of ISIS. This leads to some other possible results that future research could be carried out. Another matter worth studying is if further research could investigate current relationship based on other than government-controlled food service establishments in order to validate the results of this study.

7 CONCLUSION

In conclusion, the significant result in this study reflects the existential situation of relationship between PLE and SQ as expected. Empirically, this paper has determined that PLE is a fundamental for moral endeavor, especially for the SQ. To one extent, the SQ is gradually recognized, and at individual level, it may integrate into his or her work for a positive impact and meaningful working environment. Considering the SQ can be cultivated through questing, inquiry, and practice, food service operators should enhance the development of their PLE.

ACKNOWLEDGEMENT

The authors would like to express gratitude to RMI, Universiti Teknologi MARA for funding through the Excellent Fund (Dana Kecemerlangan 600-RMI-SSP/DANA 5/3DSP (367/2011)).

REFERENCES

Abdul Ghani, Kanesan Abdullah, Naser, Jamil alzaidiyeen, & Instsar, Turki aldarabah. (2009). Workplace spirituality and leadership effectiveness among educational managers in Malaysia. *European Journal of Social Sciences, 10*(2), 304–315.

Abu Daud, S. (2009). *Leadership theories, research and practices: Framing future leadership thinking*. Serdang, Selangor Malaysia: Universiti Putra Malaysia.

Amram, Yosi, & Dryer, D. Christopher. (2007). The development and preliminary validation of the integrated spiritual intelligence scale (ISIS) (pp. 1–45). Palo Alto, CA.

Ayranci, Evren. (2011). Effects of top Turkish managers' emotional and spiritual intelligences on their organizations' financial performance. *Business Intelligence Journal, 4*(1), 9–32.

Ayranci, Evren, & Semercioz, Fatih. (2011). The relationship between spiritual leadership and issues of spiritual and religiosity: A study of top Turkish managers. *International Journal of Business and Management, 6*(4), 136–149.

Batool, Bano Fakhra. (2013). Emotional intelligence and effective leadership. *Journal of business studies quaterly, 4*(3), 84–94.

Brown, Michael E., Treviño, Linda K., & Harrison, David A. (2005). Ethical leadership: A social learning perspective for construct development and testing. *Organizational Behavior and Human Decision Processes, 97*(2), 117–134. doi: http://dx.doi.org/10.1016/j.obhdp.2005.03.002.

Emmons, Robert. (2000). Is spirituality an intelligence: problems and prospects. *International Journal for the Psychology of Religion, 10*(1), 3–26.

Filipe, Jose Pereira Lamelas'Jose Antonio. (2011). Co-leadership in the tourism sector companies: increasing effectiveness and efficiency. *International Journal of Academic Research, 3*(4), 475–480.

Fry, Louis W. (2003). Toward a theory of spiritual leadership. *The Leadership Quarterly, 14*, 693–727.

Garcia-Zamor, J. (2003). Workplace spirituality and organizational performance. *Public Administration Review, 63*(3), 355–363.

George, J, Kong-David, CM., & Steward, K. (2007). Adherence to disease management programs in patients with COPD. *Int J Chron Obsruct Pulmon Dis, 2*(3), 253–262.

George, Mike. (2006). Practical application of spiritual intelligence in the workplace. *Human Resource Management International Digest, 14*(5), 3–5.

Gupta, Manu, Kumar, Vinod, & Singh, Mandeep. (2013). Creating satisfied employees through workplace spirituality: a study of the private insurance sector in Punjab (India). *Journal of Business Ethics*, 1–10.

Hawkshead, J. (2007). Techniques for measuring medication adherence in hypertensive patients in outpatient settings: advantages and limitations. *Dis Manag Health Out., 15*(2), 109–118.

Kinjerski, Val M., & Skrypnek, Berna J. (2004). Defining spirit at work: finding common ground. *Journal of Organizational Change Management, 17*(1), 26–42.

Lu, Chin-Shan, & Yang, Chung-Shan. (2010). Safety leadership and safety behavior in container terminal operations. *Safety Science, 48*, 123–134.

Mannion, K. (2009). Leadership for success. *Leadership & Organization Development Journal, 30*(7), 639–648.

Marques, Joan F. (2006). The spiritual worker: An examination of the ripple effect that enhances quality of life in-and-outside the work environment. *Journal of Management Development, 25*(9), 884–895.

Mazni, Saad, Toh, Poh See, & Mohamed Azam, Mohamed Adil. (2013). *Food handler's hygiene practices among government institutional training center in Northern Malaysia.* Paper presented at the Asian Journal of Environmental-Behaviour Studies.

McDermott, Aoife, Kidney, Rachell, & Flood, Patrick. (2011). Understanding leader development: learning from leaders. *Leadership & Organization, 32*(4), 358–378.

Reave, Laura. (2005). Spiritual values and practices related to leadership effectiveness. *The Leadership Quarterly, 16*, 655–687.

Robert, S. Kaplan, & David, P. Norton. (2005). The Balanced Scorecard: Measures That Drive Performance. *Harvard Business Review, 83*(7), 172.

Syed Kamran, Ali Shah, Iqbal, Jam Javed, Razaq, Adeel, Yameen, Muhammad, Sabir, Suleman, & Khan, Muhammad Asif. (2011). Influential role of culture on leadership effectiveness and organizational performance. *Information Management and Business Review, 3*(2), 127–132.

Thompson, L. (2004). Moral leadership in a postmodern world. *Journal of Leadership and Organizational Studies, 11*(1), 27–39.

Tischler, Len, Bibierman, Jerry, & McKeage, Robert. (2002). Linking emotional intelligence, spirituality and workplace performance: Definitions models and ideas for research. *Journal of Managerial Psychology, 17*(3), 203–218.

Wolf, E. (2004). Spiritual leadership: a new model. *Healthcare Executive, 19*(2), 22–26.

Yukl, Gary. (2008, April 2008). *The importance of flexible leadership.* Paper presented at the Society for Industrial-Organizational Psychology, San Francisco, CA.

Hospitality and Tourism – Sumarjan et al. (eds)
 ISBN 978-1-138-00151-0

Sensorial properties of spent hen chicken thigh marinated with nano and non-nanostructured *Zingiber officinale* Rosc (ginger) rhizome powder

N. Abdullah & N. Abdullah
Universiti Teknologi MARA, Malaysia

M.R. Mahmood
NANO, Science Tech Centre, Institute of Science, Universiti Teknologi MARA, Malaysia

ABSTRACT: The incorporation of nanoparticle in food systems for food quality has been established. The market continues to forecast that this technology has potential for commercialization. However, very limited data regarding the sensory properties and consumer acceptance of nanofood in literature. Thus, in this study the hedonic sensory analysis which consists of appearance, aroma, flavor, juiciness, tenderness and overall acceptability on spent hen chicken thigh marinated with nanostructured and non-nanostructured (micron; submicron) ginger rhizome powder were conducted. The marinated samples were compared with untreated sample (control). Nanostructured ginger rhizome marinated chicken thigh scored the highest mean score ($p < 0.05$) for all attributes tested whereas control sample, micron and submicron sample gained the significantly lower mean score. However, the sensory evaluation score was within the acceptable limits. Thus, size reduction ginger rhizome powder into nanoscale was more preferable compared to other samples. This study implicates the application of nanotechnology as a mechanism to improve the quality of food. Moreover, it also adds to the body of knowledge in the food and food-service literature.

Keywords: consumer acceptance, ginger, nanoparticle, sensory properties

1 INTRODUCTION

Consumption of poultry and poultry products has increased rapidly, resulting in a higher export value compared with other meats. The production of layer hens for egg demand has become a problem, with an excess supply of spent hens. Approximately 144 million spent hens from the poultry industry are available for processing each year in North America (Lyons & Vandepopuliere, 1996). The direct use of meat from spent hens causes problems because its texture is very tough due to high collagen content with increased age thus the toughness prevents their use in whole meat and reduces the market value (Li, 2006). However, meat from spent hen is a good protein source (Lee, et al., 2003; Rhee, et al., 1999). Globally, ginger is consumed as a food spice and obtained predominantly from the rhizome of *Zingiber officinale* Rosc. Traditionally, ginger was widely used herb and being utilized as food ingredient during food preparation and also consumed with beverages. This herb considered as a major constituent for the diet. Various in vitro and in vivo studies have been conducted on the pharmacological potential of ginger. Its medicinal properties against digestive disorder and lower the glucose level (Rani et al., 2013) have been well documented. Moreover, in Western countries, ginger being taken to treat atherosclerosis, ulcer, migraine and depression (Shukla & Singh, 2007). As an herb that's good for health, the application of ginger in variety of food also gaining an interest. There is numerous evidence demonstrating the ability of ginger to enhance the sensory properties in meat during storage. Ginger extract treated smoked buffalo showed better in appearance, color and juiciness compared to control (Anandh & Lakshmanan, 2012). Additionally it being reported that *lower shear force values were observed in ginger extract treatment yak meat and goose meat* (Gao et al., 2011).

Size reduction technology particularly nanotechnology is an emerging biomaterial processing technology that produces a narrow and uniform size distribution with greater dispersity and solubility property. On the other hand, nanotechnology improved product shelf life, increased absorption;

protect quality, freshness and safety of food. However, up to date, the data on sensory properties and consumer acceptance on nanofood relatively sparse in the scientific literature. There are also limited researches in nanotechnology that has been performed in food and food-related products and the global development of nanofood is indeed on its initial stage. Nevertheless, it is doubted whether the application of nanotechnology in ginger powder can offer greater advantages than non-nanostructured ginger powder. Therefore this study initiated to investigate the effect of nanostructured ginger rhizome on sensory properties and its acceptance. The data gained can be used as indicator on the acceptability of nano-food among consumer.

2 MATERIALS AND METHODS

Fresh ginger (*Zingiber officinale* Rosc.) rhizome variety 'Bentong' was obtained from Bentong, Pahang, Malaysia. Voucher specimens (SK 2049/12) of these plants were deposited at the Herbarium Institute of Bioscience, University Putra Malaysia. The sample weights before and after drying were noted. The dried ginger rhizome obtained was ground using a food processor (Panasonic, 176, China) for 5 min, sieved and namely as micron ginger whereas the submicron ginger was prepared by milling using a hammer mill at 2890 rpm and sieved using 250 μm sieve. On the other hand nanostructured ginger rhizome was prepared by introducing 0.5 grams of ginger powder to the planetary ball mill device (Retsch, PM 200) with 2 mm zirconia beads. The milling time was set for 4 hours at 550 rpm which was the optimum ball mill time for ginger rhizome powder to show a strong antioxidant effect according to researchers previous finding (Norhidayah et al., 2013).

Spent hen chicken thighs (aged over 69–75 weeks of laying cycle) were obtained from local market in Shah Alam area. The purchased chicken thighs were personally carried to the experimental lab in the ice box. Chicken thighs were received free from visible blood and in the ranged from 100 to 130 g of weight and 6.0 to 6.2 in pH. Any remaining visible fat was physically removed and rinsed to remove any contaminant. Meat samples were used immediately or shortly after purchase. Test marinades were based on Yusop et al. (2012) which consists of 3% marinade solution (3 gram in 100 mL of distilled water). The mixture homogenized using an ultrasonic processor (Hielscher, UP400S) for 15 minutes and kept in an airtight bottle before use. Altogether, 50 chicken thighs were allocated to five groups (10 chicken thighs per group). The first group was individually marinated with 10 mL of distilled water (control), while the other three groups were marinated with 10 mL of micron, submicron and nanostructured size solution per 100 g chicken thigh respectively. The sample was turned occasionally to ensure even marinating. The samples were subsequently grilled at 170°C in a conventional oven (Fagor Tchnology) for 30 min. All test samples were cooked at the same time and segregated to prevent any mixing.

Table 1. Sensory terms of consumer evaluation of marinated chicken thigh.

Attributes	Description
Appearance	1 = extremely dislike, 9 = extremely like
Aroma	1 = extremely dislike, 9 = extremely like
Flavor	1 = extremely dislike, 9 = extremely like
Juiciness	1 = extremely dislike, 9 = extremely like
Tenderness	1 = extremely dislike, 9 = extremely like
Overall acceptability	1 = extremely dislike, 9 = extremely like

Sensory evaluation was conducted using 50 consumers which consist of students and staff of University Teknologi MARA, Shah Alam, Malaysia. Consumer sensory analysis was conducted in the laboratory sensory panel booths university. The intensity for each attribute was rated on a 15 cm scale labelled with numbers. The numbers indicate the weak intensities on the left and stronger intensities on the right. Panelists were asked to mark the score for sensory property's base of 6 attributes (see Table 1). The cooked marinated samples were presented to the panelists along with water to clear their palates between samples. The presentation order of all samples in both sessions was randomized.

The sensorial properties were presented as mean and standard deviation. One-way analysis of variance (ANOVA) was used to analyze the data using SPSS statistical software (version 21 SPSS Inc., USA). The means were compared with Duncan's Multiple Comparison Test (DMCT) and $p < 0.05$ was considered as statistical significance.

3 RESULTS AND DISCUSSION

The effect of marination with nano and non-nanostructured ginger rhizome on sensorial properties were summarized in Table 2. Generally the mean score for appearance, aroma, flavor, juiciness, tenderness and overall acceptability ranged from 5.6 to 7.3; 5.56 to 6.86; 5.28 to 7.66; 4.82 to 7.32; 5.10 to 7.08; 5.54 to 7.6 respectively. Organoleptic parameters indicated that the samples marinated

Table 2.

	Mean scores			
Attributes	Control	Micron	Sub-micron	Nano
Appearance	5.60 ± 1.12^{c}	$5.94 \pm 1.38^{b,c}$	6.12 ± 1.18^{b}	7.30 ± 0.93^{a}
Aroma	5.56 ± 1.51^{b}	6.14 ± 1.29^{b}	6.10 ± 1.53^{b}	6.86 ± 1.31^{a}
Flavor	5.28 ± 1.31^{c}	5.90 ± 1.59^{b}	6.00 ± 1.69^{b}	7.66 ± 0.93^{a}
Juiciness	4.82 ± 1.91^{c}	5.54 ± 1.67^{b}	5.6 ± 1.55^{b}	7.32 ± 1.13^{a}
Tenderness	5.10 ± 1.93^{b}	5.72 ± 1.44^{b}	5.56 ± 1.62^{b}	7.08 ± 1.35^{a}
Overall acceptability	5.54 ± 1.76^{b}	6.14 ± 1.53^{b}	6.14 ± 1.49^{b}	7.60 ± 0.86^{a}

Values are expressed as mean ± standard deviation (n = 3) for control, spent hen thigh marinated with micron; submicron and nanostructured ginger (*Zingiber officinale* Rosc.) rhizome.
Values marked by the different superscript letters (from "a" to "c") within a column denote statistically significant differences ($p < 0.05$).

with nano and non-nanostructured ginger rhizome was in the range of dislike slightly to like moderates.

In most cases, sensory properties were found to be affected by marinade material used. Regardless the particle size of material use, it was found that higher mean score value marked for ginger marinated sample compared to control sample with significantly higher ($p < 0.05$) value showed by nanostructured marinated sample in almost all attributes tested.

Panelist rated sample marinated with nanostructured ginger rhizome having more preferable appearance and aroma whereas no difference showed by the one marinated with micron, submicron and control samples. It might be contributed by a enhanced the released of volatile compound which mainly oleoresins due to nanonization process that promoted to pleasant aroma and flavor. The nanotechnology crushed the cell membranes; therefore, most of the active components associated would be at or near the particle surface, leading to faster release. Whereas, larger particles have large cores which allow more drug to be encapsulated and slowly diffuse out (Manharaj & Chen, 2006). Essential oils and oleoresins found in ginger rhizome responsible for the characteristic flavour and spiciness of ginger (Singh et al., 2008).

Consumers tend to prefer the juiciness provided by nanostructured ginger rhizome marinated sample while the control sample gained the lowest score. It seems that marination with nanostructured ginger rhizome capable to hold better moisture content compared to control sample. It may contributed by the unique characteristic offered by nanostructured ginger rhizome itself which is larger surface area as compared to other sample. There was a significant ($p < 0.05$) increased of water holding capacity in ginger treated goose meat goose meat treated in all treated samples compared to control which lead to greater tenderization property (Gao et al., 2011).

The highest mean score obtained initiated to greater tenderness preference. The higher the juiciness, the tenderer the meat is. In this case, sample marinated with nanostructured ginger rhizomes considerably more juicy and tender compared to others which lead to highest score for overall acceptability. In contrast, due to lower rate given to control sample for almost all attributes tested, initiated to the lower acceptability rated.

4 CONCLUSION

The use of nanostructured ginger rhizome as marination ingredient enhanced the sensorial properties of spent hen chicken thigh. Nanostructured ginger rhizome marinated chicken thigh possessed higher rate of appearance, aroma, flavor, and preferable in term of juiciness and tenderness which lead to greater overall acceptability. Thus, production of nanostructured ginger rhizome could be considered in order to produce meat or meat product that has desirable sensorial quality.

ACKNOWLEDGEMENT

The authors are grateful to the Ministry of Higher Education Malaysia for the scholarship and Ministry of Science, Technology and Innovation (MOSTI) Malaysia for grant (06-01-01-SF0390).

REFERENCES

Anandh, M.A., & Lakshmanan, V. (2012). Storage stability of smoked buffalo rumen meat product treated with ginger extract. *Journal of Food Science and Technology*, 1–6.

Balestra, F., Cocci, E., Pinnavaia, G., & Romani, S. (2011). Evaluation of antioxidant, rheological and sensorial properties of wheat flour dough and bread containing ginger powder. *LWT—Food Science and Technology, 44*(3), 700–705.

Gao, H., Zhou, G., Zeng, J., Ma, H., Pan, R., & Yu, X. (2011). Tenderization of goose meat by papain, pineapple juice and ginger juice treatment. *Advanced Materials Research,* 236–238.

Lee, S.O., Min, J.S., Kim, I.S. and Lee, M. (2003). Physical evaluation of popped cereal snacks with spent hen meat. *Meat Science, 64*, 383–390.

Li, C.T. (2006). Myofibrillar protein extracts from spent hen meat to improve whole muscle processed meats. *Meat Science, 72*, 581–583.

Lyons, J.J. & Vandepopulierc, J.M. (1996). Spent Leghorn hens converted into a feedstuff. *Journal of Application Poultry Research*, 5, 18–25.

Norhidayah, A., Noriham, A., & Rusop, M. (2013). Changes in Physical and Antioxidant Properties of Nanostructured Zingiber officinale (Ginger) Rhizome as Affected by Milling Time. *Advanced Materials Research, 667*, 144–149.

Rhee, K.S. & Ziprin, Y.A. 2001. Pro-oxidative effects of NaCl in microbial growth-controlled and uncontrolled beef and chicken. *Meat Science, 57*, 105–112.

Shukla, Y., & Singh, M. (2007). Cancer preventive properties of ginger: A brief review. *Food and Chemical Toxicology, 45*(5), 683–690.

Yusop, S.M., O'Sullivan, M.G., Preuß, M., Weber, H., Kerry, J.F., & Kerry, J.P. (2012). Assessment of nanoparticle paprika oleoresin on marinating performance and sensory acceptance of poultry meat. *LWT-Food Science and Technology, 46,* 349–355.

Zieliński, H., Ciesarova, Z., Troszyńska, A., Taglińska, A., Zielińska, D., Amarowicz, R. Kukurova, K. (2012). Antioxidant Properties, Acrylamide Content and Sensory Quality of Ginger Cakes with Different Formulations. *Journal of Food and Nutrition Science, 62*(1), 41–50.

Hospitality and Tourism – Sumarjan et al. (eds)
© 2014 Taylor & Francis Group, London, ISBN 978-1-138-00151-0

Impact of nutritional information on menu towards food intake

M.F. Mohd Faiz, S.A. Talib & Z.N. Nadiah
Universiti Teknologi MARA, Malaysia

ABSTRACT: Consumer nowadays demands for nutritional information of food products before making a purchase. This leads to the importance of disclosing nutritional information by the food companies and it has been a topic of study by several researchers. Despite the importance, the restaurateurs opposed to the motion of mandatory nutritional menu labeling to be legislated. This article discussed the importance and the use of nutritional label on restaurant menus in combating obesity.

Keywords: menu nutritional labeling, nutritional knowledge, nutritional labeling, nutritional label usage

1 INTRODUCTION

The World Health Organization had stressed on the degree of dryness of obesity problems and nominated obesity as one of the top major contributors in World Health Problem (WHO, 1998). Although there had been countless campaign and awareness program to educate people to eat right, the obesity problem keep on escalating into the danger zone. Further worsen the situation, Malaysia now is the "fattest" country in the South East Asia, claims Dr Seri Liow Tiong Lai (Cheng, 2013). This scenario keep escalate even though the government and the non-government organizations violently fighting off obesity. Malaysia had shockingly 14% of the increment of obese population compared back to the 2006. This shocking revelation of statistic also includes over 2.6 million of adult while 477,000 children under 18 were categorized as obese in Malaysia in 2012. In relation with the increment of obesity cases in Malaysia, the number of Non-Communicable Disease (NCD) such as diabetes, cancers and cardiovascular were on a rise towards critical level. The health related problem always had a close connection with food intake.

1.1 *Food labeling*

One of the many means to improve the consumer food intake consumption is through the utilization of nutritional labels on food label. Food labels comprise the information about the ingredients used, expiry date, manufacturer and nutritional information (Ab-Rahman, Wan-Ahmad, Mohamad, & Ismail, 2011). Thus, nutritional label becoming a crucial tool to educate the consumers about the healthiness and the implication from consuming the foods. Nutritional Information is the disclosure of the nutritional content of the respective food product. Most of the country in the world had legislated that it is mandatory to process and packed food to disclose the information of the nutritional content on the label. In Malaysia, the Food Regulation 1985 legislated that it is compulsory to disclose the nutritional information in each proceed food. Nayga, Lipinski and Savur (1998) assert that the objective of nutritional labeling was to make the information regarding the nutrient content available to the customer so that in hope it will somehow will assist the customer to make healthier choice and dietary practices.

Nutrition labeling has emerged as an important aspect of consumers' food purchase decisions. Nutritional content in food products is considered to be credence attribute (Drichoutis, Lazaridis, & Nayga, 2006). However, if the degree of trustworthiness of the nutritional labels were high enough, it can be functioning as a search characteristic by the consumer (Grunert, F-Clemin, Wills, Bonsmann, & Nureeva, 2010). Study shows that consumers use nutritional information when making a purchasing decision that it is depending on the level health consciousness and also the awareness to consume healthily (Loureiro, Gracia, & Nayga, 2006). Plus the majority of the respondents revealed that they are willing to pay something additional for the nutritional information provided on food items.

Consumer nowadays seeks and demands nutrition information on the product for aiding them in making the purchase decision. In contrary, the restaurant industry opposed the motion for legislating mandatory menu labeling despite the importance of nutritional information to consumers is undeniable as it provides a guide for the consumer during the purchasing phase (Christina et al., 2009).

1.2 *Food habits*

Physical inactivity and poor food intake has been identified as the major factor contributing towards the prevalence of obesity. The poor food intake would lead to over consumption of calories, sodium and also fat. Food intake usually dictated by many determinants that make it very difficult for nutritionist to tackle this problem. Such factors that made up the food intake are the consumer culture, income-level, level of education and lastly yet importantly the food habits should be given the spotlight to be rectified and altered.

Food habits are determined by the social character that describes the formation of the shared character of a society. People of a society or a social class determined by their way of life and the socially typical expectations and functional requirements regarding socially adaptive behaviour. The food habits are essentially adaptive to the dominant mode of production in a society (Fromm, 1942). Interestingly, food habits are developed during the optimal growth of a normal human being which is during childhood and adolescence period.

In the recent times, Malaysian food habits had been negatively impacted by the globalization and the urbanization that happened rapidly throughout the country. Malaysian nowadays prefer to eat outside of the home.

According to Noor (2002) this phenomenon could be linked to the increasing cost of living that had forced most of Malaysian adapted to the changes. The increasing number of dual income family had been contributed to the changing of the food habits. Hence, due to the hectic lifestyle, consumer food habits shifted from eating home food to eating outside from home. This claim is supported by the report by National Restaurant Association (2008) where there is a positive relationship between the consumer spending and the rising of obesity.

In line with the notion, consumer tend to consume food that contains higher calorie, sodium and fat when dining out compared when dining at their home. Pittman (2012) wrote an article in Reuters New York about how kids and adolescences tend to consume about 200 calories more when eating out either at a fast food restaurant or full service restaurants to be compared with when they dining at their home. This event is something that need to look as something serious as the consumer would be spending a lot of their time in the restaurant or a food and beverage outlets up to 4–5 times a week.

Further supporting the above statement, Burton, Howlett and Tengari (2009) have conducted a study where they found out that very frequent restaurant diners face difficulties in maintaining or losing weight. Situation happed because the average consumer has no to very little knowledge pertaining the total calories and the nutrient content of the meal served in the restaurant. In the same study (Burton, Howlett, & Tangari, 2009) asserts that the consumer failed to consume healthier meals due to the lack of nutrient knowledge and catalysed with the absence of the information. Customarily, it publicly perceived that only fast food or quick service restaurant is the main culprit towards the inclination of the obesity cases. However on the contrary, full service restaurant also have contributed to the overconsumption of calories, sodium and fat. This is because most of the full service restaurants usually offer a menu that containing high calorie, fat and sodium.

In relation with the above issue, more consumer now more concern about the nutritional content of the food that they consume. Thus, this behavioural change leads to the creation of demand for nutritional labeling on the restaurant's menu. Although there is no legislation on disclosing the nutritional information on the food establishment's menu, the number of restaurants that includes the informational information had been increasing ever since. This shows that the demand for nutritional information keeps elevating each day and the customer aware and concern regarding the nutritional content of the food they consumed. A study done by Hwang and Lorenzen (2008) assert that with the disclosure of the information on restaurant menus would portray a healthier outlook of the restaurant compared to the restaurant that do not disclose the information on the menu.

2 NUTRITIONAL KNOWLEDGE

For the past few decades, there has been an increase in the consumer interest and their knowledge of nutritional issues (Shine et al., 1997). But sadly, the overall consumer failed to execute and apply the knowledge in daily consumption (National Consumer Council, 2003). In addition, consumer that had a good understanding of the nutritional knowledge prone to make a healthier food choice and facilitate them to follow a healthy diet (IGD, 2003).

2.1 *Types of nutritional knowledge level*

According to Blaylock et al. (1999) knowledge on nutrition can be grouped into two distinctive characteristics. The first group of nutritional knowledge is the knowledge on the general nutritional concept (fat, carbohydrate, protein). The second group of the nutritional knowledge is specific knowledge where the consumer able to demonstrate further detailed knowledge of the nutrient content of foods, such as calories, sodium and fibre. In the same study, the author had established that the nutritional knowledge had a positive relationship towards the usage of the nutritional label during the information search before pre-purchase.

3 ISSUES

As evident from past research, by disclosing the nutritional information on the menu would encourage the consumer to make a healthier and balance meal whilst dining at a restaurant or food service outlet. Although there were some research on this topic that had been conducted in other places, none of that could really reflect the Malaysian's environment. Most of the past research had been conducted in developed countries such as the United States and United Kingdom. In addition, Malaysia is a multiracial and developing country and it cannot be as certain that a similar result will occur using the Malaysian context.

Other than that, past research only looked into the quick service restaurant not the full service restaurant although full service restaurant also offers food that is high in fat, sodium and calorie. Therefore, there is a gap that exists regarding the situation in Malaysia point of view. Thus, it is crucial to conduct a study using Malaysia as the study setting.

REFERENCES

Ab-Rahman, A., Wan-Ahmad, W.I., Mohamad, M.Y., & Ismail, Z. (2011). Knowledge on Halal Food amongst Food Industry Entrepreneurs in Malaysia. *Asian Social Science*, *7*(12), 216–221.

Acumen (2012). Regulation of Nutrition Labeling & Claims in Malaysia. Retrieved from http://www.acumen.com.my/content/restriction-hazardous-substance-rohs?q=node/9.

Blaylock, J., Smallwood, D., & Kassel, K. (1999). Economics, food choices, and nutrition. *Food Policy, 24*, 269–286.

Burton, S., Howlett, E., & Tangari, A.H. (2009). Food for thought: How will the nutrition labeling of quick service restaurant menu items influence consumers' product evaluations, purchase intentions, and choices? *Journal of Retailing, 85*(3), 258–273.

Christina A., Roberto, M.S., Marlene B., Schwartz, Kelly D., & Brownell. (2009). Rationale and Evidence for Menu-Labeling Legislation. *American Journal of Preventive Medicine, 37*(6), 546–551.

Fromm, E. (1942). Character and the Social Process, *Appendix to Fear of Freedom*, Routledge.

Grunert, K.G., Fernández-Celemín, L., Wils, J.M., Bonsmann, S.S.G., & Nureeva, L. (2010). Use and understanding of nutrition information on food labels in six European countries. *Journal of Public Health, 26*, 261–277.

Hwang, J. & Lorenzen, C.L. (2008). Effective nutrition labeling of restaurant menu and pricing of healthy menu. *Journal of Foodservice, 19*(5), 270–276.

IGD (2003). "*The key to a healthier diet*." Retrieved from www.igd.com/.

Loureiro, M.L., Gracia, A., & Nayga, R.M.J. (2006). Do consumers value nutritional labels? *European Review of Agricultural Economics, 33*(2), 249–268.

National Restaurant Association (2008). Americans' Dining-out Habits, retrieved from http://www.restaurant.org/research/indglance.cfm.

Pittmen, G. (2012, November 5). *Eating in restaurants tied to higher calorie intake.* Reuters. Retrieved from www.reuters.com.

Shine, A., O'Reilly, S., & O'Sullivan, K. (1997). Consumer use of nutrition labels. *British Food Journal, 99*(8), 290–296.

Hospitality and Tourism – Sumarjan et al. (eds)
© 2014 Taylor & Francis Group, London, ISBN 978-1-138-00151-0

Factors influencing students in pursuing tourism courses: A case of Management and Science University (MSU)

S. Azaddin
Universiti Teknologi MARA, Malaysia

ABSTRACT: The high growth of inbound tourists to Malaysia has led the government, industry and students' interest in tourism for the post-secondary education level. The development of tourism industry had directed the students to believe that the tourism industry is very large and there are a lot of career opportunities and career growth. In view of the significance in this field several public and private universities, polytechnics and colleges are gearing up to offer the hospitality and tourism programs at various levels starting from Diploma, Undergraduate, Postgraduate and Doctor of Philosophy. The programmes or courses offered to students by the relevant educational institutions are very competitive and aim to produce graduates who are experts in the field of management especially in term of tourism and planning, management, development and marketing. This study will focus on factor influences students in selection of tourism courses in private university. As to seek and gather information for this study, a quantitative approach will be taken by distributing questionnaires to selected number of Management Science University students. The results of the study are useful for the running of the courses in order to increase student's intake as well as to create a great workforce for the tourism industry.

Keywords: factor influences, student, tourism courses

1 INTRODUCTION

The high growth of tourism industry in Malaysia has generated awareness to students, industries and institutions on the importance of the tourism industry that had contributed great economic and social growth to Malaysia as well as in providing high volume of career opportunity and career growth for graduates. In view of the significance of this field, several public and private universities as well as polytechnics and colleges are gearing up to offer the hospitality and tourism programs at various levels starting from Diploma, Undergraduate, Postgraduate and Doctor of Philosophy. Student selection of this program is influenced by various factors. According to Lee, Yuan, Hwang and Kim (2012), selecting a program for one's academic pursuit is an iterative, complex, and multi-factorial process that involves a wide range of influence. Once students try to make an academic or career related decision, rationality comes first. Students make decisions logically based on what they know and how they feel (Brown, 2006). Numerous studies on Tourism Education pertaining to student's motivation, preferences and satisfaction towards university and programs offered by the university were previously conducted (Lee, Kim & Lo, 2008; Nasser, Khoury & Abouchedid, 2006; Petruzzelling & Romanazzi, 2010). To date, there are a few studies pertaining to factors influencing students in choosing HTM courses in a particular university. In Malaysia, there are some articles relevant to this area of study. However there has been only limited research on factor influence student choosing tourism courses particularly in higher private institution. From the previous research they focus on motivational factors influence students choosing hospitality programs that include: 1) Self-actualization 2) job opportunity 3) field attractiveness 4) ease of study. However factor of university marketing activities that influence student choosing tourism courses is not being explored. Thus the main purpose of this research is to fill this gap.

This study is a useful source of information to the institutions that it may help in enhancing the curriculum development by putting into consideration the factor influence student to choose tourism course. Thus it will generate the quality learning outcomes and student satisfaction tourism by enrolling tourism courses and create a greater workforce for the tourism industry.

2 LITERATURE REVIEW

2.1 *Tourism courses in Management and Science University*

Tourism has been seen by both the public and private sectors as having enormous potential for growth and consequently have significant potential for employment generation. This is because the tourism sector had continuously become the forefront of Malaysia's economic development. Such sustainable and high-yield sector will continue to drive the Malaysia's economy, providing income and job opportunities to the local communities. With a strong global tourism position today, Malaysia is positioned as the 9th most visited countries in the world and received up to RM1 billion receipts per week from foreign visitors (Economic Transformation Program, 2011).

Realising the advantage and strong position in the tourism industry globally, the growing numbers of public and private higher institutions offering Tourism Program could be seen. In a way, the growth of the hospitality and tourism education can be considered as the movement effect of the growth of the tourism industry. Management and Science University (MSU) is one of the higher learning institution that offer the hospitality and tourism program which began as a university college (KUTPM) in 2001. After a full university status was conferred in October 2007, the university adopted a new name—Management and Science University (MSU). The full university status is an important milestone achieved by MSU as it demonstrates its grown to prominence in the tertiary education sector in Malaysia.

According to Prof. Dato' Wira Dr. Mohd Shukri Ab. Yajid (2012), the 30 years MSU grew from just a college to a Tier 5 Excellent rated University. This rating, accorded twice to MSU by the Ministry of Higher Education of Malaysia has helped propelled MSU to be as one of Malaysia's best. He also said that MSU will give priority to all graduates the employability; as such they will focus on the industry relevant courses and ensuring that these courses meet current trends. MSU programs offered are either approved or accredited by the Malaysian Qualifications Agency (MQA). MSU offer hospitality programs at two levels comprise of Diploma and Bachelor in Hospitality and Tourism Management under the Faculty of Business Management Professional Studies.

2.2 *Student decision consideration in program selection*

Tourism education programs in Malaysia are growing enormously. Many of higher learning institutions from both public and private institution offers various tourism programs that specialize in similar or different areas that comprise of hotel, tour and travel, event management and marketing. However student choosing the best course can be difficult decision as well as thinking about their own future career. Some students know exactly what course to pursue while others do not know which one will get them there (Korir & Wafula, 2012). Students make decisions logically based on what they know and how they feel (Brown, 2006).

Pertaining to the selection of the Tourism programs, it is coming with various factors influence students to choose the course. The selection of this course is influenced by several factors such as self-actualization, job or career opportunity, university marketing activities, ease of study and so on. According to the results of survey for study perception of hospitality and tourism students towards study motivation and preferences, the five motivational factors of student study HTM that include of self-actualization, job opportunity, field attractiveness, ease of study and scholastic achievement (Lee et al., 2009). Nevertheless (O'Mahony et al., 2001) stated that the student choice of the HTM was based on the reputation of the HTM Program and the availability of a particular course of study. They focus on three major motivational factor including: 1) student's knowledge and interest in the hospitality industry, 2) the influence of their parents, career counsellors and peers, and 3) their experiences as a customer in the hospitality industry.

2.3 *Self-actualization*

The need for self-actualization refers to the need to develop one's potentialities to their fullest (Mayo & Jarvis, 1981). Self-actualization portrays a human's quest for choice and self-determination as well as spiritual values. Putting it in personal terms, self-actualization is the developmental process of becoming or actualizing the "real you" (Wolfe & Sisodia, 2003). According to the result of a survey, among of the six motivational factor, job opportunity and self-actualization were top two motivators for student choosing Hospitality and Tourism courses. Notable themes indicated by students within self-actualization include of the perception that the hospitality and tourism industry is interesting, students enjoy serving others, student perceive their study upon graduation to be sufficient to match the demands and requirements of the hospitality and tourism industry (Lee et al., 2010).

2.4 *Job or career opportunity*

Job opportunity has been seem the most influence factor of student choosing the hospitality and tourism courses. In studies of perceptions of hospitality and tourism students towards study motivations and preferences (Lee et al., 2008) concluded that job opportunity in the hospitality and tourism industry was the most important factor for Hong Kong students. According to Berkita and Howard

(2007), students believed that the hospitality industry was growing and provided employment opportunities. Furthermore pertaining to the study concluded that African American student pursuing careers in hospitality management regardless of academic major finds career in the hospitality industry to offer opportunities for advancement. Besides that the results reported hospitality students believe the employment opportunities in this field. Overall hospitality students appeared to have a positive view of the hospitality industry and appeared to be aware of opportunities available to them within the industry.

2.5 *Field attractiveness*

The attribute of attractiveness has gained considerable controversy. Numerous researchers have contended that the attractiveness of an individual means that he or she is attributed with several positive attributes such as likable, friendly, intelligent, successful and competent (Fatt, 2000). However in a destination perspective the attractiveness reflects the feelings and opinions of its visitors about the destinations perceived ability to satisfy their needs. In other words the destination attractiveness refers as a perceived ability of the destination to deliver individual benefits. This ability is enhanced by the attributes of a destination.

2.6 *University marketing activities*

According to Kotabe and Helsen (2010), marketing is essentially the activity, set of institutions, and processes for creating, communicating, delivering and exchanging offerings that have value for customers, clients, partners, and society at large. However in the context of Hospitality marketing define as a continuous, sequential process through which management in the hospitality and travel industry plans, researches, implements, controls and evaluates activities designed to satisfy both customers' needs and wants and their own organization's objectives. To be most effective, marketing requires the efforts of everyone in an organization and can be made more or less effective by the actions of complementary organizations (Morrison, 2002). In order for the private university to achieve their sales and increase student enrolment they need marketing activities to achieve their target. The university marketing activities that for instance direct mail, online marketing, promotions, advertising, publicity, flyers and many more.

According to Russell (2005), higher educational institutions need to maintain or develop a distinct image to create a competitive advantage in an increasingly competitive market. Thus, this will encourage the university becoming more aggressive in their marketing activities and need to be clear about their positioning and the image they wish to convey to the public. An appropriate marketing mix can be developed which not only takes into account environmental factors but also an identification of student needs and the ability of the institutions to meet those needs.

3 ISSUES

Tourism has been seen by both the public and private sectors as having enormous potential for growth and consequently have significant potential for employment generation. This is because the tourism sector had continuously become the forefront of Malaysia's economic development. Such sustainable and high-yield sector will continue to drive the Malaysia's economy, providing income and job opportunities to the local communities. With a strong global tourism position today, Malaysia is positioned as the 9th most visited countries in the world and received up to RM1 billion receipts per week from foreign visitors (Economic Transformation Program, 2011). Realising the advantage and strong position in the tourism industry globally, the growing numbers of public and private higher institutions offering Tourism Program could be seen. In a way, the growth of the hospitality and tourism education can be considered as the movement effect of the growth of the tourism industry.

The growth of several public and private educational institutions that offers tourism programs increases with the restructure of their programs that are in line with the changes in the economy and government priorities. Most universities offer various categories of programs at a number of level including certification programs that specializes in hotel and travel operations, diplomas level that emphasizes the supervision of the hotel and travel operations, the bachelor level that offers managerial aspects of hotel and travel operations, masters level that present the advanced managerial aspects and doctorate level that caters to those aspire to be an academician.

However, the questions arise on how students choose these courses. What are the factors that influence student in choosing Tourism Courses in that particular university? Management and Science University offers various Hospitality and Tourism Management programs that comprises of diploma and degree level. Admission requirements for these courses is set up with minimum criteria. The issue stated here is whether students graduate from this university is really qualified. This can be seen through the minimum criteria to enter for this course either for diploma or degree programs.

4 CONCLUSION

The development of tourism education in Malaysia can be considered as the movement effect of the growth of the tourism industry. In view of the significant, there is a high growth of educational institutions offering tourism courses. Nowadays pertaining to the selection of the tourism courses, it is influenced by several factors such as self-actualization, job/career opportunity, and ease of study. Thus, the research on the factors influencing students in choosing tourism courses without any doubt will contribute to the significance in both aspects of practical and academic perspective.

The contribution for practical aspect, this study may act as a guideline to private universities and colleges on how to increase students' enrolment as well as ways to improve the services and the quality of the courses offered to students. Furthermore, this study will also provide new knowledge and information for the university itself and other higher education as well as various organizations and individuals such as the governments, corporations, privates and students.

From the academic perspective, there is a lack of research conducted in this area, particularly in investigating the factors influencing students in choosing the tourism courses offered by a particular university. Therefore, this research may contribute to a new knowledge and extending the body of literature in tourism education related study. Furthermore these studies will likely lead to other potential researchers to look more in-depth or broader scope related to tourism education.

REFERENCES

Berkita S.B., & Howard S.C. (2007). Exploring the factors that influence African American students to pursue careers in the hospitality industry, *7*(1).

Brown, R.V. (2006). Making decision research useful-not just rewarding, judgement and decision making, *1*(2), 162–173.

Buku Syarat Kemasukan UiTM (2013). http: Retrieved from www.esyarat.uitm.edu.my on Sunday 16 June 2013.

Creswell J.W. (2003). Research design. qualitative, quantitative and mixed method approaches. Thousand Oaks, CA: Sage.

ETP Handbook Chapter 10 Tourism (2011). http:www.etp.pemandu.gov.my. Retrieved on 16 June 2013.

Fatt J.P.T (2000). Attractiveness and outcomes of job interviews. *23*(12), 11–18.

Goldsmith, A. & Zahari, S. (1994). Hospitality education in Malaysia: Filling the skill gap. *International Journal of Contemporary Hospitality Management*, *6*(6), 27–31.

Insttution Profile. http:www.mohe.gov.my. Retieved on 17 June 2013.

Jr Harry, N.B & Boone, D.A. (2012). Analyzing Likertdata. *Journal of Extention,* (50) 2.

Korir, J & Wafula W. (2012). Factors that influence career choice hospitality students in Moi University, Kenya. *Journal of Education and Practice*. *13*(14), 83–90.

Kotabe, M. & Helsen, K. (2010). Global marketing management (5th Ed). John Wiley & Sons. Inc.

Lee, K.W., Yuan J., Hwang J.S., & Kim H.S. (2012). Doctoral students selection intention in Hospitality and Tourism Management (HTM) programs in the United States. *Journal of Hospitality, Leisure, Sport & Tourism Education*, *11*, 140–150.

Lee, M.J., Olds, D.A., & Lee, C. (2012). Why Students Choose A Hospitality and Tourism Program: A Pilot Study of U.S. Undergraduate Students. *Journal of Hospitality and Tourism Education*. *22*(3), 20–26.

Management and Science University (2012)."From The President Office". http: www.msu.edu.my. Retrieved on June 16 2013.

MSU Faculty of Business & Professional Study (2011). http:www.course_klass E-Learning Management and Science University. Retrieved on June 16 2013.

Malaysia tourist profile by selected market report (2010). Tourism Malaysia, 1–65.

Morrison A.M. (2002). Hospitality and travel marketing (3rd ED). Delmar. Thomson Learning.

Nasser, N.R, Khoury B., & Abouchedid, K. (2007). University students knowledge of services and programs in relation to satisfaction. *Journal of Quality Assurance in Education*, *16*(1), 80–97.

O'Mahony, G.B, Mc William's A.M., & Whitelaw, P.A. (2001). Why student choose a hospitality degree program: An Australian case study. Cornell Hotel and Restaurant Administrative Qualitative.

Petruzzellis, L. & Romarazzi, S. (2010). Educational value: How students choose university. *International Journal of Educational Management*. *24*(2), 139–158.

Russell M. (2005). Marketing Education. A review of service quality perceptions among international students. *Journal of Contemporary Hospitality Management*, *17*(1), 66–77.

Thitthongkam, T. & Wals, J. (2011). Tourism education at the tertiary level and competitive advantage: A comparison between Thailand and Malaysia.

Wolfe D.B., & Sisodia R. (2003). Marketing to the self-actualizing customer. *Journal of Consumer Marketing*, *20*(6), 555–569.

Hospitality and Tourism – Sumarjan et al. (eds)
 ISBN 978-1-138-00151-0

Assessment of food safety knowledge, attitudes and practices among food handler at childcare centre

Z.T. Muhamud & C. Tamby Chik
Universiti Teknologi MARA, Malaysia

ABSTRACT: Foodborne illness closely related to the exposure of harm resulting from consuming contaminated food and beverage. The top factors contribute to the foodborne illness include; i) poor personal hygiene, ii) time/temperature control, iii) cross contamination, and iv) contaminated sources. Foodborne illness endangers the children more than other population ages because of their early development of antibody. In Malaysia, there are increasing numbers of children being sent to childcare centre by their parent because the need of both spouses to work, due to increasing cost of living. Childcare centre acquire the responsibility to educate, provide food and taking care of the children. However, improper food handling increase risk to the production of unsafe food which can cause illness to the children. Previous study has identified food handler practices become the keys factor to foodborne illness occurrence. The knowledge, attitude and practice of the food handler at the childcare centre are the variables that wish to examine in this study.

Keywords: childcare centre, children, food safety, food handler

1 INTRODUCTION

Unsafe food endangers people's health. Symptoms such as vomiting, diarrhoea, fatigue and stomach cramp due to consuming contaminated food disrupt daily activities of the person (Scallan, Griffin, Angula, Tauxe, & Hoekstra, 2011). However, there are population that faces higher risk to food contamination such as pregnant woman, elderly, under medicine and also children (Center for Disease Control, 2013). They become risk population due to their lower level of antibody to fight with harmful bacteria that contaminated their meal. Children are one of the risk populations that highly risk suffer from unsafe food (Jahan, 2012). Children are highly exposed to the harm of foodborne illness, this is because (American Academy of Paediatrics, 2009) listed children as vulnerable to harm of unsafe food because they is still developing their immune systems which reduce the ability for the body to fight infection. Children usually have lower weight compare to adult that make only little amount of pathogens to sickens them besides that they have limited control over their diet such as related to food safety and they also have lower stomach acid production that decrease their capability to kill harmful bacteria that may contains in their food (Kathy & Jelley, 2003). This makes the children unable to fight with the harmful bacteria that contaminated their food. Besides that, children also have very lower level of knowledge about food safety (Kang, Kim, Kim, & Ha, 2010). This show that the children are highly dependable to the person who preparing the food; it can be the parent, the food caterer at school or commercial, and also the cook at the childcare centre if the children are sent to the setting during the day.

According to Malaysia Department of Welfare (2012), the total registered and hold valid licence of childcare centre in Selangor according to the districts is Shah Alam (59) centres, Gombak (30), Klang (12), Hulu Langat (52), Kuala Selangor (40), Hulu Selangor (22), Sabak Bernam (5), Petaling (47), Subang Jaya (51), and Sepang (23) (Ibrahim, 2006) Children are sent to childcare centre due to both working parents, with childcare setting facilities both parent can focus on their working commitment. (Ibrahim, 2006) also added that rising of cost of living enforce the parents to let the responsibility to raise children at early stage pass to the childcare provider, and it become a responsibility and trustworthy that childcare setting cares about. Balasundaram (2012) mentioned that there also needs for more numbers of childcare centre to support working mother and stabilize their household income due increasing cost of living. It become the reason of the need for working of both parent help the family to better funding the living expenses (Gordon & Berry, 2004). Childcare centre provider offer variety of services such as all day care, transit service and workplace based child care centres, community based childcare centre, institution based childcare centre and home based childcare centre (UNICEF, 2010). The diversification of childcare centre provide option to the parent, they are able to choose centre that fit to their needs and requirement. Working parents need to rely the trust and responsible to care; provide meal and education for their children to the childcare centre.

The responsible of preparing food at the childcare centre is the food handler. Food handler plays an important role in determining safety of food consumes, this is because (Nee & Sani, 2011) found that mistakes in food handling usually occur when serving contaminated food resulting from inadequate cooking, heating or reheating of food from unsafe sources, inadequate cooling process and also holding time longer than four hours of the food. Paez and Ortiz (2011) revealed that food handler is a critical point to determine safety of food serve. Practices of personal hygiene such as hand washing procedure contribute to the safety of food prepared. Food and Drug Administration (2009) identify risk factor and behaviour that leads to foodborne illness such as improper time and temperature management, poor personal hygiene, contaminated equipment and cross contamination. The control of those factor reduce the risk of unsafe food, the identification of the knowledge, attitudes and practices of the food handler towards the food safety first need to determine before future action can take place. The understanding on the food handler will help determine precaution and aids plan to improve and ensure the safety of food serve to the children.

2 LITERATURE REVIEW

Childcare defines as the action or skill of looking after children (Willoughby, 2006). Additionally, (Encylopedia of Children Health, 2012) also defines childcare as day-care refers to the care provided for infants and toddlers, preschoolers, and school-aged children, either in their own homes, in the home of a relative or other caregiver or in a center-based facility. Besides that, (Willoughby, 2006) distinguished several types of childcare centre such as the care of children by a day-care center, babysitter or other provider while parents are working. Referring to (Child Care Centre Act 1984) provided four types of childcare facilities such as work place based childcare centres, community based child care centres, institution based child care centres and home based child care centres.

Child Care Centre Act 1984 has stated issue on realizing the need for setting minimal standards and regulations for child care services following the report on the survey on child care (Ministry of Social Child Care in Malaysia 33 Welfare Services, 1983), the government enrolled the Child Care Centre Act (Act 308) in 1984. For example, in the act it is required for institution-based child care centres to register, but not home-based child care centres which could receive up to nine children.

Food contaminations are possible to occur either at commercial or institutional foodservice. Study by Sanlier and Konaklioglu (2010), found that contamination of food can happened from the production and if not get a serious attention can lead to foodborne illness. Foodborne illness occurs when food become contaminated and not cooking according to safety practices such as hygiene and inadequate cooking temperature. Besides that Nee & Sani (2011) also found that foodborne illness disease occurs with several factors which are improper storage or reheating, inappropriate storage system and cross contamination.

Howes, McEvan, Griffith & Harris (1996) notion that 97% of foodborne ilness outbreak cause by improper practices of food handler. This additionally happened because of food handler believe that money saving on operation is more important than the needs of proper safety practices (Griffith, 2010). To provide structure to factor contributing to foodborne illnes, (World Health Organization, 2000) identify inadequate heat treatment, inappropiate storage of foods, infected handler and cross contamination as major factor to foodborne illness.

Effect of not compliance with food safety practices has been identify by Griffith, Livesey and clayton (2010) as declining of quality, from business perspectives foodborne illness significantly brought damage towards brand identity, this situation lead to financial loses and possibility cause loses and bankruptcy. Besides the reputation and loses, children are also exposing to risk cause by mishandling and foodborne illness.

Gilling (2001) found that food safety practice only able to obtain by an organization if the employees embrace a positive behaviour with adequate knowledge and good attitudes. Therefore, Clayton and Griffith (2004) studied that is a crucial to identify and determine detailed safety practices in the catering kitchen.

3 ISSUE

The need for data collection activity to improve foodborne illness surveillance and food contamination has been highlighted by World Health Organization (2004). This enable proper documentation and evaluation of the burden of the food borne disease can be made. Study about foodborne illness among children at childcare centre also one of the branch that acquire the attention to be study.

Children are highly risk population and can be affected by unsafe food. Infants and young children have not yet fully developed their immune and digestive systems. Children younger than five years are at a higher risk of death from diarrhea than older children and adults, with infants at the highest risk of death (Paez & Ortiz, 2011) besides that Infants and children are likely to become dehydrated more quickly from diarrhea and vomiting because of their smaller body size (National Digestive Disease Information Clearing House, 2012). The food handler that prepare the food believe has significant contribution to the occurence of foodborne illness among children.

In a business organization Griffith, Livesey and Clayton (2010) stated that food safety must included in business objectives and need a proper compliance followed by an accessible documentation. In order to reduce public health risk from unsafe food. There are the arising needs for effectice food safety management control (Ramnouth, Driver, & Vial, 2008).

There still scarce of information and study about childcare centre compare to other type of foodservice establishment, (Lin & Sneed, 2007) studied about employee knowledge, attitudes and practices of food safety in University canteen, whereby (Nee & Sani, 2011) studied on the employee food safety knowledge, attitudes and practices at residential collages and canteen. There are needs for more literature on food safety knowledge, attitudes and practices at childcare centre in Malaysia. The study is important in order to contribute deeper understanding besides extending the existing literature about food at childcare for the benefit of children food safety.

4 CONCLUSION

Childcare centre become an important institutional service for Malaysian as general, it specifically help the working parents to take care of their children. The need of both parents to work is because of the increasing cost of living, besides increasing number of parent with higher education requires them to work. The responsibility of taking care of the children is passing to the childcare provider. Notice that children are highly vulnerable to harm from physical or microbial sources due to their development phases of antibody, the environment and food served at childcare centre contribute to the children's health as they spend about half of their day life at the centre.

The foods served to the children are prepared by food handler, so that the food handler's knowledge, attitudes and practices on food safety hold a significant influence to the safety of food serve. Factors such as poor personal hygiene, poor understand of foodborne illness, lack of time and temperature control, cross contamination and improper sanitizing identified contribute to unsafe food. The food handler knowledge, attitudes and practices towards the safety of food in childcare centre is very crucial to determine and become a field of interest to explore.

REFERENCES

Balasundaram, H. (2013). Family-friendly workplaces give children and business a boost. Story From the Field. University of Cambridge Press. Cambridge. Retrieved from http://www.admin.cam.ac.uk/univ/childcare/student/nonuniv1.html.

Center for Disease Control and Prevention (CDC). (2013). Health practices on cruise ship: training for employees transcripts. Foodborne Illness. Retrieve from http://www.cdc.gov/nceh/vsp/training/videos/transcripts/foodborne.pdf.

Clayton, D.A. & Griffith, C.J. (2004). Observation of Food Safety Practices in Catering Using Notation Analysis. *British Food Journal, 106*(3), 211–227.

Department of Statistic Malaysia, (2010). Food poisoning. Compendium of Environment Statistic. Retrieved from http://www.statistics.gov.my/portal/download_Environment/files/Compendium_2010/06-BAB5.pdf.

Department of Social Welfare Malaysia, (2012). Senarai taska berdaftar dibawah akta taman asuhan kanak-kanak 1984 di negeri selangor. Retrieved from http://www.jkm.gov.my/index.php?option=com_content&view=article&id=925%3Asenarai-taska-selangor-2&catid=147%3Asenarai-taska&Itemid=475&lang=en#.

Encyclopaedia of Children's Health, (2013). Day Care. Retrieved from http://www.healthofchildren.com/D/Day-Care.html.

Gilling, S.J. (2001). Is your HACCP dragging you down? A psychological perspective. *Food Science and Technology, 15*(3), 44–8.

Gordon, J.R. & Berry, K.S.W. (2004). It takes two to tango: an empirical study of perceived spousal/partner support for working women. *Women in Management View. Emerald Group Publishing, 19*(5), 260–273.

Griffith, C.J., Livesey, K.M., & Clayton, D. (2010). Food safety culture: The evolution of an emerging risk factor? *British Food Journal, 112*(4), 426–38.

Howes, M., McEwan, S., Griffiths, M., & Harris, L. (1996). Food Handler Certification by home study: Measuring changes in knowledge and behavior. *Dairy, Food and Environmental Sanitation, 16*, 737–744.

Ibrahim, Z. Malaysian National News Agency Creche, babysitter Or Housemaid? Parents' Dilemma Features. June 09, 2006, 12:38 PM. Retrieve from http://www.bernama.com/bernama/v3/printable.php?id=202481.

Jahan, S. (2012). Epidemiology of Foodborne Illness, Scientific, Health and Social Aspects of the Food Industry. Retrieved from http://www.intechopen.com/books/scientific-health-and-social-aspects-of-the-food-industry/epidemiology-offoodborne-illness.

Jappelli, R. & Marconi, N. (1997). Recommendations and prejudices in the realm of foundation engineering in Italy: A historical review. In Carlo Viggiani (ed.), *Geotechnical engineering for the preservation of monuments and historical sites; Proc. intern. symp., Napoli, 3–4 October 1996*. Rotterdam: Balkema.

Johnson, H.L. (1965). Artistic development in autistic children. *Child Development, 65*(1): 13–16.

Kang, N.E., Kim, J.H., Kim, Y.S., & Ha, A.W. (2010). Food safety knowledge and practice by the stages of change model in school children. *Nutrition Research and Practice*, *4*(6), 535–540.

Kathy, L. & Jelley, M.Ed. (2003). Health & safety for caregivers. Preventing Food Borne Illness in the Childcare Environment. OHIO State University. Retrieved from http://ohioline.osu.edu/hsc-fact/pdf/HSC5.pdf.

Lin, S.Y. & Sneed, J. (2007). University foodservice employees food safety knowledge, attitudes, practices, and training. Retrieved From http://fsmec.org/wp-content/uploads/2011/09/LinSneed22005.pdf.

Musa, M., Jusoff, K., Khalid, K., Patah, M.O.R.A., Anuar, J., & Zahari, H. (2010). Food Borne Illness Risk Factors Assessment in UiTM Shah Alam, Malaysia. Retrieved from http://www.idosi.org/wasj/wasj8%287%2910/13.pdf.

National Digestive Disease Information Clearinghouse (NDDIC). (2012). Foodborne Illness. Retrieved from http://digestive.niddk.nih.gov/ddiseases/pubs/bacteria/index.aspx.

Nee, S.O. & Sani, N.A. (2011). Assessment of knowledge, attitudes, and practices (KAP) among food handlers at residential colleges and canteen regarding food safety. *Sains Malaysiana,* 403–410.

Paez & Ortiz (2011). Food Safety Practices of Foodservice Employees.

Ramnouth, M., Driver, F., & Vial, B. (2008). Food Safety Management in the Fish Industry in Mauritus: Knowledge, *Attitude and Perception. British Food Journal, 110*(10), 989–1005.

Sanlier, N. & Konaklioglu, E. (2010). Food Safety knowledge, attitudes and food handling practices of student. *British Food Journal, 114*(4), 469–480.

Scallan, E., Griffin, P.M., Angulo, F.J., Tauxe, R.V., & Hoekstra, R.M. (2011). Foodborne illness acquired in the United States—unspecified agents. *Emerging Infectious Diseases, 17*(1), 16–22.

United Nation Children's Fund (UNICEF). (2010). Establishment of a childcare centre at workplace by corporate and business employers. Best business practice circular.

Wee, T.C. Childcare in crisis in Malaysia. Straits Times Indonesia. March 9, 2012, 12:02 PM. Retrieved from http://www.thejakartaglobe.com/archive/childcare-in-crisis-in-malaysia/.

World Health Organization (WHO). (2000). The WHO Surveillance Programme for Control of Foodborne Infections and Intoxications in Europe: 7th Report (1993–1998). Retrieved from www.who.int/foodsafety/publications/foodborne_disease/dec2000/en/.

World Health Organization (WHO). (2004). Foodborne disease monitoring and surveillance systems. FAO/WHO Regional Conference on Food Safety for Asia and Pacific. Seremban, Malaysia. Grove, A.T. 1980. Geomorphic evolution of the Sahara and the Nile. In M.A.J. Williams & H. Faure (eds), *The Sahara and the Nile*: 21–35. Rotterdam: Balkema.

Willoughby, M. (2006). A parent guide to the childcare (Pre-school services) regulations 2006. Bornados National Children's Resources Centre.

Hospitality and Tourism – Sumarjan et al. (eds)
© 2014 Taylor & Francis Group, London, ISBN 978-1-138-00151-0

Food hygiene knowledge and attitude amongst food handlers toward safe food practices in school canteens

A. Ena Arzairina, R.S. Raja Saidah & A.A. Saidatul Afzan
Universiti Teknologi MARA, Malaysia

ABSTRACT: Reducing the risk and occurrence of foodborne illness is a priority for the food service industries. This provides an overview on food hygiene knowledge and attitude amongst food handlers toward safe food practices in the school canteen. Food handlers play an important role in preventing foodborne illness and meeting the objective of serving harmless. However, food handlers were found not applying good practices of hygiene level of acceptance of the standard food hygiene procedure. Therefore, this study will provide information and guide the food handler's to practice the proper standard of food hygiene procedure on site.

Keywords: attitude, food hygiene, knowledge, practices, school canteen

1 INTRODUCTION

Foodservice is an industry which got a broad and various types of establishment. Foodservice divided into two types which is commercial and non-commercial (Palacio & Theis, 2009). Commercial is a profit oriented foodservice which is they are selling things to get a lot of profit. As for example, restaurants, supermarkets, delis, convenience store and other retail food establishment. Noncommercial foodservice is a non-profit oriented food service. It is sometimes called as institutional or on-site. This type of food service is non-profit because the business will be operated by educational, government or other organization which is not into gaining profit but more to giving or serve food to a place like military, hospital and school canteen.

Schools are key settings that will provide children with any health promotion programs that planned to make the children start healthy dietary and physical activity behaviours (Cullen & Watson, 2007). Currently, health is one of the main concerns and a key public health and food security issue. School meals are important because they are one of the most obvious instruments for policy intervention in children's diet. However, another problem which is remaining unresolved that is food poisoning at the school canteen (Hanan & Aisha, 2012). Food safety is essential for school foodservice operations serving young children to prevent foodborne illnesses.

2 PROBLEM STATEMENT

In Malaysia food borne disease has increase evidently from 2006 to 2008 (Health Facts, 2008; Ministry of Health Malaysia, 2009). Furthermore, food poisoning recorded the highest incidence rate of communicable disease in the food and water borne disease category (44.18 per 100,000 populations). The food poisoning at school still occurs until now which is the new one is in Klang (The Star Online, 2013). U.S. Food and Drug Administration (2004) found that the majority of food borne illnesses can be attributed to five risk factors: food from unsafe sources, inadequate cooking, improper holding temperatures, contaminated equipment, and poor personal hygiene. Decreased public awareness, food contamination, cross-contamination, improper temperatures during storage and food preparation, inadequate food handling and poor personal hygiene are among the most important reasons that cause food-borne diseases.

Research on food borne illness risk factors indicates that most outbreaks associated with food service establishments can be attributed to food workers' improper food preparation practices (Hertzman & Barrash, 2007). The objective of this study is to evaluate literature from various journals, article related to food safety knowledge and attitude based on the Health Belief Model (HBM) and safe food practice. Therefore, this study is to explore the food hygiene knowledge and attitude amongst food handler's practices towards school canteen and how far they can accept the food hygiene level in school canteens in order to discover the possible factors that contributes to the weakness in food handling.

3 LITERATURE REVIEW

3.1 *Food hygiene practices*

Ministry of Health Malaysia (2010) stated that there has been an increase in the incidence rate of foodborne diseases from 2005 to 2008. Although in 2009 the incidence rate are decreased but there have been a death due to foodborne disease. Root causes of these outbreaks were traced to poor food handling practices, such as cross contamination of raw and cooked products, slow cooling and inadequate refrigeration of foods, and poor worker hygiene from, for example, a failure to wash hands when handling ready-to-eat foods (Todd, Greig, Bartleson, & Michaels, 2007).

Hands are the medium or tools for every food handler to making a food. It also contains some pathogen or bacteria which can affect people who consume it. If a food handler did not practice washing hands very often, a consumer might get food-borne disease. Food handlers in schools have the responsibility of ensuring the production of safe foods, and their knowledge, attitudes and practices play a major role in the occurrence of food poisoning cases (Angelillo, Viggiani, Greco, & Rito, 2001; Patil, Cates, & Morales, 2005). In previous studies, food handlers were identified as the main cause of food contamination (Campos, Cardonha, Pinheiro, Ferreira, Azevedo, & Stamford, 2009). Bas, Ersun and Kivanc (2004), the staff employed in food and beverage services should have a clean, tidy and proper appearance, without any skin infections, good dental hygiene, have short finger nails and are not in the habit of biting nails, do not wear jewellery except wedding ring, wearing no make-up, work in clean shoes and uniform, and stick to good hygiene practices.

3.2 *Food safety knowledge*

Food safety knowledge among food handlers is vital in any food service establishment no matter what circumstances take place. It is important to expose all food handlers with knowledge and education through training where it might increase food handlers' knowledge, attitudes, and practices when handling or get in contact with food (CDC, 2011, 2009). In recent years, food poisoning cases have been increasing (Moon, 2000). Consumers can be alert with this current news through newspaper, internet and many other sources which deliberately has become serious issues nowadays (Isara & Isah, 2009). These issues should be concerned because food poisoning does not only harm the body but it is proven that it could lead to death (Gostin & Stewart, 2011). According to Abhay, Naveeta, Gargi, Ramchandra, Ajay and Vasant (2010) states that education must be considered as the only effective ways of influencing food handling practices at a working area.

Therefore, food business operators shall make sure that food handlers are supervised and instructed and are trained in any food hygiene activity (Toh & Birchenough, 2000). Most food establishments neglect the importance of training and understanding of food safety among food handlers (Chow & Mulan, 2010). Besides that, food handlers itself does not have the knowledge and zero understanding about food safety, that causes and effect of poor food safety practices towards food, and why and how the food can be contaminated (Burkitt, 2011). Promoting awareness of food safety by giving education and training can improve food handlers performance by getting the benefits from the training itself (Ministry of Education Malaysia's, 2004).

3.3 *Attitude towards behaviour*

Food safety is the responsibility of every person who is involved in foodservice to handle food. Many studies pinpoint the need for training and education of food handlers in public hygiene measures due to their lack of knowledge on microbiological food hazards, temperature ranges of refrigerators, cross contamination and personal hygiene (Bas et al., 2004; Nel, Lues, Buys, & Venter, 2004). However, in some previous studies show no differences between the staff who attended an educational course with those who did not (Angelillo et al., 2001; Askarian et al., 2004). This statement was supported by several studies (Howes, McEwen, Griffiths, & Haris, 1996; Powell, Attwell, & Massey, 1997) and it shows that although training may increase the knowledge of food safety, but it might not always produce a positive change in food handler attitudes.

Social cognition models from the realm of health psychology have been frequently posited as an important tool in improving both prediction and intervention research in safe food handling (Rennie, 1995). One such model is the Health Belief Model (HBM) (Rosenstock, 1974); which considers barriers and benefits of engaging in safe food handling as well as how severe food poisoning is seen to be and the degree of susceptibility to the illness. The HBM consists of 5 variables that interrelate that is perceived susceptibility, perceived seriousness, perceived benefits of taking action, barriers of taking actions and cues to action.

Perceived susceptibility is a belief that one is at risk of an illness that is subjective. As for example, a food handler belief that the food will not be affected if using a same cutting board for chicken and vegetables. Perceived seriousness refers to how serious the individual believes the consequences of developing the condition are (Redding, Rossi, Rossi, Velicer, & Prochaska, 2000). An individual is more likely to take action to prevent food poisoning if she/he believes that possible negative physical, psychological, and/or social effects resulting from developing the disease will give serious consequences.

Perceived benefits of taking action are a deciding on a course of action is shaped by the options accessible to the individual and the belief in their effectiveness. For example, individuals who are not convinced that there is a causal relationship between improper food handling and diarrhoea are unlikely to handle food properly because they believe that anything else can be a reason for the disease such as chemical at vegetables. Barriers of taking action are a belief that has been established by a particular course of action that may reduce a health threat, indecision may still take place. Belief alone is not enough to motivate an individual to act.

The stimulus that triggers action may be internal or external. This may be internal such as physical discomfort, or external such as a message communicating the seriousness of a disease. The external is most relevant to communications as it often relies on media and interpersonal interaction (Mattson, 1999). For example, when perceptions of susceptibility and severity are high, a very minor stimulus may be all that is needed to initiate action (Redding et al., 2000). The health belief model can observe the attitude of food handler which will become their barrier and also benefit to make a safe food handling. Individuals will demonstrate behaviour to both take action and avoid it (Lizewski & Maguire, 2010).

4 FINDINGS

In Nee and Norrakiah (2011) study shows that the knowledge level of food handlers at the two residential cafeterias and canteen of the Universiti Kebangsaan Malaysia was moderate, with a mean value of 57.8%. The respondents have good knowledge on personal hygiene where 93.9% answered are correct. Manning and Snider (1993) reported that 81% of their respondents are aware of the importance of hand washing, but only 2% observe washing their hands thoroughly. More than half of the respondents (73.9%) answered with the correct option which indicates that they realized food prepared without proper handling may contribute to the risk of food-borne illnesses (Nee & Norrakiah, 2011). They also stated that questions about time and temperature control were correctly answered by most of the respondents (61.2%), but 30.0% of them did not give answer correctly while 8.0% did not know the answer.

Noor Azira, Mohammad Faiz, Shuhaimi, Rukman and Malina (2012) reported that the majority of the food handlers have positive attitudes on most of aspects such as clean working area, hand hygiene and food preparation. Nevertheless, about 30% of the respondents either agree or are uncertain about wearing jewellery and watch while handling food. Almost 40% of the respondents disagree or are uncertain about preparing raw and cooked food separately using the same kitchen utensils like a cutting board or knife to prepare food can cause cross contamination between raw and cooked or ready to eat food. Nee and Norrakiah (2011) reported that almost 55.4% of the respondent agreed with the statement that food service staffs with abrasions and cuts on the hands should avoid from unwrapping foods. However, in the previous study by Clayton et al. (2002), food handlers might be aware of the food safety attitudes they should have, but 63.0% of their respondents admitted that they rarely practice such positive attitudes. This proves by in Noor Azira et al. (2012), about 83% of the respondents agree or are uncertain about the refreezing defrosted food.

Respondents in this study showed good practices when 75.4% of them wash hands after using the toilet and only 60.0% washed their hands thoroughly with soap and hot water before handling foods (Nee & Norrakiah, 2011). Many of the previous studies proved that it is crucial to practice self-hygiene especially hand hygiene because hand is the major agent that transmit microorganisms and intestinal parasites to foods (Aarnisalo, Tallavaara, Wirtanen, Maijala, & Raaska, 2006). Approximately, 66.5% of the respondents in study Aarnisalo et al. (2006) produced good personnel hygiene practices. Angelillo et al. (2001) and Askarian et al. (2004) stated that using of gloves is mainly influenced by employees' ages where younger workers seem to be more motivated in preventing risk practices compared with older workers. Noor Azira et al. (2012) also report that the use of gloves is also not being practiced during food preparation due to Malaysian hot climate which can cause discomfort if worn for a long period. In the aspects of food safety, answers provided by the respondents indicated that the level of their practices was average, in which the overall percentage was 59.3% (Nee & Norrakiah, 2011).

REFERENCES

Aarnisalo, K., Tallavaara, K., Wirtanen, G., Maijala, R., & Raaska, L. (2006). The hygienic working practices of maintenance personnel and equipment hygiene in the Finnish food industry. *Journal of Food Control*, 17, 1001–1011.

Abhay, B.M., Naveeta, K., GargiAbhay, M., Ramchandra C.G., Ajay, K.D., & Vasant V.W. (2010). Health status and personal hygiene among food handlers working at food establishment around a rural teaching hospital in Wardha District of Maharashtra, India. *Journal of Global Journal of Health Science*, 2(2), 190–206. Retrieved June 16, 2013 from www.ccsnet.org/gjhs.

Angelillo, I.F., Viggiani, M.A., Greco, R.M., & Rito, D. (2001). HACCP and food hygiene in hospital: Knowledge, attitudes and practices of foodservices staff in Calabria, Italy. *Infection Control Hospital Epidemiology*, 22, 1–7.

Askarian, M., Kabir, G., Aminbaig, M., Memish, Z., & Jafari, P. (2004). Knowledge, attitudes, and practices of food service staff regarding food hygienein Shiraz, Iran. *Infection Control Hospital Epidemiology*, 25, 16–20.

Bas, M., Ersun, A.S., & Kıvanc, G. (2004). The evaluation of food hygiene knowledge, attitudes and practices of food handlers in food businesses in Turkey. *Journal of Food Control*, 17, 317–322.

Burkitt, L. (2011). Why china struggles with foodsafety. Retrieved June 23, 2013 from http://www.wsj.com/chinarealtime/2011/04/25/whychina-struggles-withfood-safety.

Campos, A.K.C., Cardonha, Â.M.S., Pinheiro, L.B.G., Ferreira, N.R., Azevedo, P.R.M., & Stamford, T.L.M. (2009). Assessment of personal hygiene and practices of food handlers in municipal public schools of Natal, Brazil. *Food Control*, 20, 807–810.

Chow, S., & Mullan, B. (2010). Predicting food hygiene: An investigation of social factors and past behavior in an extended model of the Health Action Process Approach. *School of Psychology*, 54(1), 126–133.

Clayton, D.A., Griffith, D.J., Price, P., & Peters, A.C. (2002). Food handlers' beliefs and self-reported practices. *International Journal of Environmental Health Research*, 12, 25–39.

Cullen, K.W., & Watson, K. (2007). Measuring school foodservice workers perceptions of organizational culture. *The Journal of Child Nutrition and Management*.

Food and Drug Administration (FDA). (2004). FDA Report on the occurrence of foodborne illness risk factors in selected institutional foodservice, restaurant, and retail food store facility types. *Food and Drug Administration, US Department of Health and Human Services.*

Hertzman, J., & Barrash, D. (2007). An assessment of food safety knowledge and practices of catering employees. *British Food Journal*, 109(7), 562–576.

Howes, M., McEwen, S., Griffiths, M., & Haris, L. (1996). Food handler certification by home study: Measuring changes in knowledge and behaviour. *Dairy Food Environmental Sanitation*, 3, 208–214.

Isara, A.R., & Isah, E.C. (2009). Knowledge and practice of food hygiene and safety among food handlers in fast food restaurant in Benim City, Edo State. Department community health. University of Benin Teaching Hospital.

Lizewski, L., & Maguire, K. (2010). The health belief model. 1–15.

Manning, C.K., & Snider, O.S. (1993). Temporary public eating places: Food safety knowledge, attitudes and practices. *Journal of Environmental Health*, 56, 24–28.

Mattson, M. (1999). Toward a reconceptualization of communication cues to action in the Health Belief Model: HIV test. *Communication Monographs*, 66(3), 240.

Ministry of Education. (2004). Panduan pengurusan kantin sekolah. Malaysia, School Division, MOE.

Ministry of Health Malaysia. (2010). Statistik Fakta Kesihatan. Retrieved on June 10, 2013 from http://www.moh.gov.my/v/stats_si.

Moon, A. (2000). Evaluating the health promoting school. *Health Education*, 100(6), 237–241.

Nee, S.O., & Norrakiah, A.S. (2011). Assessment of Knowledge, Attitudes and Practices (KAP) among food handlers at residential colleges and canteen regarding food safety. *Sains Malaysiana*, 40(4), 403–410.

Nel, S., Lues, J.F.R., Buys, E.M., & Venter, P. (2004). The personal and general hygiene practices in the deboning room of a high throughput red meat abattoir. *Journal of Food Control*, 15, 571–578.

Noor-Azira, A., Mohammad-Faiz, A., Shuhaimi, M., Syafinaz, A., Rukman A.H., & Malina, O. (2012). Knowledge, attitude and practices regarding food hygiene and sanitation of food handlers in Kuala Pilah, Malaysia. *Food Control*, 27, 289–293.

Palacio, J.P., & Theis, M. (2009). Introduction to foodservice. USA, Pearson Prentice Hall.

Patil, S.R., Cates, S., & Morales, R. (2005). Consumer food safety knowledge, practices, and demographic differences: Findings from a meta-analysis.

Powell, S.C., Attwell, R.W., & Massey, S.J. (1997). The impact of training on knowledge and standards of food hygiene a pilot study. *International Journal of Environmental Health Research*, 7, 329–334.

Redding, C.A., Rossi, J.S., Rossi, S.R., Velicer, W.F., & Prochaska, J.O. (2000). Health behave models. *The International Electronic Journal of Health Education*, 3, 180–193.

Rennie, D.M. (1995). Health education models and food hygiene education. *Journal of the Royal Society of Health*, 115, 75–79.

Rosenstock, I.M. (1974). The health belief model and preventative health behavior. *Health Education Monographs*, 2, 354–386.

Hanan, S.A., & Aisha, A.A. (2012). Assessment of personal hygiene, knowledge and practices in Al Ain, United Arab Emirates. *Food Control*, 25, 249–253.

Todd, E.C., Greig, J.D., Bartleson, C.A., & Michaels, B.S. (2007). Outbreaks where food workers have been implicated in the spread of food-borne disease. Part 2. Description of outbreaks by size, severity and settings. *Journal of Food Protection*, 70(8), 1975–1993.

Toh, P.S. & Birchenough, A. (2000). Food safety knowledge and attitudes: Culture and environment impact on hawkers in Malaysia. Knowledge and attitudes are key attributes of concern in hawker food handling practices and outbreaks of food poisoning and their prevention. *Food Control*, 11, 447–452.

Hospitality and Tourism – Sumarjan et al. (eds)
© 2014 Taylor & Francis Group, London, ISBN 978-1-138-00151-0

Food handlers' attitude towards practicing safe food handling in kopitiam restaurants

W.H.W. Bujang, F. Ahmad Fauzi, N. Baba & N. Sumarjan
Universiti Teknologi MARA, Malaysia

ABSTRACT: There are five major factors leading to food borne illness: lack of hygiene and sanitation by the food handlers, improper cooking procedures, temperature abuse during storage, cross-contamination between fresh ready-to-eat foods and raw foods, and acquiring food from risky sources. Improper food handling by individuals was recognized as the major reason for outbreaks of food poisoning. This study was conducted to explore food handler's attitude in practicing safe food handling in kopitiam restaurants. The Theory of Planned Behavior was used to evaluate the impacts and influences of the food handler's attitude. Thirty three kopitiam outlets restaurants in Shah Alam participated in this study with a total of 188 food handlers. Data of the study were collected through self-administered questionnaires. Reliability analysis was carried out and descriptive analyses were used to answer the research questions. One main finding from this study is food safety attitude variable was not the best predictor of food handlers' behavioral intention to practice safe food handling.

Keywords: food handlers, food safety, kopitiam restaurant, theory of planned behavior

1 INTRODUCTION

Food safety problems can occur at any stage of a foodservice system and the restaurant will be the last point in the sequence, from farm to the fork, as food is processed, cooked and finally served to the customers. Ironically, restaurants have been detected as one of the most common places for food-borne illness outbreaks (Knight, Worosz, & Todd, 2007). In Malaysia, the Food Act 1983 and The Food Regulations 1985 was published to defend the public from health hazards and fraud in the preparation of foods and sale of foods by irresponsible food handlers. The Act can protect the consumers by ensuring the consumption of safe foods and beverages in the outlets and premises, which has been prepared by the food handlers (Ministry of Health, 2010), which is more to enforcement in nature rather than prevention. Thus, this research is conducted to cultivate deeper into one of the reason that influence safe food handling practices, which is attitude.

2 LITERATURE REVIEW

2.1 *Theory of planned behavior*

Theory of Planned Behaviour, often abbreviated as TPB, was created in 1980 by Ajzen and Fishbein. This theory consist of three constructs, which are attitude, subjective norm and perceived behavioural control. TPB originally is an extension of Theory of Reasoned Action (TRA), created by Ajzen and Fishbein in 1975. Previously, TRA consists only of attitude and subjective norm. Wicker (1969) has conducted a study investigating the relationship between behavior and attitudes, and came out with his conclusion and argument that attitudes possibly do not predict the behavior, only the social psychologist have improved and influenced the predictive power of attitudes.

TPB have been extensively applied to a lot of study related to consumers' behavior, such as in sexual behavior, in the field of nutrition, physical exercise, food safety as well as in other scopes of study. Most of these studies used this TPB to form the relationship between the constructs, and possibly predict human behavior. Hence, the theory could be said to measures the constructs of behavioral intentions, attitude toward behaviors, subjective norm and perceived behavioral control (Ajzen, 1991). The perceived behavioral control construct was another addition that has been made in the TRA (Ajzen & Madden, 1986), resulting in the enhanced TPB. The TPB can be used to recognize the barriers that can avoid people to perform certain behaviors. TPB can actually be used to test the factors that influence individual's behavior.

A number of studies had provided support for the use of TPB in the prediction of a wide range of behavior, including food handling (Clayton, Griffith, & Peters, 2002) and hand hygiene (Jenner et al., 2002). In another study of nutrition and food, TPB has been implemented to explain the variety of intention and behavior and healthy eating has been included in the study (Conner, 2005; Povey et al., 2000).

2.2 *Attitude*

Because people reaction varies according to their natural attitude concerning on object, all characteristics of responses are often displayed in attitude formation that include cognitions, emotions, and behavior (Chaiken, 1998). Individual attitude toward behavior was determined by the assumption that the individual holds and thinks. Generally, attitude is an evaluative response of individuals, ranging from favorable to unfavorable (Chaiken, 1998). Attitude also influence the food handlers in practicing safe food handling for each occasion, even if it fall in the last place between subjective norm and perceived behavioral control in TPB (Philip & Anita, 2010). Mullan and Wong (2009) found that attitude makes significant contribution toward the prediction of intention for hygienic food handling behavior. If a person have negative attitudes about the behavior, then he or she will be less likely to participate in the behavioras compared to individuals who has positive attitudes. Ajzen and Fishbein (1980) specified that the strength of each behavioral belief increased by the assessment and evaluation of its value, and attitude is a summation of the cross products.

2.3 *Behavior intention*

In the TPB, the basic principle showed that behavior intention is the strongest factor in predicting behavior (Fishbein, 1975). A lot of studies have found behavioral intention was the best and most powerful predictor in determining behavior of an individual, but stillhas the limitations in its capability in predicting behavior. Armitageand Conner (2001) conducted a meta-analysis of the TPB model and found that it can be used to support and predict intention, in particular action behavior.

O'Boyle, Henly and Larson (2001), suggested that the success of the TPB model will be extended if a 'situational' theory is proposed as a context for individual behavior. The same author stated that the concentration of work place activity rather than internal motivational factor had an excessive influence on the adherence to hand hygiene recommendations; however hand washing practice will be lower during the peak hour.

Mullan and Wong (2009) who investigated the relationship of food and behavior found that the model of TPB was very successful in predicting breakfast consumption, explaining 53% of variance in intention and 64% of variance in behavior. Even though the TPB model was not used directly in predicting food handling behavior, it was still significant in predicting 79% of intention and 87% of self-reported hand hygiene practice in hospitals (Jenner et al., 2002). Clayton, Griffith and Peters (2008) found the variance in hand hygiene malpractice amounting to approximately 34% in catering establishment.

Philip and Anita (2010) showed that the variance of intention to carry out safe food handling practices at every occasion was positive. This proved that TPB can be used to measure behavioral intention. This was consistent with previous studies (Armitage & Conner, 2001; Kok, 1996).

3 METHODOLOGY

This research employed correlational research design and utilized structured questionnaire as the research instrument. The population of this study was food handlers of kopitiam restaurants in Shah Alam that serve coffee as the main products along with other local favorites. *Kopitiam* or coffee house is a place serving coffee and other refreshment. This definition of *kopitiam* is established to distinguish it from other type of coffee houses. A drop off method was employed where the questionnaires were distributed to the participated food handlers and collected two weeks later for analysis.

4 RESULTS AND DISCUSSION

4.1 *Demographic*

A total of 188 questionnaires were distributed to the respondents and 100% response rate was gained as the questionnaire was personally administered by the researcher. Majority of the respondents are male (n = 133, 70.7%) while female was 29.3% (n = 55). Almost three quarter of the respondent was single (n = 139, 73.9%), while the rest were married respondents (n = 43, 22.9%), and divorced/widowed respondents (n = 6, 3.2%). Most of the respondents were from the age group of 18 to 24 years old (n = 103, 54.8%). This is followed by respondents aged from 25–34 years old (n = 55, 29.3%), 35 to 44 years old (n = 25, 13.3%) and 45 years old and above (n = 5, 2.7%).

Almost half of the respondents went to high school (n = 87, 46.3%). Sixteen percent (n = 30) and 25.5% of the respondents had a certificate and Diploma, respectively. Only 6.4% (n = 12) of the respondents had their Bachelor Degree and less than 1% (n = 1) had a Master's Degree. The remainder (5.3%, n = 10) indicated 'others' as their highest level of educational level.

The total numbers of Malaysian's respondents are 116 persons (61.7%). Bangladeshi respondents represented by 23 persons (12.2%), followed by Indonesian respondents (n = 32, 17.0%) and Vietnamese (n = 1, 0.5%), and other nationalities (n = 16, 8.6%). It can also be concluded that most of the respondents were still new with the industry because they have just been working for only one to six months period. They represented the largest group of the total respondents (n = 57, 30.3%). Forty (21.3%) respondents have been working for 7 to 12 months, 27 respondents (14.4%) have been working for 19 to 24 months, and 34 respondents (18.1%) have been working for 31 months or more. Interestingly, 98.4% (n = 185) of the respondents have heard about food safety while the rest admitted they never heard about food safety prior and while working at the restaurants.

Training course is one of the most important things that food handlers need to attend. 173 respondents (92.0%) attended food safety training course. The rest of 15 respondents (8.0%) never attend any food safety-training course. Majority of the respondent attended the food safety training once a year. Typhoid vaccination is also an important process for each of the food handler to have. In this study, 184 respondents (97.9%) have had their typhoid vaccination.

4.2 *Attitude*

The Table 1 summarizes the level of food handler's agreement on their attitude which influenced their behavioral intention. Fifty-nine% (M = 4.6, SD = 0.55) of the respondents strongly agreed that they will prepare food hygienically for every meal and occasions. Fifty-nine% (M = 4.57, SD = 0.538) of the respondents said that they wash their hands before preparing the foods. Twenty seven percent (M = 3.61, SD = 1.207) of the respondents confessed that they would carry out safe food handling if they are given more pressure by the management.

To recapitulate, the research objective of this study was to determine the influence of food safety attitude towards behavioral intention in practicing safe food handling among food handlers. The multiple regressions analysis dictated that food safety attitude (statistically) is not significant to predict the behavioral intention of food handlers to practice safe food handling as the p value is larger than 0.05.

This indicates that the food handlers do not thoroughly apply food safety procedures while working in the restaurants. One plausible explanation for this scenario is perhaps lack of knowledge and

Table 1. Food handler's level of agreement for attitude.

	Level of agreement				
Items	SD (%)	D (%)	N (%)	A (%)	SA (%)
I will prepare food hygienically for every meal and occasions	–	–	2.7	37.8	59.6
I wash my hands before prepare the foods	–	–	2.1	38.8	59.0
I wash the knife and cutting board after used	–	–	10.1	42.6	47.3
I will carry out safe food handling practices to ensure less food poisoning	–	–	2.7	43.1	54.3
I will carry out safe food handling to ensure I would gain praise from my boss	4.8	6.4	12.8	41.5	34.6
I would carry out safe food handling to increase personal satisfaction	0.5	1.1	10.6	42.0	45.7
I would carry out safe food handling to gain pay rise	6.9	8.0	17.6	31.9	35.6
I would carry out safe food handling to have better personal hygiene	2.7	0.5	6.4	42.6	47.9
I would carry out safe food handling to have a safe working environment	0.5	0.5	6.4	41.5	51.5
I would be carrying out safe food handling to achieve a good reputation responsibility from my management	1.6	5.9	12.8	39.4	40.4
I would carry out safe food handling if I'm being given more pressure	6.4	14.4	18.1	34.0	27.1
I would carry out safe food handling to ensure the customers will happy	–	2.1	4.8	41.0	52.1

Scale for items: 1 = Strongly Disagree (SD), 2 = Disagree (D), 3 = Neutral (N), 4 = Agree (A), 5 = Strongly Agree (SA), n = 188.

Table 2. Mean and standard deviation for attitude.

Items	Standard deviation	Mean
I will prepare food hygienically for every meal and occasions	0.548	4.57
I wash my hands before prepare the foods	0.538	4.57
I wash the knife and cutting board after used	0.662	4.37
I will carry out safe food handling practices to ensure less food poisoning	0.552	4.52
I will carry out safe food handling to ensure I would gain praise from my boss	1.078	3.95
I would carry out safe food handling to increase personal satisfaction	0.747	4.31
I would carry out safe food handling to gain pay rise	1.203	3.81
I would carry out safe food handling to have better personal hygiene	0.838	4.32
I would carry out safe food handling to have a safe working environment	0.685	4.42
I would be carrying out safe food handling to achieve a good reputation responsibility from my management	0.950	4.11
I would carry out safe food handling if I'm being given more pressure	1.207	3.61
I would carry out safe food handling to ensure the customers will happy	0.686	4.43

Scale for items: 1 = Strongly Disagree (SD), 2 = Disagree (D), 3 = Neutral (N), 4 = Agree (A), 5 = Strongly Agree (SA), n = 188.

exposure on the consequences of food hazards to the food handlers. The attitude toward food safety is also low, and majority confessed that their food safety knowledge is not up to the required standard. This finding is comparable to a study conducted by Byrd-bredbenner et al. (2007) who found that although 97% of their respondents ranked their own food safety knowledge as at least fair, 60% of them confessed that they did not wash their hands with soap and water after handling raw poultry, which could turn out be a fatal mistake.

From the result it can be assumed that he food handler's attitude participated in this study were quite disappointing. They were not really concerned on the importance of applying food safety procedures during their working time, even after being briefed, trained and coached at their workplace. This is parallel to the demographic result where almost majority of the total respondents (46.3%) only went to high school as their highest educational level. Food safety attitude variable in this study can't be seen as the best predictor of food handlers behavioral intention to practice safe food handling as the variance of this variable is very low. In most of the studies, attitude was established as the predictive factor compared to subjective norm (Conner & Sparks, 2005). The general Theory of Planned Behavior (TPB) stated that attitude (an individual acceptable and unacceptable evaluation towards behavior) predict intention (Ajzen, 1991).

5 CONCLUSION

This study was empirically carried out to determine the strength of attitude factor that influence the food handler's intention to practice safe food handling using TPB. Generally, it is not easy to predict how successful TPB in predicting intention to practice safe food handling in Malaysian context. This is because for most Malaysians, food safety is not necessarily a major issue in restaurant selection. Perhaps, the awareness on food safety and food hazards is still not satisfactory.

There are regulations on the issue, but lack of enforcement in relation to food safety and its necessary procedures. The results of this study showed that the food safety practices and attitude among the employees of the participating restaurants are far from satisfactory, and in need of serious attention. Even though the food safety attitude contributed minimally in the prediction of behavioral intention in TPB, this variable can still be used in other study. Most of the food handlers attended food safety training courses before they work in the industry. This at least provided them the food safety knowledge.

It is hoped that the input of this study could provide some insights about the attitude of the food handlers in providing safe food. Thus, restaurateurs could identify the appropriate actions to minimal the causes of potential outbreak and enhance the quality and safety of food serves to customers.

REFERENCES

Ajzen, I. (1991). The theory of planned behavior. *organizational behavior and human decision processes. 50*, 179–211.

Ajzen, I. & Fishbein, M. (1980). *Understanding attitude and predicting social behavior*: Prentice Hall.

Ajzen, I. & Madden (1986). Prediction of goal directed behavior: attitude, intentions, and perceived behavioral control. *Journal of Experimental Social Psychology, 22*, 453–474.

Armitage, C.J. & Conner, M. (2001). Afficacy of the theory of planned behavior: A meta-analytic review. *British Journal pf Psychology, 40*, 471–499.

Byrd-bredbenner, C., Maurer, J., Wheatley, V., Schaffer, D., Bruhn, C., & Blalock, L. (2007). Food safety self-reported behaviors and cognitions of young adults: Results of national study. *Journal of Food Protection, 70*(8), 1917–1926.

Chaiken, E.A. (1998). *Attitude Structure and Function* (4th ed. Vol. 1).

Clayton, D.A., Griffith, C., & Peters, A.C. (2002). Food handlers belief and self-reported practices. *International Journal of Environmental Health Research, 12*(1), 25–39.

Conner, M. & Sparks, P. (2005). Theory of planned behavior and health behaviour: Predicting health behaviour research and practice with social cognitions models, *170–222*.

Fishbein, A. (1975). Belief, Attitude, Intention and Behavior: An Introduction to Theory and Research Addison-Wesley.

Godin, G. & Kok, G. (1996). The Theory of Planned Behaviour: A Review of Its Applications to Health-Related Behaviour. *American Journal of Health Promotion, 11*(2), 87–98.

Jenner, L.A., Watson, P.W.B., Miller, L., Jones, F., & Scott, G.M. (2002). Explaining hand hygiene: an extended application of the theory of planned behaviour. *Psychology, Health and Medicine, 7*(3), 311–326.

Knight, A.J., Worosz, M.R., & Todd, E.C.D. (2007). Serving Food Safety: Consumer Perception of Food Safety at Restaurants. *International Journal of Contemporary Hospitality Management, 19*(6), 476–484. doi: 10.1108/09596110710775138.

Ministry of Health (2010). *Malaysian dietary guidelines*. malaysia: technical working group on nutritional guidelines.

Mullan, B.A. & Wong, C.L. (2009). Hygenic food handling behaviours: An application of the theory of planned behaviour. *Appetite, 52*, 757–761.

O'Boyle, C.A., Henly, S.J. & Larson, E. (2001). Understanding adherence to hand hygiene recommendations: The theory of planned behaviour. *American Journal of Infection Control, 31*, 415–423.

Philip, S. & Anita, E. (2010). Efficacy of the theory of planned behaviour model in predicting safe food handling practices. *Food Control, 21*, 983–987.

Povey, R., Conner, M., Aparks, P., James, R., & Sheperd, R. (2000). The theory of planned behaviour and healthy eating: examining additive and moderating effects of social influences variables. *Psychology and Health, 14*(6), 991–1006.

WHO (2006). Five Keys to Safer Food Manual. Retrieved from http://www.who.int/entity/foodsafety/publications/consumer/manual_keys.pdf.

Wicker, A.W. (1969). Attitude versus actions: The relationship of verbal and over behavioral responses to attitude objects. (25), 41–78.

Relevant areas in hospitality and tourism

Hospitality and Tourism – Sumarjan et al. (eds)
© 2014 Taylor & Francis Group, London, ISBN 978-1-138-00151-0

Assessing Knowledge, Attitude and Practice (KAP) on food safety among food handlers in Universiti Teknologi Mara (UiTM), Shah Alam

M.I. Zulkifly, M.M. Salleh, M.H. Mohd Hanafiah & M.R. Jamaluddin
Universiti Teknologi MARA, Malaysia

ABSTRACT: Food handlers' plays a major role in preventing outbreaks of foodborne illness. The purpose of this study were to assess the Knowledge Attitude and Practice (KAP) on food safety among food handlers in Universiti Teknologi MARA (UiTM), Shah Alam. The convenience sample of 112 food handlers was collected from 62 food premises (restaurants, kiosk, cafeteria and food court) in UiTM, Shah Alam. Descriptive statistic were used to analyse the data regarding food handlers' knowledge related to food safety area (personal hygiene, foodborne illness, cross contamination, time and temperature control and sanitizing), attitude and practices. In general the respondent overall knowledge was satisfactory with a good magnitude mean score of knowledge. Majority of the food handlers reported positive attitudes on most aspects such as the responsibility to obtain food safety and to ensure that food is save to serve. The attitude significantly contributed to food safety practice findings in this study. This study clearly indicates that attitude strongly determine food handlers' actual practices in their working food premises. The lack of knowledge in food safety will not influence their attitude and practices thus foodborne illness can be avoided. Monitoring and identifying food handler's level of Knowledge, Attitude and Practices (KAP) is important to assure food safety in food premises operations therefore, prevent the occurrence of foodborne illnesses, a health public problem in UiTM and Malaysia.

Keywords: attitude, food handlers, food safety, knowledge, practice

1 INTRODUCTION

Foodservice industry in Malaysia is expanding tremendously over the years. Due to the fast speed of living, most people take advantage of eating in restaurants and buying readymade food for the quick meal at home. However, food contamination by food handlers could occur and will lead to foodborne diseases if they neglect proper food handling practices in their premises (Haryani, Noorzaleha, Fatimah, Noorjahan, Patrick, Shamsinar, & Son, 2007). According to the Ministry of Health Malaysia (2010), there has been an increase in the incidence rate of foodborne diseases from 2005 to 2008. Although the trend of food poisoning declined in 2009, there was a death reported due to foodborne diseases in that year with the same mortality rate as 2008.

Previous report indicated that besides poor hand and surface hygiene, lack in personal hygiene amongst food handlers was also one of the most commonly reported practices that gave rise to foodborne illness (Nee & Sani, 2011). Food handlers may cross contaminate raw and processed foodstuffs as well as inadequate cooked and stored foods. In addition, foodborne illness has been associated with improper storage or reheating, food stored inappropriately and cross-contamination (Karabudak, Bas, & Kiziltan, 2008). This shows that if food handlers take serious note on the cleanliness of their hand, body and clothing, this will help in preventing incidence of cross-contamination from occurring (Sneed et al., 2004).

Zulkifly, Zahari and Hairi (2009) reported that food safety implementation in Malaysia is still considered very slow in term of enforcement. Proper hygiene practices and procedures on the food production line are often being ignored by some of the food handlers. Besides, the supply and the availability of raw materials are always inconsistent and consequently have caused outbreaks of food poisoning. Furthermore unhygienic food handling practices and the inadequacy of a safe water supply, as well as poor environmental sanitation were some causes of foodborne illness outbreaks in Malaysia (Siow & Norrakiah, 2011). It is understood that the risk of food getting contaminated depends largely on the health status of the food handlers, their personal hygiene, knowledge, attitude and practice of food hygiene (Soon, Singh, & Baines, 2011).

Infections can also be acquired through contaminated unwashed fingers, insects, and circulation of banknotes and by wind during dry conditions (Isara & Isah, 2009).

Based on the discussion above, it is very much clear that there is no such thing as 'zero risk' when it comes to food. With the expansion of the foodservice industry, this issue is becoming more critical than ever given that the selection of restaurants or food premises is abundant hence complicating the efforts to control food safety. However, it is important to highlight at this stage that Soon et al. (2011) reported that half of the foodborne related diseases from the early 1990s were associated with outbreaks in institutions and schools, mostly due to unhygienic food handling procedures. Having said that they suggested that some of the key agencies from the Malaysian such as Ministry of Health, academia, industries and research institutions should continue to strengthen their collaboration and networking in order to coordinate the prevention and control of foodborne diseases and thus improve public health among students of all levels.

Food handler training and education are seen as one of the main strategies to increase food safety knowledge and practices. It is very clear that good knowledge of food safety is important in order to avoid any negative consequences regarding foodborne illness. Toh and Birchenough (2000) concurred that there have been many food safety studies on domestic, commercial and retail foods but not much is known as to the extent of hygiene knowledge of food handlers in Malaysian universities and this information regarding knowledge, attitude and practice needs to be strengthened further among food handlers especially in UiTM where this study was conducted. Therefore, this study will assess the knowledge, attitude and actual practices concerning food safety among food handlers in Universiti Teknologi MARA (UiTM), Shah Alam which involved structured food premises such as at kiosks, restaurant, cafeteria and stall.

2 LITERATURE REVIEW

2.1 *Food safety knowledge*

In this study, food safety knowledge is important to prevent foodborne illness. Prevention of foodborne illnesses is one of the primary responsibilities of the foodservice industry (Cushman, Shanklin, & Niehoff, 2011). It is the manager's responsibility to ensure the safety of food prepared and served to customers. Mortlock, Peters and Griffith (2000) stated that there was a general agreement revealed from several authors as good levels of knowledge towards food safety among the food handlers and the effective practices of such knowledge in food handling were imperative in ensuring the safe production of food in any catering operations.

The majority of foodborne illnesses have been linked to foods prepared in retail foodservice operations and caused by human error at some point in the food chain. Identifying foodservice employee's knowledge, attitudes, and practices are essential to assure food safety in retail foodservice operations and to prevent the occurrence of foodborne illnesses, which is a public health problem. Recently, many studies pinpoint to the need for training and education of the food handlers in public hygiene measures due to their lack of knowledge on microbiological food hazards, temperature ranges of refrigerators, cross contamination and personal hygiene (Bas et al., 2006; Nee et al., 2011).

2.2 *Food handlers' attitude*

Knowledge on the importance of food safety should be given to all staffs in the food processing businesses so as to bring behavioral changes besides adopting of positive attitudes (Coleman & Roberts, 2005). But in some previous studies no differences were detected between the staffs who attended an educational course with those who did not (Angelillo, Viggiani, Rizzo & Bianco, 2001 and Askarian et al., 2004).

This statement was supported by several studies (Bas et al., 2006; Jevšnik et al., (2012); Nee and Sani, 2011), and it indicates that although training may increase the knowledge of food safety, this does not always produce a positive change in food handling attitudes. Meanwhile, Park, Kwak & Chang (2010) also pointed out that knowledge alone is not sufficient to promote positive attitudes and safe behaviors among the food handlers. Therefore, the food handlers need to have a confident understanding of the basic knowledge of food safety at least, to ensure they can assess relative risks and make informed decisions about food handling practices.

2.3 *Food handlers' practice*

Human handling errors have been responsible for most outbreaks of food poisoning (Jevšnik et al., 2012; Nee & Sani, 2011) for example; Hepatitis A virus can be introduced by unwashed hands of the food handlers who are themselves infected. Education is important for increasing the knowledge of the food handlers; however, increased knowledge of food safety does not always result in positive change in food handling behaviors.

There are several studies about the knowledge and practices of the food handlers conducted in various parts of the world (Bas et al., 2006; Jevsnik et al., 2008; Walker, Pritchard, & Forsythe, 2003, Annor & Baiden, 2011). Therefore, a good personal hygiene as well as sanitary handling practices at work is an essential part of any prevention programs for food safety as the food handlers themselves may be a major source of contamination and ultimate sources of health risks either as carriers of pathogens or through poor hygienic practices (Gomes-Neves, Araújo, Ramos, & Cardoso, 2007).

3 METHODOLOGY

3.1 *Research design*

Quantitative approach was chosen by utilizing the descriptive method of research which is known as a fact-finding study that involves adequate and accurate interpretation of the findings (Powell, 1997). In other words, subjectivity of judgment, which is not needed in a thesis discussion, can be avoided through quantitative methods. Thus, conclusions, discussion and experimentation involved in the process are more objective. Data were collected using self administered questionnaires that were distributed to the respondents.

3.2 *Sampling and instrument*

In evaluating the food safety knowledge, attitude and practices among food handlers within the campus of Universiti Teknologi MARA (UiTM), Shah Alam the survey was carried out from March to April 2013 at food premises such as restaurant, cafeteria, kiosk, and food court in UiTM. The respondent in this study were food handlers from all 62 structured food premises in the university which involved 240 food handlers in total. Since the food handlers population is only 240, a sample of 112 is convincing enough as a true representative and this was considered for the purpose of this study.

A self-administered structured questionnaire was designed and modified by the aid of previous research (Lin & Sneed, 2003). Questionnaire was comprised of four distinct parts; demographic characteristics of the participants, knowledge related to food safety, food handlers' attitude towards food safety actual practices. The 5-points Likert scale was used to interpret items in the questionnaire.

4 RESULTS AND DISCUSSION

4.1 *Food safety knowledge among food handlers*

The mean scores for all 10 items in the food safety knowledge variables ranges from 2.39 to 4.6, thus indicating three items with moderate scores and seven items with the high of mean scored items; which are "The most important rule of foodservice personal hygiene is that employees must wash their hands often." (M = 4.6, SD = 0.93), "Foodborne illnesses are diseases that are carried or transmitted to people by food" (M = 4.6, SD = 0.93).

It might be possible that the reason behind these high mean scores are due to the fact that based on

Table 1. Food safety knowledge among food handler.

No.	Item	Mean	S.D
1	After washing their hands, employees should avoid touching their hair.	4.41	0.97
2	The most important rule of foodservice personal hygiene is that employees must wash their hands often.	4.60	0.93
3	Foodborne illnesses are diseases that are carried or transmitted to people by food.	4.60	0.93
4	The most important factors to control the growth of bacteria are temperature and time.	3.97	1.29
5	When holding hot foods for service, it is required that internal food temperatures are taken at least every two hours.	4.02	1.22
6	The temperature danger zone for potentially hazardous foods is 41° to 140°F.	2.39	1.61
7	Under running water that is 70°F or less is an acceptable method for thawing frozen food.	4.03	1.11
8	Cross contamination is the transfer of harmful substances or micro-organisms to food from food or from a non food contact surface, such as equipment, utensils, or hands.	4.54	0.89
9	Sarah wore disposable gloves while she formed raw ground beef into patties. After she was finished, she wore the same gloves to slice roast beef for sandwich. What mistake did Sarah make? She failed to change her gloves and wash her hands after handling raw meat and before handling a ready-to-eat food item.	4.60	0.93
10	When iodine solutions (such as Microcline) are used for sanitizing, the item must be immersed in the solution for 30 seconds.	2.39	1.61

their level of education, majority of the respondent (college, diploma, bachelor and master) have been exposed to the importance of personal hygiene, foodborne illness and cross contamination.

4.2 *Food handlers' attitude*

High mean scores are reported to most of the items especially to; "I think that managers should educate employees on personal hygiene and sanitation regularly" A substantial majority of the food handler agreed that selecting a place to eat based on its reputation for good sanitation and cleanliness is important. They also reported that 'food safety knowledge not only benefits their work but also their personal life' However, the respondents disagree when provoked with statements like "tasty food more important rather than safe food" and "I think that only cooks should receive food safety training".

4.3 *Food handlers' practices*

The entire of food handlers' in UiTM claimed that they store chemicals in non food storage room (and wear a clean uniform when working in food premises. Majorities of food handlers in UiTM also reported to clean and sanitize work surfaces after each task and use gloves or utensils to handle food that is ready-to-eat in nature. The overall

Table 2. Food handlers' attitude.

No.	Item	Mean	S.D
1	I think sanitation is an important part of my job responsibilities	4.34	0.59
2	I believe that good employee hygiene can prevent foodborne illness	3.85	0.65
3	I think that it is the responsibility of all food handlers to ensure that food is safe to serve	4.54	0.55
4	I am willing to change my food handling behaviors when I know they are incorrect	4.38	0.62
5	I am willing to obtain more food safety knowledge	4.45	0.60
6	It is more important to have tasty food rather than safe food	2.55	1.11
7	I select a place to eat based on its reputation for good sanitation and cleanliness	3.48	0.56
8	I think that managers should educate employees on personal hygiene and sanitation regularly	4.88	0.38
9	I think that only cooks should receive food safety training	2.00	0.95
10	I believe that food safety knowledge not only benefits my work but also my personal life	3.49	0.48
11	I am willing to attend a food safety training course	4.29	0.46
12	I believe the food safety knowledge would make me more confident about my work	4.26	0.48

Table 3. Food handlers' practices.

No.	Item	Mean	S.D
1	I use gloves or utensils to handle food that is ready-to-eat	4.91	0.29
2	I use a separate clean utensils for each food item	4.04	0.21
3	I wash my hands with soap and water before working with food	3.80	0.58
4	I wash raw items before using it	3.81	0.61
5	I store chemicals in a non food storage room	5.00	0.00
6	I store raw food item in an area separate from cooked food	4.77	0.63
7	I wear a clean uniform, when I work in food premises	5.00	0.00
8	I wear a hair restraint (cap or helmet) when I work in foodservice	4.80	0.50
9	I wash my hands and change into a new pair of gloves after touching anything that may contaminate my hands, when I prepare or serve food	4.89	0.31
10	I drink or eat food, while I am serving or preparing food	2.36	0.48
11	I clean and sanitize work surfaces after each task	4.88	0.45
12	When I am in doubt the safety of a previously cooked food, I report it to the supervisor	4.75	0.64
13	I pay attention to expiration dates on food and do not use foods that have passed the expiration date	4.80	0.60
14	I check concentrations of sanitary solutions used for sanitizing work surfaces or items cooked in the pot and pan sink	4.29	0.97

mean score for the actual practices among food handlers' in UiTM is 4.44 and standard deviation are 0.15. This represents that majority of food handlers' in UiTM are practicing appropriate food safety in their working premises.

4.4 *The relationship between food safety knowledge and attitude among food handlers in UiTM*

On the attribute of texture, the result shows that there is relationship between overall acceptability and texture attribute. The coefficient of the relationship (R^2) 0.239 shows that 24% of overall acceptable attitude by increasing of 9%. The result also suggest that there is a significant predictor of overall acceptability.

4.5 *The relationship between knowledge and practice among food handlers in UiTM*

The result concluded that the assumption on knowledge and practice are strongly supported ($R^2 = 0.521$, $p < 0.000$). Ansari-Lari et al. (2010) also revealed a positive correlation between knowledge, attitude and practice. However this finding contradicts with the study by Bas et al. (2006) who found out that good knowledge on food safety does not necessarily leads to good handling practices.

4.6 *Mediating effect of attitude with knowledge and practice among food handlers in UiTM*

The relationship between attitude and practice is significantly positively correlated for participants among food handlers in UiTM ($r = 0.978$, $n = 112$, $p = 0.000$). So we can conclude that the attitude increase from practice among food handlers in UiTM. The strong positive correlation existed between them explained that the effect of mediating significantly contributed by increased 8 percent (R^2-0.295) to the prediction of this study.

5 CONCLUSIONS

As a conclusion, the discrepancy between the knowledge, attitude and practice needs to be studied further. Different observation techniques could also be incorporated to decrease the invasiveness of the observers and to determine the number of violations that could have been prevented. It is important to note that further research is needed to determine how food handlers' can improve their knowledge related to food safety, attitude and practices. A good knowledge and a right attitude among food handlers in their working premises would help to assure more harm to food can be avoided. In addition, the management enforcement action regarding food safety studied to determine food handlers' motivation and monitor and discipline the employees with respect to food safety in Universiti Teknologi MARA.

REFERENCES

Angelillo, I.F., Viggiani, N.M.A., Rizzo, L., & Bianco, A. (2000). Food handlers and foodborne diseases: knowledge, attitudes, and reported behavior in Italy. *Journal of Food Protection*, *63*, 381–385.

Annor, G.A. & Baiden E.A. (2011). Evaluation of food hygiene knowledge attitudes and practices of food handlers in food businesses in Accra, Ghana. *Food and Nutrition Sciences*, *02*(08), 830–836. doi:10.4236/fns.2011.28114.

Ansari-Lari, M., Soodbakhsh, S., & Lakzadeh, L. (2010). Knowledge, attitudes and practices of workers on food hygienic practices in meat processing plants in Fars, *Iran. Food Control, 21*(3), 260–263. doi:10.1016/j.foodcont. 2009.06.003.

Askarian (2004). Results of a hospital waste survey in private hospitals in Fars Province, Iran Waste Management, *24*, 347–352.

Baş, M., Şafak Ersun, A., & Kıvanç, G. (2006). The evaluation of food hygiene knowledge, attitudes, and practices of food handlers' in food businesses in Turkey. *Food Control*, *17*(4), 317–322. doi:10.1016/j.foodcont. 2004.11.006.

Coleman, P. & Roberts, A. (2005). Food hygiene training in the UK: A time for change. Food Service Technology, *5*(17)22. doi:10.1111/j.1471-5740.2005 00108.

Cushman, J.W., Shanklin, C.W., & Niehoff, B.P. (2001). Hygiene practices of part-time student employees in a university foodservice operation. *The Journal of the National Association of College and University Food Services. 23*, 37–44.

Gomes-Neves, E., Araújo, A.C., Ramos, E., & Cardoso, C.S. (2007). Food handling: comparative analysis of general knowledge and practice in three relevant groups in Portugal. *Food control*, *18*(6), 707–712.

Haryani, Y., Noorzaleha, A.S., Fatimah, A.B., Noorjahan, B.A., Patrick, G.B., Shamsinar, A.T., & Son, R. (2007). Incidence of Klebsiella Pneumonia in street foods sold in Malaysia and their characterization by antibiotic resistance, plasmid profiling, and RAPD–PCR Analysis. *Food control*, *18*(7), 847–853.

Isara & Isah (2009). Knowledge and practice of food hygiene and safety among food handlers in fast food restaurants in Benin City, Edo State, Niger. *Postgraduate Medical Journal, September*, *16*(3), 207–12.

Jevšnik, M., Ovca, A., Bauer, M., Fink, R., Oder, M., & Sevšek, F. (2012). Food safety knowledge and practices among elderly in Slovenia. *Food Control.*

Karabudak, E., Bas, M., & Kiziltan, G. (2008). Food safety in the home consumption of meat in Turkey. *Food Control*, *19*(3), 320–327.

Lin, Shu-Ying, & J. Sneed. (2003), University foodservice employees' food safety knowledge, attitudes, practices, and training. Phd Diss., Iowa State University, 2003.

Ministry of Health Malaysia. (2010). *Annual Report 2010.* Public Health Department, Ministry of Health Malaysia, Putrajaya, MOH.

Mortlock, M.P., Peters, A.C., & Griffith, C.J. (2000). A national survey of food hygiene training and qualification levels in the UK food industry. *International Journal of Environmental Health Research, 10*(2), 111–123.

Nee, S.O. & Sani, N.A. (2011). Assessment of knowledge, attitudes and practices (KAP) among food handlers at residential colleges and canteen regarding food safety. *Sains Malaysiana, 40*(4), 403–410.

Powell, R.R. (1997). *Basic research methods for librarians.* Greenwich, Conn.: Ablex Publishing Corporation.

Siow O.N. & Norrakiah A.S, (2011). Assessment of knowledge, attitudes and practices (KAP) among food handlers at residential colleges and canteen regarding food safety. *Sains Malaysiana, 40*(4), 403–410.

Soon, J.M., Singh, H., & Baines, R. (2011). Foodborne diseases in Malaysia: A Review. *Food Control, 22*(6), 823–830. doi:10.1016/j.foodcont.2010.12.011.

Toh, P.S., & Birchenough, A. (2000). Food safety knowledge and attitudes: Culture and environment impact on hawkers in Malaysia. Knowledge and attitudes are key attributes of concern in hawker food handling practices and outbreaks of food poisoning and their prevention, *11,* 447–452.

Walker, E., Pritchard, C., & Forsythe, S. (2003). Hazard analysis critical control point and prerequisite programme implementation in small and medium size food businesses. *Food Control, 14*(3), 169–174.

Zulkifly, M.I., Zahari, M.S.M., & Hairi, M. (2009). An investigative study into the hazard analysis of critical control point (Haccp) implementation in small and medium-sized food manufacturing enterprises (SMEs).

Hospitality and Tourism – Sumarjan et al. (eds)
© 2014 Taylor & Francis Group, London, ISBN 978-1-138-00151-0

The effects of tourist motivation as antecedents on revisiting heritage destination

Z.M. Rani & N. Othman
Universiti Teknologi MARA, Malaysia

ABSTRACT: Heritage tourism represents the most exciting segment in the industry, which allow tourist to experience and discover the priceless characters of the past. It is also among the oldest type of tourism segments which require comprehensive concentration for preservation and conservation. Lack of vigilance towards the heritage attributes resulted to low quality of tourist experience at the heritage destination. Thus, this issue will directly affect tourist motivation to revisit. Retaining tourists to revisit a destination presents severe challenge for most managerial level in the tourism industry. Revisit intention plays an important role to determine the success of the industry. A major concern is whether the pull factors of motivation able to influence the push factors in the context of intention to revisit. Therefore, the purpose of this study is to address the effects of tourist motivation as antecedents of intention to revisit heritage destination in Malaysia. Using quantitative method as the approach, a total of 600 questionnaires were distributed to the international and domestic tourist that is available at heritage destinations in Penang and Melaka. The data were analyzed using descriptive statistic, and factor analysis. The present study provides an understanding towards the tourist motivation to revisit heritage destination. Additionally, the study contributes to the understanding of tourist behaviors relative to heritage settings along with implications for the marketing of heritage settings.

Keywords: heritage destination, heritage tourism, intention, motivation, revisit

1 INTRODUCTION

From the standpoint of the tourism industry, tourist motivation is an amalgam of concept borrowed from the individual orientation of psychology and functional to a definite domain of human action (Pearce, Morisson, & Rutledege, 1998). Moreover, Pearce et al. (1998) also noted that to explain the function of travel motivation in the stance of tourism demand, two questions need to be addressed: (1) "why do certain groups of tourists travel?" and (2) "why do people go to a certain place?" "The first question seeks to understand the individual psychology of the traveler, whereas the second requires us to describe the important features of a tourism destination and also to assess how well these features will satisfy the potential travelers' needs" (Pearce et al., 1998).

In a similar vein, revisit intention has become one of the central and multifaceted issues in tourism industry. To date, many destinations rely on repeat tourist to stimulate the flock of tourist arrival. Researchers in service industry especially tourism ought to consider revisit intention due to the "globalization of markets, competitive pressure, brand multiplication and, consumer lifestyle change" (Flambard-Ruaud, 2005).

2 LITERATURE REVIEW

2.1 *Heritage tourism*

The emergence of culture and heritage tourism since the past few decades has led to the development of the tourism industry today. It is now one of the major sectors in the tourism industry and has experienced a significant growth (Alzua, O'Leary, & Morrison, 1998). Cultures and heritage are irreplaceable, their specific forms and means are often tangible and intangible expressions that establish the community heritage values ought to be promoted as an indispensable feature of human development (Ahmad, 2006).

Since the formation of this fragile segment (Joliffe, Rowe, & Davis, 2002; Van der Borg, Costa, & Gotti, 1996;) in the tourism industry, there are numerous definitions of heritage tourism. Although most researchers demonstrate diverse concepts and definition of heritage tourism and can be considered as problematic, yet they agreed that heritage can be considered as bridge that link between the present and the past. The definitions of heritage tourism varied due to researchers address dissimilar notion, geographical area and events. Among the earliest definition by Alzua et al. (1998) have developed a definition by associate heritage tourism with "development, education, intellectual and aesthetic training, social cooperation, and taste".

On one hand, Hitchcock, King and Parnwell (2010) have created a definition of heritage by integrating the notion of politic, economic and touristic attention. Whereas, Hitchcock et al. (2010, p. 2) refers heritage tourism as "tangible and concrete elements of the past (buildings, monuments, artifacts, sites and constructed landscapes), as well as to those aspects of culture expressed in behavior, action and performance (usually referred to as 'intangible cultural heritage') which are interpreted, valued and judged to be worthy of our attention and protection". Timothy and Nyaupane (2009) and Timothy and Boyd (2003) also agreed that heritage tourism work best on tangible and intangible resources. Hall and Zeppel (1990) defined heritage tourism as "a subset of cultural ethnic and educational forms of special interest tourism which includes many aspects of touristic behavior ranging from examination of the physical remains of the past to the experience of contemporary cultural traditions".

Meanwhile, Chhabra (2010) proposed a definition of heritage tourism in the context of sustainable marketing, which is very detailed and precise. She defines heritage tourism as "a phenomena that focuses on management of the past, inheritance, and authenticity to enhance participation and satisfy consumer motivations by evoking nostalgic emotions; its underlying purpose is to stimulate monetary benefits for its various constituencies such as the museum, historic house, festival, heritage hotels and stakeholders".

Regardless various definition created by researchers, heritage tourism has now become attraction to tourists as this priceless segment combined experience, enjoyment, information and education together which could benefit to the most tourists. It is s also considered as an established phenomena in the tourism industry (Chhabra, 2001, p. 1) and considered as an important component to most developing countries (Garrod & Fyall, 2000). Heritage tourism is a unique segment (Millar, 1989) as it offers the opportunity to portray and experience the past in the present through an endless possibility of interpretation and presentation techniques (Ahmad, 2006).

The escalation of heritage tourism can be seen as there are many tourism developments occurring in order to cope with the demand from mass tourist. The recent phenomena of no frills airline could also be among the major contributor to the growth of heritage tourism. Timothy and Boyd (2003, p. 10) pointed out that the growth of the heritage tourism arises due to the "higher level of education, more income, growing awareness of the world, globalization processes that make the world a smaller place, technology, the effects of media and telecommunication and new types of heritage attraction.

2.2 *Motivation*

Crompton (1979) is one of the pioneers for motivation study in tourism study and affirmed that motivation is one of the most important variables that contribute to behavior. Fodness (1994) in his work on motivation noted that most tourism professionals and academics agreed that motivation is a critical variable as it is the driving force behind all behavior. More specifically, overabundance of researchers has investigated travel motivations (Crompton, 1979; Fodness, 1994; Kau & Lim, 2005; Kozak, 2002).

McReynolds (1991) noted that it is not easy to coherent a precise definition of motivation due to profuse theories available. In a similar vein, the origin of study in motivation is from the phrase motive, "to denote excitement, cause, desire, want, drive, and other similar concepts, each of which is related to an activity or a disposition to act" (Furchtgott & Furchtgott, 1999).

The notion that travel motives are connected to the travelers' choice of vacation is a fascinating one. Crompton (1979) showed that travel motivations are the reasons for people to travel and such motivations will affect their decisions to choose one destination over another. A commonly acknowledged theory of travel motivation is the theory of push and pulls factors (Crompton, 1979). Crompton (1979) further highlighted that push factors are functional to explain the desire to travel while pull factors facilitate to clarify preferences for travel type or the choice of destination. Similarly, a number of researchers have endeavored to identify push factors causing a person to travel (Jang & Wu, 2006; Park & Yoon, 2009; Pearce & Lee, 2005).

Similarly, Timothy and Boyd (2003) claimed that a person visit heritage sites for reasons such as to "enhance learning, satisfy curiosity and feel nostalgia, grow spiritually, relax and get away from home". Likewise, Poria, Butler, and Airey (2004) identified three main categories of the reasons to visit heritage sites: 'heritage experience', 'learning history' and 'recreational experience' (Phosikham, 2010). Nyaupane, White and Budruk (2006) grouped heritage tourists into three categories according to their motivation to travel "culture-focused, culture-attentive, and culture-appreciation".

Meanwhile, Timothy and Boyd (2003) identified two types of heritage tourists: "passive and serious ones". They found that serious heritage tourists are motivated to obtain educational heritage experiences or personal nostalgia from their visits. Meanwhile, passive visitors are basically motivated to visit the world-famous heritage sites. In a similar setting, McKercher (2002) classified cultural tourists into five different categories: "the purposeful cultural tourists, the sightseeing cultural tourists, the casual cultural tourists, the incidental cultural tourists, and the serendipitous cultural tourists".

2.3 *Push and pull motivation*

Within every literature of motivation declared that motivation is commonly associated with "intrinsic" and "extrinsic" motivation. Intrinsic motivation is also recognized as "push" while extrinsic is identified as "pull" factors (Beldona, Morrison, & O'Leary, 2005; Chen & Chen, 2011). Dann (1977) claimed that one of the conceptual frameworks that reflect the psychological needs of tourists is the push and pull theory. The push and pull theory recognized that tourists are motivated by a push and pull factors to a destination. Dann (1977, p. 186) continues to define the concept by saying that push factors as a motivation, in which the tourist as subject and deal with those factors predisposing him to travel and the pull factors are motivators, that attract the tourist to travel. Most push factors are internal forces or intrinsic motivators (mainly social-psychological motives) that relate to the needs and wants of the tourists, for example, the desire for escape, rest and relaxation, excitement, prestige, health and fitness, relieving stress, doing different things (Correia & Valle, 2007).

On a similar vein, the pull factors come forward as an outcome of attractiveness of a destination as perceived by tourists such as recreational facilities, natural attractions, cultural attractions (Correia & Valle, 2007). According to Crompton (1979) push factors are socio-psychological motives, which are originally motivated from one's interest and desire to do a particular task rather than be affected by the external environment. Anomie and ego enhancement are said to belong to push factors (Chen & Chen, 2011). They further explained that anomie means that people's desire to escape from daily life can be fulfilled by traveling, while ego-enhancement means people need to be recognized by others.

On the contrary, extrinsic or pull motivation is basically known as pressure derived from external factors of the person. Crompton (1979) confirmed that pull factors are motives aroused by the destinations rather than emerging exclusively from within the tourist himself. In addition, Nicolau and Más, (2006) discovered that destination features are the main moderating role in nearly all of the tourist's motivation study. They examine the physical, cultural and interpersonal motivation. Given these points, to build the tourist intention to revisit, the practitioner must understand the role of motivations to enhance their behavior. Therefore, it is vital to consider both motivation and satisfaction constructs at the same time in order to access the tourists intention to revisit heritage destination.

Meanwhile, Crompton (1979) argued that, even though most researchers in tourism literature generally illustrate push motivations as those factors describing a person's desire to travel while pull factors aid to elucidate the choice of destination or preferences for travel type, it is also reasonable to predict that a person is pushed towards different destination and travel types as a result of different motivational factors.

2.4 *Revisit intention*

Revisit intention derived from the antecedent of actual behavior (Muala, 2010). As widely agreed by most researcher, repeat tourist can be considered as a major contributor towards the business opportunities and development of the tourism destination (Brida & Schuderi, 2013; Kim & Brown, 2012). Cole and Chansellor (2009) pointed out that resilient desire to revisit intention is influenced by the overall quality of experience during the visit.

In the context of travel and tourism studies, understanding customer loyalty will let the management to focus on the factors influencing customer retention (Chi & Qu, 2008). Gotlieb et al. (1994) emphasized that positive satisfaction has a positive influence on tourists' repurchase intention. Likewise, Lee et al. (2005) asserted that a person who has positive destination image is more likely to have a greater level of satisfaction and behavioral intention.

3 RESEARCH METHOD

The population of this study involved two heritage destinations namely Penang and Melaka. The reason of choosing Penang and Melaka as the preferred heritage destination was due to reputation of both states as World Heritage City. Both destinations also possess high tourist arrivals and diversity of tourists every year. Melaka and Penang own profusion of heritage buildings, houses and sites, a mixture of cultures, and rich historical diversity. Chhabra, Healy, & Sills (2003) affirmed that the built environments (such as historic homes, buildings, castles, and industrial sites) are the indicator and can be considered as the predominant attraction for heritage tourism.

In choosing appropriate respondents for data collection of the study, the researcher employed simple random sampling technique. Kumar, Talib and Ramayah (2013) and Zikmund et al. (2013) pointed out that for simple random sampling; each and every elements of the chosen population has an equal chance to be selected as sample. Thus, the international and domestic tourist has an equal chance to be chosen as a respondent.

However, it is hard to access the domestic tourists as they are possibly local people that occupy their leisure time during the weekend. Therefore, before the researcher decides to distribute questionnaire to a particular tourist, he or she was asked a few questions which entitled them to answer the questionnaire. They were orally asked the state of origin and purposes of visiting Melaka or Penang. The procedure is compulsory for every domestic person in order to differentiate the tourist from the local people in order to obtain valid respondent for the research. The push and pull motivation questionnaire were measured on the Likert Scale from 1–10, ranging from "strongly disagree" (1) to "strongly agree" (10). 600 questionnaires were distributed in two heritage tourism destination in Malaysia, namely Melaka and Penang.

4 RESULTS

A total of 510 usable questionnaires were collected from the respondents. Analyses of the data revealed that majority of the respondents were male (52.4%) and female (42.6%). The age group were fairly evenly distributed with the exception of "21–30" group consisting of 51%, and the group of "31–39" consisting of 40%. 55% of the respondents were married and 27 percent were single and the rest were divorced (18%). The respondents travel to heritage tourism destination had a high level of education 86.5% from tertiary education background, with a moderately even split of those were manager or executive, 55%. 79.4% of the respondents were from international tourist while 20.6% were from domestic tourist. Majority of the respondents (80%) indicated that this was their first trip to their heritage tourism destinations.

Findings from the assessment of fitness for the measurement model shows that all fitness indexes required in evaluating the fitness of a model are achieved. It was indicated that the result of Confirmatory Factor Analysis (CFA) was excellent. Most of the fitness indexes have achieved the required level for measurement model. The measurement model fit was great with P-value = 0.000, GFI = 0.940, CFI = 0.966, RMSEA = 0.077 and NORMEDCHISQ = 4.010 indicating a strong model fit. The result of cronbach alpha, Average Variance Extracted (AVE) and Composite Reliability (CR) for push motivation is (α = 0.763, AVE = 0.599, CR = 0.737), for pull motivation (α = 0.781, AVE = 0.573, CR = 0.714), and for revisit (α = 0.904, AVE = 0.583, CR = 0.907).

Table 1. Goodness-of fit measures for the structural equation model (N = 510).

Name of category	Name of index	Index value	Comments
1. Absolute fit	RMSEA	0.056	The required level is achieved
Absolute fit	GFI	0.968	The required level is achieved
2. Incremental fit	CFI	0.985	The required level is achieved
3. Parsimonious fit	Chisq/df	2.589	The required level is achieved

Table 1 shows that the result of the structural model fit was great with p-value = 0.000, GFI = 0.968, CFI = 0.985, RMSEA = 0.056 and NORMEDCHISQ = 2.589 indicating a strong predictive validity.

The purpose of this study is to study the effect of push motivation on revisit intention (H1), and the effect of pull motivation on revisit intention (H2). H1, H2 and H3 were supported in this study with a significant level of (p. 001). H1 is supported as the unstandard beta = 0.607, standard beta = 0.613, t-value = 9.903. It is also supported for H2 when unstandard beta = 0.212, standard beta = 0.224, t-value = 3.7133. H3: The result for H3: pull motivation positively influence push motivation to revisit intention as the unstandard beta = 0.128, standard beta = 0.148, t-value = 3.347.

5 CONCLUSION

The objective of the study was to examine and to clarify the relationship of push and pull motivation and revisit intention to heritage tourism destination. It was believed that tourists' revisit intention can be measured by push and pull motivation factor. The results of the present study indicate that motivations clearly influence tourist to revisit the same destination. Regardless of numerous studies in tourist motivation in tourism industry, the implication of the study will be valuable to the heritage tourism destination especially to the heritage site.

The findings also indicate that pull motivation (the destination attributes) has a significant relationship with push motivation (internal forces). The result proved that regardless of the push motivation that tourists have, the forces of pull motivation (listed as world heritage site, different cultures, and thrills and excitement in heritage) strengthen the tendency to revisit. It also shows that pull motivation influence the tourists to revisit the same destination.

REFERENCES

Ahmad, A.G. (2006). Cultural heritage of Southeast Asia: preservation for world recognition. *Journal of Malaysian Town Plan, 3*(1), 52–62. Retrieved from http://issuu.com/anwar_townplan/docs/townplan_04?mode=window&viewMode=doublePage.

Alzua, A., O'Leary, J.T., & Morrison, A.M. (1998). Cultural and heritage tourism: identifying niches for international travellers. *Journal of tourism studies, 9*(2), 2–13.

Beldona, S., Morrison, A.M., & O'Leary, J., (2005). Online shopping motivations and pleasure travel products: a correspondence analysis. *Tourism Management, 26*(4), 561–570.

Brida, J.G., & Scuderi, R. (2013). Determinants of tourist expenditure: A review of microeconometric models. *Tourism Management Perspectives, 6*, 28–40.

Chen, L., & Chen, J.S. (2011). The motivations and expectations of international volunteer tourists: A case study of "Chinese Village Traditions". *Tourism Management, 32*(2), 435–442. doi:10.1016/j.tourman. 2010.01.009.

Chhabra, D. (2001). Heritage Tourism: An Analysis of Perceived Authenticity and Economic Impact of the Scottish Highland Games in North Carolina (Doctoral Dissertation, North Carolina State University, North Carolina, NC). Retrieved from http://repository.lib.ncsu.edu/ir/bitstream/1840.16/4787/1/etd.pdf.

Chhabra, D. (2010). Back to the past: a sub-segment of Generation Y's perceptions of authenticity. *Journal of Sustainable Tourism, 18*(6), 793–809. doi:10.1080/09669582.2010.483280.

Chhabra, D. (2010). Sustainable marketing of cultural and heritage tourism. New York, NY: Routledge.

Chhabra, D., Healy, R., & Sills, E. (2003). Staged authenticity and heritage tourism. *Annals of Tourism Research, 30*(3), 702–719. doi:10.1016/S0160-7383(03)00044-6.

Cole, S.T., & Chancellor, H.C. (2009). Examining the festival attributes that impact visitor experience, satisfaction and re-visit intention. *Journal of Vacation Marketing, 15*(4), 323–333. doi:10.1177/1356766709335831.

Correia, A., & Valle, P.O. (2007). Why people travel to exotic places. *International Journal of Culture, Tourism and Hospitality Research, 1*(1), 45–61. doi: 10.1108/17506180710729600.

Crompton, J.L. (1979). An Assessment of the Image of Mexico as a Vacation Destination and the Influence of Geographical Location Upon That Image. *Journal of Travel Research, 17*(4), 18–23. doi:10.1177/004728757901700404.

Dann, G. (1977). Anomie, ego-enhancement and tourism. *Annals of Tourism Research, 4*(4), 184–194. doi:10.1016/0160-7383(77)90037-8.

Flambard-Ruaud, S. (2005). Relationship marketing in emerging economies: some lessons for the future. *Vikalpa, 30*(3), 53. Retrieved from http://www.vikalpa.com/pdf/articles/2005/2005_july_sep_53_63.pdf.

Fodness, D. (1994). Measuring tourist motivation. *Annals of Tourism Research, 21*(3), 555–581. doi:10.1016/0160-7383(94)90120-1.

Furchtgott, E., & Furchtgott, M.W. (1999). *Aging and human motivation*. New York, NY: Plenum Publishing Corporation.

Garrod, B., & Fyall, A. (2000). Managing heritage tourism. *Annals of Tourism Research, 27*(3), 682–708. doi:10.1016/S0160-7383(99)00094-8.

Hall, C.M., & Zeppel, H. (1990). History, Architecture, Environment: Cultural Heritage and Tourism. *Journal of Travel Research, 29*(2), 54–55. doi:10.1177/004728759002900212.

Hitchcock, M., King, V.T., & Parnwell, M. (2010). *Heritage tourism in Southeast Asia* (Eds.). Copenhagen, Denmark: NIAS Press.

Jang, S.C.S., & Wu, C.M.E. (2006). Seniors' travel motivation and the influential factors: An examination of Taiwanese seniors. *Tourism Management, 27*(2): 306–316. doi:10.1016/S0261-5177(01)00090-5.

Joliffe, L., Rowe, T., & Davis, C.H. (2002, October). *Using the WWW to develop cultural heritage destinations: An exploratory study*. Paper presented at the 2nd Tourism and Travel Research Association Canadian Conference, Ontario, Canada. Retrieved from http://www.ryerson.ca/~c5davis/publications/cultural_heritage_tourism_destinations-TTRA-Oc.PDF.

Kau, A.K., & Lim, P.S. (2005) Clustering of Chinese Tourists to Singapore: An Analysis of Their Motivation, Values and Satisfaction. *International Journal of Tourism Research* 7, 231–248. doi:10.1002/jtr.537.

Kim, A.K., & Brown, G. (2012). Understanding the relationships between perceived travel experiences, overall satisfaction, and destination loyalty. *Anatolia, 23*(3), 328–347.

Kozak, M. (2002). Comparative analysis of tourist motivations by nationality. *Tourism Management 23*(3): 221–232. doi: 10.1016/S0261-5177(01)00090-5.

Kumar, M., Talib, S.A., & Ramayah, T. (2013). *Business Research Method*. Selangor, Malaysia: Oxford University Press.

McKercher, B. (2002). Towards a classification of cultural tourists. *International Journal of Tourism Research, 4*(1), 29–39. doi:10.1002/jtr.346.

McReynolds, P. (1991). Motives and metaphors: A study in scientific creativity. In D.E. Leary (Ed.), Metaphors in the history of psychology (pp. 133–172). Cambridge, UK: Cambridge University Press.

Millar, S. (1989). Heritage management for heritage tourism. *Tourism Management, 10*(1), 9–14. doi:10.1016/0261-5177(89)90030-7.

Moscardo, G. (1996). Mindful visitors: Heritage and tourism. *Annals of Tourism Research, 23*(2), 376–397. doi:10.1016/0160-7383(95)00068-2.

Nicolau, J., & Mas, F. (2006). The influence of distance and prices on the choice of tourist destinations: The moderating role of motivations. *Tourism Management, 27*(5), 982–996. doi:10.1016/j.tourman.2005.09.009.

Nyaupane, G.P., White, D., & Budruk, M. (2006). Motive-based tourist market segmentation: An application to Native American cultural heritage site in Arizona, USA. *Journal of Heritage Tourism, 1*(2), 81–99. doi:10.2167/jht010.0.

Park, D.B., & Yoon, Y.S. (2009). Segmentation by motivation in rural tourism: A Korean case study. *Tourism Management, 30*(1), 99–108. doi:10.1016/j.tourman.2008.03.011.

Poria, Y., Butler, R., & Airey, D. (2004, May 27). Links Between Tourists, Heritage, and Reasons for Visiting Heritage Sites. *Journal of Travel Research*. Retrieved from http://epubs.surrey.ac.uk/1135/1/fulltext.pdf.

Pearce, P., & Lee, U.L. (2005). Developing the travel career approach to tourist motivation. *Journal of Travel Research,* 43, 226–237. doi:10.1177/0047287504272020.

Pearce, P.L., Morrison, A.M., & Rutledge, J.L. (1998). Tourism Bridges Across Continents. Sydney: McGraw-Hill Companies, Inc.

Phosikham, T. (2010). Heritage sites as tourist attractions: A case study of Luang Prabang, Lao People's Democratic Republic (Masters' Dissertation, Lincoln University). Retrieved from http://researcharchive.lincoln.ac.nz/dspace/bitstream/10182/2584/3/Phosikham_MTM.pdf.

Timothy, D.J., & Boyd, S.W. (2003). Heritage Tourism. Essex, England: Pearson Education.

Timothy, D.J., & Nyaupane, G.P. (2009). Cultural heritage and tourism in the developing world: A regional perspective. New York, NY: Taylor & Francis.

Van der Borg, J., Costa, P., & Gotti, G. (1996). Tourism in European heritage cities. Annals of Tourism Research, *23*(2), 306–321. doi:10.1016/0160-7383(95)00065-8.

Zikmund, W.G., Babin, B.J., Carr, J.C., & Griffin, M. (2013). Business Research Methods (9th ed.). Ohio, OH: South-Western Cengage Learning.

Hospitality and Tourism – Sumarjan et al. (eds)
© 2014 Taylor & Francis Group, London, ISBN 978-1-138-00151-0

Hospitality sponsorship program: Consumer involvement and product awareness

M.A. Hadi, S.A. Jamal & H.A. Hadi
Universiti Teknologi MARA, Malaysia

ABSTRACT: The relevance of this study lays in the search for the relationship between consumers involvement attributes in the sponsorship program, and product awareness towards the intention to purchase the sponsors product. In spite of all indicators pointing to the need for a clear understanding of the business value of sponsorships, little is known about basic aspects for the consumers' involvement in the sponsorship program, product awareness and intention to purchase the product. Using data from selected literature from relevant journals been analyzed—additional three events in different hospitality organizations categories, been evaluated by analyzing responses from participants and attendees from the respective events. This study finding has clearly revealed that consumer involvement activity plays a vital role towards the awareness of sponsor's product and in consumer purchase intention behavior. This paper helps hospitality organizations to take decisions with regard to creating consumer awareness by reviewing important issues relating to both consumers' involvement, in an event in order to reach the consumers awareness and purchase intention towards the sponsor's product.

Keywords: consumer involvement, intention to purchase, product awareness, sponsorship

1 INTRODUCTION

Sponsoring an event opens a promising marketing opportunity to any organization and institution. This is in relation to the notion that event sponsorship activities introduced by several well-known companies since year 1983 where early research on sponsorship attempted to establish the role of sponsorship in marketing communications (Meenaghan, 1983). This sponsorship program activity has gain popularity in Malaysia where some hospitality based companies take part in prestigious events, such as Marriot Hotel Putrajaya, and Kentucky Fried Chicken (KFC) in Hot Air Balloon Festival 2012, and in Frost & Sullivan 4th Annual Customer Interaction Malaysia 2012 (Malaysia Tourism Event & Festival, 2012).

At the moment where the products and services being promoted and expose through television and media, sponsorship has been another approach in promoting the product through sponsorship programs attached with the consumers involvement in an event. Some consumers' interest towards a product could be tackled through their direct involvement with the product exposure. This is where the sponsorship program related with the consumer involvement, found that when the eye-to-eye contact and personal approach gave much more result towards the consumer product awareness and will lead to their purchase intention.

This study aim to discover the effectiveness of consumer involvement in the sponsorship program could influence consumers' awareness compared to the traditional marketing approach, through several Medias in capturing the consumers' intention to purchase. In addition, from the previous studies of relevant journals, sponsorship marketing activities is under investigation whether the use of sponsorship has positive effects on consumer involvement, attitude, and awareness and purchase intention.

2 LITERATURE REVIEW

2.1 *Sponsorship program*

Sponsorship program is an orchestration and implementation of marketing activities aimed at building and communicating an association to sponsorship (Cornwell, 1995). Sponsorship program developed with intrinsically linked by dealing with the companies, firms or sponsorship by individuals. All these involve parties will have a negotiation among them and would come out with a decision in order to collaborate in the sponsorship program structure.

This sponsorship activity has become an increasingly important field in marketing research and now represents a significant part of the marketing mix. This promotional activity also has grown remarkably (Meenaghan, 2005), in the practice of promoting a company's interest and its product brand by tying them to a specific and meaningful related event, organization or charitable cause (Endogan & Kitchen, 1998).

2.2 *Consumer involvement in sponsorship program*

Consumer involvement refers to motivation to process information regarding the target object, and motivational state governed by the perceived relevance of the target object (Celci & Olson, 1988; Zaichkowsky, 1985). Accordingly, perceived personal relevance effects consumer processing of the event and the sponsorship stimuli (Johar & Pham, 1999; Pham, 1992). Attraction, centrality, and self-expression are three dimensions of involvement that consistently examined in previous studies (Alexandris & Tsiotsou 2012).

The above notion supported by McIntyre and Pigram (1992) that suggested activity involvement consists of three elements: attraction, self-expression, and centrality. Attraction is pleasure related to the activity; self-expression is the image participants want to present of themselves through participation in an activity, and centrality is the value of an activity as it relates to other aspects of a participant's life. Therefore, these dimensions applied in sponsorship event research influence sponsorship outcomes (Alexandris et al., 2007).

2.3 *Product awareness*

According to Farshchian (2001), awareness is the information that human beings exchange with their environment in order to coordinate their work with others. Consumers' product awareness and product knowledge represents the product-related experience and accumulated information (Tsai, 2007). Amaoka, Baah, Dzogbenuku, and Junior (2012) stated that in sponsorship context; product awareness consists of how consumers determine the sponsorship impacts on their product's brand preferences, their understanding of the concept of sponsorship and the message that sponsorship communicate to them and how clearly they understand the message. It is important to notice that, the message sponsorship affects their awareness of the product.

2.4 *Purchase intention*

The basis of a purchase intention is a positive and favorable attitude toward a product, and many studies have shown that sponsorship is effective in increasing purchase intentions (Lee, 2005). Firms sponsor events to elicit a variety of consumer responses, including increased interest in purchasing the firm's product. Purchase intent is the likelihood that a consumer will buy a product (Lacey, Close & Finney, 2010). Many studies have shown that sponsorship is effective in increasing purchase intentions (Lee, 2005; IEG Sponsorship Report, 1992; Pitts, 1998).

2.5 *The relationship between consumer involvement, product awareness and intention to purchase*

Hypothesis: There is a relationship between consumer involvement in the sponsorship program and product awareness and intention to purchase.

The consumer involvement level in the purchase decision-making process has received a great attention in general marketing literatures both for its encouraging function, and its influence on consumer attitude and behavior (Zaichkwosky 1985; 1987; Joel, Carmina, & Carlos, 2009). Consumer involvement has shown itself as a key variable to explain some of consumers' responses to marketing stimuli. In relation with sponsorship context, sponsorship takes learning on both sides of sponsor product and consumers' product awareness; sponsorship remains a clear avenue for companies to connect with their customers in a sincere and heartfelt way (IEG Network, 2011).

In attachment with consumer involvement in this program, sponsorship program allows companies to reach specifically targeted niche markets without any waste and offer the possibility of achieving several goals at once.

3 METHODOLOGY

3.1 *Nature of study*

In fulfilling the study objectives, a quantitative data analysis using correlation technique employed by using a structured questionnaire distributed to the target sample. A cross-sectional study method carried out for data collection where data gathered from respective events within two months period. This study has proceeded with a self-administered questionnaire feedback from the target sample.

The events held in Kuala Lumpur as well as Selangor and Putrajaya selected from Tourism Malaysia's Event and Festival 2013 calendar list. This study is focusing on the event attendees as the individual unit of analysis.

3.2 *Sampling*

The population for this study included all consumers above 18 years old participating in hospitality sponsorship program of event in

Kuala Lumpur and Selangor. The proposed sample size for this study is 384 respondents from the selected events.

3.3 *Instrumentation*

The survey instrument comprised of 41 items. It included a series of questions relating to the following five areas; Section A; Event attendees' description (Andrew, 2011; Schuweilef, 2008), Section B: Consumer involvement (Alexandris & Tsiotsou, 2012; Close, Finney, Lacey & Sneath, 2006; Gerard, Alan, Robert & James, 2004; Kyle et al., 2005; Kyle et al., 2004; Kyle et al., 2003; Lacey, Close & Finney, 2010; McIntyre, 1990; McIntyre & Pigram's, 1992; Thomas, Joseph, & Christopher, 2009), Section C: Product awareness (Amaoka, Baah, Dzogbenuku & Junior, 2012; Nelson, Keum & Yaros, 2004; Winkler & Bunker, 2010), Section D: Purchase Intention (Baker & Churchill, 1997; Yong, Kyongtae, Cathryn & Tae, 2008), Section E: Socio-demographic information (Alexandris, Douka, Bakaloumi & Tsasousi, 2008; Close, Finney, Lacey & Sneath, 2006).

3.4 *Data collection process*

The set of questionnaires forwarded to the Tourism department and the event's secretariat, and the survey questionnaires distributed to the attendees of the events. A drop-off and pick up method selected as the mode of collecting the sets of questionnaires, whereby the collection time was at the end of the event day, or maximum days of four days after the event ended.

4 RESULTS AND FINDINGS

With reference to Table 1, the coefficient alpha for the first dimension of Consumer involvement produced a high coefficient value of 0.825. It followed by the consumer involvement sub-dimensions for Activity involvement which is Centrality with 0.757 coefficient value, a high coefficient value of 0.803 for Attraction, and the coefficient value of 0.737 for Self-expression. Product knowledge represents as the second dimension under Consumer involvement has achieved 0.737 coefficient values. All the more, Product awareness dimension has achieved 0.782 coefficient values and the final dimension for this study is Purchase intention that produced a high coefficient value of 0.826. All the above coefficient results are in line with Guielford (1965), whereby the bigger in Cronbach's alpha value, the higher the internal consistency is. If the alpha value is bigger than 0.70, then it shows reliability of measurement is fairly high.

Table 1. Reliability coefficients for each section of the questionnaire.

Questionnaire	No. of items	Cronbach's alpha
Section B		
Consumer involvement:	6	0.825
a. Activity involvement		
i. Centrality	5	0.757
ii. Attraction	6	0.803
iii. Self-expression	4	0.780
b. Product knowledge	6	0.737
Section C		
Product awareness	3	0.782
Section D		
Purchase intention	5	0.826

Table 2. Summary of regression analysis H1 results.

	B	SE B	β
Constant	0.367	0.22	
CONSINV main	0.957	0.054	0.67*

Dependent variable: product awareness.
Note: R square = 0.45, $*p < 0.001$.

Table 3. Summary of regression analysis H2 results.

	B	SE B	β
Constant	1.802	0.16	
Prod. aware.	0.528	0.039	0.57*

Dependent variable: intention to purchase.
Note: R square = 0.33, $*p < 0.001$

Table 4. Summary of regression analysis H3 results.

	B	SE B	β
Constant	0.624	0.204	
CONSINV main	0.854	0.052	0.65*

Dependent variable: intention to purchase.
Note: R square = 0.42, $*p < 0.001$.

4.1 *Regression analysis*

Consumer involvement made a statistically significant contribution to the product awareness, and it explained 45% of the variation in the model (see Table 2).

Product awareness made a statistically significant contribution to the Intention to purchase, and it explained 33% of the variation in the model (see Table 3).

Consumer Involvement made a statistically significant contribution to the Intention to Purchase, and it explained 42% of the variation in the model (see Table 4).

5 DISCUSSION

From the findings, consumer involvement is a primary determinant of consumer behaviour. In this study, the influence of consumer involvement dimensions that is activity involvement and product knowledge towards the sponsor's product tested. The result of analysis supported the hypothesis that activity involvement and knowledge of the sponsor's product in a sponsorship event program has a significant relationship and does influence the awareness of product and consumer purchase intention. A significant and direct link between the consumer involvement dimensions highlighted above and purchase intention also revealed, and it has a strong significant between them. From this finding, it can be concluded that the consumer involvement (activity involvement and product knowledge) has given a significance impact towards consumer purchase decision. Once the consumer involving themselves in this type of event program, through the activity involvement and knowledge of the sponsor's product delivered directly to them, the consumers would have a better picture pertaining to any respective hospitality product sponsored and at the same time, the purchase intention to buy the sponsor's product may occur. This also supported by the descriptive result also shows that the consumers respond resulted that consumer involvement dimensions has a vital function in consumer purchase intention behaviour.

It also indicates that the respondents agree with the importance of product awareness in this situation resulted sponsorships do communicate a clear message about product information to target audience. Consumer feels that this sponsorship program activity positively influences their purchase intention when their interest toward sponsor's product by providing fruitful information of the product, and make the consumer extra aware on any stipulated product.

Regression results also shows that consumer involvement in the sponsorship program and product awareness dimensions significantly contributed in the elements of purchase intention, whereby without this post-purchase behaviour, people might have less information pertaining to certain product and might have a less confidence to choose the product. This finding has supported the hypotheses prediction.

5.1 *Study limitations/practical implication*

This study carried out with some limitations, the sample limited to the event attendees/consumers, and the feedback gathered cannot be generalized to all events. Therefore, it suggested that future study should be carried out using two samples of respondents, one from the consumers' feedback and the other one are from the sponsor company feedback. The purpose is to see the pattern of how the sponsorship program activity could improve the consumer awareness towards purchase intention and also the company sales profitability.

Secondly, future research should include the events from other area or countries outside Kuala Lumpur, Selangor and Putrajaya to compare differences in respondents of other areas.

It also suggested that future research should be conducted using sales as an additional factor to measure the effect of this sponsorship activity. It suggested that future research to measure the impact of traditional advertising (television and radio) on marketing communication performance. This will help in establishing whether sponsorship could stand on alone to contribute positively towards creating product awareness, the sponsors image and goodwill as expected by sponsorship communication.

5.2 *Conclusion/recommendations*

The analysis earlier have mentioned clearly and addresses the purpose of this study by giving a clear understanding of how consumer involvement in the sponsorship program contributes positively to marketing communication performance, in this context the hospitality product promotion. Consumer involvement in this sponsorship activity increases marketing awareness of a consumer through their direct involvement in this event program and finally, affects their behaviour intention on the purchase intention.

It found that sponsorship has a positive impact on the product awareness of the sponsor's product. This sponsorship activity also generates goodwill and product awareness and the recall of the sponsor's product by the event attendees. While consumer getting involved with sponsorship program activity, this is different from traditional advertising result although both of them have similar objectives. This approach could have change consumer perception on a product, and also projects a high persuasion of consumer. This is because sponsorship communicates the sponsor's product and its benefits. As such, there is a positive link between sponsorship and marketing communication performance.

The results show that consumer involvement in the sponsorship program affects marketing communication performance positively in attachment by having a direct involvement by the consumers, when they believe that sponsorship may permission them to have a touch to touch point with the sponsors' product. This would be different for them by only watching the advertisement and could not experience the product personally before the purchase decision made. Sponsor's product patronized by consumers as a result of the sponsorship program.

It recommended that the respective government and tourism related bodies as well as the marketing managers of the related industry should outline a strategic plan for any sponsorship program to be implemented; thus sponsorship should be a part of integrated marketing communication.

REFERENCES

Alexandris, K., Douka, S., Bakaloumi, S., & Tsasousi, E. (2008). The influence of spectators' attitudes on sponsorship awareness: A study in three different leisure events. *Managing Leisure*, *13*(1), 1–12.

Alexandris, K. & Tsiotsou, R.H. (2012). Segmenting soccer spectators by attachment levels: A psychographic profilebased on team self-expression and involvement. *European Sport Management Quarterly*, *12*(1).

Alexandris, K., Tsaousi, E., & James, J. (2007). Predicting sponsorship outcomes from attitudinal constructs: The case of a professional basketball event. *Sport Marketing Quarterly*, 16, 13–139.

Amaoka, G.K., Baah, K.D., Dzogbenuku, R.K., & Junior, S.K. (2012). The effect of Sponsorshi Marketing Communication perfomance: A case study of Airtel Ghana. *African Journal of Marketing Management, 4*(2), 65–79.

Andrew, O. (2011). Irish fair of Minnesota. Profile of Irish Fair of Minnesota Attendees.

Baker, M.J. & Churchill, G.A. Jr. (1997). The Impact of physically attractive models on advertising evaluations. *Journal of Marketing Research*, *14*(4), 538–555.

Brucks, M. (1985). The effects of product class v knowledge on information search behavior. *Journal of Consumer Research,* 12, 1–16.

Celci, R.L. & Olson, J.C. (1988). The role of involvement inattention and comprehension processes. *Journal of Cosumer Research*, *15*, 210–224.

Close, A.G., Finney, R.Z., Lacey, R.Z., & Sneath, J.Z. (2006). Engaging the consumer through event marketing: linking attendees with the sponsor, community, and brand. *Journal of Advertising Research*. doi:10.2501/S0021849906060430.

Cornwell, T.B. (1995). Sponsorship-linked-marketing development. *Sport Marketing Quarterly, 12*(4), 13–24.

Crompton, J.L. (1995). Economic impact analysis of sports facilities and events: Eleven sources of misapplication. *Journal of Sport Management*, *34*(9), 14–35.

Erdogan, B. & Kitchen, P. (1998). Managerial mindsets and the symbiotic relationship between sponsorship and advertising. *Marketing Intelligence & Planning, 16*(6).

IEG Sponsorship Report (1992). Measuring sponsorship's impact on corporate image: Finding from John Hancock, *12*(9), 4–5.

IEG Network (2011). Leading the way in sponsorship consulting, valuation, measurement and strategy. Retrieved from http://www.sponsorship.com/.

Joel, E., Carmina, F., & Carlos, F. (2009). The influence of consumer involvement on quality signals perception: An empirical investigation in the food sector. *British Food Journal, 111*(11), 1212–1236.

Johar, G.V. & Pham, M.T. (1999). Relatedness, prominence and constructive sponsor identification. *Journal of Marketing Research, 36*(3), 299–312.

Kyle, G., Graefe, A., Manning, R., & Bacon, J. (2004). Predictors of behavioural loyalty among hikers along the Appalachian Trail. *Leisure Sciences*, *26*, 99–118.

Lacey, R., Close, A.G.R., & Finney, Z. (2010). The pivotal roles of product knowledge and corporate social responsibility in event sponsorship effectiveness. *Journal of Business Research*, *63*, 1222–1228.

Lee, S. (2005). The influence of product involvement and fan identification on response to team sponsor's products. Unpublished thesis.

Malaysia Tourism Event and Festival (2012). Retrieved from http://www.ecotourismhub.com/events-and-festivalmalaysia.html.

McIntyre, N. & Pigram, J. (1992). Recreation specialisation re-examined: The case of vehicle-based campers. *Leisure Sciences*, *1*, 3–15.

McIntryre, N. (1990). Recreation involvement. The personal meaning of participation. Unpublished doctoral dissertation. University of New England, New South Wales, Australia. 183.

Meenaghan, J. (1983). Commercial Sponsorship. *European Journal of Marketing,* 5–72.

Meenaghan, T. (2005). Evaluating sponsorship effects. *Global sport sponsorship*, 243–264.

Nelson, M.R., Keum, & Yaros, R.A. (2004). Advertainment or Adcreep? Game players' attitudes toward Advertising and Product Placements in Computer Games. *Journal of Interactive Advertising*, *5*(1).

Pits, B.G. (1998). An analysis of sponsorship recall during Day Games IV. *Sport Marketing Quaterly, 7*(4), 11–18.

Schuweilef, A. (2008). International festival of owls. Profile of 2008 International Festival of Owl Attendees.

Thomas, L.B., Joseph, C.J., & Christopher, D.H. (2009). The impact of involvement on key service relationships. *Journal of Service Marketing*, 23/2, 115–124.

Tsai, W. (2007). Competitive Tension: The Awareness-Motivation-Capability Perspective. *Academy of Management Journal*, *50*(1), 101–118.

Winkler, T. & Buckner (2010). Receptiveness of gamers to embedded brand messages in advergames: Attitudes towards product placement. *Journal of Interactive Advertising, 7*(1), 24–32.

Yong, J.K., Kyongtae, K., Cathryn, L.C., & Tae, H.K. (2008). The effects of sport involvement, sponsor awareness and corporate image on intention to purchase sponsors' products. *International Journal of Sports Marketing & Sponsorship Quarterly, 7*(4), 11–18.

Zaichkowsky, J.L. (1985). Measuring the involvement construct. *Journal of Consumer Research,* 12, 341–352.

Zaichkwosky, J.L. (1987). The emotional aspect of product involvement, *Advances in Consumer Research*, 14, 32–35.

Hospitality and Tourism – Sumarjan et al. (eds)
 ISBN 978-1-138-00151-0

The role of tourism in the rural area development

A. Daneshvarinasab
Islamic Azad University, Bardsir, Iran

M. Mohd Dali
University of Malaya, Malaysia

H. Zeinizadeh
Islamic Azad University, Tabriz, Iran

ABSTRACT: Rural tourism is considered to be the kind of tourism happening around the villages. Such tourism usually follows the aims of raising the income of rural families, educating the people involved in tourism and increasing the people's participation in the tourism industry. Rural tourism includes six different types: cultural, village, agricultural, farmland, natural and ecotourism. The main reasons for paying attention to rural tourism are to increase the income of the villagers, decreasing the rate of migration from rural areas, and improving the knowledge of the inhabitants of rural areas through interaction with tourists. It seems that rural tourism plays a significant role in increasing the income of the villagers through selling their agricultural products, handicrafts, and local foods provided that the barriers on the way of this industry such as lack of managerial expertise, marketing, transportation, seasonality of tourism are eliminated or at least reduced.

Keywords: constraint to rural tourism, rural tourism, tourism

1 INTRODUCTION

Rural tourism has high significance in the countries with big rural population. Rural tourism has various and several forms. This variation and diversity which make it somehow difficult to provide a single comprehensive definition of rural tourism. According to the commission of European society (1986), rural tourism in not only confined to agricultural tourism. According to Mahdavi (2013) rural tourism refers to all the measures and activities taken by the rural inhabitants, people and government, for the purpose of recreation and relaxation of tourists in order to attract them to rural areas. These measures and activities can include agricultural tourism, farmland tourism, and cultural tourism. In general, rural tourism includes not only the rural areas and environments surrounding it, but also the provision of tourism services and products to tourists. It also includes all the activities of this industry and takes diverse forms.

2 OBJECTIVES OF RURAL TOURISM

One of the rural tourism objectives is to increase the income of rural inhabitants. Another objective of rural tourism is education. This occurs through a series of professional education which should be provided to all the people involved in this huge industry in rural tourism like villagers, personnel of active institutes, and organizations. Of course, these educational programs should be planned on the basis of the principles of sustainable development, and prepare the individuals for managerial activities (Unwin, 1996). This will not only result in the managing the time and resources but also will result in the increase of the number tourists visiting the rural areas in comparison to previous years. This will automatically result in the increase of the income of the villagers. Some of the other objectives of rural tourism are to encourage the participation of the villagers in the process of policy-making regarding provision of recreational and sport facilities, and the introduction of the natural and geographical image of their villages. Some other purposes of rural tourism are to provide the people with information about the organizations and institutions which are directly or indirectly involved in the tourism industry (Fleicher and Felsenstien, 2004).

Generally, there are four main objectives identified for tourism: training the people to provide service in rural tourism on the basis of the principles of sustainable development, to prepare the people to found and manage the small rural industries, to prepare the people to participate in the policy-making processes in rural tourism industry, training the people for developing the opportunities, activities and recreations for tourists on the basis of sustainable development (Pina, 2004).

3 TYPES OF RURAL TOURISM

Rural tourism is classified on the basis of the capacities of the local tourism and the purposes of tourists. Rural tourism includes either of the following types:

i. Cultural tourism: is related to the culture, history, cultural heritage, monuments, and the history of indigenous people (Khayati, 2002; Sharifzadeh & Moradnejad, 2002; Maghsoudi & Lashkarara, 2003)
ii. Natural tourism: has mainly recreational aspects, has the least destructive effects on ecology and the most interaction with the ecological attractions (Khayati, 2002; Sharifzadeh & Moradnejad, 2002; Maghsoudi & Lashkarara, 2003)
iii. Village tourism: in this type of tourism, tourists live in villages and take part in the social and economic activities of the villagers (Ryan, 1991; Maghsoudi and Lashkarara, 2003).
iv. Eco-tourism: it is mainly in interaction with the natural attractions. It not only preserves the natural resources, but also considers the people's welfare and social values (Khayati, 2002; Maghsoudi and Lashkarara, 2003)
v. Agricultural tourism or agro-tourism: Without having any destructive impact on nature, tourists watch the traditional agricultural activities and sometimes take part in those activities (Khayati, 2002; Maghsoudi and Lashkarara, 2003)
vi. Farmland tourism or farm-tourism: This type of tourism tries to investigate and introduce rural tourism as a special institution in the organization in order to connect them to agricultural companies (Maghsoudi and Lashkarara, 2003; FFTC, 2004).

4 FORMS OF RURAL TOURISM

The way the rural tourism is formed and organized varies one country to another. In some countries it is the government which is the main resource for the investment in rural tourism, but in some other countries private sector may also embark on investing in rural tourism (FFTC, 2004). Rural tourism might even be defined in different ways in different countries. Below, some instances of rural tourism will be mentioned. In Finland, rural tourism means providing the tourists with rental village cottage and services. In Hungary, the term village tourism is more common but in Netherland, it refers to tourists' camping in farmlands along with activities like horse riding and walking. In south Asia, where tourism industry has remarkably advanced, the local inhabitants of rural areas earn a rather significant amount of money by selling their handicrafts and services to tourists (Sharifzadeh & Moradnejad, 2002). In Indonesia, rural tourism has seriously advanced in the farmlands of Sumatra and Java where tourists usually go to see the rice plantation activities and collection of latex from rubber trees. In Japan, the common form of rural tourism is farmland tourism. In France, which is one of the absolute examples of developing rural tourism, villagers allocated some parts of their Farmlands For Tourists' Camping (FFTC, 2004).

5 NATURAL ATTRACTIONS OF RURAL TOURISM

What are the factors which motivate a person to pass a long way to stay in a rural environment even for a short while? One of the reasons which motivate rural tourism might be the rural occupations. The following reasons can be mentioned as the most important reasons: natural and geo-morphological conditions of the village, life style, traditions and culture of villages, rural house and residences (such as cottages, stone-made and mud-made houses) (Fleicher & Felsenstien, 2004) local and rural handicrafts (Rezvani, 1998), the untouched attractions surrounding the villages, economic attractions, cultural and social attractions of the village, (Moradnejad and Nazeri, 2002), indigenous fauna, rare floral cover, historical monuments of the villages, local and traditional foods of the village (FFTC, 2004), Recreation and sports (water, land, air) and health-related activities (Fleicher and Felsenstien, 2004).

6 REASONS FOR PAYING ATTENTION TO RURAL TOURISM

Tourism can be regarded as one of the components of development. It is necessary to take all the aspects of this category into the consideration. One of the reasons for this fact is the presence of the people whose income sources and residence exist in the villages. In case this issue is not properly explained for villagers, it might have irreversible and negative outcomes. For villagers, tourists are some strangers who enter their life space. Since the culture and traditions of the villagers vary from the tourists who usually come from urban areas, tourists might not be welcome to villages and villagers show no inclination to establish contact with tourists. This will result in some problems among which the followings are the most important: decrease of the economic growth of the village and the decline of the social image of villages (Fleicher & Felsenstien, 2004). On the other hand, if we had a look at the rate of income of both rural and urban residents, we would find out that the sharp gap between the rates of income of these two groups. Rural tourism can reduce this gap and result in the increase of villagers' income which is one of the objectives of rural tourism. Four groups play the roles in the development of the tourism industry: tourists, organizations and institutions involved in this industry, governments and the inhabitants of villages who form the host community (Maghsoudi & Lashkarara, 2003). Villagers and farmers are the first groups involved in rural tourism. In case the significance of this issue is taken seriously, and both villages are informed about its importance, both parties can take advantage of income increase. This will result in the increase of villagers' income and increase of foreign currency for governments through attracting tourist. On the other hand, statistics reveal that then there is a rise in the number of rural tourists during the past twenty years so that the number of the population of the villages becomes doubled during New Year and summer holidays (Pina, 2004). Below-mentioned reasons are among the ones which justify the necessity of considering rural tourism as a money-spinning industry: villagers can earn remarkable amount of money by selling their handicrafts and products, villagers can obtain new information about different issues through the tourists who come to their villages from external environments, Rural tourism will stop the increase the rate of migration of villagers towards cities, Rural tourism will result in the development and identification of villages, Rural tourism motivates the craftsmen not to let the handicrafts industry to be forgotten, Rural tourism is an answer for the friends of nature. (Fleicher & Felsenstien, 2004; Pina, 2004).

7 PROBLEMS AND CONSTRAINTS TO RURAL TOURISM

A superior plan can be beneficial for rural communities and consequently result in income increase. However, on the way to come up with such a process and to turn that type of tourism to a kind of culture, various challenges will be encountered. These challenges will directly or indirectly affect the increase or decrease of villagers' income. Solving these problems will have a remarkable contribution to the achievement of rural tourism objectives especially increase the income. Some of the challenges facing by the industry which seems to be of higher significance are as below:

i. Lack of managerial expertise: the problem that most of the countries suffer from is the lack of managerial expertise among the farmers and villagers (governmental and non-governmental authorities) who are involved in rural tourism projects. This problem will result in further problems like inability in selecting appropriate projects for developing this type of tourism and inability in evaluating the negative and positive outcomes of such projects in villages.
ii. Marketing: since the villages of a region are scattered, this may create the problem of accessing to better and more facilities for those tourists who are rich and need some marketplaces to provide their needs. This may consequently endanger the development of this industry. In addition to it, this will create a barrier on the way of selling the handicrafts and products of villagers and consequently they're not earning money.
iii. Transportation system: despite the spectacular natural scenes of the villages, most of them lack having access to appropriate roads and railroads. This way having access to these places become impossible. In some areas, the denseness of the site might create heavy road traffic. This may result in stationary traffic wasting tourists' time that has come to visit that region. The consequence of this problem is that tourists might not want to travel to those regions any more.
iv. Seasonality of the tourism industry: in some areas, farmers and villagers might not incline to receive the tourists in some certain seasons (due to the heavy load of farming activities). On the other hand tourists might be able to travel to rural areas (due to their children's schooling and their working). This results in some problems like villagers reluctance to receive tourists and to establish contact with them. Such a problem also opens the way for a villagers decrease of income.
v. Weak return of capital: in comparison to other business investments, the return of capital in rural tourism is remarkably low. Taking into consideration, villagers have no intention to invest in the facilities and new buildings for tourists. On the hand tourists prefer to travel to those areas which have better welfare facilities. Nevertheless, tourists have no intention to travel to such areas and this process results

in the decline of the number of tourists. The consequence of such decline is the decrease in the selling of products whose final results are insignificant or no growth in income (Unwin 1996; FFTC 2004; Hondale, 2004).

8 DISCUSSION AND CONCLUSION

All types of tourism industry are the source of revenue to the government if it is managed effectively. The number of the villages with significant characteristics in developing countries is not few. Particularly, most of these villages benefit from several cultural and natural attractions. These villages can pay a significant role in attracting tourists but unfortunately these areas have been neglected and have always suffered from constraints. The main reason for such a situation is authorities' inadequate management and planning, therefore authorities involved in agriculture and village affairs are recommended to investigate the various aspects of this issue and try to motivate the tourists regular travelling to those areas by providing them the welfare services and facilities. On the other hand tourism can be regarded as a source of revenue for both villagers and governments. This revenue can be obtained either through selling the products of through creating job opportunities in villages. In case we wish to benefit from this source of revenue, we should take some measures to satisfy tourists so that their presence in rural areas functions as a kind of promotion for others to travel to rural areas again and again.

One of the strategies which can flourish the rural tourism industry is to make and cast radio and television documentaries about the spectacular and attractive rural areas of the countries. Another strategy to increase the tourism-related revenues are to select some of the distinguished villages and to equip them with recreational and welfare facilities and to promote the culture of tourism in those villages. However, if rural tourism is taken into consideration by the authorities, to have developed villages, to eradicate of migration problem, and consequently decrease of the problem in the way of development will be fulfilled in short time.

REFERENCES

FFTC. (2004). "Rural Tourism." www.fftcagent.org/library/abstract/ac1998b.htm.

Fleicher, A. & Felsenstien, D. (2004). Support for Rural Tourism, does it make a difference? *Annals of Tourism Research*, 27(4), 1007–1024.

Hondale, B.W. (2004). Extension and tourism Development. *Journal of Extension*, 28(2).

Khayati, M. (2002). Rural tourism and its impact on rural society. *Jahad Quarterly,* 23(257).

Maghsoudi, T. & Lashkarara, F. (2003). Tourism, Development and Rural. *Jahad Quarterly*, 24(264).

Mahdavi, D. (2013). The role of tourism in development of rural area. Geography. Tehran, Tarbiat Modarres. M.A.

Moradnejad, H. & Nazeri, S. (2002). *Tourism and Rural Development*, 22(256).

Pina, I.P.A. (2004). Rural Tourism Demand by type of Accommodation. *Journal of Management Tourism.*

Rezvani, A. (1998). Geography and Tourism Industry. Tehran, Iran, Payamnoor university.

Ryan, C. (1991). Recreational Tourism: A Social Science per Spietive. London, Rutledge.

Sharifzadeh, A. & Moradnejad, H. (2002). Sustainable development and rural tourism. *Jahad Quarterly*, 22(250).

Unwin, J. (1996). Tourism Development in Estonia. *Journal of Management Tourism*, 17(4), 265–167.

Hospitality and Tourism – Sumarjan et al. (eds)
© 2014 Taylor & Francis Group, London, ISBN 978-1-138-00151-0

A study of food festival loyalty

Z. Mohi
Universiti Teknologi MARA, Malaysia

J.H.C. Wu & J.W.C. Wong
Macau University of Science and Technology, Avenida Wai Long, Taipa, Macau, China

ABSTRACT: Food festivals are growing in popularity and warrant in-depth studies of festival visitors. The study of food festivals is now an important and prolific area of tourism research enquiry. The aim of this study is to fill the conceptual gap by examining the interrelationships between festival loyalty and its related constructs, in order to gain an improved understanding of visitors' loyalty towards food festivals in Macau. The proposed hypothetical model in the food festival would facilitate further researches into the dynamics of tourism and offer guidelines for practitioners as they constantly strive to provide the best experience for visitors to return or revisit the food festival.

Keywords: food festival, festival loyalty, festival quality, festival satisfaction

1 INTRODUCTION

Food festivals always provide an opportunity or a stage for visitors to enjoy cuisine and experience local culture at the same time (Zhang, 2009). Cohen and Avieli (2004) indicate that food festivals play a key role in promoting a destination and attracting visitors. Lewis (1997) defines food festivals as "popular festivals that purport to center on and revolve around, food are a social phenomenon encountered in virtually all human cultures, from the ritual banquets of ancient Rome to the chili cook-offs of Texas today" (p. 73). In Macau, the food festival is an annual event, with more than 100 stalls offering tasty morsels from Europe, Malaysia, Mainland China, Taiwan, and other parts of Asia on every November (Kennedy, 2012). The live music performances, game booths, and rides are included in the Macau Food Festival (MFF) (Chen, 2011). The MFF has been attracting visitors because of its scenic venues and scrumptious dishes that can stimulate even the most discriminating taste buds. Despite the growth and popularity of food festivals in both practical and theoretical studies, there has been a little research focusing on festival loyalty. In the tourism studies, loyalty remains an important indicator of successful development of destinations (Prayag & Ryan, 2012). An understanding of the antecedents of festival loyalty, therefore, may offer service providers, practitioners and marketers additional opportunities to improve festival quality, emotion, festival image and festival satisfaction (Chen, Lee, & Lin, 2012; Drengner, Jahn, & Gaus, 2010; Huang, Li, & Cai, 2010; Kozak 2001; Razavi, Safari, Shafie, & Rezaei Vandchali, 2012; Yang, Gu, & Cen, 2011; Yoon, Lee, & Lee, 2010; Wong & Wu, 2013). Still, little research focusing on festival loyalty has been compiled to enhance the understanding of how festival quality, emotion, festival image and festival satisfaction affect festival loyalty from the perspective of food festival attendees (Yoon et al., 2010).

There are two objectives in this study. The first objective is to identify the dimensions of festival quality and their relative importance as perceived by visitors in the MFF. The second one is to examine the interrelationships among festival quality, emotion, festival image, festival satisfaction, and festival loyalty as perceived by visitors in the MFF.

This study contributes to the theoretical advancement in the field of festival management by empirically testing the interrelationships among festival quality, emotion, festival image, festival satisfaction, and festival loyalty. This study can also provide further insight for food festival organizers and exhibitors to better understand festival loyalty and provide useful information to marketing managers who are promoting festival products.

2 CONCEPTS AND OPERATIONAL DEFINITIONS

2.1 *Festival loyalty*

Behavioral intentions and (attitudinal) loyalty have been used interchangeably in the literature of marketing and tourism (Yoon et al., 2010). Loyalty has been a crucial objective of service providers since the high retention of customers or a low defection rate determines long-term profit levels (Zeithaml, Berry, & Parasuraman, 1996). Loyalty or behavioral intention has been measured by (1) positive word-of-mouth, (2) recommendation to others, (3) repurchase intention, and (4) high tolerance for price premium (Cronin & Taylor, 1992; Zeithaml et al., 1996). Understanding festival loyalty is all the more important for cultivating the establishment of a critical mass of loyal repeat visitors in the food festival. According to Yoon et al. (2010), the first three scales have been considered to be adopted to depict the traits of festival loyalty.

2.2 *Festival quality*

Service quality in tourism has been largely measured through a range of techniques that have been developed from the supplier's viewpoint (Briggs, Sutherland, & Drummond, 2007). From a provider's perspective, service quality is seen as "a global judgment or attitude relating to the superiority of the service" (Parasuraman, Zeithaml, & Berry, 1988, p. 16). Specifically, festival quality is used as predictors of satisfaction and behavioral intentions, rather than, for example, the gap between expected and perceived service quality (Parasuraman et al., 1988). The assumption is that if a festival maintains a high level of quality, there will be more satisfied visitors. For example, Baker and Crompton (2000) capture four dimensions of festival quality: generic features (festival characteristics), specific entertainment features, information sources (e.g. printed programs and information booths), and comfort amenities for festival visitors. They find that information sources and comfort amenities are hygiene factors, or a basic set of conditions, and that generic features and entertainment features strongly predict behavioral intentions. Based on the aforementioned review, festival quality has been identified to be multi-dimensional (Baker & Crompton, 2000; Chen et al., 2012; Cole & Chancello, 2009; Compton & Lee, 2007; Lee, Lee, Lee, & Babin, 2008; Ralston, Ellis, Yoon et al., 2010).

2.3 *Emotion*

According to Bagozzi, Gopinath and Nyer (1999), emotion can be defined as a psychological state of preparation that arises from cognitive appraisals of events or thoughts. For example, a visitor becomes excited when he or she is amazed by an entertainment show held in the festival. As the role of customer emotions is of high importance in such services, previous studies have demonstrated the importance of emotions experienced in defining consumption experiences and consumer reactions (Babin, Darden, & Babin, 1998; Dubé & Menon, 2000; Jang & Namkung, 2009; Lee et al., 2008).

2.4 *Festival image*

Keller (1993) refers to image as perceptions of an organization reflected in the associations held in customers' memories. A festival having social and cultural meaning holds a powerful attraction and appeal for visitors and has a major significance on creating or enhancing an image. The concept of festival image remains ambiguous and unexplored (Huang et al., 2010). According to several researchers (Alireza, Ali, & Aram, 2011; Clemes, Gan, & Kao, 2007, Clemes, Wu, Hui, & Gan, 2009; Wu, Lin, & Hsu, 2011), image is generally measured by reputation, prestige, and reputation comparison with other competitors. Therefore, those scales were adopted in this study to describe the traits of festival image.

2.5 *Festival satisfaction*

Anderson, Fornell and Lehman (1994, p. 54) define overall satisfaction as "an overall evaluation based on the total purchase and consumption experience with a good or service over time." Kim, Suh, and Eves (2010) maintain that satisfaction is based on the feeling of fulfillment that a visitor receives from a festival or event. It is hoped that visitor satisfaction with the food offered will result in the increased chance of that visitor recommending the food festival, as well as the increased chance that her or she will return to the food festival next time after the initial exposure. Festival satisfaction as operationalized as overall satisfaction since satisfaction in this study was based on overall emotions and the composite of festival quality dimensions.

3 RESEARCH MODEL AND HYPOTHESES

The conceptual research model shown in Figure 1 suggests that visitors form their overall perceptions of festival quality based on five dimensions: interaction quality, physical environment quality, outcome quality, access quality and programming quality. The festival quality and emotion perceived by visitors are expected to influence their overall festival satisfaction, which, in turn, results to festival image and festival loyalty respectively. A total of 10 hypotheses are formulated to test each path in the model.

Several researchers have indicated that overall perceived quality is influenced by five dimensions: interaction quality, physical environment quality, outcome quality, access quality and programming

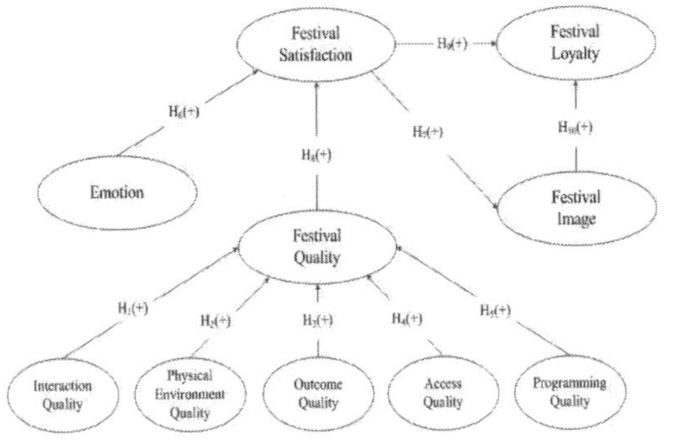

Figure 1. A conceptual research model.

quality (Brady & Cronin, 2001; Chen et al., 2012; Clemes et al., 2007, 2009, 2011a, b; Huang et al., 2010; Ko & Pastore, 2005; Shonk & Chelladurai, 2008; Wu & Hsu, 2012). The following hypotheses have been formulated to examine the effects of the dimensions on overall perceived quality based on the visitors' experiences in a food festival.

H1: Interaction quality positively affects festival quality.
H2: Physical environment quality positively affects festival quality.
H3: Outcome quality positively affects festival quality.
H4: Access quality positively affects festival quality.
H5: Programming quality positively affects festival quality.

Whereas the early models of satisfaction were centered mainly on the cognitive processes in order to understand and explain consumer satisfaction (Oliver, 1980), there is also theoretical support for linking emotions with satisfaction (Erevelles, 1998; Liljander & Strandvik, 1997; Mano & Oliver, 1993; Phillips & Baumgartner, 2002; Westbrook & Oliver, 1991). In line with previous research, the proposed model points out that, together with the cognitive component, the emotions positively impact satisfaction. Therefore, the following hypotheses are proposed:

H6: Emotions positively affect festival satisfaction.

Since customer satisfaction is depicted as a judgment made on the basis of a specific service encounter (Cronin & Taylor, 1992), satisfaction levels derived from each service encounter are believed to have an effect on image assessments (Nguyen & Leblanc, 1998). Moreover, the relationship between satisfaction and image has been reported in the previous studies (Kandampully & Hu, 2007; Lai, Griffin, & Babin, 2009; Nguyen & Leblanc, 2001). Therefore, the following hypothesis is proposed:

H7: Festival satisfaction positively affects festival image.

Several researchers (Baker & Crompton, 2000; Lee, Lee, Lee, & Babin, 2008) indicate that festival quality has been used as a precursor to festival satisfaction. Furthermore, visitors' overall satisfaction level positively affects the possibility of revisiting the destinations in the future (Hui, Wan & Ho, 2007; Kozak & Rimmington, 2000). Yoon et al. (2010) assert that the effect of festival satisfaction on festival loyalty is strong and positive. Therefore, the following hypotheses are proposed:

H8: Festival quality positively affects festival satisfaction.
H9: Festival satisfaction positively affects festival loyalty.

Image has been considered to be important factors in establishing and maintaining loyalty among people (Raj, 1985). Nguyen and Leblanc (2001) indicate that the direct of image in one organization on loyalty. Therefore, the following hypothesis is proposed:

H10: Festival image positively affects festival loyalty.

4 CONCLUSION

This study examines the theoretical concepts of festival loyalty, festival quality, emotion, festival image and festival satisfaction and their interacting relations in the food festival. The proposed conceptual model is developed for the further research on the Macau Food Festival (MFF) in 2013. MFF has been attracting visitors from all around the world by its various tastes and dishes of different Asian countries, scenic venues and good reputations due to successful experiences. A better understanding of the mechanism among festival loyalty, festival quality, emotion, festival image and festival satisfaction in the food festival may help the MFF to retain its success and hence get more contributions to the tourism industry in Macau. In this study, the literatures of the key factors in the food festival in relation to festival quality are reviewed and one proposed model is recommended based on our hypotheses.

This proposed conceptual model is still on the early developmental stage and needs to be enriched. In order to revise and apply this model in the MFF, further qualitative and quantitative research will be performed to evaluate the validity and reliability of the model. Specifically, a focus group interview focusing on visitors will be formed to generate items related to the dimensions of festival quality. Afterwards, the next step is to develop the scales by modifying the existing items and generate new items if necessary. In addition, this study will start to identify the dimensions of festival quality and their relative importance as perceived by visitors in the MFF using exploratory factor analysis and confirmatory factor analysis. This study will then examine the interrelationships among festival quality, emotion, festival image, festival satisfaction, and festival loyalty as perceived by visitors in the MFF using structural equation modeling.

REFERENCES

Alireza, F., Ali, K., & Aram, F. (2011). How quality, value, image, and satisfaction create loyalty at an Iran telecom. *International Journal of Business and Management, 6*(8), 271–279.

Anderson, E.W., Fornell, C., & Lehman, D.R. (1994). Customer satisfaction, market share, and profitability: Findings from Sweden. *Journal of Marketing, 58*(3), 53–66.

Babin, B.J., Darden, W.R., & Babin, L.A. (1998). Negative emotions in marketing research: Affect or artifact? *Journal of Business Research, 42*(3), 271–285.

Bagozzi, R., Gopinath, M., & Nyer, P. (1999). The role of emotions in marketing. *Journal of the Academy of Marketing Science, 27*(2), 184–206.

Baker, D.A., & Crompton, J.L. (2000). Quality, satisfaction and behavioral intentions. *Annals of Tourism Research, 27*(3), 785–804.

Brady, M.K., & Cronin, J.J. (2001). Some new thoughts on conceptualizing perceived service quality: A hierarchical approach. *Journal of Marketing, 65*(3), 34–49.

Briggs, S., Sutherland, J., & Drummond, S. (2007). Are hotels serving quality? An exploratory study of service quality in the Scottish hotel sector. *Tourism Management, 28*(4), 1006–1019.

Chen, S.C. (2011). Residents' perceptions of the impact of major annual tourism events in Macao: Cluster analysis. *Journal of Convention & Event Tourism, 12*(2), 106–128.

Chen, W.C., Lee, C.F., & Lin, L.Z. (2012). Investigating factors affecting festival quality: A case study of Neimen Song Jiang Jhen Battle Array, Taiwan. *African Journal of Marketing Management, 4*(2), 43–54.

Clemes, M.D., Brush, G.J., & Collins, M.J. (2011a). Analyzing the professional sport experience: A hierarchical approach. *Sport Management Review, 14*(4), 370–388.

Clemes, M.D., Gan, C., & Ren, M. (2011b). Synthesizing the effects of service quality, value and customer satisfaction on behavioral intentions in the motel industry: An empirical analysis. *Journal of Hospitality & Tourism Research, 35*(4), 530–568.

Clemes, M.D., Gan, C.E.C., & Kao, T.H. (2007). University student satisfaction: An empirical analysis. *Journal of Marketing for Higher Education, 17*(2), 292–325.

Clemes, M.D., Wu, H.C.J., Hu, B.D., & Gan, C. (2009). An empirical study of behavioral intentions in the Taiwan hotel industry. *Innovative Marketing, 5*(3), 30–50.

Cohen, E., & Avieli, N. (2004). Food in tourism: Attraction and impediment. *Annals of Tourism Research, 31*(4), 755–778.

Cole, S.T., & Chancellor, H.C. (2009). Examining the festival attributes that impact visitor experience, satisfaction and revisits intention. *Journal of Vacation Marketing, 15*(4), 323–333.

Cronin, J.J., & Taylor, S.A. (1992). Measuring service quality: A re-examination and extension. *Journal of Marketing, 56*(3), 55–68.

Drengner, J., Jahn, S., & Gaus, H. (2010). Events and loyalty formation: The role of satisfaction, felt community, emotional experience, and frequency of use. In C. Zanger, C. (Ed.), *Stand und Perspektiven der Event for schung* (pp. 151–166). Wiesbaden, Germany: Gabler.

Dubé, L., & Menon, K. (2000). Multiple roles of consumption emotions in post-purchase satisfaction with extended service transactions. *International Journal of Service Industry Management, 11*(3), 287–304.

Erevelles, S. (1998). The role of affect in marketing. *Journal of Business Research, 42*(3), 199–215.

Huang, J.Z., Li, M., & Cai, L.A. (2010). A model of community-based festival image. *International Journal of Hospitality Management, 29*(2), 254–260.

Hui, T., Wan, D., & Ho, A. (2007). Tourists' satisfaction, recommendation and revisiting Singapore. *Tourism Management, 28*(4), 965–975.

Jang, S., & Namkung, Y. (2009). Perceived quality, emotions, and behavioral intentions: Application of an extended Mehrabian–Russell model to restaurants. *Journal of Business Research, 62*(4), 451–460.

Kandamplully, J., & Hu, H.H. (2007). Do hoteliers need to manage image to retain loyal customers? *International Journal of Contemporary Hospitality Management, 19*(6), 435–443.

Keller, K.L. (1993). Conceptualization, measuring, and managing consumer-based brand equity. *Journal of Marketing, 57*(1), 1–22.

Kennedy, R. (2012). *Macau tourist spots.* Retrieved December 29, 2012, from http://traveltips.usatoday.com/macau-tourist-spots-61482.html.

Kim, Y.G., Suh, B.W., & Eves, A. (2010). The relationship between food-related personality traits, satisfaction, and loyalty among visitors attending food events and festivals. *International Journal of Hospitality Management, 29*(2), 216–226.

Ko, Y.J., & Pastore, D.L. (2005). A hierarchical model of service quality for the recreational sport industry. *Sport Marketing Quarterly, 14*(2), 84–97.

Kozak, M. (2001). Repeaters' behavior at two distinct destinations. *Annals of Tourism Research, 28*(3), 784–807.

Kozak, M., & Rimmington, M. (2000). Tourist satisfaction with Mallorca, Spain as an off-season holiday destination. *Journal of Travel Research, 38*(3), 260–269.

Lai, F., Griffin, M., & Babin, B.J. (2009). How quality, value, image, and satisfaction create loyalty at a Chinese telecom. *Journal of Business Research, 62*(10), 980–986.

Lee, Y.K., Lee, C.K., Lee, S.K., & Babin, B.J. (2008). Festivalscapes and patrons' emotions, satisfaction, and loyalty. *Journal of Business Research, 61*(1), 56–64.

Lewis, G.H. (1997). Celebrating asparagus: Community and the rationally constructed food festival. *Journal of American Culture, 20*(4), 73–78.

Liljander, V., & Strandvik, T. (1997). Emotions in service satisfaction. *International Journal of Service Industry Management, 8*(2), 148–169.

Mano, H., & Oliver, R.L. (1993). Assessing the dimensionality and structure of the consumption experience: Evaluation, feeling and satisfaction. *Journal of Consumer Research, 20*(3), 451–466.

Nguyen, N., & Leblanc, G. (1998). The mediating role of corporate image on customers' retention decisions: An investigation in financial services. *International Journal of Bank Marketing, 16*(2), 52–65.

Nguyen, N., & Leblanc, G. (2001). Corporate image and corporate reputation in customers' retention decisions in services. *Journal of Retailing and Consumer Services, 8*(4), 227–236.

Oliver, R.L. (1980). A cognitive model of the antecedents and consequences of satisfaction decisions. *Journal of Marketing Research, 17*(4), 460–469.

Parasuraman, A., Zeithaml, V.A., & Berry, L.L. (1988). SERVQUAL: A multiple-item scale for measuring consumer perceptions of service quality. *Journal of Retailing, 64*(1), 12–40.

Phillips, D.M., & Baumgartner, H. (2002). The role of consumption emotions in the satisfaction response. *Journal of Consumer Psychology, 12*(3), 243–252.

Prayag, G., & Ryan, C, (2012). Antecedents of tourists' loyalty to Mauritius: The role and influence of destination image, place attachment, personal involvement, and satisfaction. *Journal of Travel Research, 51*(3), 342–356.

Raj, S.P. (1985). Striking a balance between brand popularity and brand loyalty. *Journal of Marketing, 49*(1), 53–59.

Ralston, L.S., Ellis, G.D., Compton, D.M., & Lee, J. (2007). Staging memorable events and festivals: An integrated model of service and experience factors. *International Journal of Event Management Research, 3*(2), 24–38.

Razavi, S.M., Safari, H., shafie, H., & RezaeiVandchali, H. (2012). How customer satisfaction, corporate image and customer loyalty are related? *European Journal of Scientific Research, 78*(4), 588–596.

Shonk, D.J., & Chelladurai, P. (2008). Service quality, satisfaction, and intent to return in event sport tourism. *Journal of Sport Management, 22*(5), 587–602.

Wan, Y.K.P., & Chan, S.H.J. (in press). Factors that affect the levels of tourists' satisfaction and loyalty towards food festivals: A case study of Macau. *International Journal of Tourism Research*, doi 10.1002/jtr.1863.

Westbrook, R.A., & Oliver, R.L. (1991). The dimensionality of consumption emotion patterns and consumer satisfaction. *Journal of Consumer Research, 18*(1), 84–91.

Wong, J., & Wu, H.C.J. (2013, January 15–16). *Visitor behavioral intentions in relation to service quality and visitor satisfaction in the food festival.* Paper presented at 3rd International Conference on Applied Social Science (ICASS 2013), Taipei, Taiwan.

Wu, H.C., & Hsu, F.S. (2012). A multi-dimensional and hierarchical model of service quality in the gaming industry. *International Journal of Tourism Sciences, 12*(3), 90–118.

Wu, J.H.C., Lin, Y.C., & Hsu, F.S. (2011). An empirical analysis of synthesizing the effects of service quality, perceived value, corporate image and customer satisfaction on behavioral intentions in the transport industry: A case of Taiwan high-speed rail. *Innovative Marketing, 7*(3), 83–100.

Yang, J., Gu, Y.K., & Cen, J. (2011). Festival tourists' emotion, perceived value, and behavioral intentions: A test of the moderating effect of festivalscape. *Journal of Convention & Event Tourism, 12*(1), 25–44.

Yoon, Y.S., Lee, J.S., & Lee, C.K. (2010). Measuring festival quality and value affecting visitors' satisfaction and loyalty using a structural approach. *International Journal of Hospitality Management, 29*(2), 335–342.

Zeithaml, V.A., Berry, L.L., & Parasuraman, A. (1996). The behavioral consequences of service quality. *Journal of Marketing, 60*(2), 31–46.

Zhang, T. (2009). *The strategic use of events within food service: A case study of culinary tourism in Macao.* 2009 Paper presented at First International Conference on Information Science and Engineering, Nanjing, China.

Hospitality and Tourism – Sumarjan et al. (eds)
 ISBN 978-1-138-00151-0

The Quality of College Life (QCL) among hotel & catering students in Malaysian polytechnics

A. Ismail
Politeknik Merlimau, Malacca, Malaysia

S.M. Radzi
Universiti Teknologi MARA, Malaysia

ABSTRACT: A good Quality of College Life (QCL) is about the overall judgment of students about their satisfaction towards the academic aspects, social as well as facilities and services provided by the institutions. This study aims to examine the relationship between the quality of college life and academic as well as social aspects and also facilities and services provided. It also determines the most contributing variables to the quality of college life. This study involved 380 students from four Malaysian polytechnics who enrolled in Diploma of Hotel and Catering. Systematic sampling was used to select the sample based on the students' list as the sampling frame. Findings revealed strong supports for the importance of the facilities and services in the quality of college life. The findings also revealed the trends of college life among Hotel and Catering students in Malaysian polytechnics. The academic aspects have less impact on their quality of life which signaled that polytechnics should take proper actions regarding this matter.

Keywords: hotel & catering students, Malaysian polytechnics, Quality of College Life (QCL)

1 INTRODUCTION

The Polytechnics Education was made known in Malaysia 41 years ago. It began with the establishment of Polytechnic Ungku Omar (PUO) as the first polytechnic, in 1969 under the United Nations Development Plan. After 41 years, on 25 February 2009, the Department of Polytechnic Education launched the transformation of polytechnics with the aim of polytechnics to sustain human capital development that will produce a new capability to the polytechnics in developing human resources towards 2020.

The transformation aimed to reinforce the role of polytechnics in providing education and training. Its mission is to break boundaries for the creation of transformative and creative learning environment for an innovation—led economy (Jabatan Pengajian Politeknik, 2010). Transformation of polytechnic education in Malaysia was planned to enhance the quality of technical education in Malaysia (Jabatan Pengajian Politeknik, 2010).

According to Wahab, Zakaria and Jasmi (2010) polytechnics will have to deal with many issues in achieving the transformation agenda and indirectly providing the best quality experiences to polytechnic students. The issues were related to the changes in technology, quality of programmes, staffing, organizational culture, infrastructure, and facilities. In reality, polytechnics are facing problems which are related to insufficient and out-of-date infrastructures and lab devices. The quality of programmes, facilities, academic staff, support services, accessibility and location, and campus climate are the predictors of students' satisfactions (Abidin, 2008). However, Polytechnics will have to improve on these matters to provide excellent learning experiences.

2 LITERATURE REVIEW

Malaysian Polytechnic Education is producing minds-on, heads-on and hearts-on individuals. Therefore, polytechnics education is implemented based on the hybrid concept of learning experiences which are education and training. It is a combination of classroom and industrial experiences. It consists of theory in classroom learning and practical in the lab and training in the hospitality industry (Jabatan Pengajian Politeknik, 2010). The transformation of polytechnic education in 2009 introduced the Outcome Based Education (OBE) as required by the Malaysian Qualification Agency (MQA). The implementation of OBE in polytechnic education is new to lecturers and students (Shaari and Jusoh, 2012). Thus, the new systems and curriculum should provide a high quality of education to enhance the quality of college life among students.

Quality of College Life (QCL) can be defined as overall feelings of satisfaction experienced by a student with regards to the college life. The quality of students' life is determined by their satisfactions towards academic and social aspects as well as facilities and services (Sirgy, Grzeskowiak, & Rahtz 2007).

There were many dimensions which are researchable in relation to academic aspects. The academic aspects valued by students are faculty, teaching methods, the classroom environment, the workload, the academic reputation of the institutions and academic diversity (Sirgy et al., 2007). Based on British Columbia and College Student Outcomes (2003), the dimensions of educational experiences are curriculum, teaching, analytical skills, communication skills, social skills and personal growth.

Social aspects contributed to students' satisfaction towards their quality of college life. The quality of college life can be measured by considering campus housing, spiritual and services programmes, club and parties, collegiate and recreational activities (Sirgy et al., 2007, and Yu & Lee, 2007). According to Joyce (2009) there is a relationship between student participations in social activities in college and students satisfactions towards their life in college.

The facilities and services such as labs, library, transportation, hostels, cafeterias, student centre, clinics, sport centre, I.T. Department and workshop helps students to experience a conducive teaching and learning process. It enhances their creativity and helps them to perform better in their coursework (Wahab et al., 2010). A study by Manley (2011) related to outsourcing student residential services proved that there is no significant relationship between the outsourcing of the services and facilities and the students' satisfaction towards their college life. The study proved that students were not really concerned with facilities and service provided throughout their college life.

Satisfaction is a key element in determining the quality of the products and services provided (Hashim & Azemi, 2010). Customer satisfaction is a common issue for any business and organizations. It is important to ensure consumer confidence in the products or services provided. As well as product and services, quality of college life is also determined by students' satisfaction. Students are considered as customers in educational institutions. Students' satisfaction is also an important issue to be discussed because they are the main clients of educational institutions (Kwek, 2010).

3 RESEARCH METHODOLOGY

This study used questionnaires in identifying the relationship between satisfaction with academic and social aspects as well as facilities and services and the quality of college life. This study was conducted by utilizing self-administered questionnaires to gather information related to students' satisfaction towards the academic and social aspects as well as facilities and services. This study began with the adaptation of instruments by Sirgy et al. (2007), pilot testing and later proceeded with the actual study after the researcher is satisfied with the research instruments.

The actual study was administered to only four (4) polytechnics. There were four hundred (400) respondents involved from Diploma in Hotel and Catering students in four (4) Malaysian polytechnics in Malaysia which are offering this programme. The polytechnics are:

a. Tuanku Syed Sirajuddin Polytechnic, Perlis,
b. Ibrahim Sultan Polytechnic, Johor Bahru,
c. Kota Kinabalu Polytechnic, Sabah, and
d. Merlimau Polytechnic, Melaka.

The populations of this study were Diploma in Hotel and Catering students from semester 1 to semester 6. The researcher then analysed the data by using individual unit of analysis. The respondents were chosen through systematic sampling. Systematic sampling is efficient for a large sampling frame and the advantages are ease and quickness in sample determination (Sekaran, 2009). The total population was based on the current student cohorts. The students were selected from the sampling frame which listed every 4th targeted sample in the students' record from the Students Affairs Unit.

The questionnaire consisted of five (5) major sections. There were 76 items developed to measure the relationship between academic aspects and social aspects as well as facilities and services towards the quality of college life. The respondents were required to answer questionnaires using six points Likert Scale with six point scale represented strongly agree = 6, agree = 5, somewhat agree = 4, somewhat disagree = 3, disagree = 2 and strongly disagree = 1. The six points Likert Scale was used to eliminate the neutral response from the respondents and creating less biased measurement (Sekaran, 2009).

The reliability and validity of the instruments was tested by the implementation of pilot test. The reliability and validity of the content was based on discussions between the researcher and supervisor before distributing the questionnaire to the respondents. The pilot study was carried out involving 40 students from Sultan Idris Shah Polytechnic, Sabak Bernam.

The data analysis process was a quantitative analysis using Statistical Package for Social Science (SPSS) version 18.0. The data final were analyzed using the multiple regressions to analyze the relationship among variables and determine the most contributing variable to the quality of college life. The overall conclusions were made based on the acceptance or rejection of hypotheses on null hypotheses.

4 DATA ANALYSIS AND RESULTS

The highlight of this study was to measure the relationship between academic and social aspects, as well as facilities and services provided towards the quality of college life among hospitality respondents in polytechnics. The results provided were based on 380 sample size gathered from four polytechnics. The multiple regression analysis was performed to examine the relationship and determine the most contributing variables to the quality of college life. Regression models explained that 55.7% of the variance represented by the quality of college life. The other 44.3% represented by other predictor variables. The coefficient results reported that only facilities and services ($p = 0.00 < \alpha = 0.05$) and social aspects were significant at ($p = 0.02 < \alpha = 0.05$) level.

However, academic aspects were found not significant at ($p = 0.42 > \alpha = 0.05$) level. The beta value (β) was used to compare and evaluate the relationship between the independent and dependent variable. Among the three predictors entered into the model facilities and services ($\beta = 0.574$) contributed the largest unique determiners to the quality of college life. These findings provided strong supports for the importance of the facilities and services in the quality of college life. The findings also revealed the trends of college life among Hotel and Catering students in Malaysian polytechnics. The academic aspects have less impact on their quality of life which signaled that polytechnics should take proper actions regarding this matter. The quality of human capital produced for hospitality industry will be enhanced by understanding their satisfaction. Therefore, understanding the relationship of all predictors to the quality of college life is important to provide a good quality of hospitality education. This study helped the polytechnics to remain competitive in the higher education system and to have a reflection of their strengths and weaknesses in order to be a first choice institution.

5 DISCUSSIONS AND CONCLUSIONS

The outcomes of hospitality education are very important to determine the development of quality human capital for the hospitality industry. The issue stimulated the researcher's interest to study the quality of college life among hospitality students in Malaysian Polytechnics. The quality of college life will determine the performance of the students in the real industry (Nguyen, Shultz, & Westbrook, 2011).

This study was conducted to examine the relationship between academic and social aspects as well as facilities and services towards the quality of college life. The objective of this study was also to determine the most contributing aspects to the quality of college life. The quality of human capital produced for hospitality industry will be enhanced by understanding their satisfaction. Therefore, understanding the relationship of all redictors to the quality of college life is important to provide a good quality of hospitality education.

This study provided many implications to the polytechnics' education systems and other institutions. It helped the polytechnics to remain competitive in the higher education system by understanding their students' satisfaction towards the academic, social and facilities and services. It also developed future students by using the current students as the tools of promotion. In terms of finances, this study helped polytechnics' top management to improve their financial management. An effective financial management can be implemented by allocating the resources efficiently.

As a conclusion, the study on quality of college life is about understanding the students' satisfaction towards the academics and social life as well as facilities and services. In a very competitive situation, academic institution compete each other to offer higher quality of education. Polytechnics as a technical education provider in Malaysia must also compete with public and private universities, private institutions, and community colleges in winning the students' hearts to enrol for their future development. Therefore, this study helped the polytechnics to have a reflection of their strengths and weaknesses in order to be a first choice institution.

Based on the findings, there was a significant relationship between quality of college life and facilities and services. There was also a significant relationship between social aspects and quality of college life. However, there was no significant relationship between quality of college life and academic aspects. It can be concluded that, polytechnic students were more concerns on facilities, services and the social aspect through their life in polytechnics. The quality of their college life was not largely affected by the academic aspects. However, polytechnics must not neglect the quality of education offered. Hence, in order to generate students' satisfaction towards their quality of college life, polytechnics must have a continuous improvement on the quality of education, facilities and services.

The findings revealed the trends of college life among Hotel and Catering students in Malaysian Polytechnics. The academic aspects have less impact on their quality of life which signalled that polytechnics should take proper actions regarding this matter. This situation also showed that the students were less concerned about academic. Based on the findings, polytechnics or management need to enhance the students' awareness towards the importance of academic aspects. The main objectives of their enrolment in polytechnics need to be highlighted as a reminder.

Many activities, seminars or conferences must be organized for the students. From the findings we can also assumed that the students were satisfied with the academic aspects in polytechnics. It can be concluded that the students are satisfied with the faculty, teaching method, classroom environment, workload, academic reputation and academic diversity. Therefore, academic issues were not influencing their quality of college life.

REFERENCES

Abidin, N.Z. (2008). The relationship between perception and satisfaction of service quality among business students in polytechnics in Malaysia. *School of Graduate School*. Abstract retrieved from http://www.psasir.upm.edu.my.

British Columbia College & Institute Student Outcomes. (2003). *Issue paper on understanding student satisfaction: CIESS Research and IT Solutions, 3*(1), 1–4. Retrieved from http://admin.selkirk.bc.ca/research/documents/issue_satisfaction[1].pdf.

Department of Polytechnic Education. (2010). *Competency standards for polytechnic graduates*. Ministry of Higher education Malaysia. Seri Kembangan, Selangor.

Hashim, F.M.A. & Azemi, N.C. (2010). Tahap kepuasan pelajar terhadap kemudahan dan perkhidmatan yang disediakan di Politeknik Port Dickson bagi sesi Disember 2010. Retrieved from http://www.polipd.edu.my.

Jabatan Pengajian Politeknik. (2010). *Competency Standards for Polytechnics Graduates.* (1st Edition). Selangor, Malaysia.

Jabatan Pengajian Politeknik. (2010). *Fakta ringkas.* (April Edition). Selangor, Malaysia.

Joyce, G.J. (2009). *Student satisfaction with academic experiences, faculty, campus life, and mission as predictors of retention at a Catholic University* (Doctoral Dissertation). Retrieved from Proquest Dissertation and Theses. (UMI No: 3348455).

Krejcie, R.V. & Morgan, D.W. (1970). Determining sample size for research activity: *Educational and Psycological Measurement, 30*, 607–610.

Kwek, Choon Ling (2010). The 'inside-out' and 'utside-in' approaches on students' perceived service quality: An empirical evaluation. *Management Science and Engineering*, *4*(2), 01–26.

Manley, J.H. (2011). *A study of effects of outsourcing residence life programmes on students life satisfaction.* (Doctoral Dissertation). Retrieved from Proquest Dissertation and Theses. (UMI No: 3473211).

Nguyen, T.D., Shultz, C.J., & Westbrook, M.D. (2011). Psychological hardiness in learning and quality of college life of business students: Evidence from Vietnam. *Journal of Happiness Studies*, doi:10.1007/s10902-011-9308-0.

Sekaran, U. & Bougie, R. (2009). *Research methods for business: A skill building approach.* Fifth Edition, John Wiley & Son, Inc.

Shaari, B.M. & Jusoh, Y.Y. (2012). The Outcome-Based Education (OBE) curriculum for electrical engineering departments at Malaysian Polytechnics. *Education Research Journal, 2*(3), 66–74. Retrived from http://www.resjournal.com/ERJ.

Sirgy, M.J., Grzeskowiak, S., & Rahtz, D. (2007). Quality of College Life (QCL) of students: Developing and validating a measure. *Social Indicator Research, 80* (2), 343–360. Retrieved from http://www.springerlink.com.

Wahab, S.H.A., Zakaria, M.A., & Jasmi, M.A. (2010). Transformational of Malaysian's polytechnic into university college in 2015: Issues and challenges for Malaysian technical and vocational education. *Proceedings of the 1st UPI International Conference on Technical and Vocational Education and Training* (pp. 570–578). Bandung, Indonesia.

Yu, G.B. & Lee, D.J. (2007). A model of Quality of College Life (QCL) of students in Korea. *Social Indicator Research*, 87, 269–285. doi:10.1007/s11205-007-9172-9.

Hospitality and Tourism – Sumarjan et al. (eds)
© 2014 Taylor & Francis Group, London, ISBN 978-1-138-00151-0

Community participation toward states tourism development: A case of Melaka Baba Nyonya and Portuguese minority

N.S. Mior Shariffuddin & M.S. Mohd Zahari
Universiti Teknologi MARA, Malaysia

ABSTRACT: The continuous development of tourism sectors worldwide not only for the purpose to sustain the traditional resources but expanding the contemporary tourism products and cultural values. The combination of these two elements directly boosts up the economy of the country, situational contexts and the local communities. Tourism development besides others is closely associated with communal efforts of all parties. Recognition of the tourism destinations could not be achieved by solidly depends on the government initiatives alone without participation of the private sectors, industry players, NGO's and the local communities. The involvement or participation of the community in tourism development is crucial in ensuring the development is meeting the economy and social objectives. This research note discusses the community capacity building toward state tourism development.

Keywords: minority ethnic group, participation, state tourism developments

1 INTRODUCTION

The continuous development of tourism sectors worldwide not only for the purpose to sustain the traditional resources but expanding the contemporary tourism products and cultural values (UNWTO, 2008). In this sense, tourism development is closely associated with communal efforts of all parties. According to Vourc'h (2003) recognition of the tourism destinations could not be achieved by solidly depend on the government initiatives alone without participation of the private sectors, industry players, NGO's and the local communities.

With regard to the local communities, their involvement significantly contributes to the development of the tourism industry. Feighery (2002) stated that the key successful element in tourism development, tourism planning and other aspects concerning tourism destination without doubt depending broadly on community support. Similarly, to achieve sustainable tourism there is always in need of support from the host community (STCRC, 2010).

An approach called Community-Based Tourism (CBT) was being introduced and practiced in many third world, developing and developed countries in accelerating of the tourism industry. Hitchens and Highstead (2005) reported that areas that successfully practiced of CBT are usually in the prime areas that have good proximity in establishing tourism accessibility, links to the private sector as well as local communities although in many instances the Community-Based Tourism (CBT) was not successfully achieved due to the limited involvement of the community in the main tourism development process (Cooper, 2004). Therefore, it is crucial for any tourism destination to allow the engagement of their local communities towards tourism development.

Sustainable Tourism Cooperative Research Centre (STCRC) (2010) argued that the best way to recognize a destination is by identifying the communities itself because they are part of host destination and familiar with the resources of destination that can be used as the tourism products. On the other hand, the participation of the local community in tourism destination through community capacity buildings cannot be ignored as this could contribute to a significant tourism development. This research note is therefore conceptually discussing the relationship between community participation and state tourism development and relates the issue to the researcher proposed study.

2 LITERATURE REVIEW

2.1 *Community participation toward tourism development*

Generally, community participation is concerned on the involvement of individuals and communities in decision making that may affect their lives

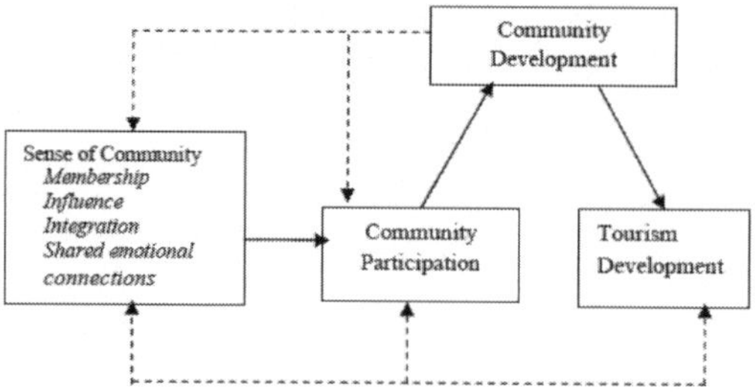

Figure 1. Tourism development chain (Phillips & Pittman, 2008).

(Burns et al., 2004). The community participation is also known as the mechanism for active community involvement in partnership working, decision making and representation in community structures (Chapman & Kirk, 2001). Phillips and Pittman (2008) came out with a model that relates to how the community participation in tourism development contributes to the whole progress of the outcome of the development or they called tourism development chain.

This model explains the relationship how sense of community which comprises the attributes (membership, influence, integration, shared emotional, connections) contribute to community participation and development and at the end lead tourism development. Aref et al. (2010) suggested that sense of community as one of the traits on their study on the concept and dimensions of community capacity building with respect to tourism development. The parties involved in this development such as community leaders, stakeholders and tourism development practitioners should encourage the concept of the sense of community as to improve and allow community participation.

This relationship has also proven by Aref (2011) as he adapted the model to investigate the relationship between the sense of community and level of participation in tourism development in Shiraz, Iran. This is owing to the fact that the elements of the sense of community are a motivation for the participation of the tourism activities of the destination. The findings resulted that there are significant relationship between the community participation and sense of community in the tourism development.

In contrast, even though it is being urged that the local communities participate in the tourism development but studies revealed that some dimensions limit the participation of local community's tourism development. According to Tosun (2000) there are three main dimensions of limitations for participation of local communities in developing countries towards tourism development which are classified as (1) Operational Limitations (2) Structural Limitations and (3) Cultural Limitations. Under the operational level include (a) centralization of public administration of tourism, (b) lack of co-ordination, and (c) lack of information. As for the structural limitations include (a) attitudes of professionals; (b) lack of expertise; (c) elite domination; (d) lack of appropriate legal system; (e) lack of trained human resources; (f) relatively high cost of community participation and lack of financial resources. Last dimension is cultural limitations consisting of (a) limited capacity of poor people and (b) apathy and low level of awareness in the local community.

Tosun (2000) explained the lack of participation of the local communities is not a simple matter but it involves different ideological beliefs, political forces, administrative arrangements and re-distribution of wealth as well as power in developing countries. Dukeshire & Thurlow (2002) identified the barriers and challenges that may hinder the progress of the tourism development in the rural Canada areas which contribute to the government further understanding towards this issue. Seven of the reasons found were (1) Lack of understanding of the policy process which allows the individuals empowerment and community-based organizations to make decisions, (2) Lack of community resources which relate to funding, government training programs, education, leaders and volunteers to support rural causes and initiatives.

3 MINORITY GROUP COMMUNITY PARTICIPATION AND MELAKA STATE TOURISM

Known by their developed trading cultural exchanges between East and West for over the past 500 years (UNESCO, 2008), Melaka is one of the oldest heritage cities. Owing to many tourism products ranging from traditional heritage to modern products Melaka is increasingly receiving a substantial number of local and international tourists' arrival. In 2012 alone, more than four million tourists are reported of visiting Melaka with China, Singapore tourists in the top notches follow by Indonesia, Taiwan, Hong Kong, European and Middle Eastern countries (Syahrin, 2012).

In hastening the tourism development, the state government through the department of tourism also strongly encourages the participation of NGOs, international organizations, private agencies, tourism suppliers, tourism private sectors and the local communities (NEAC, 2010). The collaborations between all of these tourism key players have resulted in many profitable and constructive tourist attractions. Many opportunities for the development of tourism allow the partnership between the key players especially with local communities. Participation of the local communities with the state tourism can be seen through the selling of handicrafts, homestay, food and traditional costumes and others in many tourism spots in the state (Fernandis, 2003). In fact, the state head minister stated that the multiracial of the local communities in Melaka could create a positive force in the tourism development.

Despite the urging from the government, the participation of certain minority ethnic groups in Melaka tourism especially among the Baba Nyonya and Portuguese is slightly slow compare to Malays, Chinese and Indian. For instance, record from the Melaka tourism department revealed the participation of these two minority ethnic groups is not aggressive (Melaka State Tourism, 2012). This scenario is receiving a greater concern to state tourism department and public as these two minority ethnics have long been part of Melaka heritage. This issue raises critical questions as to why less participation among the minority ethnic groups and what are the underlying reasons of this causation. In other word, are these problems related to lack of community capacity building among those ethnic groups? In this sense, community capacity building is related to the support of communities in developing their skills and knowledge as to allow them to work together and give positive change within their communities (Department for Social development, n.d). The community capacity building consists of the influence by the community's commitment, resources and skills that build the community strengths and address on the community problems and opportunities (Aspen Institute, 1996). Therefore, it is worth to investigate and revealed the causation which the empirical investigation is still undertaking.

4 CONCLUSION

The study on Baba Nyonya and Portuguese community capacity building is believed to have significantly influenced towards the tourism development of the state.

From the academic perspective, this study will create new literature in Malaysia as well as expanding the existing body of literature existed. Consequently this research will contribute in adding knowledge to the future researcher to further explore in various areas in Malaysia for the benefit of tourism Malaysia.

In practical perspective, the findings will benefit the government, especially the Melaka State Ministry of Tourism and other non-government related agencies. The result of the research will also help the industry provider in making improvements and also giving information to come out with a guideline for future planning.

REFERENCES

Alpeni, L. (2013). A missed opportunity: Community participation in tourism in South Africa.

Aref, F., Redzuan, M., & Gill, S. (2010). Dimensions of community capacity building: A review of its implications in tourism development. *Journal of American Science*, *6*(1).

Aref, F. (2011). Sense of community and participation for tourism development. *Life Science Journal*, *8*(1).

Aspen Institute (1996). *Measuring Community Capacity Building*. Queenstown, MD.

Breed of Life to Tourism in Melaka. *Travel Daily News* [Asia Pacific]. Retrieved from http://www.traveldailynews.asia/news/article/49888/unesco-gives-a-new-breed.

Burns, D., Heywood, F., Taylor, M., Wilde, P., & Wilson, M. (2004). *Making community participation meaningful: A handbook for development and assessment*. Policy Press.

Chapman, M. & Kirk, K. (2001). *Lessons for Community Capacity Building: A summary of the research evidence*. Retrieved from http://www.scothomes.gov.uk/pdfs/pubs/260.pdf.

Cooper, G. (2004). Community based tourism experiences in the Caribbean: Lessons and key consideration. Paper presented at the *Caribbean tourism organization Sixth Annual Caribbean Conference on Sustainable Tourism Development: Keeping the right balance—Land and Sea Encounters*.

Darus, M. (2008, March 11). Growth in Malaccatourism. *The Star Online*. Retrieved from http://thestar.com.my/metro/story.asp?file=/2008/3/11/southneast/20579177&sec=southneast.

Department for Social Development (n.d.). *Voluntary and Community Sector: Community Capacity Building*.

Dukeshire, S. & Thurlow, J. (2002). *Challenges and barriers to community participation in policy development*. Canada: Rural communities impacting policy project.

Ertuna, B. & Kirbas, G. (2012). Local community involvement in rural tourism development: The case of Kastamonu, Turkey. *Revista de Turismo y Partrimonio Cultural*, *10*(2), 17–24.

Feighery, W. (2002). Community participation in rural tourism development: A social representations approach. *Tourism Research 2002: An interdisciplinary conference in Wales*.

Fernandis, G. (2003). The Portuguese community at the periphery: A Minority report on the Portuguese quest for Bumipatera status. Kajian Malaysia, XXI (1), 285–301.

Ghimire, S. & Upreti, B. (2011). Community Participation For Environment-friendly Tourism: The avenue for local peace. *The Journal of Tourism and Peace Research*, *2*(1), 55–69.

Havana, Cuba, (2004) Citrinot, L. (2012, July 4). UNESCO Gives a New.

Hitchens, R. & Highstead, J. (2005). Community-based tourism in Namibia–A Discussion Paper, Brighton: University of Sussex.
Ismail, M. (1992). Tourism of historic city Of Melaka in Peninsular Malaysia. Retrieved from http://eprints.uitm.edu.my/3346/1/MOHD_YUSOFF_ISMAIL_92_24.pdf.
NEAC (2010). *Transforming the tourism sector into a high income industry*. Retrieved from National economic advisory council website: http://www.neac.gov.my/files/Transforming_the_Tourism_Sector_into_A_High_Income_Industry.pdf.
Phillips, R. & Pittman, R. (Eds.). (2008). *An introduction to community development,* Taylor & Francis.
Rahman, S. (2013). Tourism is here to Stay. *Buletin Melaka Taming Sari Berhad.*
Sustainable Tourism Cooperative Research Centre (STCRC) (2010). *Sustainable Tourism–Destinations and Communities*. Retrieved from Sustainable Tourism Online website: http://www.sustainabletourismonline.com.
Tosun, C. (2000). Limits to community participation in the tourism development process in developing countries. *Tourism Management*, 613–633.

Hospitality and Tourism – Sumarjan et al. (eds)
© 2014 Taylor & Francis Group, London, ISBN 978-1-138-00151-0

Residents' perceived social impacts and desire to support Kenyir Lake International Triathlon event in Terengganu

S.K.B. Mohd Supar
Universiti Teknologi MARA, Malaysia

ABSTRACT: Sports tourism is not something new in Malaysia and it comprises the whole part of Malaysia which not only in the urban area but also in the rural area. Hosting the sport event not just involve the participant of the event and the host community residents', but also the spectators that willing to travel to watch the event. No matter where the event is, organizing an international sports event more or less will influence and affect the behaviour of local people especially with massive amount of media attention, political involvement and also money. This study will empirically investigate whether perceived socio impacts will affect residents' desire to support the Kenyir Lake International Triathlon (KLIT) event. In this study, self-administered questionnaire will be adapted from past researchers and the issues of interest will be obtained. The result from this study will enable the Malaysia government to take immediate action to ensure that the negative social impacts will not outweigh the positive impacts in hosting the sport event.

Keywords: residents' desire, rural area, social impact, sport tourism

1 INTRODUCTION

Sports tourism is a phenomena that ascending from the unique contact of people, place and activity through the cultural, social and economic (Weed and Bull, 2004). The activity is referring to people's travel motives either to watch or to participate in the sport events. Regardless of the size of the sport events, events that have known in the international level are an important element of sport tourism which may strengthen the element of tourism destination marketing and growth. In line with the statement, the sport activities are able to attract large number of tourists even from outside of the region, and it have high potential on attracting the non-residents of host community such as the coaches, athletes, media and other sport event officers.

In line with above notion, Malaysia is one of the countries that had hosted various types of sports events local and internationally. One of the international sports events that hosted by Malaysia is the Kenyir Lake International Triathlon. According to the organizer of the event, the Lembaga Kemajuan Terengganu Tengah (KETENGAH) (2013) in collaboration with the state government of Teregganu, it is a sport tourism event with the objective of promoting Kenyir Lake as Aqua-eco tourism to the local and international tourist. This international event has been listed as the annual event in the tourism calendar of Terengganu. The event of Kenyir Lake International Triathlon used the Olympic distance and regulations which the participants have to complete their event starting with 1.6 km of swimming, 45 km of cycling and 11 km of running (KETENGAH, 2013). KETENGAH (2013) proudly stated that this is their ninth year of organizing the Kenyir Lake International Triathlon with at least more than 28 countries participated in this event including country from Australia, Canada, Spainand France.

Based on the statement above, according to Turco et al. (2003) the bidding process for a sport event and hosting a sport event has become essential parts of the tourism strategy of a destination. As for that, the relations effects between sport event and destination have also been found (Xing & Chalip, 2006).

2 LITERATURE REVIEW

At this proposal stage, the review of the literature specifically touches on the most significance variables related to the study such as the sport tourism events, the positive and negative of social impacts and the residents' perception.

2.1 *The sport tourism events*

The definition of sport tourism given by Weed and Bull (2004, p. 37): "Sports tourism is a social, economic and cultural phenomenon arising from the unique interaction of activity, people and place". The activity is always linked to the motives of travelling either to participate or watching the events. Based on the statement, Nogawa et al. (1996) stated that the possibilities for the sport tourists to take part in more activities at the destination is higher if they stay more than one night as compared to those who stay less than 24 hours. In line with that statement, more or less it will affect the host community of the sport events both positively and negatively. For example, the positive impact exist when the destination resident support the sport event and give great hospitality to the tourist. Nevertheless, the negative impact may also arise due to hooligans or security problem whichever to the tourist or the destination residents (Smith, 2006).

Based on the sport tourism definition (Wedd & Bull, 2004), the sport tourists might perceive different types of impact from the sport events that they attend either to participate or to watch. In line with the statement, this is because they have different travel motives which lead to the various activities throughout the stay.

2.2 *Positive and negative of social impacts*

The residents of hosting events both local and national, plays an important role in the economies sector to the country. According to Preuss and Sulberg (2006), it is an important role because welcoming a particular events to the area by the residents, the politician are more willing to allocate monies to fund the event. As for that, at any discussion of events, it is important to consider the residents of hosting community and pay attention on the impact of their quality of life by hosting an event at a destination (Fredline, 2005). Based on Crompton and Ap (1994), the residents' views are important and must not be neglected by the organizer of the events. They are the main supporters for the event which able to help promote the event, willingness to volunteer at the event and industry tourism and most important factor is hospitality to the tourists. Preuss and Solberg (2006) noted that both positive and negative social impacts were produced by hosting a sport event.

Based the above notion, the social impacts consequences of sport tourism events that may arise have been identified by Higham (1999), Fredline (2005), Preuss and Solberg (2006), Ohman et al. (2006) and Bull and Lovell (2007). From all the social impact consequences, Higham (1999) is the only one who stated that hosting sport tourism event mostly gives negative impacts without present any of positive impacts. This is in contrast with all other author which mentioned that even though there are negative impacts, the positive impacts to the host community also exist. In line with the statement, it is clearly shows that understanding the social cultural impacts that perceived by the residents' host community is important to avoid the negatives impacts outweigh the positives impacts.

2.3 *The residents' perception*

The attitudes of residents' towards the development of tourism event are important for the organizer to understand either the residents' are supportive or vice versa, and by understanding the attitudes, suitable action can be taken to avoid the negative influences that affect the residents' perceptions that arise from the tourism development process (Sharma & Dyer, 2009). Based on Andrereck et al. (2005), there is a positive relationship between the strong contact of tourists with the residents and the residents' knowledge on tourism. This relationship is supported by Davis et al. (1988) and Lankford and Howard (1994), which based on their studies on the level of residents knowledge may influence their perception on tourism development.

In a study by Liu and Var (1986), they underlined that the resident who derive income and dependent on tourism industry are more supportive on the tourism development compared to those who didn't. In line with the statement, Gursoy et al. (2002) stated that the residents' perceptions can be positive towards the tourism industry if they enjoy the improvement of the facilities in their area or the increases number of opportunities to the resident activity. Alternatively, O'leary (1976) mentioned that the residents' perceptions may able to be negative if they believe that the tourism industry may affect their traditional activities.

3 ISSUES

It is generally understood that events can impacts the host communities on the environmental, economic and social either positive and/or negative impact (Bull & Lovell, 2007). According to Otmann et al. (2006), the research on social impacts empirically under-research, compared to the economic effects on the sport tourism events. In recent years, knowing the inequity of studies on the social impact of sport tourism especially in mega events have grab the attention of researchers (Bob & Swart, 2009). To date, the researchers have concentrating most social impact on sport tourism events in developed country (Bull & Lovell, 2007), while research on event in rural area is scarce (Ljudevit, 2012). Additionally, based on Nunkoo and Ramkissoon (2011), there are many studies which have adopted the Social Exchange Theory (SET) to enlighten the tourism benefits, it appear to provide an understanding based on benefits of the economic only, while the non-economic benefits are neglected.

Based the above notion, since hosting international sport events can affect the resident of host community a range of social impacts either positive or negative impacts (Preuss & Solberg, 2006), by understanding the both positive and negative social impacts and the perceptions of the resident may enable action that can be taken to ensure that negative social impact will not outweigh the positive impacts (Delamere, 2001). Furthermore, according to Ohmann et al. (2006), to organize an international event, the supports of the residents' are reliant on their perceptions on the events. Their perceptions are important as the organizer of the sport event need continuous support from the residents' of the host community.

4 CONCLUSION

Sport events are believed to have significantly contributed to the tourism development industry in Malaysia especially in Terengganu. Therefore, by investigating the social impacts to the host community in rural area will contribute to the consequence in two aspects which is the practical and academic aspects.

The contribution for practical aspect, it will allow the state government especially the state of Terengganu to ensure that negative social impacts will not outweigh the positive impacts of hosting the sport events. Not only that, by knowing the social impact of hosting the sport events, it will enable the government, specifically the Ministry of Tourism and other related agencies to take immediate action of the negative impacts. Besides that, with less negative social impacts towards the host community, it will encourage them to cooperate with Ministry of Tourism and Culture and will be able to organize better event for the purpose of revisiting process among the local and international tourist to Kenyir Lake, Terengganu. The revisiting process will create more business opportunities for many sectors such as accommodation, transportation, and food service operators. Hence, directly or indirectly, these will contribute to the economic development.

From the academic perspective, there is little research conducted on this area, specifically towards the relationship of sports events and social impacts and the impacts towards the host community in rural area. Therefore, this research will in some way contribute to the new body of knowledge in Malaysia and extending the literature in Malaysian sport tourism related study.

REFERENCES

Andereck, K.L., Valentine, K.M., Knopf, R.C., & Vogt, C.A. (2005). Residents' perceptions of community tourism impacts. *Annals of Tourism Research, 32*(4), 1056–76.

Bull, C. & Lovell, J. (2007). The impact of hosting major sporting events on local residents: An analysis of the views and perceptions of Canterbury residents in relation to the Tour de France 200. *Journal of Sport & Tourism, 12*(3–4), 229–48.

Crompton, J.L. & Ap, J. (1994). Development of a tourism impact scale in host resident context research enhancement program: Final Report, Texas A&M University, College Station, TX.

Davis, D., Allen, J., & Cosenza, R.M. (1988). Segmenting local residents by their attitudes, interests, and opinions towards tourism. *Journal of Travel Research, 27*(2), 2–8.

Delamere, T.A. (2001). Development of a scale to measure resident attitudes toward the social impacts of community festivals, part 2: Verification of the scale. *Event Management, 7*(1), 25–38.

Fredline, E. (2005). Host and guest relations and sport tourism. *Sport in Society, 8*(2), 263–79.

Gursoy, D., Jurowskiand, C., & Uysal, M. (2002). Resident attitudes: A structural modelling approach. *Annals of Tourism Research, 29*(1), 79–105.

Higham, J. (1999). Sport as an avenue of tourism development: An analysis of the positive and negative impacts of sport tourism. *Current Issues in Tourism, 2*(1), 82–90.

Liu, J. & Var, T. (1986). Resident attitudes towards tourism impacts in Hawaii. *Annals of Tourism Research, 13*, 193–214.

Ohmann, S., Jones, I., & Wilkes, K. (2006). The perceived social impacts of the 2006 Football World Cup on Munich residents. *Journal of Sport & Tourism, 11*(2), 129–52.

Preuss, H. & Solberg, H.A. (2006). Attracting major sporting events: The role of local residents. *European Sport Management Quarterly, 6*(4), 391–411.

Sharma, B. & Dyer, P. (2009). Residents' involvement in tourism and their perceptions of tourism impacts. *Benchmarking: An International Journal, 16*(3), 351–71.

Turco, D.M., Swart, K., Bob, U., & Moodley, V. (2003). Socio-economic impacts of sport tourism in the Durban Unicity, South Africa. *Journal of Sport Tourism, 8*(4), 223–39.

Zhou, Y. & Ap, J. (2009). Residents' perceptions towards the impacts of the Beijing 2008 Olympic games. *Journal of Travel Research, 48*(1), 78–91

Hospitality and Tourism – Sumarjan et al. (eds)
© 2014 Taylor & Francis Group, London, ISBN 978-1-138-00151-0

The influence of graduate attributes towards their employability: A case in the Faculty of Hotel and Tourism Management, Universiti Teknologi MARA (UiTM)

E.S. Mohd Salleh & M.F.S. Bakhtiar
Universiti Teknologi MARA, Malaysia

ABSTRACT: Local hospitality and tourism sectors are growing particularly over the past decade while continues to contribute largely to Malaysia economy. The growth of the industry is apparent through development of new hospitality related products and services each year; even thousand of new jobs were created. Unfortunately, unemployment rate among graduates including those from the hospitality and tourism fields remain high. Debates and studies were unveiling the reason behind such unemployment issue. Nevertheless, an approach to look at the success factors among graduates who secured employment was minimally covered. Having said that, this study, examines the influence of graduate attributes contributing towards their employability in the Faculty Hotel and Tourism Management, UiTM. Through self reported questionnaire, 283 valid respondents were analyzed following necessary filtering procedure. The result shows that skills are the key factor contributes to graduate employability.

Keywords: employability, graduates, hospitality and tourism, Malaysia

1 INTRODUCTION

Local hospitality and tourism sectors are growing particularly over the past decade while continues to contribute largely to Malaysia economy. Increased in tourist arrival and spending pattern are reflected through expansion of revenue earning yearly. In 2011, these combined two sub-sectors of service industry has brought over RM37 billion into the national gross income (ETP Annual Report, 2011). The growth of the industry is apparent through development of new hospitality related products and services each year; for example the accommodation sector alone has seen additional 99 new hotels built and opened between 2005 and 2012 making up a total of 2355 establishments operating by end of that year. In terms of human capital, 55,565 new jobs were created in 2011 and it stretches up to 90,542 by the following year (The Star, May 2012). Notwithstanding substantial opportunities available, unemployment rate among graduates remain high including from those with hospitality and tourism background. According to Graduates Tracer Study report established by the Ministry of Education in 2011 alone, a total of 44,391 or 24 percent of graduates remain unemployed 6 months after graduation. An in-depth review of the report revealed that Universiti Teknologi MARA (UiTM) produced the largest number of graduates in hospitality and tourism fields with total of 664 in 2011 as compared to other public and private higher learning institutions. From 664 graduated, 28% of them remain jobless 6 months after graduation. It represents 5.8% increment from 23 percent in 2010 as it was also 4% above national graduate unemployment record for that year (UiTM Centre for Strategic Planning, 2012).

Following such perturbing occurrence for the past few years; ongoing debates and studies focusing to graduate employability were carried at national level to unravel this issue. A study by Nordin (2011) in addressing this matter revealed that graduates were lacking critical key traits needed by employers. Graduates unimpressive personality and lacking of soft skills were among the reasons why they were not preferred by employers. Other influencing factors would include personal attitude of being too selective with task performance, frequent job-hopping, inability to work as a team, less willing to learn, self-centered and demanding for workplace with higher pay while maintaining work-life balance (Bernama, 2011; Blumenfeld et al., 1987; Domonte & d Vaden, 1987; Kusluvan & Kusluvan, 2000; National Higher Education Research Institute, 2003; Price water house Coopers, 2009).

Despite all discouraging reasons made to uncover causes of unemployment among graduates, it is equally significant to note that another

72% from the 664 graduates in hospitality and tourism fields at UiTM in 2011 were doing differently; they could either being employed or pursuing higher level education. According to Cooperrider and Whitney (2005) through the principle of appreciative inquiry, it is noteworthy to look upon individual's core competencies rather than soliciting of problem or issue hindering a success. By understanding the means, alternative focus towards graduate unemployment issue may be shifted from observing the unfortunate causes alone to a new dimension of appreciating and learning from those who were successful. Hence, this paper aims to examine the influence of graduate attributes contributing towards their employability; a case in the Faculty Hotel and Tourism Management, UiTM.

2 LITERATURE REVIEW

Researchers have continuously unveiling variation of definitions and concepts with regard to employability; however most of them agree that employability constitutes of graduates' skills, attitudes, capabilities and competencies which valuable for getting a job with sustainable employments.

Many factors have been identified as being the reasons why the graduates are unemployed. Based on the ongoing debate with regard to the attitude of the graduates toward seeking employment (Blumenfeld et al., 1987; Domonte and Vaden, 1987; Kusluvan, & Kusluvan, 2000; Price water house Coopers, 2009; The National Higher Education Research Institute, 2003; The Star, 2012), further with the skills and competencies demanded by industries (Johanson, 2010; Nordin, 2011; Zopiatis, 2010); considerable attention further has been given to these three factors—attitude, skills and competencies in this study.

3 METHODOLOGY

In assessing factors affecting employability among graduates, a quantitative research approach was employed using self administered questionnaire.

All 2011 bachelor degree graduates from the Faculty Hotel and Tourism Management, UiTM were approached for the study. Initial screening was performed on all returned responses; only those who declare themselves as being employed were chosen for analysis. Following such guideline, only 283 from 303 responses received were further analyzed.

4 FINDINGS AND DISCUSSION

4.1 *Respondents' profile*

Prior to commencing into statistical analysis, all data were checked for irregularities, missing item, or unrealistic responses. Frequency test was used against all demographic variables. Of the 283 respondents, 41.2% were graduates of Tourism Management, 23.1% in Foodservice Management, another 17.6% in Hotel Management while the remaining graduated in Culinary Arts Management. Moving to respondents' employment profile, 56.3% stated that they were able to secure a job within 3 to 7 months; another 32.8% succeeded within 7 to 11 months and only 10.9% spend more than 11 months in search for a job.

4.2 *Reliability and validity of the instrument*

The reliability of data was validated using Cronbach's Alpha (α) coefficient analysis. According to George and Mallery's (2003) rule of thumb, a minimum value of 0.70 may be considered acceptable reflecting high internal consistency reliability. A sample of 30 respondents differ from actual sampling frame was used during preliminary questionnaire testing. From the analysis, Cronbach's Alpha coefficient value for attitude was 0.911, whereby skills 0.944 and competencies 0.943. Since all Cronbach's Alpha coefficient values were above 0.70, existing questionnaire may be considered fit for actual data collection.

4.3 *Relationship between attitude, skills and competencies towards employability*

The result of correlation analysis between two variables is shown in Table 2. With significant level of 0.05, attitude was considered significant but, with very minimal positive relationship to graduates employability ($r = 0.138$ and $p = 0.033$); at significant level 0.01 both skills and competencies shown significant but, with low relationship to employability where $r = 0.400$ and $p = 0.000$ and $r = 0.367$ and $p = 0.00$ respectively. This finding reveals that skill had the highest association to graduate employability.

Table 1. Reliability test of pilot study.

Variables	Pilot test (α)	Actual study (α)	Items
Attitudes	0.903	0.911	9
Skills	0.884	0.944	6
Competencies	0.921	0.943	5

Table 2. Result of correlation analysis.

	Correlation coefficient	
Variables	*r*	Sig.
Attitudes	0.138*	0.033
Skills	0.400**	0.000
Competencies	0.367**	0.000

*Correlation is significant at the 0.05 level (2-tailed).
**Correlation is significant at the 0.01 level (2-tailed).

Table 3. Result of multiple regression.

	Multiple regression		
Variables	b	β	Sig.
Attitudes	−0.186	−0.193	0.006
Skills	0.663	0.605	0.000
Competencies	0.112	0.110	0.430

The multiple regression model with three predictors produces R square (R^2), $F = 43.505$, $p < 0.05$ ($p = 0.000$). As shown in Table 3, skill had significant positive regression weights, indicating graduates with high skills were expected to achieve high level of employability, after controlling for the other variables in the model. The Attitude ($b = -0.164$) was not significant ($p = 0.006$); further with the coefficient being negative ($\beta = -0.193$), it could be indicated that possessing positive attitude alone is nota guarantee to become highly employable. Finally, the competencies ($b = 0.110$, $p = 0.430$) seems to be unrelated to graduate employability. This would seem as if the level of competency is not an important factor in predicting graduate employability; interestingly, this result was relatively unexpected.

This study found that there is general agreement among the graduates that the skills is the most influential and become key factors contribute to the employability. This is not surprising since the nature of hospitality and tourism industry is highly exposed to the human relations and "moment of truth" (Calzon, 1987); which requires specific "technical or hand-on skills" and interpersonal skills in operational areas and industry segment (Johanson et al., 2010). From the findings, it can be assumed that competencies can be developed from time to time as result from the work-disciplined experience. Meanwhile, the attitude actually is very subjective and it represents evaluating statement (Poezter, 1987). Therefore, it can be changing accordingly based on experience and environmental setting rather than behaviors or values (Maio et al., 2000). However, the graduates need to adjust their employment expectation according to their actual capability to ensure they can get placement in the job markets.

5 CONCLUSION

Work environment in the 21st century is absolutely challenging especially for graduates. Organizations today strived to respond to the dynamic-changing market as result from globalization and rapid technological advancement. Hence, to be employable graduates are expected to have interactive personal with highly skilled in job disciplined (Harvey, 2000) able to show evidence of having the ability to cope with uncertainty, a readiness to explore and create opportunities, self-confidence and willingness to learn (Raybould & Sheedy, 2005).

The result of the finding, gives an empirical evidence that it is important for the graduates to prepare themselves with the skills that most demanded by industries' players. These elements are required for flourishing career in the hospitality industry (Kay & Moncarz, 2004; Raybould & Wilkins, Jauhari, 2006; Tas, 1988;). Having an excellent academic qualification or straights 'As' is not sufficient to secure a job or to give one a competitive edge for being employed (Yorke, 2006; Purcell et al., 2000).

REFERENCES

Blumenfeld, W.S., Kent, W.E., Shock, P.J., & Jourdan, L.F. (1987). Job attributes preferences of a group of potential hospitality industry managers: What makes a job good or bad. *Hospitality Education and Research Journal, 10*(2), 79–93.

Coakes, S.J., & Clara, O. (2010). SPSS Version 18.0 For Windows: Analysis Without Anguish. Wiley.

Confederation of British Industry (CBI), (1999). Making Employability Work: An Agenda for Action. London: CBI.

Confederation of British Industry (CBI), (2007). Time well spent: Embedding employability in work experience. London: CBI.

Confederation of British Industry (CBI), (2009) Future fit: Preparing graduates for the world of work. London: CBI.

Dileep Kumar, M. & Jain, Vishal. (2010). Survival skills of business management graduates: A study with reference to retail and banking. *Far East Journal of Psychology and Business, 1*(1), 49–77.

Domonte, T. & Vaden, A.G. (1987). Career decisions in hospitality management. *Hospitality Education and Research Journal, 11*, 51–63.

Economic Transformation Programme (2011). Annual report five reasons; why graduates are unemployed. (2010, February 2). The Star Online. Retrieved from http://thestar.com.my/news/story.asp?file=/2010/2/27/nation/20100227211559&sec=nation.

George, D. & Mallery, P. (2003) SPSS for Windows step by step: A simple guide and reference. 11.0 update (4th Edition). Boston: Allyn & Bacon.

Great Place to Work Institute. (2009). Executive briefing, motivating generation Y: Across the great divide. Retrieved from http://www.greatplacetowork.co.uk.

Gursoy, D. & Swanger, N. (2005). An industry-driven model of hospitality curriculum for programs housed in accredited colleges of business: Part 2. *Journal of Hospitality and Tourism Education, 17*(2), 46–54.

Hair, J.F., Anderson, R.E., Tatham, R.L., & Black, W.C. (1998). Multivariate data analysis. New Jersey: Prentice Hall.

Harvey, L. (2000). New realities: The relationship between higher education and employment. *Tertiary Education and Management*, *6*(1), 3–17.

Hillage, J. & Pollard, E. (1998). Employability: Developing a frame work for policy analysis. Research Brief No. 85, Department for Education and Employment, London. Retrieved from: http://www.employmentstudies.co.uk/summary/summary.php?id=emplblty.

Hinchliffe, Geoffrey., & Jolly, Adrienne. (2009). Employer concepts of graduate employability. Retrieved from http://dera.ioe.ac.uk/13064/.

Ismail, Rahmah., Yusoff, Ishak., & Lai, Wei Sieng. (2011). Employers' perceptions on graduates in Malaysian services sector. *International Business Management*, *5*(3), 184–193.

Jauhari, V. (2006). Competencies for a career in the hospitality industry: An Indian perspective. *International Journal of Contemporary Hospitality Management, 18*(2), 123–134.

Johanson, Misty, Ghiselli, Richard., Shea, Linda J., & Roberts, Chris. (2010). Revealing key competencies of hospitality graduates demanded by industry: A 25-year review. International CHRIE Conference-Refereed Track, Paper 5. Retrieved from http://scholarworks.umass.edu/refereed/CHRIE_2010/Saturday/5.

Kay, C. & Moncarz, E. (2004). Knowledge, skills and abilities for lodging management success. *Cornell Hotel and Restaurant Administration Quarterly, 45*(3), 285–298.

Kusluvan, S. & Kusluvan, Z. (2000). Perceptions and attitudes of undergraduate tourism students towards working in the tourism industry in Turkey. *Tourism Management, 21*, 251–269.

Lai, W.S. (2011). Unemployment among graduates: Study of employers' perception on graduates. *Proceeding of the International Conference on Social Science, Economics and Art 2011*, 15–18.

Lowden, Kevin., Hall, Stuart., Elliot, Dr Delly., & Lewin, John. (2011). Employers' perceptions of the employability skills of new graduates; Research commissioned by the Edge Foundation. University of Glasgow SCRE Centre and Edge Foundation, London.

Mason, Geoff., Williams, Gareth. & Cranmer, Sue. (2006). Employability skills initiatives in higher education: What effects do they have on graduate labor outcomes? National Institute of Economic and Social Research, London. Retrieved from http://citeseerx.ist.psu.edu/viewdoc/summary?doi=10.1.1.143.3726.

McQuaid, Ronald W. & Lindsay, Colin. (2005). The concept of employability. *Urban Studies*, *42*(2), 197–219.

Ministry of Higher Education Malaysia, (2006). Development of soft skills for Institutions of Higher Learning. Universiti Putra Malaysia, Selangor DarulEhsan.

Moreland, N. (2006). Entreprenuership & Higher Education: An employability perspective, learning & employability series. ESECT, York.

National Higher Education Research Institute. (2003). Research on unemployment among graduates.

Noe, R.A. & Hollenbeck, J. (2007). Fundamental of human resource management, McGraw-Hill, New York, NY.

Nordin, K. (2009). Enhancing graduate employability: Issues, concerns and the way forward. Retrieved from http://khalednordin.com/wp-content/uploads/2009/07/july-21st-2009-seminar-on-enhancing-graduate-employability-issues-concerns-and-the-way-forward.pdf.

Pavesic D.V. (1991). Another view of the future of hospitality education. *Cornell Hotel and Restaurant Administration Quarterly*, *32*(4), 8–9.

Pizam, A. (2011). The domains of tourism and hospitality management. Paper presented in the Plenary Section at the First International Conference on Tourism & Management Studies, Faro.

Poczter, A. (1987). Attitude development hierarchy and segmentation. *Review of Business, 9*(1), 17.

Price water house Coopers. (2009). Malaysia's Gen Y unplugged, Price water house Coopers Advisory Services Sdn Bhd. Retrieved from http://www.pwc.com/en_my/my/assets/publications/gen-y.pdf.

Raybould, M. & Wilkins, H. (2005). Over-qualified and under-experienced: turning graduates into hospitality managers. *International Journal of Contemporary Hospitality Management, 17*(3), 203–216.

Robbins, S.P. & Judge, T.A. (2009). Organizational Behaviour. 13th edition. Pearson Education, Upper Saddle River, NJ.

Rychen, D.S. & Salganik, L.H. (2001). Defining and Selecting Key Competencies. Retrieved from http://www.oecd.org/pisa/35070367.pdf.

Sekaran, Uma. & Bougie, Roger. (2010). Research method for business: A skill building approach. 5th edition, United Kingdom: John Wiley & Son Ltd.

Statistics of Graduates in the Labor Forces Malaysia. (2011). Department of Statistic Malaysia.

Suarta, I Made. (2012). Employability skills needed by higher vocational education graduates to be successful in the workplace. *Chinese Business Review*. *5*(7), 639–648.

Tas, F.R., (1988). Teaching future managers. Cornell Hotel and Restaurant Administration Quarterly *29*(2), 41–43.

Trochim, W.M. (2006). *Research Method Knowledge Base.* Retrieved from http://www.socialresearchmethods.net/kb/reliablt.php.

UiTM's Centre for Strategic Planning (2012); Graduate tracer study, UiTM Shah Alam.

Undergraduates need to Enhance Soft Skills-Khaled (8th April 2011). Retrieved from http://my.news.yahoo.com/undergraduates-enhance-soft-skills-khaled.

Woodley, A. & J. Brennan, (2000). Higher education and graduate employment United Kingdom. *European J. Education, Research, Development and Policies*, *35*(2), 239–249.

Yorke, M. (2006). Employability in higher education: What it is—what it is not; Learning and employability series 1. York, United Kingdom: The Higher Education Academy.
Yorke, Mantz & Knight, Peter T. (2006). Embedding employability into the curriculum; Learning and employability series one. York, United Kingdom: The Higher Education Academy.
Yorke, M. & Harvey, L. (2005). Graduate attributes and their development. New Directions for Institutional Research, 128, 41–58.
Zopiatis, Anastasios. (2010). Is it art or science? Chef Competencies for Success. *International Journal of Hospitality Management*, *29*(3), 459–467.

Hospitality and Tourism – Sumarjan et al. (eds)
 ISBN 978-1-138-00151-0

Culinary arts students' readiness towards the blended learning activities

U.A. Khalid, A. Ismail, W.S.N.A. Ab Aziz, A.A. Azdel & M.S.Y. Kamaruddin
Universiti Teknologi MARA, Malaysia

ABSTRACT: This study investigated the Culinary Arts students' perception on the blended learning activities challenges, willingness, self-efficacy and usefulness. Data were collected using self-administered survey. Proportionate stratified random sampling design used based on the Culinary Arts students stratum. Data obtained from the questionnaire was analysed using the Statistical Percentage for Social Science (SPSS). The results shown that the students perceived blended learning activities as cost saving, promoting bilateral communication, enhance autonomy and quality of learning as well as education accessibility. However, the students still believe that the face to face interactions are better for their learning process. This study also highlighted numerous factors affecting the adoption of blended learning system.

Keywords: blended learning, culinary arts students, E-learning, readiness

1 INTRODUCTION

E-learning (or blended learning) is also known as internet-based learning. It has emerged as one of the fastest moving trends in today's education that provides an exceptional opportunity to increase students access to education (Kamarulzaman, Madun & Ghani, 2011; Palmer et al., 2001).

Blended Learning is the process incorporating many different ways in which people learn (learning style) through the use of "blended" virtual and physical resources that contains a variety of learning activities with the use of technology, lecturer and peer interaction. A typical example of the delivery method of blended learning would be a combination of technology-based materials along with face-to-face instruction used in the content delivery. A lecturer can begin a course with a well-structured lesson in the classroom, and then proceed with follow-up tutorial online (i-Learn Centre, UiTM Shah Alam, 2012).

The blended learning was introduced in the Faculty of Hotel and Tourism Management in March 2013 where only 2 hours a week is spent in the classroom and the other one hour will be in the virtual classroom known as e-learning.

1.1 *Problem of the study*

Basically, Culinary Arts students are more involved in practical classes which includes cooking, baking and other means of preparing dishes. The word "culinary" is defined as something related to, or connected with, cooking or kitchens.

The nature of Culinary Arts does make students a bit shocked as they were comfortable with the traditional F2F learning environments that have been around for centuries (Graham, 2004). Use of technology to facilitate learning is accepted as a value across educational institutions. However, the focus is still largely on getting the infrastructure and creating the e-learning content. It is necessary to consider the individual factors that play an important role in the adoption of e-learning (Siavash et al., 2011). Therefore the objective of this paper is to reveal the perception of culinary arts students regarding the usefulness of blended learning and their readiness to this respective teaching and learning approach.

2 LITERATURE REVIEW

Based on the previous research, Tayerbinik (2012) noted that e-learning environments present some disadvantages such as inhibiting the socialization process of individuals resulting in a lack of face-to-face communication; a new environment has surfaced. This new environment combines the e-learning and the classical learning environments. It has been termed as blended learning, hybrid or mixed learning *(ibid)*.

Furthermore, Omidinia, Masrom and Selamat, (2010), postulated that it is necessary to consider the individual factors that play an important role in the adoption of e-learning. For example, attitude of students and teachers towards e-learning may affect their acceptance of the technology in the teaching and learning process. The four determinants of e-learning acceptance are performance expectancy, effort expectancy, social influence and facilitating conditions (Umrani-Khan & Iyer, 2009). One of the important challenges of most developing countries is not to have high speed internet access, due to a myriad of factors including but not limited to intermittent electricity, use of expensive low bandwidth satellite technology and inadequately trained personnel (ibid).

Therefore, Shraim and Khlaif (2010) stated that the students' opinion (strongly agree or disagree) based on challenges; (language, distract time & lack of motivation, lack of online learning experiences and technical problems), willingness; (prefer face to face method, adopt e-learning in the future, recommended to colleagues, willing to take another VCR), self-efficacy; (ability to learn from online tutorial, easy to use e-learning elements, no problem in access and log on, IT experience sufficient to use VCR), and usefulness; (saving cost, promoting communication, enhanced autonomy, enhanced quality of learning, and enhanced accessibility to education) reflect their information, and readiness.

3 METHODOLOGY

The sample population for this study was drawn from the Culinary Arts Students of the Faculty of Hotel and Tourism Management (FPHP), University Technology Mara (UiTM), Shah Alam, Selangor. A proportionate stratified random sampling design was used for this study where only 20% for each stratum of each semester from the Culinary Arts students were solicited.

For the research instrument, a set of questionnaire was designed and divided into 7 sections which includes the profile of the respondent, student's background or knowledge of the Blended Learning activities and factors that will show the Culinary Arts Students' readiness towards the Blended Learning activities. The questionnaire also sought information on students perception regarding the usefulness and readiness (self-efficacy and willingness) to adapt the learning process of Blended Learning and the challenges that the students will face.

The researchers personally distributed the questionnaire so that it would be rightly distributed and collected for a reliable result. Each collected questionnaire was marked and keyed for the records.

4 FINDINGS

Based on the information obtained from the study, 78% of the respondents were female and 23% were male. The findings also revealed that 40% of the students accessed the Blended Learning activities at home while the rest accessed it at their hostel and at the faculty with 22% and 14% respectively. Usefulness. Significant number of students accessed the Blended Learning System during their free time.

Based on Table 1, majority of the respondent agreed that Blended learning activities is a useful aspect in Learning System. Furthermore, the respondents perceived Blended Learning as cost saving (M = 3.00) which did not require any cost involved such as entrance fees or membership fees. Blended Learning indeed a beneficial facets as students are able to express themselves openly and it does help promoting communication skills (M = 3.00) and enhanced learning quality (M = 3.00). As blended learning activities enable the students to express themselves freely without any control of the lecturers (M = 4.00), it does enhance autonomy. Without a doubt, majority of the respondents agreed that the Blended learning activities do enhance the accessibility to education resources (M = 4.00).

4.1 *Willingness*

According to the findings above, despite being technology-savvy, the students prefer face to face interaction between lecturers (M = 4.00). This finding coherent with the previous studies (Aydin & Tasci, 2005; Shraim & Khlaif, 2010; So & Swatman, 2006; Watkins et al., 2004). Majority agreed that blended learning can be adopted (M = 3.00) and willing to attend any familirazition workshop (M = 3.00).

4.2 *Self efficacy*

Based on Table 2, most of the respondents are technology-savvy (M = 4.00). Malaysia is a well known as a developed country whereby every citizen

Table 1. Usefulness of blended learning among culinary students.

No.	Items	Mean	Std. dev.
1.	Saving cost	3.00	0.996
2.	Promoting communication	3.00	0.752
3.	Enhance autonomy (freedom)	4.00	0.753
4.	Enhance quality of learning	3.00	0.808
5.	Enhance accessibility to education	4.00	0.773

Table 2. Willingness of using blended learning among culinary students.

No.	Items	Mean	Std. dev.
1.	I would prefer face to face method then online method in my learning process	4.00	1.147
2.	I would adopt blended learning for next semester	3.00	0.998
3.	I am willing to attend any workshop for blended learning	3.00	1.102
4.	I am willing to use blended learning if appropriate support provided	4.00	1.032

Table 3. Blended learning self efficacy of culinary students using blended learning.

No.	Items	Mean	Std. dev.
1.	Ability to learn from online tutorial	4.00	0.808
2.	Easy to use e-learning elements	4.00	0.965
3.	No problem in accessing and longing in	3.00	1.178
4.	Have IT experiences that are sufficient enough to handle blended learning	3.00	0.829

Table 4. Challenges in using blended learning among culinary arts students.

No.	Items	Means	Std. dev.
1.	Languages (English as the medium of the instrument)	4.00	0.985
2.	Time distraction	4.00	0.779
3.	Lack of motivation	3.00	0.762
4.	Lack of online experiences uploading and downloading file	3.00	1.149
5.	Technical problem e.g. server down	4.00	0.988

is well equipped with basic IT knowledge. This is inline with their respons on ability to access Blended Learning system (M = 3.00) the ability to learn from the online tutorial availability (M = 4.00).

4.3 *Challenges*

Based on the findings above, students faced problem with of languages (M = 4.00; Std Dev = 0.985), time distraction (M = 4.00; Std Dev = 0.779), lack of motivation (M = 3.00; Std Dev = 0.762), lack of online experiences (M = 3.00; Std Dev = 1.149) and technical problem (M = 4.00; Std Dev = 0.988).

In every system there are always challenges that the system faces that include the Blended Learning System. Challenges such as languages, time distraction, lack of motivation, lack of experiences and technical problems (Robbins & Judge, 2013).

5 CONCLUSIONS

As a conclusion, this study highlighted the students' challenges, willingness, self efficacy and usefulness to adapt blended learning as a new tools system in learning activities and their attitude shows nonchalant result towards adopting the blended learning, but they have positively perceived it as part of their learning tools. Furthermore, the blended learning activity is not the replacement for the all process learning but only as a supplement in traditional learning which is face to face teaching method which is the best learning tool. Academician need to prepare themselves promoting the students with different academic background to adapt this new learning methodology.

REFERENCES

Bonk, C.J., & Graham, C.R. (2004). Handbook of blended learning: Global perspectives. *Local Designs*. San Francisco, CA: Pfeiffer Publishing.

I-Learn (UITM). (2012). Blended learning: For lecturers, students and admin guideline, i-learn.uitm.edu.my/v2/wp-content/.../01/BL_Booklet_New2012.pdf.

Omidinia, S., Masrom, M., & Selamat, H. (2011). Review of e-learning and Ict infrastructure in developing countries (case study of Iran). *American Journal of Economics and Business Administration*, *3*(1), 120–125.

Robbins, S.P., & Judge, T.A. (2013), *Organizational Behavior*. England: Pearson Education Limited.

Shraim, K., & Khlaif, Z. (2010). Students' readiness toowards e-learning. A case study of virtual classroom for secondary education in Palestine: *3rd Annual Forum on e-learning Excellence Bringing Global Quality to a Local Context*, Dubai U.A.E.

Soydal, I., Alır, G., & Ünal, Y. (2011). Are Turkish universities ready for e-learning: A case of Hacettepe University Faculty of Letters. *Information Services and Use*, *31*(3), 281–291.

Tayebinik, M., & Puteh, M. (2012). Blended Learning or E-learning?. *International Magazine on Advances in Computer Science and Telecommunications*, *3*(1), 103.

Umrani-Khan, Farida., & Iyer, Sridhar. (2009). ELAM: A Model for Acceptance and Use of E-learning by Teachers and Students. *Indian Institute of Technology Bombay*, Mumbai, India.

Yusniza, K., Azian, M., & Farinda, A.G. (2011). Attitude towards e-learning among students: Evidence from a Malaysian Public University: British. *Journal of Arts and Social Sciences*, *3*(2), 2046–9578.

Author index

An environmentally friendly book printed and bound in England by www.printondemand-worldwide.com

PEFC Certified
This product is from sustainably managed forests and controlled sources
www.pefc.org
PEFC
PEFC/16-33-415

FSC
www.fsc.org
MIX
Paper from responsible sources
FSC® C004959

This book is made entirely of sustainable materials; FSC paper for the cover and PEFC paper for the text pages.

#0136 - 030614 - C0 - 246/174/34 [36] - CB